Inquiry into Life

Inquiry into Life
Fourth Edition

Sylvia S. Mader

wcb
Wm. C. Brown Publishers
Dubuque, Iowa

Cover Image
The beautifully colored insect on the cover is not a
butterfly; it is an atlas moth *(Attacus atlas),* a giant among
moths having a wingspan of up to ten inches. You can tell
it is a moth because of its stout body and feathery
antennae. Many unique creatures inhabit the earth—all
products of the evolutionary process, just as we are. A
major decision facing us today is whether we care enough
about other forms of life to help them survive. Photo by
Gunter Ziesler/Peter Arnold, Inc.

Book Team

John Stout   *Senior Editor*
Mary J. Porter   *Assistant Editor*
Eugenia M. Collins   *Senior Production Editor*
Geri Wolfe   *Designer*
Mary Sailer   *Design Layout Assistant*
Vicki Krug   *Permissions Editor*
Faye M. Schilling   *Photo Research Editor*

**wcb
group**

Wm. C. Brown   Chairman of the Board
Mark C. Falb   *President and Chief Executive Officer*

wcb

Wm. C. Brown Publishers, College Division

James L. Romig   *Vice-President, Product Development*
David A. Corona   *Vice-President, Production and Design*
E. F. Jogerst   *Vice-President, Cost Analyst*
Bob McLaughlin   *National Sales Manager*
Marcia H. Stout   *Marketing Manager*
Craig S. Marty   *Director of Marketing Research*
Eugenia M. Collins   *Production Editorial Manager*
Marilyn A. Phelps   *Manager of Design*
Mary M. Heller   *Photo Research Manager*

Copyright © 1976, 1979, 1982, and 1985 by Wm. C. Brown
Publishers. All rights reserved

Library of Congress Catalog Card Number: 84–72409

ISBN 0–697–04798–9

Printed in the United States of America
10 9 8 7 6 5 4

For my children

Brief Contents

Contents

Part Three
Human Anatomy and Physiology

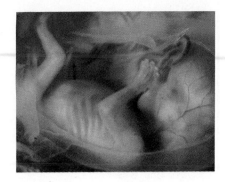

Part Four

Reproduction, Development, and Inheritance

Part Five
Evolution and Diversity

Part Six
Behavior and Ecology

Tables

Readings

*These readings were written by the author to highlight and expand upon topics discussed in the text.

INQUIRY INTO LIFE is written for the young adult who would like a working knowledge of biology. Educational theory tells us that adults are most interested in knowledge of immediate practical application. This book attempts to remain true to this idea, and its basic theme is knowledge about and understanding of human biology. Plants and other animals are included, however, because human beings cannot understand themselves unless they understand other living things. All organisms face the same problems, but through the evolutionary process, they solve these problems differently. Human beings can better understand themselves when they see the unity of life while at the same time seeing the diversity. Concerned citizens need to realize that humans are not the pivot point, nor even the culmination, of life but are a part of a great, overall, interrelated network.

While this text covers the whole field of basic biology, it emphasizes the application of this knowledge to aspects regarding humans themselves and the relation of humans to other organisms. Students will thereby be able to see that biology is truly relevant. Concurrent with this approach, the student is led to a discovery of biological concepts and principles. Detailed, high-level scientific data and terminology are not included because the author believes that true knowledge consists of working concepts rather than technical facility. The latter can always be added by more advanced study after acquisition of a core of knowledge upon which to build.

The fourth edition of *Inquiry into Life* has the same format and style as the first three editions. Each chapter presents the topic clearly, simply, and distinctly so that the student will feel capable of achieving an adult level of understanding. While the book was being revised, it was computerized for the first time, thus providing the author an opportunity to review each and every word. Although it is quite possible for succeeding editions of a text to become more lengthy and erudite, this tendency has been strenuously avoided, and adopters will find the fourth edition to the point and straightforward like the previous editions.

Preface

Organization of the Text

An introductory chapter precedes the thirty-three chapters of *Inquiry,* which are grouped into six parts.

Introduction
The introductory chapter reviews the characteristics of life. This discussion is designed so students can see that humans are but one kind of living thing and that all types of organisms share the same characteristics. There follows an overview of the book that indicates the manner in which its various parts are related to one another. The overview provides students with an opportunity to see the whole before beginning their study of the individual parts.

A reading that outlines the scientific method illustrates that the biological view of life is a product of this method. Scientific information is subject to change because new studies bring forth further clarification.

Part One, The Cell, a Unit of Life
The first part introduces basic biological principles and serves as a foundation for the parts that follow. The organization and content remain essentially as they were in the third edition. The chemistry chapter has been simplified somewhat, and new findings on the structure and function of the cell have been incorporated. In particular, recent information regarding the cytoplasm is included. Presentation of cellular metabolism has been shortened, and instructors should find it even more feasible to give only an overview of this topic if they so wish.

Part Two, Plant Biology

Coverage of photosynthesis precedes that of plant structure and function. Terminology relating to photosynthesis has been revamped and the treatment is simpler than before. Again, there is a choice as to depth of presentation. The chapter devoted to plant structure and function was completely rewritten and much care was taken to keep the terminology correct but uncluttered.

Part Three, Human Anatomy and Physiology

Just as the flowering plant is used as the basis for discussing plant physiology, humans are used as the basis for discussing animal physiology. Updating and clarification characterize the revision of this part. The latest information on diet and cancer, immunity, muscle contraction, and brain research, for example, is included. Throughout, however, the emphasis is placed on presenting these topics at the proper level.

Part Four, Reproduction, Development, and Inheritance

Human reproduction, development, and inheritance are the topics for this part. Instructor feedback has helped the author improve the pedagogy of Mendelian genetics and rewrite the molecular genetics chapter, which now includes a reading about Watson and Crick's discovery of DNA structure. The illustrations continue to improve, and this should make it easier for instructors to present these topics. As before, genetic engineering, regulation of genes, and the related topic, cancer, are covered. New findings regarding oncogenes are included.

Part Five, Evolution and Diversity

Chapter 23, "Evolution," and chapter 24, "Origin of Life," precede a survey of living organisms. The evolution chapter now presents evidences for evolution and contains a reading regarding Darwin. The **five kingdom** system of classification has replaced the four kingdom system used in the previous editions. Anatomy and physiology of the various groups are stressed rather than classification, however. There is now only one chapter devoted to the animal kingdom rather than two chapters; this was achieved without sacrificing significant sections. New findings regarding human evolution have been incorporated into the last chapter of this part.

Part Six, Behavior and Ecology

This last part required little revision. As before, it features two chapters devoted to behavior before ecosystems, biomes, and human population are discussed. Chapter 29, "Behavior within Species" discusses the why and how of animal behavior, including the formation of societies. Chapter 30, "Behavior between Species" contains topics such as competition and predation, which are also important components of ecological interactions. Chapter 31, "Ecosystems," explains ecological principles and also examines ecological problems such as air and water pollution. Chapter 32, "The Biosphere," examines natural communities within the biosphere while emphasizing how they have been altered by humans. By studying the effects of human intervention, students may become aware of the need for conservation. Chapter 33, "Human Population Concerns," uses the latest data and includes two related topics, energy sources and food production.

Comparison of Texts

Biology: Evolution, Diversity, and the Environment

As many may know, I am also the author of a new general biology text, *Biology: Evolution, Diversity, and the Environment*. This text approaches the study of life from the perspective of the three themes in the title—evolution, diversity, and environment. In contrast to *Inquiry, Biology* explains general principles primarily in reference to forms of life other than the human form and secondarily in reference to the human organism. For example, the animal physiology section emphasizes comparative animal physiology in an evolutionary context. Even so, the discussion includes all aspects of human physiology. Ecology is a constant theme that runs throughout the text. Several readings highlight various modern ecological problems and they, along with suggested solutions, are also discussed at length in the last few chapters.

Inquiry into Life

Inquiry into Life, as has been discussed, focuses on human biology. The human biology focus is an emphasis that immediately appeals to many students. General principles are explained primarily in reference to the human organism and secondarily in reference to other forms of life. A significant portion of *Inquiry into Life* is devoted to human physiology. The text also includes plant and animal diversity and ecological principles and their application to modern ecological problems.

Both *Inquiry into Life* and *Biology: Evolution, Diversity, and the Environment* cover all aspects of general biology, giving nonmajors a review of the entire field of biology and majors a firm basis on which to build. The availability of both of these texts gives instructors a choice of emphasis, according to their own preference and knowledge of the students at their particular institution.

Aids to the Reader

Inquiry into Life includes a number of aids that have helped students study biology successfully and enjoyably.

Text Introduction

The introductory chapter discusses the characteristics of life and presents an overview of the book. This chapter surveys the field of biology as a whole and prepares the reader for an examination of its individual portions.

Part Introductions

An introduction for each part highlights the central ideas of that part and specifically tells the reader how the topics within each part contribute to biological knowledge.

Chapter Concepts

Each chapter begins with a list of concepts stressed in the chapter. This listing introduces the reader to the chapter by organizing its content into a few meaningful sentences. The concepts provide a framework for the content of each chapter.

Boldfaced Words

New terms appear in boldface print as they are introduced within the text and are immediately defined in context. If any of these terms are reintroduced in later chapters, they are italicized. Many of these terms are also defined in the glossary, where a phonetic pronunciation is also given.

Readings

Two types of boxed readings are included in the text. Readings chosen from popular magazines illustrate the applications of concepts to modern concerns. These spark interest by illustrating that biology is an important part of everyday life. The second type of reading is designed to expand, in an interesting way, on the core information presented in each chapter. Topics such as plant growth regulators and U.S. population problems are addressed in these readings.

Chapter Summaries

Chapter summaries offer a concise review of material in each chapter. Students may read them before beginning the chapter to preview the topics of importance, and they may also use them to refresh their memories after they have a firm grasp of the concepts presented in each chapter.

Study Questions

Study questions at the end of each chapter allow students to test their knowledge of the material presented in the chapter. Following each question is the page number where the answer can be found.

Selected Key Terms

Important terms and their phonetic pronunciation now appear at the end of each chapter. These terms represent major ideas in the chapter and can serve as a mechanism for review.

Further Readings

For those readers who would like more information about a particular topic or are seeking references for a research paper, the list of articles and books under further readings will help them get started. For the most part, the entries are *Scientific American* articles and specialty books that expand on the topics covered in the chapter.

Tables and Illustrations

Numerous tables and illustrations appear in each chapter and are placed near their related textual discussion. The tables clarify complex ideas and summarize sections of the narrative. Once students have achieved an understanding of the subject matter by examining the chapter concepts and the text, these tables can be used as an important review tool. The photographs and drawings have been selectively chosen and designed to help students visualize structures and processes.

Appendix and Glossary

The appendix contains optional information for student referral. It includes a discussion of the light microscope versus the electron microscope, an expanded Periodic Table of the Elements, and a review of the metric system. An important part of the appendix is the Classification System of Organisms used in this text.

The glossary defines the terms most necessary for making the study of biology successful. By using this tool, students can review the definitions of the most frequently used terms.

Additional Aids

Student Study Guide

To ensure close coordination with the text, the author wrote the *Student Study Guide* that accompanies this text. Each text chapter has a corresponding study guide chapter that includes a listing of behavioral objectives, a pretest, study exercises, and a posttest. Answers to study guide questions appear at the end of each study guide chapter, giving students immediate feedback.

Instructor's Manual

The *Instructor's Manual* is designed to assist instructors as they plan and prepare for classes using *Inquiry into Life*. Possible course organizations for semester and quarter systems are suggested, along with alternate suggestions for sequencing the chapters. An outline and a general discussion are provided for each chapter; together these give the overall rationale for the chapter. A large number of objective test questions and several essay questions are provided for each chapter. A list of suggested films for the various topics and a list of film suppliers are included at the end of the *Instructor's Manual*.

Laboratory Manual

The author also wrote the *Laboratory Manual* that accompanies *Inquiry into Life*. With few exceptions, each chapter in the text has an accompanying laboratory exercise in the manual (some chapters have more than one accompanying exercise). In this way, instructors will be better able to emphasize particular portions of the curriculum if they wish. The thirty-three laboratory sessions in the manual were designed to further help students appreciate the scientific method and learn the fundamental concepts of biology and the specific content of each chapter. All exercises have been tested for student interest, preparation time, and feasibility.

Acknowledgments

Many instructors have contributed not only to this edition of *Inquiry* but to previous editions. The author is extremely thankful to each one, for we have all worked diligently to remain true to our calling and to provide a product that will be the most useful to our students.

In particular, it seems proper to acknowledge the help of the following individuals:

Warren Smith, *Central State University,* Edmund, OK;
Roy B. Clarkson, *West Virginia University,* Morgantown, WV;
Russell Davis, *University of Arizona,* Tucson, AZ;
Ken Abbott, *Yavapai College,* Prescott, AZ;
Steven N. Murray, *California State University,* Fullerton, CA;
Mark Levinthal, *Purdue University;*
Kathy Burt-Utley, *University of New Orleans;*
Joyce Maxwell, *California State University* at Northridge

These people contributed valuable suggestions for this edition by responding to our telephone survey, and we gratefully acknowledge their help:

David Sulter, *College of the Desert,* CA; Don Collier, *John C. Calhoun State Community College,* AL; J. Rosko, *St. Thomas Aquinas College,* NY; Jerry Henderson, *St. Louis Community College,* MO; Ann Robinson, *Carnegie-Mellon University,* PA; John Biehl, *Riverside City College,* CA; Lynne Osborn, *Middlesex Community College,* MA; Gary Fungle, *Santa Barbara City College,* CA; James McIver, *Idaho State College,* ID; Shirley Crawford, *SUNY Agricultural & Technical College,* NY; Gordon Ashcroft, *San Juan College;* Ellen McGloflin, *Sanford University,* CA;

William Carden, *Grossmont College,* CA; Ted Sherill, *Eastfield College,* TX; Jean Monier, *College of Notre Dame,* MD; Roger McPherson, *Clarion University of Pennsylvania;* Maurice Sweatt, *Canada College,* CA; John Sharp, *John Tyler Community College,* VA; Dick Birkholz, *Sheridan College;* Gary Wrinkle, *West Los Angeles College,* CA; Milton Nathanson, *CUNY Queens College,* NY; Donald Butler, *Duquesne University,* PA; Glen Drews, *Chaffey College;* George DeHullu, *Northern Essex Community College,* MA; Barbara Clarke, *American University,* DC; William Whittaker, *Greenville Technical College,* SC; William Carr, *Alexander City State Junior College;* Doris Tingle, *North Greenville College,* SC; Jim Royce, *Iowa Central Community College,* IA; Kathy Fergen, *Miami-Dade Community College,* FL; Fred First, *Wytheville Community College,* VA; Patrick Daley, *Lewis & Clark Community College,* IL; William Neilson, *Harrisburg Area Community College,* PA; George Mason, *East Central Junior College;* James Kane, *Muskegon Community College,* MI; Carlos Estol, *New York University;* Billy Williams, *Dyersburg State Community College;* Jim Milek, *Casper College;* Michael Bell, *Richland College,* TX; George Washington, *Jackson State University,* MS; Al Chiscon, *Purdue University,* IN; Jean Shuemaker, *Gallaudet College,* DC; Judith Slon, *Hilbert College,* NY; Barbara Berkley, *St. Paul's College,* VA; Michael Willig, *Texas Tech University;* Carol Gerding, *Cuyahoga Community College,* West Campus; Sheila Brown, *University of S. Mississippi;* Eldon Carins, *Auburn University at Montgomery;* David Cherney, *Azusa Pacific College;* Frank Johnson, *Massasoit Community College,* MA; Sadako Houghten, *Los Angeles Pierce College,* CA; Virginia Teagarden, *Lees-McRae College;* Joseph Cancannon, *St. Johns University,* NY; Neil Johnson, *N. Dakota State School of Science;* Marie Fitzgerald, *Holyoke Community College,* MA; James Sandoval, *LA City College;* Benjamin Lowenhaupt, *Edinboro State College,* PA; William R. Hawkins, *Mt. St. Antonio College,* CA; Stephen Wheeler, *Alvin Junior College,* TX; Laird Hartman, *University of South Dakota-Springfield;* Hossam A. Negm, *Grambling State University,* LA; Robert Schodorf, *Lake Michigan College,* MI; Charles Cottingham, *South Carolina State College,* SC; Madhu Narayanan Mahadeva, *University of Wisconsin-Oshkosh;* Wesley Thompson, *Rio Hondo College,* CA; Robert Neher, *University of LaVerne,* CA; John M. Chapin, *St. Petersburgh Junior College,* FL; Hessel Bouma III, *Calvin College,* MI; Ivan Huber, *Farleigh Dickenson University-Madison,* NJ; W. Brooke Yeager, *Luzerne Community College,* PA; A. Floyd Scott, *Austin Peay State University,* TN; Jack Dennis McCullough, *Stephen F. Austin State University,* TX; Harry S. McDonald, *Stephen F. Austin State University,* TX; John Hoagland, *Danville Junior College,* IL; Edwin Lane Netherland, *Cameron University,* OK; Arthur A. Cohen, *Massachusetts Bay Community College,* MA; Karen Belcher, *Lake City Community College,* FL; Vincent A. Kissel, *CUNY-Bronx Community College,* NY; Ben Van Wagner, *Bethany Bible College,* CA; Edward Lucier, *Becker Junior College,* MA; George Klee, *Kent State University-Stark,* OH; Mario A. Vecchiarelli, *Housatonic Regional Community College,* CT; Robert Dahm, *William Wright College;* Mary Lou Longo, *American International College,* MA; Lewis Grey, *Truman City College-Chicago;* Jerry A. Clonts, *Anderson College,* SC; James Smith, *Lawson State Community College,* AL

The personnel at Wm. C. Brown Publishers have always lent their talents to the success of *Inquiry into Life,* and I especially want to thank John Stout, biology editor, for guiding the book from its inception to its completion. During production, my illustrators, Kathleen Hagelston and Anne Greene, not only drew but helped develop many fine drawings, and Faye Schilling selected just the right photographs. Finally, my secretarial assistant, Adelle Robinson, helped with many tedious tasks, and Gene Collins, production editor, and Geri Wolfe, designer, orchestrated the input of all.

As always, I am thankful for the patience of my family and friends. Preparing the manuscript would have been most difficult without the constant encouragement of my sister, Rhetta McMeans, and the steadfast understanding and helpfulness of my children, Karen and Eric.

Inquiry into Life

Introductory concepts

1 Although life is difficult to define, it can be recognized by certain common characteristics.

2 There are various approaches to the study of biology. (The approach chosen in this text stresses human biology.)

3 The information in this book has been gathered by using the scientific method, a process based on certain identifiable procedures.

Introduction

Figure I.1

Diversity of life. *a.* These hippos are engaged in a territorial battle. When they open wide, oversize canine teeth up to two feet long are exposed. The ensuing combat lasts until one gives way, usually without harm to either one. *b.* Goldfinch feeding on a thistle while another approaches. Both types of organisms are interacting with the environment in numerous ways.

a.

b.

This book is about living things (fig. I.1) and therefore it is appropriate to first define life. Unfortunately this is not so easily done—life cannot be given a simple one-line definition. Since this is the case, it is customary to discuss life in terms of its characteristics; the following five characteristics are commonly attributed to living things:

1. Living things have a structure that is ultimately made up of cells.
2. Living things grow and maintain their structure by taking chemicals and energy from the environment.
3. Living things respond to the external environment.
4. Living things reproduce and pass on their organization to their offspring.
5. Living things evolve and adapt to the environment.

Living things have a structure that is ultimately made up of cells. Figure I.2 shows that both plants and animals are organisms, which are composed of organs, which in turn are composed of tissues, and tissues are composed of cells. Cells are made up of chemicals (molecules), which are nonliving substances that contain atoms, the smallest units of matter nondivisible by chemical means.

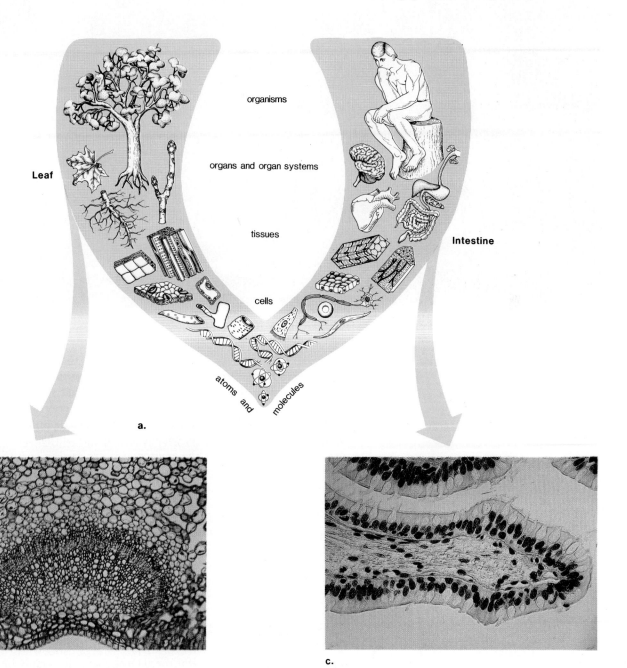

Figure I.2
Organization of plants and animals. *a.* Cells are composed of chemicals; tissues are made up of cells; organs are composed of tissues; and organisms contain organs. *b.* Cross section of plant (privet) leaf showing individual cells. *c.* Longitudinal section of villi from small intestine showing individual cells. Some are goblet cells which are discharging mucus into the gut cavity.

Leaf

organisms

organs and organ systems

tissues

Intestine

cells

atoms and molecules

a.

b.

c.

Figure I.3
Grasshopper feeding from a flower of the mimosa
family in the state of Veracruz, Mexico. Plants carry
on photosynthesis, a process which provides them
with organic nutrients, but animals must either eat
plants or other animals to acquire nutrients.

*Living things grow and maintain their structure by taking chemicals
and energy from the environment.* Only plants and plantlike organisms are
capable of utilizing inorganic chemicals and the energy of the sun in the process of making their food. Other organisms, such as animals, must take in
preformed food as a source of chemicals and energy (fig. I.3). The chemicals
and energy obtained from the environment are used in part to maintain the
organization of living organisms. Each living organism has its own particular
organization that is maintained only because energy can be acquired from the
environment.

Living things respond to the external environment. This characteristic
is commonly recognized by people who acknowledge a living thing by its ability
to move. A multicellular animal can move because it possesses a nervous system,
but other living things, including plants, possess a variety of mechanisms by
which they respond to the physical and biological environment. Responses to
the environment constitute the behavior of organisms. The reading on page 8
discusses the behavior of the hornbill, a bird in the tropical forests of Africa
and Southeast Asia, and of the behavior of people that interact with this bird.

*Living things reproduce and pass on their organization to their off-
spring.* A unicellular organism reproduces asexually simply by dividing into
two new organisms. Because the new organisms have the same hereditary factors, or genes, as the original organism, they also possess the same structure
and function. In contrast, multicellular organisms often reproduce by means
of sexual reproduction. Male and female organisms each contribute one-half
the total number of genes to the new organism, which often resembles the
parents even as it is maturing (fig. I.4).

Figure I.4
Offspring resemble their parents because of their genetic inheritance.
a. These Emperor penguins breed on floating ice in Antarctica. After the female lays one egg in the winter, she passes it to the male who keeps it warm by draping a fold of skin over it while carrying it on his feet. The chick hatches in time to feed on the plentiful food of the short summer. *b*. Aside from their biological inheritance, humans also have a unique cultural inheritance. Culture is a set of ideas that determines how the person views and relates to the world.

a.

b.

Hornbill

The beak is the hornbill's most distinctive attribute. Superficially, it resembles that of a New World toucan, though the birds are not related. There are at least 44 species of hornbills, which range from Africa to the Solomon Islands, and many of them sport a peculiar protrusion atop the beak. It's called a casque, and in some forms, it sticks up like a 1959 DeSoto tail fin with an overdose of hormones. Much lighter in weight than it looks, it is a horny shell supported internally by spongelike tissue. . . .

The helmeted hornbill is the only hornbill with a solid casque, and this feature has caused it no end of trouble. Malays of Perak State make poison-detecting spoons out of its casque, the idea being that the spoon will cause a lethal brew to bubble and simmer. Across the Malay Peninsula, traditional healers of Kelantan State mix the casque with elephant tusk, dugong bone and white hibiscus to make an antidote for cyanide of potassium poisoning. The casque of this magnificent four-foot bird also has long been sought by Chinese artisans who carve the precious substance called *ho-ting* into belt buckles, plume holders, decorative miniatures and snuff bottles. Centuries ago, Chinese traders traveled to Borneo and exchanged glass beads and iron tools for the golden orange casque, which was valued more highly than jade. Like all hornbills, this species is still intensively hunted wherever it is found, and its prospects are closely linked to the future of the tropical rain forests—which are being cut and burned at a rapid rate.

While the Chinese were trading trinkets for the helmeted hornbill, the Ibans of Borneo were worshiping its relative, the rhinoceros hornbill. This is the sacred Iban bird of war, the *kenyalang,* which is perhaps the most glorious hornbill of them all with its huge, upturned casque the size of a banana. Birds are important omens throughout Borneo, but this impressive creature, which is one of the eight hornbill species found on the islands, is the only one that has a festival named after it: the *gawai kenyalang.* . . .

The hornbill is at its noisiest when flying; it has air spaces at the bases of the primary wing quills, which produce a bellowslike sound while flapping through the air. All hornbill species have distinctive calls, but the most interesting is that of the male helmeted hornbill. Local myth suggests that its call replicates the sound of an angry son-in-law chopping down his father-in-law's house. It starts softly with a series of "tooks," builds in tempo and volume over ten minutes or more, and ends in an amplified climax of frenzied laughter which defies description.

If you were to follow the call of a hornbill, you might notice orangutans, the great apes of Borneo and Sumatra, doing the same. Jan Wind, an ornithologist with the World Wildlife Fund in Bogor, Indonesia, says that hornbills and orangutans both favor large fruits, particularly figs, and often the red apes follow flocks of up to 50 of the big birds in search of fruiting trees. But fruit isn't the only component of the hornbill's menu. Some species, particularly the ground hornbills of Africa, enjoy insects, lizards and snakes, along with an occasional bird.

Noisy as the hornbill is in its daily activities, it draws a curtain of privacy around its unusual nesting habits. A naturalist friend at Jakarta's Ragunan Zoo told me how, in many species, the female imprisons herself during the four- to six-week incubation period and subsequent six-week rearing period. She first seeks out a large, hollow tree, dropping sticks and debris into the cavity to raise its floor to a comfortable level. The female hornbill then spends several days plastering up the entrance with her own droppings, which harden into an impenetrable claylike substance. The male helps from the outside, using the flattened sides of his bill like a trowel until only a narrow slit remains.

Thus protected from predators, the female will stay in this cell while her mate feeds her fruits. One observer recorded 18 visits by the male in ten hours, and another bird-watcher, no doubt bleary-eyed and semicatatonic from looking up a tree for 3½ months, recorded that

one male brought 24,000 fruits to his mate during that time. A male hornbill takes his responsibility seriously. One bird was seen repeatedly bringing his imprisoned consort flowers, which are not part of the hornbill diet, but certainly add a nice touch.

To catch a female hornbill and her young, all you need to do is find a nest and haul them out, since the female quickly molts and becomes flightless and very fat during part of her imprisonment. Wallace, the British naturalist who, together with Darwin, helped formulate the theory of evolution, wrote of coming across a hornbill nest in Sumatra: "After a while, we heard the harsh cry of a bird inside. . . . I offered a rupee to anyone who would go up and get out the bird . . . but they all declared it was too difficult, and they were afraid to try. . . . In about an hour afterward, much to my surprise, a tremendous loud, hoarse screaming was heard and the bird was brought to me, together with a young one which had been found in the hole. This was a most curious object, as large as a pigeon, but without a particle of plumage on any part of it. It was exceedingly plump and soft, and with a semitransparent skin, so that it looked more like a bag of jelly, with head and feet stuck on, than like a real bird."

The adults are equally intriguing. One afternoon at a zoo, I was entranced by a rufous hornbill from the Philippines. It hopped around with both feet leaving the ground at the same time. Although it is dangerous to anthropomorphize, I got the feeling that the bird knew just how silly it looked, and tried to make *me* feel bad by staring me down with its big eyes.

Suddenly, it made an astonishingly delicate maneuver. It bent down to pick up a berry, but it couldn't simply swallow the fruit, since, as in all hornbills, its tongue is shorter than its beak. It had to flip up its food like a piece of popcorn and gulp it down. Then the bird made a most unusual burp—a cross between the scream of a gagged mezzo-soprano in distress and the not-very-polite sound of an obese diner in urgent need of Alka-Seltzer. . . .

A hornbill feeding with its ponderous saw-toothed beak. Since the tongue is smaller than the beak, the hornbill has to flip the food in before it can gulp it down.

The rhinoceros hornbill is a war god in Borneo. As its tropical forest habitat dwindles, its chances for survival are not good.

Figure I.5
Extinct saber-toothed cat. Some believe this animal became extinct because its canine teeth became too long to use. But its real problem seems to have been a lack of speed. When its large, slow-moving prey became extinct, it could not adjust to catching smaller, more speedy prey.

Figure I.6
Fishing bat. Perfectly adapted for nighttime hunting, this fishing bat locates its prey with sonar and catches it with oversized claws. The bones of a bat's wing are essentially the same as those in the human arm and hand, but the "fingers" are elongated and a double membrane stretches between them. Many people don't realize that bats rarely cause any harm to humans and feed on a variety of substances. Only one-third of 1 percent drink blood.

Living things evolve, or change, and thereby become adapted to the environment. While the preceding characteristics of life pertain to individual living things, this fifth characteristic is concerned with the **species,** a group of similarly constructed organisms that share common genes. It is the species, rather than the individual, that evolves. Evolution begins when certain organisms happen to inherit a genetic change that causes them to be better suited to a particular environment. These organisms, which are said to be better adapted, tend to survive and have more offspring than those that are not as well suited. In this way, evolution produces successive generations of organisms that are better adapted to the environment. If the environment should change, they may no longer possess the genetic capability of adapting to the new environment, and extinction can follow (fig. I.5).

The evolutionary process causes life to have a history. The belief today is that a chemical evolution produced the first cell or cells and from these all other forms of life have evolved. There is a variety of life forms on this planet because they are adapted to various ways of living in specific environments (fig. I.6).

Inquiry into Life

There are many different ways to approach the study of life. This text, while covering all aspects of biology, focuses on human biology. For example, human anatomy and physiology is studied as representing vertebrate anatomy and physiology; and the chief environment of humans, the country and city, is studied as an example of an ecosystem.

A brief introduction to the many topics discussed in the six units of this text follows.

Introduction

The Cell, a Unit of Life

Multicellular organisms, including humans, are composed of many cells (fig. I.2). In order to understand how multicellular organisms function, it is necessary to understand how the cell functions. Since cells are made up of chemicals, we begin our study by considering some basic chemistry essential to the cell. Our study of cells also includes how cells grow and reproduce, two vital functions in the life of the cell.

Plant Biology

There are two major types of higher organisms—plants and animals. Flowering plants are the most recently evolved of the plants, and they serve as our basis for the study of plant biology in this unit. Plant cells carry on photosynthesis, the process by which they make their own organic food after capturing energy from sunlight (fig. I.7). Photosynthesis is extremely important because it ultimately provides food for all living things.

Human Physiology

As in our study of plant biology, our study of animal biology centers on the most recently evolved animal—humans. The human body is composed of many systems, each designed to perform a particular life function (table I.1). The unit on human physiology discusses the anatomy and, in particular, the physiology of each of these systems while it also highlights related areas of interest, such as dieting, drugs, smoking, and important illnesses. It is hoped that the knowledge you gain about the human body will assist you in understanding the workings of your own body.

Reproduction, Development, and Inheritance

The unit on human reproduction and inheritance explores topics of extreme interest to young people, who are just beginning the reproductive years of their lives. The anatomy and physiology of the reproductive system is considered, as well as birth control, infertility, and sexually transmitted diseases. The stages of development are reviewed, giving you an opportunity to see how a fertilized egg becomes the newborn infant. Inheritance is an extremely important topic today for several reasons. Much more is known today about human genetic diseases than was known even as little as a decade ago. Science has made it possible to sometimes predict the chances of a child being born with a genetic disease. Also, we are entering an era of genetic engineering, that may make it possible to correct genetic abnormalities.

Figure I.7
Sunlight streaming through the trees of a forest. Plants use the energy of the sun to carry on photosynthesis, a process that produces food not only for themselves but for most living things.

Table I.1	Animal organ systems
Name	**Function**
Digestive	Convert food particles to nutrient molecules
Circulatory	Transport of molecules to and from cells
Immune	Defense against disease
Respiratory	Exchange of gases with environment
Excretory	Elimination of metabolic wastes
Nervous and sensory	Regulation of systems and response to environment
Muscular and skeletal	Support and movement of organism
Hormonal	Regulation of internal environment

Name of Kingdom	Representative Organisms	
	Drawings	Descriptions
Monera		Bacteria and cyanobacteria
Protista		Protozoans, unicellular algae of various types
Fungi		Molds and mushrooms
Plants		Green algae, mosses, ferns, various trees and flowering plants
Animals		Sponges, worms, insects, fishes, amphibians, reptiles, birds and mammals

Evolution and Diversity

After an introductory chapter that presents the principles of evolution, the unit on evolution and diversity surveys living things from the origin of life to human evolution. Particularly today, it is important for you to become acquainted with the diversity of life not only because it illustrates our relationship with other living things but because it may also enhance your awareness of the need to preserve and protect all forms of life.

Biologists classify living things according to their evolutionary relationships. Since these relationships are not known exactly, various classification systems are in use. This text classifies organisms into five major groups, or kingdoms, as outlined in figure I.8.

Behavior and Ecology

The behavior of organisms can be studied just as their anatomy and physiology can be. In fact, much of an organism's behavior can be attributed to its genetic inheritance. Certain behavior patterns are particularly applicable to the study of ecology, which is defined as the interactions of organisms with each other and with the physical environment. In many cases, humans have drastically altered the environment (fig. I.9) and are just now coming to realize their potential for destroying nature.

The future existence of human beings is dependent on preserving the natural world, and it is the goal of this unit to make you aware of this dependence and what should be done to protect the balance of nature. The last chapter in this unit on behavior and ecology considers the growth and size of the human population. Only humans have increased their number to the extent that they dominate the planet. Our use of fossil fuel energy (coal, natural gas, petroleum), especially as it is used to grow food, has made this inordinate increase possible. This last chapter in the book also considers energy resources and food production problems.

a.

b.

Figure I.9
Natural area versus one developed by humans. Humans need both types of places. *a.* They often benefit psychologically from visiting natural areas that provide a home for many plants and animals. This is a pond in Maine in the fall. *b.* Humans carry on most of their activities in developed areas. This aerial view of a Miami beach illustrates the degree to which humans can profoundly affect the nature of the environment.

Scientific Knowledge

The information presented in this book was acquired by utilizing the scientific method. This method is explained on page 14. As you read the text, keep in mind the mechanisms by which the information presented here has been gathered and also keep in mind the limitations of science. Science deals with supported hypotheses, not absolute truths; therefore its conclusions are subject to change. Also, it is not the role of science to answer ethical questions and to dictate the direction of the future. While scientists can inform and even recommend, it is up to individuals to decide to what degree and in what manner the conclusions of science should be implemented. For this reason it is extremely important for you to possess a knowledge of human biology, a knowledge that will allow you to contribute intelligently to shaping the future destiny of human beings.

The Scientific Method

The scientific method is as varied as scientists themselves, but even so there are certain processes that are typically characteristic of science. First of all, scientists seek to understand the material world (that which can be observed). When doing this, scientists ask only **causality questions**, such as what caused this or how does this occur, rather than **teleological questions** that ask for what purpose something occurs. For example, a scientist might address the question, What causes this particular illness? but would not ask the more philosophical question, Why should humans get sick?

Scientists answer questions by collecting data, or evidence, using simple **observation** or carefully conducted **experiments.** The observations and experiments must be repeatable; that is, others must report the same observations under the same circumstances. Otherwise, the observations and experiments are considered invalid.

Observations permit scientists to formulate **hypotheses**, tentative explanations of observed phenomena. To arrive at a hypothesis, scientists use various methods of reasoning, especially inductive reasoning. **Inductive reasoning** allows scientists to arrive at a generalization after observing specific facts. For example, in 1976, delegates to an American Legion convention in Philadelphia became ill with what came to be called Legionnaire's disease. After observing that all ill persons had spent time in the same vicinity and that their symptoms included fever and pneumonia, certain scientists suggested that the causative agent was an infectious airborne organism.

After a **hypothesis** such as this has been stated, a second type of reasoning, **deductive reasoning**, comes into play. Deductive reasoning begins with a general statement that infers a specific conclusion. It often takes the form of an if . . . then statement. In science, deductive reasoning allows a scientist to determine the type of experiment and/or observation necessary to support or refute a hypothesis. For example, it follows that *if* Legionnaire's disease is caused by an infectious organism, *then* it should be possible to isolate and identify the organism. Scientists made this deduction, and they began to examine specimens from all ill persons. They also inoculated guinea pigs with tissues obtained from persons who had died from the condition.

Summary

All living things display certain characteristics: they are made up of cells, maintain their structure by taking chemicals and energy from the environment, respond to external stimuli, and reproduce. As species, living things evolve and change. Evolution accounts for the diversity of life we see about us.

This text, which covers the whole scope of biology, the study of life, has a human orientation. Humans are made up of cells, and the first unit of the book covers cell biology. Because flowering plants are the most advanced of the plants and provide most of the food for all other living things, the unit on plant biology is based on the flowering plant. Human beings are the most advanced of the animals and a significant portion of this book is concerned with human organ systems. The knowledge gained should help you understand your own body. Unit 5 is about evolution and diversity. Humans can only understand their place in nature by studying all living things. The final unit of the book concerns behavior and ecology. Only when we realize our dependence on the natural world will we be able to understand the importance of trying to preserve it.

Science does not answer ethical questions; we must do this for ourselves. Knowledge provided by science can assist you in making decisions that will influence the destiny of human beings and other living things.

Just as in this case, scientists may first make observations outside the laboratory and then proceed to perform experiments inside the laboratory, where conditions can be controlled. The laboratory environment protected the guinea pigs from exposure to other infective agents, for example. To ensure that illness in the guinea pigs was due to the inoculum and not some other factor involved in the experiment, scientists devised a **control sample** (a sample that undergoes all the steps in the experiment except the one being tested). The control sample consisted of a group of guinea pigs that received treatment identical to the others except that the inoculum given them did not contain the tissue in question. In this way, when only the experimental group became ill, it allowed scientists to conclude that the illness was indeed caused by the inoculum.

As it happened, when scientists observed the spleen of the experimental guinea pigs, they found small rod-shaped organisms that were able to grow outside living things on specially prepared media suitable for bacteria. In this example, the **data** (pertinent facts) collected does not lend itself to mathematical interpretation, except perhaps if it were found that only a certain percentage of the guinea pigs had infected spleens. In many instances, however, scientists rely greatly on measurements and mathematical data. *Mathematical data* is highly desirable because it helps define variables (components of experiments) and enables scientists to see relationships that might not otherwise be obvious.

The data collected from experiments either supports or fails to support the hypothesis. However, data that supports the hypothesis does not prove the hypothesis true. After all, it is possible that the original hypothesis was misleading or that the observations were inconclusive. While a hypothesis can never be proven true, it can sometimes be proven false. Suppose that a scientist had formulated the hypothesis that Legionnaire's disease could be contracted only by Orientals. Obviously the first Caucasian to contract the disease would prove the hypothesis false.

If the hypothesis is constantly supported, the confidence of certainty becomes greater and scientists may then present the conclusions as if they are factual. For example, Legionnaire's disease has been given the name Legionellosis, and it is stated that this disease is caused by a bacteria given the name *L. pneumophila*. Dealing with conclusions as if they are factual facilitates communication. However, scientists are always aware that the present body of information represents the truest available at the moment and that further observations and experiments could lead to changes in this information.

The ultimate goal of science is to understand the natural world in terms of **concepts,** interpretations that take into account the results of many experiments and many observations. For example, the theory of evolution is one such conceptual scheme. It allows scientists to understand the history of life, the variety of living things, the anatomy and physiology of organisms, embryological development, and so forth. When the designation **theory** is used in science it means that scientists have the utmost confidence in a concept whose broad scope gives it fundamental importance. This is contrary to the way in which the word *theory* is used in every day language.

Study Questions

1. Name the five characteristics of life and discuss each one. (pp. 4–10)
2. Support the statement "All living things are organized." (p. 4)
3. Food provides what two necessities for living things? (p. 6)
4. Explain the process by which living things become increasingly more adapted to the environment. (p. 10)
5. Give two reasons why humans should study the biology of other living things. (p. 12)

Further Readings

Asimov, I. 1960. *Wellsprings of life*. New York: Abelard-Schuman.
Baker, J., and Allen, G. 1971. *Hypothesis, prediction, and implication in biology*. Reading, Mass.: Addison-Wesley.
Grobstein, C. 1965. *The strategy of life*. San Francisco: W. H. Freeman.
Luria, S. E. 1973. *Life: The unfinished experiment*. New York: Charles Scribner's Sons.
Szent-Gyorgyi, A. 1972. *The living state*. New York: Academic Press.

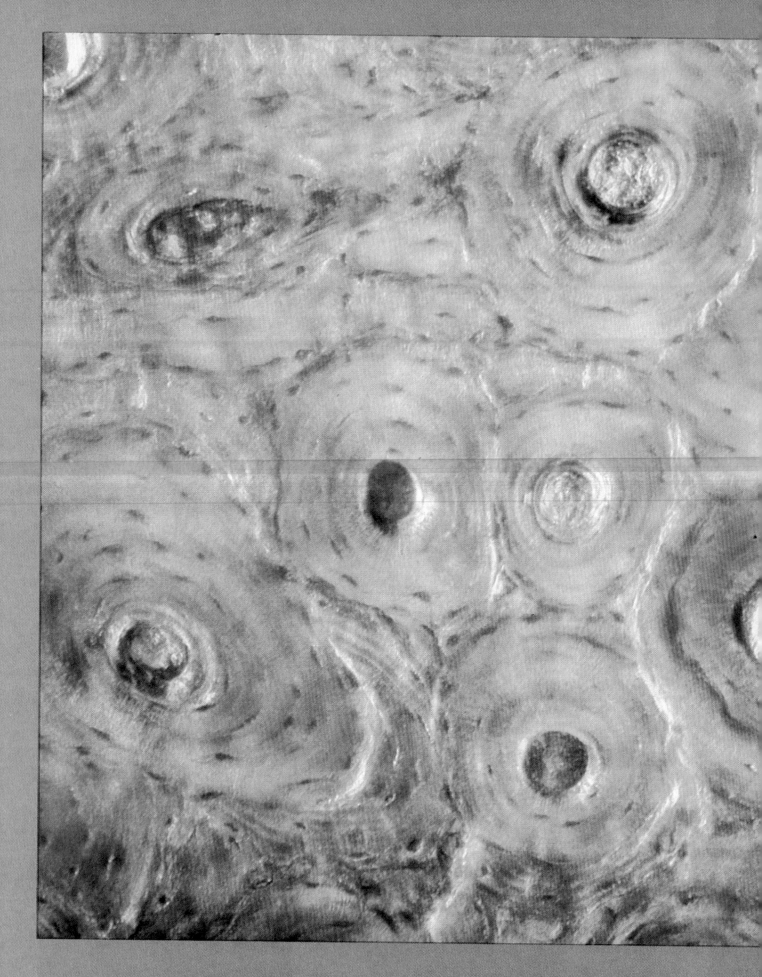

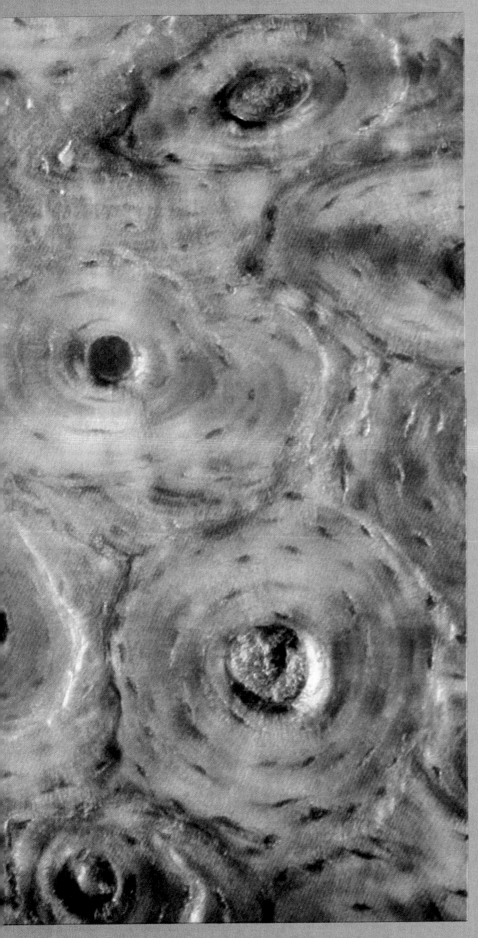

The Cell,
A Unit of Life

The cell is the smallest of living things, and all the characteristics of life are found here. An understanding of cell structure, physiology, and biochemistry serves as a foundation to an understanding of multicellular forms.

Part 1 studies each aspect of cellular biology in detail and thereby covers the fundamental concepts of biology.

Principles of inorganic and organic chemistry are discussed before a study of cell structure is undertaken. The cell is bounded by a membrane and contains organelles, many of which are also membranous. It is membrane that regulates the entrance and exit of molecules and determines how cellular organelles carry out their functions.

Cell reproduction is dependent upon mitotic cell division; whereas animal and plant reproduction are dependent upon meiotic cell division. Both types of cell division are introduced in part 1. Cells require energy for growth and reproduction, and the biochemical means by which this energy is provided is considered in chapter 5.

1

Chemistry and Life

Chapter concepts

1 All living organisms are composed only of inorganic and organic chemicals.

2 Atoms, the smallest units of matter, react with one another to form chemicals.

3 Some important inorganic chemicals in living organisms are acids, bases, and salts.

4 Some important organic molecules in living organisms are proteins, carbohydrates, fats, and nucleic acids, each of which is composed of smaller molecules joined together.

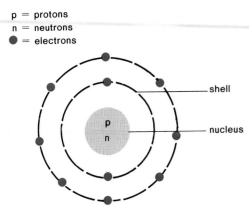

p = protons
n = neutrons
● = electrons

shell

nucleus

Figure 1.1.
Representation of an atom. The nucleus contains protons and neutrons; the shells contain electrons. The first shell is complete with two electrons, and every shell thereafter may contain as many as eight electrons.

It is not easy to accept and understand that living things are chemical and physical machines, especially since it is not possible to see the chemicals that make up an organism's body. However, a few minutes' reflection regarding modern advances in biology and medicine usually convinces one that humans are indeed made of chemicals. Various genetic diseases, some of which cause an early death if untreated, can now be chemically controlled. Drugs are often given to correct abnormal physical conditions: adrenalin is given for a heart attack; tranquilizers reduce nervous tension; and iron is used in the treatment of anemia. Good nutrition is based on our knowledge of the everyday chemical requirements necessary to keep the body in good running order. For example, the reading on page 40 suggests that dietary calcium may be important in preventing heart disease, especially in persons drinking chlorinated water.

Since living things, including humans, are composed only of chemicals, it is absolutely essential for a biology student to have a basic understanding of chemistry.

Atoms

The matter that makes up the substance of living things is composed of chemicals, and chemicals are made of atoms. The word **atom** is merely a term meaning the smallest unit of matter, nondivisible by chemical means. While it is possible to split an atom by physical means, an atom is the smallest unit to enter into chemical reactions.

Investigators tell us that it is permissible to think of an atom (fig. 1.1) as having a center, called the **nucleus,** and energy levels about the nucleus, called **shells.** While several various types of subatomic particles have been discovered, only a few are of interest to us. The subatomic particles called **protons** and **neutrons** are located within the nucleus; the subatomic particles called **electrons** are located in the shells. The shells are actually an approximate path within which the electrons circle or move about the nucleus. It has been shown that electrons have varying amounts of energy; those with the greatest amount of energy are located in shells farthest from the nucleus. Another important feature of protons, neutrons, and electrons is their weight and/or charge, which is indicated in table 1.1.

Table 1.1	Subatomic particles	
Name	**Charge**	**Weight**
Electron	One negative unit	Almost no weight
Proton	One positive unit	One atomic unit
Neutron	No charge	One atomic unit

Figure 1.2

Periodic Table of the Elements (simplified). See the Appendix for a complete table. Each element has an atomic number, atomic symbol, and atomic weight.

I	II	III	IV	V	VI	VII	VIII
1 H hydrogen 1.0							2 He helium 4.0
3 Li lithium 7.0	4 Be beryllium 9.0	5 B boron 11.0	6 C carbon 12.0	7 N nitrogen 14.0	8 O oxygen 16.0	9 F fluorine 19.0	10 Ne neon 20.2
11 Na sodium 23.0	12 Mg magnesium 24.3	13 Al aluminum 27.0	14 Si silicon 28.1	15 P phosphorus 31.0	16 S sulfur 32.1	17 Cl chlorine 35.5	18 Ar argon 40.0
19 K potassium 39.1	20 Ca calcium 40.1						

Atomic Number — *Atomic Symbol* — *Atomic Weight* (labels pointing to element 6, C)

The Periodic Table of the Elements in the Appendix (fig. A.2, p. A-2) shows the 92 naturally occurring atoms. The word **element** refers to a basic substance that makes up the material world. Each type of element is composed of just one kind of atom. Since in the following discussion we will be concerned only with the first few atoms, a simplified Table of the Elements is given in figure 1.2. Notice that in the Periodic Table each specific atom

1. has been given a **symbol**—for example, C = carbon and N = nitrogen;
2. has an **atomic number**—carbon is #6 and nitrogen is #7. *The atomic number equals the number of protons;*
3. has an **atomic weight** or mass—carbon has an atomic weight of 12[1] and nitrogen has an atomic weight of 14. *The atomic weight equals the number of protons plus the number of neutrons* (table 1.1).

With this knowledge in mind, it is now possible to diagram a specific electrically neutral atom. In an **electrically neutral** atom the number of protons (+) is equal to the number of electrons (−). The first shell of any atom contains

[1]Atomic weights are relative weights. The most common isotope of carbon has been assigned a weight of 12, and the other atoms are either lighter or heavier than carbon.

The Cell, A Unit of Life

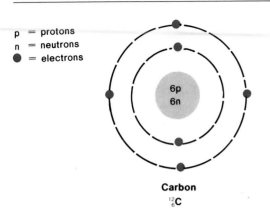

p = protons
n = neutrons
● = electrons

Carbon
$^{12}_{6}$C

Figure 1.3
Carbon atom. The diagram of the atom shows that the number of protons (the atomic number) equals the number of electrons when the atom is electrically neutral. Carbon may also be written in the manner shown below the diagram. The subscript is the atomic number, and the superscript is the weight.

only two electrons; thereafter each shell may contain eight.[2] Thus carbon can be diagrammed as in figure 1.3. The representations of some other atoms can be found in figure 1.4 so that you may study additional configurations.

It is readily apparent that the elements in the Periodic Table are linearly arranged in order of increasing atomic number and weight, but they are also vertically arranged according to similar chemical properties. For example, the atoms in the first column all have one electron in the outermost shell, and we shall see that they give up this electron in chemical reactions.

Isotopes

The atomic weights listed in the Periodic Table are actually the average weight of each type atom. Individual atoms can vary as to weight. When they do, they are called **isotopes** of one another. The isotopes of carbon may be written in the following manner in which the subscript stands for the atomic number and the superscript stands for the weight:

$$^{12}_{6}C \qquad ^{13}_{6}C \qquad ^{14}_{6}C* \qquad ^{15}_{6}C$$

The number of protons in these isotopes does not vary, but the weight does; this indicates that the number of neutrons must be responsible for the weight difference since electrons have almost no weight.

Certain isotopes, called **radioactive** isotopes, are unstable and emit radiation, which may be detected with a Geiger counter. Carbon-14 is radioactive, as the asterisk indicates. Radioactive isotopes are widely used in biological research because it is possible to trace their presence in various chemical substances and tissues.

Reactions between Atoms

Usually reactions between atoms involve the electrons in their outer shell. The **octet rule,** based on chemical findings, states that *atoms react with one another in order to achieve eight electrons in their outer shells.* An exception to this rule occurs when the outer shell is the first shell, which is complete with two electrons only.

---

[2]The maximum number of electrons for any shell except the outer shell is $2n^2$ where n is the shell number. However we will consider only atoms 1–20 in which all shells may have eight electrons except the first shell, which has two electrons.

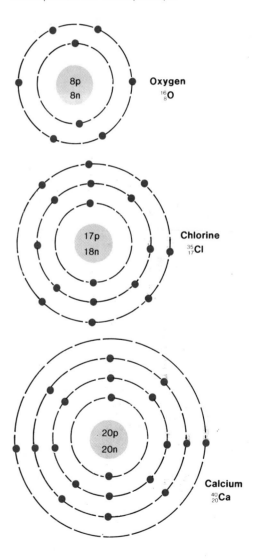

Figure 1.4
Representative atoms. How do you know that these atoms are electrically neutral? The atomic weight of each atom is equal to the number of protons plus the number of neutrons. In the symbols of the side of the figure, what does the subscript stand for? The superscript?

Oxygen
$^{16}_{8}$O

Chlorine
$^{35}_{17}$Cl

Calcium
$^{40}_{20}$Ca

Figure 1.5

In this ionic reaction, chlorine takes an electron from sodium; thereafter each atom has eight electrons in the outer shell.

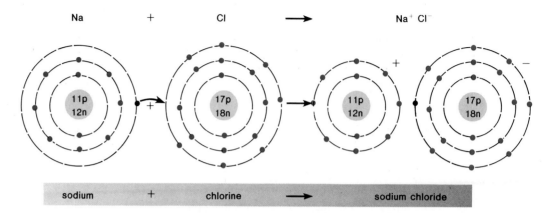

Ionic Reactions

In one type of reaction, atoms give up or take on electrons in order to achieve a completed outer shell. Such atoms, which thereafter carry a charge, are called **ions** and the reaction is called an ionic reaction. In **ionic reactions,** atoms lose or gain electrons to produce a compound that contains ions in a fixed ratio to one another. The **formula** for the compound shows the proper ratio of atoms in the compound.

For example, figure 1.5 depicts a reaction between sodium (Na) and chlorine (Cl) in which chlorine takes an electron from sodium. The resulting ions in the compound sodium chloride, Na^+Cl^-, have eight electrons each in the outer shell. Notice that when sodium gives up an electron, the second shell with eight electrons becomes the completed outer shell. Chlorine, on the other hand, needs an electron to achieve a total of eight electrons in the outer shell.

It is easy to understand the charge of any ion by realizing that while the number of electrons can vary, the number of protons must stay constant. A minus charge indicates that there are more electrons ($-$) than protons ($+$), and a plus charge indicates that there are more protons ($+$) than electrons ($-$) in the ion.

In ionic compounds, the ions are attracted to one another because of their opposite charges. This attraction is referred to as an **ionic bond.** Sodium chloride is a crystalline compound in which the sodium ions and the chlorine ions are stacked up in a characteristic pattern (fig. 1.6).

Ionic bond formation occurs when a **metal** reacts with a **nonmetal.** Metals are those atoms that appear to the left of the dark line in the simplified Periodic Table (fig. 1.2), and nonmetals appear to the right of the dark line. Metals lose electrons and become positively charged. In contrast, nonmetals gain electrons and become negatively charged. Two additional ionic reactions are shown in figure 1.7.

Covalent Reactions

When nonmetals react with nonmetals, a **covalent bond** is formed and the atoms share electrons instead of losing or gaining them. Sharing is usually equal; each atom contributes one electron to each pair that is shared. These electrons spend part of their time in the outer shell of each atom; therefore they may be counted as belonging to both atoms. When this is done, each atom will have eight electrons in the outermost shell.

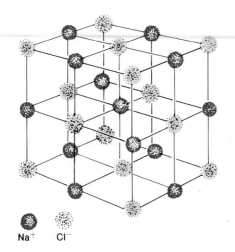

Na^+ Cl^-

Figure 1.7
Additional ionic reactions. In ionic reactions, nonmetals take electrons from metals and, in that way, each atom achieves eight electrons in the outer shell. The unlike charges on the resulting ions produce the ionic bond.

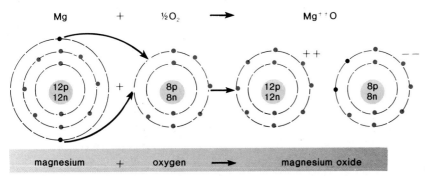

Mg   +   ½O_2 ⟶ $Mg^{++}O$

a. **magnesium + oxygen ⟶ magnesium oxide**

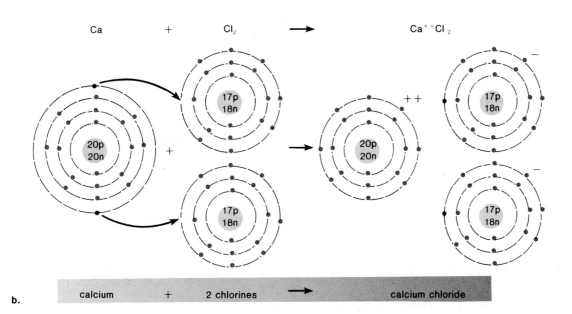

Ca   + Cl_2 ⟶ $Ca^{++}Cl_2$

b. **calcium + 2 chlorines ⟶ calcium chloride**

Figure 1.8

Covalent reaction. Each chlorine atom requires an electron to complete its outer shell. After the covalent reaction, the chlorine atoms share a pair of electrons and in this way, each atom has eight electrons in its outer shell.

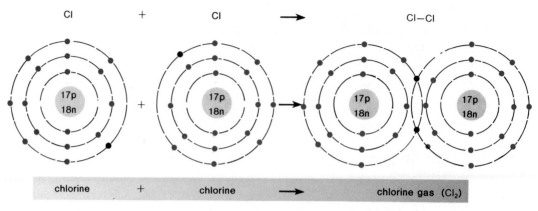

Cl   +   Cl   ⟶   Cl—Cl

chlorine   +   chlorine   ⟶   chlorine gas (Cl_2)

Figure 1.9

Additional covalent reactions. After a covalent reaction, atoms are covalently bonded to each other. In the covalent bond, atoms share electrons; and the shared electrons should be counted as belonging to both atoms.

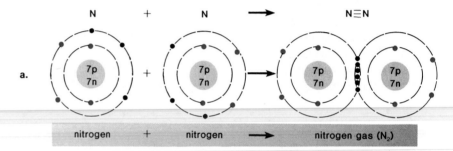

N   +   N   ⟶   N≡N

a.

nitrogen   +   nitrogen   ⟶   nitrogen gas (N_2)

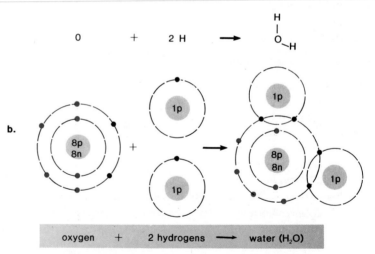

b.

oxygen   +   2 hydrogens   ⟶   water (H_2O)

An example of a reaction between two nonmetals that results in a **molecule** is shown in figure 1.8. The term *molecule* may be used to refer to a chemical in which the atoms are covalently bonded.[3] Notice that the overlapping outer shells in the molecule indicate that the atoms are sharing in order to achieve eight electrons in their outer shells. When atoms share electrons equally, neither atom carries a charge. Two additional examples of covalent reactions are given in figure 1.9.

[3]The term compound refers to a *combination* of two or more different atoms. The term *molecule* can refer to combinations of like or different atoms. A compound also means a large number of atoms held together in fixed proportion, while a molecule is one unit of atoms in correct proportion.

The Cell, A Unit of Life

Since it is inconvenient to show complete diagrams of atoms, electron dot diagrams are sometimes used instead. In these diagrams the atom is indicated by its symbol, and only the electrons in the outer shell are designated. The shared electrons are placed between the two sharing atoms, as shown here:

$$:\ddot{C}l\cdot\ +\ \cdot\ddot{C}l: \qquad\qquad :\ddot{C}l:\ddot{C}l:$$

Electron dot diagrams are a bit cumbersome, and covalent bonds are often indicated simply by straight line structural formulas.[4] At times, even the lines are omitted, and molecular formulas that indicate only the number of each type atom are given:

$$Cl-Cl \quad or \quad Cl_2$$

Even if the molecule is written in this way, it is easy to tell that the two chlorines are sharing electrons because (1) they are both nonmetals and (2) no charge is indicated. Additional examples of electron dot diagrams versus structural formulas versus molecular formulas are shown in figure 1.10.

Double Bonds

Besides a single bond, like that between the two chlorine atoms, a double bond or triple bond may form in order for two atoms to complete their octets. In a triple bond, two atoms share three pairs of electrons between them. For example, in figure 1.9a the reaction between two nitrogen atoms results in a triple bond because each nitrogen needs three electrons in order to achieve a total of eight electrons in the outermost shell. Notice that in the diagrammatic representation six electrons are placed in the outer overlapping shells and that three straight lines are indicated in the structural formula for nitrogen gas.

Oxidation-Reduction

When oxygen combines with a metal, oxygen receives electrons and becomes negatively charged; the metal loses electrons and becomes positively charged. For example, consider the reaction that is illustrated in figure 1.7:

$$Mg + \tfrac{1}{2}O_2 \longrightarrow Mg^{++}O^{--}$$

In such cases, it is obviously appropriate to say that the metal has been oxidized and that because of oxidation, the metal has lost electrons. Then we need only admit that the oxygen has been reduced because it has gained electrons, or minus charges.

Today, the terms *oxidation* and *reduction* are applied to many ionic reactions whether or not oxygen is involved. Very simply, *oxidation refers to the loss of electrons, and reduction refers to the gain of electrons*. In our previous ionic reaction, $Na + Cl \longrightarrow Na^+ Cl^-$, the sodium has been oxidized (loss of electron) and the chlorine has been reduced (gain of electron).

The terms *oxidation* and *reduction* are also applied to certain covalent reactions. In this case, however, oxidation is the loss of hydrogen atoms, and reduction is the gain of hydrogen atoms. A hydrogen atom contains one proton and one electron; therefore when a molecule loses a hydrogen atom, it has lost an electron; when a molecule gains a hydrogen atom, it has gained an electron. We will have occasion to refer to this form of oxidation-reduction again in chapter 5.

---

[4]Structural formulas show the orientation of the atoms to one another and try to reflect how the atoms are arranged in space.

Figure 1.10
Electron dot, structural, and molecular formulas. In the electron dot formula, only the atoms in the outermost shell are designated. In the structural formula, the lines represent a pair of electrons that are being shared between two atoms. The molecular formula indicates only the number of atoms found within a molecule.

Electron Dot Formula	Structural Formula	Molecular Formula
$:\ddot{O}::C::\ddot{O}:$ carbon dioxide	$O=C=O$ carbon dioxide	CO_2 carbon dioxide
H $:\ddot{N}:H$ H ammonia	H | N—H | H ammonia	NH_3 ammonia
H $:\ddot{O}:H$ water	H | O—H water	H_2O water

Table 1.2 Inorganic versus organic chemistry

Inorganic compounds	Organic compounds
Usually contain metals and nonmetals	Always contain carbon and hydrogen
Usually ionic bonding	Always covalent bonding
Always contain a small number of atoms	May be quite large with many atoms
Often associated with nonliving elements	Often associated with living organisms

Inorganic versus Organic

There are two types of chemistry that are pertinent to our study, **inorganic chemistry** and **organic chemistry.** Ionic reactions are common to inorganic chemistry, while covalent reactions are common to organic chemistry. Table 1.2 lists the important differences between inorganic and organic compounds. As will be apparent in the following discussion, both types of compounds are necessary to the proper functioning of the living organism.

Some Important Inorganic Compounds

The inorganic compounds discussed in the following paragraphs affect the well-being of all organisms.

Water

Structure

Water, or H_2O, is not an organic molecule because it does not contain carbon; but as figure 1.10 shows, water is covalently bonded. Also within the molecule, as indicated in figure 1.11, there is a partial negative charge (δ^-) on the oxygen and a partial positive charge (δ^+) on the hydrogen atoms. This partial charge comes about because the larger oxygen is capable of holding onto electrons to a greater extent than hydrogen can, and shared electrons, therefore, spend more time circling oxygen than circling hydrogen. When this situation arises between two atoms, the bond between them is called a **polar bond** and the molecule itself is called a **polar molecule** because it carries charges. In polar bonding, a partial negative charge exists on the atom that has the electron pair more often, and a partial positive charge exists on the atom that has the pair less time. Polar bonds are not only found in water but they also occur whenever there is unequal sharing of an electron pair between two atoms.

In polar bonds, the atom that has the electron pair most of the time and carries the partial negative charge is called the **electronegative atom.** When hydrogen is bonded to an electronegative atom, its partial positive charge enables it to be attracted to still another electronegative atom. The latter bond, called a **hydrogen bond,** is represented by a dotted line in figure 1.11 because it is a weak bond that is easily broken.

Hydrogen Bonds   Hydrogen bonding will occur whenever hydrogen is bonded to a highly electronegative atom such as oxygen or nitrogen and there is another atom of this type in the vicinity. For example, in water two hydrogen atoms are covalently bonded to one oxygen atom, and each, in turn, is hydrogen bonded to another oxygen atom (fig. 1.11).

Figure 1.11
Water molecules are polar; each hydrogen carries a partial positive charge and each oxygen carries a partial negative charge. The polarity of the water molecules brings about hydrogen bonding between the molecules in the manner shown. The dotted lines represent hydrogen bonds. δ = partial.

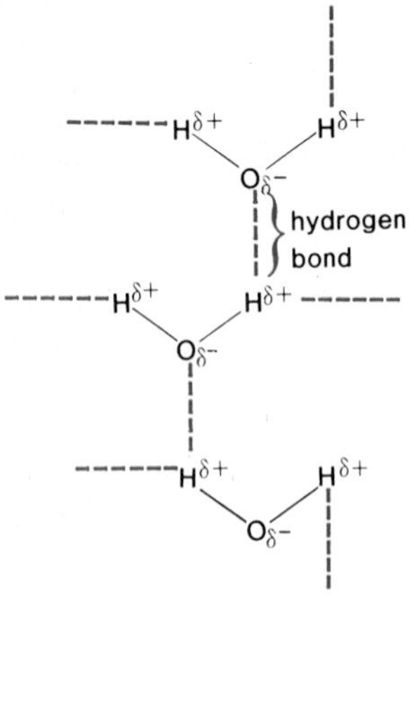

The Cell, A Unit of Life

Characteristics

The bonding properties of water account for some of its characteristics that are very important to living things. Water absorbs a great deal of heat before it becomes warm and evaporates; on the other hand, it gives off this heat as it cools down and freezes. This property allows great bodies of water, such as the oceans, to maintain a relatively constant temperature.

Life is believed to have come into existence in the ocean, and all living organisms are composed mostly of water. A large amount of water within the body is particularly helpful to cold-blooded terrestrial animals. The water content of the body can protect them from drastic changes in external temperatures. For example, if the environment suddenly becomes hot, the water in their bodies can absorb the heat; on the other hand, if the environment becomes cold, heat will be given off by this internal water. The heat will be distributed throughout the organism because heat moves through a liquid until it is uniformly distributed.

Also because of hydrogen bonding, liquid water is more dense than ice. Therefore ice floats on liquid water, and bodies of water always freeze from the top down. Furthermore, the layer of ice protects the organisms below and helps them survive the winter (fig. 1.12).

Water acts as a solvent and dissolves various chemical substances, particularly other polar molecules. The fact that water is capable of dissolving so many chemicals greatly facilitates chemical reactions and the movement of chemicals within living organisms.

Figure 1.12
The presence of hydrogen bonding causes ice to be less dense than liquid water. Therefore lakes freeze from the top down and this is a protection to the organisms that survive in the water beneath the layer of ice. It also permits humans to enjoy ice skating.

Figure 1.13
In water there is always a very few water molecules that have dissociated. Dissociation produces an equal number of hydrogen and hydroxide ions.

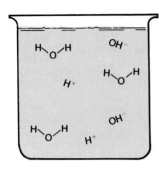

Figure 1.14
HCl is an acid that releases hydrogen ions as it dissociates in water. Notice that the addition of HCl to this beaker has caused it to have more hydrogen ions than hydroxide ions.

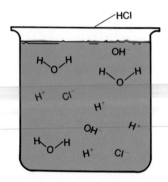

Figure 1.15
NaOH is a base that releases hydroxide ions as it dissociates. Notice that the addition of NaOH to this beaker has caused it to have more hydroxide ions than hydrogen ions.

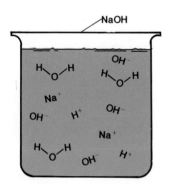

Dissociation

The presence of polarity also causes water molecules to tend to **dissociate,** or split up, in this manner:

$$H - O - H \longrightarrow H^+ + OH^-$$

The hydrogen ion (H^+) has lost an electron; the hydroxide ion (OH^-) has gained the electron. Very few molecules actually dissociate; therefore few hydrogen ions and hydroxide ions result. As an aid in grasping the concept of dissociation, envision a beaker of water, as illustrated in figure 1.13. In this beaker most of the water molecules are intact, but a very few have dissociated into H^+ and OH^-.[5]

Acids

Acids are compounds that dissociate in water and release hydrogen ions (or protons).[6] For example, an important inorganic acid is hydrochloric acid (HCl), which dissociates in this manner:

$$HCl \longrightarrow H^+ + Cl^-$$

Dissociation is almost complete, and this acid is called a strong acid. Envision that HCl has been added to a beaker of water (fig. 1.14). It is easy to see that acids increase the number of hydrogen ions when added to water.

Bases

Bases are compounds that dissociate in water and release hydroxide ions (OH^-). For example, an important inorganic base is sodium hydroxide (NaOH), which dissociates in this manner:

$$NaOH \longrightarrow Na^+ + OH^-$$

Dissociation is complete, and sodium hydroxide is called a strong base. Envision that NaOH has been added to a beaker of water (fig. 1.15). It is easy to see that bases increase the number of hydroxide ions when added to water.

pH

The pH scale[7] ranges from 0–14. Any pH value below 7 is acid, with ever-increasing acidity toward the lower numbers. Any pH value above 7 is basic (or alkaline), with ever-increasing basicity toward the higher numbers. *A pH of exactly 7 is neutral.* Water has an equal number of H^+ and OH^- ions, and thus one of each is formed when water dissociates. The fraction of water molecules that dissociate is 10^{-7} (or 0.0000001), which is the source of the pH value for neutral solutions. The pH scale was devised to simplify discussion of the hydrogen ion concentration [H^+], without using cumbersome numbers. For example:

    a. 1×10^{-7} [H^+]=pH7
    b. 1×10^{-2} [H^+]=pH2
    c. 1×10^{-9} [H^+]=pH9

Acids add hydrogen ions to solutions and increase the H^+ concentration of water. Which of the items (*a,b,c*) preceding indicates a higher concentration of hydrogen ions and therefore refers to an acid? The numbers with the

[5]The diagram is for illustration purposes and is not mathematically accurate.
[6]A hydrogen atom contains one electron and one proton. A hydrogen ion is only one proton and is often called a proton.
[7]pH is defined as the negative logarithm of the hydrogen ion concentration.

The Cell, A Unit of Life

Figure 1.16
The pH scale. The proportionate amount of hydrogen ions (H⁺) to hydroxide ions (OH⁻) is indicated by the diagonal line. Any pH above 7 is basic, while any pH below 7 is acidic.

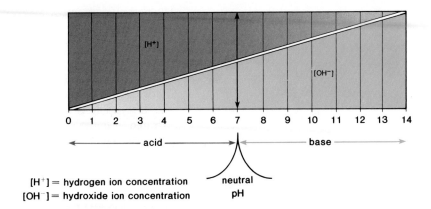

[H⁺] = hydrogen ion concentration
[OH⁻] = hydroxide ion concentration

neutral
pH

smaller negative exponents indicate a greater quantity than those with a larger negative exponent. Therefore, *b* refers to an acid. In an acidic solution there are more H⁺ ions than OH⁻ ions. Look again at the beaker of acid solution in figure 1.14; notice that there are more H⁺ ions than OH⁻ ions in the beaker after acid is added. Acidic solutions have the greater concentration of H⁺ but have the less pH numbers. Each lower pH unit has ten times the amount of H⁺ as the next higher unit.

Bases add hydroxide ions to solutions and increase the OH⁻ ion concentration of water. Look again at the beaker of basic solution, and notice that there are more OH⁻ ions in the beaker than H⁺ ions. Basic solutions, then, have fewer H⁺ ions compared to OH⁻ ions. In the preceding list (p. 28), *c* refers to a base because it indicates a lesser concentration of H⁺ than OH⁻ ions compared to water. Basic solutions have the lesser concentrations of H⁺ but have the bigger pH numbers. Figure 1.16 gives the complete pH scale with proper notations.

The concept of pH is important in biology because living organisms are very sensitive to hydrogen ion concentration. For example, in humans the pH of the blood must be maintained at about 7.4 or we become ill. All living things need to maintain the hydrogen ion concentration, or pH, at a constant level. They do this by the presence of buffers. A **buffer** is a chemical or a combination of chemicals that can take up excess hydrogen ions or excess hydroxide ions. When an acid is added to a solution, a buffer takes up excess hydrogen ions, and when a base is added to a solution, a buffer takes up excess hydroxide ions. Thus in the presence of a buffer the pH value changes less when either H⁺ or OH⁻ ions are added.

Salts
When a strong acid reacts with a strong base, a salt and water results:

$$HCl + NaOH \longrightarrow \underset{salt}{Na^+Cl^-} + \underset{water}{HOH}$$

If an equal quantity of strong acid and strong base take part in this reaction, **neutralization** occurs; the solution will be neither acid nor base as the salt and water form. In the neutralization process, the H⁺ from the acid and the OH⁻ from the base combine to form water. The salt consists of the positive ion of the base and the negative ion of the acid.

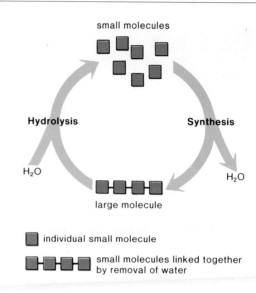

Figure 1.17
Synthesis and hydrolysis of an organic polymer. When the unit molecules join together to form the polymer (synthesis), water is released, and then when the polymer is broken down (hydrolysis), water is added.

small molecules

Hydrolysis   Synthesis

H_2O H_2O

large molecule

■ individual small molecule

■■■■ small molecules linked together by removal of water

Some Important Organic Compounds

Unit Molecules

The unique properties of carbon account for the formation of the very large number of organic compounds we associate with living organisms. Carbon is a nonmetal with four electrons in the outer shell. In order to achieve eight electrons in the outer shell, it must share with other nonmetals. It may share with as many as four other atoms. Many times, carbon atoms share with each other to form rings or chains of carbon atoms. These act as a skeleton for the unit molecules found in the life compounds—proteins, carbohydrates, fats, and nucleic acids. Thus the properties of carbon are essential to life as we know it.

Synthesis and Hydrolysis

Figure 1.17 diagrammatically illustrates that large organic molecules are synthesized or made when small unit molecules join together. A bond that joins two unit molecules together is created after the removal of H^+ from one molecule and OH^- from the next molecule. As water forms, **dehydration synthesis** occurs.

Large organic molecules are often **polymers,** or chains of unit molecules joined together. They can be broken down in a manner opposite to synthesis: the addition of water leads to the disruption of the bonds linking the unit molecules together. During this process, called **hydrolysis,** one molecule takes on H^+ and the next takes on OH^-.

Proteins

Functions

Proteins are large, complex macromolecules that sometimes have mainly a structural function. For example, in humans, *keratin* is a protein that makes up hair and nails, while *collagen* is a protein found in all type of connective tissue, including ligaments, cartilage, bone, and tendons. The muscles (fig. 1.18) contain proteins that account for their ability to contract.

Enzymatic Action   Some proteins function as **enzymes,** necessary contributors to the chemical workings of the cell and thus of the body. Enzymes are organic catalysts that speed up chemical reactions. They work so quickly that

Figure 1.18
Most athletes have well-developed muscle cells containing many protein molecules. A major controversy today is the taking of steroids to promote the buildup of muscles.

The Cell, A Unit of Life

Figure 1.20
Notice that as the peptide bond forms, water is given off. In other words, the water molecule on the right-hand side of the equation is derived from components removed from the amino acids on the left-hand side.

a reaction that might normally take several hours or days will take only a fraction of a second when an enzyme is present. Enzymes are discussed more fully in chapter 5.

Figure 1.21
The peptide bond is a polar bond; the oxygen is partially negative, and the hydrogen is partially positive. (δ = partial.)

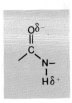

Structure

Amino Acids   The unit molecules found in proteins are called **amino acids.** Most amino acids have the structural formula shown in figure 1.19. The name *amino acid* refers to the fact that the molecule has two functional groups:

the **amino group** $H - \overset{\overset{\text{H}}{|}}{N} -$ or $H_2N -$

and the **acid group** $-\overset{\overset{\text{O}}{\parallel}}{\underset{\text{OH}}{C}}$ or   COOH

Amino acids differ from one another by their **R groups.** The letter *R* is used in organic chemistry to stand for the *R*emainder of the molecule. In amino acids, the R group varies from being a single hydrogen atom to those that are complicated rings. There are about 20 different common amino acids found in the proteins of living things. Therefore there are about 20 different types of R groups.

Peptides   The bond that joins two amino acids together is called a **peptide bond.** As you can see in figure 1.20, when synthesis occurs, the acid group of one amino acid reacts with the amino group of another amino acid and water is given off. The atoms associated with the resulting peptide bond, namely oxygen, carbon, nitrogen, and hydrogen, share the electrons in such a way that the oxygen carries a partial negative charge (fig. 1.21).

Figure 1.22

Polypeptide chain has a backbone. The R groups project to the sides. The polarity of the peptide bond (fig. 1.21) permits hydrogen bonding between members of the bonds. Only a few such hydrogen bonds are represented in the figure by dotted lines.

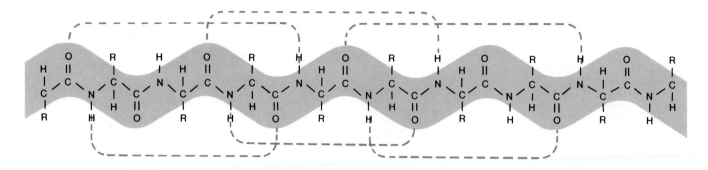

A **dipeptide** contains only two amino acids, but when up to 10 or 20 amino acids have joined together, the resulting chain is called a **polypeptide** (fig. 1.22). A polypeptide has a repeating sequence of N-C-C-N-C-C-N, which is called the backbone of the chain. The R groups extend from the backbone. A very long polymer of approximately 75 amino acids is called a protein.

Levels of Structure

Proteins are said to have at least three levels of structure: primary, secondary, and tertiary. The **primary structure** is simply the sequence or order of the different amino acids. Any number of the 20 different amino acids may be joined in various sequences and each type protein has its own particular sequence. The resulting chain is like a necklace comprised of 20 different types of beads that reoccur and are linked in a set way. The **secondary structure** is the usual orientation of the amino acid chain. This arrangement in space is locked in place by hydrogen bonding between members of the various peptide bonds. One common arrangement of the chain is the **alpha helix,** or right-handed spiral, with 3.6 amino acid residues per turn. In figure 1.22, a dotted line is indicated between a hydrogen attached to a nitrogen and a double-bonded oxygen four peptide bonds away. It is this bonding that causes the twisting characteristic of the alpha helix.

The **tertiary structure** of a protein refers to its final three-dimensional shape. In a structural protein like collagen, the helical chains lie parallel to one another. But enzymes are **globular proteins** in which the helix bends and twists in different ways. The final shape of a protein is maintained by various types of bonding between the R groups. Covalent, ionic, and hydrogen bonding are all seen.

Figure 1.23 illustrates the main features of protein chemistry. At the far left of the diagram is the final tertiary shape of a globular protein. But within this shape lies the alpha helix, as is apparent when the protein is stretched out. Finally, we see that the helix itself contains a particular sequence of amino acids. Each of these levels of organization is dependent on a particular type of bonding, as listed in table 1.3.

The final tertiary shape of a protein is very important to its function, as will be emphasized again when discussing enzyme activity. Proteins are very sensitive to both temperature and pH because a change in these conditions causes them to change their shape. For example, we are all aware that the addition of acid to milk causes curdling; heating causes egg white, a protein

The Cell, A Unit of Life

Figure 1.23
Proteins have at least three levels of structure.
Primary structure is the order of the amino acids;
secondary structure is often an alpha helix; and in
globular proteins, the tertiary structure is the
twisting and turning of the helix that takes place
because of bonding between the R groups.
Enzymes are globular proteins.

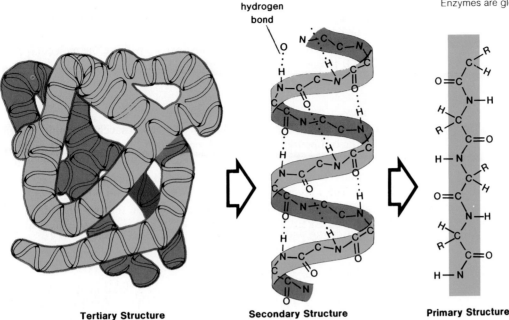

hydrogen
bond

Tertiary Structure   **Secondary Structure**   **Primary Structure**

Table 1.3   Types of bonding in a protein molecule

Location of bond	Level of structure	Type of bond
Between amino acids	Primary	Peptide bond
Between members of peptide bond	Secondary	Hydrogen bond
Between R groups	Tertiary	Various bonds

Figure 1.24
Plant cells have walls containing cellulose. The
rigidity of the cell walls permits nonwoody plants to
stand upright as long as they receive an adequate
supply of water.

called albumin, to coagulate. When a protein loses its normal configuration, it is said to be **denatured.** Denaturation occurs because the normal bonding between the R groups has been disturbed. For example, acids and bases disrupt the normal noncovalent bonding between the R groups, and this allows the protein to unfold. Once a protein loses its normal shape, it is no longer able to perform its usual function.

Carbohydrates

Carbohydrates are characterized by the presence of the atomic grouping H-C-OH in which the ratio of hydrogen atoms to oxygen atoms is approximately 2:1. Since this is the same as the ratio in water, it accounts for their name, which means *hydrates of carbon.* If the number of carbon atoms in the compound is low (from about three to seven), then the carbohydrate is a simple sugar, or **monosaccharide.** Thereafter, larger carbohydrates are created by joining together monosaccharides in the manner described in figure 1.17 for the synthesis of organic compounds.

Monosaccharides are used by all organisms as an energy source. Energy is released when they are broken down, and this energy is used by the organism to do work. The polymer cellulose is found in plant cell walls and accounts in part for the strong nature of these walls. In fact, it may be said that cellulose is the primary structural component of plants (fig. 1.24).

Monosaccharides

As their name implies, monosaccharides are simple sugars having only one unit. These compounds are often designated by the number of carbons they contain; for example, **pentose** sugars have five carbons and **hexose** sugars have six carbons. **Glucose** is a six-carbon sugar, with the structural formula shown in figure 1.25. Glucose is the primary energy source of the body, and most carbohydrate polymers can be broken down into monosaccharides that either are or can be converted to glucose. All these monosaccharides have the molecular formula $C_6H_{12}O_6$. However, in this text we will use the molecular formula $C_6H_{12}O_6$ to mean glucose since glucose is the most common six-carbon monosaccharide found in cells.

Disaccharides

The term **disaccharide** tells us that there are two monosaccharide units joined together in the compound. When two glucose molecules join together, **maltose** (fig. 1.26) is formed. The chemical equation for this reaction indicates that the forward direction is a dehydration synthesis and the backward reaction is a hydrolysis. You may also be interested in knowing that when *glucose* and another monosaccharide, *fructose,* are joined together, the disaccharide called **sucrose** is formed. Sucrose is derived from plants and is commonly used at the table to sweeten foods.

Polysaccharides

A **polysaccharide** is a carbohydrate that contains a large number of monosaccharide molecules. There are three polysaccharides that are common in organisms: starch, glycogen, and cellulose. All of these are polymers, or chains, of glucose, just as a necklace might be made up of only one type bead. Even though all three polysaccharides contain only glucose, they are distinguishable from one another.

As figure 1.27 shows, **starch** has few side branches; that is, chains of glucose that go off from the main chain. Starch is the storage form of glucose in plants. Just as we store orange juice as a concentrate, plants store starch as a concentrate of glucose. This analogy is particularly apt because, like the synthetic reaction described in figure 1.17, water is removed when glucose molecules are joined together to form starch. The following equation also represents the synthesis of starch:

$$n \text{ glucose} \underset{\text{hydrolysis}}{\overset{\text{synthesis}}{\rightleftharpoons}} \text{starch} + (n - 1)H_2O$$

n = some large number

Glycogen is characterized by the presence of side chains of glucose. Glycogen is the storage form of glucose in animals. After an animal eats, the liver stores glucose as glycogen; then in between eating, the liver releases glucose so that the blood concentration of glucose is always 0.1 percent.

In **cellulose,** a primary component of plant cell walls, the glucose units are joined by a slightly different type of linkage compared to that of starch and glycogen, as illustrated in figure 1.27. While this might seem to be a technicality, actually it is important because, for example, we are unable to digest foods containing this type of linkage; therefore cellulose passes through our digestive tract as roughage. Recently it has been suggested that the presence of roughage in the diet is necessary to good health.

The Cell, A Unit of Life

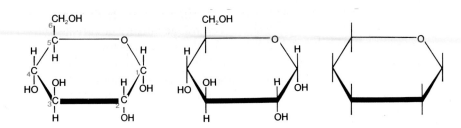

Figure 1.25
Three ways to represent the structure of glucose, a six-carbon sugar. At the far left, the small numbers count the number of carbon atoms present. In the center, the carbon atoms have not been indicated but are assumed to be present. To the right, all notations of atoms except the oxygen member of the ring have been omitted. Regardless of how it is drawn, glucose has six carbon, 12 hydrogen, and six oxygen atoms.

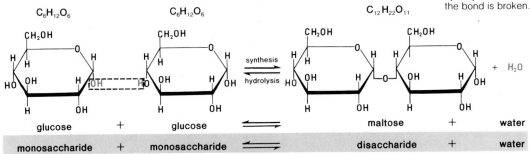

Figure 1.26
Synthesis and hydrolysis of maltose, a disaccharide containing two glucose units. During synthesis, a bond forms between the two glucose molecules as the components of water are removed. During hydrolysis, the components of water are added as the bond is broken.

Figure 1.27
Polysaccharides are composed of glucose. Starch is a relatively straight chain of glucose units. In comparison, glycogen is highly branched, and cellulose contains a slightly different type of linkage between glucose molecules.

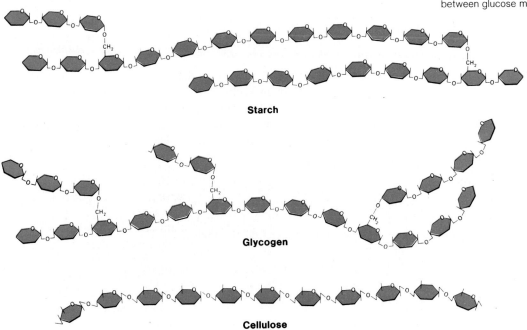

Figure 1.28

Fatty acids are either saturated (having no double bonds) or unsaturated (having double bonds).
a. Palmitic acid is a saturated fatty acid that can be represented in either way shown here. b.Linolenic acid is an unsaturated fatty acid that can be represented in either way shown.

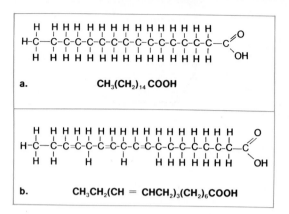

a. $CH_3(CH_2)_{14}COOH$

b. $CH_3CH_2(CH = CHCH_2)_3(CH_2)_6COOH$

Table 1.4 Common acids derived from fats

Name	Molecular formula	Structural formula
Lauric acid	$C_{11}H_{23}COOH$	$CH_3(CH_2)_{10}COOH$
Myristic acid	$C_{13}H_{27}COOH$	$CH_3(CH_2)_{12}COOH$
Palmitic acid	$C_{15}H_{31}COOH$	$CH_3(CH_2)_{14}COOH$
Stearic acid	$C_{17}H_{35}COOH$	$CH_3(CH_2)_{16}COOH$
Oleic acid	$C_{17}H_{33}COOH$	$CH_3(CH_2)_7CH=CH(CH_2)_7COOH$
Linoleic acid	$C_{17}H_{31}COOH$	$CH_3(CH_2)_4CH=CHCH_2CH=CH(CH_2)_7COOH$
Linolenic acid	$C_{17}H_{29}COOH$	$CH_3CH_2(CH=CHCH_2)_3(CH_2)_6COOH$

Lipids

Lipids are organic molecules that are insoluble in water. The most familiar lipids are the neutral fats such as lard, butter, and oil, which are used in cooking or at the table. In the body, fats serve as long-term energy sources. Adipose tissue is composed of cells that contain many molecules of neutral fat.

Neutral Fats (Triglycerides)

A neutral fat contains two types of unit molecules: **glycerol** and **fatty acids.**

Fatty Acids   Each fatty acid has a long chain of carbon atoms, with hydrogens attached, and ends in an acid group (fig. 1.28). Most of the fatty acids in cells here contain 16 or 18 carbon atoms per molecule, although smaller ones are also found. Fatty acids are either saturated or unsaturated. **Saturated** fatty acids, such as palmitic acid (fig. 1.28a), have no double bonds between the carbon atoms. The carbon chain is saturated, so to speak, with all the hydrogens that can be held. **Unsaturated** fatty acids, such as linolenic acid (fig. 1.28b), have double bonds in the carbon chain wherever the number of hydrogens is less than two per carbon atom. Unsaturated fatty acids are most often found in vegetable oils and account for the liquid nature of these oils. Vegetable oils are hydrogenated to make margarine. Polyunsaturated margarine still contains a large number of unsaturated, or double, bonds.

Table 1.4 gives other examples of saturated and unsaturated fatty acids. Notice that each fatty acid has a molecular formula and a structural formula that indicates the double bonds, if present.

Synthesis   Glycerol is a compound with three hydrates of carbon. Notice in figure 1.29 that glycerol has three —OH groups. When fat is formed, the acid portions of three fatty acids react with these groups so that fat and three molecules of water are formed. Again, the larger fat molecule is formed by dehydration synthesis in the forward direction. The backward direction represents how fat can be hydrolyzed to its components.

The Cell, A Unit of Life

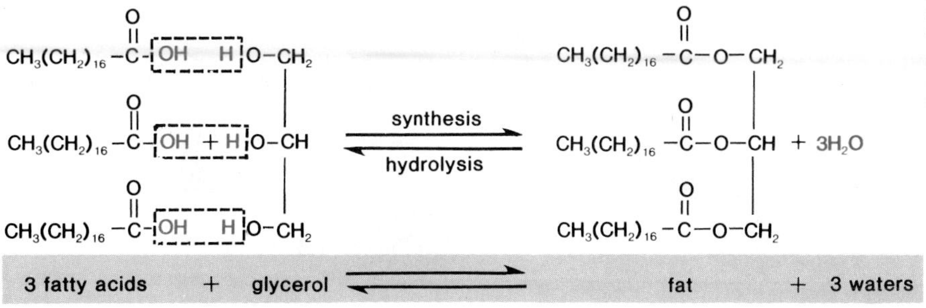

Figure 1.29
Synthesis and hydrolysis of a neutral fat. Three fatty acids plus glycerol react to produce a fat molecule and three water molecules. A fat molecule plus three water molecules react to produce three fatty acids and glycerol.

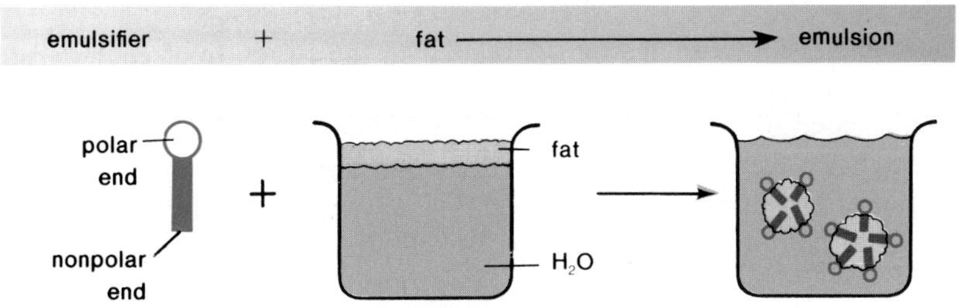

Figure 1.30
Fat molecules, being nonpolar, will not disperse in water. An emulsifier contains molecules that have a polar end and nonpolar end. When an emulsifier is added to a beaker containing a layer of nondispersed fat molecules, the nonpolar ends are attracted to the nonpolar fat, and the polar ends are attracted to the water. This causes droplets of fat molecules to become dispersed.

Soaps

A soap is a salt formed by a fatty acid and inorganic base, for example:

$$NaOH + RCOOH \longrightarrow RCOO^- Na^+$$

sodium hydroxide   fatty acid   soap

Whereas fats do not mix with water because they are nonpolar, a soap, being polar, will mix with water. When soaps are added to oils, then oils too will mix with water. Figure 1.30 shows how a soap positions itself about an oil droplet so that the polar ends project outward. Now the droplet will be soluble in water. This process of causing an oil to disperse in water is called **emulsification,** and it is said that an emulsion has been formed. Emulsification occurs when dirty clothes are washed with soaps and detergents. Also, prior to the digestion of fatty foods, fats are emulsified by bile. Usually a person who has had the gallbladder removed has trouble digesting fatty foods because the gallbladder stores bile for use at the proper time during the digestive process.

Phospholipids

Phospholipids contain a phosphate group and this accounts for their name:

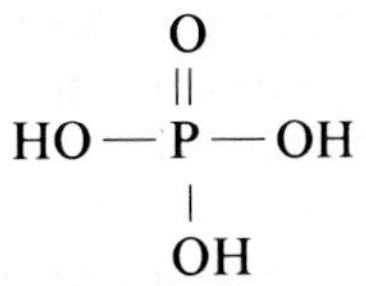

Essentially, phospholipids are constructed as neutral fats are, except that in place of the third fatty acid there is a phosphate group or a grouping that contains both phosphate and nitrogen (fig. 1.31). These molecules are not electrically neutral as are the fats because the phosphate group can ionize. Notice, then, that phospholipids have both a nonpolar (uncharged) region and a polar (charged) region. Thus, phospholipids are soluble in water. This latter property makes them very useful compounds in the body, as we will see in the next chapter.

Figure 1.31
Phospholipids are constructed similarly to fats except that they contain a phosphate group. Lecithin, shown here, has a side chain that contains both a phosphate group and a nitrogen containing group called choline.

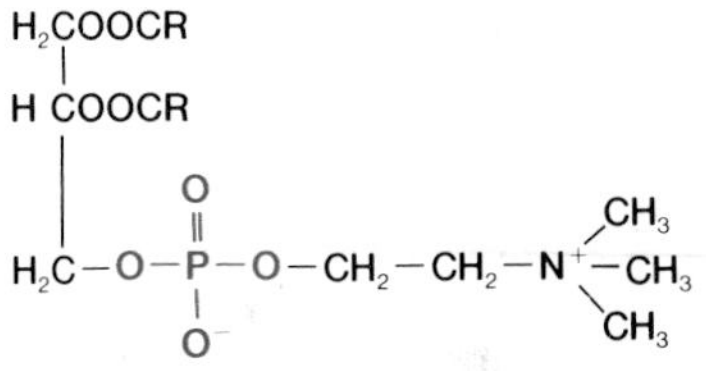

Figure 1.32
Like cholesterol (a), a steroid molecule (b) has four adjoining rings but steroid molecules differ significantly in the type of chain attached in the location indicated.

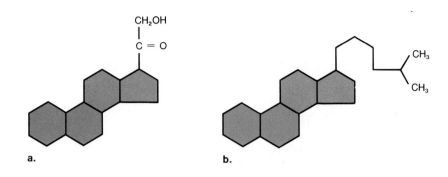

a.

b.

Figure 1.33
Generalized nucleotides. All nucleotides contain a phosphate molecule, a pentose sugar, and a base. The two types of nucleotides differ as to whether the base has *(a.)* two rings or *(b.)* one ring.

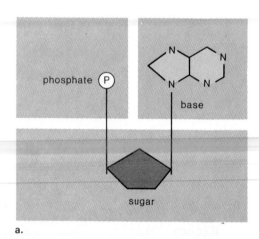

a.

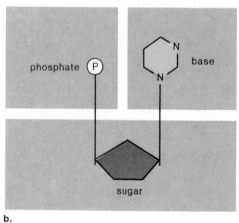

b.

Steroids

The **steroids** have a structure that is related to the structure of **cholesterol.** They are constructed of four fused rings of carbon atoms to which is usually attached a chain of varying length (fig. 1.32). For many years, it has been suggested that a diet high in saturated fats and cholesterol can lead to circulatory disorders due to reduced blood flow caused by the deposit of fatty materials on the inner linings of blood vessels. In a 10-year study of 3,806 men conducted by the National Heart, Lung and Blood Institute it was found that when, by the daily dose of a drug, the cholesterol level of the blood was reduced by 25 percent, the risk of heart disease was cut by 50 percent.

Despite these findings, the steroids are very important compounds in the body; for example, the sex hormones are steroids.

Nucleic Acids

Nucleic acids are huge, macromolecular compounds with very specific functions in cells; for example, the genes are composed of a nucleic acid called **DNA** (deoxyribonucleic acid). Another important nucleic acid, **RNA** (ribonucleic acid), works in conjunction with DNA to bring about protein synthesis.

Both DNA and RNA are *polymers of nucleotides* and therefore are chains of nucleotides joined together. Just like the other synthetic reactions we have studied in this section, these units are joined together to form nucleic acids by the removal of water molecules:

$$\text{n nucleotides} \underset{\text{hydrolysis}}{\overset{\text{synthesis}}{\rightleftharpoons}} \text{nucleic acid} + (n-1)\ H_2O$$

n = some large number

Nucleotides

Every **nucleotide** is a molecular complex of three types of unit molecules: phosphoric acid (phosphate), a pentose sugar, and a nitrogen base. DNA is composed of nucleotides that contain the sugar deoxyribose, while RNA has nucleotides having the sugar ribose. The bases in both DNA and RNA have either a single ring or double ring. Figure 1.33 shows generalized nucleotides because the specific type of base is not designated; the phosphate is simply represented as ⓟ. When nucleotides join together, they form a polymer in which the backbone is made up of phosphate-sugar-phosphate-sugar, with the bases projecting to one side of the backbone (fig. 1.34). Such a polymer is

The Cell, A Unit of Life

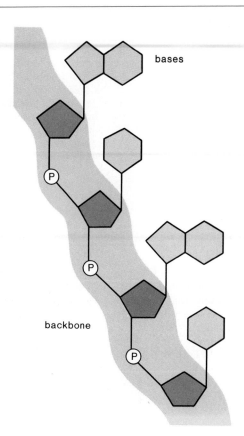

bases

backbone

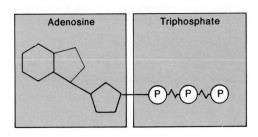

Adenosine	Triphosphate

Figure 1.35
ATP is a nucleotide with three phosphate units; two of the phosphate bonds are high energy bonds, indicated by wavy lines.

called a strand. RNA is single stranded and DNA is double stranded, the two strands being held together by hydrogen bonding between the bases (fig. 22.4). This is only a brief description of the structure and function of DNA and RNA; they are considered again in more detail in chapter 22.

Adenosine triphosphate (ATP) is a nucleotide that is used as a carrier of energy in cells. The structure of ATP is similar to that shown in figure 1.33a. Adenine is the base; the sugar is ribose; and there are three phosphate groups instead of one. It is customary to draw the molecule as shown so that the three phosphate groups appear on the right. ATP is known as the energy molecule because the triphosphate unit contains two high-energy bonds, represented in figure 1.35 by wavy lines.

Calcium, Chlorine, and Heart Disease

Many persons require more dietary calcium. Drinking milk can help supply this calcium.

It is possible that there is more reason to drink milk than to build strong bones and teeth. Recent studies conducted by Richard Bull, who heads the Environmental Protection Agency's toxicology and microbiological division in Cincinnati, suggest that it may protect against heart disease in persons who drink chlorinated water.[8]

In one of Bull's studies, pigeons were fed a diet that was normal except that it contained only 80 percent of the recommended daily allowance of calcium [for pigeons]. (The recommended daily allowance [for humans] is 800 mg of calcium.) Half the birds drank unchlorinated water, and the other half drank water containing 10 milligrams of chlorine per liter. After three months, serum cholesterol levels of the birds that drank chlorinated water were 50 percent higher than those that drank chlorine-free water. However, no such contrast was noted in a control group of birds that had received the recommended daily allowance of calcium.

In another study, birds were fed a diet that was both calcium deficient and contained 10 percent lard. Birds that drank chlorine-free water had only slightly elevated cholesterol levels, but those that drank chlorinated water had three times as much cholesterol. Autopsies of the high-cholesterol birds also suggested they had developed more plaques that could eventually block blood passages. Such deposits have been implicated in strokes, heart attacks, and other forms of human heart disease.

These studies are significant because at least half the population in the United States drinks chlorinated water and according to National Center for Health Statistics, the diet of many adults in this country is deficient in calcium. The most recent survey shows that especially women, of all ages, are apt to be below the recommended daily intake of calcium. Men usually exceed the recommended daily allowance until age 65 when their calcium intake drops below this amount.

[8]Findings in bird studies cannot be related directly to humans; they only provide ancillary information on human physiology.

Summary

All matter is made up of atoms, which are arranged in the Periodic Table of the Elements according to increasing weight and chemical properties. The weight of an atom is dependent upon the number of protons and neutrons in the nucleus, while the chemical properties are dependent on the number of electrons in the outermost shell.

Atoms react with one another in order to acquire eight electrons in the outermost shell. In one major type of reaction, metals give electrons to non-metals, forming ionic compounds in which ions are attracted to one another because of their opposite charges. The formula for such a compound shows the correct ratio of ions in relationship to one another. In the other major type of reaction, nonmetals share electrons to form covalent compounds. The formula for such a compound may indicate the bonds as lines between atoms (structural formula) or may simply indicate the number of each kind of atom (molecular formula). Ionic reactions are of primary importance in inorganic chemistry, while covalent reactions are of primary importance in organic chemistry. Other differences between these two types of chemistry are listed in table 1.2. Oxidation-reduction reactions occur in both types of chemistry; the manner in which oxidation (loss of electrons) and reduction (gain of electrons) occurs is different.

Water, acids, and bases are important inorganic compounds. While water has a neutral pH, acids have a pH less than seven and bases have a pH greater than seven. Thus acids increase the hydrogen ion concentration of water while bases decrease this concentration. When an equal amount of strong acid is added to an equal amount of strong base, a salt and water will result.

The carbon atom is the basis for organic chemistry. Carbon atoms share electrons with each other to form the unit molecules that make up the macromolecular life compounds. In this chapter, we studied the macromolecular molecules and their unit molecules listed in table 1.5. The unit molecules listed in the second column in this table join together by dehydration synthesis to form the macromolecules in the first column of the table, and the macromolecules can be decomposed by hydrolytic degradation to release the unit molecules.

Proteins are macromolecules that are structurally and otherwise functionally important; in all cells they function as enzymes that speed up chemical reactions. Proteins contain 20 different types of amino acids that differ from one another only in their respective R groups and are joined together by peptide bonds. Proteins have at least three levels of structure: primary, secondary, and tertiary. A weak type of bond, called the hydrogen bond, is important to the secondary structure of a protein.

Carbohydrates include the simple sugars and polymers of these sugars. Ribose and deoxyribose are five-carbon sugars known as pentoses, and glucose is a six-carbon sugar known as a hexose. Glucose is the common energy source in the body. Maltose is a disaccharide made up of two glucose units. Starch, glycogen, and cellulose are long polymers of glucose, or polysaccharides.

Lipids include the neutral fats, phospholipids, and steroids. Neutral fats contain glycerol and three fatty acids joined in such a way that three molecules of water are released. Fatty acids are saturated when they contain as much hydrogen as possible and unsaturated when they do not. Phospholipids are like fats except that in place of the third fatty acid they have a phosphate or a grouping that contains both phosphate and nitrogen. Steroids are chemically related to cholesterol. They also serve important chemical functions in humans. The sex hormones are steroids.

Nucleic acids are of two types, DNA and RNA. Both of these are polymers of nucleotides, complex molecules containing phosphate, a sugar, and a base. DNA is composed of nucleotides having the sugar deoxyribose, while RNA nucleotides contain the sugar ribose. RNA is a single-stranded polymer and DNA is a double-stranded polymer. Both nucleic acids are involved in protein synthesis, a process discussed in detail in chapter 22.

Table 1.5 Organic compounds of life

Macromolecules	Unit molecule	Usual atoms
Protein	Amino acid	C, H, O, N
Carbohydrate, e.g., starch	Glucose	C, H, O
Lipid	Glycerol and fatty acids	C, H, O
Nucleic acid	Nucleotide	C, H, O, N, P

Study Questions

1. Name the subatomic particles of an atom; describe their charge and weight and their location in the atom. (p. 19)
2. Draw the atomic diagram for calcium. (p. 21)
3. State the octet rule and explain how it relates to chemical reactions. (p. 21)
4. Give an example of an ionic reaction and explain it. Mention in your explanation: compound, ion, formula, and ionic bond. (p. 22)
5. Give an example of a covalent reaction and explain it. (pp. 22–24)
6. Explain oxidation-reduction in terms of loss or gain of electrons. (p. 25)
7. Name five possible differences between inorganic and organic compounds. (p. 26)

8. On the pH scale, which numbers indicate a basic solution? An acidic solution? Why? (pp. 28–29)
9. What are buffers and why are they important to life? (p. 29)
10. Explain dehydration synthesis of organic compounds and hydrolytic degradation of organic compounds. (p. 30)
11. What are some functions of proteins? What is the unit molecule of protein? What is a peptide bond, a dipeptide, a polypeptide? (pp. 31–32)
12. Discuss the primary, secondary, and tertiary structure of a protein and state the type of bonding associated with each of these. (pp. 32–33)
13. Name some monosaccharides, disaccharides, and polysaccharides, and state appropriate functions. (pp. 33–35) What is the most common unit molecule for these?(p. 34)
14. Name some important lipids and state their function. (p. 36) What is a saturated fatty acid? (p. 36) An unsaturated fatty acid? (p. 36) How is fat formed? (p. 36)
15. What are the two important nucleic acids? (p. 38) What is the unit molecule? (p. 38)

Selected Key Terms

The accent marks used in the pronunciation guides are derived from a simplified system of phonetics standard in medical usage. The single accent (') denotes the major stress. Emphasis is placed on the most heavily pronounced syllable in the word. The double accent ('') indicates secondary stress. A syllable marked with a double accent receives less emphasis than the syllable that carries the main stress, but more emphasis than neighboring unstressed syllables.

atom (at'om)
isotopes (i'so-tōps)
ion (i'on)
bond energy (bond en'er-je)
dissociation (dis-so''she-a'shun)
acid (as'id)
base (bās)
polymer (pol'ĭ-mer)
hydrolysis (hi-drol'ĭ-sis)
protein (pro'te-in)

peptide (pep'tīd)
denaturation (de-na-tur-a'shun)
glycogen (gli'ko-jen)
cellulose (sel'u-lōs)
lipid (lip'id)
triglyceride (tri-glis'-er-īd)
emulsification (e-mul''sĭ-fĭ-ka'shun)
steroid (ste'roid)
nucleic acid (nu-kle'ik as'id)

Further Readings

Asimov, I. 1966. *The world of carbon.* Rev. ed. New York: Collier Co.
———. 1962. *The world of nitrogen.* New York: Collier Co.
Baker, J. J., and Allen, G. E. 1981. *Matter, energy, and life.* 4th ed. Reading, Mass.: Addison-Wesley.
Crick, F. H. C. 1957. Nucleic acids. *Scientific American* 197(3):62.
Kendrew, J. C. 1961. The three-dimensional structure of a protein molecule. *Scientific American* 205(6):34.
Roberts, J. D. 1957. Organic chemical reactions. *Scientific American* 197(5):38.
Seaborg, G. T. 1980. The new elements. *American Scientist* 68(3):279.
Stein, W. H., and Moore, S. 1961. Chemical structure of proteins. *Scientific American* 204(2):28.
Stryer, L. 1981. *Biochemistry.* 2d ed. San Francisco: W. H. Freeman.
Thompson, E. O. P. 1955. The insulin molecule. *Scientific American* 192(5):18.

The Cell, A Unit of Life

2

Chapter concepts

1 The fundamental unit of life is the cell, which is highly organized and contains organelles that carry out specific functions.

2 The organelles have been divided into four groups: (*a*) the nucleus, (*b*) membranous canals and vacuoles, (*c*) energy-related organelles, and (*d*) centrioles and related organelles.

3 The nucleus, a centrally located organelle, controls the metabolic functioning and structural characteristics of the cell.

4 Endoplasmic reticulum, Golgi apparatus, vacuoles, and lysosomes are all membranous tubules or vesicles concerned with the entrance, production, digestion, excretion, or transportation of molecules.

5 Mitochondria in both plant and animal cells are organelles concerned with the production of a form of energy usable by cells; they are the powerhouses of the cell. Chloroplasts are unique to plant cells and absorb the energy of the sun in order to produce glucose.

6 Centrioles and related organelles are structures concerned with the shape and/or movement of the cell. Therefore both microfilaments and microtubular structures are included in this category. Centrioles, cilia, flagella, and spindle fibers all contain microtubules.

Cell Structure and Function

Figure 2.1

Animal and plant cells. These generalized representations are based on electron micrographs. *a.* Animal cell. *b.* Plant cell (p. 45). In the average mature plant cell, the vacuole actually occupies a greater percentage of the cell's volume than it does in this drawing.

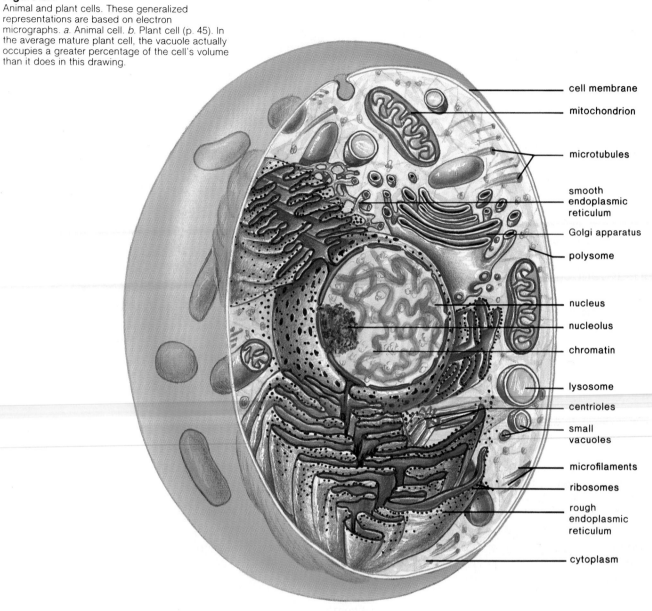

cell membrane

mitochondrion

microtubules

smooth endoplasmic reticulum

Golgi apparatus

polysome

nucleus

nucleolus

chromatin

lysosome

centrioles

small vacuoles

microfilaments

ribosomes

rough endoplasmic reticulum

cytoplasm

a. Animal cell

Cell Theory

All living things are made up of **cells,** which are the smallest units of life. Cells come in many different shapes and sizes; but no matter what the shape or size, each one carries on the functions associated with life, interacting with the environment, growing, and reproducing.

The cell marks the boundary between the nonliving and the living. The molecules that serve as food for a cell and the organic molecules that make up a cell are not alive and yet the cell is alive.

The answer to what life is will have to be found within the cell, because the smallest living organisms are single cells while larger organisms are **multicellular** and composed of many cells. The statement that all living things are composed of cells is called the **cell theory.**

The Cell, A Unit of Life

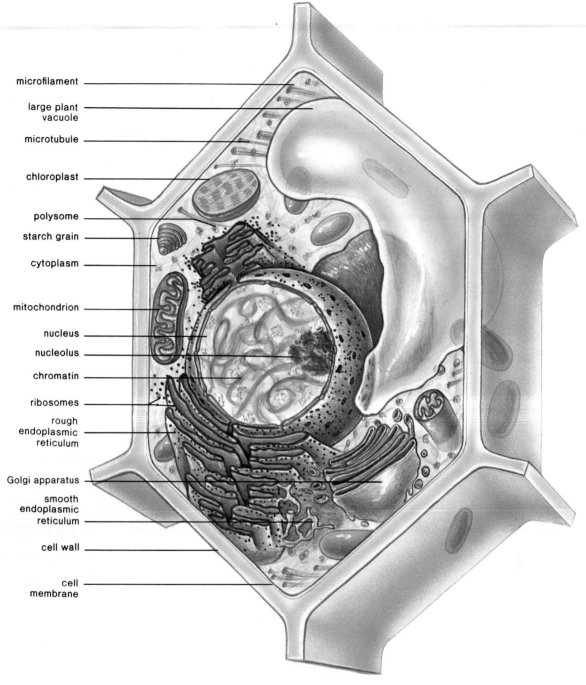

microfilament

large plant
vacuole

microtubule

chloroplast

polysome

starch grain

cytoplasm

mitochondrion

nucleus

nucleolus

chromatin

ribosomes

rough
endoplasmic
reticulum

Golgi apparatus

smooth
endoplasmic
reticulum

cell wall

cell
membrane

b. Plant cell

Generalized Cells

Cells are usually divided into two main groups called **prokaryotic cells** and **eukaryotic cells.** We will begin with an examination of eukaryotic cells and will then compare them to prokaryotic cells. Even though there is a great variety of eukaryotic cells, differing in regard to specific structure and function, they all have the same basic organization. This chapter will stress the generalized animal cell depicted in figure 2.1a and the generalized plant cell depicted in figure 2.1b.

Figure 2.2
Under this type of light microscope *(a.)*, the mitochondria in a mouse liver cell *(b.)* appear as dark spots.

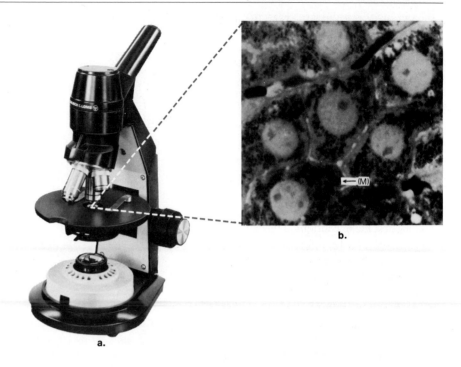

b.

a.

Figure 2.3
Under this type of transmission electron microscope *(a.)*, the details of a single mitochondrion *(b.)* are clear.

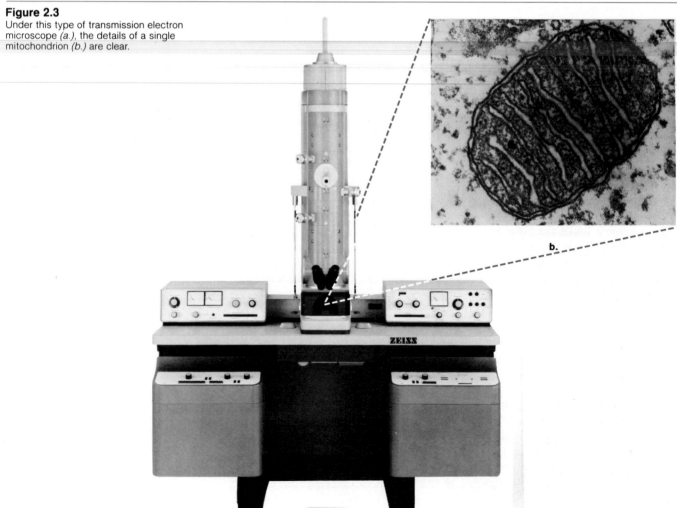

b.

a.

The Cell, A Unit of Life

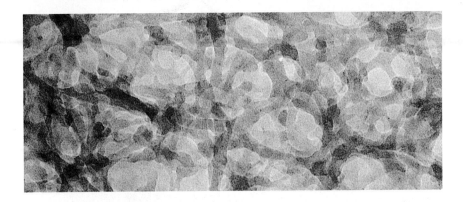

Figure 2.4
A high power electron micrograph that shows the
fine filaments found within a lattice that pervades
the cytoplasm. Figure 2.18 shows how this lattice
helps suspend the organelles.

With some exceptions, such as various types of eggs, cells are not readily visible to the eye; therefore a microscope is needed to view them. **Light microscopes,** such as the one shown in figure 2.2, utilize light to view the object. Light microscopes do not provide as much detail as do the **electron microscopes,** such as the one shown in figure 2.3, which utilize electrons to view the object. A comparison of these two types of microscopes (see the appendix, p. A-1) shows that the most important difference between them is not the degree to which they magnify but is instead their **resolving power,** the capacity to distinguish between two points. If two points are seen as separate, then the image appears more detailed than if the two points are seen as one point. Resolving power of a microscope is improved as the wavelength of the illumination becomes shorter, and an electron beam has a much shorter wavelength than a visible light ray. At the very best, a light microscope can distinguish two points separated by 200 nm (nanometer = 1×10^{-6} mm), but the electron microscope can distinguish two points separated by only .5 nm.[1] Thus the electron microscope gives a much more detailed image.

Pictures obtained by using the light microscope are sometimes called **photomicrographs,** and pictures resulting from the use of the electron microscope are called **electron micrographs.** The drawings in figures 2.1a and 2.1b are based on electron micrographs of animal and plant cells. Actual electron micrographs are found throughout the chapter.

Parts of a Eukaryotic Cell

Eukaryotic cells are surrounded by an outer membrane, or **cell membrane,** within which is found the **cytoplasm,** the substance of the cell outside the nucleus in which there are various **organelles,** small bodies with specific structures and functions. In addition to a cell membrane, plant cells have an exterior **cell wall** (fig. 2.1b). Detailed consideration of the cell wall and cell membrane is delayed until the next chapter so that these topics can be discussed in depth.

Cytoplasm

At one time, the cytoplasm was thought to be a homogenous substance, but now it is known to contain several types of filamentous protein structures that together form a **cytoskeleton,** which extends throughout the cytoplasm suspending the organelles. The cytoskeleton includes the microtubules and microfilaments discussed on page 57 and also a newly discovered irregular, three-dimensional network or lattice (fig. 2.4) (termed the *microtrabecular lattice*) made up of slender strands that interconnect with all of the other structures in the cytoplasm. The lattice is dynamic and its exact shape is constantly changing. While the lattice itself is protein rich, the spaces between the strands are water rich. Together these two phases give the cytoplasm a gel-like consistency.

---

[1]For a review of linear metric units see the Appendix, p. A-2.

Table 2.1 Sizes of organelles

Organelles	Size
Nucleus	3–25 μm
Membranous canals and vacuoles	
Endoplasmic reticulum	Variable
Ribosomes	15–25 nm
Vacuoles	Variable
Golgi apparatus	0.5–1.0 μm
Lysosomes	0.5 μm
Energy-related organelles	
Mitochondria	1–10 μm $\times$ 0.3–1.0 μm
Chloroplasts	1–10 μm $\times$ 2–4 μm
Centrioles and related organelles	
Centrioles	0.16–5.6 μm $\times$ 0.16–0.23 μm
Microtubules	Variable $\times$ 20–25 nm
Microfilaments	Variable $\times$ 4–7 nm
Cilia	10–20 μm $\times$ 0.5 μm
Flagella	100–200 μm $\times$ 0.5 μm

*As noted in the Appendix (p. A-2), nm = nanometers; μm = micrometers; sizes are length $\times$ diameter.

Organelles

The various organelles are specialized for particular functions. For the sake of discussion, we will divide the organelles into four categories: (1) the nucleus, (2) membranous canals and vacuoles, (3) energy-related organelles, and (4) centrioles and related organelles.

Organelles vary in size, but most are very small and only the electron microscope can make out their inner detail. Table 2.1 gives the size of the organelles in terms of the metric system, which is explained in the appendix. Notice that while the nucleus, chloroplast, and mitochondrion are within the range of the light microscope, the electron microscope is required to distinguish the others.

Many organelles are composed of membrane and therefore we will now discuss the structure of membrane.

Intracellular Membrane

Biochemical assay, the use of chemical tests to identify compounds, has shown that membrane is composed of protein and phospholipid molecules. Since electron micrographs (fig. 2.5) show membrane as two dense lines separated by a clear space, it was at first believed that phospholipid molecules are always sandwiched between a top and bottom layer of protein molecules. Most investigators today, however, believe that this is an oversimplification. They point out that we should expect the structure of membrane to vary in the same manner that the function of membrane varies. While it is believed that phospholipid is always a component of membrane, the protein portion most likely varies in regard to both specific makeup and placement. Depending on the membrane, the protein molecules may reside above or below the phospholipids, extend from top to bottom of the membrane, or simply penetrate a short

The Cell, A Unit of Life

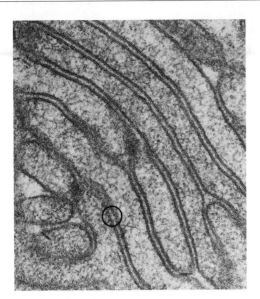

Figure 2.5
Electron micrograph of intracellular membranes.
The circle surrounds one section of membrane,
which appears to be composed of two outer
electron-dense layers separated by a less dense
central zone.

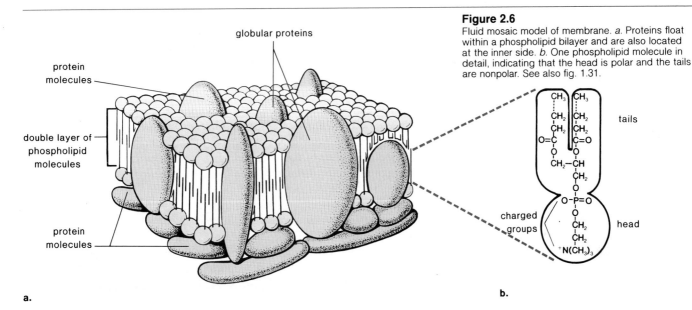

globular proteins

protein
molecules

double layer of
phospholipid
molecules

protein
molecules

a.

Figure 2.6
Fluid mosaic model of membrane. *a.* Proteins float
within a phospholipid bilayer and are also located
at the inner side. *b.* One phospholipid molecule in
detail, indicating that the head is polar and the tails
are nonpolar. See also fig. 1.31.

tails

charged
groups

head

b.

distance. This model of the membrane (fig. 2.6) is called the **fluid mosaic model**
because the protein molecules form a pattern within the lipid bilayer, which
is in a liquid state having the consistency of light oil. Notice the manner in
which the phospholipid molecules arrange themselves. Their structure, discussed in chapter 1, causes each molecule to have a polar head and nonpolar
tails. Within the phospholipid bilayer, the tails face inward and the heads face
outward where they are likely to encounter a watery environment.

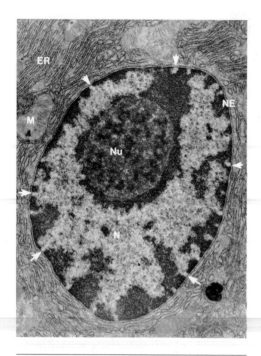

Figure 2.8
Experiments with the large single-celled alga, Acetabularia. a. In this experiment, the stalk and cap without a nucleus die, but the base with a nucleus regenerates. b. In this experiment, the regenerated cap resembles that of the species of the nucleus and not that of the cytoplasm. These experiments show that the nucleus controls the rest of the cell.

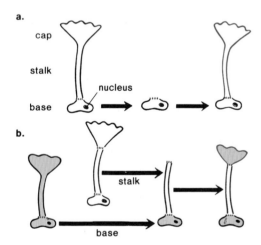

Nucleus

The **nucleus** (fig. 2.7), the largest organelle within the cell, is enclosed by a double-layered membrane (fig. 2.1), called the **nuclear envelope.** There are pores, or openings, in this envelope through which large molecules pass from the nucleoplasm, the fluid portion of the nucleus, to the cytoplasm.

The nucleus is of primary importance in the cell because it is the control center that oversees the metabolic functioning of the cell and ultimately determines the cell's characteristics, as experimentation has shown. In one type of green alga, *Acetabularia,* the organism consists of a base, stalk, and cap (fig. 2.8). If the stalk and cap are removed from the base as in figure 2.8a, the stalk and cap die but the nucleus-containing base develops into a new organism. The importance of the nucleus is further exemplified in figure 2.8b when the base is combined with the stalk of a different species, and the cap regenerated is appropriate to the species of the nucleus rather than to the cytoplasm of the stalk. This demonstrates that the nucleus controls both the function and structure of the cell.

Within the nucleus there are masses of threads called **chromatin,** so called because they take up appropriate stains and become colored. Chromatin is indistinct in the nondividing cell, but it condenses to rodlike structures called **chromosomes** (fig. 4.5) at the time of cell division. Chemical analysis shows that chromatin, and thus chromosomes, contain the chemical DNA (deoxyribonucleic acid) along with certain proteins and some RNA (ribonucleic acid). It is now known that DNA, with the help of RNA, *controls protein synthesis* within the cytoplasm and that it is this function that allows DNA to control the cell.

Nucleoli

One or more **nucleoli** are present in the nucleus. These dark-staining bodies are actually specialized parts of chromosomes in which a special type of RNA called ribosomal RNA (rRNA) is produced. Ribosomal RNA joins with proteins before migrating to the cytoplasm where it becomes part of the ribosomes, organelles to be discussed in the following sections.

Membranous Canals and Vacuoles

Endoplasmic reticulum, the Golgi apparatus, vacuoles, and lysosomes (fig. 2.1) are structurally and functionally related membranous structures. Ribosomes are not composed of membrane but are included in this category because they are often intimately associated with the endoplasmic reticulum.

Endoplasmic Reticulum

The **endoplasmic reticulum** (ER) forms a membranous system of tubular canals that begins at the nuclear envelope and branches throughout the cytoplasm. Small granules, called ribosomes, are attached to some portions of the endoplasmic reticulum. If they are present, the reticulum is called **rough endoplasmic reticulum;** if they are not present, it is called **smooth endoplasmic reticulum.** Figure 2.9 illustrates rough endoplasmic reticulum and figure 2.10 illustrates the smooth. Apparently, smooth endoplasmic reticulum contains, within its membrane, enzymes that synthesize lipids. Thus smooth endoplasmic reticulum is abundant in cells of the testes and adrenal cortex, both of which produce steroid hormones. Also, it is known that the administration of drugs increases the amount of smooth endoplasmic reticulum in the liver. It would seem then that the reticulum has enzymes that detoxify drugs.

The Cell, A Unit of Life

Figure 2.9

Rough endoplasmic reticulum *a.* Electron micrograph showing that in some cells the cytoplasm is packed with this organelle. *b.* A three-dimensional drawing gives a better idea of the organelle's actual shape. *c.* This three-dimensional model of a ribosome is based on high-power electron microscopy studies that indicate that a ribosome is composed of a small and a large subunit.

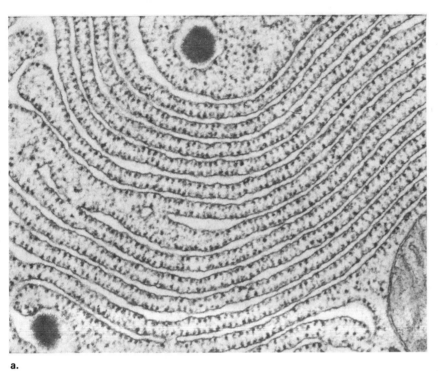

a.

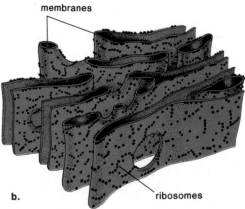

membranes

ribosomes

b.

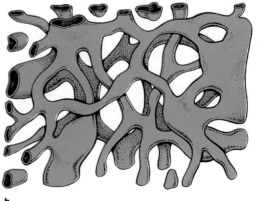

c.

Figure 2.10

Smooth endoplasmic reticulum. *a.* Electron micrograph showing the abundance of this organelle in cells that secrete steroids. *b.* A three-dimensional drawing illustrates that this organelle lacks ribosomes.

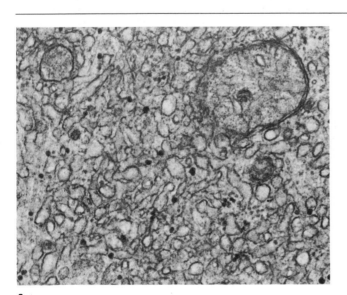

a.

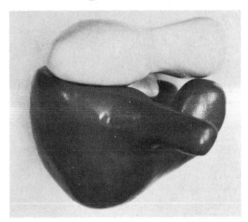

b.

Cell Structure and Function

51

Ribosomes

Ribosomes look like small, dense granules in low-power electron micrographs (fig. 2.9a), but a higher resolution shows that each one contains two subunits (fig. 2.9c). As their name implies, ribosomes contain RNA (ribonucleic acid), but they also contain proteins. The larger of the two subunits contains at least thirty different proteins, and the smaller unit contains at least twenty different proteins.

Ribosomal RNA produced in the nucleolus joins with proteins before migrating to the cytoplasm. Once the ribosomes are fully assembled within the cytoplasm, they function in the process of protein synthesis. Synthesis, as discussed in chapter 1, refers to the joining together of small organic molecules to make larger ones. In this case, amino acids are joined together to make a protein.

Ribosomes are very often attached to endoplasmic reticulum (fig. 2.9) but are also found unattached within the cytoplasm. In these instances, several ribosomes, each of which is producing the same type protein, are arranged in a functional group called a **polysome.** Most likely these proteins are for use inside the cell. But when the ribosomes are attached, it most likely means that the protein is for export outside the cell. For example, certain pancreatic cells that produce digestive enzymes for use in the small intestine have extensive amounts of rough endoplasmic reticulum.

Protein that is destined for export outside the cell is prepared at the ribosomes and temporarily stored in the channels of the reticulum. Small portions of the endoplasmic reticulum then break away to form membrane-enclosed vesicles (small vacuoles) that migrate to the Golgi apparatus, where the product is received and repackaged for export.

Golgi Apparatus

The **Golgi apparatus** (fig. 2.11) is named for the person who first discovered its presence in cells. It is composed of a stack of about a half-dozen or more *saccules* that look like flattened vacuoles. One side of the stack faces the nucleus and the other side faces the cell membrane. Vesicles are seen especially at the rims of the saccules, but vesicles also occur along the length of the saccules at either face of the stack.

These observations along with biochemical evidence suggest that the Golgi apparatus receives protein-filled vesicles from the endoplasmic reticulum at its inner face. After the proteins are sorted out, they are packaged into vesicles at the rims of the saccules and at the outer face. Thereafter, the vesicles move to different locations in the cell (fig. 2.12).

Vacuoles

A cross section of a **vacuole,** which is called a **vesicle** when small in size, most often shows a clear area bounded by a membrane. Vacuoles are storage sites for various kinds of molecules in solution or suspension. Vesicles can be made by the Golgi apparatus, as previously described, or they can arise by an infolding of the cell membrane. The large, central plant vacuole (fig. 2.1b) is attached to the endoplasmic reticulum, and it seems that this particular vacuole might be a part of this system. The central vacuole helps support a plant cell when it is full of water. When a plant wilts, its cells' vacuoles are not filled with water.

The Cell, A Unit of Life

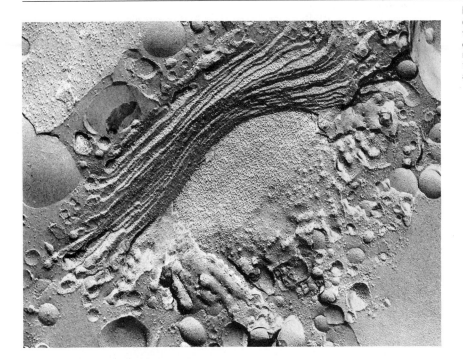

Figure 2.11
Golgi apparatus structure. A frozen cell was fractured (split) and then subjected to scanning electron microscopy. In this case, the treatment revealed the interior of the saccules and shows that vesicles appear along both sides of the Golgi apparatus and at the rims of the saccules.

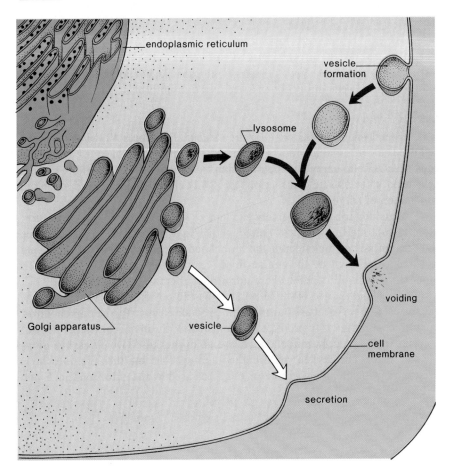

Figure 2.12
Golgi apparatus function. The Golgi apparatus receives vesicles from the endoplasmic reticulum and thereafter forms at least two types of vesicles, lysosomes and secretory vesicles. Lysosomes contain hydrolytic enzymes that can break down large molecules. Sometimes lysosomes join with vesicles, bringing large molecules into the cell. Thereafter, any nondigested residue is voided at the cell membrane. The secretory vesicles formed at the Golgi apparatus also discharge their contents at the cell membrane.

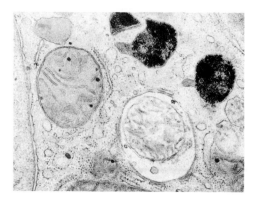

Lysosomes

Lysosomes are a special type of vesicle (fig. 2.12), most likely formed by the Golgi apparatus. All lysosomes are concerned with intracellular digestion and contain powerful enzymes called *hydrolytic enzymes* (p. 183). Following formation the lysosome may fuse with a vesicle that contains a substance to be digested. While the products of digestion enter the cytoplasm, the nondigested residue is expelled from the cell.

Occasionally, a person is unable to manufacture an enzyme normally found within the lysosome. In these cases, the lysosome fills to capacity as the substrate for that enzyme accumulates. The cells may become so filled with lysosomes of this type that it brings about the death of the individual.

Lysosomes also carry out autodigestion, or the disposal of worn-out or damaged cell components such as mitochondria, which have a short life span in the cell. This is an essential part of the normal process of cytoplasmic maintenance and turnover. By turnover it is meant that the cell is constantly breaking down and remaking its parts. The lysosomes no doubt play a role in this process, and electron micrographs sometimes show mitochondria enclosed within these structures (fig. 2.13).

When a cell dies, or for some other reason is destined for total destruction, the lysosome sometimes releases its contents. The term *suicide bag* has been used in connection with the lysosome because of its ability to bring about a complete breakdown of the cell itself. Normally, complete cell destruction occurs when a change of shape is required by the organism. For example, the disappearance of a tadpole's tail may be brought about by lysosome action.

Energy-related Organelles

The energy-related organelles, mitochondria and chloroplasts, are both transformers, changing one form of energy into another. While chloroplasts are unique to plant cells, mitochondria are found in both plant and animal cells.

Mitochondria

A **mitochondrion** is a rather complex organelle that produces ATP energy. Every cell needs a certain amount of ATP energy to synthesize molecules, but many cells need ATP to carry out their specialized functions. For example, muscle cells need it for muscle contraction (fig. 2.14) and nerve cells need it for the conduction of nerve impulses.

Mitochondria are extremely efficient at what they do. One investigator, after calculating the weight of mitochondria in a horse's legs, suggested that the "magnitude of the effect of the mitochondria per weight unit is the same as the one delivered by the engines in a jet plane in vertical ascent. . . . In other words, the mitochondria are admirably effective machines."[2]

Mitochondria are often referred to as the powerhouses of the cell because, just as a powerhouse burns fuel to produce electricity, the mitochondria burn glucose products to produce ATP molecules, the chemical energy needed by cells. In the process, mitochondria use up oxygen and give off carbon dioxide and water. The oxygen you breathe in enters cells and then the mitochondria; the carbon dioxide you breathe out is released by the mitochondria. Since

[2]Bjorn Afzelius, *Anatomy of the Cell* (Chicago: University of Chicago Press, 1966), p. 11.

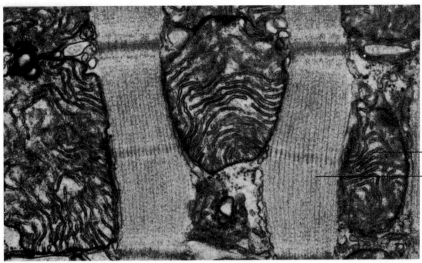

Figure 2.14
Mitochondria are especially abundant in heart muscle. The heart requires a continual supply of ATP in order to keep pumping the blood to the body.

_____ mitochondrion

_____ muscle filaments

Figure 2.15
Mitochondrion structure. _a._ An electron micrograph of a mitochondrion surrounded by rough endoplasmic reticulum. Note the shelflike cristae formed by the inner membrane. _b._ Diagrammatic drawing shows outer and inner structure more clearly.

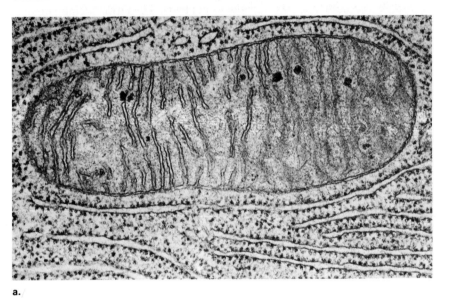

a.

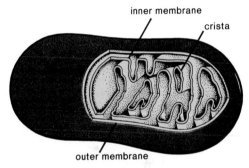

inner membrane

crista

outer membrane

b.

gas exchange is involved, it is said that mitochondria carry on **cellular respiration.** A shorthand way to indicate the chemical transformation associated with cellular respiration is:

carbohydrate + oxygen ⟶ carbon dioxide + water + energy

Each mitochondrion is composed of two membranes, an outer membrane and an inner membrane (fig. 2.15a, 2.15b). The inner membrane is convoluted into shelflike projections called **cristae.** The respiratory enzymes that aid in the production of energy are located in an assembly-line fashion on these membranous shelves. The membrane is divided into functional units, and a very small area of each crista contains one complete set of enzymes. The inner membrane lends itself to this type of arrangement and thus we see that structure aids function.

Chloroplasts

Plastids are membranous structures that often contain pigments and give plant cells their color. Some plastids, however, are colorless and act as storage bodies for starch, protein, or oils.

The most familiar and abundant plastid is the **chloroplast** (fig. 2.16). In higher plants, chloroplasts have an ovoid, or disklike, shape and may be even larger than mitochondria (table 2.1).

Within a chloroplast there are stacks of membranous sacs called **grana.** The green pigment, chlorophyll, is found within the grana, and thus chloroplasts are green. The inner portion of a chloroplast is called the **stroma.**

Chlorophyll is a chemical molecule that can absorb the energy of the sun. This energy allows photosynthesis to take place. **Photosynthesis** (synthesis by means of light energy) refers to the production of food molecules, glucose for example. The glucose molecules may later be joined together to form starch. Chloroplasts take in carbon dioxide, water, and radiant energy from the sun in order to produce glucose. They give off oxygen, which leaves the plant as a gas. Again, we can use the shorthand method to describe what has been said:

$$\text{energy} + \text{carbon dioxide} + \text{water} \longrightarrow \text{carbohydrate} + \text{oxygen}$$

The equation for photosynthesis is the opposite of cellular respiration, as you can see by comparing the shorthand statements for each.

Chloroplasts also illustrate that structure facilitates the function of an organelle. Within the grana, light energy is used to form ATP molecules, the type of energy that can be used by the enzymes within the stroma to make food molecules. It is believed that the grana are highly organized, just as the cristae of the mitochondria are organized. Electron micrographs show particles that are believed to be photosynthetic units.

Figure 2.16
Chloroplast structure. *a.* An electron micrograph. The darker regions are the grana joined by lamellae. *b.* Diagrammatic drawing shows outer and inner structure more clearly.

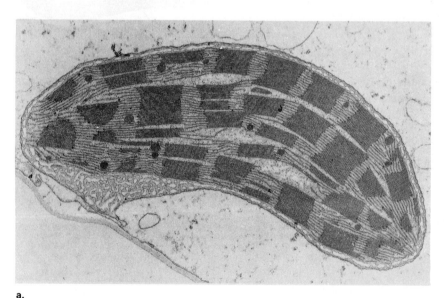

a.

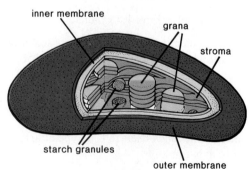

b.

The Cell, A Unit of Life

Centrioles and Related Organelles

All the organelles in this category are associated with maintaining cell shape or allowing the cell (fig. 2.17) or its components to move about.

Microfilaments and Microtubules

Microfilaments and microtubules assist cellular movement and help give a cell an internal cytoskeleton (or framework) (fig. 2.18). This cytoskeleton helps maintain the cell's shape, anchors the organelles, or allows them to move as appropriate. **Microfilaments** are extremely thin fibers (table 2.1) that usually occur in bundles or other groupings. Microfilaments have been isolated from a number of cells. When analyzed chemically, their composition is very similar to that of either *actin* or *myosin,* the two proteins responsible for the contraction of muscle.

Microtubules look like thin cylinders and are several times larger than microfilaments (table 2.1). Each cylinder contains thirteen rows of a globular protein, *tubulin,* arranged in a helical fashion. Aside from existing independently in the cytoplasm, microtubules are also found in certain organelles such as cilia, flagella, and centrioles.

Remarkably, both microfilaments and microtubules assemble and disassemble within the cell. When they are assembled, the protein molecules are bonded together and when they are disassembled, the protein molecules are not attached to one another. When microfilaments and microtubules are assembled, the cell has a particular shape and when they disassemble, the cell can change shape.

Figure 2.17
A scanning electron micrograph of an individual cell in a tissue culture. Notice the fingerlike projections on the "ruffle," which marks the leading edge of the cell. Microfilaments and microtubules are most likely present in this ruffle.

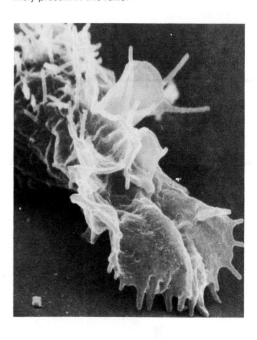

Figure 2.18
Cytoskeleton of cell. Notice that the various organelles are suspended by a cytoplasm that includes microtubules, microfilaments, and a lattice composed of very fine protein strands. *a.* Electron micrograph. *b.* Drawing of microtubules and microfilaments. *c.* Detailed structure of a microtubule. *d.* Detailed structure of a microfilament.

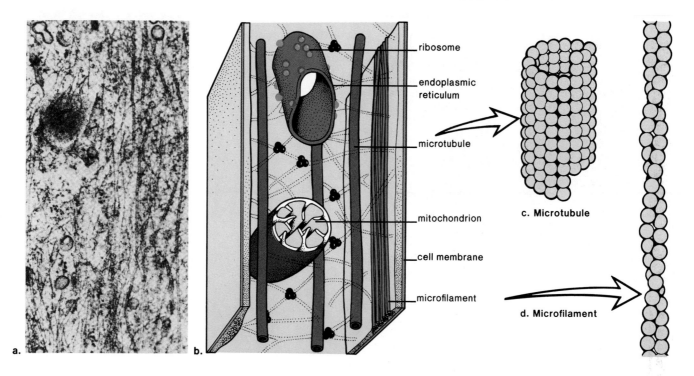

ribosome

endoplasmic reticulum

microtubule

mitochondrion

cell membrane

microfilament

c. Microtubule

d. Microfilament

Figure 2.19

Anatomy of cilia and flagella. *a*. Drawing showing
that the 9 + 2 pattern of microtubules within a
cilium or flagellum is derived (in some unknown
way) from the 9 + 0 pattern in a basal body.
b. Cross-section drawing of the 9 + 2 pattern
shows the exact arrangement of microtubules.
Notice the clawlike arms of the outer doublets and
the spokes that connect them to the central pair.
c. Electron micrograph of the cross section of a
basal body.

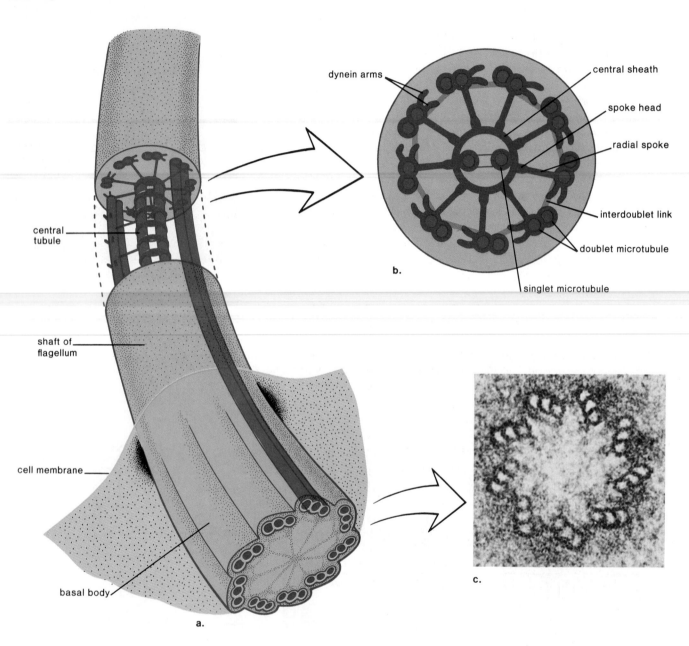

Cilia and Flagella

Cilia and **flagella** are hairlike projections of cells that can move either in an undulating fashion, like a whip, or stiffly like an oar. Cells that have these organelles are capable of movement. For example, a single-celled paramecium (p. 547) moves by means of cilia. Sperm, on the other hand, which carry genetic material to the egg, move by means of flagella. In our bodies, the cells that line the upper respiratory tract are ciliated (p. 265). The action of the cilia sweeps debris trapped within mucus back up into the throat and in this way helps keep the lungs clean.

Cilia are much shorter than flagella (table 2.1), but even so they both are constructed similarly (fig. 2.19). They are membrane-bound cylinders enclosing a matrix area. In the matrix are nine microtubule doublets arranged in a circle around two central microtubules (fig. 2.19b). This is called the 9 + 2 pattern of microtubules. Each doublet also has pairs of arms projecting toward a neighboring doublet and spokes extending toward the central pair of microtubules. Recent evidence indicates that cilia and flagella move when the microtubule doublets slide along one another. The claw-like arms and spokes seem to be involved in causing this sliding action, which requires ATP energy.

Each cilium and flagellum has a basal body (fig. 2.19c) lying in the cytoplasm at its base. **Basal bodies,** which are short cylinders with a circular arrangement of nine microtubule triplets called the 9 + 0 pattern, are believed to organize the structure of cilia and flagella.

Centrioles

Centrioles are short cylinders, with a 9 + 0 pattern of microtubule triplets. There are usually two pairs (fig. 2.1a) lying to one side of the nucleus in certain eukaryotic cells such as lower fungi, lower plants, and animal cells. The members of each pair are at right angles to one another.

Centrioles give rise to basal bodies that direct the formation of cilia and flagella. Centrioles may also be involved in other cellular processes that use microtubules, such as the movement of material throughout the cell or the appearance and disappearance of the spindle apparatus (p. 84) in animal cells. Their exact role in these processes is uncertain however.

Cellular Comparisons

Prokaryotic versus Eukaryotic Cells

Thus far in this chapter we have been discussing **eukaryotic cells,** a term that means that the cell has a membrane-bound, or "true," nucleus (karyote). **Prokaryotic cells** evolved before eukaryotic cells and they lack a true nucleus.

Table 2.2 compares prokaryotic cells to the two types of eukaryotic cells studied in this chapter. You'll notice that prokaryotic cells, represented only

Table 2.2 Comparison of prokaryotic and eukaryotic cells

	Prokaryotic	Eukaryotic	
		Animal	**Plant**
Cell membrane	yes	yes	yes
Cell wall	yes	no	yes
Nuclear membrane	no	yes	yes
Mitochondria	no	yes	yes
Chloroplasts	no	no	yes
Endoplasmic reticulum	no	yes	yes
Ribosomes	yes, small	yes, large	yes, large
Vacuoles	no	yes, small	yes, large, central
Lysosomes	no	yes, usually	no, usually
Centrioles	no	yes	no

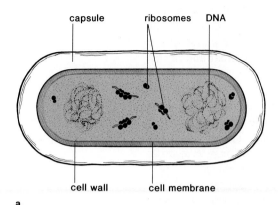

a.

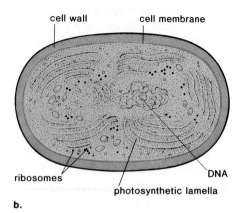

b.

by bacteria and cyanobacteria (formerly termed blue-green algae), also lack most of the other types of organelles we have been discussing (fig. 2.20). This does not mean, however, that their cells do not carry on the functions performed by organelles. The functions simply occur within the cytoplasm of these much smaller cells. Thus they have chromosomes but not nuclei; respiratory enzymes but no mitochondria; and when they have chlorophyll, there are no chloroplasts, although the chlorophyll is associated with lamellae.

Prokaryotes do have a cell wall and if motile, they possess flagella. However, the structure of these cell features differs from those found in eukaryotic cells.

Plant versus Animal Cells

Table 2.2 also compares plant and animal cells. It is wise to keep in mind that while all cells have a cell membrane, plant cells have a cell wall in addition; and that while all eukaryotic cells have mitochondria, plant cells also have chloroplasts. However, higher plant cells typically lack centrioles and thus the cilia and flagella of animal cells. Even so, plant cells do have a mitotic spindle as will be discussed in chapter 4.

Summary

Figure 2.21 summarizes the animal cell organelles we have studied in this chapter. To simplify their study, organelles are divided into four categories: (1) the nucleus, (2) membranous vacuoles and canals, (3) energy-related organelles, and (4) centrioles and related organelles.

The nucleus is a large centrally located organelle of primary importance since it controls the rest of the cell. Within the nucleus lies the chromatin material, which condenses to become chromosomes during cell division. Chromosomes contain DNA that, with the help of RNA, directs protein synthesis in the cytoplasm. Another type of RNA, called ribosomal RNA, is made within the nucleolus before migrating to the cytoplasm and becoming incorporated into ribosomes.

Membrane, which is abundant in eukaryotic cells but limited in prokaryotic cells, is composed of phospholipid molecules and protein molecules.

The Cell, A Unit of Life

The endoplasmic reticulum is a membranous system of tubular canals that ramify through the cell and may have attached ribosomes. Proteins synthesized at the ribosomes are sometimes packaged at the Golgi apparatus for export. The Golgi apparatus is a stack of saccules with associated vesicles. One type of vesicle moves toward the cell membrane where it discharges its contents. Other vesicles are lysosomes that contain digestive enzymes capable of breaking down the large molecules that form the cell itself. Usually, lysosomes fuse with incoming vacuoles to digest any material enclosed therein, but lysosomes sometimes carry out autodigestion of old parts of the cell.

The energy-related organelles include mitochondria and chloroplasts. Mitochondria are composed of two layers of membrane; the inner layer forms shelves called cristae. Here are found enzymes that convert carbohydrate energy to ATP energy; mitochondria are called the powerhouses of the cell for this reason. Since the breakdown of carbohydrates requires gas exchange, the mitochondria are said to carry out cellular respiration.

Chloroplasts are a type of plastid composed of two sets of membrane. The inner membrane is arranged in stacks of grana, and chlorophyll is located within these stacks. When photosynthesis occurs, chlorophyll absorbs the energy of the sun and subsequently it is transformed to ATP, a molecule that may be used by enzymes located in the stroma, the inner portion of the chloroplast where carbohydrates are produced.

The centrioles and related organelles are associated with maintaining cell shape and with the movement of either cell contents or of the cell itself. Microfilaments and microtubules are found within a cytoskeleton that maintains shape and also directs the movement of cell parts. Microfilaments are thin actin or myosin strands, but microtubules contain 13 rows of tubulin protein molecules arranged to form a hollow cylinder. Cilia and flagella contain microtubules arranged in a 9 + 2 pattern. The sliding of these microtubules past one another is believed to produce the motion of these organelles. Cilia and flagella arise from basal bodies that contain microtubules in a 9 + 0 pattern, also seen in centrioles that give rise to basal bodies. Centrioles are in some way connected to the origination of microtubules and the spindle fibers that are seen during cell division.

Prokaryotic cells lack the organelles typically found in eukaryotic cells. Nevertheless, they carry on all the same functions as eukaryotic cells.

Study Questions

1. Briefly define the cell theory. (p. 44)
2. Describe the structure and biochemical makeup of membrane. (pp. 48–49)
3. Describe the nucleus and its contents, including the terms *DNA* and *RNA* in your description. (p. 50)
4. How does the experiment with *Acetabularia* illustrate the function of the nucleus in the cell? (p. 50)
5. Describe the structure and function of endoplasmic reticulum. Include the terms *rough* and *smooth ER, ribosomes, Golgi apparatus, vacuoles,* and *lysosomes* in your description. (pp. 50–54)
6. Describe the structure and function of mitochondria and chloroplasts. (pp. 54–56)
7. Describe the structure and function of microfilaments, microtubules, centrioles, cilia, and flagella. (pp. 57–59)
8. What are the two main types of cells and how do they differ structurally? (p. 59)
9. What are the structural differences between animal and plant cells? (p. 60)

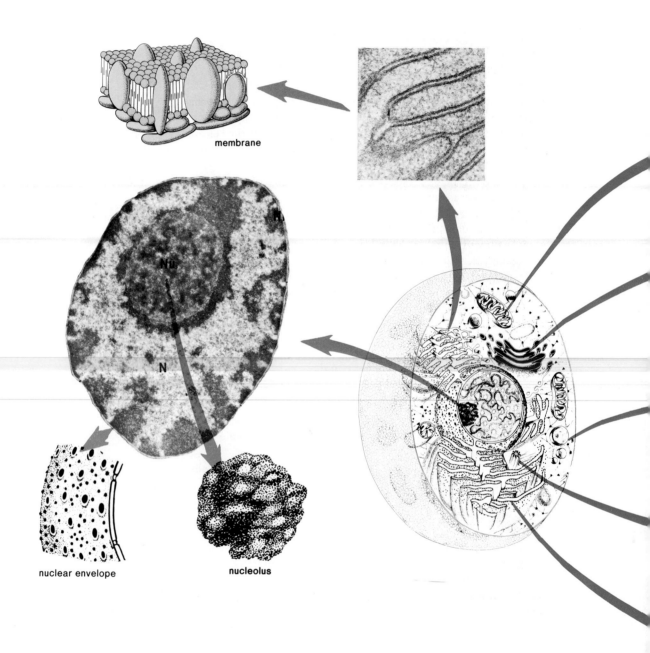

membrane

nuclear envelope

nucleolus

The Cell, A Unit of Life

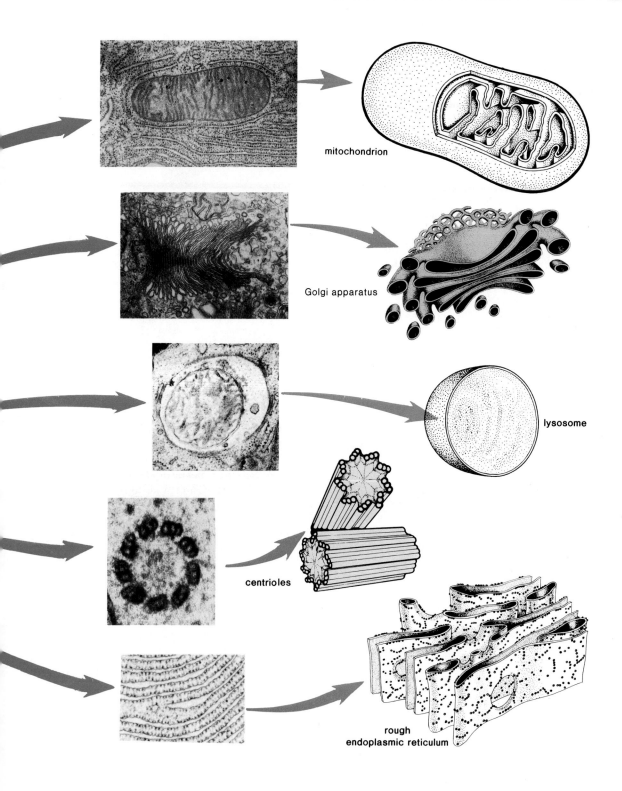

mitochondrion

Golgi apparatus

lysosome

centrioles

rough
endoplasmic reticulum

Selected Key Terms

prokaryotic (pro''kar-e-ot'ik)
eukaryotic (u''kar-e-ot'ik)
cytoplasm (si'to-plazm)
organelle (or''gah-nel')
nucleus (nu'kle-us)
chromatin (kro'mah-tin)
chromosomes (kro'mo-sōmz)
nucleolus (nu-kle'o-lus)
endoplasmic reticulum (en-do-plaz'mic rĕ-tik'u-lum)
ribosomes (ri'bō-sōmz)
polysome (pol'e-sōm)
Golgi apparatus (gol'ge ap''ah-ra'tus)
lysosome (lī'so-sōm)
mitochondrion (mi''to-kon'dre-on)
chloroplast (klo'ro-plast)
microfilament (mi''kro-fil'ah-ment)
microtubule (mi''kro-tu'būl)
cilia (sil'e-ah)
flagella (flah-jel'ah)
centriole (sen'tre-ōl)

Further Readings

Avers, C. J. 1981. *Cell biology.* 2d ed. New York: D. Van Nostrand Co.

Berns, M. W. 1983. *Cells.* 2d ed. New York: Holt, Rinehart & Winston.

Brachet, J. 1961. The living cell. *Scientific American* 205(3):39.

Capaldi, R. A. 1974. A dynamic model of cell membranes. *Scientific American* 230(3):26.

Dustin, P. 1980. Microtubules. *Scientific American* 243(2):66.

Engelman, D. M., and Moore, P. B. 1976. Neutron-scattering studies of the ribosome. *Scientific American* 235(4):44.

Fox, C. F. 1972. The structure of cell membranes. *Scientific American* 226(2):30.

Grivell, L. A. 1983. Mitochondrial DNA. *Scientific American* 248(3):78.

Jensen, W. A. 1970. *The plant cell.* 2d ed. Belmont, Calif.: Wadsworth.

Lake, J. A. 1981. The ribosome. *Scientific American* 245(2):84.

Lazarides, E., and Revel, J. P. 1979. The molecular basis of cell movement. *Scientific American* 240(5):100.

Loewy, A. G., and Siekevitz, P. 1970. *Cell structure and function.* 2d ed. New York: Holt, Rinehart & Winston.

Neutra, M., and Leblond, C. P. 1969. The Golgi apparatus. *Scientific American* 220(2):100.

Novikoff, A. B., and Holtzman, E. 1976. *Cells and organelles.* 2d ed. Modern Biology Series. New York: Holt, Rinehart & Winston.

Porter, K. R., and Bonneville, M. A. 1973. *Fine structure of cells and tissues.* 4th ed. Philadelphia: Lea & Febiger.

Porter, K. R., and Tucker, J. B. 1981. The ground substance of the living cell. *Scientific American* 244(3):56.

Satir, P. 1975. The final step in secretion. *Scientific American* 233(4):28.

Unwin, N., and Henderson, R. 1984. The structure of proteins in biological membranes. Scientific American 250(2):78.

Wessells, N. K. 1971. How living cells change shape. *Scientific American* 225(4):12.

3

Cell Membrane and Cell Wall Function

Chapter concepts

1 The cell membrane is selectively permeable, allowing some substances to pass through freely and restricting the passage of other substances.

2 Small molecules can pass through a cell membrane along a concentration gradient either by diffusion or facilitated transport, the latter requiring protein carriers.

3 Osmosis is the diffusion of water across a selectively permeable membrane. The phenomenon of osmosis can affect the concentration of substances within cells.

4 Active transport is the passage of molecules against a concentration gradient requiring protein carriers and an expenditure of cellular energy.

5 Endocytosis is vesicle formation to take large molecules or matter into the cell, and exocytosis is vesicle formation to discharge substances from the cell.

6 Within tissues, cells form junctions and one of these, the gap junction, has cell-to-cell channels that allow exchange of molecules.

Cell Membrane

The fluid mosaic model of membrane structure, discussed in chapter 2, also applies to the **cell membrane.** There is a double layer of phospholipid molecules, having the consistency of light oil, in which protein molecules are either partially or wholly embedded. Since the proteins are scattered throughout the membrane, they form a mosaic pattern. The presence of proteins in the membrane is revealed when the membranes of red blood cells are first frozen and then fractured before they are subjected to electron microscopy (fig. 3.1).

Figure 3.1
Fluid mosaic model of a cell membrane. *a.* Protein molecules are embedded in and project to either side of a double layer of phospholipid molecules. *b.* Enlargement of a section of the membrane to show glycoproteins and glycolipids. *c.* A frozen cell that is fractured before being viewed by a scanning electron microscope reveals proteins that are embedded, along the fracture line, in the membrane. *d.* Electron micrograph of red blood cell membranes prepared by the freeze-fracture technique.

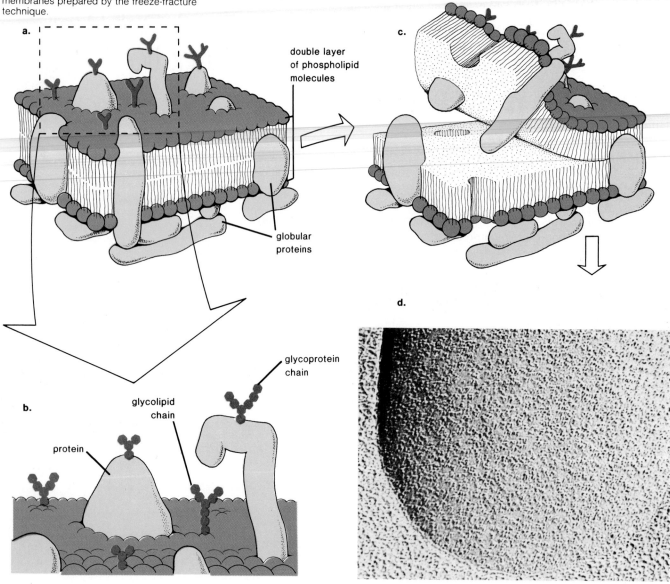

The Cell, A Unit of Life

Table 3.1 Passage of molecules into and out of cells

Name	Direction	Requirements	Examples
Diffusion	Toward lesser concentration	———	Lipid-soluble molecules Water Gases
Transport			
Facilitated	Toward lesser concentration	Carrier	Sugars and amino acids
Active	Toward greater concentration	Carrier plus energy	Sugars, amino acids, and ions
Endocytosis and exocytosis			
Pinocytosis	Either	Vacuole formation plus energy	Macromolecules
Phagocytosis	Either	Vacuole formation plus energy	Cells or subcellular material

The cell membrane, unlike the intracellular membrane discussed previously, also contains carbohydrate. Simple sugars are strung together in chains that are attached to proteins and lipids. The chains are always on the outside of the membrane and probably function as markers to identify the cell. Cells that are not recognized by the body are apt to be rejected, as sometimes happens when individuals receive organ transplants. On the other hand, cell recognition processes may be faulty when cancerous cells are permitted to grow and reproduce.

The cell membrane serves as a boundary for the cell and regulates the entrance and exit of molecules into and out of the cell. This property of the cell membrane is most likely, at least in part, dependent on the proper orientation of proteins within the membrane. The manner in which proteins assist the transport of molecules is discussed in detail in the following paragraphs.

Cell Membrane Permeability

The cell membrane is not freely permeable; some molecules enter cells and others do not. In this regard, then, the cell membrane can be called semipermeable. A permeable membrane allows all molecules to pass through; an impermeable membrane allows no molecules to pass through; and a semipermeable membrane allows some molecules to pass through. Certain small molecules can cross a cell membrane, while large molecules cannot cross a membrane. However, some small molecules pass through the cell membrane quickly, while others have difficulty in passing through or fail to pass through at all. Therefore, a cell membrane is often regarded as **selectively permeable** rather than semipermeable.

As listed in table 3.1, there are three general means by which substances can enter and exit cells from the surrounding medium: (1) diffusion, (2) transport by carriers, and (3) endocytosis and exocytosis.

Diffusion

Diffusion is a physical process that can be observed with any type of particle. Diffusion is such a universal phenomenon that there is a physical law called the law of diffusion, which states that *particles move from the area of greater concentration to the area of lesser concentration until equally distributed.* To illustrate diffusion, imagine opening a perfume bottle in the corner of a room (fig. 3.2a). The smell of the perfume soon permeates the room because the molecules that make up the perfume have drifted to all parts of the room. Another example is putting a tablet of dye into water (fig. 3.2b). The water eventually takes on the color of the dye as the tablet dissolves.

Figure 3.2
Diffusion occurs when *(a.)* a perfume bottle is opened and the scent fills a room because the molecules have moved away from the bottle and *(b.)* a tablet of dye is placed in a beaker and water becomes colored because the molecules have moved away from the original area of the tablet.

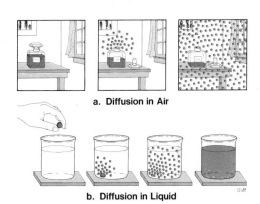

a. Diffusion in Air

b. Diffusion in Liquid

Figure 3.3
Certain molecules move freely across the
membrane by means of diffusion. Notice that a
carrier protein is not utilized for diffusion.

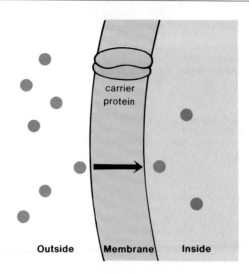

Figure 3.4
Oxygen *(dots)* diffuses into the capillaries of the
lungs because there is a greater concentration of
oxygen in the air sacs than in the capillaries.

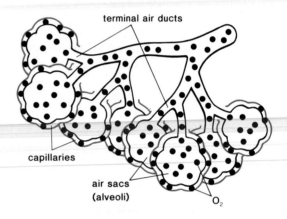

The movement of molecules by diffusion alone is a slow process. The
rate of diffusion is affected by the **concentration gradient** (the difference in
concentration of the diffusing molecules between the two regions), the size
and shape of the molecule, and the temperature. Also, diffusion in a liquid
medium is slower than in a gaseous medium; however, distribution of molecules in cytoplasm is speeded up by an ever-constant flow of the cytoplasm
that is called cytoplasmic streaming.

The chemical and physical properties of the cell membrane allow few
types of molecules to enter and exit by means of diffusion (fig. 3.3). Lipid-
soluble molecules, such as alcohols, can diffuse through the membrane simply
because phospholipids are the membrane's main structural components.

Gases can also diffuse through the membrane; this is the mechanism by
which oxygen enters cells and carbon dioxide exits cells. As an example, consider the movement of oxygen from the air sacs (alveoli) of the lungs to the
blood in lung capillaries (fig. 3.4). After inspiration (breathing in), the concentration of oxygen in the alveoli is greater than the concentration of oxygen
in the blood; therefore oxygen diffuses into the blood.

Water passes into and out of cells with relative ease. Since water is not
lipid soluble, it has been hypothesized that the membrane contains protein-
lined pores large enough to allow the passage of water and perhaps some ions
(fig. 3.5). Other molecules cannot utilize these pores because they are either
too large or they carry a charge that prevents passage. The fact that water
can penetrate a membrane has important biological consequences, as described in the discussion that follows.

The Cell, A Unit of Life

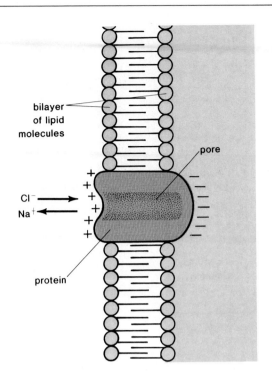

bilayer of lipid molecules

pore

Cl^- →

Na^+ ←

protein

Figure 3.5

Water is presumed by some investigators to enter the cell by way of protein-lined pores that are charged similarly to the membrane itself. Ions would have difficulty passing through the pore. For example, positive ions trying to enter the cell would be repelled by the like charge, and negative ions would be at first attracted but then would tend to be held at the surface.

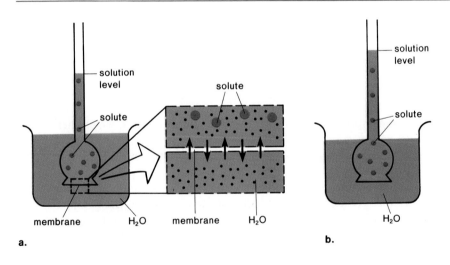

solution level

solute

solute

membrane

H_2O

membrane

H_2O

solution level

solute

H_2O

a.

b.

Figure 3.6

Osmosis demonstration. *a.* A thistle tube, covered at the broad end by a membrane, contains a solute *(large circles)* in addition to a solvent *(small circles).* The beaker contains only solvent. The solute is unable to pass through the membrane, but the solvent passes through in both directions. *b.* There is a net movement of solvent toward the inside of the thistle tube. This causes the solution to rise in the thistle tube until a back pressure develops that prevents any further net gain of solvent.

Osmosis

The diffusion of water across a selectively permeable membrane has been given a special term: it is called **osmosis.** Osmosis is defined as the *net movement of water molecules from the area of greater concentration of water to the area of lesser concentration of water across a selectively permeable membrane.*

To illustrate osmosis, a thistle tube containing a protein solution and covered at one end by a membrane is placed in a beaker of distilled water. Inside the tube is a solution containing both a **solute** (particles) and a **solvent** (water), while outside the tube is pure water (fig. 3.6a). Obviously, the greater concentration of water is outside the tube, and therefore there will be a net movement of water from outside the tube to inside the tube across the membrane. The solute (protein) is unable to pass out of the tube because the membrane is impermeable to it; therefore the solution within the tube rises (fig. 3.6b).

Figure 3.7

Effect of tonicity on an animal cell. In an isotonic solution, there is no net movement of water and the appearance of a red blood cell remains the same. In a hypertonic solution, there is a net movement of water *(arrow)* to the outside of the cell and the cell shrinks. In a hypotonic solution, there is a net movement of water *(arrow)* to inside the cell and the cell swells to bursting. (Circles = solute; stipples = solvent)

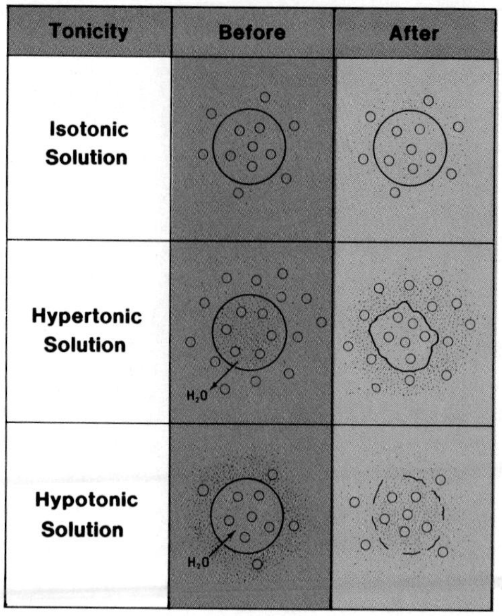

The pressure due to the flow of water from the area of greater concentration to the area of lesser concentration is called **osmotic pressure.** In this case, osmotic pressure will eventually be counterbalanced by a hydrostatic pressure (pressure exerted by liquids) caused by the height of the solution. When hydrostatic pressure equals osmotic pressure, **equilibrium** is reached, the net flow of water ceases, and the solution rises no higher in figure 3.6b because as many water molecules now exit the tube as enter the tube.

Notice that in this illustration of osmosis

1. a membrane separates two liquids;
2. the membrane is impermeable to the solute particles, which therefore do not move through the membrane;
3. a difference in water concentration exists on the two sides of the membrane;
4. the membrane is permeable to water, which therefore moves from the area of greater concentration to the area of lesser concentration;
5. due to the process of osmosis or, if you prefer, due to osmotic pressure, the amount of liquid increases on the solution side of the membrane.

These considerations will be important as we discuss osmosis in relation to cells placed in different solutions. Recall that solutions are made up of two parts: the solute and the solvent. The solute is the solid substance dissolved or suspended in the solvent, which is usually water. The concentrations of solutions are usually described in terms of percentages of solute; for example, a 10 percent salt solution. This solution would contain 90 percent water.

Cells may be placed in solutions that contain the same number of water molecules per volume, a lesser number of water molecules per volume, or a greater number of water molecules per volume than does the cell. These solutions are called isotonic, hypertonic, and hypotonic, respectively. Figure 3.7 depicts and table 3.2 describes the effects of these solutions on animal and plant cells.

The Cell, A Unit of Life

Table 3.2 Effect of osmosis on cells

Tonicity of solution	Description	Results Animal	Plant
Isotonic	Equal concentration of water on both sides of cell membrane	No change	No change
Hypertonic	Lesser water and greater solute concentration outside the cell than inside	Shrink	Plasmolysis
Hypotonic	Greater water and lesser solute concentration outside the cell than inside	Swell to bursting	Turgor pressure

Isotonic Solutions

In the laboratory, cells are normally placed in solutions that cause them neither to gain nor to lose water. Such solutions are said to be **isotonic;** that is, the number of water molecules per volume is the same on both sides of the membrane, and thus there is no net gain or loss. The term *iso* means "the same as" and the term *tonicity* refers to the strength of the solution.

It is possible to determine, for example, that a 0.9 percent solution of the salt sodium chloride (Na^+Cl^-) is isotonic to red blood cells because the cells neither swell nor shrink when placed in such a solution (fig. 3.8).

Hypertonic Solutions

Solutions that cause cells to shrink or shrivel due to a loss of water are said to be **hypertonic** solutions. The prefix *hyper* means "greater than" and refers to a solution with a *greater concentration of solute* (lesser concentration of water) than the cell. If a cell is placed in a hypertonic solution, water will leave the cell—the net movement of water is from the inside to the outside of the cell (fig. 3.7).

A 10 percent solution of sodium chloride is hypertonic to red blood cells. In fact, any solution with a concentration higher than 0.9 percent NaCl is hypertonic to red blood cells. If red blood cells are placed in this solution, they will shrink (fig. 3.9). The term **crenated** is used to refer to red cells in this condition.

Hypotonic Solutions

Solutions that cause cells to swell or even burst due to an intake of water are said to be **hypotonic** solutions. The prefix *hypo* means "less than" and refers to a solution with a *lesser concentration of solute* (greater concentration of water) than the cell. If a cell is placed in a hypotonic solution, water will enter the cell—the net movement of water is from the outside to inside of the cell (fig. 3.7).

Any salt solution less than 0.9 percent is a hypotonic solution to red blood cells. Red blood cells placed in such a solution will expand (fig. 3.10) and even burst due to the buildup of osmotic pressure. The term **lysis** is used to refer to disrupted cells.

Importance

Osmosis occurs constantly in living organisms. For example, due to osmosis, water is absorbed from the human large intestine, retained by the kidneys, and taken up by the blood. Since living things contain a very high percentage of water, osmosis is an extremely important physical process for their continued good health.

Figure 3.8
Scanning electron micrograph of normal-appearing red blood cells. If red blood cells are placed in an isotonic solution, they have this appearance.

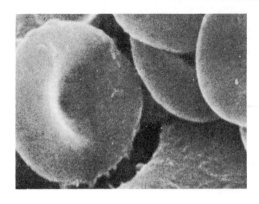

Figure 3.9
Scanning electron micrograph of red blood cells that were placed in a hypertonic solution. The cells are crenated due to the loss of water.

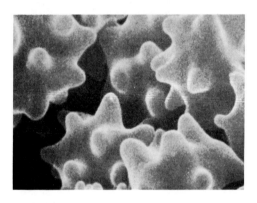

Figure 3.10
Scanning electron micrograph of red blood cells that were placed in a hypotonic solution. The cells are swollen due to the uptake of water.

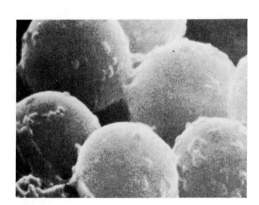

Figure 3.11

All plant cells are surrounded by a primary cell wall in addition to the cell membrane. Some, as illustrated here, also have a secondary cell wall that forms inside the primary wall. The middle lamella is a complex organic substance that cements two adjacent cell walls together.

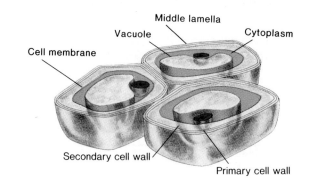

Figure 3.12

Effect of tonicity on a plant cell. In an isotonic solution, the appearance of a plant cell remains the same because there is no net movement of water. In a hypertonic solution, the cell membrane withdraws from the cell wall as the plant vacuole shrinks because of a net movement of water to outside the cell *(arrow)*. In a hypotonic solution, the cell membrane is pressed up against the cell wall and the vacuole expands due to a net movement of water to inside the cell *(arrow)*.

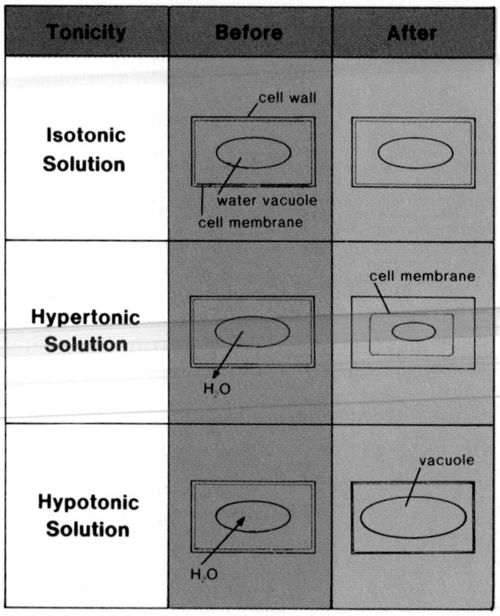

Tonicity	Before	After
Isotonic Solution	cell wall / water vacuole / cell membrane	
Hypertonic Solution	H_2O	cell membrane
Hypotonic Solution	H_2O	vacuole

Figure 3.13

Effect of tonicity on a plant cell. *a.* Plasmolysis occurs when plant cells are placed in a hypertonic solution. *b.* Plant cells develop turgor pressure when they are placed in a hypotonic solution.

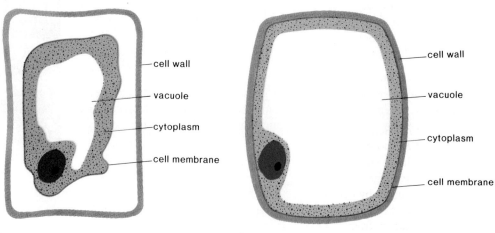

a.   b.

The Cell, A Unit of Life

Cell Wall

Plant Cells

In addition to the cell membrane, plant cells are surrounded by rigid, boxlike layers of cellulose that collectively are called the **cell wall** (fig. 3.11). This inflexible wall lies external to the cell membrane and maintains the shape of the plant cell. It also gives support to the plant as a whole. Sometimes in woody plants, the cells produce a secondary cell wall that lies inside the first primary wall. The secondary wall is impregnated with lignin, a hardening material.

The cell wall is porous, allowing all types of molecules to pass through freely; the rest of the plant cell, however, responds to hypertonic and hypotonic solutions, as described in figure 3.12 and in the following paragraphs.

Plasmolysis

When a plant cell is placed in a hypertonic solution, the large central vacuole loses water, and the cell membrane pulls away from the wall (fig. 3.13a). In this instance, the plant cell is said to have undergone **plasmolysis.**

Turgor Pressure

When a plant cell is placed in a hypotonic solution, the cytoplasm and large central vacuole gain water, and the cell membrane pushes against the rigid cell wall (fig. 3.13b). The hydrostatic pressure increases until it is equal to the osmotic pressure; the plant cell does not burst because the cell wall does not give way. Under these circumstances, the plant cells are said to have developed **turgor pressure.** Turgor pressure in plant cells is extremely important to maintaining the erect position of the plant.

Transport by Carriers

Facilitated Transport

The concept of **facilitated transport** explains the passage of molecules, such as glucose and amino acids, which are observed to cross the cell membrane even though they are lipid insoluble. There is evidence that the passage of these molecules is facilitated by a reversible combination with proteins that in some manner transport the smaller molecules through the cell membrane. These proteins, called **carriers,** are highly specific and combine only with one particular type of molecule. For example, various sugar molecules of identical size might be present inside or outside the cell, but certain ones cross the membrane hundreds of times faster than others. As stated earlier, this is the reason that the membrane may be called selectively permeable.

A model of a facilitated transport system (fig. 3.14) shows that after a carrier has assisted the movement of a molecule to the other side of the membrane, it is free to assist the passage of other similar molecules. Neither simple diffusion, explained previously, nor facilitated transport by means of protein carriers requires an expenditure of energy by the cell because the molecules are moving down a concentration gradient in the same direction they would tend to move anyway. Sometimes, therefore, facilitated transport is called facilitated diffusion, or passive transport. In passive transport, a substance moves only from the area of higher concentration to the area of lower concentration.

Active Transport

Due to **active transport,** molecules or ions move through the cell membrane against a concentration gradient, accumulating either inside or outside the cell in the region of *higher* concentration. For example, iodine collects in the cells of the thyroid gland; sugar is completely absorbed from the gut by the cells lining the digestive tract; and sodium (Na^+) is sometimes almost completely withdrawn from urine by cells lining the kidney tubules.

Both protein carriers and an expenditure of energy (fig. 3.15) are needed to transport substances from an area of lesser concentration to an area of higher

Figure 3.14
Facilitated transport is apparent when certain molecules easily cross a cell membrane toward the area of lesser concentration. A protein carrier is presumed to transport these molecules across the membrane.

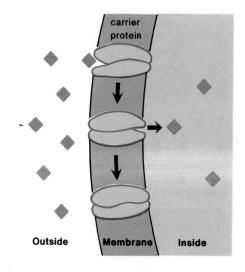

Figure 3.15
Active transport is apparent when a molecule crosses the cell membrane toward the area of greater concentration. An expenditure of ATP energy is required, presumably to allow a protein carrier to transport molecules across the cell membrane.

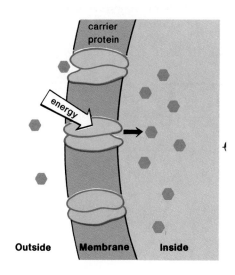

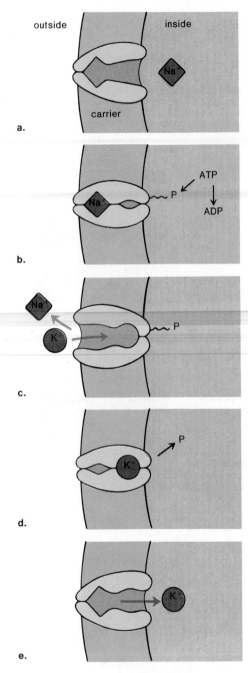

concentration. In this case, energy (ATP molecules) is required to cause the carrier to combine with the substance to be transported. Thus it is not surprising that cells primarily involved in active transport, such as kidney cells, have a large number of mitochondria near the membrane where active transport is occurring (fig. 14.12).

One type of active transport present in all cells, but especially important in nerve and muscle cells, is the active accumulation of sodium ions outside the cell and a corresponding accumulation of potassium ions within the cell. These two events are presumed to be linked, and a **sodium/potassium pump** has been hypothesized. There is evidence that the "pump" is a protein molecule capable of combining with both sodium and potassium ions. Externally the protein combines with potassium, assisting its passage to the inside of the cell; internally the protein combines with sodium, assisting its passage to the outside of the cell. Presumably the shape of the carrier changes so that it can alternately move potassium to the inside of the cell and move sodium to the outside of the cell (fig. 3.16). ATP energy released by an ATPase (an enzyme that breaks down ATP) is believed to be necessary to bring about the necessary change in shape. A recent study discussed in the reading on page 75 has shown that obese individuals have a lower amount of ATPase than do persons with normal weight. The investigators speculate that these individuals are obese because of a malfunctioning of the sodium/potassium pump.

Endocytosis and Exocytosis

Endocytosis
At times, large molecules or other matter become incorporated into cells by the process of **endocytosis** (fig. 3.17), which requires the formation of a vesicle. Endocytosis, even when not moving substances against a concentration, requires energy.

When the material taken in by the process is quite large, the process is called **phagocytosis** (cell eating). Phagocytosis is common to amoeboid-type cells, such as human white blood cells. These cells phagocytize bacteria, helping the body fight infections.

Vesicles also form around large-sized molecules such as proteins; this is called **pinocytosis** (cell drinking). Whereas phagocytosis can be seen with the light microscope, pinocytosis requires the use of the electron microscope.

Once formed, vesicles or vacuoles contain a substance enclosed by membrane. In order that these substances might be broken down and incorporated into the cytoplasm, digestion is required. Therefore, it is believed that lysosomes probably fuse with these bodies in order that digestive enzymes may begin to break down the molecules they contain.

Exocytosis
Exocytosis (fig. 3.18) is the reverse of endocytosis and requires that a vesicle fuse with the membrane, thereby discharging its contents. As we saw in chapter 2, vesicles formed at the Golgi apparatus transport cell products out of the cell; this entire process is called secretion. Also, residues remaining after digestion by lysosome enzymes may be discharged from the cell by fusion of a vesicle with the cell membrane.

Fat folks have long made the familiar claim, "My metabolism is off." Doctors have long answered that they just eat too much. Lately, though, scientists have been more willing to entertain the notion that biochemistry may indeed play a critical role in weight control. How else to explain thin people who gorge but never gain a pound and hefties who eat little? Now three researchers have announced evidence of a link between obesity and a biochemical abnormality. Drs. Jeffrey Flier, Mario De Luise and George Blackburn at Boston's Beth Israel Hospital compared blood samples taken from 28 normal adults with those from 21 obese individuals. They found that the levels of the enzyme adenosine triphosphatase, or ATPase, were 22% lower in the red blood cells of the fat subjects, and that the heavier the person the lower the enzyme level. ATPase is critical to a basic process, the pumping of sodium and potassium across cell membranes. The exchange generates an estimated 20% to 50% of the body's total heat production and consumes considerable calories. Low levels of ATPase, speculates Dr. Flier, may "predispose people to be overweight by causing fewer calories to be burned up as heat and more to be stored as fat."

Figure 3.17

Endocytosis. In both phagocytosis and pinocytosis, the cell membrane forms a vesicle around the substance to be taken in.

Figure 3.18

Exocytosis. A substance within a vacuole is deposited outside the cell when the vesicle fuses with the cell membrane.

cell membrane
substance

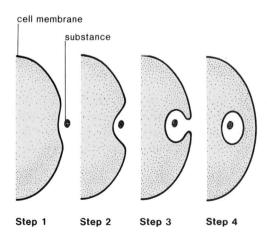

Step 1   Step 2   Step 3   Step 4

cell membrane
substance

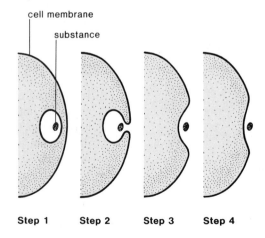

Step 1   Step 2   Step 3   Step 4

Figure 3.19
The same three types of intercellular junctions are
shown by means of (a.) transmission electron
micrograph and (b.) a scanning electron
micrograph. To the far right, a tight junction (TJ)
joins the two adjoining cell membranes. Midcenter,
there is a band desmosome (BD) and to the far left
there is a spot desmosome (SP).

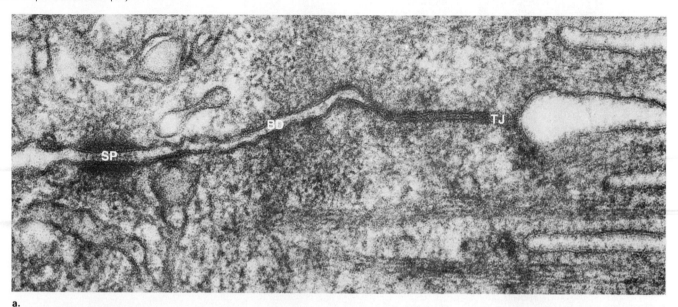

a.

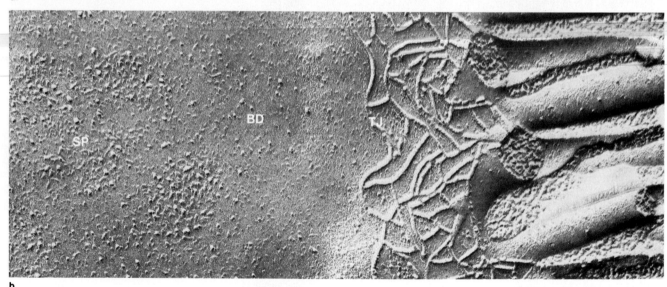

b.

Junctions between Cells

In multicellular organisms, cells often form tissues in which the cells function
and work together. Electron micrographs of animal tissues that line cavities
have shown that cell membranes do not merely touch—they are joined by
three types of specialized junctions: tight junctions, desmosomes, and gap
junctions.

The first two types of junctions mentioned bind tissue cells together. In
a **tight junction** (fig. 3.19) cell membranes from two cells interlock in a zip-
perlike fashion. If many cells are joined by tight junctions there can be no
leakage of molecules into or out of a tissue. The cells that line the intestines

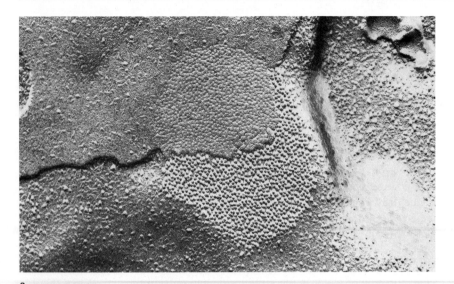

a.

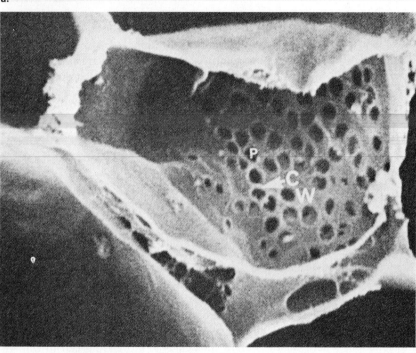

b.

Figure 3.20
Exchange of molecules between cells is sometimes possible. *a*. The closely packed particles shown in this micrograph are channels between animal cells in the region of a gap junction. *b*. Molecules can pass from one sieve tube plant cell to another by means of the pores shown in the end wall of one such cell.

are joined by tight junctions and this prevents intestinal contents from leaking into this tissue. In a **desmosome** (fig. 3.19), a fibrous plaque just within each cell is joined by a dense mat of thin filaments and this gives the two cells greater stability.

A **gap junction** is different from the first two because it allows an exchange of molecules between cells. In a gap junction (fig. 3.20a), protein molecules from each cell membrane aggregate to form cell-to-cell channels. By utilizing tracer molecules, it has been possible to show that ions, most sugars, amino acids, nucleotides, and vitamins are able to move from cell to cell through the channels of a gap junction. Thus gap junctions represent another way in which molecules enter and exit cells.

Certain plant cell tissues also have connections between cells. For example, organic nutrients translocate in tube-shaped plant cells that are joined by channels at their end walls. These end walls are appropriately called **sieve plates** (fig. 3.20b).

Summary

The fluid mosaic model of membrane structure is also appropriate for the cell membrane. Unlike the intracellular membrane, however, it has chains of sugar molecules attached extracellularly to lipid and protein molecules.

A principal function of the cell membrane is the regulation of molecules into and out of the cell from the surrounding medium. The membrane allows certain small molecules to pass through, while restricting the passage of large molecules. Therefore the cell membrane is semipermeable or, preferably, selectively permeable.

There are three mechanisms by which molecules can move into and out of cells from the surrounding medium: diffusion, facilitated transport, and active transport. Diffusion is the passage of molecules from the area of greater concentration to the area of lesser concentration. Lipid-soluble molecules, as well as gases and water, can diffuse through the membrane. It is hypothesized that water crosses the membrane by way of pores.

The diffusion of water across a membrane is called osmosis. When a cell is placed in an isotonic solution, it neither gains nor loses water because the concentration of water is the same inside and outside the cell. When there is an unequal distribution of nonpermeable solutes, and thus a difference in concentration of water on either side of the membrane, water alone moves from the area of greater to the area of lesser concentration. In hypertonic solutions, the cell loses water; in hypotonic solutions, it gains water. Plant cells that have strong cell walls in addition to cell membranes react somewhat differently in these solutions than do animal cells. Table 3.2 summarizes the effect of these solutions on both animal and plant cells.

Facilitated transport, by which small molecules and ions selectively pass through the membrane, requires the presence of protein carriers in the membrane. Cellular energy is not expended because the molecules are moving down a concentration gradient.

Active transport, which accounts for the passage of molecules across the membrane against a concentration gradient, requires protein carriers and an expenditure of cellular energy.

Vesicle formation during endocytosis permits large molecules (pinocytosis) and even smaller cells or parts of cells (phagocytosis) to be taken into larger cells. Exocytosis, or the discharge of substances by vesicle formation and subsequent fusion with the cell membrane, describes the process of secretion.

In multicellular organisms, cells often form tissues in which they function and work together. In animal tissues that line cavities, cells are joined at gap junctions where channels allow molecules to move from cell to cell. In plants, organic nutrients translocate in a tissue having cells that are joined end to end by means of sieve plates.

Study Questions

1. How does the cell membrane differ from intracellular membrane? (p. 66)
2. Why can a cell membrane be called semipermeable or selectively permeable? (p. 67)
3. What are the three mechanisms by which substances enter and exit cells? (p. 67)
4. Define diffusion and give an example. (p. 67)
5. Define osmosis. (p. 69) Define isotonic, hypertonic, and hypotonic solutions, and give examples of these concentrations for red blood cells. (p. 71)

The Cell, A Unit of Life

6. Draw a simplified diagram of a red blood cell before and after being placed in these solutions. What terms are used to refer to the condition of the red blood cell in a hypertonic and hypotonic solution? (pp. 70–71)

7. Draw a simplified diagram of a plant cell before and after being placed in these solutions. (p. 72) What terms are used to refer to the condition of a plant cell in a hypertonic and a hypotonic solution? (p. 73)

8. How does facilitated transport differ from simple diffusion across the cell membrane? (p. 73)

9. How does active transport differ from facilitated transport? Give an example. (p. 73)

10. Draw diagrams that show endocytosis and exocytosis. Give an example for each of these. (pp. 74–75)

11. Describe a junction by which molecules can pass from cell to cell in animal cells; in plant cells. (p. 77)

Selected Key Terms

permeable (per′me-ah-b′l)
diffusion (dĭ-fu′zhun)
osmosis (oz-mo′sis)
solute (sol′ūt)
solvent (sol′vent)
osmotic pressure (oz-mot′ik presh′ur)
isotonic solution (i′′so-ton′ik so-lu′shun)
hypertonic solution (hi′′per-ton′ik so-lu′shun)
crenated (kre′nāt-ed)
hypotonic solution (hi′′po-ton′ik so-lu′shun)
lysis (li′sis)
plasmolysis (plaz-mol′ĭ-sis)
turgor pressure (tur′gor presh′ur)
facilitated transport (fah-sil′ĭ-tāt-ed trans′port)
active transport (ak′tiv trans′port)
endocytosis (en′′do-si-to′sis)
phagocytosis (fag′′o-si-to′sis)
pinocytosis (pin′′o-si-to′sis)
exocytosis (eks′′o-si-to′sis)

Further Readings

Capaldi, R. A. 1974. A dynamic model of cell membranes. *Scientific American* 230(3):26.

Dautry-Varsat and Lodish, H. F. 1984. How receptors bring proteins and particles into cells. *Scientific American* 250(5):52.

Fox, C. F. 1972. The structure of cell membranes. *Scientific American* 226(2):30.

Kennedy, D., ed. 1974. *Cellular and organismal biology: Readings from Scientific American*. San Francisco: W. H. Freeman.

Keyes, R. D. 1979. Ion channels in the nerve cell membrane. *Scientific American* 240(3):126.

Lodish, H. F., and Rothman, J. E. 1979. The assembly of cell membranes. *Scientific American* 240(1):48.

Solomon, A. K. 1960. Pores in the cell membrane. *Scientific American* 203(6):40.

———. 1962. Pumps in the living cell. *Scientific American* 207(2):22.

Staehelin, L. A., and Hull, B. E. 1978. Junctions between living cells. *Scientific American* 238(5):140.

Unwin, N. 1983. The structure of proteins in biological membranes. *Scientific American* 250(2):78.

4

Cell Division

1 Cell division includes the division of both the cytoplasm and the nucleus.

2 The nucleus contains the gene-bearing chromosomes. Ordinary cell division in plants and animals is carried out in such a way that each daughter cell receives a full complement of chromosomes and genes.

3 In animals, reduction division is required to produce the sex cells, which contain one-half the full number of chromosomes. This means that when the sperm (male sex cell) fertilizes the egg (female sex cell), the full number of chromosomes is restored.

4 In plants, reduction division produces spore structures that precede the formation of the sex cells in the life cycle of these organisms.

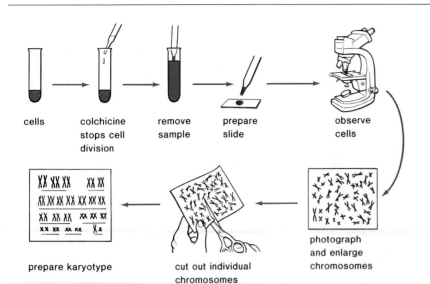

Figure 4.1
Preparation of a karyotype. Cell division is halted by chemical means and then the cells are microscopically magnified and photographed. An enlargement of the photograph permits the chromosomes to be cut out and arranged by pairs.

cells

colchicine stops cell division

remove sample

prepare slide

observe cells

photograph and enlarge chromosomes

cut out individual chromosomes

prepare karyotype

Cell division is necessary for growth and repair of multicellular organisms and for reproduction of all organisms. Cell division requires that not only the nucleus but also the cytoplasm be divided. An examination of the body cells of a multicellular organism shows that all the nuclei have the same number of chromosomes. This number is characteristic of the organism—corn plants have 20 chromosomes, fruit flies have 8, and humans have 46. In order to clearly view the chromosomes so that they can be counted, a cell is treated and photographed just prior to division, as described in figure 4.1. The chromosomes may then be cut out of the photograph and arranged by pairs. (Pairs of chromosomes are recognizable in that the chromosomes of the pair have the same size and general appearance.) The resulting display of chromosome pairs is called a **karyotype.**

Human Karyotype

A human karyotype is shown in figure 4.2. Although both male and female have 23 pairs of chromosomes, in the male one of these pairs is of unequal length. The larger chromosome of this pair is called the X and the smaller is called the Y. Females have two X chromosomes in their karyotype. The X and Y chromosomes are called the **sex chromosomes** because they carry genes that determine sex in humans and other organisms. The other chromosomes, known as **autosomes,** include all the pairs of chromosomes except the X and Y chromosomes.

Notice, as further illustrated in figure 4.2b, that each chromosome prior to division is composed of two identical parts, called **chromatids.** These two sister chromatids are genetically identical and contain the same **genes,** the units of heredity that control the cell. The chromatids are held together at a region called the **centromere.**

Figure 4.2

a. Karyotype in a male. Note pairs of autosomes, numbered from 1 to 22, and one pair of sex chromosomes, X and Y. The chromosomes are also grouped (by letters) according to their length and the location of the centromere. *b.* Enlargement of one chromosome.

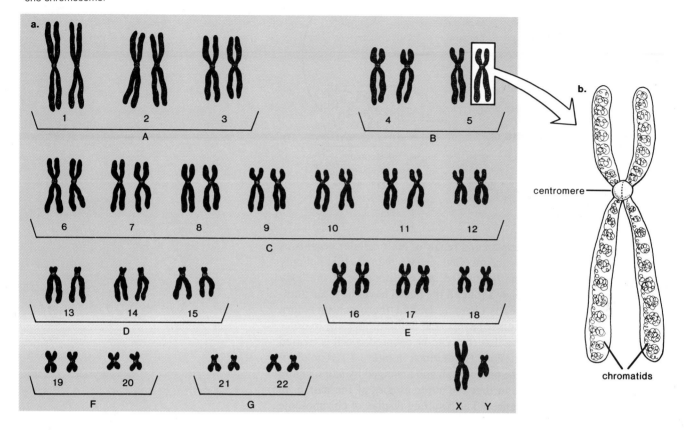

Life Cycle of Animals

Advanced multicellular animals, including humans, typically have a life cycle (fig. 4.3) that requires two types of cell division: **meiosis** and **mitosis.**

Meiosis occurs during the production of the **sperm** and **egg,** which are the sex cells, or **gametes.** A new individual comes into existence when the sperm of the male fertilizes the egg of the female. In humans, the resulting **zygote** contains 46 chromosomes, and as the zygote grows to become the adult, *mitosis* occurs so that each and every cell has 46 chromosomes. In this way each body cell contains the full complement of chromosomes and genes. The full complement of chromosomes is called the **2N,** or **diploid,** number of chromosomes.

Meiosis occurs only in the sex organs, or **gonads**—the testes in males and the ovaries in females. Here diploid germ cells develop into the gametes, which have half the total number, called the **N,** or **haploid,** number of chromosomes. *The haploid number always has one from each of the pairs of chromosomes. Thus the haploid number has one of each kind of chromosome.* For example, in humans the germ cells have 46 chromosomes, or 23 pairs, but gametes contain only 23 chromosomes—one of each of the pairs.

When a haploid sperm fertilizes a haploid egg, the new individual has the diploid number of chromosomes, half of which came from the father and half of which came from the mother. Thus, each parent contributes one of each of the pairs of chromosomes possessed by the new individual. Table 4.1 summarizes the major differences between mitosis and meiosis in multicellular animals.

The Cell, A Unit of Life

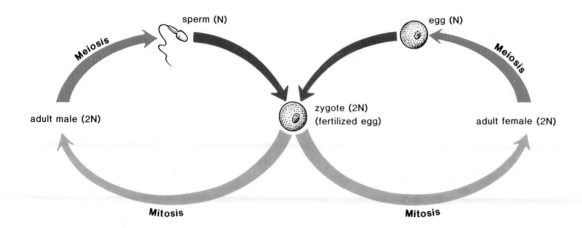

Table 4.1	Mitosis versus meiosis in animals		
Cell type	**Cell division**	**Description**	**Result**
Somatic or body cells	Mitosis	2N (diploid)→2N(diploid)	More body cells = growth
Germ cells in gonads of animals	Meiosis	2N(diploid)→N(haploid)	Gamete or sex cell production

Figure 4.4
Mitosis overview. Following replication of the genes each chromosome in the mother cell contains two sister chromatids. During mitotic division, the sister chromatids separate so that daughter cells have the same number and kinds of chromosomes as the mother cell.

Mitosis

Overview

Mitosis is *cell division in which the daughter cells retain the same number and kinds of chromosomes as the mother cell.*[1] Therefore the mother cell and the daughter cells are genetically identical. The **mother cell** is the cell that divides, and the **daughter cells** are the resulting cells. Figure 4.4 is an overview of mitosis; each cell in the diagram contains four chromosomes. (In determining the number of chromosomes it is necessary to count only the number of independent centromeres.) Figure 4.4 points out that a cell prepares for mitosis by replication of the genetic material contained within each chromosome. **Replication** is the process by which DNA makes a copy of itself as is described in detail in chapter 22. Because of replication, each chromosome in the mother cell contains two sister chromatids, sometimes called a chromatid pair, or **dyad.** During mitosis, sister chromatids separate and go to newly forming daughter cells, ensuring that each daughter cell receives a copy of each chromosome rather than two copies of one chromosome and none of another. Different genes are on different chromosomes, and it is necessary for each daughter cell to receive a copy of each chromosome in order to have a full complement of genes.

As an aid in describing the events of mitosis, the process has been divided into four phases: prophase, metaphase, anaphase, and telophase (fig. 4.5). In between cell divisions, the cell is said to be in interphase.

2N = 4

replication of genes

2N = 4

cell division

2N = 4   2N = 4

---

[1]The term *mitosis* technically refers only to nuclear division but for convenience is used here to refer to division of the entire cell.

Figure 4.5

Mitosis has four stages, not including interphase and daughter cells. Daughter cells are shown here in late interphase.

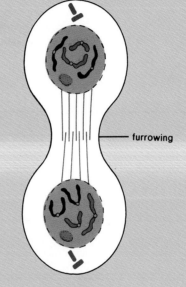

Interphase — centrioles, chromatin, nucleolus, nuclear membrane

Prophase — spindle fibers, chromosome, centromere

Metaphase — spindle, equator, aster

Anaphase

Telophase — furrowing

Daughter Cells

Figure 4.6

Microtubules can assemble at one end and disassemble at the other. During assembly, protein dimers join together, and during disassembly the protein dimers separate from one another.

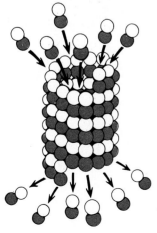

assembly end

disassembly end

Stages

Interphase

During **interphase** an animal cell resembles figure 2.1a. The nuclear envelope and the nucleoli are visible. The chromosomes, however, are not visible because the chromosomal material is dispersed into fine threads called **chromatin.** In animal cells there are two pairs of centrioles just outside the nucleus.

It used to be said that interphase was a resting stage, but we now know that this is not the case. During interphase the organelles are metabolically active and are carrying on their normal functions. Also during this phase, DNA is replicating so that as mitosis begins each chromosome consists of sister chromatids.

Prophase

It is apparent during **prophase** that cell division is about to occur. The chromatin material shortens and thickens so that the chromosomes are readily visible. The pairs of centrioles begin separating and moving toward opposite ends of the nucleus. **Spindle fibers** appear between the separating pairs of centrioles. As the spindle appears, the nuclear envelope and nucleolus begin to disappear.

Spindle fibers are now known to be bundles of microtubules. As discussed previously, microtubules can assemble and disassemble (fig. 4.6).

The Cell, A Unit of Life

Whether the separating pairs of centrioles aid the assembly process that presumably leads to the formation of the spindle or are simply pushed apart by the newly assembling microtubules is not known.

Figure 4.7 is a micrograph of an animal cell at the time of late prophase. The chromosomes are randomly placed even though the spindle appears to be fully formed. The entire spindle apparatus can be removed from the cell and examined, as shown in figure 4.8. Notice the **asters,** short lengths of microtubules that radiate out from the centrioles, located at the **poles** (ends) of the spindle. The presence of asters indicates that the cell is an animal cell because plant cells, since they lack centrioles, don't have asters during cell division.

Figure 4.7
Micrograph of late prophase. The chromosomes are moving toward the equator of the spindle located midway between the asters that are found at the poles of the spindle.

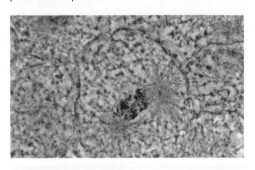

Figure 4.8
The spindle apparatus from an animal cell consists of the structures shown. Each spindle fiber is a bundle of microtubules. Some believe that the centrioles are involved in the production of microtubules and thus of the spindle.

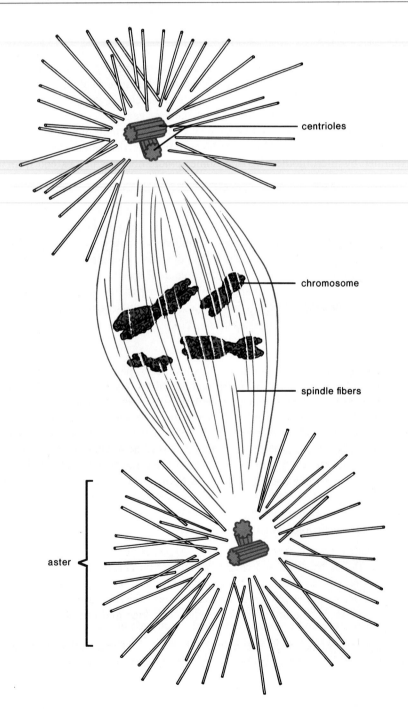

centrioles

chromosome

spindle fibers

aster

Figure 4.9
Micrograph of metaphase. The chromosomes are now lined up along the equator of the spindle.

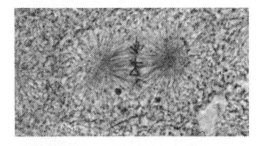

Figure 4.10
Micrograph of anaphase. Separation of sister chromatids results in chromosomes that are pulled by spindle fibers to opposite poles of the spindle.

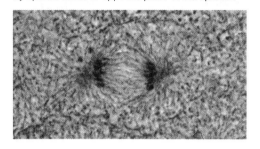

Figure 4.11
Micrograph of late telophase. Furrowing has resulted in two daughter cells separated by membrane. Remnants of the spindle are still seen, but these will disappear as the daughter nuclei re-form.

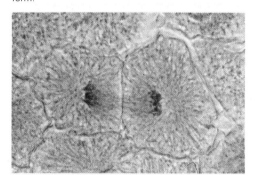

Figure 4.12
Electron micrograph of isolated human cell in advanced stage of cytokinesis. In animal cells, furrowing divides the cytoplasm.

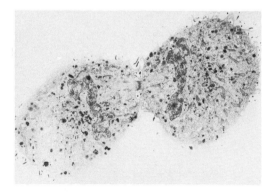

Table 4.2	Stages of mitosis
Stage	**Events**
Prophase	Replication has occurred and each chromosome is composed of a pair of sister chromatids
Metaphase	Chromatid pairs (dyads) are at the equator of the cell
Anaphase	Chromatids separate and each one is now termed a chromosome
Telophase	At each pole there is a diploid number of chromosomes, the same number and kinds of chromosomes as the mother cell

Metaphase

As **metaphase** begins, the nuclear envelope has disappeared and the spindle now occupies the region formerly occupied by the nucleus. Each chromosome attaches to a fiber and moves to the equator (center) of the spindle. Metaphase is characterized by a fully formed spindle, with the chromosomes, each composed of two chromatids, arranged across the equator (fig. 4.9).

Anaphase

During **anaphase** each centromere divides. Now the sister chromatids separate and each moves toward an opposite pole of the spindle (fig. 4.10). *Once separated, the chromatids are called chromosomes.* Separation of the sister chromatids ensures that each daughter cell will receive a copy of each type chromosome and thus have a full complement of genes.

Notice in figure 4.5 that as the newly formed chromosomes move to opposite poles, the entire cell elongates. Perhaps both these two events can be explained by microtubule behavior. Some spindle fibers, which extend from a pole to the equator, are attached to chromosomes, and some, which extend the length of the spindle apparatus, are not. Those that go from pole to pole appear to lengthen, and thus the cell elongates during cell division. The microtubules that are attached to chromosomes, however, appear to shorten, and thus the newly formed chromosomes are pulled to opposite poles. Microtubule assembly may cause the pole-to-pole spindle fibers to lengthen, and microtubule disassembly may cause the pole-to-equator spindle fibers to shorten. It is also possible that the short spindle fibers slide past the long spindle fibers with the help of contractile proteins that are known to be present in the spindle apparatus.

Telophase

During **telophase,** (fig. 4.11) the spindle disappears, possibly due to disassembly of the microtubules making up the spindle fibers. As the nuclear envelopes re-form and the nucleoli reappear in each daughter cell, the chromosomes become indistinct chromatin again. Following nuclear division, cytoplasmic division, sometimes called **cytokinesis,** usually occurs. In animal cells, **furrowing,** or an indentation of the membrane between the two daughter cells, divides the cytoplasm (fig. 4.12). Furrowing is complete when each daughter cell has a complete membrane enclosing it. Microfilaments are believed to take part in the furrowing process since they are always present in the vicinity. In each newly formed daughter cell, an immature centriole forms at a right angle to each mature centriole. Thus each daughter cell now has two pairs of centrioles.

Table 4.2 is a summary of the stages of mitosis.

Cell Cycle

As mentioned earlier, interphase follows telophase. At this point, some daughter cells may mature and become specialized. Each type of specialized cell has a characteristic life span. For example, red blood cells live about 120 days while many nerve cells live as long as the individual does.

The Cell, A Unit of Life

Other daughter cells may enter the cell cycle (fig. 4.13) and prepare to divide again. First, these cells usually increase to normal size before replication of DNA takes place. Following replication, each chromosome consists of sister chromatids. The chromosomes, however, are greatly elongated and do not condense until cell division is imminent. Before dividing, the cell becomes slightly larger than normal size.

Interphase is the longest of the phases in the cell cycle. The length of time required for the entire cycle varies according to the organism and even the type of cell within the organism, but 18 to 24 hours is typical for animal cells. Mitosis is usually the shortest portion of the cycle, lasting from less than an hour to slightly more than two hours.

There is a limit to the number of times an animal cell will divide before death follows degenerative changes. Most will divide about 50 times and only cancer cells retain the ability to divide repeatedly. Aging of cells in the individual appears to be a normal process most likely controlled by the nucleus. Cancer cells have abnormal chromosomes and other abnormalities of cell structure. In the laboratory, aging of cells can be delayed by environmental circumstances such as reduced temperature, but it cannot be postponed indefinitely.

Plant Mitosis

There are two main differences between plant and animal cell mitosis. First of all, in higher plant cells, centrioles and asters are not seen during mitosis. However spindle fibers do appear (fig. 4.14). Interesting is the fact that animal cells deprived of centrioles will also form a spindle. It may be, therefore, that centrioles do *not* contribute to spindle formation.

Figure 4.13
The cell cycle consists of mitosis and interphase. During interphase, there is growth before and after DNA synthesis. DNA synthesis is required for the process of replication by which DNA makes a copy of itself. Some daughter cells "break out" of the cell cycle and become specialized cells performing a specific function.

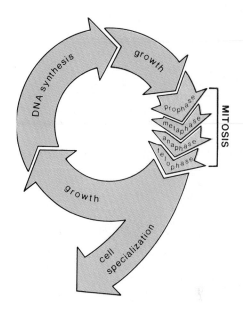

Figure 4.14
Plant cell mitosis. Notice the absence of asters and the square shape of the cells. In telophase, a cell plate develops in between the two daughter cells. The cell plate marks the boundary of the new daughter cells where new cell membrane and new cell wall will form for each cell.

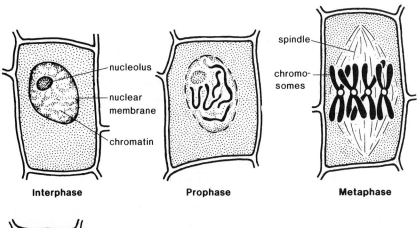

Interphase   Prophase   Metaphase

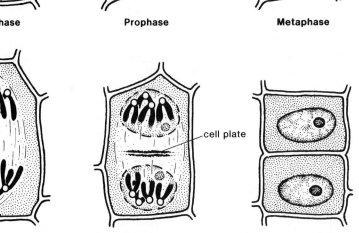

Anaphase   Telophase   Daughter Cells

Figure 4.15
Electron micrograph of a plant cell during
cytokinesis. Nuclei have re-formed. Some spindle
fibers *(SpF)* are still seen. The cell plate *(CP)* marks
the location of the formation of new plasma
membrane *(Pl)* and cell wall *(CW)*.

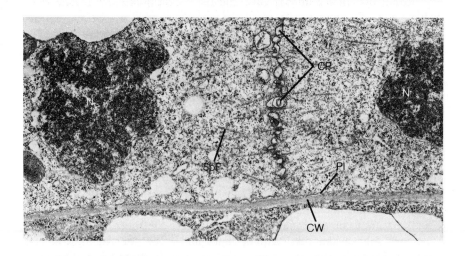

The second difference occurs during telophase when a **cell plate** (fig. 4.15) develops in the equatorial plane. The cell plate, which initially develops in the center and spreads to the sides, marks the boundary of the two daughter cells. It forms the middle lamella, a gummy substance that lies between two adjacent plant cells (fig. 3.11). Each daughter cell is complete once it forms a new cell membrane and cell wall next to the cell plate. Thus, no furrowing is observed in plant cells.

Protist Mitosis
Single-celled organisms, such as some protozoans and algae, normally reproduce by means of mitosis. For example, when an amoeba (fig. 25.10) undergoes mitosis, it divides into two amoebas, whereas before there was only one. Since an adult amoeba has only a haploid number of chromosomes, it is obvious that mitosis, on occasion, can be represented by N ———►N, rather than 2N ———►2N.

Importance of Mitosis
Mitosis assures that each daughter cell receives the same number and kinds of chromosomes as the mother cell, and thus mitosis assures that each daughter cell is *genetically identical* to the mother cell.

Mitosis is important to the growth and repair of multicellular organisms. When a baby develops in its mother's womb, mitosis occurs as a component of growth. As a wound heals, mitosis occurs to repair the damage.

Mitosis also occurs during the process of asexual reproduction. In protists, one division results in two organisms whereas before there was only one. Some lower animals (fig. 27.10) and many plants (fig. 7.21) are also capable of reproducing asexually. In these instances, several divisions are required to produce a replica of the organism.

Meiosis

Overview of Animal Meiosis
In animals, meiosis occurs during the production of the sex cells or gametes. The gametes are the egg and sperm. Meiosis, which requires two cell divisions, results in *four daughter cells, each having one of each kind of chromosome and thus half the number of chromosomes as the mother cell.*[2] The mother cell has the diploid number of chromosomes, while the daughter cells have the haploid number of chromosomes.

---

[2]The term *meiosis* technically refers only to nuclear division but for convenience is used here to refer to division of the entire cell.

The Cell, A Unit of Life

Recall that in the diploid condition the chromosomes are paired. Humans possess 46 chromosomes, or 23 pairs of chromosomes. These pairs are called **homologous chromosomes.** During meiosis, the homologous chromosomes separate. In this way, each daughter cell will receive half the total number of chromosomes but one of each kind. Halving the chromosome number is necessary to keep the chromosome number constant from generation to generation. Each sperm and egg has only 23 chromosomes, so that after fertilization the new individual has 46 chromosomes. By way of the gametes, each parent contributes one chromosome of each homologous pair of chromosomes in the new individual.

Figure 4.16 presents an overview of meiosis, indicating the two cell divisions, **meiosis I** and **meiosis II.** Prior to meiosis I, replication has occurred and each chromosome consists of sister chromatids. During meiosis I, the homologous chromosomes come together and line up side by side, due to a means of attraction still unknown. This so-called **synapsis** results in **tetrads,** an association of four chromatids that stay in close proximity during the first two phases of meiosis I. During synapsis, the chromatids exchange genetic material as illustrated in figure 4.17. The exchange of genetic material between chromatids is called **crossing over** and is an additional means by which new combinations of genes occur so that an offspring will have a different genetic makeup than either of its parents.

Following synapsis, the homologous chromosomes separate during meiosis I. This separation means that one chromosome of every homologous pair will reach each gamete.[3] There are no restrictions to the separation process; either chromosome of a homologous pair may occur in a gamete with either chromosome of any other pair.[4]

Notice that at the completion of meiosis I (fig. 4.16), the chromosomes still consist of sister chromatids. During meiosis II, the sister chromatids separate, resulting in four daughter cells, each of which has the haploid number of chromosomes. Although two cell divisions have taken place, replication occurred only once. Remembering that one counts only the number of independent centromeres verifies that the mother cell has the diploid number of chromosomes, while each of the four daughter cells has the haploid number.

[3]See Mendel's law of segregation, p. 442.
[4]See Mendel's law of independent assortment, p. 446.

Figure 4.16
Overview of meiosis. Following replication of genes, the mother cell undergoes two divisions, meiosis I and meiosis II. During meiosis I, homologous chromosomes separate, and during meiosis II, chromatids separate. The final daughter cells are haploid.

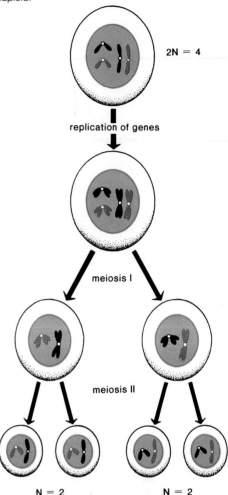

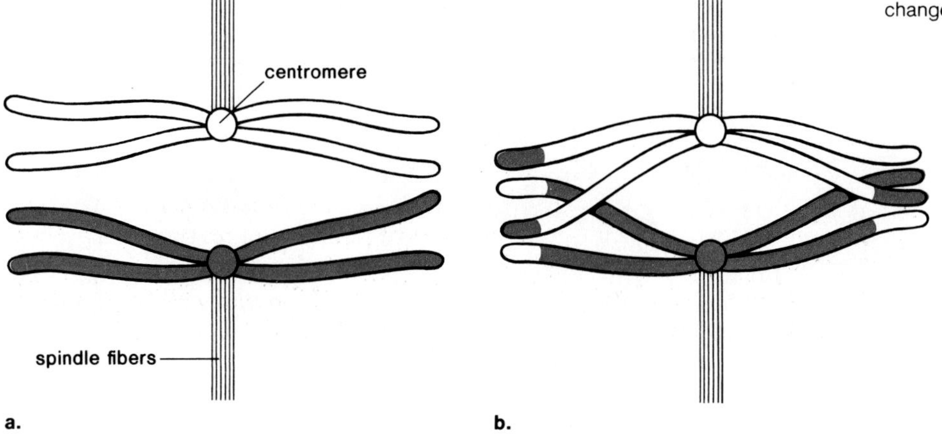

Figure 4.17
During crossing over, pieces of chromosomes are exchanged between chromatid pairs. *a.* Chromatid pairs before crossing over has occured. *b.* Chromatid pairs after crossing over. Notice the change in chromosome structure.

centromere

spindle fibers

a.

b.

Figure 4.18
Meiosis I requires a complete set of stages.

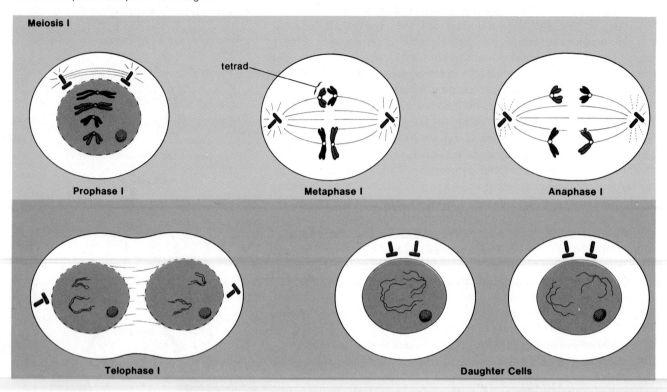

Figure 4.19
Meiosis II requires a complete set of stages. In this illustration, Meiosis II is shown for only one daughter cell from Figure 4.18. Please note, at the end of Meiosis II there will be four cells. (See figure 4.21).

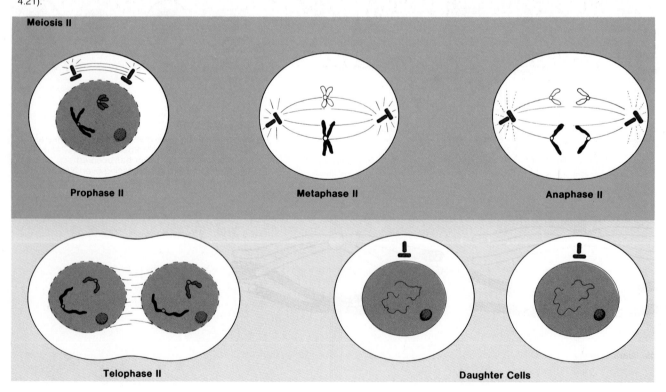

The Cell, A Unit of Life

Table 4.3 Stages of meiosis I and meiosis II

	Meiosis I	Meiosis II
Prophase	Homologous chromosomes, each one composed of chromatids synapse	Each chromosome is still composed of sister chromatids
Metaphase	Tetrads (chromosome pair = four chromatids) are at the equator	Chromatid pairs (dyads) are at the equator
Anaphase	Homologous chromosomes separate	Sister chromatids separate and each one is now termed a chromosome
Telophase	At each pole there is one from each pair of homologous chromosomes	At each pole there is the haploid number and one of each kind of chromosome

Stages

First Division

The stages of meiosis I are diagrammed in figure 4.18. During **prophase I,** the spindle appears while the nuclear envelope and the nucleolus disappear. Homologous chromosomes undergo synapsis, forming tetrads. At **metaphase I,** tetrads line up at the equator of the spindle. During **anaphase I,** homologous chromosomes separate and the chromosomes (still composed of two chromatids) move to the poles of the spindle. Each pole receives one-half the total number of chromosomes. In **telophase I,** the nuclear envelope and the nucleolus reappear as the spindle disappears. In certain species, the cell membrane furrows to give two daughter cells and in others, the second division begins without benefit of complete furrowing. Regardless, each daughter nucleus now contains only one from each homologous pair of chromosomes.

Second Division

The stages of meiosis II are diagrammed in figure 4.19. At the beginning of **prophase II,** a spindle appears while the nuclear envelope and nucleolus disappear. Each chromosome with its two chromatids attaches to the spindle independently. During **metaphase II,** the chromosomes are lined up at the equator. During **anaphase II,** the centromeres divide and the chromatids separate and move toward the poles. Each pole receives the same number of chromosomes. In **telophase II,** the spindle disappears as the nuclear envelope reappears. The cell membrane furrows to give two complete cells, each of which has the haploid, or N, number of chromosomes. Since each daughter cell from meiosis I undergoes meiosis II, there are four daughter cells altogether. Figure 4.19 shows only two of these.

Table 4.3 is a summary of the stages of meiosis I and meiosis II. This summary is appropriate to both plants and animals. Our preceding discussion refers, in general, to animal cell meiosis because centrioles are present in the figures, the cells are rounded without cell walls, and furrowing occurs to divide the cells. In animals, meiosis is specifically involved in either spermatogenesis or oogenesis.

Figure 4.20
Spermatogenesis produces four viable sperm, whereas oogenesis produces one egg and at least two polar bodies. In humans, both sperm and egg have 23 chromosomes each; therefore, following fertilization the zygote has 46 chromosomes.

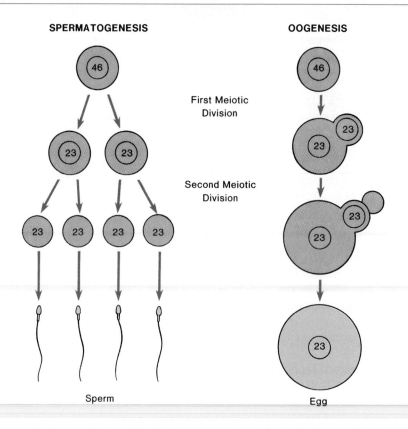

SPERMATOGENESIS   OOGENESIS

First Meiotic Division

Second Meiotic Division

Sperm   Egg

Spermatogenesis and Oogenesis

During **spermatogenesis** in males, sperm are produced, and during **oogenesis** in females, eggs are produced. The final product of meiosis, as well as the process itself, is different in the two sexes as you can see in figure 4.20. In males, meiosis results in four viable sperm from each original cell. In females, the first meiotic division produces two cells, but one is much larger than the other. The smaller nonfunctional cell is called a **polar body.** The second division also results in two cells of unequal size, but again one is much larger than the other which is another nonfunctional polar body. Thus meiosis in females produces only one functional egg that contains a rich supply of cytoplasmic components. The polar bodies are simply a way of discarding unnecessary chromosomes. Figure 4.20 shows how the sperm and egg are adapted to their function. The sperm is a tiny, flagellated cell that is adapted for swimming to the maturing egg, a large cell that contributes most of the cytoplasm and nutrients to the new individual.

There is another difference between spermatogenesis and oogenesis. Spermatogenesis, once started, goes to completion and mature sperm result. In contrast, oogenesis does not necessarily go to completion. Only if a sperm fertilizes the maturing egg does it undergo meiosis II, otherwise it simply disintegrates. Regardless of this complication, however, both the sperm and egg contribute the haploid number of chromosomes to the new individual. In humans each contributes 23 chromosomes.

Plant Meiosis

While the events of meiosis are essentially the same in plants, the resulting cells called spores have a different function, which will be brought out in the discussion of plant life histories in chapter 26.

Importance of Meiosis

Meiosis is nature's way of keeping the chromosome number constant from generation to generation. In animals, it occurs prior to maturation of the egg and sperm. In higher plants, it occurs when spores are formed.

Meiosis assures that the next generation will have a different genetic makeup than that of the previous generation. As a result of crossing over, the chromosomes carry a new combination of genes. In animals, the egg carries one-half the genes from the female parent and the sperm carries one-half the genes from the male parent. When the sperm fertilizes the egg, the zygote has a different combination of genes than either parent. In this way, meiosis assures *genetic variation,* generation after generation.

Summary

Each organism has a characteristic number of chromosomes; humans have 46. A human karyotype shows 22 pairs of autosomes and one pair of sex chromosomes. The sex pair is an X and a Y chromosome in males and two X chromosomes in females. Each chromosome in a karyotype is composed of two sister chromatids, held together at the centromere.

The life cycle of humans requires two types of cell divisions: mitosis and meiosis. Mitosis is responsible for growth and repair, while meiosis is required for gamete production.

Cell division is made up of four stages: prophase, metaphase, anaphase, and telophase. The cell cycle includes an additional stage termed interphase. During interphase, DNA replication causes each chromosome to have sister chromatids. These same stages take place in animal and plant cells, but there are no centrioles or asters (although there is a spindle) in higher plant cells, and division of the cytoplasm takes place by formation of a cell plate instead of by furrowing, as in animal cells.

Mitosis ensures that each body cell will have the full diploid, or 2N, number of chromosomes and will be genetically identical to the mother cell. This comes about because each chromosome within the mother cell consists of sister chromatids. When the centromeres divide and the sister chromatids separate, each newly forming daughter cell receives the same number and kinds of chromosomes as the mother cell. It is important to remember that when determining the number of chromosomes it is necessary to count only the number of centromeres.

Meiosis involves two cell divisions. During meiosis I, homologous chromosomes come to lie side by side during synapsis. The chromatids making up the resulting tetrad exchange chromosome pieces; this is called crossing over. When the homologous chromosomes separate during meiosis I, each daughter cell receives one from each pair of chromosomes. Separation of sister chromatids during meiosis II then produces a total of four daughter cells, each with the haploid number of chromosomes.

Spermatogenesis in males produces four viable sperm, while oogenesis in females produces one egg and at least two polar bodies. Each gamete is specialized for the job it does; the sperm is a tiny, flagellated cell that swims to the cytoplasm-laden egg.

Figure 4.21 contrasts the process of mitosis with the process of meiosis.

Figure 4.21
Mitosis compared to meiosis.

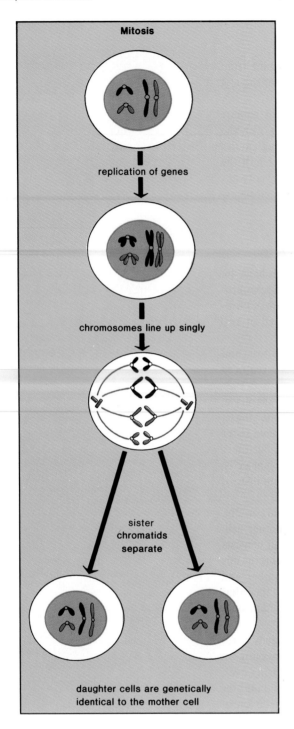

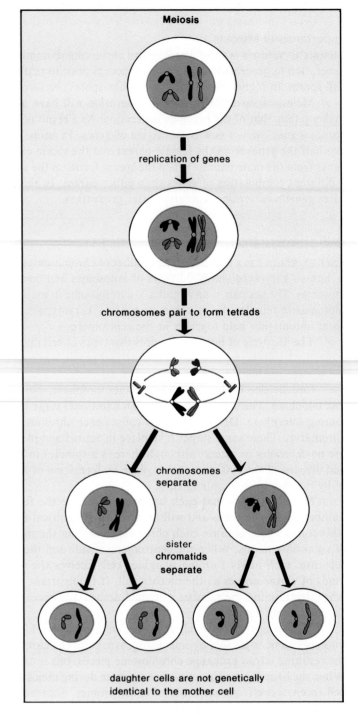

Study Questions

1. Describe the normal karyotype of a human being. What is the difference between a male and female karyotype? (pp. 81–82)
2. Relate the terms *diploid (2N)* and *haploid (N)* to mitosis in somatic cells and meiosis in germ cells. (pp. 82, 88)
3. Explain the makeup of a chromosome prior to cell division. (pp. 81–82)
4. Describe the stages of animal mitosis, including in your description the terms *centriole, nucleolus, spindle,* and *furrowing.* (pp. 83–86)
5. Name two differences between plant cell mitosis and animal cell mitosis. (pp. 87–88)
6. Give several instances when mitosis occurs in humans. (p. 88)
7. Describe the stages of meiosis I, including in your description the term *tetrad.* (p. 91)
8. Compare the second series of stages of meiosis to a mitotic division. (pp. 86, 91)
9. Explain the fact that oogenesis produces one mature egg, but spermatogenesis results in four sperm. (p. 92)
10. What is the importance of meiosis in the life cycle of any organism? (p. 93)
11. Give several differences between mitosis and meiosis. (pp. 82, 94)

Selected Key Terms

karyotype (kar′e-o-tīp)   **dyad** (di′ad)
chromatids (kro′mah-tidz)   **spindle** (spin′d′l)
centromere (sen′tro-mēr)   **aster** (as′ter)
meiosis (mi-o′sis)   **cytokinesis** (si″to-ki-ne′sis)
mitosis (mi-to′sis)   **furrowing** (fur′o-ing)
gamete (gam′ēt)   **cell plate** (sel plāt)
zygote (zi′gōt)   **tetrad** (tet′rad)
diploid (dip′loid)   **crossing over** (kros′ing o′ver)
gonad (go′nad)   **spermatogenesis** (sper″mah-to-jen′e-sis)
haploid (hap′loid)   **oogenesis** (o″o-jen′e-sis)
replication (re″plĭ-ka′shun)   **polar bodies** (po′lar bod′es)

Further Readings

Dustin, P. 1980. Microtubules. *Scientific American* 243(2):66.
Hayflick, L. 1980. The cell biology of human aging. *Scientific American* 242(1):58.
Kornberg, R. D., and Klug, A. 1981. The nucleosome. *Scientific American* 244(2):52.
Maiza, D. 1974. The cell cycle. *Scientific American* 230(1):54.
———. 1961. How cells divide. *Scientific American* 205(3):100.
Sloboda, R. D. 1980. The role of microtubules in cell structure and cell division. *American Scientist* 68(3):290.
Swanson, C. P. 1977. *The cell.* 4th ed. Englewood Cliffs, N.J.: Prentice-Hall.
Taylor, J. H. 1958. The duplication of chromosomes. *Scientific American* 198(6):36.
Thomas, L. 1974. *Lives of a cell: Notes of a biology watcher.* New York: Viking Press.

5

Cellular Metabolism

Chapter concepts

1 A metabolic pathway is a series of reactions controlled by enzymes.

2 Enzymes are protein molecules that only function properly when they retain their normal tertiary, or three-dimensional, shapes.

3 Cells require the energy molecule ATP to drive forward synthetic reactions and for various other functions.

4 Cellular respiration, or the metabolic pathway for the breakdown of glucose and related molecules, provides the necessary energy to form ATP molecules. A study of cellular respiration shows that glucose is oxidized by the removal of hydrogen atoms and that the resulting energy of oxidation is used to form ATP from ADP + (P) molecules.

Metabolism

Cells are not static; they are dynamic. Drawings of cells and even microscopic slides of cells give us the impression that cells are inactive; actually, cells are constantly active. Pinocytotic and phagocytotic vesicles are constantly being formed, organelles are moving about, and division may be taking place. A vital part of this activity is constantly occurring chemical reactions, which collectively are termed the **metabolism** of the cell.

Metabolic Pathways

Reactions do not occur haphazardly in cells; they are usually a part of a metabolic pathway. **Metabolic pathways** begin with (a) particular reactant(s) and terminate with (an) end product(s). While it is possible to write overall equations for metabolic pathways, the actual pathway itself proceeds by many minute steps. One reaction leads to the next reaction, which leads to the next reaction, and so forth in an organized, highly structured manner. This system makes it possible for one pathway to lead to several others, especially since various pathways have several substances in common. Also, metabolic energy is more easily captured and utilized if it is released in small increments rather than all at once.

Metabolic pathways can be represented by the following diagram, as long as we realize that side branches may occur at any juncture:

$$A \xrightarrow{1} B \xrightarrow{2} C \xrightarrow{3} D \xrightarrow{4} E \xrightarrow{5} F \xrightarrow{6} G$$

In the pathway represented, the letters are products of the previous reaction and the reactants for the next reaction. *A* is the beginning substance(s) and *G* is the end product(s). The numbers in the pathway refer to different enzymes. *Every reaction in a cell requires a specific enzyme.* Enzymes are protein molecules (p. 32) that speed up chemical reactions, and their concentration controls the rate at which reactions occur. In effect, no reaction occurs in a cell unless its enzyme is present. For example, if enzyme number 2 in the diagram is missing, the pathway cannot function; it will stop at *B*. Since enzymes are so necessary in cells, their mechanism of action has been studied extensively.

Enzymes

Each and every enzyme typically speeds up only one particular reaction and therefore is said to be *specific* for that reaction. The specificity of enzymes is explained by the **lock-and-key theory** of enzymatic action, which is often discussed in terms of the reaction $E + S \longrightarrow ES \longrightarrow E + P$. In this reaction, E = enzyme, S = substrate, ES = enzyme − substrate complex, and P = product.

Enzymatic Action

Substrate

As shown in figure 5.1, the substrate(s) is (are) the reactant(s) in an enzyme's reaction. Table 5.1 indicates that the name of an enzyme is often formed by adding *ase* to the name of its substrate. Some enzymes are named for the action they perform, as, for example, a dehydrogenase is an enzyme that removes hydrogen atoms from its substrate.

Table 5.1 Enzymes named for their substrates

Substrate	Enzyme
Lipid	Lipase
Urea	Urease
Maltose	Maltase
Ribonucleic acid	Ribonuclease
Lactose	Lactase

Figure 5.1

Lock-and-key theory of enzymatic action. An enzyme has an active site where the substrates and enzyme fit together in such a way that the substrates are oriented to react. Following the reaction, the products are released and the enzyme assumes its prior shape.

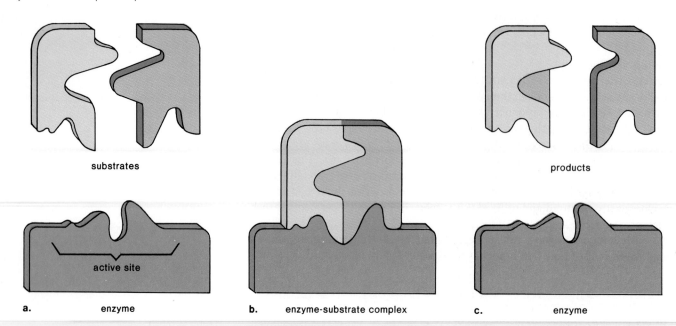

substrates

active site

a.   enzyme

b.   enzyme-substrate complex

products

c.   enzyme

Enzyme

Notice in figure 5.1 that the enzyme is unaltered by the reaction. Only a small amount of enzyme is actually needed in a cell because enzymes are used over and over again.

Enzyme-Substrate Complex

For a reaction to occur, the reactants must be brought into close proximity. Thus in figure 5.1, substrates come together on the surface of an enzyme because their shape fits the shape of their enzyme as a key fits a lock. The region where the substrate(s) attach(es) is called the **active site,** and it is here that the reaction takes place. The substrate(s) fit(s) onto the active site in such a way that they become orientated in the proper manner for the reaction to take place. Sometimes the active site assumes a different configuration as the reaction occurs. After the reaction has been completed, the product(s) is (are) released and the active site returns to its original state.

Product

Only (a) certain product(s) can be produced by any particular reactant(s), thus an enzyme cannot bring about impossible products. However the presence of an enzyme can determine whether a reaction takes place or not. For example, if substance A can react to either B or C, whichever enzyme is present, 1 or 2, will determine which product is produced.

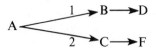

Since enzymes cannot bring about impossible reactions or products, it is usually emphasized that enzymes only speed up their particular reaction. They do this by lowering the energy of activation.

The Cell, A Unit of Life

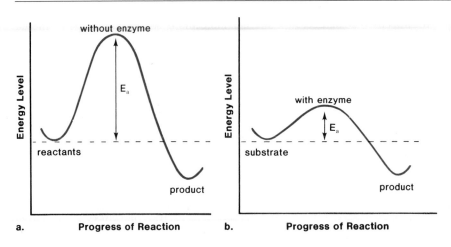

Figure 5.2
a. Energy of activation (E_a) that is required for a reaction to occur when an enzyme is not available.
b. Required energy of activation (E_a) is much lower when an enzyme is available. Notice that the energy level of the entire system is always lower following a reaction.

Energy of Activation

Organic molecules frequently will not react with one another unless they are activated in some way. For example, wood does not burn unless it is heated to a moderately high temperature. In the laboratory, too, activation is very often achieved by heating the reaction flask so that the number of effective collisions between molecules increases, allowing the molecules to react with one another.

The energy that must be supplied to cause molecules to react with one another is called the **energy of activation.** Figure 5.2 compares the energy level of nonreactive molecules, reactive ones, and the energy level of the products after the reaction. The difference between the nonreactive energy level and the reactive energy level is the necessary energy of activation. Enzymes lower the necessary energy of activation. For example, the following data are available for the hydrolysis of casein, the protein found in milk.

	Energy of Activation cal/mole[1]
Without an enzyme	>20,600
With an enzyme	12,000

This means that when an enzyme is present, less energy, and thus less heat, is needed to bring about a reaction. The lock-and-key theory of enzymatic action indicates that enzymes lower the energy of activation by bringing the substrates together so that a reaction can occur.

Conditions Affecting Yield

The model for the lock-and-key theory of enzymatic action can also be used to explain why certain factors influence the yield of enzymatic reactions, such as the factors discussed in the following sections.

Concentrations

If the concentration of the substrate or enzyme is increased, the amount of product increases; that is, the more S or E available, the more P there is within a certain amount of time. In many instances, the substrate is plentiful within the cell, but the enzyme is present only in small amounts. The amount of enzyme limits the overall rate of the reaction. In other words, if there is only a small amount of enzyme present, there will be less product in a given unit of time.

---

[1]Cal/mole = Calories per mole. A Calorie is a common method of measuring heat, and a mole is 6.02×10^{23} molecules of the substance being considered.

Competitive Inhibition

It happens that on occasion another molecule is so close in shape to the enzyme's substrate that this molecule can compete with the true substrate for the active site of the enzyme. However, since this molecule is not a reactant in the enzyme's reaction, a product is never realized. Such a molecule is designated as *I* for inhibitor in the lock-and-key reaction:

$$I + E \rightarrow EI \rightarrow \text{no further reaction}$$
$$I + E \rightarrow EI \rightarrow E + I$$

In the first reaction, the enzyme and inhibitor never become unlocked. This situation is very serious since the enzyme is thereafter incapacitated. In the second reaction, the enzyme is only temporarily out of action.

Hydrogen cyanide is an inhibitor for a very important enzyme (cytochrome oxidase) in all cells, and this accounts for its lethal effect on the human body. However, the phenomenon of inhibition can be used to advantage at times. The sulfonamide drugs are inhibitors of an enzyme unique to bacteria. When sulfa drugs are taken, bacteria die but the human body is unaffected.

Denaturation

A denatured protein is one that has lost its normal configuration and therefore its ability to form an enzyme-substrate complex. Three environmental factors that can cause denaturation are:

1. *Heavy metals.* Metals, such as Pb^{++} (lead) and Hg^{++} (mercury), bond with proteins and thereby inactivate them.
2. *Temperature.* Cold temperatures slow down chemical reactions, and warm temperatures speed up chemical reactions (including enzymatic reactions). High temperatures, however, affect hydrogen bonding and can cause the enzyme to restructure in a way that inactivates it.
3. *pH.* Each enzyme has a preferred pH at which the speed of the reaction is optimum. Presumably, any other pH affects the ionic bonding between the side chains of the molecule, leading to denaturation.

Coenzymes

Enzymes are very often composed of two portions: a protein called an **apoenzyme** and a nonprotein group called a **coenzyme.** The protein portion (apoenzyme) of the enzyme accounts for its specificity; that is, the ability of the enzyme to speed up only one particular reaction. The coenzyme portion of the enzyme may actually participate in the reaction by accepting or contributing atoms to the reaction:

Coenzyme	Apoenzyme
nonprotein	protein
helper	specificity

Coenzymes are generally large molecules that the body may be incapable of synthesizing. Many are vitamins that are required organic molecules in our diet. Vitamins are needed in only small amounts for efficient metabolism. Niacin (or nicotinic acid), thiamine (or vitamin B_1), riboflavin, folic acid, and biotin are all well-known vitamins that are parts of coenzymes.

NAD Cycle

Presently we wish to consider a coenzyme, known as **NAD**[2], that contains the vitamin niacin. NAD is part of a **dehydrogenase,** a type of enzyme that removes hydrogen atoms from substrates and also at times passes them to other substances.

[2]Nicotinamide adenine dinucleotide

Figure 5.3
The NAD cycle. NAD is reduced and becomes $NADH_2$ when it accepts hydrogen atoms and becomes oxidized to NAD again when the hydrogens are passed to another acceptor.

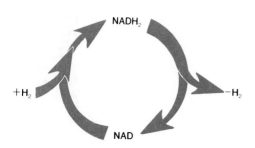

The Cell, A Unit of Life

Figure 5.4
ATP, the energy molecule in cells, has two high-energy phosphate bonds (indicated in figure by wavy lines). When cells require energy, the last phosphate bond is broken and a phosphate molecule is released.

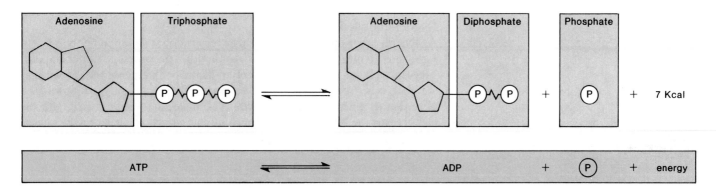

Therefore dehydrogenases are agents of *oxidation* (removal of hydrogen atoms) and *reduction* (addition of hydrogen atoms). Only a small amount of a coenzyme, like NAD, is present in a cell because the same coenzyme molecule is used over and over again, just as an enzyme is used over and over again. Figure 5.3 illustrates this point. After NAD accepts hydrogen atoms and becomes NADH$_2$, NADH$_2$ is apt to turn around and pass the hydrogen atoms to another acceptor, becoming NAD again.

Energy

When cells require energy, a certain kind of energy must be available. Electricity is the type of energy we use to light our homes and run our electric appliances. Cells, through the process of evolution, have come to depend on the molecule **ATP** (adenosine triphosphate) whenever they require energy, just as our society has come to depend on electricity.

ATP Reaction

ATP (fig. 5.4) is a nucleotide composed of the base adenine and the sugar ribose (together called adenosine) and three phosphate groups. The wavy lines in the formula for ATP indicate high-energy phosphate bonds; when these bonds are broken, an unusually large amount of energy is released. Because of this property, ATP is the energy currency of cells; when cells "need" something, they "spend" ATP.

ATP is used in body cells for synthetic reactions, active transport, nervous conduction, and muscle contraction. When energy is required for these processes, the end phosphate group is removed from ATP, breaking down the molecule to ADP (adenosine diphosphate) and (P) (phosphate) (fig. 5.4).

ATP Cycle

The reaction shown in figure 5.4 occurs in both directions; not only is ATP broken down it is also built up when ADP joins with (P). Since ATP breakdown is constantly occurring, there is always a ready supply of ADP and (P) to rebuild ATP again.

Figure 5.5 illustrates the ATP cycle in a diagrammatic way. Notice that when ATP is broken down, energy is released and when it is built up, energy is required. We shall see that aerobic cellular respiration, a metabolic pathway that takes place largely within mitochondria, produces energy needed for ATP build-up.

Figure 5.5
The ATP cycle. When ADP joins with a P group, energy is required; but when ATP breaks down to ADP and a P group, energy is given off.

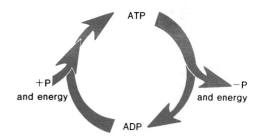

Aerobic Cellular Respiration (Simplified)

Both plants and animals, whether they reside in the water or on land, carry on aerobic cellular respiration (fig. 5.6). During this process, organic molecules are oxidized by the removal of hydrogen atoms. Oxidation releases the energy needed to cause ATP buildup (fig. 5.7).

Overall Equation

An overall equation for aerobic cellular respiration is shown in figure 5.8. The term **aerobic** means that oxygen is required for the process. As the equation suggests, cellular respiration most often begins with glucose but, as we shall see later, other molecules can also be used. Most importantly, we want to realize that as glucose is broken down, ATP molecules are produced, and this is the reason the ATP reaction is drawn using a curved arrow above the glucose reaction arrow.

Figure 5.6
Almost all organisms, whether they reside in the water or on land, take in oxygen and carry on aerobic cellular respiration. It is sometimes hard to detect that photosynthetic organisms are carrying on aerobic cellular respiration because this process can make use of the oxygen given off by photosynthesis.

Figure 5.7
Removal of hydrogen atoms ($H^+ + e^-$) results in oxidation of an organic molecule. During oxidation, bonds are broken and energy is released. In the opposite direction, acceptance of hydrogen atoms results in reduction of an organic molecule. During reduction, energy is required because bonds are formed.

The Cell, A Unit of Life

Subpathways

During cellular respiration, glucose is oxidized to carbon dioxide and water (fig. 5.8). The oxidation of glucose does not actually occur in one step, however. The entire process requires three subpathways (*glycolysis, the Krebs cycle, and respiratory chain*) and the transition reaction, as illustrated in figure 5.9. In the figure, each arrow represents a different enzyme, and the letters represent the product of the previous reaction and the substrate for the next. Notice how each pathway resembles a conveyor belt in which a beginning substrate continuously enters at the start and, after a series of reactions, end products leave at the termination of the belt. It is important to realize, too, that all three pathways are going on at the same time. They can be compared to the inner workings of a watch in which all parts are synchronized.

It is possible to relate the reactants and products of the overall reaction (fig. 5.8) to the subpathways in figure 5.9:

1. Glucose, $C_6H_{12}O_6$, is to be associated with **glycolysis,** the breakdown of glucose to two molecules of pyruvic acid (PYR). Oxidation by removal of hydrogen atoms provides enough energy for the buildup of two ATP molecules.
2. Carbon dioxide, CO_2, is to be associated with the transition reaction and the Krebs cycle. During the **transition reaction,** PYR is oxidized to active acetate (AA). AA enters the **Krebs cycle,** a cyclical series of oxidation reactions that give off CO_2 and produce one ATP molecule. Notice that since glycolysis ends with two molecules of PYR, the transition reaction and the Krebs cycle occur twice per glucose molecule. Altogether, then, the Krebs cycle accounts for two ATP molecules per glucose molecule.
3. Oxygen, O_2, and water, H_2O, are to be associated with the respiratory chain. The respiratory chain begins with $NADH_2$, the coenzyme that carries most of the hydrogens to the chain. The **respiratory chain** is a series of molecules that pass hydrogen and/or electrons from one to the other until they are finally received by oxygen, which is then reduced to water. As the electrons pass from one molecule to the next, oxidation occurs and this releases the energy needed for ATP buildup.
4. ATP is to be associated with glycolysis, the Krebs cycle, and the respiratory chain. Most ATP, however, is produced by the respiratory chain. It is usually said that the chain produces 34 ATP molecules per glucose molecule.

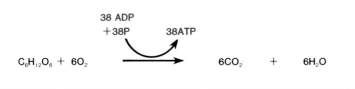

$$C_6H_{12}O_6 + 6O_2 \xrightarrow[+38P]{38 ADP \quad 38ATP} 6CO_2 \quad + \quad 6H_2O$$

glucose + oxygen ⟶ carbon dioxide + water

Figure 5.8
The overall equation for aerobic cellular respiration. ATP buildup is indicated with a curved arrow above the reaction arrow because it indicates that *as glucose is oxidized, ATP is produced.*

Figure 5.9

Aerobic cellular respiration contains three subpathways: glycolysis, Krebs cycle, and the respiratory chain which are discussed on page 103. As the reactions occur, a number of hydrogen atoms (nH_2) and carbon dioxide molecules are removed from the various substrates. Oxygen acts as the final acceptor for the hydrogen atoms.

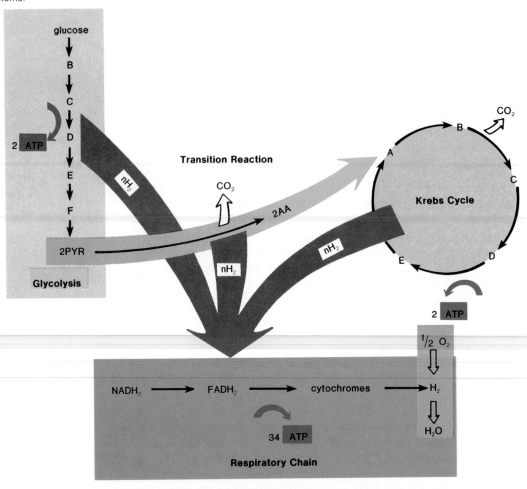

Table 5.2 Overview of aerobic cellular respiration

Name of pathway	Result
Glycolysis	Removal of H_2 from substrates Produces 2 ATP molecules
Transition reaction	Removal of H_2 from substrates Releases CO_2
Krebs cycle	Removal of H_2 from substrates Releases CO_2 Produces 2 ATP after two turns
Respiratory chain	Accepts H_2 from other pathways and passes them on to O_2 producing H_2O Produces 34 ATP

Table 5.2 summarizes our discussion of aerobic cellular respiration. This chart assumes that aerobic cellular respiration produces 38 ATP per glucose molecule.

Mitochondria

A significant portion of aerobic cellular respiration takes place in mitochondria (fig. 5.10). Glycolysis occurs outside the mitochondria, but the transition reaction, Krebs cycle, and respiratory chain occur within the mitochondria.

The Cell, A Unit of Life

Figure 5.10

Two-dimensional view of a mitochondrion based on electron micrographs (fig. 2.15) shows the location of the matrix and cristae. The enzymes responsible for the Krebs cycle are located in the matrix and the enzymes of the respiratory chain are located along the cristae.

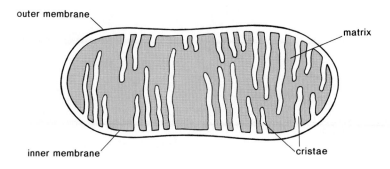

Evidence suggests that the enzymes of the Krebs cycle are located in the matrix, while the enzymes for the respiratory chain are located along the cristae of the inner membrane.

It is interesting to think about how our bodies provide the reactants for aerobic cellular respiration and how they dispose of the products. The air we breathe contains *oxygen,* and the food we eat contains *glucose.* These enter the bloodstream that carries them to the body's cells, where they diffuse into each and every cell. In mitochondria, glucose products are broken down to carbon dioxide and water as ATP is produced. All three of these leave the mitochondria. The *ATP* is utilized inside the cell for energy requiring processes. Carbon dioxide diffuses out of the mitochondria and out of the cell into the bloodstream. The bloodstream takes the *carbon dioxide* to the lungs, where it is exhaled. The *water* molecules produced, called **metabolic water,** only become important if by chance this is the organism's only supply of water. In these cases, it can help prevent dehydration of the organism.

Aerobic Cellular Respiration (in Detail)

In the detailed discussion that follows, cellular respiration is divided into two stages. The first stage takes place outside the mitochondria and the second stage takes place inside the mitochondria.

Stage One

Glycolysis

Glycolysis is that portion of cellular respiration that takes place outside the mitochondria. **Glycolysis** is the breakdown of glucose to the end product **pyruvic acid** (pyruvate). During these reactions, hydrogen atoms are removed from the substrates of the pathway and are picked up by NAD. When the hydrogen atoms are removed, oxidation of the substrates has occurred. Oxidation releases energy and, in this case, the energy is first captured within the substrate molecules and then used to supply enough energy to form two ATP molecules. The net result of glycolysis is the formation of two $NADH_2$ molecules and two ATP molecules. This can be better appreciated by examining the series of reactions that comprise glycolysis (fig. 5.11). In this figure, each of the molecules involved is represented by the number of carbon atoms it contains (each molecule also contains hydrogen and oxygen atoms). Table 5.3 summarizes the reactions of glycolysis.

As we shall see on page 111, glycolysis is also a part of anaerobic respiration, a process that is often called fermentation.

Figure 5.11

Glycolysis is a metabolic pathway that begins with glucose and ends with pyruvic acid. Net gain of two ATP molecules can be calculated by substracting those expended from those produced. Print in the boxes explains each reaction.

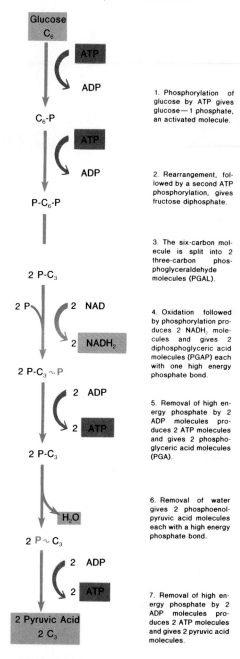

1. Phosphorylation of glucose by ATP gives glucose—1 phosphate, an activated molecule.

2. Rearrangement, followed by a second ATP phosphorylation, gives fructose diphosphate.

3. The six-carbon molecule is split into 2 three-carbon phosphoglyceraldehyde molecules (PGAL).

4. Oxidation followed by phosphorylation produces 2 $NADH_2$ molecules and gives 2 diphosphoglyceric acid molecules (PGAP) each with one high energy phosphate bond.

5. Removal of high energy phosphate by 2 ADP molecules produces 2 ATP molecules and gives 2 phosphoglyceric acid molecules (PGA).

6. Removal of water gives 2 phosphoenolpyruvic acid molecules each with a high energy phosphate bond.

7. Removal of high energy phosphate by 2 ADP molecules produces 2 ATP molecules and gives 2 pyruvic acid molecules.

Table 5.3 Glycolysis summary

Reaction	$NADH_2$	ATP
1		−1 ATP
2		−1 ATP
3		
4	2 $NADH_2$	
5		+2 ATP
6		
7		+2 ATP
Net gain	2 $NADH_2$	+2 ATP

Alcoholic Beverages

Wine, beer, and whiskey production all require yeast fermentation. To produce wine, grape juice is allowed to ferment. After the grapes are picked, they are crushed in order that the juice may be collected. In the old days, wine makers simply relied on spontaneous fermentation by yeasts that were on the grape skins, but now many add specially selected cultures of yeast. Also, it is common practice to maintain the temperature at about 20°C for white wines and 28°C for red wines. Fermentation ceases after most of the sugar has been converted to alcohol. Various methods are used to clarify the wine, that is, the removal of any suspended materials. Also many fine wines improve when they are allowed to "age" during barrel or bottle storage.

Brewing beer is more complicated than wine production. Usually grains of barley are first *malted;* that is, allowed to germinate for a short time so that amylase enzymes are produced that will break down the starch content of the grain. After the germinated grains have been crushed and mixed with water, the *malt wort* is separated from the spent grains and traditionally boiled with hops (an herb derived from the hop plant) to give flavor to the beer. Now the *hop wort* is seeded with a strain of yeast that converts the sugars in the wort to alcohol and carbon dioxide. At the end of fermentation, the yeast is separated from the beer, which is then allowed to mature for an appropriate period. After filtration and pasteurization, the beer is packaged.

The production of whiskey (from grains), brandy (from grapes), and rum (from molasses) differs from wine and beer production chiefly in that the alcohol is removed from the fermented substance by distillation. Most often, in the United States, corn or rye is used in the production of whiskey. These grains are ground up and mashed to release their starch content. Then amylase enzymes are added to convert the starch to fermentable sugars. Now yeast is added so that fermentation can occur. Following fermentation, the alcohol is concentrated by distillation. A warm temperature causes the alcohol to become gaseous and rise in a column where it condenses to a liquid before entering a collecting vessel. The alcohol content of the collecting vessel is much higher following this distillation process. The distillate is usually stored, quite often in an oak barrel, to improve the aroma and taste of the final product.

Fermentation of grapes (above left), barley (above right), and corn produces wine, beer, and whiskey respectively.

Stage Two

This portion of aerobic cellular respiration occurs within the mitochondria (fig. 5.10) and consists of a transition reaction, the Krebs cycle, and the respiratory chain.

Transition Reaction

The transition reaction connects glycolysis to the Krebs cycle, as is apparent in figure 5.9. In this reaction, pyruvic acid (PYR) is converted to a molecule called **active acetate** (AA), or acetyl coenzyme A, and carbon dioxide is given off in the process.

As figure 5.12 shows more clearly, this is an oxidation reaction in which hydrogen atoms are removed from pyruvic acid by NAD. Also, active acetate is actually an acetyl group attached to a coenzyme called **coenzyme A.** This coenzyme activates the acetyl group, hence the name *active acetate*. Notice that since glycolysis results in two pyruvic acid molecules, the transition reaction occurs twice per glucose molecule. Altogether, the reaction produces two molecules of carbon dioxide, two molecules of active acetate, and 2 $NADH_2$ per glucose molecule.

Krebs Cycle

The Krebs cycle, represented in detail in figure 5.13, is named for the person who discovered it. It is called a cycle because it is a series of reactions, or steps, that begin and end with citric acid. Sometimes the cycle is called the **citric acid cycle.** During this series of reactions, *oxidative decarboxylation* occurs.

Figure 5.12
During the transition reaction, pyruvic acid is oxidized to a two-carbon acetyl group attached to coenzyme A.

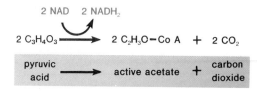

$$2\ NAD \quad 2\ NADH_2$$

$$2\ C_3H_4O_3 \longrightarrow 2\ C_2H_3O{-}Co\ A \ + \ 2\ CO_2$$

pyruvic acid $\longrightarrow$ active acetate $+$ carbon dioxide

Figure 5.13
The Krebs cycle is a metabolic pathway that begins and ends with citric acid. Print in the boxes explains each reaction.

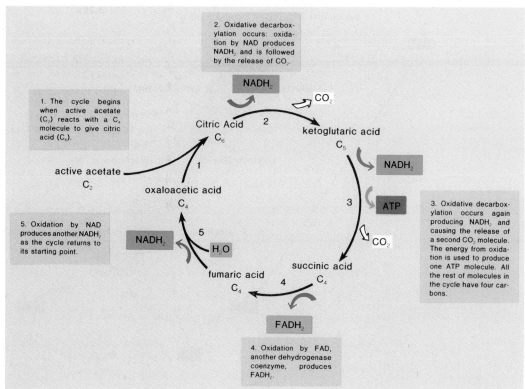

1. The cycle begins when active acetate (C_2) reacts with a C_4 molecule to give citric acid (C_6).

2. Oxidative decarboxylation occurs: oxidation by NAD produces $NADH_2$ and is followed by the release of CO_2.

3. Oxidative decarboxylation occurs again producing $NADH_2$ and causing the release of a second CO_2 molecule. The energy from oxidation is used to produce one ATP molecule. All the rest of molecules in the cycle have four carbons.

4. Oxidation by FAD, another dehydrogenase coenzyme, produces $FADH_2$.

5. Oxidation by NAD produces another $NADH_2$ as the cycle returns to its starting point.

active acetate C_2

Citric Acid C_6

ketoglutaric acid C_5

oxaloacetic acid C_4

succinic acid C_4

fumaric acid C_4

$NADH_2$ CO_2 $NADH_2$ ATP CO_2 FADH$_2$ $NADH_2$ H_2O

Table 5.4 Summary of Krebs cycle*

Step	CO₂	NADH₂	FADH₂	ATP
1				
2	CO_2	$NADH_2$		
3	CO_2	$NADH_2$		ATP
4			$FADH_2$	
5		$NADH_2$		
	2 CO_2	3 $NADH_2$	$FADH_2$	ATP

*Per turn. Cycle turns twice per glucose molecule.

Oxidative decarboxylation means that hydrogen atoms and carbon dioxide are removed from the substrates at the same time. Oxidation results in energy that is partially captured by the molecules within the cycle and this energy is used to form 1 ATP per turn. The carbon dioxide given off is a metabolic waste and is excreted by cells. Most of the hydrogen atoms are picked up by NAD, but a few are taken by FAD. **FAD** is another coenzyme of oxidation-reduction, which is used infrequently compared to NAD. Altogether, the Krebs cycle turns twice per glucose molecule and produces 4 CO_2, 6 $NADH_2$, 2 $FADH_2$, and 2 ATP. Table 5.4 summarizes the reactions of the Krebs cycle.

Respiratory Chain

The respiratory chain (figs. 5.9 and 5.14) is a series of molecules that pass electrons from one to the other. The electrons are at first a part of the hydrogen atoms ($H^+ + e^-$) attached to NAD or FAD. These are the same hydrogen atoms that were removed from the molecules of glycolysis and the Krebs cycle.

Figure 5.14

Respiratory chain. Hydrogen atoms enter the chain attached to either NAD or FAD. Thereafter, certain molecules of the chain accept only electrons and deposit hydrogen ions between the inner and outer membrane of the mitochondrion (see fig. 5.15). This creates an electrochemical gradient that supplies the energy necessary for ATP synthesis.

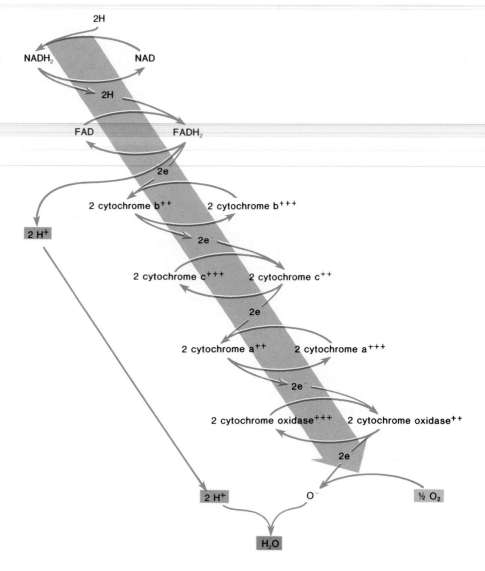

The Cell, A Unit of Life

Figure 5.15

Mitochondrion structure in more detail. The Krebs cycle occurs in the matrix background substance. The respiratory chain is located along the cristae of the inner membrane and ATP formation occurs at the F_1 factor particles located similarly.
a. Longitudinal section of mitochondrion.
b. Micrograph showing inner membrane. *c.* Drawing of F_1 factor particles.

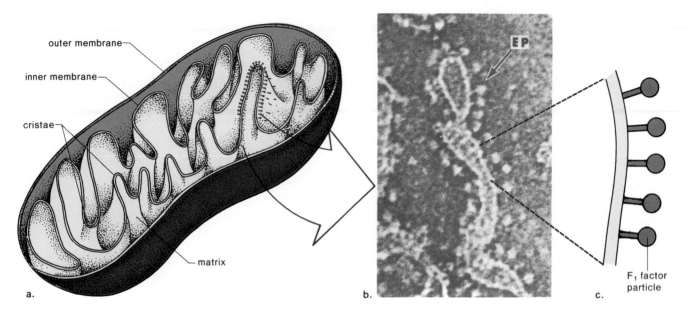

When the electrons enter the chain, they are at a high-energy level, but as they are passed "downhill" from one molecule to another, they lose energy as oxidation occurs. This energy of oxidation is indirectly used to make ATP molecules. If the electrons enter attached to NAD, there is enough energy to produce 3 ATP for every two hydrogen atoms. However, if they enter by way of $FADH_2$ only 2 ATP molecules are produced.

$FADH_2$ passes electrons to a series of cytochrome molecules. Because of this, another name for the respiratory chain is the **cytochrome system.** It happens that the cytochromes, and other respiratory chain molecules not mentioned, accept only electrons and deposit the hydrogen ions (H^+) outside the inner mitochondrial membrane (fig. 5.15). This uneven distribution of hydrogen ions is believed to be highly significant since it creates an electrochemical gradient that supplies the necessary energy for ATP production.[3]

The passage of electrons down the respiratory chain gives it still another name, the **electron transport system.** Oxygen is the final acceptor for both electrons and hydrogen ions. Therefore, water is an end product of cellular respiration.

As table 5.5 indicates, the respiratory chain accounts for most of the ATP produced during aerobic cellular respiration and since the chain is present in mitochondria, this makes them the powerhouses of the cell. For the sake of discussion, it is usually calculated that the chain produces 34 of a possible total 38 ATP molecules per glucose molecule. These numbers are for discussion purposes because the chain produces a varying amount of ATP that is usually less than 34 ATP per glucose molecule.

[3]This is known as the chemiosmotic theory.

Table 5.5 Summary of ATP produced by cellular respiration

	Direct	By way of respiratory chain
Glycolysis	2 ATP	2 $NADH_2$ = 6 ATP*
Transition reaction		2 $NADH_2$ = 6 ATP
Krebs cycle	2 ATP	6 $NADH_2$ = 18 ATP
		2 $FADH_2$ = 4 ATP
Subtotal	4 ATP	34 ATP
Grand total		38 ATP*

*The numbers in this column and the total number of ATP are usually less because the chain does not always produce the maximum possible number of ATP per $NADH_2$.

Figure 5.16
The major metabolic pathways within a cell are
interrelated as shown.

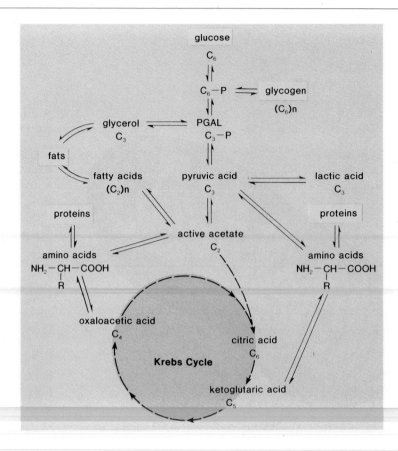

Other Metabolites

Other molecules besides glucose can be oxidized by the cell to release energy. When fats are used as an energy source they break down to glycerol and three fatty acids. As figure 5.16 indicates, glycerol is easily converted to PGAL, a metabolite in the glycolytic pathway. The fatty acids are converted to active acetate, which enters the Krebs cycle. A fatty acid that contains 18 carbons will result in nine active acetates. Calculation shows that respiration of these can produce a total of 216 ATP. For this reason, fats are an efficient form of stored energy because there are three long fatty acid chains per fat molecule.

As figure 5.16 shows, certain amino acids can enter the Krebs cycle. Before they enter, they undergo **deamination,** or the removal of the amino group. This group is converted to ammonia (NH_3), an excretory product of cells. Just where the amino acids enter the Krebs cycle is dependent on the length of the R group since this determines the number of carbons left following deamination.

The substrates that make up these pathways can also be used as starting molecules for various synthetic reactions. As you know, it is possible to gain weight if large amounts of carbohydrates are consumed. The connection between the metabolism of carbohydrates and fats is by way of active acetate. In figure 5.16 notice that polysaccharides, such as glycogen, lead directly to glucose phosphate molecules, each of which is then broken down to active acetate. A number of active acetate units may then be joined together to form a fatty acid. It is possible, too, for the body to produce some amino acids; the starting molecules of these are very often oxaloacetic and ketoglutaric acids. Thus it can be seen that the reactions of cellular respiration do more than produce energy. These reactions form the center of a "metabolic mill" that enables the cell to convert one kind of organic molecule to another.

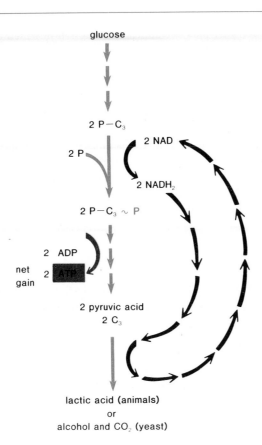

glucose

2 P—C₃

2 P

2 NAD

2 NADH₂

2 P—C₃ ~ P

2 ADP

net gain 2 ATP

2 pyruvic acid
2 C₃

lactic acid (animals)
or
alcohol and CO_2 (yeast)

Figure 5.17
Fermentation consists of the glycolytic pathway plus one additional reaction in which the end product of the glycolytic pathway, pyruvic acid, accepts hydrogens and becomes reduced. This "frees" NAD so that it may return to pick up more hydrogen atoms.

Anaerobic Respiration

If oxygen is not available to cells, the respiratory chain soon becomes inoperative because electrons plug up the system when the final acceptor, oxygen, is not present. In this case, most cells have a safety valve so that some ATP can still be produced.

The glycolytic pathway will run as long as it is supplied with "free" NAD; that is, NAD that can pick up hydrogen atoms. Normally, $NADH_2$ passes hydrogens to the respiratory chain and thereby becomes "free" of hydrogen atoms. However if the chain is not working due to the lack of oxygen, $NADH_2$ passes its hydrogen atoms to pyruvic acid, as shown in the following reactions:

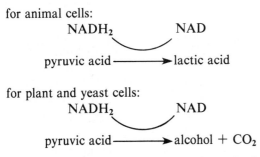

for animal cells:

$NADH_2$ → NAD

pyruvic acid ——→ lactic acid

for plant and yeast cells:

$NADH_2$ → NAD

pyruvic acid ——→ alcohol + CO_2

The whole purpose of this reaction, as shown in figure 5.17, is to keep the limited amount of NAD, available in the cell, free of hydrogen atoms so NAD can continue to oxidize PGAL (designated as 2P—C₃ in the diagram). Without this reaction the glycolytic pathway would also stop.

Notice that anaerobic respiration, also called **fermentation,** consists of the glycolytic pathway plus one additional reaction in which the end product of the glycolytic pathway, pyruvic acid, accepts hydrogens and becomes reduced. The Krebs cycle and respiratory chain do not function as part of fermentation, but when oxygen is available again, lactic acid can be converted back to pyruvic acid and metabolism can proceed as usual.

Fermentation in animals is an impractical process for two reasons. First, it produces only 2 ATP per glucose molecule compared to a much larger number for aerobic respiration. Second, it results in lactic acid buildup. Lactic acid is toxic to cells and causes muscles to cramp and fatigue (p. 339). If fermentation continues for any length of time, death will follow.

Yeast fermentation is useful to humans in at least two ways: the baking of bread and the production of alcohol. The reading on page 106 discusses the latter process.

Summary

Metabolic pathways are a series of reactions that proceed in an orderly step-by-step manner. Each reaction within a cell requires a specific enzyme. The substrate and enzyme for the reaction fit together as a key fits a lock. The formation of an enzyme-substrate complex allows the reaction to proceed in a regulated manner and lowers the energy of activation. Thus it is that enzymes speed up chemical reactions within cells. Enzymes are protein molecules that have a shape appropriate to their substrate. Any environmental factor that affects the shape of a protein also affects the ability of an enzyme to do its job. Among the environmental factors that affect enzymes are temperature, pH, poisons, and inhibitors. Most enzymes require a coenzyme. In this case, the protein portion of the enzyme, called the apoenzyme, provides specificity; the coenzyme, or nonprotein, portion of the enzyme participates in the reaction.

The breakdown of glucose to carbon dioxide and water requires three subpathways: glycolysis, or the breakdown of glucose to pyruvic acid; the Krebs cycle, a cycle of reactions that begins and ends with citric acid; and the respiratory chain, a chain of molecules that pass electrons from one to the other until oxygen acts as the final acceptor. Passage of the electrons received from each molecule of glucose through the respiratory chain produces 34 ATP (in addition to the 4 ATP produced directly by glycolysis and the Krebs cycle). Various types of molecules, such as amino acids and fatty acids, can be converted to molecules that can enter the Krebs cycle. Thus other types of molecules besides glucose can be used as an energy source in cells.

If oxygen is not available in cells, the respiratory chain is inoperative and fermentation, or anaerobic respiration, occurs. Fermentation is glycolysis with the addition of just one reaction. Glycolysis is completed with the production of pyruvic acid, but in fermentation, pyruvic acid is reduced by $NADH_2$ to lactic acid (animal cells) or alcohol and CO_2 (yeast cells). This reaction frees NAD so that it may return to an earlier reaction of glycolysis and pick up more hydrogen atoms. In this manner, it is possible for the process of glycolysis to continue. Fermentation results in lactic acid buildup in animal cells. Therefore as soon as oxygen becomes available again, lactic acid is oxidized to pyruvic acid and metabolism proceeds further.

Study Questions

1. Discuss and draw a diagram to describe a metabolic pathway. (p. 97)
2. Discuss and give a reaction to describe the lock-and-key theory of enzymatic action. (pp. 97–98)

The Cell, A Unit of Life

3. Name several factors that affect the yield of enzymatic reactions and explain the effect of these factors. (pp. 99–100)
4. Define coenzyme and give several examples. (p. 100)
5. Which molecule is known as the energy molecule of cells? (p. 101) Why is this molecule appropriate to the task? (p. 101)
6. Give the overall equation for cellular respiration and discuss the equation in general. (p. 103)
7. Name and describe the events within the three subpathways and transition reaction that make up aerobic cellular respiration. (p. 103)
8. How do our body cells obtain glucose and oxygen? What happens to the carbon dioxide given off by these cells? (p. 105)
9. Calculate the number of ATP that are formed as a *result* of glycolysis and the Krebs cycle. (pp. 105, 108)
10. Explain the term *oxidative decarboxylation*, which occurs in the Krebs cycle. (p. 107)
11. The respiratory chain is composed of what type molecules? (p. 109)
12. Calculate the number of ATP that are actually produced as hydrogens pass down the respiratory chain. (p. 109)
13. Why is fermentation wasteful and potentially harmful to the human body? (p. 112)

Selected Key Terms

metabolism (mĕ-tab′o-lizm)
metabolic pathway (met″ah-bol′ik path′wa)
substrate (sub′strāt)
enzyme (en′zīm)
coenzyme (ko-en-zīm)
denaturation (de-na-tur-a′shun)
NAD (en a de)
dehydrogenase (de-hi′dro-jen-as)
ATP (a te pe)
aerobic (a″er-ōb′ik)
glycolysis (gli-kol′ĭ-sis)
pyruvate (pi′roo-vāt)
Krebs cycle (krebz si′kl)
oxidative decarboxylation (ok″sĭ-da′tiv de″kar-bok″sĭ-la′shun)
respiratory chain (re-spi′rah-to″re chān)
cytochrome (si′to-krom)
deamination (de-am″ĭ-na′shun)
anaerobic (an-a″er-ōb′ik)
fermentation (fer″men-ta′shun)

Further Readings

Baker, J. W., and Allen, G. E. 1981. *Matter, energy, and life*. 4th ed. Reading, Mass.: Addison-Wesley.

Brock, T. D., and Brock, K. M. 1978. *Basic microbiology*. 2d ed. Englewood Cliffs, N.J.: Prentice-Hall.

Hinkle, P. C., and McCarty, R. E. 1978. How cells make ATP. *Scientific American* 238 (3):104.

Lehninger, A. L. 1961. How cells transform energy. *Scientific American* 205(3):39.

Margaria, R. 1972. The sources of muscular energy. *Scientific American* 226(3):84.

Rose, A. H. 1981. The microbiological production of food and drink. *Scientific American* 245(3):126.

Part Two

Plant Biology

Plants capture radiant energy from
the sun and use this energy to
convert inorganic chemicals to organic
chemicals. Photosynthesis is therefore a
two-step process. The first step, which
requires light energy, produces ATP and
$NADPH_2$. The second step uses these
molecules to reduce CO_2 molecules to
carbohydrate molecules.

Plant form and function can be related
to a plant's ability to carry out
photosynthesis. Flowering plants,
representative of plants in general, are
composed of a root system and a shoot
system; the latter includes the stems,
leaves, and flowers. Photosynthesis
occurs primarily in the leaves, which
contain the green pigment chlorophyll,
which is capable of absorbing the
energy of the sun.

Photosynthesis is essential to the
continued existence of all living things.
Its by-product, oxygen, makes cellular
respiration possible, and its primary
product, carbohydrates, are food for the
entire biosphere.

6

Photosynthesis

Chapter concepts

1 Photosynthetic organisms produce the organic molecules that are used as a source of food and chemical energy by all living things.

2 Photosynthesis takes place primarily within chloroplasts. Recent scanning electron micrographs suggest the location of photosynthetic subpathways.

3 Photosynthesis has two subpathways. The first drives the second by providing the energy (ATP) and the hydrogen atoms ($NADPH_2$) needed to reduce carbon dioxide (CO_2).

4 Photosynthesis can be compared to cellular respiration; both similarities and differences exist.

| Autotrophs | Herbivores | Carnivores |

Figure 6.1
a. Autotrophs represented by green plants, algae, and a few bacteria (the latter two represented in circles) produce food that feeds both (b.) herbivores which feed directly on plants or plant products, and (c.) carnivores, which feed on herbivores or other carnivores.

Only plants, algae, and a few bacteria carry on photosynthesis. As the name implies, photosynthesis refers to the ability of these organisms to make their own food in the presence of sunlight. The food produced in this manner eventually becomes the food for the rest of the living world (fig. 6.1). For example, humans and all other animals either eat plants directly or eat animals that have eaten plants, and so forth. Another way to express this idea is to say that **autotrophs,** which have the ability to synthesize organic molecules from inorganic raw materials, feed **heterotrophs,** which must take in preformed organic molecules as food. Thus autotrophs produce organic food, while heterotrophs consume it.

After food has been eaten and digested, the resulting small molecules may be used either as building blocks for growth or as a source of energy (for the production of ATP). Therefore the food provided by autotrophs supplies energy for metabolism.

Plants also supply energy in another sense because their bodies became the fossil fuel coal that is still in abundant supply. For several reasons, then, it is correct to say that all life is ultimately dependent upon the energy of the sun:

1. Sunlight supplies the energy needed for photosynthesis.
2. The food made by photosynthetic organisms becomes the food of the biosphere. This food may be used not only for growth but also for metabolic energy purposes.
3. The bodies of plants became the fossil fuel coal, upon which we are still dependent today.
4. Sunlight can be an energy source, called solar energy, for private and industrial use.

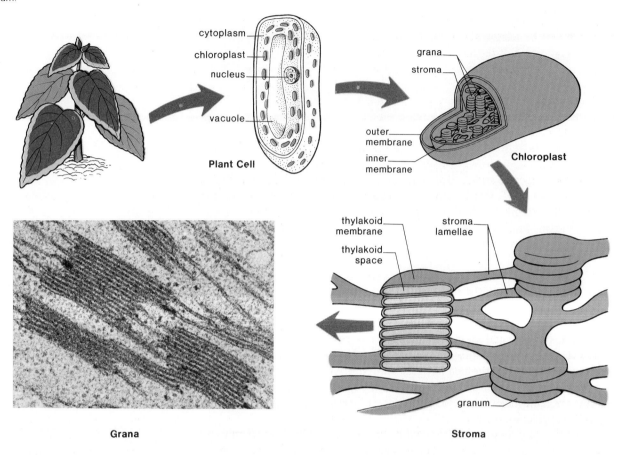

Plant Cell — cytoplasm, chloroplast, nucleus, vacuole

Chloroplast — grana, stroma, outer membrane, inner membrane

Grana

Stroma — thylakoid membrane, thylakoid space, stroma lamellae, granum

Chloroplasts

Chloroplasts are the organelles found in a plant cell (fig. 6.2) in which photosynthesis takes place.

Physiology

The overall equation for photosynthesis may be written in this manner to indicate that carbohydrates in general are the end product of photosynthesis:

$$CO_2 + H_2O + \text{light energy} \longrightarrow (CH_2O) + O_2$$

Multiplying this equation by 6 shows that glucose is often an end product of photosynthesis:

$$6\,CO_2 + 6\,H_2O + \text{light energy} \longrightarrow C_6H_{12}O_6 + 6\,O_2$$

However, we now know that the oxygen molecules that appear on the right-hand side of the equation were originally a part of the water molecules. This was proven experimentally by exposing plants first to carbon dioxide and then to water that contained an isotope of oxygen, O^{18}, called heavy oxygen. Only in the latter instance did this isotope appear as molecular oxygen given off by the plant. Thus it was shown that oxygen released by chloroplasts comes from water and not from carbon dioxide. To indicate this, the overall equation for

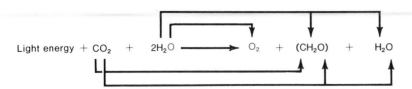

$$\text{Light energy} + CO_2 + 2H_2O \longrightarrow O_2 + (CH_2O) + H_2O$$

Figure 6.3
Equation for photosynthesis showing the relationship between reactants and products. Oxygen given off by photosynthesis is derived from water.

photosynthesis may be written as shown in figure 6.3. The arrows in the equation indicate the relationship between the molecules on the left and those on the right. If we analyze this equation, we see that photosynthesis provides a link between the nonliving and living world. The small inorganic molecules (on the left side of the equation) are converted to a much larger organic molecule (on the right side of the equation). The carbohydrate molecule represents the food that is produced by photosynthesis.

The equation (fig. 6.3) also indicates that photosynthesis is an energy requiring synthetic reaction. As with some synthetic reactions, photosynthesis involves reduction: hydrogen atoms ($H^+ + e^-$) are added to a molecule. Notice that hydrogen is added to the carbon-oxygen combination when carbohydrate is formed. This addition of hydrogen requires the formation of new bonds and therefore requires energy. Thus, it is not surprising to learn that carbon dioxide and water are low-energy molecules and that carbohydrates are high-energy molecules.

Two Subreactions

It was discovered some years ago that the process of photosynthesis involves two subreactions. It was proposed that these subreactions be called the "light" reaction, which requires that light be present, and the "dark" reaction, which does not require light. These two subreactions were detected because when light is being maximally absorbed by a photosynthetic system, a rise in temperature still increases the rate of photosynthesis (fig. 6.4). This indicates that the first subreaction (or pathway) is primarily dependent on light, while the second subpathway is dependent primarily on temperature. Today the terms "light" and "dark" reaction are being phased out because they do not adequately point out the most important difference between the two subpathways of photosynthesis. The first pathway represented by the reaction given in figure 6.5a captures the energy of sunlight and uses this energy to produce $NADPH_2$ and ATP, and the second pathway represented by the reaction given in figure 6.5b uses these molecules to reduce carbon dioxide to a carbohydrate.

Structure

Four concentric "layers" of a chloroplast can be distinguished (fig. 6.2). The first layer is the chloroplast envelope, a double membrane that separates the chloroplast from the rest of the cell. The raw materials for photosynthesis enter the chloroplast across this envelope. Similarly, the photosynthetic products exit across the envelope when they leave the chloroplast and enter the cytoplasm.

The second "layer" is the area within the chloroplast, called the **stroma.** There is a fluid within the stroma that contains a number of enzymes that catalyze various reactions responsible for the second subreaction (fig. 6.5b).

Grana comprise the third "layer." They are called grana because they appear as dark "grains" under the light microscope. The electron microscope shows that grana are composed of flattened sacs, or discs, called **thylakoids.** There are membranous connections between some adjacent thylakoids and these are called **stroma lamellae.** Chlorophyll and other pigments, as well as numerous proteins, are contained in the membranes making up the thylakoids. This is where the energy of sunlight is captured during photosynthesis and the subreaction represented in figure 6.5a occurs.

Figure 6.4
At lower light intensities *(shaded area)*, the amount of light limits photosynthetic rate. Above maximum light intensity *(broken line)*, the photosynthetic rate increases if the temperature is increased. On the basis of these results it was proposed that there are separate "light" and "dark" reactions in photosynthesis.

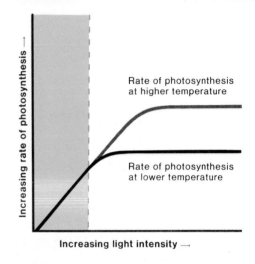

Figure 6.5
Photosynthesis involves two pathways. *a.* The overall reaction for the pathway that occurs in the thylakoid membranes: ATP and $NADH_2$ are formed as water is split, releasing oxygen. *b.* The overall reaction for the pathway that occurs in the stroma: ATP and $NADH_2$ from the first reaction is used to reduce carbon dioxide. (R = remainder of the molecule).

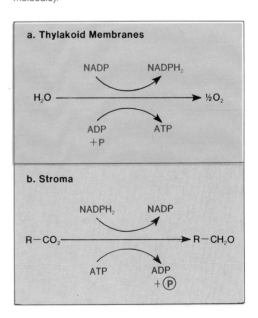

Figure 6.6

Anatomy of a granum. Each granum is composed of thylakoids. *b.* Drawing of an enlarged thylakoid membrane to show location of particles in the inner half and outer half of the membrane. *c.* Drawing is based on electron micrographs, such as this one that was made following freeze-fracture of thylakoid membrane. The smaller particles are believed to be the location of Photosystem I and the larger particles are believed to be the location of Photosystem II. Light energy is captured by the photosystems.

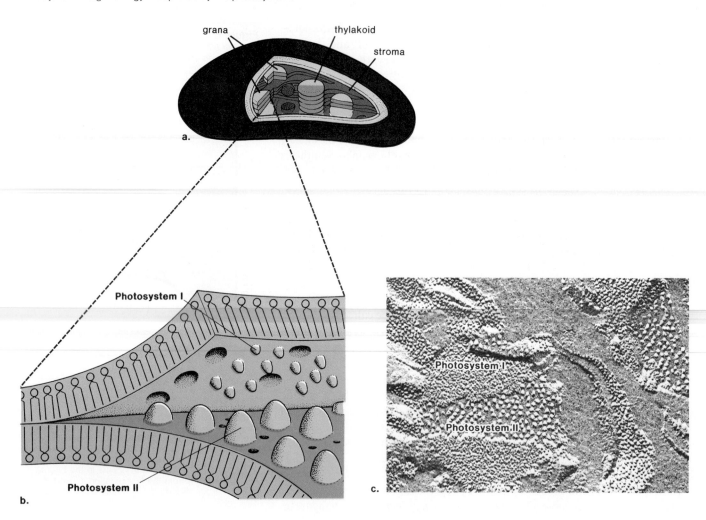

The freeze-fracture method of preparing thylakoid membrane shows two types of particles (fig. 6.6b and 6.6c). The smaller particles that appear in the outer half of the membrane are believed to be locations for *Photosystem I*, while the larger particles that appear in the inner half of the membrane are believed to be locations for *Photosystem II*. Both photosystems contain pigments and carrier molecules that help capture the energy of sunlight.

The spaces inside the thylakoids make up the fourth "layer" of the chloroplast. As the first subreaction occurs, hydrogen ions collect in this space and then later move out of it across the thylakoid membrane into the stroma. Movement of these ions is believed to supply the energy for ATP production.

Capturing the Energy of Sunlight

Sunlight

Radiant energy from the sun can be described in terms of its wavelength and its energy content. Figure 6.7 lists the different types of radiant energy from the shortest wavelength, gamma rays, to the longest, radio waves. The shorter

Plant Biology

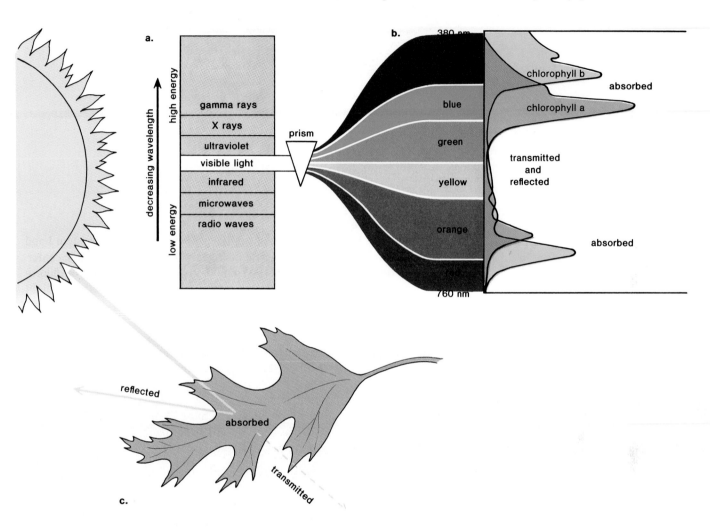

Figure 6.7
a. Energy from the sun is categorized into the types listed in the gray portion. Gamma rays have the shortest wavelength and the most energy content. Radio waves have the longest wavelength and the least energy content. Visible (white light) actually contains many colors of light as can be detected when the light is passed through a prism. *b.* Absorption spectrum of chlorophyll *a* and *b* shows that these two pigments absorb violet and red light the best. *c.* Leaves appear green to us because the color green is reflected or transmitted by chlorophyll.

wavelengths contain more energy than the longer ones. **White,** or **visible, light** is only a small portion of this spectrum. Visible light itself contains various wavelengths of light, as can be proven by passing it through a prism; then we see all the different colors that make up visible light. (Actually, of course, it is our eyes that interpret these wavelengths as colors.) The colors range from violet (the shortest wavelength) to blue, green, yellow, orange, and red (the longest wavelength). The energy content is highest for violet light and lowest for red light.

The pigments found within photosynthesizing cells are capable of absorbing various portions of visible light. The absorption spectrum for chlorophyll *a* and chlorophyll *b* is shown in figure 6.7b. Both chlorophyll *a* and chlorophyll *b* absorb violet, blue, and red better than the other colors. Because the color green is only minimally absorbed and is primarily reflected (fig. 6.7c), leaves appear green to us.

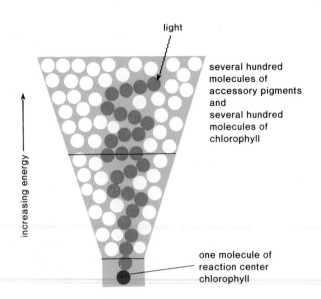

light

several hundred
molecules of
accessory pigments
and
several hundred
molecules of
chlorophyll

increasing energy

one molecule of
reaction center
chlorophyll

The Photosystems

Photosynthesis begins when pigment complexes within both Photosystem I and Photosystem II (fig. 6.8) absorb light energy and funnel it to a reaction center. The **reaction center** of each photosystem contains a different type of chlorophyll *a*. **Photosystem I** has a reaction-center chlorophyll *a* that absorbs light maximally at a wavelength around 700 nm and is therefore called **P700** (P stands for pigment). **Photosystem II** has a reaction-center chlorophyll *a* that absorbs slightly shorter wavelengths maximally and is called **P680.**

Most often the two photosystems are at work at the same time, but there is also evidence that at times Photosystem I can work alone.

Noncyclic Photophosphorylation

As figure 6.9 indicates, both photosystems most likely absorb sunlight at the same time. However, it is easier to describe the manner in which energy is usually captured by beginning with Photosystem II:

1. As energy is received by P680, its electrons become so highly charged that a few actually leave the chlorophyll molecule. P680 would soon disintegrate if it did not receive replacement electrons. It receives these electrons from water, which splits in the following manner:

$$H_2O \longrightarrow 2\,H^+ + 2e^- + \tfrac{1}{2}O_2$$

This freed oxygen is the oxygen gas given off during photosynthesis.

2. The electrons that leave P680 pass down an **electron transport system** that, like the one in mitochondria, is a series of carriers, some of which are cytochrome molecules. For this reason, the system is sometimes referred to as the **cytochrome system.** As the electrons pass "downhill" from one cytochrome to another, energy is made available for ATP formation:

$$ADP + \text{P} + energy \longrightarrow ATP$$

Just how this occurs is still under investigation. It is believed that some of the members of the electron transport system deposit hydrogen ions within the thylakoid space and this creates an electrochemical gradient that serves as a source of energy for ATP formation when these ions flow from the thylakoid space to the stroma.[1]

[1]This is known as the chemiosmotic theory.

Figure 6.9
Photophosphorylation (ATP formation) occurs in thylakoid membranes. In noncyclic photophosphorylation, electrons move from water to P680, to an acceptor molecule that passes them down a transport system, to P700, which sends them to another acceptor molecule, before they are finally sent to NADP. In cyclic photophosphorylation *(dotted lines)*, electrons pass from P700 to an acceptor molecule that sends them down the electron transport system before they return to P700.

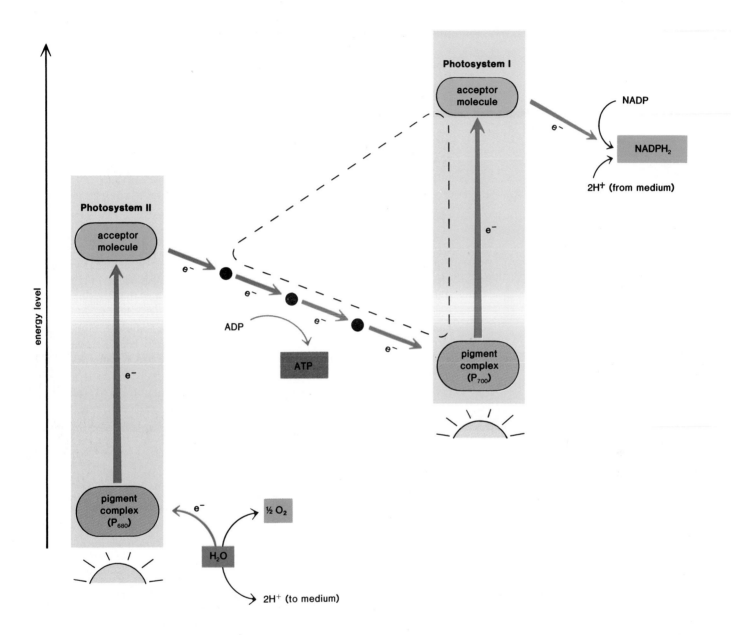

3. After the electrons leave the cytochrome system, they are in a lower energy state and are received by P700, the reaction center chlorophyll *a*, in Photosystem I. These electrons replace some that previously left P700. Prior to their arrival the pigment complex of Photosystem I has absorbed light energy and funneled it to P700. A few of its electrons were so highly energized that they left the chlorophyll molecule to be eventually taken up by NADP, as shown in the following reaction:

$$NADP + 2e^- + H^+ \longrightarrow NADPH_2$$

NADP is similar in structure to NAD except that it carries an extra phosphate group, as indicated by the additional *P*.

This series of three events just described is called **noncyclic photophosphorylation** because it is possible to trace electrons in a one-way direction from water to NADP and because sunlight energy (*photo*) has caused ATP production (*phosphorylation*).

Cyclic Photophosphorylation

At times, electrons leave P700 and then return to it instead of being taken up by NADP (dotted line in figure 6.9). Before arriving, they pass down the electron transport system, mentioned earlier, and ATP is produced.

This is called **cyclic photophosphorylation** because it is possible to trace electrons in a cycle (from P700 to P700) and because sunlight energy (*photo*) has caused ATP production (*phosphorylation*). Cyclic photophosphorylation is believed to occur whenever CO_2 is in such limited supply that the dark reaction is not occurring.

Cyclic photophosphorylation provides an independent means by which ATP can be generated. But it does not result in the release of oxygen or in NADPH production. Perhaps this form of photophosphorylation evolved before noncyclic, simply as a means to make ATP.

Reducing Carbon Dioxide

C$_3$ Photosynthesis

Figure 6.10 is a simplified diagram of the reactions that occur as carbon dioxide is taken up and reduced so that carbohydrate synthesis occurs. Figure 6.11 illustrates the same reactions in more detail. Altogether, the reactions make up a cycle called the Calvin cycle, named for the man who discovered it. Carbon dioxide is combined with the five-carbon sugar, ribulose biphosphate (RuBP)[2]. The resulting six-carbon molecule immediately breaks down to two molecules of PGA (phosphoglyceric acid). This is called **C$_3$ photosynthesis** because the first molecule that can be detected, namely PGA, is a three-carbon molecule. PGA immediately undergoes reduction to PGAL (phosphoglyceraldehyde):

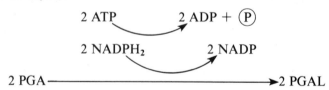

[2]Formerly called ribulose diphosphate.

Figure 6.10
The Calvin cycle (simplified). RuBP accepts carbon dioxide (CO_2), forming a six-carbon molecule (C_6), which immediately breaks down to two PGA molecules that are reduced to 2 PGAL, one of which represents the net gain of the cycle. PGAL can be combined with another PGAL to give glucose-6 phosphate that can be metabolized to other organic molecules.

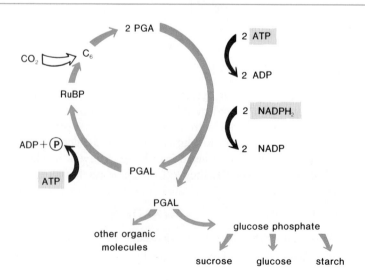

Plant Biology

This reaction, which takes place in the stroma, uses the $NADPH_2$ and ATP formed previously in the thylakoid membranes. The NADP and the $ADP + \textcircled{P}$ that remain after the reaction is completed return to the thylakoid membrane (fig. 6.12). This is the reaction that represents the reduction of carbon dioxide and the conversion of a low-energy molecule to a high-energy molecule. PGAL is the immediate photosynthetic product of the Calvin cycle.

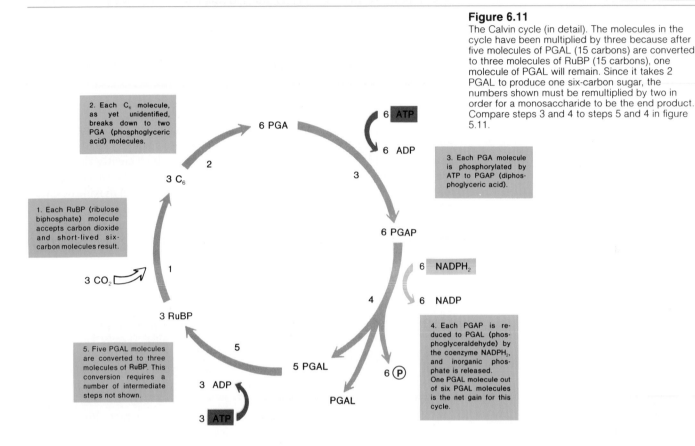

Figure 6.11
The Calvin cycle (in detail). The molecules in the cycle have been multiplied by three because after five molecules of PGAL (15 carbons) are converted to three molecules of RuBP (15 carbons), one molecule of PGAL will remain. Since it takes 2 PGAL to produce one six-carbon sugar, the numbers shown must be remultiplied by two in order for a monosaccharide to be the end product. Compare steps 3 and 4 to steps 5 and 4 in figure 5.11.

2. Each C_6 molecule, as yet unidentified, breaks down to two PGA (phosphoglyceric acid) molecules.

3. Each PGA molecule is phosphorylated by ATP to PGAP (diphosphoglyceric acid).

1. Each RuBP (ribulose biphosphate) molecule accepts carbon dioxide and short-lived six-carbon molecules result.

4. Each PGAP is reduced to PGAL (phosphoglyceraldehyde) by the coenzyme $NADPH_2$, and inorganic phosphate is released. One PGAL molecule out of six PGAL molecules is the net gain for this cycle.

5. Five PGAL molecules are converted to three molecules of RuBP. This conversion requires a number of intermediate steps not shown.

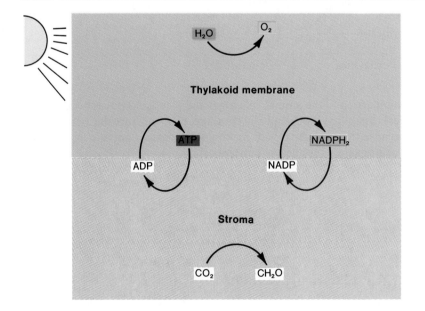

Figure 6.12
Diagram illustrating that the metabolic pathway located in the thylakoid membranes produces the ATP and $NADPH_2$ that is used by the pathway located in the stroma. The ADP and P resulting from ATP breakdown and the NADP (minus hydrogen atoms) return to the thylakoid membrane.

Some PGAL is used within the chloroplast to re-form ribulose biphosphate.

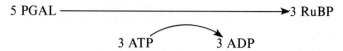

This ATP is also produced by the light reactions. Altogether the ATP and $NADPH_2$ consumed in converting CO_2 to the level of a hexose (a six-carbon sugar) results in about a 30 percent energy-efficiency rate for photosynthesis.

PGAL, the product of the Calvin cycle, can combine with another PGAL to form glucose phosphate, which can be converted to glucose. The overall equation for photosynthesis often shows glucose as the end product for the process. However, glucose phosphate is converted to sucrose for transport in phloem or to starch for storage.

C_4 Photosynthesis

The more the Calvin cycle functions, the greater the amount of food produced. Recently, it has been discovered that plants adapted to a hot, dry environment, having so called *C_4 photosynthesis,* produce food more efficiently than those adapted to a temperate climate having so-called *C_3 photosynthesis.* The reason for this is based on their relative abilities to fix (reduce) carbon dioxide despite its low concentration in the atmosphere. This is discussed in the reading for this chapter on page 127. Whereas the concentration of CO_2 in the atmosphere is a **limiting factor** (a factor that determines whether growth occurs or not) for temperate zone plants, it is not a limiting factor for many tropical zone plants.

Plant Metabolism

The carbohydrates (PGAL, glucose phosphate, sucrose, and so forth) synthesized by means of photosynthesis in the leaves are transported to other parts of the plant, usually in the form of sucrose. Although it may be temporarily stored in the form of starch, much of this carbohydrate will eventually be broken down by means of cellular respiration (just as in animals) to produce the ATP needed for cell metabolism.

Plants not only have the enzymatic capability to produce sugars from carbon dioxide and water but they also have the ability, by way of the metabolic pathways diagrammed in figure 5.16 and by even more complex pathways, to produce all the various types of organic molecules they require:

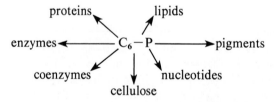

In comparison to the animal cell, algal and plant cells possess enormous biochemical capability. Biochemically speaking, plant cells are due our utmost respect because they are capable of making all necessary organic molecules, using only inorganic molecules as nutrients.

The world's population continues to increase and, as it does so, encroaches on agricultural land. In the United States where there were once fields of grain, there are now towns, suburbs, and shopping malls. More food must be produced on less land, therefore agricultural yields must be constantly increased. Despite advanced technology and the development of mutant plants, the average yield per acre has not significantly increased in recent years. New ideas are needed and one may have been discovered.

Observation has shown that temperate zone plants, such as wheat, alfalfa, and potato, do not take up CO_2 as efficiently as do tropical zone plants, such as corn. The fault lies with the first enzyme of the Calvin cycle. This enzyme can catalyze one of two reactions:

1. $CO_2 + RuBP \rightarrow \rightarrow C_3$
2. $O_2 + RuBP \rightarrow \rightarrow CO_2 + H_2O$

The first reaction is a normal part of photosynthesis. The second reaction is called **photorespiration** because oxygen is taken up and CO_2 is given off to the environment. Photorespiration, which competes with the photosynthetic reaction, accounts for the lower yield of temperate zone plants because it obviously does not lead to carbohydrate synthesis.

How is photorespiration avoided in tropical zone plants? First, photosynthesis only occurs in certain cells called the **bundle-sheath cells** (p. 144). Second, other cells specialize in capturing CO_2 and passing it to the bundle-sheath cells. CO_2 is captured by a reaction that occurs despite the low concentration of CO_2 in the atmosphere (0.4%):

3. $CO_2 + PEP \rightarrow C_4$

These C_4 molecules are transported into the photosynthesizing bundle-sheath cells where CO_2 is released.

Notice that it is possible to make a distinction between the two groups of plants according to the first molecule detected following CO_2 uptake. In temperate zone plants, C_3 is always the first molecule detected following CO_2 uptake (see reaction #1), and in the tropical zone plants under discussion, C_4 is the first molecule detected (see reaction #3). Therefore, it is now customary to speak of C_3 versus C_4 photosynthesis—C_4 being much more efficient. Scientists are now exploring the possibility of transforming temperate zone plants into C_4 photosynthesizers. If this can be accomplished, agricultural yields will be greatly increased.

Increasing Crop Yields: C_3 versus C_4 Photosynthesis

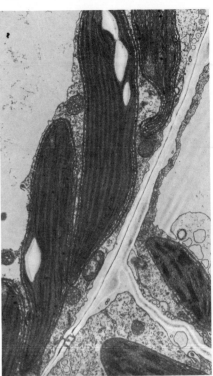

Two corn cells at bottom right capture CO_2 and pass it to the bundle sheath cells on the left. Notice that the chloroplasts in the bundle sheath cells have stroma lamellae but lack grana. Could this be related to their ability to carry on C_4 photosynthesis?

Autotrophic Bacteria

Bacterial Photosynthesis

Some bacteria are photosynthetic, but unlike cyanobacteria (formerly called blue-green algae), eukaryotic algae, and plants, they do not release oxygen when they photosynthesize. This indicates that they do not use water as a source of electrons. Rather, depending on the type of bacteria, they utilize hydrogen gas (H_2), hydrogen sulfide (H_2S), or organic compounds for this purpose. It is possible that early prokaryotes utilized this type of photosynthesis. Later, algae and plants developed photosynthesis that could make use of water.

Figure 6.13
Giant tube worms are a part of the community of
organisms found near the deep ocean ridges and
vents. These communities have been given such
names as Dandelions, the Rose Garden, and the
Garden of Eden.

Chemosynthesis

Some bacteria can oxidize inorganic compounds, such as ammonia, nitrates, and sulfides, and can trap the small amount of energy released to synthesize carbohydrates. While this capability is not believed to contribute greatly to the support of life on land, it has recently been found capable of supporting an entire community a mile and a half below sea level. By means of deep-diving research submarines, scientists have been examining the mid-ocean ridge system where hot minerals spew out from the inner earth. Here, bacteria in the vent water oxidize hydrogen sulfide and carry on chemosynthesis. Living off the resulting organic molecules are giant tube worms (fig. 6.13), clams, and crabs. There is evidence that such bacteria may also reside within the bodies of the worms. Prior to finding these organisms it was not thought that such communities could exist on the ocean floor where light never penetrates.

Comparison of Cellular Respiration and Photosynthesis

Differences

Whereas only certain cells carry on photosynthesis, all cells, including photosynthesizing cells, carry on cellular respiration, either aerobic or anaerobic. The cellular organelle for cellular respiration is the mitochondrion, while the cellular organelle for photosynthesis is the chloroplast.

The overall equation for aerobic cellular respiration is the opposite of that for photosynthesis:

$$\text{energy} + \text{carbon dioxide} + \text{water} \rightleftharpoons \text{carbohydrate} + \text{oxygen}$$

The reaction in the forward direction represents photosynthesis, and the energy is the energy of the sun. The reaction in the opposite direction represents cellular respiration, and the energy then stands for ATP.

Obviously, photosynthesis is the building up of glucose, while cellular respiration is the breaking down of glucose. See table 6.1 for a summarized comparison of these processes.

Table 6.1 Cellular respiration and photosynthesis

Cellular respiration	Photosynthesis
Oxidation	Reduction
Releases energy	Requires energy
Requires oxygen	Releases oxygen
Releases carbon dioxide	Requires carbon dioxide

Figure 6.14

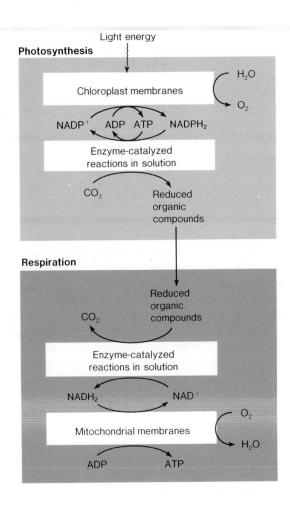

Photosynthesis

Light energy

Chloroplast membranes

H_2O

O_2

$NADP^+$ ADP ATP $NADPH_2$

Enzyme-catalyzed
reactions in solution

CO_2

Reduced
organic
compounds

Respiration

Reduced
organic
compounds

CO_2

Enzyme-catalyzed
reactions in solution

$NADH_2$ NAD^+

O_2

Mitochondrial membranes

H_2O

ADP ATP

Diagram illustrating similarities and differences between photosynthesis that takes place in chloroplasts and cellular respiration that takes place in mitochondria. Both have a cytochrome system located within membrane where ATP is produced. Both have enzyme-catalyzed reactions in solution. The coenzyme NAD(P) cycles between the membrane and the solution. However, whereas photosynthesis releases oxygen and reduces carbon dioxide into carbohydrates, cellular respiration reduces oxygen and releases carbon dioxide.

Similarities

Both photosynthesis and cellular respiration are metabolic pathways within cells and therefore consist of a series of reactions that the overall reaction does not indicate. Within the pathways, both make use of a cytochrome system located in membrane (fig. 6.14) to generate a supply of ATP, and both make use of a hydrogen carrier: cellular respiration uses NAD and photosynthesis uses NADP.

Both pathways utilize this overall reaction but in opposite directions. For photosynthesis, read from left to right in the following diagram, and for cellular respiration, read from right to left.

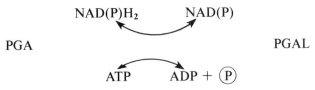

$NAD(P)H_2$ $NAD(P)$

PGA PGAL

ATP ADP + P

Both photosynthesis and cellular respiration occur in plant cells. While both of these occur during the daylight hours, only cellular respiration occurs at night. During daylight hours, the rate of photosynthesis exceeds the rate of cellular respiration, resulting in a net increase and storage of glucose. The stored glucose is used to support cellular metabolism that continues during the night.

Summary

Photosynthesis is absolutely essential for the continuance of life because it supplies the biosphere with food and energy. Chloroplasts carry on photosynthesis. The overall equation for photosynthesis shows that it is an energy-requiring synthetic reaction that changes inorganic substances to an organic substance, namely a carbohydrate. Photosynthesis has been found to consist of two subreactions. The subreaction located in the stroma uses $NADPH_2$ and ATP to reduce carbon dioxide. These molecules are produced by the other subreaction, located in thylakoid membranes of the grana, after chlorophyll captures the energy of sunlight. Violet, blue, and red rays of visible light are especially effective in photosynthesis because these are the colors absorbed by chlorophyll.

Small particles located on thylakoid membrane are thought to be the location for Photosystem I and Photosystem II. Photosynthesis begins when pigment complexes within Photosystem I and Photosystem II absorb radiant energy and pass it to a reaction center chlorophyll *a* molecule. In noncyclic photophosphorylation, electrons move from water to Photosystem II and then to Photosystem I before finally arriving at NADP. Negatively charged NADP then combines with hydrogen ions, forming $NADPH_2$. As the electrons move between the photosystems, they pass down a cytochrome system that produces ATP. If CO_2 is in limited supply, cyclic photophosphorylation occurs; electrons circle from Photosystem I to the cytochrome system and then back again.

The ATP and $NADPH_2$ made in thylakoid membranes pass into the stroma where carbon dioxide is reduced. Carbon dioxide is picked up by RuBP, forming a six-carbon sugar that immediately breaks down to two PGA molecules. The PGA molecules are reduced to PGAL, and this represents the reduction of carbon dioxide and the formation of a high-energy molecule from a low-energy molecule. Two PGAL molecules join to form glucose phosphate, which can be converted to sucrose for transport and starch for storage. Table 6.2 summarizes the roles played by the various participants in photosynthesis.

A plant can use the products of photosynthesis to synthesize all the organic molecules it needs by utilizing the pathways described in figure 5.16 as well as additional ones. Aside from photosynthetic algae and plants, some bacteria are also autotrophic. Photosynthetic bacteria do not release oxygen, and chemosynthetic bacteria remove hydrogen atoms from a number of substances in order to capture energy. A surprising find has been deep ocean communities supported by chemosynthetic bacteria.

Table 6.2 Participants in photosynthesis

	Participant	Role
Thylakoid membranes	Sunlight	Provides energy
	Chlorophyll	Absorbs energy
	Water	Donates electrons and releases oxygen
	ADP + (P)	Forms ATP
	NADP	Accepts electrons and H^+ and becomes $NADPH_2$
Stroma	RuBP	Takes up CO_2
	CO_2	Reduced to PGAL
	ATP	Provides energy for reduction
	$NADPH_2$	Provides electrons for reduction
	2 PGAL	Becomes glucose phosphate

Plant Biology

Study Questions

1. Why are all living things dependent upon the process of photosynthesis and the energy of the sun? (p. 117)
2. Give the overall equation for photosynthesis and the equations for the two subpathways involved in the process. (p. 119)
3. Which rays of light are most important for photosynthesis and why? (p. 121)
4. Describe the structure of a chloroplast, and indicate where photophosphorylation and the Calvin cycle occur. In what way are these two processes connected? (p. 119)
5. Trace the path of electrons during noncyclic photophosphorylation. (pp. 122–24)
6. Give the primary steps of the Calvin cycle, indicating the reaction that represents the reduction of carbon dioxide. (pp. 124–26)
7. Describe the role of each participant in photosynthesis. (p. 150)
8. Why would it be correct to say that a plant cell is more biochemically competent than an animal cell? (p. 126)
9. How is bacterial photosynthesis different from plant photosynthesis? (p. 127)
10. Contrast cellular respiration and photosynthesis in at least five ways. How are the two cellular processes similar? (pp. 128–29)

Selected Key Terms

photosynthesis (fo″to-sin′thĕ-sis)
autotroph (aw′to-trōf)
heterotroph (het′er-o-trōf)
stroma (stro′mah)
stroma lamellae (stro′mah lah-mel′ e)
grana (gra′nah)
chlorophyll (klo′ro-fil)
thylakoid (thi′lah-koid)
photosystem (fo′to-sis″tem)
noncyclic photophosphorylation (non-sik′lik fo″to-fos″for-i-la′shun)
cyclic photophosphorylation (sik′lik fo″to-fos″for-i-la′shun)
NADP (en a de pe)
cytochrome system (si′to-krōm sis′tem)
Calvin cycle (kal′vin si′kl)
RuBP (ar u be pe)
C₃ photosynthesis (se thre fo″to-sin′thĕ-sis)
C₄ photosynthesis (se for fo″to-sin′thĕ-sis)
chemosynthesis (ke″mo-sin′the-sis)

Further Readings

Arnon, D. I. 1960. The role of light in photosynthesis. *Scientific American* 203(5):50.
Bassham, J. A. 1962. The path of carbon dioxide in photosynthesis. *Scientific American* 206(6):40.
Edmond, J. M., and Von Damm, K. 1983. Hot springs on the ocean floor. *Scientific American* 248(4):78.
Govindjee, and Govindjee, R. 1974. The absorption of light in photosynthesis. *Scientific American* 231(6):68.
Levine, R. P. 1969. The mechanism of photosynthesis. *Scientific American* 221(6):15.
Miller, K. R. 1979. The photosynthetic membrane. *Scientific American* 241(4):102.
Ray, P. M. 1972. *The living plant.* 2d ed. New York: Holt, Rinehart & Winston.

7

Plant Form and Function

Chapter concepts

1 Plant anatomy and physiology, exemplified by a flowering plant, can be correlated with the process of photosynthesis by which plants make their own organic food.

2 The plant body, which includes the roots, stems, leaves, and flowers, is made up of a few major types of specialized cells and tissues derived from an ever-present source of embryonic cells.

3 Theories have been formulated to explain the transport of water and minerals from the roots to the leaves and the transport of organic nutrients in the reverse direction.

4 The sex organs of a flowering plant are located in the flower. Flowering is controlled by the length of daylight (photoperiod) in some plants.

5 Seeds within fruits are the products of sexual reproduction in flowering plants. Germination of a seed results in another plant.

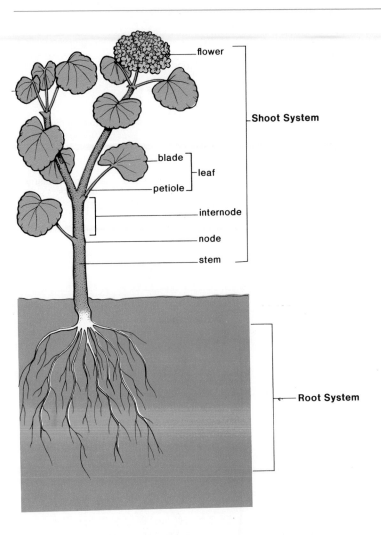

flower

Shoot System

blade ─┐
       ├─ leaf
petiole ─┘

internode

node

stem

Root System

The body of a flowering plant is divided into two portions, the **root system** and the **shoot system** (fig. 7.1). The roots, which lie below ground level, anchor the plant and absorb water and minerals. Within the shoot system, the stem lifts the leaves to catch the rays of the sun. The leaves receive water and minerals that are sent from the roots up through the stem and take in carbon dioxide from the air. Thus the leaves are provided with the light energy, water, and carbon dioxide they need to carry on photosynthesis.

This chapter will begin with the basic anatomy of the roots, stems, and leaves of a flowering plant. These are vegetative organs, meaning they are not concerned with reproduction. Some of the tissues found within each organ are listed in table 7.1. In addition to the tissue types listed, plants also contain embryonic tissue called **meristem tissue,** which is unspecialized and continually capable of cell division. Cell division is followed by differentiation into the cell types depicted in figure 7.2. Parenchyma and sclerenchyma cells are found in most tissues. Parenchyma cells are relatively unspecialized and correspond best to the generalized cell of a plant (fig. 2.1b). Sclerenchyma cells are hollow, nonliving cells with extremely strong walls that give support to plant tissues and organs. Two tissue types, epidermis and endodermis, do not have parenchyma and sclerenchyma cells. **Epidermis** covers the entire body of

Table 7.1 Vegetative organs and major tissues

	Roots	Stems	Leaves
Function	Absorb water and minerals	Transport water and nutrients	Carry on photosynthesis
	Anchor plant	Support leaves and flowers	
Tissue			
Epidermis	Root hairs absorb water and minerals	Protect inner tissues	Stomata carry on gas exchange
Cortex	Store products of photosynthesis	Carry on photosynthesis, if green	————
Vascular	Transport water and nutrients	Transport water and nutrients	Transport water and nutrients
Pith	————	Store products of photosynthesis	————
Mesophyll	————	————	
Spongy layer			Gas exchange
Palisade layer			Photosynthesis

Figure 7.2
Differentiation of plant cells. The meristem cell is nonspecialized and, by a process of maturation, may become the specialized cells shown here, or others. Meristem cells also constantly divide to produce new meristem cells. Parenchyma and sclerenchyma cells are found in most plant tissues. Parenchyma cells are nonspecialized vegetative cells but the sclerenchyma cells are highly specialized as support cells. They lack cytoplasm and have very strong walls.

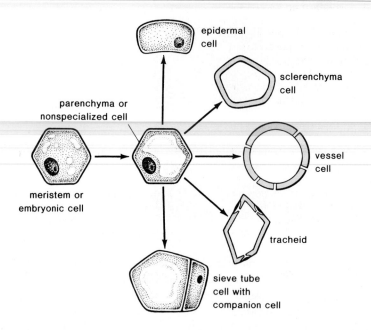

a nonwoody and young woody plants. **Epidermis** functions to protect inner body parts and to prevent the plant from drying out. In addition, the epidermis has specialized structures and functions that are discussed when each organ is considered. In contrast to epidermis, **endodermis,** which is nearly universally found in roots, contains only endodermal cells.

The other cell types in figure 7.2 are found in vascular (transport) tissue, which is considered in detail on page 145. **Sclerenchyma cells** lack cytoplasm and are nonliving support cells with strong walls.

Plants need only inorganic nutrients in order to produce all the organic molecules that make up their bodies. Aside from carbon, hydrogen, and oxygen (obtained from carbon dioxide and water), the mineral elements listed in table 7.A are required nutrients for plants. The mineral elements are classified as **macronutrients** when they are used by plants in greater amounts and **micronutrients** when they are needed by plants in very small amounts. Both types of minerals are found in the soil, but in low concentrations; not only must a plant be able to take them up, it must also be able to concentrate them. Fortunately, the root system of a plant is designed for just this purpose. As the root system grows, it branches and branches again so that the roots are exposed to a tremendous amount of soil. It has been estimated that a rye plant has roots totaling about 900 kilometers (more than 650 miles) in length. Further, because of the extensive number of root hairs, the total surface area is about 635 square meters, or more than 7,000 square feet! Water, and possibly minerals too, enters root hairs by diffusion, but eventually active transport is used to concentrate the minerals within the organs of a plant. A plant uses a great deal of ATP for active transport.

It is lucky for us that plants have the capability of concentrating minerals for we are often dependent on them for our basic supply of such ions as sodium to maintain blood pressure, calcium to build bones and teeth, and iron to help carry oxygen to our cells. Once plants have taken the minerals up, they are incorporated into proteins, fats, and vitamins when these substances are formed from carbohydrates, the product of photosynthesis. When we eat plants, we are supplied with minerals and all types of organic molecules, some of which become building blocks for our own cells and some of which are used as an energy source.

Table 7.B lists examples of foods that humans consume directly from plants. Each type of food is associated with a particular organ of a flowering plant: the root, stem, leaf, or flower.

Plants as Food

Table 7.A Inorganic nutrients necessary for plant life

Compound	Element supplied
Macronutrients	
KNO_3	K;N
$CaNO_3$	Ca;N
$NH_4H_2PO_4$	N;P
$MgSO_4$	Mg;S
Micronutrients	
KCl	Cl
H_3BO_3	B
$MnSO_4$	Mn
$ZnSO_4$	Zn
$CuSO_4$	Cu
H_2MoO_4	Mo
Fe-EDTA	Fe

Table 7.B Food from plants

Plant part	Foods
Roots	Sweet potato, beets, radish, carrot, turnip, parsnip
Stems	White potato, sugar cane, asparagus
Leaves	Cabbage, kale, spinach, lettuce, tea leaves
Petioles*	Celery, rhubarb
Seeds†	Peas, navy beans, lima beans, nuts, coffee beans
Fruits†	Wheat, rice, corn, oats, rye, string beans, apple, orange, peach, tomato, squash

*Part of a leaf
†Derived from flower parts

Figure 7.3
Root tip. *a.* Root tip is divided into four zones best seen in a longitudinal section such as this. *b.* The tissues of a root are best seen in a cross section. *c.* Vascular cylinder of a root contains the vascular tissue; xylem is typically star shaped, and the phloem lies between the points of the star.

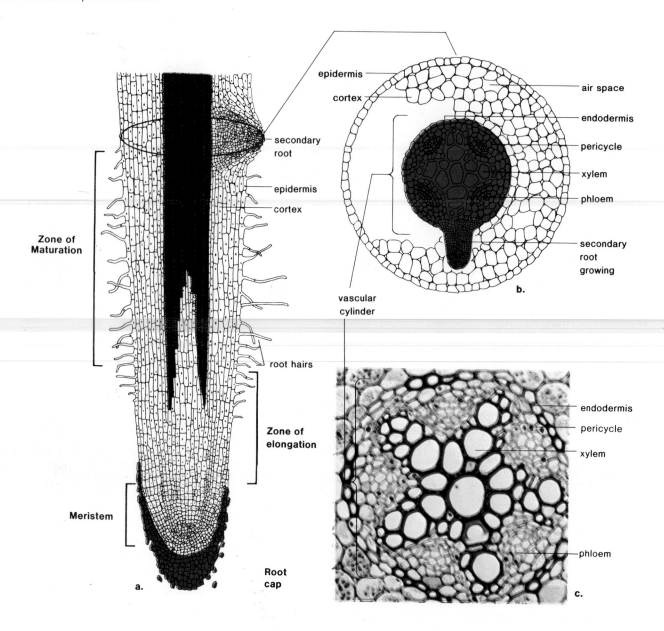

Root System

A longitudinal section of a root (fig. 7.3a) shows apical meristem located at the tip, protected by a **root cap** composed of a thimble-shaped mass of parenchyma cells. Outer root cap cells contain a slimy substance that assists movement through the soil. Above the meristem are nonspecialized cells that eventually elongate and develop into specialized tissues. Specifically, the root has four zones: (1) root cap, (2) meristem, or zone of cell division, (3) zone of elongation where the cells increase in length, and (4) zone of maturation where the cells are differentiated into specialized tissues.

Plant Biology

In a cross section (fig. 7.3b), the outer epidermal cells display **root hairs** (fig. 7.4), which tremendously increase the absorptive surface of the root. The absorbed water and minerals pass through the cortex, a tissue composed largely of parenchyma cells. The water and minerals must then cross through a layer of endodermis before entering the **vascular cylinder,** where the vascular tissue is arranged in a star-shaped pattern. Within the vascular tissue, water and minerals are transported upward into the stem by way of the **xylem,** and the products of photosynthesis are transported downward by way of the **phloem** for storage in the cortex. Lying between the endodermis and the vascular tissue, the **pericycle,** composed of parenchyma cells, retains the ability to undergo cell division and it on occasion produces secondary roots.

Types of Roots

Root systems are classified into two types according to their growth pattern: the fibrous root system and the taproot system (fig. 7.5). The **fibrous root system** is typical of grasses, such as the cereals. Each root is approximately the same size, and the many branches of the root intertwine about the soil particles. The **taproot system** is typical of many vegetables, such as carrots, beets, and turnips, in which there is one large, principle root that grows straight down, deep into the soil. Taproots are often fleshy and store the products of photosynthesis.

Shoot System

The **shoot system** consists of stems and their numerous appendages, the leaves. Although leaves have a shape and arrangement on the stem that is typical of the particular species, leaves are always arranged so that each one is exposed to the rays of the sun. The area or region of a stem where a leaf or leaves are attached is called a **node,** and stem regions between nodes are called **internodes** (fig. 7.1).

The shoot has apical meristem but no cap. Instead, the apical meristem produces leaves that grow up and around it, forming the apical bud (fig. 7.6). At the junction of a leaf and the stem is a group of meristem cells that may become branches or may develop into next year's flowers. This group of cells is called an **axillary,** or **lateral bud.**

Beneath the apical and lateral bud meristem, newly formed cells gradually elongate and differentiate into the various kinds of cells characteristic of the mature tissues of the plant (table 7.1). Plant growth is regulated by hormones, chemicals produced by one part of a plant that affect another part. As the reading on page 139 indicates, it is hoped that appropriate use of natural and synthetic plant hormones will contribute much to crop yield.

Stems

In nonwoody stems (fig. 7.6), the epidermis is covered by a waxlike substance, the **cuticle,** that prevents water loss. Beneath the epidermis a small circular region of cortex often surrounds the pith. Both cortex and pith are typically composed of parenchyma cells, but the former may carry on photosynthesis, while the latter stores the products of photosynthesis. Instead of a vascular cylinder, stems have **vascular bundles** that contain xylem and phloem. Between the *primary xylem and phloem* produced by the apical meristem, there may be a layer of meristematic tissue called the **vascular cambium.** Cells produced by the vascular cambium become *secondary xylem* to the inside of the meristem and to the outside they become *secondary phloem.* These tissues, which have the same function as their primary counterparts, contribute to the increased width of plants.

Figure 7.4
Scanning electron micrograph of root hairs, which increase the absorptive surface of roots. A root hair is an extension of a root epidermal cell.

Figure 7.5
Two types of roots. *a.* Fibrous root has many secondary roots with no one main root. *b.* Taproot has secondary roots in addition to a main root.

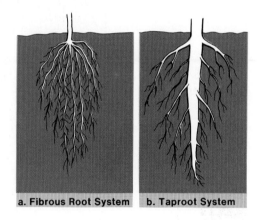

a. Fibrous Root System   b. Taproot System

Figure 7.6

Dicot herbaceous shoot system. *a*. Portion of shoot system in which the location of vascular tissue is shown diagrammatically. *b*. Apical bud contains the apical meristem protected by many immature leaves. *c*. Micrograph of cross section of the stem shows that the vascular bundles occur in a ring.

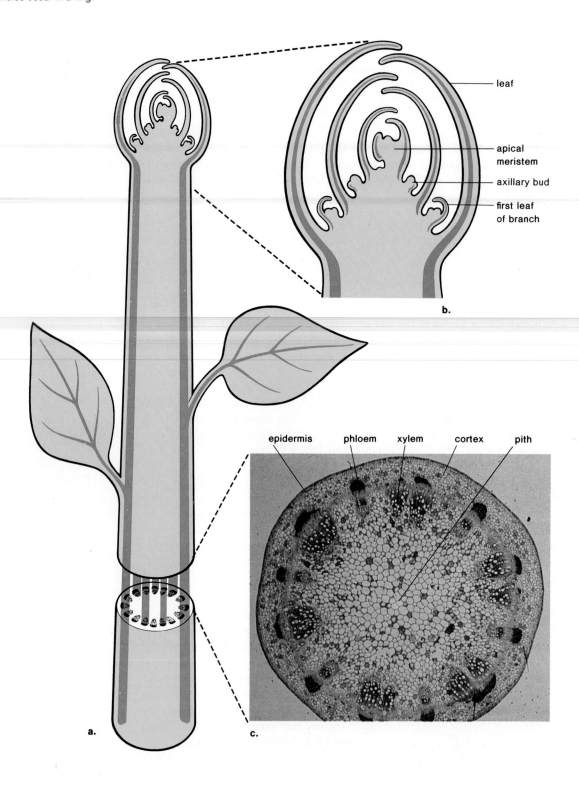

Plant Biology

Plant growth involves production of cells by means of cell division, enlargement of these cells, and finally, differentiation as the cells take on specific functions. Three types of hormones are known to promote plant growth: the **cytokinins** stimulate cell division, the **auxins** cause enlargement of plant cells, and the **gibberellins** promote both cell division and enlargement. A fourth class of plant hormones, termed **inhibitors,** retard or prevent growth in general. Plant growth regulators include natural hormones and related synthetic hormones. It is hoped by many that plant growth regulators will bring about an increase in crop yields just as fertilizers, irrigation, and pesticides have done in the past.

Plants bend toward light, and experiments with oat seedlings have shown that bending occurs because auxin is transported to the shady side of the shoot. This can be proven by removing the tip of a shoot and placing an auxin-containing agar block on one side of the stump. The cells on this side elongate causing bending to occur. Since the time these experiments were first performed, many commercial uses for auxins have been discovered. Auxins can cause the base of a shoot to form new roots so that new plants can be started from cuttings. When sprayed on trees, auxins can prevent fruit from dropping too soon. Auxins also inhibit the growth of lateral buds; potatoes sprayed with auxin will not sprout and thus can be stored longer. In high concentrations, auxins are used as herbicides that prevent the growth of broad-leaved plants. The synthetic auxins known as 2,4D and 2,4,5T were used as defoliants during the Vietnam war.

Gibberellins cause the entire plant, including all its parts, to grow larger. Before World War II, the Japanese studied a disease they called ''foolish seedling disease'' because the young plants grew rapidly, became spindly, and fell over. They found that this disease was caused by gibberellins secreted by a fungus that had infected the plants. Since this time it has been discovered that the application of gibberellins can cause seeds to germinate and plants, such as cabbages, to bolt (meaning rapid stem elongation) and flower. Gibberellins are used commercially to increase the size of plants. Treatment of sugarcane with as little as two ounces per acre increases the yield of cane by more than five tons.

Cytokinins were discovered when mature carrot and tobacco plant cells began to divide when grown in coconut milk. Testing revealed the presence of cytokinins. Later, scientists were able to grow entire plants from single cells in test tubes when various plant hormones were present in correct proportions. This has encouraged some to believe that plant growth regulators will soon permit tissue culture of important food plants. This means that isolated tissues of these plants could be grown in a test tube, a procedure that might assist the process of developing varieties of food plants with particular characteristics, such as tolerance to heat, cold, toxins, and drought.

The hormone ethylene, which is classified as an inhibitor, causes fruit to ripen. Fruits are commonly kept in cold storage to prevent the release of ethylene. Many synthetic inhibitors simply oppose the action of the natural stimulatory hormones (auxins, gibberellins, and cytokinins). The application of synthetic inhibitors can cause leaf and fruit drop. Removal of the leaves of cotton plants by chemical means aids harvesting; thinning the fruit of young fruit trees produces larger fruit from the trees as they mature; and retarding the growth of some plants increases their hardiness. For example, an inhibitor has been used to reduce stem length in wheat plants so that they do not fall over in heavy winds and rain. Other synthetic inhibitors mimic the action of ethylene and cause ripening of fruit and other crops. Fields and orchards are now sprayed with synthetic growth regulators just as they are sprayed with pesticides.

Plant Growth Regulators

Figure 7.7
Monocot shoot system. *a.* Portion of stem and leaves in which the location
of vascular tissue is shown diagrammatically. *b.* Micrograph of cross
section of several vascular bundles shows that they are scattered about in
the stem. *c.* Drawing of a vascular bundle shows the location of the xylem and phloem.

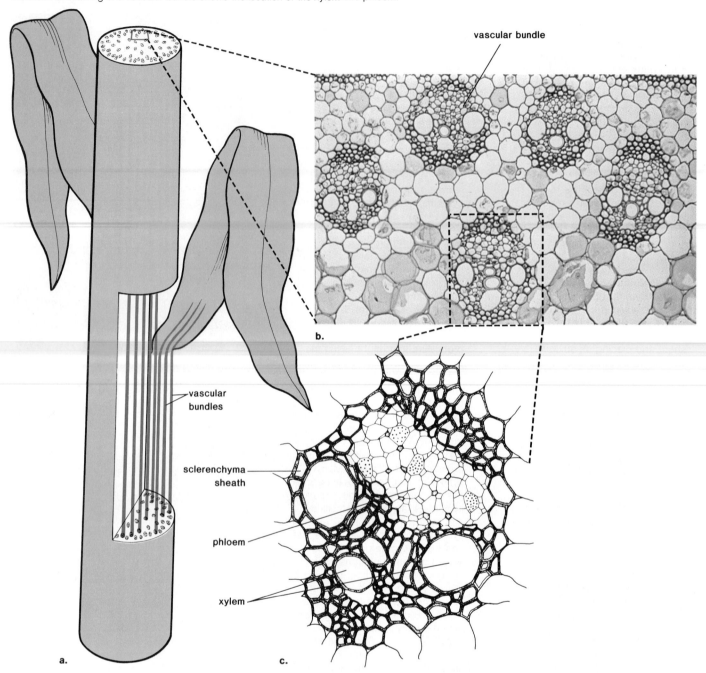

vascular bundle

b.

vascular
bundles

sclerenchyma
sheath

phloem

xylem

a.   c.

The arrangement of the vascular bundles varies for **monocot** and **dicot** plants. In dicots (fig. 7.6), the bundles are usually arranged in a circular pattern around the central pith. In monocots (fig. 7.7), the bundles are randomly scattered, without a definite pattern. This is only one difference between the two major types of flowering plants; figure 7.8 illustrates other comparisons.

So far we have been discussing **nonwoody herbaceous stems,** such as those of common garden plants; trees have **woody stems,** of course. Woody stems also increase in diameter due to secondary growth produced by vascular cambium. In trees, the vascular cambium, which lies between rings of xylem and phloem, produces new secondary xylem and phloem each succeeding year. Secondary xylem builds up and forms the **annual rings** (fig. 7.9), which can

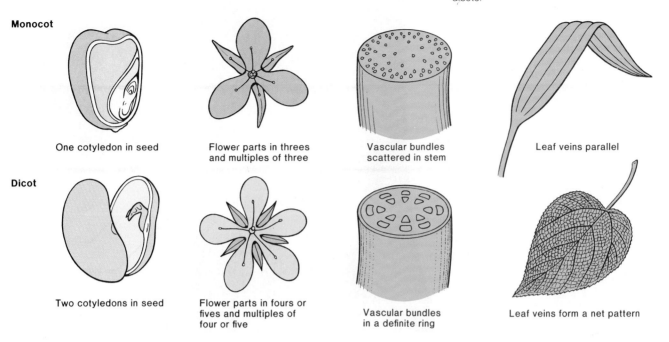

Monocot

One cotyledon in seed

Flower parts in threes and multiples of three

Vascular bundles scattered in stem

Leaf veins parallel

Dicot

Two cotyledons in seed

Flower parts in fours or fives and multiples of four or five

Vascular bundles in a definite ring

Leaf veins form a net pattern

Figure 7.8
Comparison of monocots and dicots. These four features are used to distinguish monocots from dicots.

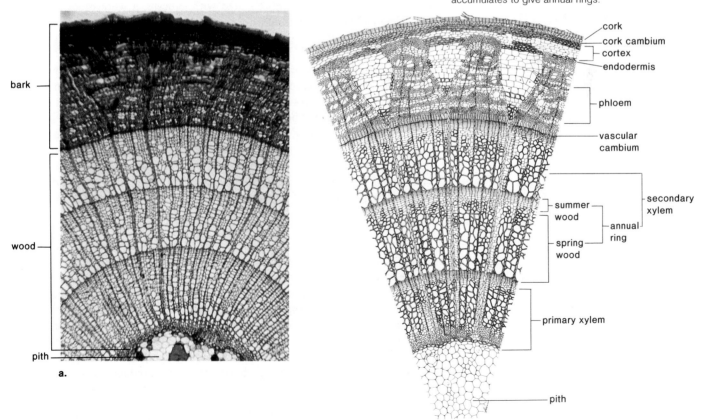

Figure 7.9
Woody stem. *a.* Photomicrograph of a cross section shows the three parts of a woody stem: bark, wood, and pith. *b.* Drawing indicating location of cork, phloem, vascular cambium, and xylem that accumulates to give annual rings.

bark

wood

pith

a.

cork
cork cambium
cortex
endodermis

phloem

vascular cambium

summer wood

spring wood

secondary xylem

annual ring

primary xylem

pith

b.

Figure 7.10
Tree trunk drawn to show its various tissues. The bark contains the cork and the phloem. The vascular cambium gives rise to new xylem and phloem every year. The sapwood contains the functional layers of xylem. The heartwood is dead and contains nonfunctional xylem tissue.

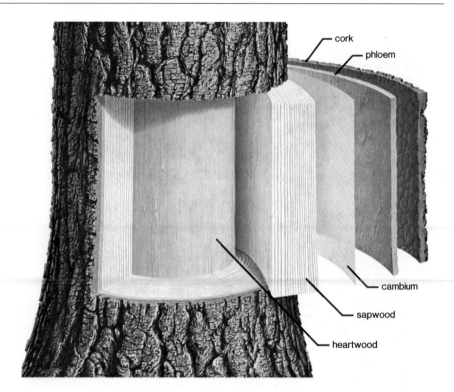

cork

phloem

cambium

sapwood

heartwood

be counted to tell the age of a tree. It is easy to tell where one ring begins and another ends; in the spring, when moisture is plentiful, the xylem cells are much larger than in the late summer, when moisture is scarcer.

As the girth of a woody stem increases, the epidermis is replaced by cork produced by **cork cambium,** a meristem tissue derived from the cortex. **Cork** is made up of dead cells impregnated with suberin, a waterproof material. Occasionally the cork cambium forms pockets of loosely arranged cells that do not become impregnated with suberin. These are the **lenticels** that function in gas exchange.

The stem of a tree can be divided into three parts: bark, wood, and pith (fig. 7.9). The **bark** contains cork, cork cambium, cortex, and phloem. Since the phloem is in the bark of a tree, even partial removal of the bark can seriously damage a tree. The **wood** contains the annual rings of xylem. In older stems, most of the xylem is nonliving inner *heartwood* that no longer transports water but does function to support the tree (fig. 7.10). The most recent annual rings contain functioning cells that do transport xylem sap, the watery contents of vessel cells, and are therefore collectively called *sapwood.* Wood is used to make all sorts of products. For example, lumber, furniture, and paper are made from wood.

Leaves

A leaf is usually composed of a **blade** and a **petiole,** which connects the blade to the stem. In between the upper and lower epidermis, the blade contains mesophyll tissue and veins. In the leaves of plants adapted to the temperate zone (fig. 7.11a) the **mesophyll** contains two layers of cells: the **palisade layer,** composed of compact but elongated cells, and the **spongy layer,** with irregular cells bounded by air spaces. The parenchyma cells of the palisade layer have many chloroplasts and carry on most of the photosynthesis for the plant. Water needed in the process of photosynthesis is transported to these cells by the **leaf veins,** the final extensions of vascular tissue. Leaf veins have a *net pattern* in

Plant Biology

Figure 7.11
Leaf anatomy. *a.* Drawing of a leaf adapted to temperate climate. Notice that the bundle sheath cells do not contain chloroplasts. The mesophyll contains two distinct layers of cells, the palisade layer and the spongy layer. *b.* Scanning electron micrograph of the mesophyll tissue in particular. Notice the distinct difference in the shape of the cells of the palisade layer and the spongy layer (as identified in *a.*). *c.* Drawing of a leaf adapted to a hot, dry climate. Notice that the bundle sheath cells contain large chloroplasts. The cells analogous to the palisade layer encircle the bundle sheath cells and contain small chloroplasts. Cells analogous to the spongy layer have no chloroplasts.

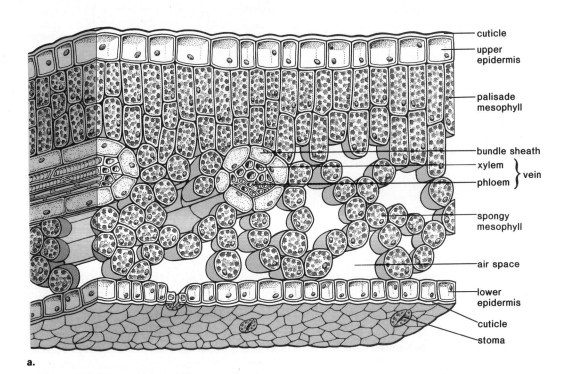

cuticle

upper epidermis

palisade mesophyll

bundle sheath
xylem ⎱ vein
phloem ⎰

spongy mesophyll

air space

lower epidermis

cuticle

stoma

a.

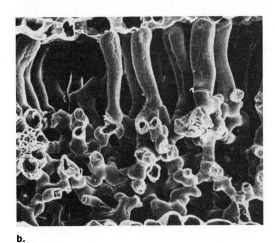

b.

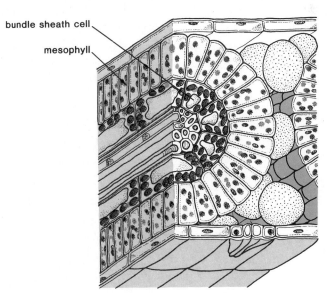

bundle sheath cell

mesophyll

c.

Figure 7.12

Stomata. *a.* Drawing of lower layer of epidermis of a leaf showing the presence of several stomata. Each stoma is surrounded by two guard cells that regulate whether the stoma is open or closed. *b.* Scanning electron micrograph of one stoma.

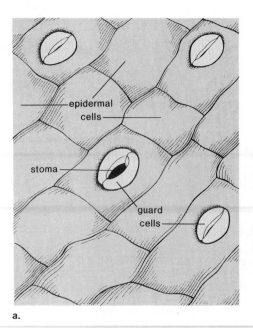

a.

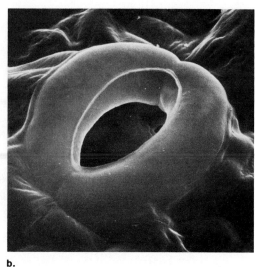

b.

Figure 7.13

Opening of a stoma requires a sequence of events. *a.* K^+ enters guard cells and this creates an osmotic pressure that causes *(b.)* water to enter the guard cells. *c.* Guard cells have thick inner walls but weak outer walls. Therefore the entrance of water creates a pressure that causes the cells to bulge outward, opening the stomata.

a.

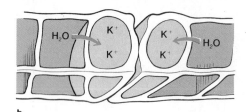

b.

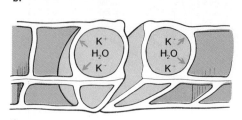

c.

dicot leaves and a *parallel pattern* in monocot leaves (fig. 7.8). The larger leaf veins are enclosed by a circular layer of tightly packed parenchyma cells known as the **bundle sheath.** Notice in figure 7.11a that these cells do not contain chloroplasts.

The anatomy of a leaf in a plant adapted to a hot, dry, sunny environment is slightly different from the one just described. As shown in figure 7.11c, the mesophyll cells that correspond to the palisade layer are arranged circularly about the bundle sheath cells and both types of cells contain chloroplasts. However, the chloroplasts in the bundle sheath cells are much larger than those in the mesophyll cells. As discussed in the reading on page 127, the mesophyll cells are specialized to capture carbon dioxide after which they pass it to the bundle sheath cells where carbohydrate is synthesized. Apparently this arrangement evolved as a way to overcome the limited supply of carbon dioxide in the atmosphere.

Stomata

Carbon dioxide, which is necessary for photosynthesis, enters the air spaces of a spongy layer by diffusion through openings, called **stomata,** present particularly in the lower epidermis. When a leaf receives a plentiful amount of water and carbon dioxide and is carrying on photosynthesis, it gives off oxygen, which also exits by way of the stomata. Stomata are tiny pores, each surrounded by two specialized epidermal cells called **guard cells** (fig. 7.12). Due to thickened inner cell walls, guard cells expand outward only when they fill with water. This outward expansion thereafter causes the stomata to open. When guard cells lose water, the loss of turgor causes the stomata to close. Water has been shown to enter the guard cells by osmosis because of a high K^+ (potassium) concentration (fig. 7.13). Since the stomata are open when a plant is photosynthesizing, it is assumed that photosynthesis in some way triggers a potassium pump that actively transports K^+ into the guard cells, creating an osmotic pressure that eventually opens the stomata. Another possibility

Plant Biology

is that light directly initiates opening of the stomata but, as you might predict, stomata will not open unless there is a plentiful supply of water.

Much of the water that is transported from the roots to the leaves evaporates and escapes from the leaf by way of the stomata. The evaporation of water from a leaf is called **transpiration;** table 7.2 gives the transpiration rates for a number of plants. The amount of water that is lost in this manner is truly phenomenal, but it is believed that evaporation of water keeps the leaf cool enough to function in the bright sun and also aids the transport of water. Stomata close at night to prevent water loss.

Transport

Vascular tissue, which is composed of xylem and phloem, is responsible for transport in plants. Xylem, which is present in all parts of a plant, transports water and minerals primarily from the roots to the leaves, and phloem, which is also present in all parts of a plant, transports the products of photosynthesis from the leaves to the roots during the growing season.

Xylem
Xylem contains two types of conducting cells: vessel cells and tracheids, each of which is a cell specialized for transport (fig. 7.14). The nonliving **vessel cells** are sometimes called vessel elements because piled one on top of the other

Table 7.2 Transpiration rates

Per day midsummer

Ragweed	6–7 quarts
10 foot apple tree	10–20 quarts
12 foot cactus	0.02 quarts
Coconut palm	70–80 quarts
Date palm	400–500 quarts

Per growing season

Tomato	100 days	30 gallons
Sunflower	90 days	125 gallons
Apple tree	188 days	1800 gallons
Coconut	365 days	4200 gallons
Date palm	365 days	35,000 gallons

From BOTANY: A HUMAN CONCERN by David L. Rayle and Hale L. Wedberg. Copyright © 1980 by Saunders College/Holt, Rinehart and Winston. Reprinted by permission of CBS College Publishing.

Figure 7.14
Xylem structure. General organization of xylem at far left followed by an external view of vessel elements stacked one on top the other and a longitudinal view of several tracheids. Tracheids and vessel elements usually conduct water in the direction shown.

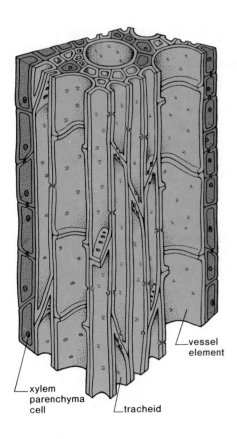

xylem parenchyma cell

vessel element

tracheid

vessel elements

tracheids

primary direction of transport

Figure 7.15

Scanning electron micrograph of xylem vessel elements showing how the cells are stacked one on top of the other to form a pipeline that stretches from the roots to the leaves.

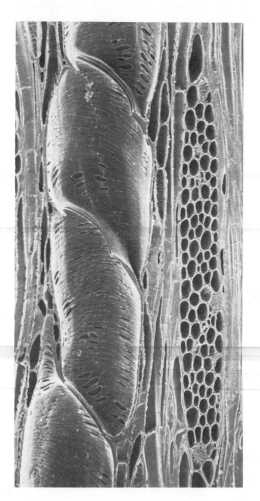

(fig. 7.15) they form a continuous pipeline that stretches from the root to the leaves. It is an open pipeline because the elements have no end walls separating one from the other. The elongated **tracheids** are also dead at maturity, but they have tapered end walls that are unperforated. Both types of cells have secondary walls that contain lignin, an organic substance that makes the walls tough and hard. Even so, water can move between the lateral walls of both types of cells and the end walls of tracheids because of the presence of pits, depressions where the secondary wall does not form.

Water absorbed by root hairs must be transported from roots to the leaves. While this may not seem to be very difficult for short garden plants, one would think it might pose a problem for very tall trees. Botanists who have studied this phenomenon have arrived at the cohesion-tension theory of water transport.

Cohesion-Tension Theory of Water Transport

As figure 7.16 shows, water (and minerals) primarily enter the root of a flowering plant at the region of the root hairs. In order to reach the xylem in the vascular cylinder, the water must pass through the cortex. Water may follow one of two pathways; it may move between the cells via the porous cell walls or directly into a root hair cell and then progress from cell to cell across the cortex. With either of these routes, however, water (and minerals) must eventually enter the endodermal cells if they are to finally reach the xylem. This is because endodermal cells have a band of fatty material called the **Casparian strip** that prevents substances from moving between them. Thus the endodermis regulates the entrance and exit of substances to and from the vascular cylinder.

To understand the transport of water upward in the vascular cylinder we have to realize that water molecules have a great tendency to cling together because they are polar molecules (p. 26). This **cohesion** property of water means that a column of water can be pulled without breaking. In fact, it is easier to pull apart the molecules in fine wires made of some common metals than it is to pull apart a column of water in a small-diameter, airtight tube. Secondly, we must remember that water is continuously lost at the leaves because of transpiration. When water evaporates from the mesophyll cells in a leaf, these cells become less turgid and this creates an osmotic pressure that causes water to move out of xylem into these cells. Therefore, transpiration creates a **tension** that in effect pulls water upward in the xylem.

An experiment can be done to show that transpiration aids the movement of water in xylem (fig. 7.17). A pan filled with mercury is placed under a tube that has been evacuated and the mercury rises 76 centimeters into the tube due to a force provided by atmospheric pressure. This shows that atmospheric pressure alone could not raise water to treetops. But when a transpirational pull is added at the top of the tube, either in the form of a twig with leaves or a porous clay bulb, the mercury column rises to a height of 100 centimeters or more. If this height is expressed in terms of water rather than of mercury, it shows that transpiration must definitely be a factor in pulling water along in xylem vessels to the tops of tall trees.

Plant Biology

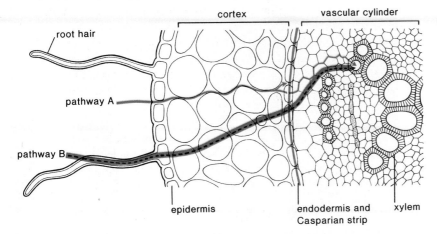

Figure 7.16
Water and minerals enter a root at the region of the root hairs and then move across the cortex. Pathway A: Water can move between cells by way of porous cell walls until it reaches the endodermis. Because endodermal cells are bordered by a waxy substance, water and minerals must pass through these cells in order to reach the vascular cylinder. Pathway B: Water and minerals enter a root hair and thereafter pass from cell to cell to reach the vascular cylinder.

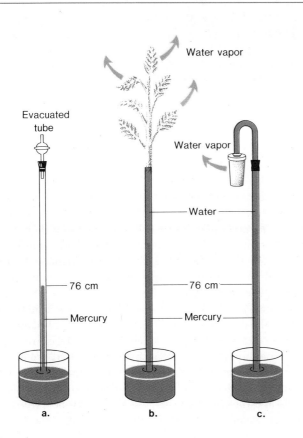

Figure 7.17
Measurement of the forces involved in transpirational pull. *a.* Atmospheric pressure raises liquid mercury only 76 cm in a tube evacuated of air. *b.* Transpirational pull at leaves pulls on water and raises liquid mercury 100 cm or more. *c.* Transpirational pull at a porous clay cup also exerts a pulling force that is greater than that of *a.*

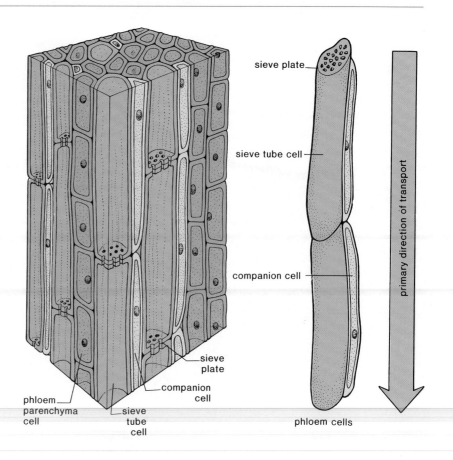

sieve plate

sieve tube cell

companion cell

sieve plate

companion cell

phloem parenchyma cell

sieve tube cell

phloem cells

primary direction of transport

Phloem

The conducting cells in phloem are sieve-tube cells, each of which typically has a companion cell (fig. 7.18). **Sieve-tube cells** contain cytoplasm but no nucleus; as their name implies, these cells have pores in their end walls that make these walls resemble a sieve. Through these pores, strands of cytoplasm extend from one cell to the other. The smaller **companion cells** are more generalized cells and have nuclei. It is speculated that the nucleus may control and maintain the life of both cells.

Chemical analysis of phloem sap shows that it is composed chiefly of sucrose and that the concentration of nutrients is 10 to 13 percent by volume. Interesting is the fact that samples for chemical analysis are most often obtained by the use of aphids (fig. 7.19), small insects that are phloem feeders. The aphid drives its stylet, a short mouthpart that functions like a hypodermic needle, between the epidermal cells and withdraws sap from a sieve-tube cell. If the aphid is anesthetized by ether, the body may be carefully cut away, leaving the stylet, which exudes the phloem contents that can be collected and analyzed.

Pressure-Flow Theory of Phloem Transport

The most widely accepted theory of phloem transport is called the **pressure-flow theory,** which is demonstrated by an experiment described in figure 7.20. The first thistle tube contains sucrose (colored dots in figure 7.20) at a higher concentration than in the second thistle tube. Water will tend to enter both tubes, but it will enter the first tube with more force than it enters the second tube. In this way a pressure difference is created that causes water to flow from the first tube to the second and even to exit from the second tube. As the water flows, it carries sucrose with it from the first to the second container.

Plant Biology

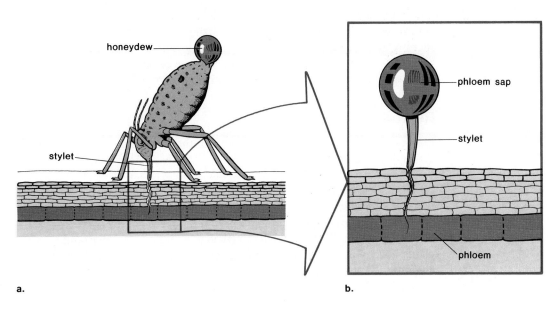

Figure 7.19
Aphids are small insects that withdraw organic nutrients from phloem by means of a hypodermic-like mouthpart called a stylet. *a.* Aphid with stylet in place. *b.* When the aphid's body is removed, phloem sap can be collected from the stylet in a syringe.

honeydew

stylet

phloem sap

stylet

phloem

a.

b.

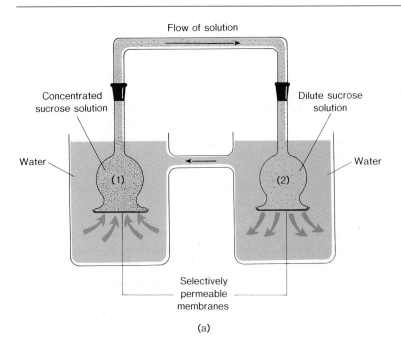

Flow of solution

Concentrated sucrose solution

Dilute sucrose solution

Water

Water

(1)

(2)

Selectively permeable membranes

(a)

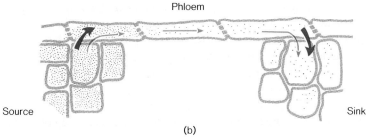

Phloem

Source

Sink

(b)

Figure 7.20
Phloem transport. *a.* Model system in which pressure flow occurs. Thistle tube (1) contains more sucrose *(dots)* than thistle tube (2). Therefore more water enters (1) than (2). This creates a pressure that causes water to flow from (1) to (2) and even to exit from (2). As the water flows, sucrose also moves from (1) to (2). *b.* In the plant, sucrose is actively transported into phloem at a source (leaves in growing season and roots prior to growing season). Sucrose is transported out of phloem at a sink (roots during growing season and leaves prior to growing season). So much water enters the source end of the phloem by osmosis that water is forced out the sink end. This results in flow of water that sweeps sucrose along with it.

An analogous situation is believed to take place in a plant. When a plant is photosynthesizing, sucrose is actively transported into the sieve-tube cells of a leaf and water will therefore follow by osmosis. This creates a pressure that causes water, along with sucrose, to flow toward the roots where the cells remove sucrose by means of active transport.

The pressure-flow hypothesis can account for the observed reversal of flow in the sieve-tube system; for example, in the spring before the leaves are out. When the plant is not photosynthesizing, the roots serve as a source of sucrose, and the other parts of the plant remove it from the sieve-tube cells. Therefore water will tend to enter the sieve-tube cells in the roots with a greater force and will flow toward the leaves, carrying sucrose with it.

Reproduction

Plants can reproduce both asexually, without the need for gametes, or sexually, which does require gametes.

Asexual Reproduction

Asexual reproduction, also known as **vegetative propagation,** is common in plants. In vegetative propagation, a portion of one plant gives rise to a completely new plant. Both plants now have identical genes. As an example, some plants have above-ground horizontal stems, called *runners,* and others have underground stems, called *rhizomes,* that produce new plants. To take a concrete example, strawberry plants grow from the nodes of runners and violets grow from the nodes of rhizomes. White potatoes can be propagated in a similar manner. White potatoes are actually portions of underground stems, and each eye is a node that will produce a new potato plant. Sweet potatoes are modified roots and may be propagated by planting sections of the root. You may have noticed that the roots of some fruit trees, such as cherry and apple trees, produce "suckers," small plants that can be used to grow new trees.

Asexual reproduction has a great deal of commercial importance. Once a plant variety with desired characteristics has been developed through vegetative propagation, new plants can be supplied to gardeners and farmers. Cuttings can be taken from the plant, and the cut end can be treated to encourage it to grow roots or a cutting can be grafted to the stem of a plant that has a root. Budding is a form of grafting most often used commercially. In this procedure, just the axillary buds are grafted onto the stem of another plant. Today, entire plants can also be produced by tissue culture (fig. 7.21), a technique that may eventually replace the older methods thus far discussed. Usually an embryonic *tissue* is removed from a plant and placed in a special *culture* medium. After the tissue has grown for a while, it is subdivided so that many identical plants are produced from a very small amount of starting cells. Nurseries now culture all sorts of plants with assembly-line efficiency. Plant breeders are extremely interested in utilizing a modification of the tissue culture technique in which most often leaf cells are treated to produce *protoplasts,* cells that have been chemically stripped of their outer wall. (A single protoplast will give rise to a new plant, identical in nature to the original plant.) It is quicker and easier to test protoplasts instead of entire plants for desired characteristics such as resistance to bacteria and fungi, high temperatures, and drought. Also, protoplasts can be made to fuse together. In one experiment, a potato and a tomato protoplast were fused and a hybrid plant was eventually grown. Perhaps protoplast fusion will eventually allow botanists to alter the genetic makeup of a variety of plants.

Figure 7.21
Each dish shows several shoots of Douglas Fir growing from a single cotyledon. Such cultures are part of a research project for cloning genetically improved trees. Subsequently, the shoots will be cut, rooted, and planted in the forest. Tissue culture propagation is expected to play an important part in bringing forest yields toward their theoretical maximum.

Plant Biology

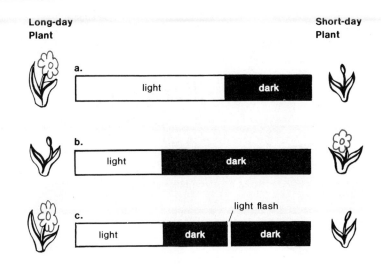

Long-day Plant ... Short-day Plant

a.
light | dark

b.
light | dark

c.
light flash
light | dark | dark

Figure 7.22
Daylength effects flowering in some plants. *a.* Long-day plants flower when the days are long and the nights are short. *b.* Short-days plants flower when the nights are long and the days are short.

Sexual Reproduction

Plants also reproduce sexually. This may come as a surprise to those who never thought of plants as being male and female. Sexual reproduction is properly defined as reproduction that requires gametes, often an egg and a sperm. In flowering plants, the sex organs are located in the flower, and therefore we will digress just now to discuss flowering, an event that occurs only at a particular time of year in certain plants.

Flowering

It has been observed that some plants seem to favor a particular length of day in order to flower. Such plants are said to exhibit **photoperiodism.** Short-day plants flower only when the days are shorter than some critical length, and others, called long-day plants, flower only when the days are longer than some critical length (fig. 7.22). This terminology is unfortunate because it is actually the length of darkness that is critical—interrupting the dark period with a flash of white light prevents flowering in a short-day plant and induces flowering in a long-day plant. Therefore, short-day plants are really long-night plants, and long-day plants are really short-night plants.

Plants do not flower if the leaves are removed or if they are placed in total darkness. Based on these two facts, scientists have suggested that flowering is dependent on a light-sensitive pigment called **phytochrome,** which is present in the leaves. The biologically active form of phytochrome is called P730 because it absorbs red light of this wavelength. It is now believed that P730 communicates, in an undetermined manner, by means of a biological clock, an internal system that controls the timing of certain physiological and behavioral responses; in this case, flowering. (Biological clocks are discussed in more detail in chapter 29.) The presence of P730 during a critical time period causes the clock to promote flowering in long-day plants and inhibit flowering in short-day plants.

Flowering was formerly thought to be dependent on the presence of a hormone termed **florigen.** Since this hormone has never been isolated, it is generally believed that the biological clock may instead control the balance between stimulatory and inhibitory plant hormones, such as those discussed in the reading on page 139. Evidence for this comes from the fact that when a day-neutral plant is grafted to a long-day plant, appropriate lighting for the long-day plant promotes flowering in the day-neutral plant but inappropriate lighting for the long-day plant inhibits flowering in the day-neutral plant (fig. 7.23).

Figure 7.23
A day-neutral plant is grafted to a long-day plant. When the plants receive long-day lighting *(a.)* both plants flower; but when the plants receive short-day lighting *(b.)* neither plant flowers. Since the amount of light has no effect on the flowering of day-neutral plants, it is assumed that inhibiting hormones passed from the long-day plant to the day-neutral plant, overcoming stimulatory hormones for flowering.

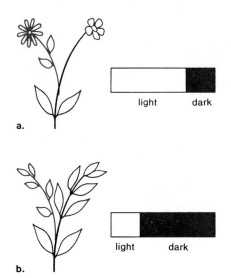

a.
light | dark

b.
light | dark

Figure 7.24

Diagram illustrating flower parts. The mature pollen grain contains two sperms that travel down a pollen tube to the ovule. If conditions are correct, one sperm fertilizes the egg cell within this structure.

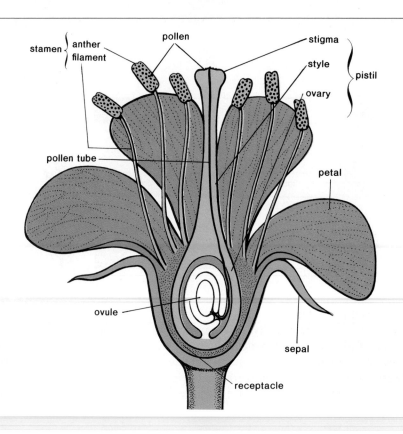

Flower

Figure 7.24 shows the parts of a typical flower. The **sepals,** most often green, form a whorl about the **petals,** whose color accounts for the attractiveness of many flowers. In the center of the flower is a small vaselike structure, the **pistil,** which usually has three parts: the **stigma,** an enlarged sticky knob; the **style,** a slender stalk; and the **ovary,** an enlarged base. The ovary contains a number of ovules that play a significant role in reproduction. Grouped about the pistil are a number of **stamens,** each of which has two parts: the **anther,** a saclike container, and the **filament,** a slender stalk.

Flowering plants have a life cycle called **alternation of generations** (p. 561) because it contains two generations: the sporophyte generation and the gametophyte generation. The sporophyte generation is a diploid (2N) generation that produces haploid spores by meiosis. The spores give rise to a gametophyte (N) generation. The gametophyte generation produces gametes that join to give rise to the sporophyte generation once again.

A plant that flowers is actually a **sporophyte generation** that produces spores. Within the ovary, each **ovule** contains a megaspore (*mega* means large) mother cell (fig. 7.25) that undergoes meiosis to produce four haploid megaspores. Three of these disintegrate, leaving one functional megaspore that divides mitotically. The result is the **female gametophyte generation,** which typically consists of eight nuclei embedded in a mass of cytoplasm that is partly differentiated into cells. One of these cells is an egg.

The anther contains numerous microspore (*micro* means small) mother cells each of which undergoes meiosis to produce four haploid cells called microspores. The microspores usually separate and each one becomes a **pollen**

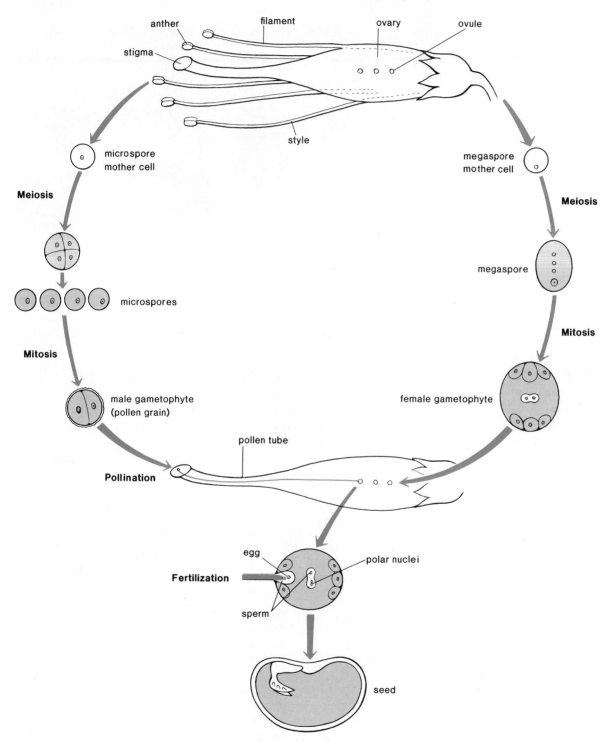

Figure 7.25

Events leading to fertilization and seed formation in a flowering plant. There is a megaspore mother cell in each ovule found in the ovary. The megaspore mother cell undergoes meiosis to give one functional megaspore that undergoes mitosis. The result is the female gametophyte generation, which contains an egg cell. The anther contains microspore mother cells that undergo meiosis to give four functional microspores. Following mitosis, each is a pollen grain containing two cells. At about the time of pollination, one of these cells divides to give two sperms that travel down a pollen tube to the ovule. During double fertilization, one sperm nucleus joins with the egg nucleus and the other joins with two polar nuclei. The ovule now matures and becomes a seed that contains an embryonic plant and stored food. The seed shown here is a dicot seed with two large cotyledons.

anther

stigma

filament

ovary

ovule

style

microspore mother cell

megaspore mother cell

Meiosis

Meiosis

microspores

megaspore

Mitosis

Mitosis

male gametophyte (pollen grain)

female gametophyte

pollen tube

Pollination

egg

polar nuclei

Fertilization

sperm

seed

grain (fig. 7.26), or male gametophyte generation. At this point, the young male gametophyte generation contains two nuclei, the *generative nucleus* and the *tube nucleus*.

Pollination (fig. 7.27) occurs when pollen is windblown or carried by insects, birds, or bats to the stigma of the same type plant. Only then does a pollen grain germinate and produce a long pollen tube that grows within the style until it reaches an ovule in the ovary. Before fertilization occurs, the generative nucleus divides, producing two sperms that have no flagellae. On reaching the ovule, the pollen tube discharges the sperms. One of the two sperms migrates to and fertilizes the egg, forming a zygote; the other sperm migrates to and unites with the polar nuclei, producing a 3N (triploid) endosperm nucleus. The endosperm nucleus divides to form endosperm, food for the developing plant. Note that flowering plants have a **double fertilization.** One fertilization produces the zygote, the other produces endosperm.

Seed

Following fertilization, each ovule becomes a mature seed containing an embryo that has at least three parts: the **cotyledon(s),** or seed leaf (leaves); the **epicotyl,** that portion of the embryo above the attachment of the cotyledon(s); and the **hypocotyl,** which lies below the attachment of the cotyledons. The epicotyl contains the apical meristem of the shoot and sometimes bears young leaves, in which case it is called a **plumule.** The hypocotyl is sometimes prolonged at its base into the **radicle,** which becomes the primary root of the seedling.

Monocots have seeds with one cotyledon and dicots have seeds with two cotyledons (fig. 7.28). Cotyledons typically provide nutrient molecules for the growing embryo. In monocot embryos, the cotyledon rarely stores food; rather, it absorbs food molecules from the endosperm and passes them on to the embryo. In many dicot embryos, the cotyledons replace the endosperm, which typically has already transferred its nutrients to the cotyledons.

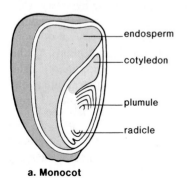

a. Monocot

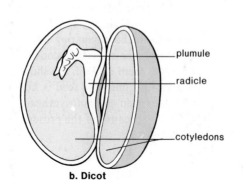

b. Dicot

Figure 7.28
Seed anatomy. *a.* Longitudinal section of monocot seed shows a large amount of endosperm (stored food) in addition to the cotyledon and parts of the embryo that will produce the root, stems, and leaves. *b.* Longitudinal section of a dicot seed shows two large cotyledons, one on either side of the embryo which is easily recognized by its leafy plumule.

Figure 7.29
Development of fruit from a flower. Once seed formation begins, the ovary and accessory parts begin to enlarge and grow larger until only remnants of the other flower parts remain and finally disappear entirely leaving only the mature fruit.

In addition to the embryo and stored food either within cotyledon(s) or within endosperm, a seed is covered by a seed coat. In flowering plants, all seeds are also enclosed within a fruit that develops from the ovary (fig. 7.29) and, at times, from other accessory parts. Although peas, beans, tomatoes, and cucumbers are commonly called vegetables by laymen, botanists categorize them as fruits. Fruits protect seeds, sometimes provide extra nourishment, and also sometimes aid dispersal. For example, winged dry fruits like that of a maple tree are adapted to distribution by the wind while fleshy fruits like that of a cherry are eaten by birds and the seeds are deposited some distance away.

Seed Germination A seed will not normally germinate unless environmental factors are favorable. Water and oxygen are essential to the completion of germination in nearly all seeds, but light is not necessarily a requirement because some prefer darkness. Most seeds need a temperature that is above freezing but below 45°C. The first event normally observed in a germinating seed is the emergence of the root formed from the base of the hypocotyl (fig. 7.30). This is followed shortly by the appearance and expansion of the seedling shoot formed from the elongating epicotyl. In dicots, the cotyledons degenerate after their nourishment is consumed by the developing plant. As the seedling emerges from the soil, the shoot may be hook shaped, protecting the delicate leaves. But once the seed is above ground, the stem straightens out and the leaves expand as photosynthesis begins. The plant continues to grow as long as it lives because of the presence of meristem tissue.

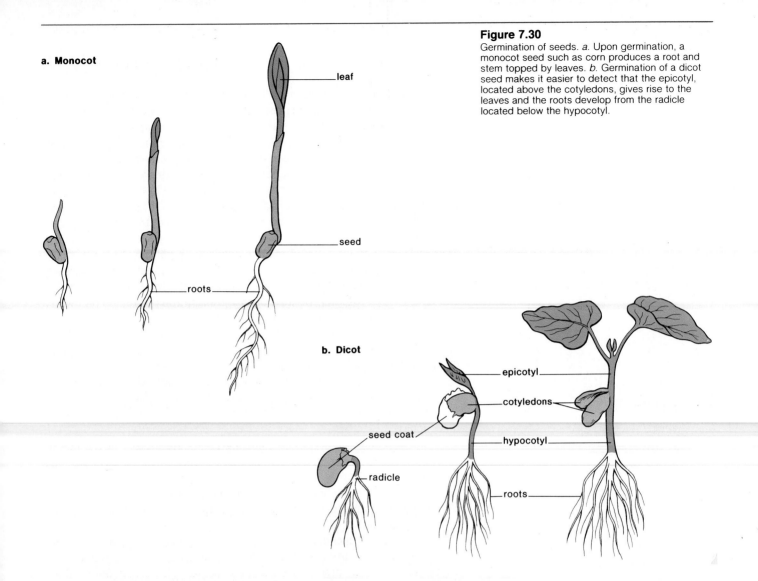

a. Monocot

leaf

seed

roots

b. Dicot

epicotyl

cotyledons

seed coat

hypocotyl

radicle

roots

Figure 7.30
Germination of seeds. *a.* Upon germination, a monocot seed such as corn produces a root and stem topped by leaves. *b.* Germination of a dicot seed makes it easier to detect that the epicotyl, located above the cotyledons, gives rise to the leaves and the roots develop from the radicle located below the hypocotyl.

Summary

Flowering plants carry on photosynthesis and this is reflected in their anatomy and physiology. The root system of a plant (taproot or fibrous root) absorbs water and minerals. In longitudinal section, a root has four zones: (1) root cap, (2) meristem tissue, (3) zone of elongation, and (4) zone of maturation. In a cross section of the latter zone, the cortex lies between the epidermis and the vascular cylinder. It is bounded by a layer of endodermis, within which is the pericycle that produces secondary roots, and the vascular tissue, which consists of xylem, for water and mineral transport upward, and phloem, for the transport of organic molecules, usually downward.

The shoot system of a plant consists of stems and leaves. A stem can be herbaceous or woody. Herbaceous stems are either dicots, with vascular bundles arranged in a ring, or monocots, with vascular bundles randomly arranged. In cross section, a dicot stem has an outer epidermis and an inner pith; a narrow band of cortex lies between the epidermis and pith.

In woody dicot stems the layer of vascular cambium that lies between the phloem and xylem produces new secondary vascular tissue each year. This

accounts for the increase in diameter of tree trunks. Secondary phloem disintegrates, but secondary xylem builds up, forming the annual rings. A woody stem has three parts: the bark, which contains cork, cork cambium, cortex, and phloem; wood, or xylem; and pith.

In most plants of the temperate zone, leaves carry on C_3 photosynthesis, particularly within the palisade cells of the mesophyll. Reactants for photosynthesis are water absorbed by the roots and transported upward in the xylem to its final extension, the leaf veins (having a net pattern in dicots and a parallel pattern in monocots), and carbon dioxide, which passes through the stomata by the process of diffusion, before entering the air spaces of the spongy layer of mesophyll. Stomata, located particularly in the lower epidermis, open in the presence of light. Potassium ions are pumped into the guard cells, which then take up water by osmosis and expand, opening the stomata.

The process of transport in plants is twofold: transport of water and minerals in xylem and transport of organic nutrients in phloem. Xylem is made up of two types of nonliving cells: tracheids and vessel cells. The cohesion-tension theory of water transport points out that water molecules are cohesive and tend to stick together. Because of this, transpiration is able to pull water as a continuous column from the roots. According to the pressure-flow theory of phloem transport, sugar and then water molecules enter the phloem at a source (e.g., the leaves) and exit the phloem at a sink (e.g., the roots). This creates a difference in water pressure that causes phloem sap to flow.

Plants reproduce both asexually and sexually. Asexual reproduction occurs when a portion of one plant gives rise to an entirely new plant. Grafting, particularly budding, and tissue culture propagation have commercial importance today.

The reproductive organs of flowering plants are located in the flower. Some plants exhibit photoperiodism and flower only when the days are long and others only when the days are short. Interrupting the night with a flash of light supports the contention that it is actually the amount of darkness that affects flowering and therefore it is accurate to call them short-night plants and long-night plants.

Flowering is dependent on the presence of phytochrome P730, a pigment that, when present during a critical time period, communicates by means of a biological clock and promotes flowering in long-day plants and inhibits flowering in short-day plants. It is believed that a balance between stimulatory and inhibitory hormones controls flowering.

A plant that flowers is the sporophyte generation because it produces haploid spores. In ovules within the ovary, located at the base of the pistil, megaspore mother cells undergo meiosis to give one viable haploid nucleus that divides to become the female gametophyte generation, composed of eight cells. One of these cells is the egg cell. In the anther, located at the top of the stamen, microspore mother cells undergo meiosis to give four viable haploid cells, each of which becomes a pollen grain. At this point the young pollen grain contains a generative cell and a tube cell. At about the time of pollination, the generative cell divides, giving two sperms. As the pollen grain forms a long pollen tube, the sperms travel down it to the ovule. During fertilization, one sperm combines with the egg, forming a zygote, and the other combines with two polar nuclei, forming the triploid endosperm. The ovule now becomes the seed. In flowering plants, seeds are enclosed by a fruit that develops from the ovary. Thus, all flowering plants produce fruits.

The mature seed contains an embryo consisting of the cotyledon(s), epicotyl, and hypocotyl. In monocots, the endosperm typically remains, but in dicots the endosperm is typically absorbed by the cotyledons. Upon germination, the epicotyl gives rise to the stem and leaves while the roots develop from the radicle.

Study Questions

1. Name the two major divisions of a plant and the organs that each contains. (p. 133)
2. Discuss the anatomy of a root, stem (nonwoody and woody), and leaf. (pp. 136, 137, 142)
3. What are the two types of roots? Give examples. (p. 137)
4. Explain how stomata open and close. (p. 144)
5. Describe the structure of xylem, and discuss water and mineral transport. (pp. 145–46)
6. Describe the structure of phloem, and discuss organic nutrient transport. (pp. 148–50)
7. Give examples of asexual reproduction in plants. (p. 150)
8. Explain periodic flowering in some plants. (p. 151)
9. Trace the production of seeds in flowering plants, starting with a megaspore mother cell in the ovule and microspore mother cells in the anther. (pp. 152–54)
10. Describe the structure and germination of a monocot and dicot seed. (pp. 154–56)
11. Name four differences between dicots and monocots. (p. 141)

Selected Key Terms

meristem (mer'ĭ-stem)
epidermis (ep''ĭ-der-mis)
endodermis (en''do-der'mis)
vascular cylinder (vas'ku-lar sil'in-der)
vascular cambium (vas'ku-lar kam'be-um)
herbaceous (her-ba'shus)
mesophyll (mes'o-fil)
stomata (sto'mah-tah)
guard cell (gahrd sel)
vessel cells (ves'el selz)
tracheids (tra'ke-idz)
transpiration (tran''spĭ-ra'shun)
sieve tube cell (siv tūb sel)
companion cell (kom-pan'yun sel)
photoperiodism (fo''to-pe're-od-izm)
phytochrome (fi'to-krōm)

ovary (o'var-e)
anther (an'ther)
pollen (pol'en)
cotyledon (kot''ĭ-le'don)

Further Readings

Biddulph, O., and Biddulph, S. 1959. The circulatory system of plants. *Scientific American* 200(2):26.

Bold, H. C. 1980. *Morphology of plants.* 4th ed. New York: Harper & Row.

Butler, W. L., and Downs, R. J. 1960. Light and plant development. *Scientific American* 203(6):38

Jacobs, W. P. 1955. What makes leaves fall? *Scientific American* 192(5):20.

Muller, W. H. 1979. *Botany: A functional approach.* 4th ed. New York: Macmillan.

Ray, P. M. 1972. *The living plant.* 2d ed. New York: Holt, Rinehart & Winston.

Rayle, D., and Wedberg, L. 1980. *Botany: A human concern.* 2d ed. Boston: Houghton Mifflin.

Salisbury, F. B. 1958. The flowering process. *Scientific American* 198(4):108.

Shepard, J. F. 1982. The regeneration of potato plants from leaf-cell protoplasts. *Scientific American* 246(5):154.

van Overbeek, J. 1968. The control of plant growth. *Scientific American* 219(2):17.

Zimmerman, M. H. 1963. How sap moves in trees. *Scientific American* 208(3):132.

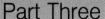

Human Anatomy and Physiology

The study of human anatomy and physiology serves as a guide to an understanding of the vertebrate body. A limited number of tissues make up organs, which form systems to carry out the functions assumed by the cell in less complex animals. All body systems help maintain a relatively constant internal environment so that the proper physical conditions exist for each cell.

The digestive system provides nutrients, and the excretory system rids the body of metabolic wastes. The respiratory system supplies oxygen but also eliminates carbon dioxide. The circulatory system carries nutrients and oxygen to and wastes from the cells so that tissue fluid composition remains constant. The immune system helps protect the body from disease. The nervous and hormonal systems control body functions. The nervous system directs body movements, allowing the organism to manipulate the external environment, an important life-sustaining function.

8

Human Organization

Chapter concepts

1 Animal tissues, including human tissues, can be categorized into four major types: epithelial, connective, muscle, and nervous tissues.

2 Organs usually contain several types of tissues. For example, although skin is primarily composed of epithelial and connective tissue, it also contains muscle and nerve fibers.

3 Organs are grouped into organ systems; the location of organ systems in the body determines whether an animal is a vertebrate or invertebrate.

4 Mammals, including humans, exhibit a marked ability to maintain a relatively constant internal environment. All organ systems contribute to homeostasis.

In the chapters to follow, human physiology is studied as representative of vertebrate physiology. Our study will be more meaningful if we first review human organization. Figure I.2 shows that the human body, like that of other organisms, has levels of organization. Cells of the same type are joined together to form a tissue. Different tissues are found in an organ, and various types of organs are arranged into an organ system. Finally, the organ systems comprise the organism.

Tissues

The tissues of the human body can be categorized into four major types: epithelial tissue that covers body surfaces and lines body cavities, connective tissue that binds and supports body parts, muscular tissue that causes parts to move, and nervous tissue that responds to stimuli and transmits impulses from one body part to another (fig. 8.1).

Epithelial Tissue

Epithelial tissue forms a continuous layer, or sheet, over the entire body surface and most of the body's inner cavities. On the external surface, it forms a covering that, like the epidermis in plants, protects the animal from injury and drying out. On internal surfaces, this tissue may be specialized for other functions in addition to protection; for example, it secretes mucus along the digestive tract, it sweeps up impurities from the lungs by means of hairlike extensions called cilia, and it efficiently absorbs molecules from kidney tubules because of fine cellular extensions called microvilli.

There are three types of epithelial tissue. **Squamous epithelium** (fig. 8.2) is composed of flat cells and is found lining the lungs and blood vessels. **Cuboidal epithelium** (fig. 8.3) contains cube-shaped cells and is found lining the kidney tubules. In **columnar epithelium** (fig. 8.4), the cells resemble pillars or columns, and nuclei are usually located near the bottom of each cell. This epithelium is found lining the digestive tract. Each type of epithelium may have microvilli or cilia as appropriate for its particular function.

Each of these types of epithelium may be stratified. **Stratified** means to exist as layers piled one over the other. The skin is stratified squamous epithelium. Some epithelium cells are **pseudostratified** and appear to be layered, but actually each cell touches the base line and thus true layers do not exist. The lining of the windpipe, or trachea, is called pseudostratified ciliated columnar epithelium (fig. 8.5).

Epithelial cells sometimes secrete a product, in which case they are described as glandular. A gland can be a single cell, as in the case of the mucus-secreting goblet cells found within the columnar epithelium lining the digestive tract, or a gland can contain numerous cells. Glands that secrete their products into ducts are called **exocrine glands,** and those that secrete directly into the bloodstream are called **endocrine glands.**

Since epithelium covers and lines body parts, it always has an exposed outer surface and an inner surface that adjoins connective tissue. The inner surface is anchored to connective tissue by a thin, noncellular layer called the **basement membrane.**

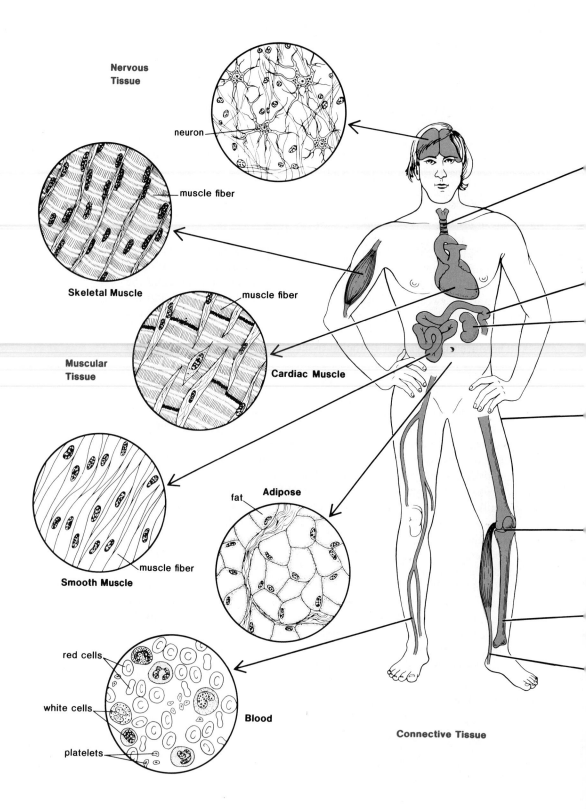

Nervous
Tissue

neuron

muscle fiber

Skeletal Muscle

muscle fiber

Muscular
Tissue

Cardiac Muscle

Adipose

fat

muscle fiber

Smooth Muscle

red cells

white cells

Blood

platelets

Connective Tissue

Figure 8.1
The major tissues in the human body. Reading counterclockwise, observe that nervous tissue contains specialized cells called neurons. Muscular tissue is of three types: skeletal, cardiac, and smooth. Connective tissue includes adipose tissue, blood, fibrous tissue, bone, and cartilage. Epithelial tissue includes squamous, cuboidal, columnar and ciliated columnar epithelium.

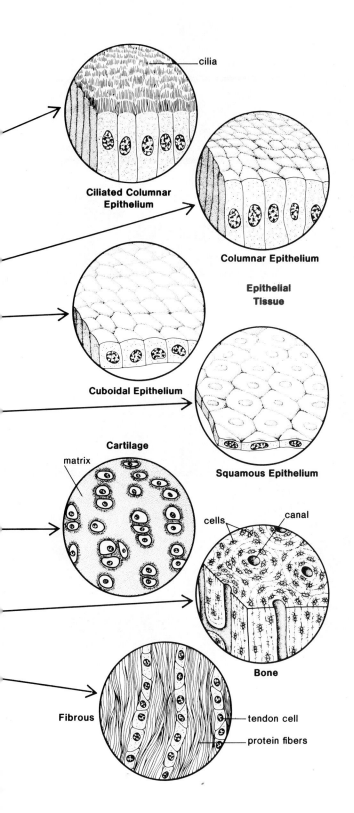

cilia

Ciliated Columnar Epithelium

Columnar Epithelium

Epithelial Tissue

Cuboidal Epithelium

Cartilage

matrix

Squamous Epithelium

cells   canal

Bone

Fibrous

tendon cell

protein fibers

Figure 8.2
Simple squamous epithelial cells form the lining of
the heart. Dome-shaped areas (*) of each cell
indicate the location of underlying nuclei. Surface
folds (arrows) are frequently visible between
adjoining cells. Notice how flat the cells are.
From: TISSUES AND ORGANS: A TEXT-ATLAS OF
SCANNING ELECTRON MICROSCOPY by R. G. Kessel
and R. Kardon. W. H. Freeman and Company, © 1979.

Figure 8.3
Cuboidal epithelial cells with microvilli (Mv) form the
lining of kidney tubules. Each cell resides on a
basal lamina (BL), or basal membrane. Notice the
cubelike shape of the cells.
From: TISSUES AND ORGANS: A TEXT-ATLAS OF
SCANNING ELECTRON MICROSCOPY by R. G. Kessel
and R. Kardon. W. H. Freeman and Company, © 1979.

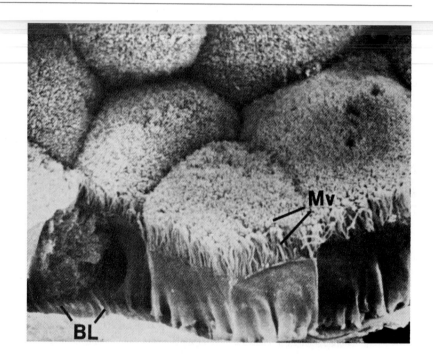

Human Anatomy and Physiology

Figure 8.4
Columnar epithelial cells from the lining of the
oviduct. Some cells have ciliated surfaces *(Ci)* and
others have microvilli *(Mv)*. Notice that the cells
have a columnlike appearance.
From: TISSUES AND ORGANS: A TEXT-ATLAS OF
SCANNING ELECTRON MICROSCOPY by R. G. Kessel
and R. Kardon. W. H. Freeman and Company, © 1979.

Figure 8.5
Pseudostratified ciliated columnar epithelium from
lining of the trachea. *Pseudostratified* means that
although there appears to be more than one layer
of cells, the cells are not layered since each cell
touches the basal membrane.
From: TISSUES AND ORGANS: A TEXT-ATLAS OF
SCANNING ELECTRON MICROSCOPY by R. G. Kessel
and R. Kardon. W. H. Freeman and Company, © 1979.

Figure 8.6
Loose connective tissue. Fibroblasts *(Fi)* vary in
shape but can be compared in size to a nearby red
blood cell (erythrocyte, *Er*). Collagen fibers *(CF)*
surround the cells. The term *loose* is used because
the cells are not tightly packed together.
From: TISSUES AND ORGANS: A TEXT-ATLAS OF
SCANNING ELECTRON MICROSCOPY by R. G. Kessel
and R. Kardon. W. H. Freeman and Company, © 1979.

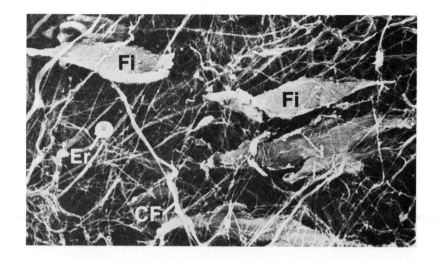

Figure 8.7
Hyaline cartilage. The cells, called chondrocytes
(Ch), are located in the many spaces found within
the matrix they synthesize. Because the matrix is
composed of fibers and proteins, cartilage is a
flexible tissue.
From: TISSUES AND ORGANS: A TEXT-ATLAS OF
SCANNING ELECTRON MICROSCOPY by R. G. Kessel
and R. Kardon. W. H. Freeman and Company, © 1979.

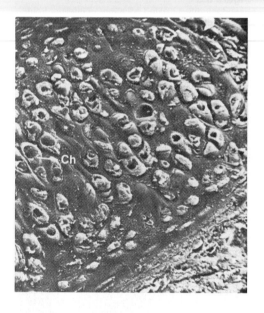

Connective Tissue

Connective tissues bind structures together, provide support and protection, serve as a framework, fill spaces, store fat, and form blood cells. As a rule, connective tissue cells have an abundance of intercellular material (matrix) between them.

Loose Connective Tissue

Loose connective tissue forms delicate, thin membranes throughout the body (fig. 8.6). The cells of this tissue, which are mainly **fibroblasts,** are located some distance apart from one another and are separated by a jellylike intercellular material that contains many white (collagen) and yellow (elastin) fibers. The white fibers occur in bundles and are strong and flexible. The yellow fibers form networks that are highly elastic—when stretched they return to their original length. Loose connective tissue commonly lies beneath epithelium; it binds skin to underlying organs and fills spaces between muscles. **Adipose,** or fat, tissue is a type of loose connective tissue in which the fibroblasts enlarge and store fat and the intercellular matrix is reduced.

Fibrous Connective Tissue

Fibrous connective tissue contains large numbers of collagenous fibers that are closely packed together. This type of tissue often functions to bind body parts and is found, for example, in **tendons,** which connect muscles to bones, and **ligaments,** which connect bones to other bones at joints. Tendons and ligaments take a long time to heal following an injury because their blood supply is relatively poor.

Cartilage

Cartilage (fig. 8.7) is a fairly rigid connective tissue whose cells occupy small chambers called **lacunae,** which are separated by a matrix composed largely of fibers and protein. An important feature of cartilage is its ability to increase in size; the developing embryonic skeleton is composed of cartilage. Later, the cartilaginous skeleton is replaced by bone, and cartilage remains only in certain locations, such as the soft portion of the nose, the outer ear flaps, and the ends of the bones.

Bone

Bone (fig. 8.8) is the most rigid of the connective tissues. It consists of an extremely hard matrix of calcium salts deposited around protein fibers. The minerals give it rigidity, while the protein fibers provide elasticity and strength, much as steel rods do in reinforced concrete. Bone cells are located in lacunae

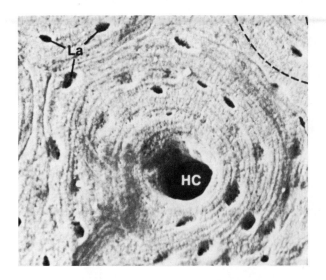

Figure 8.8
A ground, polished specimen of compact bone from which the cells and other organic constituents have been removed. The Haversian canal (HC) formerly contained blood vessels and nerves, while the lacunae (La) formerly contained bone cells. Notice how the lacunae are arranged in concentric circles around the canal.
From: TISSUES AND ORGANS: A TEXT-ATLAS OF SCANNING ELECTRON MICROSCOPY by R. G. Kessel and R. Kardon. W. H. Freeman and Company, © 1979.

Table 8.1 Blood plasma

Water	92% water
Inorganic ions (salts)	Na^+, Ca^{++}, K^+, Mg^{++}; Cl^-, HCO_3^-, HPO_4^{--}, SO_4^{--}
Gases	O_2 and CO_2
Plasma proteins	Albumin, globulin, fibrinogen
Organic nutrients	Glucose, fats, phospholipids, amino acids, etc.
Nitrogenous waste products	Urea, ammonia, uric acid
Regulatory substances	Hormones, enzymes

Table 8.2 Types of muscle

Name	Fiber appearance	Location	Control
Skeletal	Striated	Attached to skeleton	Voluntary
Smooth	Spindle shaped	Internal organs	Involuntary
Cardiac	Striated and branched	Heart	Involuntary

Figure 8.9
Biconcave red blood cells and rounded white cells within an arteriole. Red blood cells are scientifically termed erythrocytes and white blood cells are scientifically termed leukocytes (Le).
From: TISSUES AND ORGANS: A TEXT-ATLAS OF SCANNING ELECTRON MICROSCOPY by R. G. Kessel and R. Kardon. W. H. Freeman and Company, © 1979.

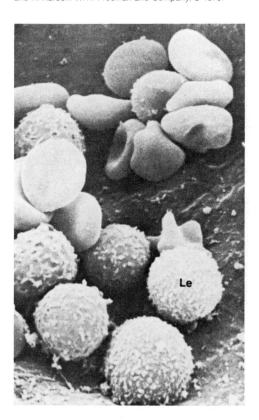

that are arranged in concentric circles around tiny tubes called **Haversian canals,** which contain blood vessels and nerve fibers. Bone cells deposit or liberate calcium from the bone, depending on the present nutritional state of the body. Branching out in all directions from the lacunae are minute canals (canaliculi) containing thin processes that connect the bone cells with one another and with the Haversian canals. Bone has a more adequate blood supply than does cartilage; therefore bone injuries heal faster than cartilage injuries.

Blood
Blood (fig. 8.9) is a connective tissue in which the matrix is a liquid called **plasma,** the contents of which are listed in table 8.1. The cells are of two types: **red** (erythrocytes), which carry oxygen, and **white** (leukocytes), which aid in fighting infection.

Also present in blood are **platelets,** which are important to the initiation of blood clotting. Platelets are not complete cells; rather, they are fragments of giant cells found in the bone marrow.

Muscular Tissue
The cells that make up muscle tissue contain microfilaments whose interaction accounts for the movements we associate with animals. There are three types of vertebrate muscles: skeletal, smooth, and cardiac (table 8.2).

Figure 8.10

Surface view of three muscle fibers (cells). It is possible to observe the transverse striations *(St)* in areas where the external membrane, called sarcolemma *(Sa)*, has been sheared away *(arrows)*. Connective tissue (called endomysium, *En*) surrounds each fiber.

From: TISSUES AND ORGANS: A TEXT-ATLAS OF SCANNING ELECTRON MICROSCOPY by R. G. Kessel and R. Kardon. W. H. Freeman and Company, © 1979.

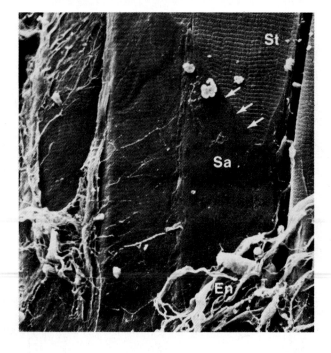

Figure 8.11

Nerve anatomy. When a portion of the connective tissue sheath surrounding a nerve is removed, underlying nerve fibers *(NF)* can be observed.

From: TISSUES AND ORGANS: A TEXT-ATLAS OF SCANNING ELECTRON MICROSCOPY by R. G. Kessel and R. Kardon. W. H. Freeman and Company, © 1979.

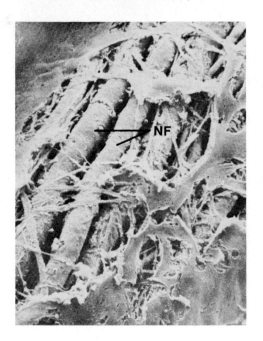

Skeletal muscle is attached to the bones of the skeleton and functions to move body parts. It is under our *voluntary control* and has the fastest contraction of all the muscle types. The cylindrical cells of this muscle have characteristic light and dark bands perpendicular to the length of the cell or fiber. These bands give the muscle a **striated** appearance (fig. 8.10).

Smooth muscle is so named because it lacks striations. The spindle-shaped cells that make up smooth muscle are not under voluntary control and are said to be *involuntary*. Smooth muscle, which is found in the viscera (intestine, stomach, and so on) and blood vessels, contracts more slowly than skeletal muscle but can remain contracted for a longer time. The cells tend to form layers in which the thick middle portion of one cell is opposite the thin ends of adjacent cells. Consequently, the nuclei form an irregular pattern in the tissue.

Cardiac muscle seems to combine features of both smooth and skeletal muscle. It has *striations* like those of skeletal muscle, but the contraction of the heart is involuntary for the most part. Heart muscle cells also differ from skeletal muscle cells in that they are branched and seemingly fused, one with the other, so that the heart appears to be composed of one large, interconnecting mass of muscle cells. Actually, however, cardiac muscle cells are separate and individual.

Nervous Tissue

The brain and nerve cord (also called the spinal cord) contain conducting cells called neurons. A **neuron** is a specialized cell that has three parts: (1) **dendrites** conduct impulses to the cell body; (2) the **cell body** contains most of the cytoplasm and the nucleus of the neuron; (3) the **axon** conducts impulses away from the cell body.

Axons and dendrites, when long, are called **nerve fibers** (fig. 8.11). Outside the brain and spinal cord, nerve fibers are bound together by connective tissue to form **nerves.** Nerves conduct impulses from sense organs to the spinal cord and brain, where the phenomenon called **sensation** occurs, and they also conduct nerve impulses away from the spinal cord and brain to the muscles, causing them to contract.

Human Anatomy and Physiology

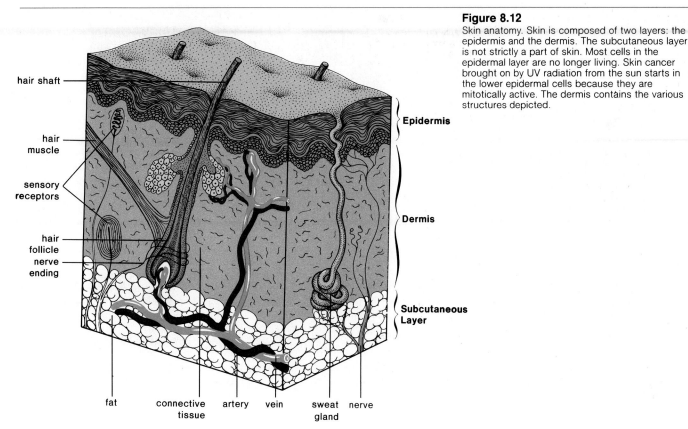

hair shaft

hair muscle

sensory receptors

hair follicle nerve ending

Epidermis

Dermis

Subcutaneous Layer

fat   connective tissue   artery   vein   sweat gland   nerve

Figure 8.12
Skin anatomy. Skin is composed of two layers: the epidermis and the dermis. The subcutaneous layer is not strictly a part of skin. Most cells in the epidermal layer are no longer living. Skin cancer brought on by UV radiation from the sun starts in the lower epidermal cells because they are mitotically active. The dermis contains the various structures depicted.

Organs

Some organs in the body are composed largely of one type of cell. For example, muscles are made up of muscle cells; and glands are most often made up of epithelial cells. Many organs, however, are a composite of different types of tissues. We will consider skin as an example.

Skin

Skin (fig. 8.12) covers the body, protecting it from loss of water and detering invasion by microorganisms. It also helps regulate body temperature and contains sense organs for touch, pressure, temperature, and pain. Skin has two tissue layers, the outer epidermis and the inner dermis. Beneath the dermis there is a subcutaneous layer that binds the skin to underlying organs.

Epidermis

The epidermis is made up of stratified squamous epithelial cells that are continually produced by a germinal layer that lies next to the dermis. Here, constant cell division of *basal cells* produces many new cells that gradually rise to the surface when outer layers of cells are worn off. Specialized cells in this layer called melanocytes produce **melanin,** the pigment responsible for skin color. As the reading on page 172 describes, a suntan is caused by the appearance of melanin granules in the outer surface of the skin. Unfortunately, UV (ultraviolet) radiation from the sun also causes the cells that are undergoing mitosis to become cancerous. There has been such a high increase in recent years of skin cancers due to sunbathing that physicians are strongly recommending that everyone stay out of the sun—or at least, use suncreen lotions to protect the skin.

The cells produced by the germinal layer gradually become flattened and hardened as they are pushed toward the surface. Hardening is caused by **keratinization,** cellular production of a fibrous, waterproof protein called keratin. Over much of the body, keratinization occurs only minimally. In these

Bring Back the Parasol

With summer just around the corner, the pale of face throughout the Northern Hemisphere will soon be hitting beaches in pursuit of a deep, dark and sexy tan. The Victorian ideal of delicate, camellia-white skin has long since been supplanted by the bronzed-god look. But the trend has taken a mortal toll. Sun-related skin cancer is rapidly on the rise in the U.S. and Europe, and afflicting younger and younger people. The incidence of the most lethal form, malignant melanoma, though less directly linked to sunshine, has jumped tenfold in the past 20 years. Last week some 300 dermatologists and others gathered in Manhattan to discuss the problem at the first World Congress on Cancers of the Skin. Their message: bring back the parasol.

Doctors have long known that ultraviolet (UV) radiation from the sun produces profound changes in human skin. "Even one day's exposure can cause damage," says Dermatologist Fred Urbach of Temple University in Philadelphia. The most insidious rays are the short wave-length UVB, which prevail during the peak sun hours (between 11 a.m. and 3 p.m.). But new research has shown that even longer UVA waves, which are present all day, can promote skin cancer. The damage caused by these invisible rays ranges from ordinary sunburn, to the wrinkles and liver spots caused by years of sunbathing, to the precancerous dark patches known as actinic keratosis and, finally, cancer. Each of these is part of the same process, says Urbach. "First you look old, then if you've had a lot more sun, you get keratosis, and after that skin cancer. If we all lived long enough, we would all get skin cancer."

The process begins when solar UV damages basal cells near the surface of the skin, causing them to swell. The pain and redness, which appear a few hours after exposure, are caused by the dilation of blood vessels in the damaged area. The ensuing tan is the body's desperate effort to save its skin from further injury. Tiny granules of melanin, a brownish pigment made in specialized skin cells, rise to the surface in response to UV radiation and act as sunlight deflectors. Over the years, however, the beachgoer pays for this glamorous natural shield. The buildup of melanin, combined with UV damage to the elastic fibers in underlying layers, gives the skin the texture of an old baseball mitt.

Like X rays, UV radiation can alter cell DNA, producing the mutations associated with cancer. "Both UVA and UVB are carcinogenic," says Harvard Photobiologist Madhu Pathak. UV also appears to suppress the body's immune system. This may explain why certain viral infections, such as chicken pox and fever blisters, become more severe in the sun. And since the immune system is believed to play a role in preventing tumor growth, its

---

regions, patches of overly keratinized cells are called spots of keratosis, which are early indications of possible skin cancer. In contrast, both the palm of the hand and the sole of the foot normally have a particularly thick outer layer of dead keratinized cells arranged in spiral and concentric patterns. These patterns are unique to each person and are known on the fingers as fingerprints. Hair is not present here but is found wherever the skin is thinner. Nails and hair are also composed of tightly packed keratinized cells.

Dermis

The dermis is largely fibrous connective tissue with many inclusions. The reading on this page tells us that UV light contributes to the aging process because it causes this tissue to lose its elasticity, producing "wrinkles" in the skin.

The epidermis from above regularly dips down into the dermis, serving as a location for sweat glands and hair follicles, with their associated sebaceous glands. The sweat glands help regulate body temperature. They become active and release fluid onto the skin when the body temperature rises. The

Human Anatomy and Physiology

suppression "may also be an aggravating factor in the development of skin cancer," says Dr. Margaret Kripke of the National Cancer Institute.

About 80% of the skin cancers caused by the sun are basal-cell carcinoma. Usually occurring on the head or neck, they are the most common and curable form of cancer in the U.S. Nancy Reagan was one of 400,000 Americans treated for this disease last year [1982]; she has more recently had several spots of keratosis removed from her face to prevent a recurrence. Skin cancers that appear elsewhere on the body are usually squamous-cell carcinoma, also easily cured by surgery.

Far more lethal are the darkly pigmented spots of malignant melanoma, which strikes more than 15,000 Americans a year, killing 45% of them. Though melanoma tends to occur on such sun-exposed areas as the chests of men and legs of women, its relationship to the sun remains unclear. A history of severe sunburns may play a role; pregnancy and birth control pills have also been implicated.

The evils of ultraviolet are easily escaped. Mrs. Reagan had the right idea when she resolved last New Year's Day to stay out of the sun. For those unwilling to make that sacrifice, the doctors at last week's conference urged the use of sunscreen lotions, some of which are now formulated to block both UVA and UVB. The Food and Drug

Administration has published guidelines recommending a sunscreen of certain strength for each type of skin. For fair-skinned people who never tan and always burn (Type I skin), sunscreens labeled with the number 15 are best. The number indicates that it will take at least 15 times longer to burn when the product is used than when the skin is unprotected. People who sometimes burn and never tan should use sunscreens in the 6-to-8 range; those who occasionally burn but tan well need only 4-to-8 protection; olive-complected Type IVs, who never get scorched, can manage with a factor of 2 to 4. In the future, however, a sunscreen pill being developed in Pathak's lab may make the lotions obsolete.

evaporation of this fluid cools off the body. The sebaceous glands lubricate the adjoining hairs and skin. At the time of puberty, they are also responsible for acne.

Blood vessels and nerve endings are also present in the dermis. Some of the latter control a small muscle attached to each hair follicle. When this muscle contracts, the hair stands on end, causing what is commonly known as "goose bumps." Encapsulated (membrane-surrounded) nerve endings are found in the microscopic sense organs for touch, pressure, and temperature. Stimulation of free nerve endings produces pain. Regulation of the size of the arteries in skin helps maintain a constant internal body temperature. When the arteries increase in size, more blood is brought to the surface of the skin where it is cooled off. At the same time, sweat glands become active and the person perspires.

Subcutaneous Layer

The **subcutaneous layer** is a layer of loose connective tissue containing adipose cells. A well-developed subcutaneous layer gives a rounded appearance to the body. Excessive development accompanies obesity.

Figure 8.13
Organization of the human body. Like other mammals, humans have a dorsal nervous system and well-developed coelom that contains the internal organs. The coelom is divided by the diaphragm into the thoracic and abdominal cavity.

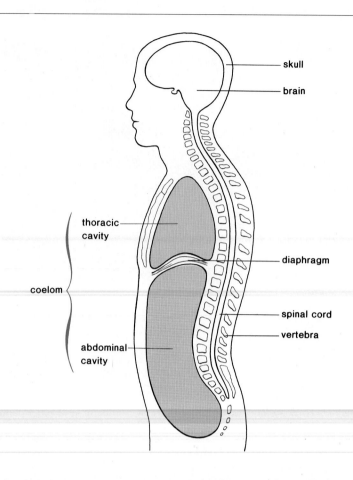

Table 8.3	Human organ systems
Name	**Function**
Digestive	Convert food particles to nutrient molecules
Circulatory	Transport of molecules to and from cells
Immune	Defense against disease
Respiratory	Exchange of gases with environment
Excretory	Elimination of metabolic wastes
Nervous and sensory	Regulation of systems and response to environment
Muscular and skeletal	Support and movement of organism
Hormonal	Regulation of internal environment

Organ Systems

In this part of the text, we are going to study the organ systems listed in table 8.3. Each of these systems has a specific location within the body. The central nervous system is dorsally located; the brain is protected by the skull, and the spinal cord, which gives off spinal nerves, is protected by the vertebrae (fig. 8.13). The repeating units of vertebrae and spinal nerves show that humans are segmented animals, meaning that body parts reoccur at regular intervals.

Within the musculoskeletal system, the skeleton provides the surface area for attachment of striated muscles that are well developed and powerful. The musculoskeletal system makes up most of the body weight and is specialized for locomotion.

The other internal organs are found within a body cavity called the **coelom.** In humans and other mammals, the coelom is divided by a muscular diaphragm that assists breathing. The heart, a pump for the closed circulatory system, and the lungs are located in the upper thoracic (chest) cavity. The major portion of the digestive system, the entire excretory system, and much of the reproductive system are located in the lower abdominal cavity. The major organs of the excretory system are the paired kidneys, and the accessory organs of the digestive system are the liver and pancreas. Each sex has characteristic sex organs.

The preceding are vertebrate characteristics and, in this text, human physiology is studied as representative of vertebrates in general. Of all the vertebrates, mammals (animals who nourish their young by means of mammary glands) and birds are best able to maintain constancy of internal body conditions. No doubt this has contributed greatly to their success.

Human Anatomy and Physiology

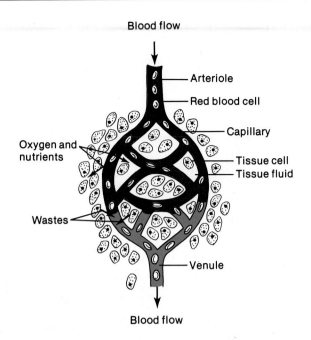

Blood flow

Arteriole

Red blood cell

Capillary

Oxygen and
nutrients

Tissue cell
Tissue fluid

Wastes

Venule

Blood flow

Figure 8.14
The internal environment of the body is the blood
and tissue fluid. Tissue cells are surrounded by
tissue fluid which is continually refreshed because
nutrient molecules constantly exit from and waste
molecules continually enter the blood stream as
shown.

Homeostasis

Homeostasis means that the internal environment remains relatively constant regardless of the conditions in the external environment. In humans, for example:

1. Blood glucose concentration remains at about 0.1%.
2. The pH of the blood is always near 7.4.
3. Blood pressure in the brachial artery averages near 120/80.
4. Blood temperature averages around 37°C (98.6°F).

The ability of the body to keep the internal environment within a certain range allows humans to live in a variety of habitats, such as the arctic regions, deserts, or the tropics.

The internal environment includes a tissue fluid that bathes all the tissues of the body. The composition of tissue fluid must remain constant if cells are to remain alive and healthy. Tissue fluid is created when water, oxygen (O_2), and nutrient molecules leave a capillary (the smallest of the blood vessels), and it is purified when water, carbon dioxide (CO_2), and other waste molecules enter a capillary from the fluid (fig. 8.14). Tissue fluid remains constant only as long as blood composition remains constant. Although we are accustomed to using the word *environment* to mean the external environment of the body, it is important to realize that it is the internal environment of tissues that is ultimately responsible for our health and well-being.

Most systems of the body contribute to maintaining a constant internal environment. The digestive system takes in and digests food providing nutrient molecules, which enter the blood and replace the nutrients that are constantly being used up by the body cells. The respiratory system adds oxygen to the blood and removes carbon dioxide. The amount of oxygen taken in and carbon dioxide given off can be increased to meet body needs. The chief regulators of blood composition, however, are the liver and the kidneys. They monitor the chemical composition of plasma (table 8.1) and alter it as required. Immediately after glucose enters the blood, it can be removed by the liver for

Figure 8.15

Diagram illustrating the principle of feedback control. A receptor (sense organ) responds to a stimulus such as high or low temperature and notifies a regulator center that directs an adaptive response such as sweating. Once normalcy such as a normal temperature is achieved, the receptor is no longer stimulated.

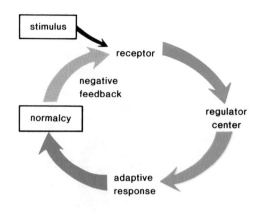

Figure 8.16

Temperature control. When the body temperature rises, the blood vessels dilate and the sweat glands become active. When the body temperature lowers, the blood vessels constrict and shivering may occur. In between these extremes the receptor is not stimulated and thus body temperature fluctuates above and below normal.

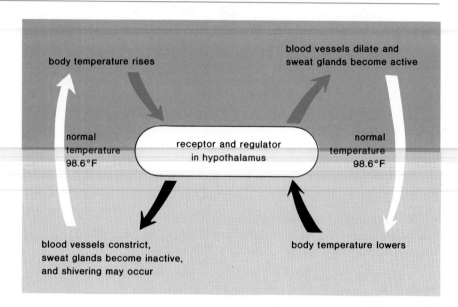

storage as glycogen. Later the glycogen can be broken down to replace the glucose used by the body cells; thus the glucose composition remains constant. The hormone insulin secreted by the pancreas regulates glycogen storage. The liver also removes toxic chemicals, such as ingested alcohol and drugs and nitrogenous wastes given off by the cells. These are converted to molecules that can be excreted by the kidneys. The kidneys are also under hormonal control as they excrete wastes and salts, substances that can affect the pH level of the blood.

Although homeostatis is, to a degree, controlled by hormones, it is ultimately controlled by the nervous system. The brain contains centers that regulate such factors as temperature and blood pressure. Maintaining proper temperature and blood pressure levels requires sensors that detect unacceptable levels and signal a control center. If a correction is required, the center then directs an adaptive response (fig. 8.15). Once normalcy is obtained, the receptor no longer stimulates the center. This is called control by **negative feedback** because the control center shuts down until it is stimulated to be active once again. This type of homeostatic regulation results in fluctuation between two levels, as illustrated for temperature control in figure 8.16.

Human Anatomy and Physiology

Summary

Human tissues are categorized into four groups. Epithelial tissue covers the body and lines its cavities. The different types of tissue (squamous, cuboidal, and columnar) can be stratified and have cilia or microvilli. Also, columnar cells can be pseudostratified. Epithelial cells sometimes form glands that secrete either into ducts or into blood.

Connective tissues in which cells are separated by a matrix often bind body parts together. Loose connective tissue has both white and yellow fibers and may also have fat (adipose) cells. Fibrous connective tissue such as tendons and ligaments contains closely packed white fibers. Both cartilage and bone have cells within lacunae, but the matrix for cartilage is more flexible than that for bone, which contains calcium salts. In bone, the lacunae lie in concentric circles about a Haversian canal. Blood is a connective tissue in which the matrix is a liquid called plasma.

Muscular tissue is of three types. Both skeletal and cardiac muscle are striated; both cardiac and smooth are involuntary. Skeletal muscle is found in muscles attached to bones, and smooth is found in internal organs. Cardiac muscle makes up the heart.

Nervous tissue has one main type of cell, the neuron, that possesses two kinds of fibers, dendrites and axons. The brain and spinal cord contain complete neurons, while the nerves contain only fibers. Neurons and their fibers are specialized to conduct nerve impulses.

Tissues are joined together to form organs, each one having a specific function. Skin is a two-layered organ that waterproofs and protects the body. The epidermis contains a germinal layer that produces new epithelial cells that become keratinized as they move toward the surface. The dermis, a largely fibrous connective tissue, contains epidermally derived glands and hair follicles, nerve endings, and blood vessels. Encapsulated nerve endings form sense organs for touch, pressure, and temperature; free nerve endings register pain. Sweat glands and blood vessels help control body temperature. A subcutaneous layer, which is made up of loose connective tissue containing adipose cells, lies beneath the skin.

Organs are grouped into organ systems. The location of the organ systems indicates whether an animal is a vertebrate or an invertebrate. In vertebrates, the well-developed brain is protected by the skull, and the dorsal spinal cord is protected by vertebrae. Other internal organs are located in a cavity called the coelom. In humans, as in other mammals, this cavity is divided by the diaphragm. Also, like other vertebrates, the skeleton is internal and the muscles are large. In this text, human physiology is studied as representative of vertebrate and mammalian physiology.

Homeostasis is characteristic of mammals. All organ systems contribute to the constancy of tissue fluid and blood. Special contributions are made by the liver, which keeps blood glucose constant, and the kidneys, which regulate the pH. The nervous and hormonal systems regulate the other systems. Both of these are controlled by a feedback mechanism, which results in fluctuation above and below the desired level.

Study Questions

1. Name the four major groups of tissues. (p. 163)
2. What are the functions of epithelial tissues? Name the different kinds and give a location for each. (pp. 163–67)
3. What are the functions of connective tissue? Name the different kinds and give a location for each. (pp. 168–69)

4. What are the functions of muscular tissue? Name the different kinds and give a location for each. (pp. 109–110)
5. Nervous tissue contains what type cell? Which organs in the body are made up of nervous tissue? (p. 110)
6. Describe the structure of skin and state at least two functions of this organ. (pp. 171–73)
7. In general terms, describe the location of the human organ systems. (p. 174)
8. List at least four vertebrate characteristics of humans. (p. 174)
9. What is homeostasis and how is it achieved in the human body? (pp. 175–76)

Selected Key Terms

epithelial (ep″ĭ-the′le-al)
stratified (strat′ĭ-fīd)
tendon (ten′don)
ligament (lig′ah-ment)
cartilage (kar′tĭ-lij)
lacuna (lah-ku′nah)
Haversian canal (ha-ver′shan kah-nal′)
skeletal muscle (skel′ĕ-tal mus′el)
striated (stri′āt-ed)
smooth muscle (smōoth mus′el)
cardiac muscle (kar′de-ak mus′el)
neuron (nu′ron)
dendrite (den′drīt)
axon (ak′son)
epidermis (ep″ĭ-der′mis)
keratinization (ker″ah-tin″i-za′shun)
dermis (der′mis)
subcutaneous (sub″ku-ta′ne-us)
coelom (se′lom)
homeostasis (ho″me-o-sta′sis)

Further Readings

Caplan, A. I. 1984. Cartilage. *Scientific American* 251(4):84.
Hickman, C. P. et al. 1982. *Biology of animals*. 3d ed. Saint Louis: C. V. Mosby.
Hole, J. W. 1984. *Human anatomy and physiology*. 3d ed. Dubuque, Iowa: Wm. C. Brown Publishers.
Kessel, R. G., and Kardon, R. H. 1979. *Tissues and organs: a text-atlas of scanning electron microscopy*. San Francisco: W. H. Freeman.

Human Anatomy and Physiology

9

Chapter concepts

1 Small molecules, such as amino acids, glucose, and fatty acids, that can cross cell membranes are the products of digestion that nourish the body.

2 Regions of the digestive tract are specialized to carry on specific functions; for example, the mouth is specialized to receive and chew food, and the small intestine is specialized to absorb the products of digestion.

3 Digestion requires enzymes that function according to the lock-and-key theory. They are specific and have a preferred pH.

4 Proper nutrition requires that the energy needs of the body be met and that the diet be balanced so that all vitamins and essential amino acids and fatty acids are included.

Digestion

igestion takes place within a tube, often called the gut, that begins with the mouth and ends with the anus (table 9.1 and fig. 9.1). Digestion of food in humans is an extracellular process. Digestive enzymes are secreted into the gut by glands that reside in the lining or lie nearby. Food is never found within these *accessory glands,* only within the gut itself.

Table 9.1 Path of food

Organ	Special features	Functions
Mouth	Teeth	Chewing of food; digestion of starch to maltose
Esophagus		Passageway
Stomach	Gastric glands	Digestion of protein to peptides
Small intestine	Intestinal glands Villi	Digestion of all foods and absorption of unit molecules
Large intestine		Absorption of water
Anus		Defecation

Figure 9.1
The human digestive system. The liver is drawn smaller than normal size and moved back to show the gallbladder and to expose the stomach and duodenum.

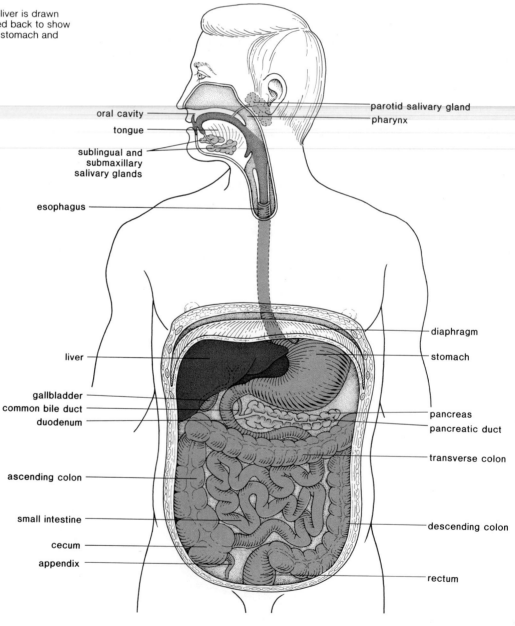

oral cavity

tongue

sublingual and submaxillary salivary glands

esophagus

liver

gallbladder

common bile duct

duodenum

ascending colon

small intestine

cecum

appendix

parotid salivary gland

pharynx

diaphragm

stomach

pancreas

pancreatic duct

transverse colon

descending colon

rectum

Human Anatomy and Physiology

While the term **digestion,** strictly speaking, means the breakdown of food by enzymatic action, we will expand the term to include both the physical and chemical processes that reduce food to small soluble molecules. Only small molecules can cross cell membranes and be absorbed by the gut lining. Too often we are inclined to think that since we eat meat (protein), potatoes (carbohydrate), and butter (fat), these are the substances that nourish our bodies. Instead, it is the amino acids from the protein, the sugars from the carbohydrate, and the glycerol and fatty acids from the fat that actually enter the blood and are transported about the body to nourish our cells. Any component of food, such as cellulose, that is incapable of being digested to small molecules leaves the gut as waste material.

Digestion of food requires a cooperative effort between different parts of the body. We shall see that the production of hormones and the nervous system achieve the coordination of parts needed to achieve cooperation of body parts.

Digestive System

Mouth

The **mouth** receives the food in humans. Most people enjoy eating because of the combined sensations of smelling and tasting food. The olfactory receptors located in the nose are responsible for smelling; tasting is, of course, a function of the taste buds located on the tongue. (See chapter 17 for a description of these sense organs.)

Normally, everyone has 32 teeth (fig. 9.2) that chew the food into pieces convenient to swallow. One-half of each jaw has teeth of four different types: two chisel-shaped incisors for biting; one pointed canine for tearing; two fairly flat premolars for grinding; and three molars, well flattened for crushing. The last molar, or wisdom tooth, may fail to erupt, or if it does it is sometimes crooked and useless.

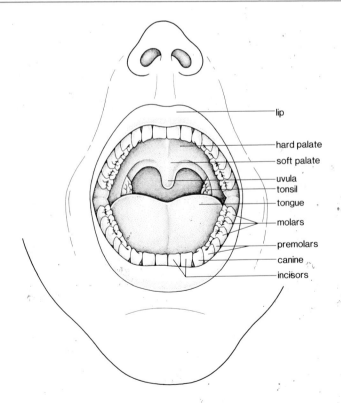

Figure 9.2
Diagram of the mouth showing the permanent teeth. The sizes and shapes of teeth correlate with their functions.

lip
hard palate
soft palate
uvula
tonsil
tongue
molars
premolars
canine
incisors

Figure 9.3
Longitudinal section of a molar. A tooth contains
nerves and blood vessels within the pulp.

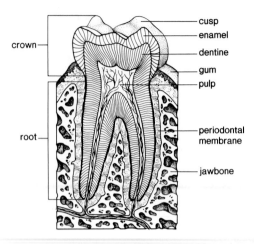

Each **tooth** (fig. 9.3) has a layer of enamel, an extremely hard outer covering of calcium compounds; dentine, a thick layer of bonelike material; and an inner pulp that contains the nerves and blood vessels. Tooth decay or **caries,** commonly called cavities, occurs when the bacteria within the mouth metabolize sugar and give off acids that corrode the teeth. Two measures may prevent tooth decay: eating a limited amount of sweets and daily brushing and flossing of teeth. It has also been found that fluoride treatments can make the enamel stronger and more resistant to decay. Gum disease is more apt to occur as one ages. Inflammation of the gums (gingivitis) may spread to the periodontal membrane (fig. 9.3) that lines the tooth socket. The individual then has **periodontitis,** characterized by a loss of bone and loosening of the teeth so that they may have to be pulled. Stimulation of the gums in a manner advised by dentists has been found helpful in controlling this condition.

In humans, the roof of the mouth separates the air passages from the mouth cavity. The roof has two parts: an anterior **hard palate** and a posterior **soft palate** (fig. 9.2). The hard palate contains several bones, but the soft palate is merely muscular. The soft palate ends in the **uvula,** a suspended process often mistaken by the layman for the tonsils. But as figure 9.2 shows, the tonsils lie to the sides of the throat.

There are three pairs of **salivary glands,** which send their juices by way of ducts to the mouth. The parotid glands lie at the sides of the face immediately below and in front of the ears. They become swollen when a person has the mumps, a viral infection most often seen in children. Each parotid gland has a duct that opens on the inner surface of the cheek just at the location of the second upper molar. The sublingual glands lie beneath the tongue, and the submaxillary glands lie beneath the lower jaw. The ducts from these glands open into the mouth under the tongue. You can locate all these openings if you use your tongue to feel for small flaps on the inside of the cheek and under the tongue.

Saliva, secreted by the salivary glands, contains mostly water, mucus, and the digestive enzyme **salivary amylase.** This enzyme acts on starch. Like all the digestive enzymes, salivary amylase is a **hydrolytic enzyme.** This means that its substrate is broken down upon the addition of water:

$$\text{starch} + (n-1)\text{H}_2\text{O} \xrightarrow{\text{salivary amylase}} n \text{ maltose}$$

In this equation, *salivary amylase* is written above the arrow to indicate that it is neither a reactant nor a product in the reaction. It merely speeds up the reaction in which its substrate, starch, is digested to many molecules of maltose. The *n* in the equation stands for some large number of molecules of maltose. Maltose is a disaccharide whose chemical structure is shown in figure 9.4. Maltose is not one of the small molecules that can be absorbed by the gut lining. Additional digestive action is required to convert maltose to glucose. This occurs farther along the digestive tract.

No other digestive process occurs in the mouth. The tongue takes the chewed food and forms it into a mass called a **bolus,** in preparation for swallowing.

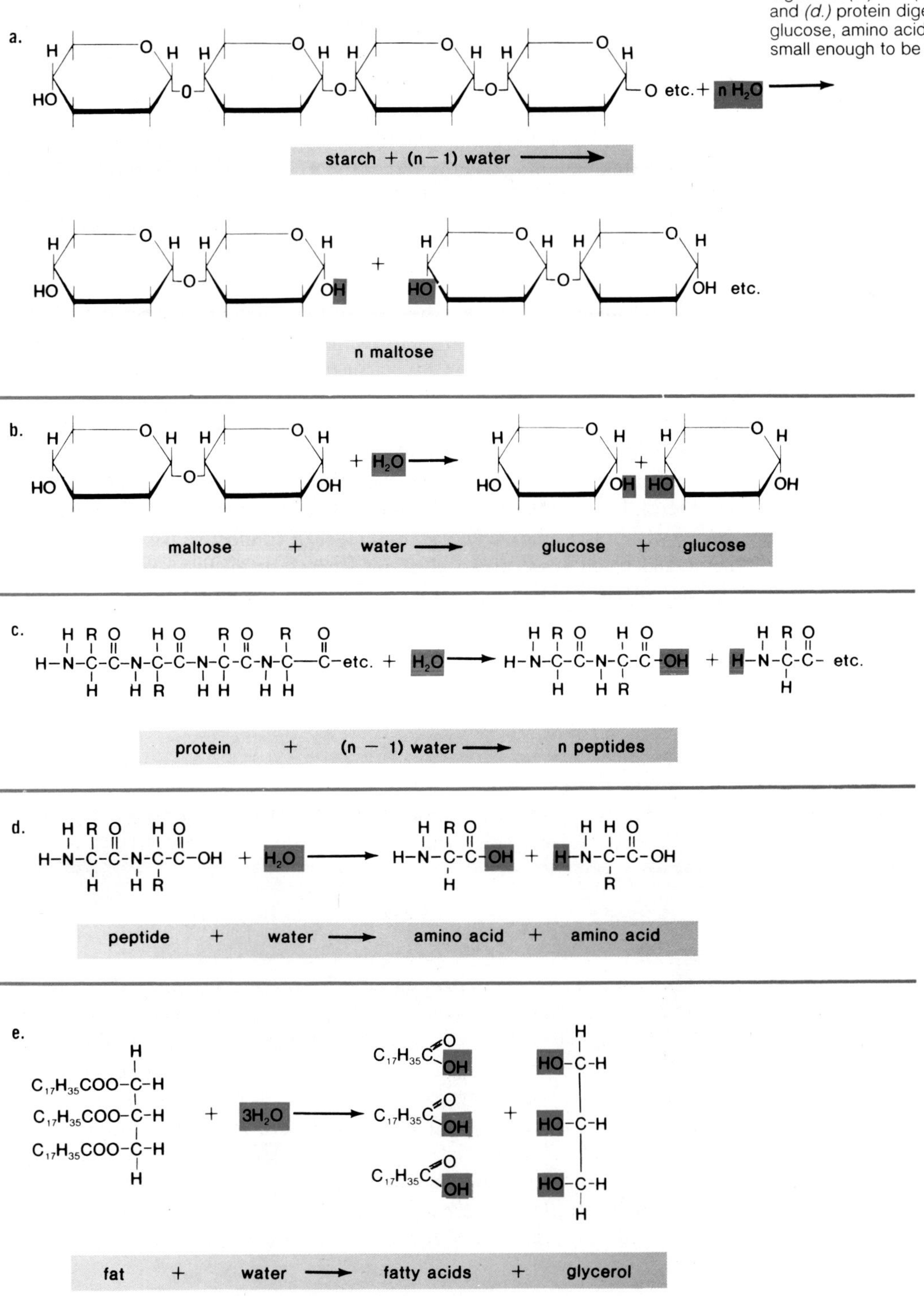

Figure 9.4
Five hydrolytic reactions occur when food is digested: (a.) and (b.) carbohydrate digestion; (c.) and (d.) protein digestion; (e.) fat digestion. Only glucose, amino acids, fatty acids, and glycerol are small enough to be absorbed by the gut lining.

Figure 9.5

When food is swallowed, the soft palate covers the nasopharyngeal openings and the epiglottis covers the glottis so that the food bolus must pass down the esophagus. Therefore one does not breathe during swallowing.

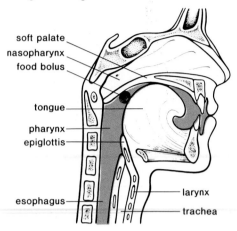

soft palate
nasopharynx
food bolus
tongue
pharynx
epiglottis
esophagus
larynx
trachea

Pharynx

Swallowing (fig. 9.5) occurs in the **pharynx,** a region between the mouth and the esophagus, a long muscular tube that leads to the stomach. Swallowing is a **reflex action,** which means the action is usually performed automatically and does not require conscious thought. Normally, during swallowing food enters the esophagus because the air passages are blocked. Unfortunately, we have all had the unpleasant experience of having food "go the wrong way." The wrong way may be either into the nose or into the windpipe (trachea). If it is the latter, coughing usually forces the food up out of the trachea into the pharynx again. Food usually goes into the esophagus because the openings to the nose, called the **nasopharyngeal openings,** are covered when the soft palate moves back. The opening to the larynx (voice box) at the top of the trachea, called the **glottis,** is covered when the trachea moves up under a flap of tissue, called the **epiglottis.** It is easy to observe the up-and-down movement of the **Adam's apple,** a part of the larynx, when a person eats. Notice that breathing does not occur during swallowing because air passages are closed off.

Esophagus

After swallowing occurs, the **esophagus** conducts the bolus through the thoracic cavity. The wall (fig. 9.6) of the esophagus is representative of the gut in general. It contains epithelial tissue, called the **mucosa,** that lines the lumen (space within the tube); a **submucosal** layer of connective tissue that contains nerve and blood vessels; two layers of smooth or involuntary muscle, called the **muscularis;** and finally, a **serosa** that usually has a thin outer layer of squamous epithelium.

A rhythmical contraction of the esophageal wall, called **peristalsis,** (fig. 9.7) pushes the food along. Occasionally peristalsis begins even though there is no food in the esophagus. This produces the sensation of a lump in the throat.

Figure 9.6

Wall of esophagus. Like the rest of the digestive tract, several different types of tissues are found in the wall of the esophagus. *Lu* = central lumen; *Mu* = mucosal lining; *Su* = submucosa; *Me* = muscularis externa; and *Ad* = adventitia, or serosa.
From: TISSUES AND ORGANS: A TEXT-ATLAS OF SCANNING ELECTRON MICROSCOPY by R. G. Kessel and R. Kardon. W. H. Freeman and Company, © 1979.

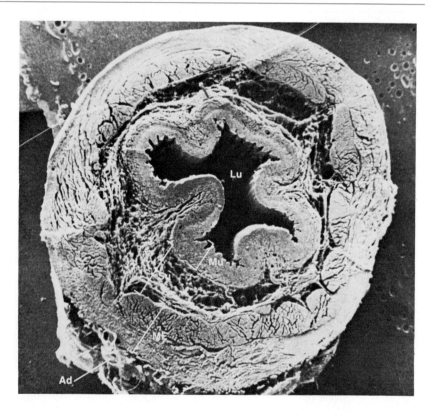

Human Anatomy and Physiology

There is a thickened region, or constrictor, where the esophagus enters the stomach, and this constrictor relaxes to allow the bolus to pass into the stomach. Normally, the constrictor prevents food from moving up out of the stomach, but when vomiting occurs, a reverse peristaltic wave causes the constrictor to relax and the contents of the stomach are propelled upwards through the esophagus.

Stomach

The **stomach** is a thick-walled, J-shaped organ that lies on the left side of the body beneath the diaphragm. It is an enlarged portion of the gut that can stretch to hold about a half gallon of liquids and solids. The walls (fig. 9.8) contain three layers of muscle instead of two layers, and contraction of these muscles causes the stomach to churn and mix its contents. Hunger pangs are felt when an empty stomach churns.

The mucosa lining of the stomach contains millions of microscopic digestive glands called **gastric glands** (the term *gastric* always refers to the stomach). The gastric glands produce a gastric juice. **Gastric juice** contains pepsinogen and hydrochloric acid (HCl). When **pepsinogen** is exposed to hydrochloric acid within the stomach, it becomes the digestive enzyme pepsin. **Pepsin** is a hydrolytic enzyme that acts on protein to produce peptides:

$$\text{protein} + (n - 1)\ H_2O \xrightarrow{\text{pepsin}} n \text{ peptides}$$

Peptides vary in length, but always consist of a number of amino acids joined together (fig. 9.4). Peptides are too large to be absorbed by the gut lining. However, they are later broken down to amino acids in another part of the digestive tract.

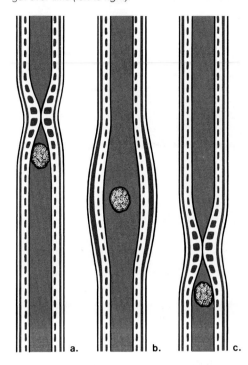

Figure 9.7
Peristalsis in the digestive tract. Rhythmic waves of muscle contraction move material along the digestive tract. The three drawings show how a peristaltic wave moves through a single section of gut over time *(left to right).*

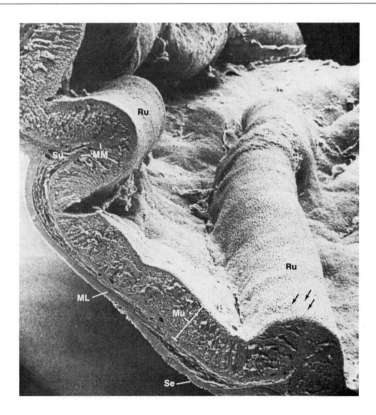

Figure 9.8
Wall of stomach. *Mu* = mucosa; *MM* = muscularis mucosa; *Su* = submucosa; *ML* = muscularis layer; *Se* = outer serosa. The wall of the stomach has folds called rugae *(Ru)* that disappear when the stomach is full. The arrows indicate openings to the gastric glands.
From: TISSUES AND ORGANS: A TEXT-ATLAS OF SCANNING ELECTRON MICROSCOPY by R. G. Kessel and R. Kardon. W. H. Freeman and Company, © 1979.

Figure 9.9

Control of gastric secretions. Especially after eating a protein-rich meal, (a.) gastrin produced by the lower part of the stomach enters the bloodstream and later, (b.) it stimulates the upper part of the stomach to produce more digestive juices.

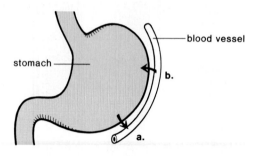

Figure 9.10

Peptic ulcer in the stomach. Normally the digestive tract produces enough mucus to protect itself from digestive juices. An ulcer often begins when an excessive amount of gastric juices is produced due to an increased amount of nervous stimulation.

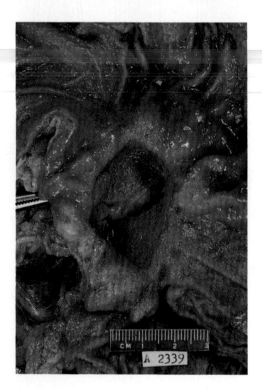

When a person has eaten a meal particularly rich in protein, the gastric glands secrete an added amount of pepsinogen; the hormone **gastrin** brings this about. **A hormone** is often a substance produced by one set of cells that affects a different set of cells, the so-called target cells. Hormones are transported by the bloodstream. Gastrin produced by the lower part of the stomach enters the bloodstream and thereby reaches the upper part of the stomach, where it causes the gastric glands to secrete more pepsinogen (fig. 9.9).

The presence of hydrochloric acid causes the contents of the stomach to have an acid pH of about 3. This low pH is beneficial in that it kills most bacteria present in food. However, HCl, although it is not an enzyme, still has a corrosive action on food and, at times, on the stomach wall itself. Normally, the wall is protected by a thick layer of mucus, but if by chance HCl does penetrate this mucus, pepsin starts to digest the stomach lining and an ulcer results. An **ulcer** (fig. 9.10) is an open sore in the wall caused by the gradual disintegration of tissues. It is believed that the most frequent cause of an ulcer is oversecretion of gastric juice due to too much nervous stimulation. Persons under stress tend to have a greater incidence of ulcers.

Normally, the stomach empties in about two to six hours. By this time, the bolus of food has become a semiliquid food mass called acid chyme. Acid chyme leaves the stomach and enters the small intestine by way of a sphincter. **Sphincters** are muscles that encircle tubes and act as valves; tubes close when sphincters contract and they open when sphincters relax.

Small Intestine

Digestion

The small intestine gets its name from its small diameter (compared to that of the large intestine). But perhaps it should be called the long intestine because it averages about 20 feet in length compared to about 5 feet for the large intestine. The first few inches of the small intestine are called the **duodenum.** Duodenal ulcers sometimes occur because the acid and pepsin within the acid chyme from the stomach may corrode and digest the internal wall in this region.

Liver and Pancreas   Two very important accessory glands, the **liver** and the **pancreas,** send secretions to the duodenum (fig. 9.1). The liver produces **bile,** which is stored in the **gallbladder** and sent by way of the bile duct to the duodenum. Bile looks green because it contains pigments that are products of hemoglobin breakdown. This green color is familiar to anyone who has observed the color changes of a bruise. Within a bruise, the hemoglobin is breaking down into the same type of pigments found in bile. However, bile also contains bile salts, which are emulsifying agents that break up fat into fat droplets:

$$\text{fat} \xrightarrow{\text{bile salts}} \text{fat droplets}$$

When fat is physically broken apart in this way and caused to mix with water, it is said to have been emulsified and an emulsion is present (fig. 1.30). These fat droplets are now ready for chemical digestion.

The pancreas sends **pancreatic juice** into the duodenum by way of the pancreatic duct (fig. 9.1). You may be more familiar with the pancreas as the source of the hormone insulin. But some other pancreatic cells produce a juice

that contains digestive enzymes and **sodium bicarbonate** ($NaHCO_3$). The latter makes pancreatic juice alkaline (pH 8.5). The alkaline pancreatic juice neutralizes the acid of the acid chyme and causes the pH of the small intestine to be basic.

Pancreatic juice contains enzymes that act on every major component of food. There is a pancreatic enzyme called **pancreatic amylase** that digests starch:

$$starch + (n - 1) H_2O \xrightarrow{\text{pancreatic amylase}} n \text{ maltose}$$

Trypsin is an example of a pancreatic enzyme that digests protein:

$$protein + (n - 1) H_2O \xrightarrow{\text{trypsin}} n \text{ peptides}$$

Trypsin is secreted as trypsinogen and changes to trypsin in the gut.

Lipase digests the fat droplets:

$$fat\ droplets + 3 H_2O \xrightarrow{\text{lipase}} n \text{ glycerol} + 3n \text{ fatty acids}$$

The digestion of fat is now complete and the molecules of glycerol and fatty acids (fig. 9.4) are small enough to be absorbed by the lining of the small intestine.

Experimental evidence has shown that the gallbladder will secrete bile and the pancreas will secrete pancreatic juice even if all nerves in the immediate area are cut. This indicates that these secretions are at least in part controlled by hormones. The duodenal wall produces hormones, the most important of which are **secretin** and **CCK (cholecystokinin)**, in response to the presence of acid chyme. Acid, especially HCl, stimulates the release of secretin, while partially digested protein and fat stimulate the release of CCK. The hormones enter the blood stream (fig. 9.11) and signal the pancreas and the gallbladder to send secretions to the duodenum.

Intestinal Glands   The wall of the small intestine contains millions of digestive glands that produce an intestinal juice. The enzymes in this juice complete the digestion of protein and carbohydrate.

Peptides, which result from the first step in protein digestion, are digested to amino acids:

$$peptides + H_2O \xrightarrow{\text{peptidases}} amino\ acids$$

Maltose, which results from the first step in starch digestion, is digested to glucose:

$$maltose + H_2O \xrightarrow{\text{maltase}} 2 \text{ glucose}$$

Other disaccharides, each of which has its own enzyme, are digested in the small intestine. The absence of any one of these enzymes may cause illness. For example, many people, including as many as 75 percent of American blacks, cannot digest lactose, the sugar found in milk, because they lack the enzyme lactase. Drinking milk often gives these individuals severe diarrhea. However, specially prepared milk containing lactose which has been enzymatically converted to glucose and lactose is now available in many stores.

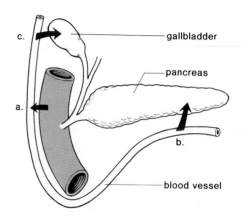

Figure 9.11
Control of intestinal secretions. *a.* Secretin and CCK produced by the duodenum enter the bloodstream. *b.* Secretin stimulates the pancreas to release digestive enzymes. *c.* CCK stimulates the gallbladder to release bile.

gallbladder

pancreas

blood vessel

Figure 9.12

Anatomy of intestinal lining. The products of digestion are absorbed by villi *(a.)*, fingerlike projections of the intestinal wall, each of which contains blood vessels and a lacteal *(b.)*. *c.* The scanning electron micrograph shows that the villi themselves are covered with microvilli *(Mv)*. *d.* A transmission electron micrograph shows that the microvilli contain microfilaments *(Mf)*. These allow limited motion of the microvilli and extend to the terminal web *(TW)*. Adjacent epithelial cells are joined by tight junctions termed zonula occludens *(ZO)*.

(c. and *d.)* From: TISSUES AND ORGANS: A TEXT-ATLAS OF SCANNING ELECTRON MICROSCOPY by R. G. Kessel and R. Kardon. W. H. Freeman and Company, © 1979.

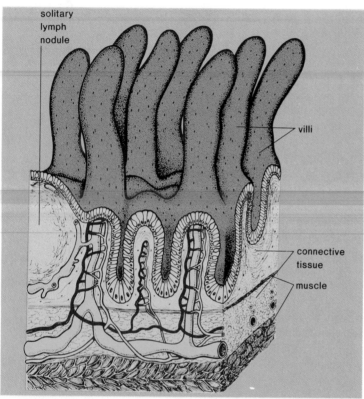

a.

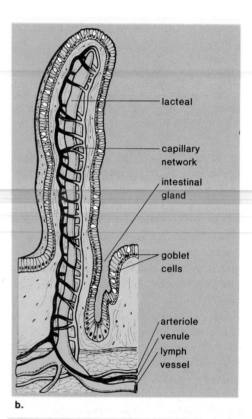

b.

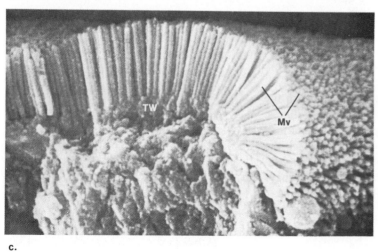

c.

d.

Human Anatomy and Physiology

Absorption

The small intestine is specialized for absorption. First, it is quite long with convoluted walls. Secondly, the absorptive surface is increased by the presence of fingerlike projections called **villi,** and the villi themselves have tiny microvilli (fig. 9.12). The huge number of villi that cover the entire surface of the small intestine give it a soft, velvety appearance. Each villus (fig. 9.12b) has an outer layer of columnar cells and contains blood vessels and a small lymph vessel called a **lacteal.** The lymphatic system is an adjunct to the circulatory system and returns fluid to the veins.

Absorption takes place across the wall of each villus, continuing until all small molecules have been absorbed. Thus absorption is an active process involving active transport of molecules across cell membranes and requiring an expenditure of cellular energy (p. 73). Sugars and amino acids cross the columnar cells to enter the blood, but glycerol and fatty acids enter the lacteals.[1]

Liver

Blood vessels from the villi merge to form the **hepatic portal vein,** which leads to the liver (fig. 9.13). The liver acts in some ways as the gatekeeper to the blood; it removes poisonous substances from the blood and works to keep the

[1]For the fate of these molecules see also pages 196–98.

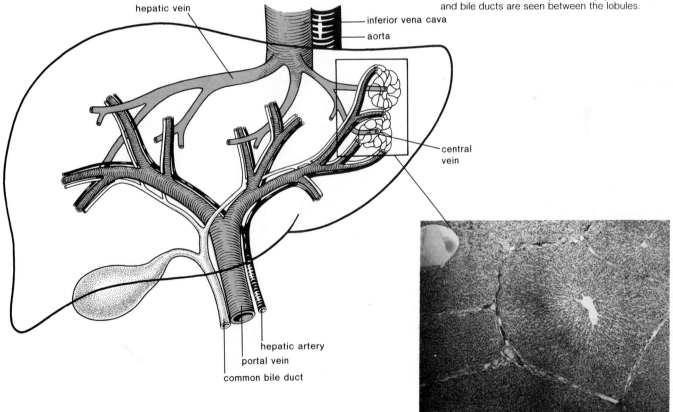

Figure 9.13
Liver anatomy. Drawing shows the blood vessels and bile ducts that serve the liver, which is composed of lobules. Illustration insert is a low-power micrograph of one lobule, surrounded by several partial lobules. The central cavity contains the central vein, a branch from the hepatic vein. Cross sections of the hepatic artery, portal vein, and bile ducts are seen between the lobules.

hepatic vein

inferior vena cava

aorta

central vein

hepatic artery

portal vein

common bile duct

contents of the blood constant. In particular, we may note that the glucose level of the blood is always about 0.1 percent even though we eat intermittently. Any excess glucose that is present in the hepatic portal vein is removed and stored by the liver as glycogen:

$$n \text{ glucose} \longrightarrow \text{glycogen} + (n-1) \text{ H}_2\text{O}$$

Between eating periods, glycogen is broken down to glucose, which enters the hepatic vein, and in this way the glucose content of the blood remains constant. It is interesting to note that glycogen is sometimes called animal starch because both starch and glycogen are made up of glucose molecules joined together (p. 55).

If by chance the supply of glycogen or glucose runs short, the liver will convert amino acids to glucose molecules:

$$\text{amino acids} \longrightarrow \text{glucose} + \text{amino groups}$$

You will recall that amino acids contain nitrogen in the form of amino groups, whereas glucose contains only carbon, oxygen, and hydrogen. Thus before amino acids can be converted to glucose molecules, **deamination,** or the removal of amino groups from the amino acids, must take place. By an involved metabolic pathway, the liver converts these amino groups to urea:

$$\underset{\text{H}_2\text{N}-\text{C}-\text{NH}_2}{\overset{\overset{\text{O}}{\|}}{}}$$

Urea is the common nitrogen waste product of humans; and after its formation in the liver, it is transported to the kidneys for excretion.

The liver also makes blood proteins from amino acids. These proteins are not used as food for cells; rather, they serve important functions within the blood itself.

Altogether we have mentioned the following functions of the liver:

1. Destroys old red blood cells and converts hemoglobin to the breakdown products in bile (bilirubin and biliverdin).
2. Produces bile that is stored in the gallbladder before entering the small intestine where it emulsifies fats.
3. Stores glucose as glycogen after eating and breaks down glycogen to glucose between eating to maintain the glucose concentration of the blood constant.
4. Produces urea from the breakdown of amino acids.
5. Makes the blood proteins.
6. Detoxifies the blood by removing poisonous substances and metabolizing them.

Liver Disorders

Jaundice   When a person is jaundiced there is a yellowish tint to the skin due to an abnormally large amount of bilirubin in the blood. In one type of jaundice called hemolytic jaundice, red blood cells are broken down in such quantity that the liver cannot excrete the bilirubin fast enough, and an extra amount spills over into the bloodstream. In obstructive type jaundice, there is an obstruction of the bile duct or damage to the liver cells and this causes an increased amount of bilirubin to enter the bloodstream.

Obstructive type jaundice often occurs when crystals of cholesterol precipitate out of bile and form gallstones, which on occasion also contain calcium carbonate. The stones may be so numerous that passage of bile along the bile duct is blocked and the gallbladder must be removed. In the meantime, the bile leaves the liver by way of the blood and a jaundiced appearance results.

Jaundice is also frequently caused by liver damage due to viral hepatitis, a term that includes two separate but similar diseases known as **infectious hepatitis** and **serum hepatitis.** Contact with fecal matter is probably the most common mode of contracting infectious hepatitis, although persons have been known to acquire the disease after eating shellfish from polluted waters, for example. Serum hepatitis is commonly spread by means of blood transfusions, kidney dialysis, and injection with inadequately sterilized needles. For this reason, drug addicts are particularly susceptible to serum hepatitis. To recover from hepatitis, a long recuperation period is commonly required, during which time the patient is in a very weakened condition. To prevent the possibility of passing on the disease, a person who has had serum hepatitis cannot give blood.

Cirrhosis   This is a chronic disease of the liver in which the organ first becomes fatty and then liver tissue is replaced by inactive fibrous scar tissue. This condition is common among alcoholics in which case it is most likely caused by the need for the liver to break down excessive amounts of alcohol. When alcohol, a two-carbon compound, is metabolized, active acetate forms. As figure 5.16 shows, active acetate molecules can be synthesized to fatty acids. To accomplish this synthesis, smooth endoplasmic reticulum increases dramatically in the liver—this may be the first step toward cirrhosis.

Large Intestine

The large intestine includes the colon and rectum. The **colon** has three parts: the ascending colon goes up the right side of the body to the level of the liver; the transverse colon crosses the abdominal cavity just below the liver and stomach; and the descending colon passes down the left side of the body to the rectum. The **rectum** is the last 6 to 8 inches of the intestinal tract. The opening of the rectum to the exterior is called the **anus.**

About 1.5 liters of water enter the digestive tract daily as a result of eating and drinking. An additional 8.5 liters enter as secretions from the various glands. About 95 percent of this water is reabsorbed by the cells that line the colon. If too little water is absorbed, diarrhea results and if too much water is absorbed, constipation occurs. These conditions are discussed in the following.

Indigestible remains finally enter the rectum. As the rectum fills, it distends until it is sufficiently stimulated to give rise to a nervous reflex called the defecation reflex (fig. 9.14). In addition to indigestible remains, **feces** also contains certain excretory substances such as bile pigments and heavy metals and large quantities of the noninfectious bacterium *E. coli.*

The large intestine normally contains a large population of bacteria that live off any substances that were not digested earlier. When they break this material down, they give off odorous molecules that cause the characteristic odor of feces. Some of the vitamins, amino acids, and other growth factors produced by these bacteria spill out into the gut and are absorbed by the gut lining. In this way *E. coli* performs a service for us.

Water is considered unsafe for swimming when the *E. coli* count reaches a certain level. This is not because *E. coli* causes disease but because a high count is an indication of the amount of fecal material that has entered the water. The more fecal material present, the greater the possibility that pathogenic or disease-causing organisms are also present.

Diarrhea and Constipation

Two common everyday complaints associated with the large intestine are diarrhea and constipation.

The major causes of **diarrhea** are infection of the lower tract and nervous stimulation. In the case of infection, the intestinal wall becomes irritated and peristalsis increases. Lack of absorption of water is a protective measure, and

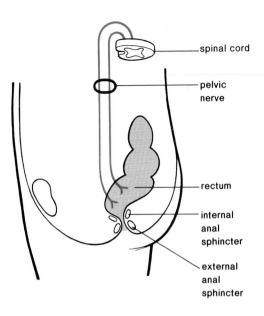

Figure 9.14
Defecation reflex. The accumulation of feces in the rectum causes it to stretch, which initiates a reflex action resulting in rectal contraction and expulsion of the fecal material.

spinal cord

pelvic nerve

rectum

internal anal sphincter

external anal sphincter

the diarrhea that results serves to rid the body of the infectious organisms. In nervous diarrhea, the nervous system stimulates the intestinal wall and diarrhea results. Loss of water due to diarrhea may lead to dehydration, a serious condition in which the body tissues lose their normal water content.

When a person is **constipated,** the feces are dry and hard. One cause of this condition is that socialized persons have learned to inhibit defecation to the point that the normal reflexes are often ignored. Two components of the diet can help to prevent constipation: water and roughage. A proper intake of water prevents the drying out of the feces. Roughage, or nondigestible plant substances, provides the bulk needed for proper elimination.

Appendicitis and Colectomy

Two other usually more serious medical conditions to be associated with the large intestine are appendicitis and colectomy.

Appendicitis The small intestine joins the large intestine in such a way that there is a blind end to one side (fig. 9.1). This blind sac, or cecum, has a small projection about the size of the little finger, called the **appendix.** In humans, the appendix is quite small, has no known function, and is vestigial, meaning that the organ is underdeveloped in humans but developed in other closely related animals. (In cows and sheep, it is considerably larger and is used as a place where cellulose is subjected to prolonged digestion.) Unfortunately, the appendix can become infected. When this happens, the individual has appendicitis, a very painful condition. It is possible for the fluid content of the appendix to rise to the point that it bursts. It is better to have the appendix removed before this occurs because it may lead to a generalized infection of the lining of the body wall.

Colectomy Cancer of the colon sometimes requires that the entire colon be removed. If so, the end of the small intestine is brought out the abdominal wall and the digestive remains are collected in a plastic bag fastened around the opening. The patient can remain healthy if vitamins that are synthesized within the large intestine are administered, and if the fluid and salt intake are regulated.

Control of Digestive Gland Secretion

The flow of digestive juices appears to be controlled by at least three types of mechanisms: reflex nervous action, conditioned reflex nervous action, and hormonal production and stimulation.

When food is present in the mouth, stomach, and small intestine, nervous impulses reach the brain, which in turn stimulates the digestive glands to secrete. This is the path of a **simple nervous reflex.**

But it is also possible for an external stimulus to influence the production of digestive juices. In such a case, it is said that the person is *conditioned* to associate this stimulus with food in the tract. Pavlov demonstrated this so-called **conditioned reflex** when he fed a dog and rang a bell at the same time. After some time, simply ringing the bell caused the dog to salivate because the dog associated the ringing of the bell with food.

We have already given examples pertaining to control of secretion by the production of hormones. Previously, we saw that the production of gastrin by the lower part of the stomach causes the gastric glands to produce additional pepsin, and that the production of secretin and CCK by the wall of the small intestine causes the pancreas and gallbladder to send their juices to the small intestine. Other hormones, not discussed here, are also involved in the control of digestive secretions.

Digestive Enzymes

The digestive process in each part of the digestive tract has been described on the preceding pages. However, it is also possible to take each type of food—protein, carbohydrate, and fat—and discuss the digestion of each. Table 9.2 lists the enzymes that are needed for each of these major components of food. As you can see, there are two enzymes for the digestion of starch to maltose: salivary amylase and pancreatic amylase. Once starch has been converted to maltose, an intestinal enzyme breaks down maltose to glucose. Glucose is absorbed into the blood by the intestinal villi.

Protein is broken down to peptides by both pepsin, an enzyme produced by the gastric glands of the stomach, and trypsin, an enzyme produced by the pancreas. Peptides must be further digested by peptidases, produced by the intestinal glands. After digestion, the released amino acids can be absorbed into the blood of the intestinal villi.

Fats are first emulsified by bile to fat droplets and then these are digested by pancreatic lipase to glycerol and fatty acids. Glycerol and fatty acids are absorbed into the lacteals of the intestinal villi.

Digestive enzymes are hydrolytic enzymes that catalyze degradation by the introduction of water at specific bonds (fig. 9.4). Digestive enzymes are no different from any other enzyme of the body. For example, they are proteins with a particular shape that fits their substrate. The lock-and-key theory of enzymatic action is described on page 97; digestive enzymes are included in this theory. Enzymes have a preferred pH because this pH maintains their shape in order that they may speed the reaction for which they are specific. Table 9.3 lists the enzymes and their preferred pH.

Table 9.2   Digestive enzymes

Reaction	Enzyme	Gland	Site of occurrence
starch + $H_2O \longrightarrow$ maltose	Salivary amylase	Salivary	Mouth
	Pancreatic amylase	Pancreas	Small intestine
maltose + $H_2O \longrightarrow$ glucose	Maltase	Intestinal	Small intestine
protein + $H_2O \longrightarrow$ peptides	Pepsin	Gastric	Stomach
	Trypsin	Pancreas	Small intestine
peptides + $H_2O \longrightarrow$ amino acids*	Peptidases	Intestinal	Small intestine
fat + $H_2O \longrightarrow$ glycerol + fatty acids*	Lipase	Pancreas	Small intestine

*Absorbed by villi.

Food is largely made up of carbohydrate (starch), protein, and fat. These very large macromolecules are broken down by digestive enzymes to small molecules that can be absorbed by intestinal villi. This chart indicates the steps needed for carbohydrate digestion (starch and maltose), protein digestion (protein and peptides), and fat digestion (fat) and shows that they are all hydrolytic reactions.

Table 9.3   Comparison of enzymes

Enzyme	Source	Optimum pH	Type of food digested	Product
Salivary amylase	Saliva	Neutral	Starch	Maltose
Pepsin	Stomach	Acid	Protein	Peptides
Pancreatic amylase	Pancreas	Alkaline	Starch	Maltose
Lipase	Pancreas	Alkaline	Fat	Glycerol; fatty acids
Trypsin	Pancreas	Alkaline	Protein	Peptides
Nucleases	Pancreas	Alkaline	RNA, DNA	Nucleotides
Peptidases	Intestine	Alkaline	Peptides	Amino acids
Maltase	Intestine	Alkaline	Maltose	Glucose

All enzymes have a preferred pH that maintains their proper shape to do their job. This table indicates the pH for each of the enzymes in table 9.2.

Figure 9.15

An experiment to demonstrate that enzymes digest food when the environmental conditions are correct. #1 lacks enzyme and no digestion occurs; #2 has too high a pH and little or no digestion occurs; #3 has the proper pH because of the presence of HCl, but still no digestion occurs because the enzyme is missing; #4 contains enzyme and the environmental conditions are correct for digestion.

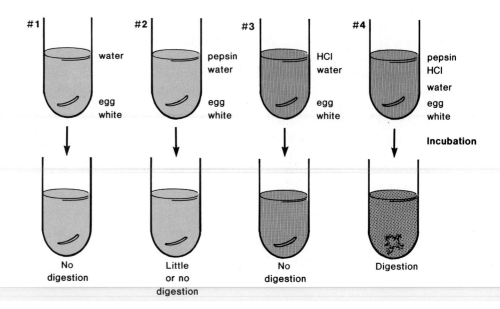

Simple laboratory experiments (fig. 9.15) can be done to show that it is enzymes that bring about the breakdown of food and not some other substance, such as hydrochloric acid (HCl) or bile. For example, the following four test tubes may be prepared in which egg white, or the protein called albumin, is to be tested for digestion:

1. Water + a small sliver of egg white
2. Pepsin + water + a small sliver of egg white
3. HCl + water + a small sliver of egg white
4. Pepsin + HCl + water + a small sliver of egg white

All tubes are now placed in an incubator at body temperature for at least an hour. At the end of this time, we can predict that tube 4 will show the best digestive action. Tube 3 does not contain the enzyme and tube 2 has too high a pH, so these two tubes will show little or no digestion. Tube 1 is a control tube and there will be no digestion in it.

This experiment can be expanded and three test tubes similar to test tube 4 can be prepared. One is kept in the cold, one at body temperature, and the last is boiled. The body temperature tube will, of course, show the best action. Like any chemical reaction, enzymatic reactions speed up if warmed; however, boiling destroys enzymes and so the boiled tube will show no digestion.

Nutrition

The four major classes of nutrients are proteins; carbohydrates; lipids; vitamins and minerals. Proteins, carbohydrates, and lipids are all utilized by the body as a source of energy.

Human Anatomy and Physiology

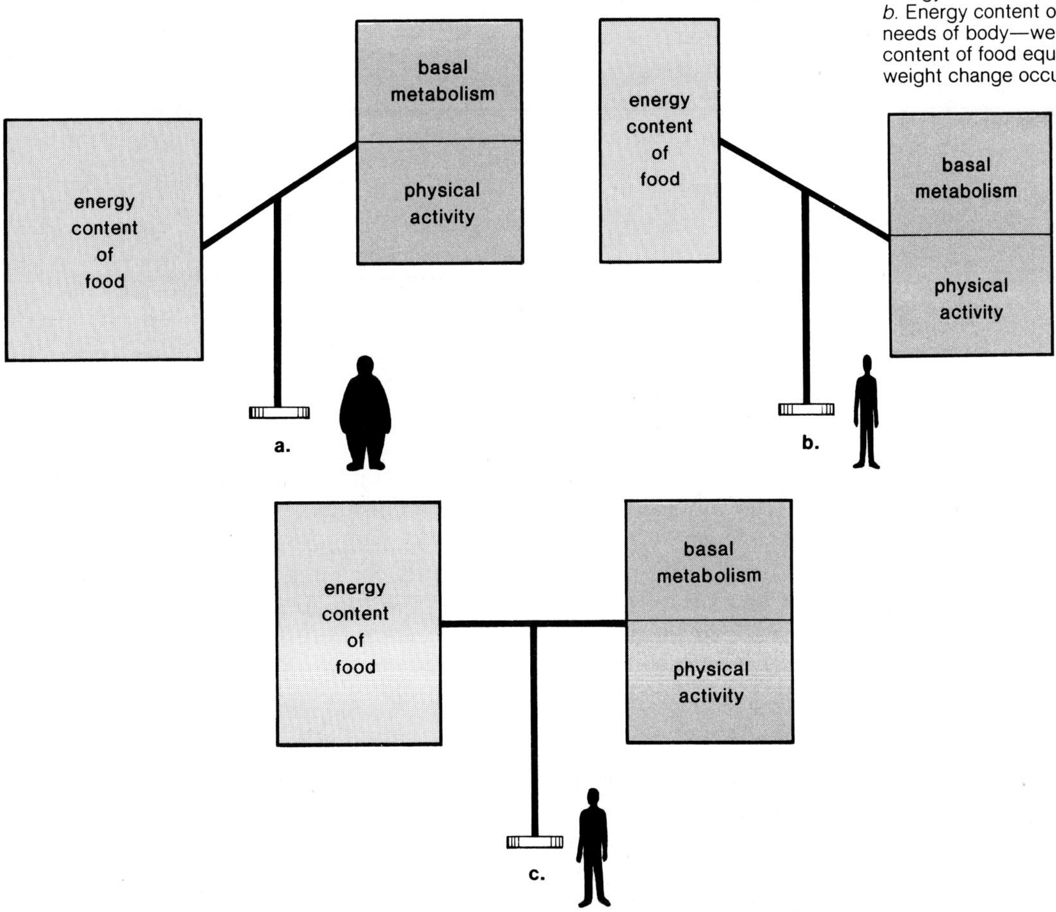

Figure 9.16
A diagram illustrating the relationship between caloric intake and weight gain or loss. In each instance, energy needs are divided between basal metabolism (basal metabolic rate) and physical activity. *a.* Energy content of food is greater than energy needs of body—weight gain occurs. *b.* Energy content of food is less than energy needs of body—weight loss occurs. *c.* Energy content of food equals energy needs of body—no weight change occurs.

Energy

Humans need energy first for basal metabolism and then for physical activities. Each individual's **basal metabolic rate** (BMR) is the amount of energy measured in calories[2] needed to maintain the body while at rest. BMR is best measured 14 hours after the last meal (because absorption of digestive products requires energy) with the subject lying down at complete physical and mental rest. BMR is usually lower for women than for men and, in general, is affected by size, shape, weight, age, activity of endocrine glands, and so forth. In addition, any type of exercise requires still more energy, which is shown by the values in table 9.4. The recommended daily intake of calories for a woman 19 to 22 years of age, height 64 inches, who is only doing light exercise each day, is 21,000C. The recommended daily intake of calories for a man 19 to 22 years of age, height 70 inches, who is only doing light exercise each day, is 29,000C.

If you use up as many calories a day as you consume in food, your weight will remain the same. If you use more calories a day than you eat, your body will break down stored fat and you will lose weight. If you use fewer calories than you eat, your body will store the excess as fat (fig. 9.16) in adipose tissue.

[2]One Calorie (capital C) is the amount of heat required to raise the temperature of 1,000 grams of water one degree centigrade, while one calorie (lowercase c) is the amount needed to raise the temperature of one gram of water one degree.

Table 9.4 Number of calories utilized by various actions

Kinds of activity	Calories (per hour)*
Walking up stairs	1100
Running (a jog)	570
Swimming	500
Vigorous exercise	450
Slow walking	200
Dressing and undressing	118
Sitting at rest	100

*Includes BMR.

Table 9.5 Suggested desirable weights for heights and ranges for adult males and females

Men

Height		Small frame	Medium frame	Large frame
Feet	Inches			
5	2	128–134	131–141	138–150
5	3	130–136	133–143	140–153
5	4	132–138	135–145	142–156
5	5	134–140	137–148	144–160
5	6	136–142	139–151	146–164
5	7	138–145	142–154	149–168
5	8	140–148	145–157	152–172
5	9	142–151	148–160	155–176
5	10	144–154	151–163	158–180
5	11	146–157	154–166	161–184
6	0	149–160	157–170	164–188
6	1	152–164	160–174	168–192
6	2	155–168	164–178	172–197
6	3	158–172	167–182	176–202
6	4	162–176	171–187	181–207

Weights at ages 25–59 based on lowest mortality. Weight in pounds according to frame (in indoor clothing weighing 5 lbs., shoes with 1″ heels).

Women

Height		Small frame	Medium frame	Large frame
Feet	Inches			
4	10	102–111	109–121	118–131
4	11	103–113	111–123	120–134
5	0	104–115	113–126	122–137
5	1	106–118	115–129	125–140
5	2	108–121	118–132	128–143
5	3	111–124	121–135	131–147
5	4	114–127	124–138	134–151
5	5	117–130	127–141	137–155
5	6	120–133	130–144	140–159
5	7	123–136	133–147	143–163
5	8	126–139	136–150	146–167
5	9	129–142	139–153	149–170
5	10	132–145	142–156	152–173
5	11	135–148	145–159	155–176
6	0	138–151	148–162	158–179

Weights at ages 25–59 based on lowest mortality. Weight in pounds according to frame (in indoor clothing weighing 3 lbs., shoes with 1″ heels).

Table 9.6 Caloric energy release

	Calories/gram
Carbohydrate	4.1
Fat	9.3
Protein	4.1

An obese person is commonly, although not always, one who weighs 20 percent more than the suggested weight for his or her height and build (table 9.5). Obesity indicates poor nutrition because the intake of food is inappropriate to the body's needs.

Carbohydrates, fats, and proteins all contribute to human energy needs. The amount of energy released upon full oxidation of 1 gram of each of these is shown in table 9.6. For the average American, carbohydrates supply about 40 percent of energy needs, fats supply about 45 percent, and protein supplies about 15 percent. Among poorer people, either in this country or other countries, carbohydrates may supply as much as 80 percent of energy needs.

Carbohydrates

The quickest, most easily available source of energy for the body is carbohydrates. Starchy foods, such as bread and potatoes, provide the largest quantity of carbohydrates, but so do meat and seafood because they contain glycogen.

As mentioned previously, all dietary carbohydrates are digested to glucose, which is stored by the liver in the form of glycogen. In between eating, the liver attempts to maintain the blood glucose level at 0.1 percent, either by breaking down glycogen or by converting amino acids to glucose (p. 190). If necessary, these amino acids are taken from the muscles, even from the heart muscle. This is how people who starve waste away. A constant supply of glucose is necessary because the brain utilizes only glucose as an energy source. Other organs can metabolize fatty acids for energy, but unfortunately this results in **acidosis,** an acid blood pH. In order to avoid this situation, it is suggested that the diet contain at least 100 grams of carbohydrate daily.[3]

Even so, foods such as candy, ice cream, sugar-coated cereals, soft drinks, and alcohol, which are rich in simple sugars, are labeled "empty calories" by some because they contribute to energy needs and weight gain without supplying any other nutritional requirements. Government agencies charged with advising the public about their dietary needs suggest that we limit our sugar intake (table 9.7).

Fats

Fats are present not only in butter and margarine but also in meat, eggs, milk, nuts, and a variety of vegetable oils. Fats from an animal origin tend to have saturated fatty acids and those from plants tend to have unsaturated fatty

[3]A slice of bread contains approximately 14 grams of carbohydrate.

Table 9.7 Dietary recommendations

The less-fat recommendations:
1. Choose as protein foods lean meat, poultry, fish, dry beans and peas. Trim fat off before you eat.
2. Eat eggs and such organ meats as liver in moderation. (Actually, these are high in cholesterol rather than fat.)
3. Broil, boil, or bake, rather than fry.
4. Limit your intake of butter, cream, hydrogenated oils, shortenings, coconut oil.

The less-salt recommendations:
1. Learn to enjoy unsalted food flavors.
2. Add little or no salt to foods at the table and add only small amounts of salt when you cook.
3. Limit your intake of salty prepared foods such as pickles, pretzels, and potato chips.

The less-sugar recommendations:
1. Eat less sweets such as candy, soft drinks, ice cream, and pastry.
2. Eat fruit which is fresh or canned fruit without heavy syrup.
3. Use less sugar—white, brown, raw—and less honey and syrups.

American Dietetic Association based on *Dietary Guidelines for Americans* 1980, U.S. Department of Agriculture and Department of Health, Education, and Welfare.

Human Anatomy and Physiology

acids. An increase in the amount of fats in the diet can greatly increase the number of calories consumed. The pad of butter or margarine on a potato contains almost as many calories as the potato. This is understandable when you compare the amount of calories derived from a gram of fat to the amount derived from a gram of carbohydrate (table 9.6).

After being absorbed, the products of fat digestion are transported by the lymph and blood to the tissues. The liver can alter ingested fats to suit the body's needs, except it is unable to produce the fatty acids linolenic and linoleic acids. Since these are required for phospholipid production, they are considered the **essential fatty acids.** Essential molecules must be present in our food because the body is unable to manufacture them.

Dietary lipids, especially saturated fatty acids and cholesterol have been found to cause circulatory difficulties, such as hypertension and heart attack due to hardening of the arteries. In a recent study, conducted by the National Heart, Lung and Blood Institute, of 3,806 men who had blood cholesterol levels of 250 milligrams per deciliter or above, the risk of heart disease was cut by 50 percent for those who, by the daily dose of a drug, lowered their blood cholesterol level by 25 percent. Not only does the American Heart Association and the governmental agencies mentioned earlier recommend that we limit our fat intake (table 9.7), they also suggest, as described in the reading on page 200, that in doing so we may be protecting ourselves from certain types of cancer.

Proteins

Foods rich in protein include meat, fish, poultry, cheese, nuts, milk, eggs, and cereals. Various legumes such as beans and peas also contain lesser amounts of protein. Following digestion of protein, amino acids enter the blood and are transported to the tissues. Some are incorporated into structural proteins and some are used to synthesize such proteins as hemoglobin, plasma proteins, enzymes, and hormones.

Except for nine amino acids, cells have no difficulty in changing one type of amino acid into another type. These nine amino acids are called the **essential amino acids** because they must be provided in the diet. Some protein sources, which include those available from animal sources, are **complete proteins**—they contain adequate amounts of the amino acids essential to maintaining body tissues and promoting normal growth and development. Other protein sources, such as those from plant sources, are **incomplete proteins** and are not able to maintain body tissue or promote normal growth. Since all of the essential amino acids must be present before protein synthesis is possible, every culture seems to have evolved its own mixture of complementary foods. For example in the Middle East, wheat bread, which lacks adequate levels of the amino acid lysine, is eaten with cheese, which has a high lysine content. Mexicans eat beans and rice, Jamaicans eat rice and peas, and Americans eat breakfast cereals with milk. In less prosperous countries, the people usually subsist on diets primarily composed of grains and vegetables. Even so, it is often difficult to provide all persons with sufficient dietary protein. In such instances, small children are observed to suffer a protein deficiency, especially after weaning. The malady, known as **kwashiorkor** (fig. 9.17), develops due to a lack of protein even if caloric intake is adequate. Unfortunately, even when a complete protein source is given, recovery is often marked by mental retardation.

Figure 9.17
Child with kwashiorkor. The swollen abdomen is caused by edema due to the lack of plasma proteins in the blood. Protein deficiency is the most common form of malnutrition in poorer countries.

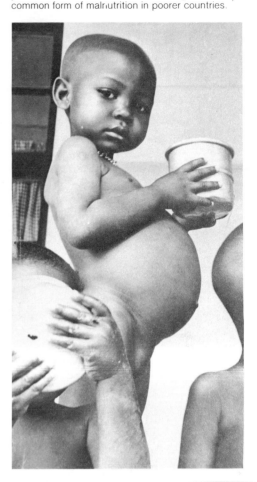

Table 9.8 Vitamins and minerals—their best food sources

Fat-soluble vitamins	Best food sources
A	Liver, milk and milk products, sweet potatoes, carrot, spinach, cantaloupe, squash, broccoli, apricots.
D	Sun on skin, fish liver oils, sardines, salmon, milk and milk products, egg yolk.
E	Vegetable oils, margarine.
K	Green vegetables, tomato.
Water-soluble vitamins	
C	Citrus, strawberry, and other fruits, broccoli, potato.
Thiamin (B-1)	Liver, pork products, peas, legumes, whole grain and enriched breads and cereals.
Riboflavin (B-2)	Liver, milk, beef, pork, chicken, egg, cheese, broccoli, salmon, whole grain and enriched breads and cereals.
Niacin (B-3)	Enriched flour, red meat, poultry, peanut butter, whole grain and enriched breads and cereals.
Pyridoxine (B-6)	Liver, pork, red meat, whole grains, vegetables.
Folacin	Liver, kidney, yeast, green leafy vegetables.
B-12	Organ meats, egg, fish, milk.
Biotin	Yeast, organ meats, egg yolk, whole grains.
Pantothenic acid	Organ meats, red meat, eggs.
Inositol	Liver, yeast, cereals, fruit.
Choline	Egg yolk, cereals, vegetables.
Minerals	
Calcium	Milk and milk products, salmon, legumes, broccoli, orange, sweet potato, lettuce.
Phosphorus	Milk and milk products, whole grains.
Magnesium	Seeds, nuts, whole grains, milk, egg, fish, red meat.
Iron	Liver, red meat, chicken, legumes, spinach, peas, prune, apricot, tomato juice, whole grain enriched breads and cereals.
Zinc	Meat, liver, eggs, oysters.
Iodine	Seafood, iodized salt.
Copper	Oysters, nuts, organ meats, corn oil, legumes, raisins.
Manganese	Nuts, whole grains.
Fluoride	Drinking water, fish, legumes.
Chromium	Yeast, red meat, cheese, whole grains.
Selenium	Seafood, organ meats, red meat.
Molybdenum	Meat, grain, legumes.
Sodium	Salt, cured meats and vegetables.
Potassium	Bananas, citrus fruits.
Chloride	Salt.
Sulfur	Protein foods.
Cobalt	Red meat.

The combination of grains and vegetables just mentioned are efficient means of providing dietary protein because, as a rule of thumb, humans store only about 10 percent of the calories available in food. Thus, 100 calories from grain results in the storage of 10 human calories, but if this grain is fed to cattle it ultimately provides only one human calorie for storage. Nevertheless, in this country the practice of fattening cattle in feedlots where they are fed a rich diet of grain is common because the cattle fatten quickly and the resulting high-cholesterol, fatty beef has been preferred by consumers. This practice may change, however; first, because it requires more fossil fuel energy and, second, because the public is becoming more sensitive to their health needs.

Vitamins and Minerals

Vitamins are organic compounds (other than carbohydrates, lipids, and proteins) that the body is unable to produce and therefore must be present in the diet. Table 9.8 lists some of the more important vitamins and minerals and the best sources of food for each one. Various symptoms develop when vitamins are lacking in the diet (fig. 9.18).

Human Anatomy and Physiology

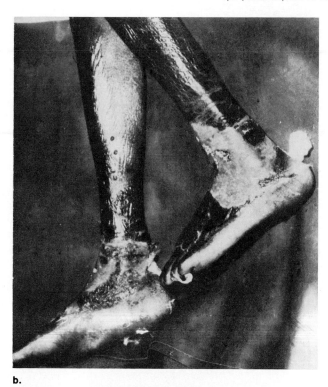

Figure 9.18
Illnesses due to vitamin deficiency. *a.* Bowing of bones (rickets) due to vitamin D deficiency. *b.* Dermatitis of areas exposed to light (pellagra) due to niacin deficiency. *c.* Bleeding of gums (scurvy) due to vitamin C deficiency. *d.* Fissures of lips (cheilosis) due to riboflavin deficiency.

a.

b.

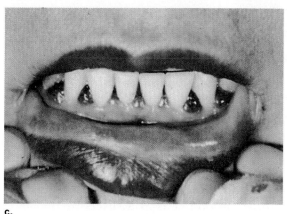

c.

d.

Although vitamins are an important part of a balanced diet, they are required only in very small amounts. As was discussed in chapter 5, many vitamins are portions of coenzymes, or enzyme helpers. For example, niacin is part of the coenzyme NAD, and riboflavin is part of FAD. Coenzymes are needed in only small amounts because each one can be used over and over again. This means that the daily requirement for vitamins is low and a properly balanced diet usually provides the amount needed. However, synthetic vitamins perform the same functions in the body as do those present in food since they are identical chemicals.

The intake of excess vitamins can possibly lead to illness. For example, excess vitamin C is converted to a product, oxalic acid, that is toxic to the body. Vitamin A taken in excess over long periods can cause such effects as

Cancer Diet

Researchers have found that fresh fruits and vegetables seem to guard against cancer.

Health food faddists have been saying it for years: eat right and you may reduce the risk of getting cancer. At a press conference in Washington last week, the National Academy of Sciences not only signaled its agreement with that view, it issued a 500–page report specifying just what is meant by eating right. "Our committee's recommendations should not be regarded as assuring a cancer-free life," said Dr. Clifford Grobstein, University of California biologist and chairman of the NAS panel that released the two-year study, but "by controlling what we eat we may prevent diet-sensitive cancer." Among the recommendations:

Cut fat consumption by 25% or more, both the saturated variety of fat found in meat and whole milk products and unsaturated lipids like those in vegetable oils. Animal tests and human population studies have shown a strong correlation between fat intake and rates of cancer of the breast, colon and prostate.

Eat less smoked, pickled and salt-cured foods, including sausages, smoked fish and bacon. In Japan, China and Iceland, where such foods are frequently consumed, there is a higher incidence of cancers of the stomach and esophagus. These foods also tend to contain nitrosamines and polycyclic aromatic hydrocarbons, chemicals known to cause cancer in animals.

Eat more fruits and vegetables containing vitamin C (oranges, broccoli and tomatoes) and beta-carotene, a precursor of vitamin A found in squash, carrots and other yellow and green vegetables. Both substances inhibit the formation of chemically induced cancers in laboratory tests; both are associated with lower cancer rates in human populations. The committee counseled against high-dose vitamin pills because of insufficient evidence about their health benefits. High doses of vitamin A, it added, can be toxic.

The committee advised Americans to drink only "moderate" amounts of alcohol, although it did not specify how much. Alcohol consumption, particularly when combined with smoking, has been linked to mouth, larynx, liver and lung cancers. Panel members were not, however, able to confirm reports that dietary fiber reduces the risk of bowel cancer. Nor was the evidence sufficient to convince them of the prophylactic benefits of vitamin E or the perils of preservatives, food dyes and other chemical additives.

The report drew criticism from the National Cattlemen's Association, which labeled it "inconclusive and premature," and from the American Meat Institute, which said it was based on "insufficient evidence." Grobstein acknowledged that his panel was "exploring a relationship between two still largely unknowns," but he added: "I don't think we're disseminating unproven theories."

loss of hair, bone and joint pains, and loss of appetite. Excessive vitamin D can cause an overload of calcium in the blood, which in children leads to loss of appetite and retarded growth. Megavitamin therapy should always be supervised by a physician.

In addition to vitamins, various **minerals** are also required by the body. Minerals are divided into the macrominerals that are needed in gram amounts per day and the trace elements that are needed in only microgram amounts per day. The macrominerals sodium, magnesium, phosphorus, chlorine, potassium, and calcium serve as constituents of cells and body fluids and as structural components of tissues. For example, calcium is needed for the construction of bones and teeth and also for nerve conduction and muscle contraction. The trace elements seem to have very specific functions. For example,

Human Anatomy and Physiology

iron is needed for the production of hemoglobin, and iodine is used in the production of thyroxin, a hormone produced by the thyroid gland. As research continues, more and more elements have been added to the list of those considered essential. During the past three decades, molybdenum, selenium, chromium, nickel, vanadium, silicon, and even arsenic have been found to be essential to good health in very small amounts.

Food Preparation

As the American public has come to depend more and more on processed and convenience foods, the use of **food additives** has increased until now it is estimated that 26,000 different chemicals are deliberately added to foods. Some of these are natural substances, such as salt and spices, while many others are artificial, such as synthetically produced preservatives, sweetners, and dyes. All additives must be approved by the Food and Drug Administration (FDA), which is responsible for determining the safety and effectiveness of such products. The FDA has taken some additives off the market because they have been found to produce cancer in animals. About half of the food additives banned by the FDA have been food colorings made from coal-tar dye.

It has been very difficult to establish a standardized method by which to deal with suspect chemicals. The Food Safety Council, a private industry/consumer group, after meeting for six years, finally issued some suggestions in 1982. The council felt that tests should be run for each chemical in order to establish a dose that was most likely safe for that chemical; then a safety factor should be subtracted from this in order to determine the maximum dose considered acceptable for human exposure. The council acknowledged that it would sometimes be necessary to make a risk-benefit determination. For example, there is evidence to suggest that some preservatives are not altogether safe and yet their use protects the public from very dangerous bacterial infections. In such instances it may be more beneficial in the long run to continue to use the preservative.

There is one additive that people themselves add to food that they might be well advised to eliminate or reduce. This additive is salt, or sodium chloride. The nutritional requirement for sodium is in the range of .1–.2 grams per day, which is equivalent to .25–.5 grams of salt per day. The average intake of salt is twenty times the nutritional requirement. There is evidence to suggest that excess salt in the diet promotes high blood pressure in susceptible individuals. Hypertension is absent in some nonindustrialized populations, such as those of the Solomon Islands, the Amazon basin, and the Coco Islands of Polynesia where the salt intake is about 2 grams per day. Americans receive about 3 grams of salt per day even if they add no salt at all to their food either during cooking or at the table; this is due to modern methods of commercial food processing. It is recommended that we try to eliminate salty foods from our diets (table 9.7).

Summary

The human gut consists of the mouth, pharynx, esophagus, stomach, small intestine, and large intestine. Only these structures actually contain food, but the salivary glands, liver, and pancreas supply substances that aid in the digestion of food. In the mouth, food is chewed and acted upon by salivary amylase before it is swallowed. Peristaltic action then moves the food along the esophagus to the stomach, a J-shaped muscular organ. Here the gastric glands produce HCl, an acid, and precursor of pepsin, an enzyme that breaks down

protein. Both the stomach and the first part of the small intestine are subject to ulcers because of the caustic effect of HCl on the gut lining. In the duodenum, acid chyme from the stomach stimulates the production of secretin and CCK. Following this, both pancreatic juice and bile, which is made by the liver and stored in the gallbladder, enter the small intestine.

In the small intestine, fat is emulsified by bile salts to fat droplets before being acted upon by pancreatic lipase. Protein is digested by pancreatic trypsin, and starch is digested by pancreatic amylase. The intestinal wall contains digestive glands that secrete intestinal enzymes that finish the digestion of proteins and carbohydrates. The action of digestive enzymes is summarized in table 9.1.

The walls of the small intestine have fingerlike projections called villi, within which are blood capillaries and a lymphatic lacteal. Amino acids and glucose enter the blood; glycerol and fatty acids enter the lymph. The blood from the small intestine moves into the hepatic portal vein, which goes to the liver, an organ that monitors blood composition and, for example, maintains a constant level of glucose in the blood.

The only material that passes from the small intestine to the large intestine is nondigestible material. The large intestine absorbs water from this material and contains a large population of bacteria that can use it as food. In the process, the bacteria produce vitamins that can be absorbed and used by our bodies. They also produce the odor characteristic of feces, which pass out of the body during defecation, a process controlled by reflex action. Feces contain nondigestible substances and bacteria. Diarrhea occurs when water is not absorbed, due to infection or nervousness; constipation results when too much water is absorbed by the large intestine.

Control of the secretion of digestive enzymes involves three mechanisms: (1) simple reflex action takes place when food is present in the mouth or gut; (2) conditioned reflex action occurs once we associate food with a particular stimulus other than food; and (3) hormonal production and stimulation is exemplified by the production of gastrin by the lower part of the stomach and the production of secretin and CCK by the wall of the duodenum.

Digestive enzymes function according to the lock-and-key theory of enzymatic action and have the usual enzymatic properties. They are specific to their substrate and speed up at body temperature and proper pH. They are destroyed by boiling.

A balanced diet is required for good health. Food should provide us with all necessary vitamins, amino acids, fatty acids, and an adequate amount of energy. Both the energy needs of the body and the energy value of food are expressed in terms of Calories. A comparison of these two helps one decide if weight loss or weight gain will occur. If the caloric intake is greater than the amount expended for BMR and exercise, weight gain will occur. If the intake is less than expended, weight loss will occur.

Study Questions

1. Name the parts of the digestive tract; anatomically describe them and state the contribution of each to the digestive process. (pp. 180–91)
2. Name the accessory glands and describe the part that they play in the digestion of food. (pp. 186–87)
3. Name six functions of the liver. (p. 190) How does the liver maintain a constant glucose level in the blood? What is jaundice? Cirrhosis of the liver? (pp. 190–91)

Human Anatomy and Physiology

4. What is the common intestinal bacterium? What do these bacteria do for us? (p. 191)
5. Discuss the three ways in which secretion of digestive enzymes is controlled. (p. 192)
6. What are gastrin, secretin, and CCK? Where are they produced? What are their functions? (pp. 186–87)
7. Discuss the digestion of starch, protein, and fat, listing all the steps that occur to bring about digestion of each of these. (p. 193)
8. Describe an experiment that shows that pepsin digests protein and that it has a preferred pH. Describe an experiment that would show that trypsin digests protein and that it has a preferred pH. (p. 194)
9. What factors determine how many calories should be ingested? (p. 195)
10. Give reasons why carbohydrates, fats, proteins, vitamins, and minerals are all necessary to good nutrition. (pp. 196–200)

Selected Key Terms

palate (pal′at)
amylase (am′i-lās)
pharynx (far′ingks)
glottis (glot′is)
esophagus (ĕ-sof′ah-gus)
peristalsis (per″ĭ-stal′sis)
gastric (gas′trik)
pepsin (pep′sin)
duodenum (du″o-de′num)
bile (bīl)
trypsin (trip′sin)
lipase (li′pās)
secretin (se-kre′tin)
villi (vil′i)
lacteal (lak′te-al)
hepatic (hĕ-pat′ik)
urea (u-re′ah)
colon (ko′lon)
rectum (rek′tum)
vitamin (vi′tah-min)

Further Readings

Davenport, H. W. 1981. *Physiology of the digestive tract.* 5th ed. Chicago: Yearbook Medical Publishers.
———. 1972. Why the stomach does not digest itself. *Scientific American* 226(1):86.
Fitch, K., and Johnson, P. 1977. *Human life science.* New York: Holt, Rinehart and Winston.
Human nutrition: Readings from Scientific American. 1978. San Francisco: W. H. Freeman.
Kappas, A., and Alvares, A. P. 1975. How the liver metabolizes foreign substances. *Scientific American* 232(6):22.
Lieber, C. S. 1976. The metabolism of alcohol. *Scientific American* 234(3):25.
Robinson, C. H., and Weigley, E. S. 1978. *Fundamentals of normal nutrition.* 3d ed. New York: Macmillan.
Scrimshaw, N. S., and Young, V. R. 1976. The requirements of human nutrition. *Scientific American* 235(3):50.

10

Circulation

Chapter concepts

1 In human beings, the blood, kept in motion by the pumping of the heart, circulates through a series of vessels.

2 The heart is actually a double pump: the right side pumps blood to the lungs and the left to the rest of the body.

3 The lymph vessels form a one-way lymphatic system that transports lymph from the tissues to certain cardiovascular veins.

4 While the circulatory system is very efficient, it is still subject to various degenerative illnesses.

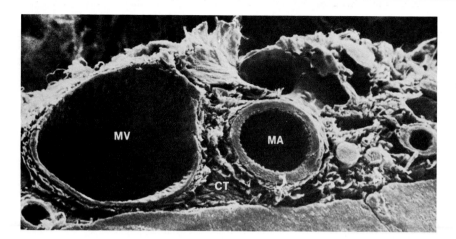

Figure 10.1
Scanning electron micrograph of a major artery
(MA) and a major vein *(MV)* separated by
connective tissue *(CT)*. Notice that the wall of the
artery is much thicker than that of the vein.
From: TISSUES AND ORGANS: A TEXT-ATLAS OF
SCANNING ELECTRON MICROSCOPY by R. G. Kessel
and R. Kardon. W. H. Freeman and Company, © 1979.

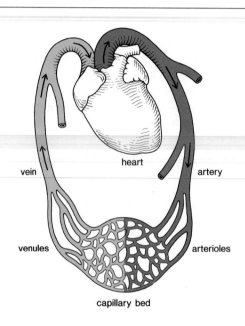

Figure 10.2
Diagram illustrating the path of blood. Blood
leaving the heart moves from an artery to arterioles
to capillaries to venules and then returns to the
heart by way of a vein. Thus arteries are vessels
that take blood away from the heart and veins are
vessels that return blood to the heart.

The heart and blood vessels form a closed circulatory system, meaning that
the blood is always contained within a series of vessels and never runs free.
The heart pumps the blood and keeps it moving within this closed system.
Circulation of the blood is so important that if the heart discontinues beating
for only a few minutes, death will result. Although the amount of blood pumped
with each beat is small (2 to 4 ounces), the heart beats about 70 to 75 times
a minute; within a minute, it pumps 10 to 12 pints of blood, or the entire
amount of blood contained within the blood vessels of an adult's body.

Cardiovascular System

Blood Vessels
The blood vessels are arranged so that they continually carry blood from the
heart to the tissues and then return it from the tissues to the heart. Blood
vessels (fig. 10.1) are of three types: the **arteries** (and **arterioles**) carry blood
away from the heart; the **capillaries** exchange material with the tissues; and
the **veins** (and **venules** return blood to the heart [fig. 10.2]).

Arteries and Arterioles

Arteries have thick walls (fig. 10.3) because, in addition to an inner endothelium layer and an outer connective tissue layer, they have a thick middle layer of elastic and muscle fibers. The elastic fibers enable an artery to expand and accommodate the sudden increase in blood volume that results after each heartbeat. Arterial walls are so thick that the walls themselves are supplied with blood vessels. The **arterioles** are small arteries just visible to the naked eye. The middle layer of these vessels has some elastic tissue but is composed mostly of smooth muscle whose fibers encircle the arteriole. The contraction of the smooth muscle cells is under the control of the sympathetic portion of the autonomic nervous system. If the muscle fibers contract, the bore of the arteriole gets smaller; if the fibers relax, the bore of the arteriole enlarges. Whether arterioles are constricted or dilated affects blood pressure. The greater the number of vessels dilated, the lower the blood pressure.

Capillaries

Arterioles branch into small vessels called capillaries. Each one is an extremely narrow, microscopic tube with a wall composed of only one layer of endothelial cells (fig. 10.3). **Capillary beds** (a network of many capillaries) are present in all regions of the body; consequently, a cut to any body tissue draws blood. The capillaries are the most important part of a closed circulatory system because exchange of nutrient and waste molecules takes place across their thin walls. Oxygen and glucose diffuse out of a capillary into the tissue fluid that surrounds cells, while carbon dioxide and ammonia diffuse into the capillary (fig. 11.8). Since it is the capillaries that serve the needs of the cells, the heart and other vessels of the circulatory system can be thought of as a means by which blood is conducted to and from the capillaries.

Not all capillary beds (fig. 10.4) are open or in use at the same time. After eating, the capillary beds of the digestive tract are usually open; during muscular exercise, the capillary beds of the skeletal muscles are open. Most capillary beds have a "thoroughfare channel" that allows blood to move directly from arteriole to venule when the capillary bed is closed. There are sphincter muslces that encircle the entrance to each capillary. These are constricted, preventing blood from entering the capillaries, when the bed is closed and are relaxed when the bed is open. As would be expected, the larger the number of capillary beds open, the lower the blood pressure.

Veins and Venules

Veins and venules take blood from the capillary beds to the heart (fig. 10.2). First, the **venules** drain the blood from the capillaries and then join together to form a vein. The wall of a venule is much thinner than that of an arteriole or artery because the middle layer of muscle and elastic fibers is poorly developed (fig. 10.3). Within some veins, especially in the major veins of the arms and legs, there are **valves** (fig. 10.14) that allow blood to flow only toward the heart when they are open and prevent the backward flow of blood when they are closed.

At any given time, more than half of the total blood volume is found in the veins and venules. If a loss of blood occurs, for example due to hemorrhaging, nervous stimulation causes the veins to constrict, providing more blood to the rest of the body. In this way, the veins act as a blood reservoir.

Human Anatomy and Physiology

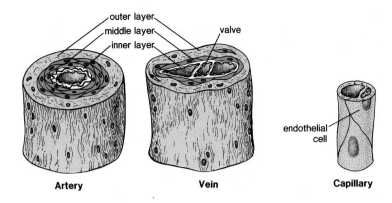

outer layer
middle layer
inner layer
valve
endothelial cell

Artery   **Vein**   **Capillary**

Figure 10.3
A comparison of artery, vein, and capillary structures shows that arteries have strong walls while veins have weak walls, largely due to the difference in size of the middle layer, which is composed of smooth muscle and connective tissue. Capillaries are much smaller, with walls one cell thick.

Figure 10.4
Anatomy of a capillary bed. Capillary beds form a matrix of vessels that lie between an arteriole and a venule. *a*. Sphincter muscles are found at the junctions between an arteriole and capillaries. When these are contracted, the capillary bed is closed. Blood moves from the arteriole to the venule by way of a thoroughfare channel. *b*. When a capillary bed is open, blood moves freely in the matrix of vessels making up the bed. If all capillary beds were open at the same time, an individual would suffer very severe low blood pressure.

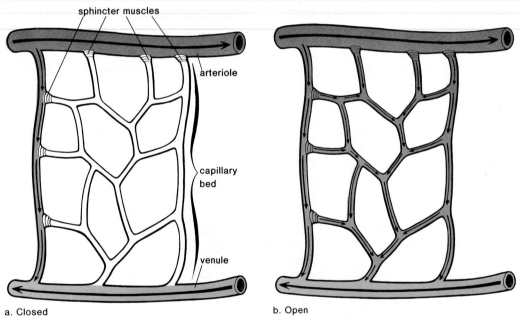

sphincter muscles

arteriole

capillary bed

venule

a. Closed   b. Open

Figure 10.5

External heart anatomy. The coronary arteries bring oxygen and nutrients to the heart muscle. The individual suffers a heart attack should they fail to do so.

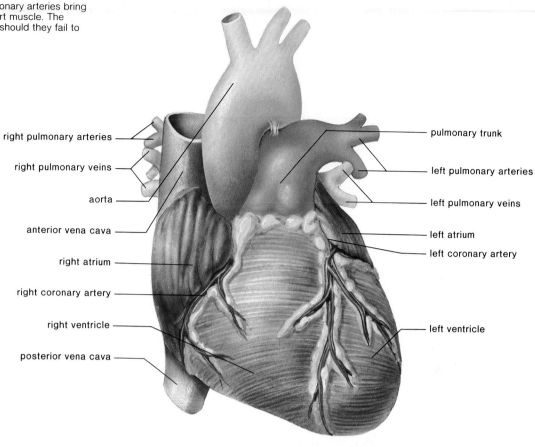

right pulmonary arteries

right pulmonary veins

aorta

anterior vena cava

right atrium

right coronary artery

right ventricle

posterior vena cava

pulmonary trunk

left pulmonary arteries

left pulmonary veins

left atrium

left coronary artery

left ventricle

Figure 10.6

Internal view of heart. The chordae tendineae support the atrioventricular valves and keep them from inverting during ventricle contraction.

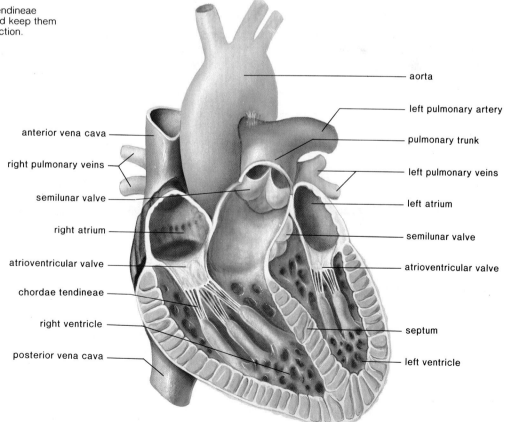

anterior vena cava

right pulmonary veins

semilunar valve

right atrium

atrioventricular valve

chordae tendineae

right ventricle

posterior vena cava

aorta

left pulmonary artery

pulmonary trunk

left pulmonary veins

left atrium

semilunar valve

atrioventricular valve

septum

left ventricle

Heart

The **heart** (fig. 10.5) is a cone-shaped, muscular organ, about the size of a fist. It is located between the lungs, directly behind the sternum, and is tilted so that the apex is directed to the left. The major portion of the heart is called the **myocardium,** which consists largely of **cardiac muscle** tissue. The muscle fibers within the myocardium are branched and joined to one another so tightly that, prior to studies with the electron microscope, it was thought that they formed one continuous muscle; now it is known that there are individual fibers. The inner surface of the heart is lined with endothelial tissue called **endocardium,** which resembles squamous epithelium. The outside of the heart is covered with an epithelial tissue called **pericardium,** which forms a sac called the pericardial sac, within which the heart is located. Normally, this sac contains a small quantity of liquid to lubricate the heart.

Internally (fig. 10.6), the heart has a right and left side, separated by the **septum.** The heart has four chambers: two upper, thin-walled **atria** (singular, atrium), sometimes called auricles, and two lower, thick-walled **ventricles.** The atria are much smaller than the strong, muscular ventricles.

The heart also has valves that direct the flow of blood and prevent a backflow. The valves that lie between the atria and ventricles are called the **atrioventricular valves.** The valves are supported by strong fibrous strings called **chordae tendineae.** These cords, which are attached to muscular projections of the ventricular walls, support the valves and prevent them from inverting. The atrioventricular valve on the right side is called the tricuspid valve because it has three cups, or flaps; and the valve on the left side is called the bicuspid, or mitral, because it has two flaps. There are also **semilunar valves,** which resemble half moons, between the ventricles and their attached vessels.

Double Pump

Figure 10.7 indicates that the right side of the heart sends blood through the lungs, and the left side sends blood throughout the body. Thus there are actually two circular paths of the blood: (1) from the heart to the lungs and back to the heart and (2) from the heart to the body and back to the heart. The right side of the heart is a pump for the first of these circuits, and the left side of the heart is a pump for the second. Thus the heart is a double pump. Since the left ventricle has the harder job because it pumps blood to all the body, its walls are much thicker than those of the right ventricle.

Heartbeat and Heart Sounds

From this description of the path of blood through the heart, it might seem that the right and left side of the heart beat independently of one another, but actually they contract together. First, the two atria contract simultaneously; then the two ventricles contract at the same time. The word **systole** refers to contraction of heart muscle, and the word **diastole** refers to relaxation of heart muscle; thus, atrial systole is followed by ventricular systole. The heart contracts, or beats, about 70 times a minute and each heartbeat lasts about 0.85 second. Each heartbeat, or **cardiac cycle** (fig. 10.8), consists of the following elements:

Time	Atria	Ventricles
0.15 sec.	systole	diastole
0.30 sec.	diastole	systole
0.40 sec.	diastole	diastole

This shows that while the atria contract, the ventricles relax, and vice versa, and that all chambers rest at the same time for 0.40 second. The short systole of the atria is appropriate since the atria send blood only into the ventricles.

Figure 10.7
Diagram of pulmonary and systemic systems. The blue-colored vessels carry deoxygenated blood, while the red-colored vessels carry oxygenated blood. Notice that the blood cannot move from the right side of the heart to the left side without passing through the lungs.

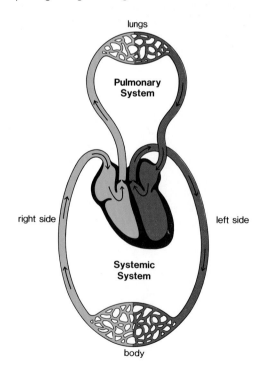

lungs

Pulmonary System

right side   left side

Systemic System

body

Figure 10.8

Stages in the cardiac cycle. *a.* When the heart is relaxed, both atria and ventricles are filling with blood. *b.* When the atria contract, the ventricles are relaxed and filling with blood. *c.* When the ventricles contract, the atrioventricular valves are closed, the semilunar valves are open, and blood is pumped into the pulmonary artery and aorta.

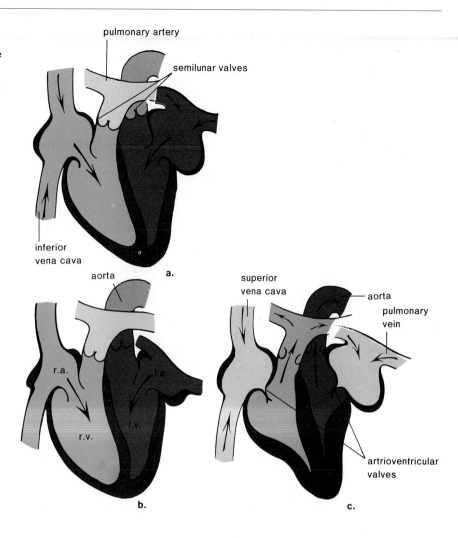

It is the muscular ventricles that actually pump blood out into the circulatory system proper. When the word *systole* is used alone, it usually refers to the left ventricular systole.

When the heart beats, the familiar lub-DUPP sound may be heard as the valves of the heart close. The *lub* is caused by vibrations of the heart when the atrioventricular valves close, and the *DUPP* is heard when vibrations occur due to the closing of the semilunar valves. Heart murmurs, or a slight slush sound after the lub, are often due to ineffective valves that allow blood to pass back into the atria after the atrioventricular valves have closed. Rheumatic fever resulting from a strep infection is one cause of a faulty valve, particularly the mitral valve. If operative procedures are unable to restructure the valve, it may be replaced by an artificial valve.

The beat of the heart is *intrinsic,* meaning the heart will beat independently of any nervous stimulation. In fact, it is possible to remove the heart of a small animal, such as a frog's heart, and watch it undergo contraction in a petri dish. The reason for this lies in the fact that there is a unique type of tissue called **nodal** tissue, with both muscular and nervous characteristics, located in two regions of the heart. The first of these, the **S-A (sinoatrial) node,** is found in the upper dorsal wall of the right atrium; the other, the **A-V (atrioventricular) node,** is found in the base of the right atrium very near the septum (fig. 10.9). The S-A node, or the **pacemaker,** initiates the heartbeat and automatically sends out an excitation impulse every 0.85 second to cause the

Human Anatomy and Physiology

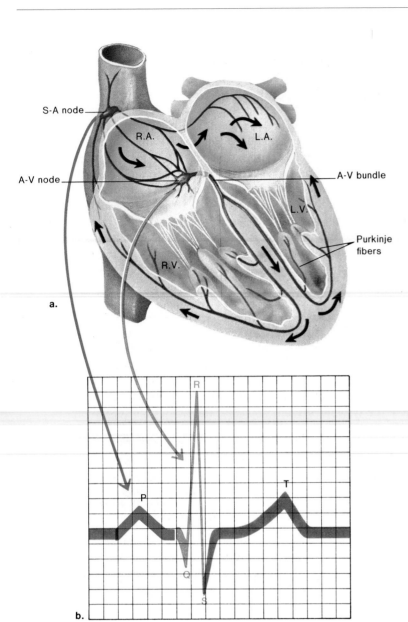

S-A node

R.A.

L.A.

A-V node

A-V bundle

L.V.

Purkinje fibers

R.V.

a.

R

P

T

Q

S

b.

Figure 10.9
Control of the heart cycle. *a*. The *S-A* node sends out a stimulus that causes the atria to contract. When this stimulus reaches the *A-V* node it signals the ventricles to contract by way of the Purkinje fibers. *b*. A normal EKG indicates that the heart is functioning properly. The *P* wave indicates that the atria have contracted; the *QRS* wave indicates that the ventricles have contracted; and the *T* wave indicates that the ventricles are recovering from contraction.

atria to contract. When the impulse reaches the A-V node, it signals the ventricles to contract by way of specialized fibers called Purkinje fibers. The S-A node is called the pacemaker because it usually keeps the heartbeat regular. If the S-A node fails to work properly, the heart will still beat, but irregularly. To correct this condition, it is possible to implant in the body an artificial pacemaker that automatically gives an electric shock to the heart every 0.85 second. This causes the heart to beat regularly again.

The rate of the heartbeat is also under nervous control. There is a heart-rate center in the medulla oblongata (p. 316) of the brain that can alter the beat of the heart by way of the autonomic nervous system (p. 313). The latter is made up of two divisions: the parasympathetic, which promotes those functions we tend to associate with normal activities, and the sympathetic system, which brings about those responses we associate with times of stress. For example, the parasympathetic system causes the heartbeat to slow down and the sympathetic system increases the heartbeat. Various factors, such as the relative need for oxygen or the blood pressure level, determine which of these systems becomes activated.

Vascular System

The vascular system, which is represented in figure 10.10, can be divided into two systems: the **pulmonary system,** which circulates blood through the lungs, and the **systemic system,** which serves the needs of the body's tissues.

Pulmonary System

The path of blood through the lungs can be traced as follows. Blood from all regions of the body first collects in the right atrium and then passes into the right ventricle, which pumps it into the pulmonary trunk. The pulmonary trunk divides into the pulmonary arteries, which divide up into the arterioles of the lungs. The arterioles take blood to the pulmonary capillaries where carbon dioxide and oxygen are exchanged. The blood then enters the pulmonary venules that lead back through the pulmonary veins to the left atrium. Since the blood in the **pulmonary arteries** is low in oxygen while the blood in the **pulmonary veins** is high in oxygen, it is not correct to say that all arteries carry blood high in oxygen and all veins carry blood low in oxygen. It is just the reverse in the pulmonary system.

Systemic System

The systemic system includes all the other arteries and veins shown in figure 10.10. The largest artery in the systemic system is the **aorta,** and the largest veins are the **anterior** (superior) and **posterior** (inferior) **vena cava.** The anterior vena cava collects blood from the head, chest, and arms, and the posterior vena cava collects blood from the lower body regions. Both enter the right atrium. The aorta and the vena cavae serve as the major pathways for blood in the systemic system.

The path of systemic blood to any organ in the body begins in the left ventricle, which pumps blood into the aorta. Branches from the aorta go to the major body regions and organs. For example, the path of blood to the kidneys may be traced as follows:

> Left ventricle—aorta—renal artery—renal arterioles, capillaries, venules—renal vein—vena cava—right atrium.

To trace the path of blood to any organ in the body, you need only mention the aorta, the proper branch of the aorta, the organ, and the returning vein to the vena cava. Figure 10.10 shows that in most instances the artery and vein that serve the same organ are given the same name. In the systemic system, unlike the pulmonary system, arteries contain oxygenated blood and appear a bright red, while veins contain deoxygenated blood and appear a purplish color.

The **coronary arteries** (fig. 10.5), which are a part of the systemic system, are extremely important arteries because they serve the heart muscle itself. (The heart is not nourished by the blood in its chambers.) The coronary arteries arise from the aorta just above the semilunar valve. They lie on the exterior surface of the heart, where they branch off in various directions into the arterioles. The coronary capillary beds join to form venules. The venules converge into the coronary vein, which empties into the right atrium. Although the coronary arteries receive blood under high pressure, they have a very small diameter and may become blocked for one reason or another (p. 220).

Human Anatomy and Physiology

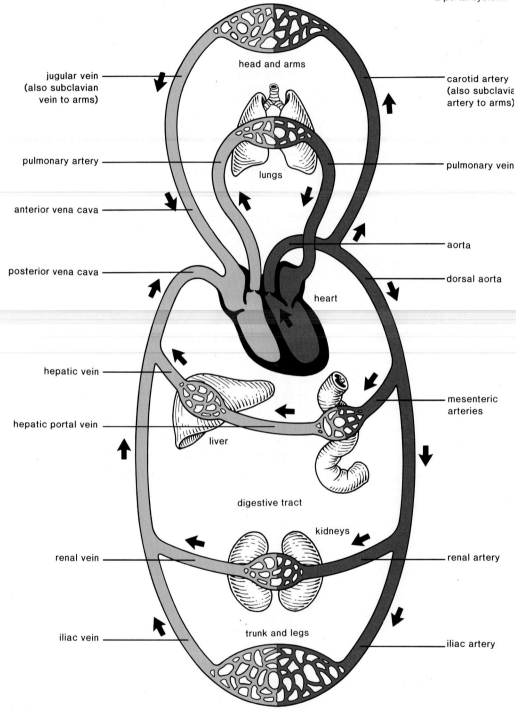

Figure 10.10
Blood vessels in the pulmonary and systemic circulatory systems. The blue-colored vessels carry deoxygenated blood and the red-colored vessels carry oxygenated blood; the arrows indicate the flow of blood. The hepatic portal vein lies between the intestine and the liver. What is the definition for a portal system?

jugular vein (also subclavian vein to arms)

head and arms

carotid artery (also subclavia artery to arms)

pulmonary artery

lungs

pulmonary vein

anterior vena cava

posterior vena cava

aorta

dorsal aorta

heart

hepatic vein

hepatic portal vein

liver

mesenteric arteries

digestive tract

kidneys

renal vein

renal artery

iliac vein

trunk and legs

iliac artery

Figure 10.11
Human circulatory system. A more realistic representation of the major blood vessels in the body shows that arteries and veins go to all parts of the body. The anterior and posterior vena cava take their name from their relationship to what organ?

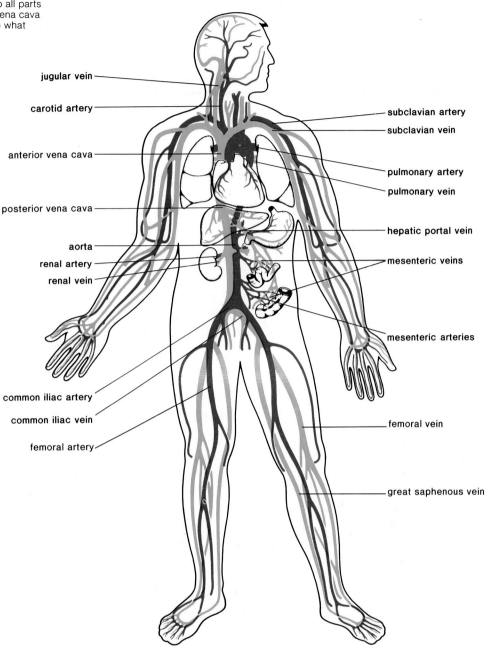

jugular vein

carotid artery

anterior vena cava

posterior vena cava

aorta

renal artery

renal vein

common iliac artery

common iliac vein

femoral artery

subclavian artery

subclavian vein

pulmonary artery

pulmonary vein

hepatic portal vein

mesenteric veins

mesenteric arteries

femoral vein

great saphenous vein

The body has only one important portal system, the **hepatic portal system** (fig. 10.10). A portal system is one that begins and ends in capillaries; the first set of capillaries occurs at the villi of the small intestine and the second occurs in the liver. Blood passes from the capillaries of the villi into venules that join to form the hepatic portal vein, a vessel that leaves the liver to enter the vena cava.

While figure 10.10 is helpful in tracing the path of the blood, it must be remembered that all parts of the body receive both arteries and veins, as illustrated in figure 10.11.

Human Anatomy and Physiology

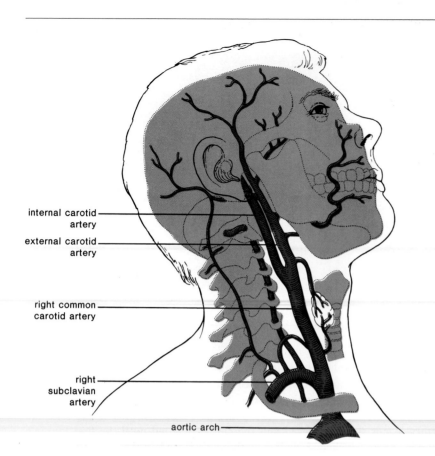

internal carotid artery

external carotid artery

right common carotid artery

right subclavian artery

aortic arch

Standard Medical Procedures

Knowledge of the cardiac cycle is helpful in understanding three medical procedures.

Electrocardiogram (EKG)

With contraction of any muscle, including the myocardium, ionic changes occur that can be detected by electrical recording devices. Thus it is possible to study the heartbeat by recording voltage changes that occur when the heart contracts. (Voltage, which in this case is measured in millivolts, is the difference in polarity between two electrodes attached to the body.) The record that results is called an **electrocardiogram** (fig. 10.9b), which clearly shows an atrial phase and a ventricular phase. The first wave in the electrocardiogram, the so-called *P* wave, represents the excitation and contraction of the atria. The second wave, or the *QRS* wave, occurs during ventricular excitation and contraction. The third, or *T* wave, is caused by the recovery of the ventricles. An examination of the electrocardiogram indicates whether the heartbeat has a normal or irregular pattern.

Pulse

When the left ventricle contracts and sends blood out into the aorta, the elastic walls of the arteries swell, but then almost immediately recoil. This alternate expanding and recoiling of an arterial wall can be felt as a **pulse** in any artery that runs close to the surface. It is customary to feel the pulse by placing several fingers on the radial artery, which lies near the outer border of the palm side of the wrist. The carotid artery is another good location (fig. 10.12). Normally the pulse rate indicates the rate of the heartbeat because the arterial walls pulse whenever the left ventricle contracts.

Figure 10.13

Determination of blood pressure by the use of a sphygmomanometer. The technician inflates the cuff with air and then as she gradually reduces the pressure, she listens by means of a stethoscope for the sounds that indicate the blood is moving past the cuff. A pressure gauge on the cuff is used to tell the systolic and diastolic blood pressure.

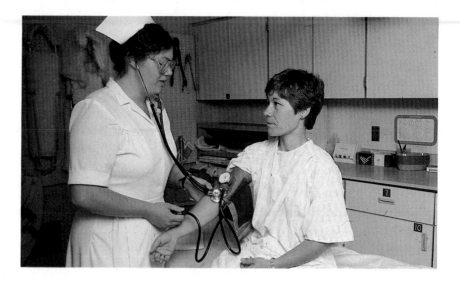

Blood Pressure

Blood pressure, the pressure of the blood against the wall of a vessel, is created by the pumping action of the heart.

To measure blood pressure (fig. 10.13), a sphygmomanometer is used. This consists of a hollow cuff attached to a pressure gauge. The cuff is placed about the upper arm over the brachial artery and inflated with air until there is no pulse felt in the wrist. At this point, no blood is flowing in the brachial artery. Air is slowly released and the cuff is deflated until the first tapping sounds can be detected through a stethoscope placed on the arm just beneath the cuff. The examiner glances at the manometer, or pressure gauge, and notes the pressure at this point. This is the value to be assigned to **systolic blood pressure,** the highest arterial pressure, reached during ejection of blood from the heart. The systolic pressure has overcome the pressure exerted by the cuff and has caused the blood to flow in the artery. The cuff is further deflated while the examiner continues to listen. The tapping sounds become louder as the pressure is lowered. Finally, the sounds become dull and muffled just before there are no sounds at all. Now the examiner again notes the pressure. This is **diastolic blood pressure,** the lowest arterial pressure. Diastolic pressure occurs while the heart ventricles are relaxing.

Normal blood pressure for a young adult is said to be 120 millimeters of mercury (Hg) over 80 millimeters, or simply 120/80.[1] Actually, this is the expected blood pressure in the brachial artery of the arm; blood pressure varies in different parts of the body (fig. 10.17). Blood pressure also varies with activity, being higher, naturally, during vigorous activity. When the blood pressure reading for a person at rest is higher than expected, the person is said to have **hypertension,** and when the reading is lower than expected the person is said to have **hypotension.** Hypertension, in particular, is often associated with cardiovascular disease, p. 220.

Lymphatic System

The lymphatic system is closely associated with the cardiovascular system because it consists of vessels that take up excess tissue fluid and transport it to the bloodstream. **Tissue fluid** is the fluid that surrounds cells. Localized swelling due to excess tissue fluid not collected by the lymphatic system is called **edema.**

---

[1] To say that the pressure is 120 millimeters Hg means that the force exerted would be sufficient to push a column of mercury up to a level of 120 millimeters. Pressure is also sometimes measured in terms of centimeters of water, but mercury is used more frequently because it is 13.6 times heavier than water.

Human Anatomy and Physiology

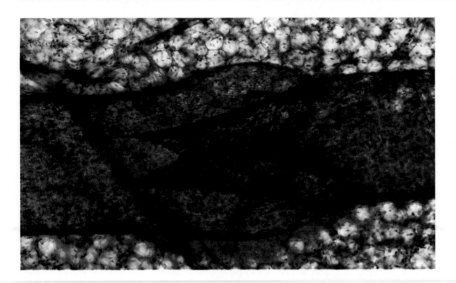

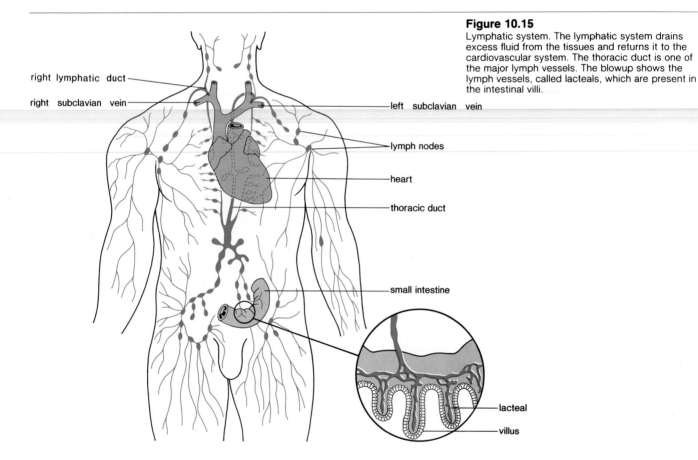

Figure 10.15
Lymphatic system. The lymphatic system drains excess fluid from the tissues and returns it to the cardiovascular system. The thoracic duct is one of the major lymph vessels. The blowup shows the lymph vessels, called lacteals, which are present in the intestinal villi.

right lymphatic duct

right subclavian vein

left subclavian vein

lymph nodes

heart

thoracic duct

small intestine

lacteal

villus

Lymph Vessels

Lymph vessels consist of **lymph capillaries** and **lymph veins.** The latter have a construction similar to cardiovascular veins, including the presence of valves (fig. 10.14).

The lymphatic system is a one-way system rather than a circulatory system (fig. 10.15). The system begins with lymph capillaries that lie near blood capillaries and take up fluid that has left the cardiovascular system but has not been reabsorbed by it. Once tissue fluid enters the lymph vessels, it is called **lymph.**

Lymph is collected in vessels that join to form two main trunks: the right lymphatic duct, which drains the upper right portion of the body, and the thoracic duct, which drains the rest of the body. The former empties into the right subclavian vein and the latter into the left subclavian vein.

The **lacteals** are blind ends of lymph vessels found in the villi. As previously mentioned, the products of fat digestion enter the lacteals. These products eventually enter the cardiovascular system when the lymph ducts join the subclavian veins.

Lymph Organs

At certain strategic points along medium-sized lymph vessels, there occur small ovoid, or round, structures called **lymph nodes** that are composed of lymphoid tissue (fig. 12.1). Lymph nodes produce lymphocytes, a type of white blood cell, and these are found packed into the spaces of a lymph node. Some lymphocytes produce antibodies, proteins that are capable of combining with foreign proteins called antigens. At times, the foreign proteins are disease-causing agents and thus lymph nodes help fight infection.

Lymph nodes also filter and trap bacteria and other debris, helping purify the blood. When a local infection is present, such as a sore throat, the lymph nodes in that region swell and become painful. Lymph nodes may be removed in cancer operations because they may help spread cancer cells if allowed to remain in the body.

Lymphoid tissue is also present in four specific organs of the body that help fight infection. The *tonsils,* located at the back of the mouth and throat (fig. 9.2), and the *appendix,* a projection from the cecum of the large intestine, are sometimes removed because they tend to become infected. The *spleen,* the largest mass of lymphoid tissue in the body, is located in the abdominal cavity below the stomach. Not only does the spleen produce white cells, it also stores blood, contracting when the blood pressure drops. The *thymus* gland is a bilobed mass of lymphoid tissue in the upper thoracic cavity that becomes progressively smaller with age. The thymus has an important function in the maturation of some lymphocytes, and its decrease in size may be important in the aging process.

Features of the Circulatory System

The circulatory system is a system of vessels that become progressively smaller and then progressively larger again. Such a system is desirable because it permits a large surface area for exchange between the tissues and the blood in the region of the capillaries. Nevertheless, this does have an effect on certain physiological features of the circulatory system as discussed in the following sections.

Blood Pressure

As figure 10.16 indicates, blood pressure decreases with distance from the left ventricle, the chamber that pumps blood into the aorta. Blood pressure is therefore higher in the arteries than in the arterioles. Further, there is a sharp drop in blood pressure when the arterioles reach the capillaries. The decrease in blood pressure is largely attributed to the fact that the total cross-sectional area of the vessels increases as blood moves through arteries, arterioles, and then into capillaries. There are more arterioles than arteries, and many more capillaries than arterioles.

As discussed earlier, there is a fluctuation in blood pressure in arteries and arterioles due to ventricular systole and diastole. As blood approaches the capillaries, there is a steeper decline in systolic pressure than diastolic pressure. Can you give an explanation for this based on the preceding information?

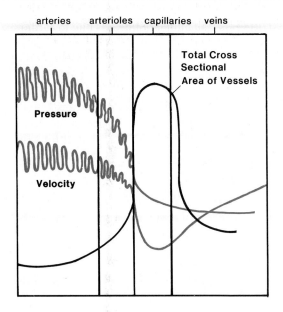

arteries   arterioles   capillaries   veins

Pressure

Velocity

Total Cross Sectional Area of Vessels

Figure 10.16
Diagram illustrating how velocity and blood pressure are related to the total cross-sectional area of blood vessels. Capillaries have the greatest cross-sectional area and the least pressure and velocity. Skeletal muscle contraction, not blood pressure, accounts for velocity of blood in the veins.

Velocity of Blood Flow

The velocity of blood varies in different parts of the circulatory system (fig. 10.16) and this can also be related to the total cross-sectional area of the various vessels.

Notice that the velocity of the blood in the arteries decreases as blood pressure decreases. In other words, blood pressure accounts for the velocity of the blood flow in the arterial system. Therefore, as blood pressure decreases due to the increased cross-sectional area of the arterial system, so does velocity. The sum total of the cross-sectional area of the capillaries is 600 to 800 times greater than that of the aorta, therefore the blood moves much slower through the capillaries than it does in the aorta. This is important because the slow progress allows time for the exchange of molecules between the blood and the tissues.

Notice that the velocity of the blood in the arterial system varies with the phases of the heart cycle. During systole, the velocity of the arterial blood rises to a maximum, and during diastole, the velocity of the blood decreases to a minimum.

Blood pressure cannot account for the movement of blood through the venules and veins since they lie on the other side of the capillaries. Instead, movement of the blood through the venous system is due to skeletal muscle contraction. When the skeletal muscles contract, they press against the weak walls of the veins and this causes the blood to move past a valve (fig. 10.14). Once past the valve, the blood will not fall back. The importance of muscle contraction in moving blood in the venous system may be demonstrated by forcing a person to stand rigidly still for a number of hours. Frequently, fainting will occur because the blood collects in the limbs, robbing the brain of oxygen. In this case, fainting is beneficial because the resulting horizontal position aids in getting blood to the head.

Blood flow gradually increases in the venous system (fig. 10.16) due to a progressive reduction in the cross-sectional area as small venules join to form veins. The two vena cavae together have a cross-sectional area of only about double that of the aorta. The chest cavity experiences a lower pressure whenever the chest expands during inspiration. This also aids the flow of venous blood into the chest area because blood flows in the dirction of reduced pressure.

Figure 10.17
Atherosclerosis. Accumulation of plaque on the inner wall of an arteriole can lead to either a stroke or heart attack, depending on where the arteriole is located. *a.* Plaque buildup has begun. *b.* Plaque buildup has almost occluded vessel.

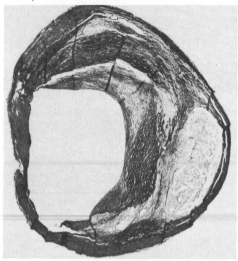

a.

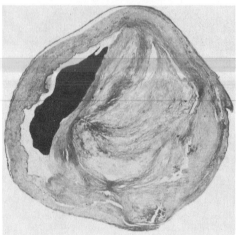

b.

Figure 10.18
Coronary bypass operation. During this operation, the surgeon grafts segments of a leg vein between the aorta and the coronary vessels, bypassing areas of blockage. Patients who are ill enough to require surgery often receive two or three bypasses in a single operation.

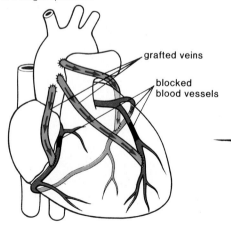

grafted veins

blocked blood vessels

Circulatory Disorders

During the past 30 years, the number of deaths due to cardiovascular disease has declined more than 30 percent. Even so, more than 50 percent of all deaths in the United States are still attributable to cardiovascular disease. The number of deaths due to hypertension, stroke, and heart attack are more than those due to cancer and accidents combined.

It is estimated that about 35 million Americans suffer from hypertension, or high blood pressure. Even though hypertension is easily detected by blood pressure readings (p. 216) it is believed that at least one-third of these people are unaware they have this condition that can lead to failure of the cardiovascular system.

Atherosclerosis and Arteriosclerosis

Hypertension is often seen in individuals who also have atherosclerosis and arteriosclerosis. **Atherosclerosis** (fig. 10.17) is the accumulation of soft masses of fatty materials, particularly cholesterol, beneath the inner linings of arteries. Such deposits are called **plaque,** and as they develop they tend to protrude into the vessel and interfere with the flow of blood. **Arteriosclerosis** becomes apparent when the atherosclerotic arteries undergo still more degenerative changes, during which they lose their elasticity and become hardened by calcium deposits. Arteriosclerosis is often termed *hardening of the arteries.*

The occurrence of plaque can cause a clot to form on the hardened arterial wall. As long as the clot stays stationary it is called a **thrombus,** but when and if it dislodges and moves along with the blood, it is called an **embolus.** Thrombi reduce the flow of blood, particularly in small arteries. An embolus causes **embolism** when it comes to a standstill and entirely blocks the flow of blood in a small vessel. If an embolism is not treated, the tissue fed by the vessel dies from lack of blood.

Stroke and Heart Attack

Hypertension can lead to stroke and heart attack. A **stroke** occurs when a portion of the brain dies due to a lack of oxygen, and a **heart attack** occurs when a portion of the heart muscle dies due to a lack of oxygen. A stroke, characterized by paralysis or death, often results when a small arteriole bursts or becomes blocked by an embolism.

When a person has hypertension, cardiac muscles require more oxygen because they must work harder to pump the blood against the increased arterial pressure. However, the coronary arteries may be hardened and unable to provide an adequate blood supply. A thrombus may be present, termed **coronary thrombosis,** or an embolus may have moved into a coronary arteriole. At first, the individual may suffer **angina pectoris,** characterized by a radiating pain in the left arm. Then if circulation to a significant portion of the heart is blocked entirely, a heart attack occurs.

Surgical Treatment

Surgical treatments are now available for coronary thrombosis. As many as 100,000 persons a year have **coronary bypass** surgery. During this operation, surgeons take segments from another blood vessel, often a large vein in the leg, and stitch one end to the aorta and the other end to a coronary artery past the point of obstruction (fig. 10.18). Between 75 and 90 percent of those who have had bypass surgery say their angina pain has been relieved.

Recently, new technologies have been developed that do away with the obstruction instead of bypassing it. In both of these, a plastic tube is threaded into an artery of an arm or leg and guided through a major blood vessel toward the heart. In one procedure, when the tube reaches the clot, a balloon attached to the end of the tube inflates and breaks up the clot. In the other procedure, a drug called streptokinase is injected to dissolve the clot. Then the artery

Human Anatomy and Physiology

opens and the blood begins to flow again (fig. 10.19). The latter procedure has been done even while a person is suffering a heart attack. If the clot is removed within six hours after a heart attack begins, there is usually no damage to the heart muscle.

Persons with weakened hearts may eventually suffer from **congestive heart failure,** meaning that the heart is no longer able to pump blood adequately. These individuals are candidates for a heart transplant, or even implantation of an artificial heart. The difficulty with a heart transplant is, first, one of availability and, second, the tendency of the body to reject foreign organs. On December 2, 1982, Barney Clark was the first person to receive an artificial heart. The heart's two polyurethane ventricles were attached to Clark's own atria and blood vessels by way of dacron fittings. Two long tubes stretched between the artificial heart and an external machine that periodically sent bursts of air into the ventricles, forcing the blood out into the aorta and pulmonary trunk. Clark died on March 23, 1983, but his death was not due to failure of the artificial heart; it was due to complications caused by his having had heart trouble for so long. Whereas Clark's body was permanently attached to an external machine, it is hoped that eventually an artificial heart can be powered by batteries so that the patient will be completely mobile (fig. 10.20).

Medical Treatment

There are several different types of drugs that can be taken to control hypertension. A drug is now available to stop the sequence of events by which a hormone produced by the kidney called **renin** leads to hypertension in some individuals. Diuretics, which have been used for some time, cause the kidneys to excrete excess salts and fluids from the body. Vasodilators lower blood pressure by relaxing the muscles of the peripheral arteries. Sympatholytics interfere with the functioning of the sympathetic nervous system, thereby preventing arterial constriction and/or slowing the heartbeat so that its oxygen needs can be more easily met. There are other drugs that also affect the heartbeat. For example, digitalis slows the heartbeat while at the same time increasing its contractile power.

Varicose Veins and Phlebitis

Varicose veins are abnormal and irregular dilations in superficial (near the surface) veins, particularly those in the lower legs. Varicose veins in the rectum, however, are commonly called piles, or more properly, **hemorrhoids.** Varicose veins develop when the valves of the veins become weak and ineffective due to a backward pressure of the blood. The problem can be aggravated when venous blood flow is obstructed by crossing the legs or by sitting in a chair so that its edge presses against the back of the knees.

Phlebitis, or inflammation of a vein, is a more serious condition, particularly when a deep vein is involved. Blood in the inflamed vessel may clot, in which case *thromboembolism* has occurred. An embolus that originates in a systemic vein may eventually come to rest in a pulmonary arteriole, blocking circulation through the lungs. This condition, termed **pulmonary embolism,** can result in death.

Prevention of Cardiovascular Disease

Investigators have been able to identify certain factors as contributing to cardiovascular disease. It is hoped that recognition of these factors by all will help prevent the occurrence of hypertension and heart disease.

Heredity

It appears that persons who have parents with cardiovascular disease are more susceptible to this condition. These are the individuals, in particular, that should

Figure 10.19
Successful thrombolytic therapy. The patient had a blockage in the coronary artery that was dissolved by infusing it with streptokinase by means of a plastic tube threaded all the way from an arm or leg vessel. *a.* Blocked coronary artery. *b.* Artery is open and blood is flowing.

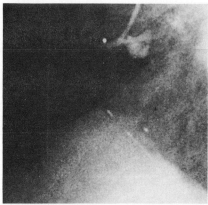
a.

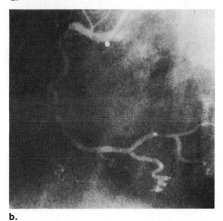
b.

Figure 10.20
Artificial heart. It is hoped that some day the artificial heart will be powered by batteries the patient can wear about the waist. A microcomputer would control the signals that enable the artificial heart to pump blood.

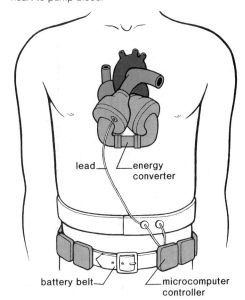

lead

energy converter

battery belt

microcomputer controller

Although psychological stress has long appeared to contribute to high blood pressure, the means by which it does so haven't been clear. Now, however, the mechanism seems to have been pinpointed: Sympathetic nervous system reaction to stress causes the kidneys to retain salt, and this salt retention upsets blood pressure regulation.

Studies with rats have already shown that psychological stress is capable of inducing kidney retention of salt and that the amount of such retention is greater in rats that develop high blood pressure than in those that do not. The rat studies also showed stress-induced sodium retention to be mediated by the sympathetic nervous system—the part of the involuntary nervous system that makes blood vessels contract and prevents smooth muscle from contracting.

These findings prompted Kathleen C. Light and colleagues of the University of North Carolina School of Medicine in Chapel Hill to see whether psychological stress can cause sodium retention in humans, and, if so, whether the effect is greater among persons with a predisposition to high blood pressure or whose sympathetic nervous systems are especially responsive to psychological stress.

They studied 13 subjects with normal blood pressure but at risk of developing high pressure either because of family history or because their own blood pressure was in the "high-normal" range. Another 11 subjects had normal blood pressure and did not possess these high-risk factors. The 24 subjects were given a psychological stress test consisting of various competitive mental tasks and further subdivided into high and low sympathetic nervous system responders. Six low-risk subjects and six high-risk subjects

be certain to have their blood pressure and heart functioning checked regularly and should be careful to attend to all the suggestions for keeping the heart functioning normal and the blood pressure under control.

Race

Blacks are more likely to develop hypertension. This may be due to a genetic defect, but it may also be due to their diet and/or exposure to stress. Careful attention to these factors and the others listed may help reduce the incidence of hypertension among blacks.

Smoking

Smoking contributes to high blood pressure and a malfunctioning heart. Nicotine constricts blood vessels, cutting off circulation to the extremities and causing the blood pressure to rise. It also triggers the brain to release extra amounts of adrenalin, a hormone that speeds the heart rate. At the same time, however, the carbon monoxide content of cigarette smoke prevents the blood from delivering the normal amount of oxygen to the heart.

Obesity

All tissues require a supply of blood carried by blood vessels. When weight is gained, the circulatory system also increases in size and the heart must increase blood pressure in order to have the blood travel further. Excess fatty tissue can collect around the heart.

Exercise

Regular moderate exercise contributes to cardiovascular health. However, it is not good to alternate between long periods of rest and overexertion. A call for a rapid response to an immediate need can cause the sudden exposure of a weakness, particularly in the coronary circulation, without giving the body a chance to repair itself gradually.

were low responders, and five low-risk subjects and seven high-risk subjects were high responders.

Sodium excretion levels measured before and during the test showed the seven high-risk, high sympathetic response subjects retained sodium during the test. The other three subgroups did not. Thus, psychological stress can cause sodium retention in humans, but only among persons at high risk of developing high blood pressure who also have a high sympathetic nervous system response to psychological stress, Light and her team conclude.

The next challenge, . . . will be to see whether individuals vulnerable to stress-induced sodium retention actually get high blood pressure. If so, it would constitute further evidence that psychological stress causes high blood pressure via sympathetic nervous system stimulation of the kidneys and salt retention.

Researchers are discovering why stress such as taking a test causes high blood pressure.

Diet

Increased salt (sodium chloride) intake causes fluid to be retained in the blood vessels, and this can cause high blood pressure since there is more fluid to press against arterial walls. There is evidence to suggest that a low-salt diet can actually prevent the occurrence of high blood pressure.

Fatty foods and foods high in cholesterol contribute to atherosclerosis and arteriosclerosis and therefore to high blood pressure. Investigators have found that cholesterol is carried in the bloodstream mainly by low-density lipoproteins (LDLs), which possibly contribute to the development of plaque, while high-density lipoproteins (HDLs) most likely encourage the breakdown of cholesterol in the liver and prevent the development of plaque. Eggs are a rich source of cholesterol; animal fat found in meat and butter contain saturated fatty acids that promote the formation of LDLs. Fatty acids found in oils and vegetables are polyunsaturated and these promote the formation of HDLs.

Both alcohol and nicotine apparently cause an increased level of fatty acids in the bloodstream and, in this way, may contribute to cardiovascular disease. In addition, alcohol seems to cause deterioration of the heart muscle itself.

Stress

Blood pressure normally rises with excitement or alarm due to the involvement of the sympathetic nervous system, which causes arterioles to constrict and the heart to beat faster. Learning to control these effects of stress can contribute to cardiovascular health.

Some investigators have collected data, as described in the reading on this page, indicating that stress also causes the kidneys to excrete less sodium. Perhaps this is the mechanism by which stress eventually leads to permanent high blood pressure.

Summary

Blood vessels include arteries (and arterioles) that take blood away from the heart; capillaries, where exchange of molecules with the tissues occurs; and veins (and venules) that take blood to the heart.

The heart is a double pump that keeps blood moving in the closed circulatory system of humans. The beat of the heart is intrinsic. During the cardiac cycle, the S-A node, called the pacemaker, initiates the beat and causes the atria to undergo contraction, or systole. The A-V node picks up the stimulus and stimulates the ventricles to contract. Thus, for the first 0.15 second both atria contract, then for 0.30 second both ventricles contract, and finally all chambers rest for 0.40 second. The heart sounds, lub-DUPP, are due to the closing of the atrioventricular valves followed by the closing of the semilunar valves. The activity of the cardiovascular system can be monitored by taking the pulse, observing the electrical activity of the heart (EKG), and by taking the blood pressure.

The circulatory system is divided into two parts: the pulmonary system and the systemic system. In the pulmonary system, blood circulates through the lungs. The pulmonary artery takes blood to the lungs, and the pulmonary veins return blood to the heart.

To trace the path of blood in the systemic system, start with the aorta and follow its path until it branches to the specific organ in question. It may be assumed that this artery will divide into arterioles and capillaries, and that the capillaries will lead to venules. The vein that joins the vena cava to return the blood to the heart most likely will have the same name as the artery. The names of some of the different arteries and veins in the systemic system are given in figure 10.13. In the adult systemic system, but not in the pulmonary system, the arteries carry oxygenated blood and the veins carry deoxygenated blood.

Lymph vessels, or veins, are constructed similarly to cardiovascular veins and contain valves to keep lymph moving from the tissues to the veins. The lymphatic system is a one-way system taking excess tissue fluid to the subclavian veins. The lacteals, which absorb the products of fat digestion, are a part of the lymphatic system. Also, lymph nodes are placed in strategic places along the length of the lymph vessels, and these filter the lymph and produce lymphocytes to fight infection.

The movement of blood in the arteries is closely related to blood pressure. As blood pressure decreases due to an increase in cross-sectional area, so does velocity. Velocity is slowest in the capillaries; this is beneficial because the slow movement of blood aids in the exchange of molecules between the blood and the tissues. The movement of blood in the veins is largely due to skeletal muscle contraction.

Cardiovascular disease is the leading cause of death. Hypertension, atherosclerosis, arteriosclerosis, and thromboembolism all contribute to circulatory failure. If the flow of blood to the brain is blocked, a stroke occurs; if the flow of blood to the heart is blocked, a heart attack occurs.

Study Questions

1. What types of blood vessels are there? Discuss their structure and function. (pp. 205–7)
2. Trace the path of blood in the pulmonary system as it travels from and returns to the heart, naming as you do the anatomical structures through which the blood passes. (p. 212)

3. Describe the cardiac cycle (using the terms *systole* and *diastole*) and explain the heart sounds. (pp. 209–10)
4. Trace the path of blood from the mesenteric arteries to the aorta, indicating which of the vessels are in the systemic system and which are in the pulmonary system. (pp. 212–14)
5. Describe an EKG and tell how its components are related to the cardiac cycle. (pp. 211, 215)
6. What is blood pressure, and why is the average normal arterial blood pressure said to be 120/80? (p. 216)
7. What is a lymph vessel? (p. 217) Give three functions of the lymphatic system and tell how these functions are carried out. (pp. 216–18)
8. In which type of vessel is blood pressure highest? Lowest? (p. 218) Velocity is lowest in which type vessel and why is it lowest? Why is this beneficial? (p. 219) What factors assist venous return of the blood? (p. 219)
9. What are atherosclerosis and arteriosclerosis? (p. 220) Name two illnesses associated with hypertension and thromboembolism. (p. 220) Discuss the treatment and prevention of cardiovascular disease. (pp. 220–23)

Selected Key Terms

artery (ar′ter-e)
capillaries (kap′i-ler″es)
vein (vān)
valves (valvz)
septum (sep′tum)
atria (a′tre-ah)
ventricle (ven′tri-k′l)
atrioventricular (a″tre-o-ven-trik′u-lar)
semilunar (sem″e-lu′nar)
systole (sis′to-le)
diastole (di-as′to-le)
S-A node (es a nōd)
A-V node (a-ve nōd)
pulmonary system (pul′mo-ner″e sis′tem)
systemic system (sis-tem′ik sis′tem)
aorta (a-or′tah)
vena cava (ve′nah ka′vah)
coronary artery (kor′ŏ-na-re ar′ter-e)
tissue fluid (tish′u floo′id)
lymph (limf)

Further Readings

Adolph, E. 1967. The heart's pacemaker. *Scientific American* 216(3):32.
Crouch, J. E. 1978. *Functional human anatomy.* 3d ed. Philadelphia: Lea and Febiger.
Jarvik, R. K. 1981. The total artificial heart. *Scientific American* 244(1):74.
Mayerson, H. S. 1963. The lymphatic system. *Scientific American* 208(6):80.
Spain, D. M. 1966. Atherosclerosis. *Scientific American* 215(2):48.
Wiggers, C. J. 1957. The heart. *Scientific American* 196(5):74.
Winfree, A. T. 1983. Sudden cardiac death: a problem in topology. *Scientific American* 248(5):144.
Wood, J. E. 1968. The venous system. *Scientific American* 218(1):86.

11

Blood

Chapter concepts

1 Blood, which is composed of cells and a fluid containing many inorganic and organic molecules, has three primary functions: transport, clotting, and fighting infection.

2 Blood transports nutrients to and wastes from the tissue capillaries. Exchange of molecules with tissue fluid takes place across capillary walls.

3 The process of clotting requires a complicated series of reactions and normally prevents the loss of blood after an injury.

4 White cells, some of which phagocytize invading microbes, and gamma globulins, which combine with foreign proteins in a specific manner, are the body's defense against infection and disease.

5 All of the functions of blood may be correlated with the ability of the body to maintain a constant internal environment.

It is a curious fact that more than half the body is water; the total quantity of water is around 70 percent of the body's weight. By far, most of this water (50 percent) is found within the cells. A smaller amount (20 percent) lies outside the cells. This water is found in (1) the **tissue fluid** that surrounds the cells, (2) lymph contained within lymph vessels, and (3) in blood vessels.

If blood is transferred from a person's vein to a test tube and prevented from clotting, it separates into two layers (fig. 11.1). The lower layer consists of red blood cells (erythrocytes), white cells (leukocytes), and blood platelets (thrombocytes). Collectively, these are called the **formed elements** (fig. 11.2) that take up about 45 percent of the volume of whole blood. The upper layer, called **plasma,** contains a variety of inorganic and organic substances dissolved or suspended in water. Plasma accounts for about 55 percent of the volume of whole blood. Table 11.1 lists the components of blood, which we will discuss

Figure 11.1
Volume relationship of plasma and formed elements (cells) in blood. Red cells are by far the most prevalent blood cell and this accounts for the color of blood.

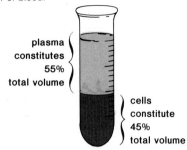

plasma constitutes 55% total volume

cells constitute 45% total volume

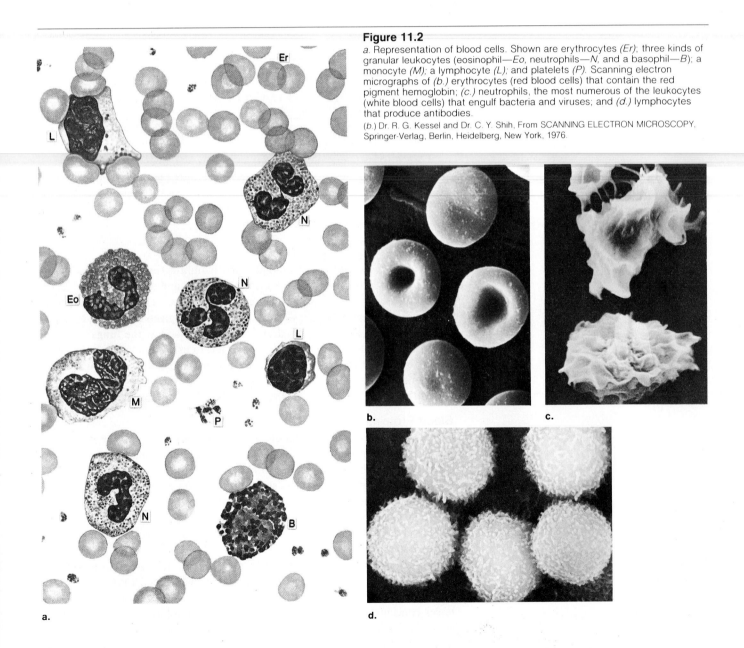

Figure 11.2
a. Representation of blood cells. Shown are erythrocytes *(Er)*; three kinds of granular leukocytes (eosinophil—*Eo*, neutrophils—*N*, and a basophil—*B*); a monocyte *(M)*; a lymphocyte *(L)*; and platelets *(P)*. Scanning electron micrographs of *(b.)* erythrocytes (red blood cells) that contain the red pigment hemoglobin; *(c.)* neutrophils, the most numerous of the leukocytes (white blood cells) that engulf bacteria and viruses; and *(d.)* lymphocytes that produce antibodies.
(b.) Dr. R. G. Kessel and Dr. C. Y. Shih, From SCANNING ELECTRON MICROSCOPY, Springer-Verlag, Berlin, Heidelberg, New York, 1976.

Table 11.1 Components of blood

Blood	Function	Source
I. Formed elements		
Red cells	Transport oxygen	Bone marrow
Platelets	Clotting	Bone marrow
White cells	Fight infection	Bone marrow and lymphoid tissue
II. Plasma*		
Water	Maintains blood volume and transports molecules	Absorbed from intestine
Plasma proteins	All maintain blood osmotic pressure and pH	
Albumin	Transport	Liver
Fibrinogen	Clotting	Liver
Globulins	Fight infection	Lymphocytes
Gases		
Oxygen	Cellular respiration	Lungs
Carbon dioxide	End product of metabolism	Tissues
Nutrients		
Fats, glucose, amino acids, etc.	Food for cells	Absorbed from intestinal villi
Salts	Maintain blood osmotic pressure and pH; aid metabolism	Absorbed from intestinal villi
Wastes		
Urea and ammonia	End products of metabolism	Tissues
Hormones, vitamins, etc.	Aid metabolism	Varied

*Plasma is 90–92 percent water, 7–8 percent plasma proteins, not quite 1 percent salts, and all other components are present in even smaller amounts.

in terms of three functions: *transport, clotting,* and *infection fighting.* All of these can be related to blood's primary function of maintaining a constant internal environment, or homeostasis.

Transport

The transport function of the blood helps maintain the constancy of tissue fluid. The blood transports oxygen from the lungs and nutrients from the intestine to the capillaries where they enter tissue fluid. Here, it also takes up carbon dioxide and nitrogen waste (i.e., ammonia) given off by the cells and transports them away. Carbon dioxide exits the blood at the lungs, and ammonia exits at the liver where it is converted to urea, a substance that later travels by way of the bloodstream to the kidneys and is excreted. Figure 11.3 diagrams the major transport functions of blood, indicating the manner in which this function helps keep the internal environment relatively constant.

Blood Proteins

While small organic molecules such as glucose and urea simply dissolve in plasma, large organic molecules such as hormones, vitamins, fatty acids, and other lipids combine with proteins for transport.

The blood proteins also assist the transport function of the blood by contributing to blood pressure and osmotic pressure. The **viscosity,** or thickness of the blood, is largely dependent on the presence of plasma proteins and on red blood cells. Reduction in the amount of protein and red blood cells results in low blood pressure.

Plasma proteins, together with salts, create an osmotic pressure that maintains the water content of the blood. You will recall that water moves across cell membranes from the area of greater concentration to the area of lesser concentration of water. Since proteins are too large to pass through or across a capillary wall, the fluid within the capillaries is always the area of lesser concentration of water, and water will therefore tend to pass into the capillaries.

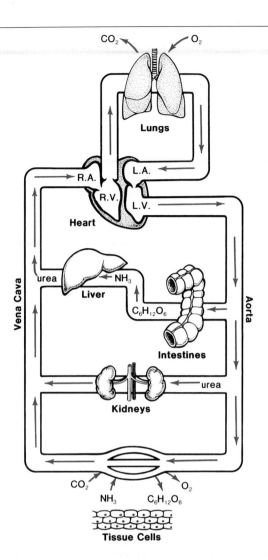

CO₂ ← → O₂

Lungs

R.A.   L.A.
R.V.   L.V.

Heart

Vena Cava

urea   NH₃

Liver

C₆H₁₂O₆

Intestines

Aorta

urea

Kidneys

CO₂   O₂
NH₃   C₆H₁₂O₆

Tissue Cells

Figure 11.3

Diagram illustrating the transport function of blood. Oxygen (O_2) is transported from the lungs to the tissues, and carbon dioxide (CO_2) is transported from the tissues to the lungs. Ammonia (NH_3) is transported from the tissues to the liver where it is converted to urea, a molecule excreted by the kidneys. Glucose ($C_6H_{12}O_6$) is absorbed by the gut and may be temporarily stored in the liver as glycogen before it is transported to the tissues.

The blood proteins we will be discussing in this chapter are listed in table 11.2. All of these except albumin have a special function. Albumin, globulin, and fibrinogen are present in the plasma, while hemoglobin is present in the red cells. Hemoglobin is a conjugated protein because it is composed not only of the protein globin but also the nonprotein group, heme.

Transport of Oxygen

Since humans are active warm-blooded animals, the brain and then the muscle cells require much oxygen within a short period of time. The use of the respiratory pigment, hemoglobin, allows the blood to carry much more oxygen than would otherwise be possible. Plasma carries only about 0.3 milliliters of oxygen per 100 milliliters, but whole blood carries 20 milliliters of oxygen per 100 milliliters. This shows that the presence of hemoglobin increases the carrying capacity of blood 60 times.

Although the iron portion of hemoglobin carries oxygen, the equation for oxygenation of hemoglobin is usually written as:

$$Hb + O_2 \underset{\text{tissues}}{\overset{\text{lungs}}{\rightleftharpoons}} HbO_2$$

The hemoglobin on the right, which is combined with oxygen, is called oxyhemoglobin. **Oxyhemoglobin** forms in the lungs and is a bright red color. The hemoglobin on the left, which has given up oxygen to tissue fluid, is called **reduced hemoglobin** and is a dark purple color.

Table 11.2	Blood proteins	
Name	**Location**	**Special function**
Albumin	Plasma	———
Globulin	Plasma	Antibodies to fight infection
Fibrinogen	Plasma	Blood clotting
Hemoglobin	Red cells	Carries gases (oxygen and carbon dioxide)

Hemoglobin is an excellent carrier for oxygen because it forms a loose association with oxygen in the cool, neutral conditions of the lungs and readily gives it up under the warm and more acidic conditions of the tissues.

Carbon monoxide, present in automobile exhaust, combines with hemoglobin more readily than does oxygen, and it stays combined for several hours, regardless of the environmental conditions. Accidental death or suicide from carbon monoxide poisoning occurs because the hemoglobin of the blood is not available for oxygen transport. This transport function of blood is so important that life can be temporarily sustained by giving a patient a hemoglobin substitute transfusion when whole blood is not available or cannot be given. The reading on this page discusses the possible benefits of this "artificial blood."

Hemoglobin does not float free within the plasma; it is enclosed within cells. Since hemoglobin is a red pigment, the cells appear red and their color also makes the blood red. There are 5 million red cells (fig. 11.4) per cubic millimeter of whole blood, and each of these cells contains about 200 million hemoglobin molecules. If this much hemoglobin were suspended within the plasma rather than being enclosed within the cells, the blood would be so thick the heart would have difficulty pumping it.

Blood without Donors

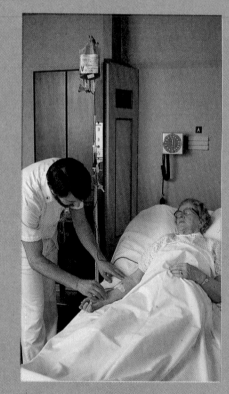

Haldor Mickelson lay dying. He suffered from severe anemia and infection, complications of an operation performed weeks before. Mickelson had refused a potentially life-saving blood transfusion because he is a Jehovah's Witness, a sect whose literal interpretation of Leviticus 17:10–12 does not permit blood transfusions: "Therefore I said unto the children of Israel, no soul of you shall eat blood."

As a last resort Mickelson's surgeon, Robert Anderson, petitioned the University of Minnesota Hospital, the U.S. Food and Drug Administration, and the Japanese-owned Alpha Therapeutic Corporation for permission to use Japan's Fluosol-DA, an experimental substitute for the oxygen-carrying hemoglobin of the blood. On humanitarian grounds Mickelson was granted a transfusion of two liters of the milky-white substance.

"Within a couple of hours he began to communicate," Anderson said. The Fluosol-DA, an oxygenated fluoro-chemical emulsion, had taken over the task, normally performed by red blood cells, of transporting oxygen. Within days Mickelson's own bone marrow began producing new red blood cells. The Fluosol-DA, having done its job, transpired out of the body through the lungs and skin. . . .

Blood is such a complex substance that scientists may never find a substitute to perform all its many functions. As Thomas Zuck, past president of the American Association of Blood Banks, points out, the synthetic has "no clotting factors, no platelets, no immunoglobulins, no antibodies, no hormones, no enzymatic activity." For this reason, human blood donations are as necessary as ever. Nonetheless, synthetic hemoglobin can save lives in emergency situations where the oxygen-carrying capability of blood is the most important order of business. . . .

One major drawback of the Japanese formula is that it must be frozen and, once thawed, used immediately. Furthermore, in the current primitive state of the art, oxygen must be delivered to

Human Anatomy and Physiology

Figure 11.4
a. Photomicrograph of red blood cells contained within a blood vessel. When one views arterioles in the living animal, it is possible to see the red cells scurrying along. *b.* Electron micrograph of red blood cells contained within each of two adjacent capillaries. Capillaries are just large enough for the small red cells to squeeze through.

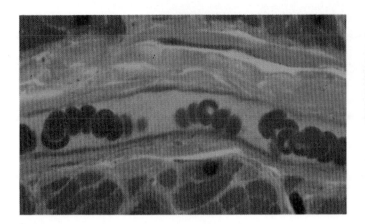

a.

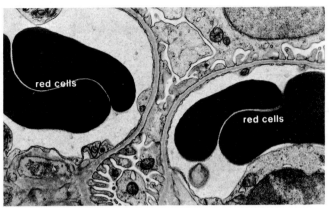

red cells

red cells

b.

patients separately through a mask while the PFC emulsion is dripped into the veins.

"A more serious problem," says George Nemo, a project officer at the National Heart, Lung, and Blood Institute just outside Washington, D.C., is the amount of time PFCs reside in the body. "Maybe there's no harm in that," says Nemo, "but maybe they're carcinogenic. We don't know. At this point we just don't have enough information about the chemical and biological properties of the compounds to say how the body will react to them."

Critical as he is of the current hype for artificial hemoglobin, Nemo is enthusiastic about the future and believes "the possibilities are endless." . . .

One gauge of success is the length of time organs can be kept alive in a PFC bath. Geyer [Robert Geyer of the Harvard School of Public Health] has kept a heart beating for more than 10 hours; livers continue to manufacture bile for up to 20 hours. It may be possible one day to transport organs destined for transplantation

halfway around the world in a bath of PFCs.

PFCs can be sterilized, thus preventing hepatitis and other infections associated with blood transfusions. In addition, they can be used by anyone, anywhere, without time-consuming blood typing. PFCs are good emergency substitutes because there is rarely enough whole blood on hand. "PFC emulsions could prove a real benefit to the Red Cross and other blood banks," Clark [Leland C. Clark, physiologist at Children's Hospital in Cincinnati] adds. Whole blood deteriorates rapidly. If a substantial supply of oxygen carriers were available, whole blood could be separated immediately into its valuable components such as platelets, immunoglobulins, and protein fractions. . . .

In carbon monoxide poisoning, death occurs because the hemoglobin, saturated with carbon monoxide, has lost much of its ability to carry oxygen. Administering oxygen through a mask is ineffective because the hemoglobin is tied up by the

carbon monoxide. An immediate transfusion with a PFC solution not only brings the needed oxygen to the brain and other vital organs but allows the carbon monoxide to dissolve in the solution and transpire out through the lungs.

Clark hopes that one day PFC emulsions will come to the rescue of stroke and heart attack victims. He believes that PFC droplets have a better chance than red blood cells do of traveling through a blocked blood vessel. If a heart attack victim were given an injection of PFC emulsion in the critical half hour to hour following the attack, significant amounts of oxygen-deprived heart muscle could be saved.

Far in the future, PFC emulsions may be used in therapy for burns, in treating sickle cell anemia, and in better methods of radiography and chemotherapy. First, questions of safety and efficacy must be answered. Meanwhile, says Robert Moore, a chemist who synthesizes PFC formulations, "We're on the front burner now. And we intend to stay there."

Figure 11.5

The hemoglobin molecule is a globular protein that contains four polypeptide chains, two of which are alpha (α) and two of which are beta chains (β). The rectangle in the center of each chain represents an iron-containing heme group.

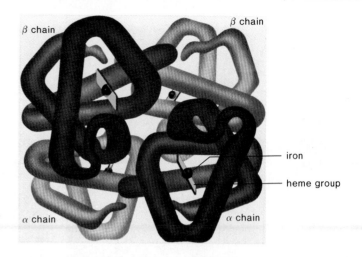

Each hemoglobin molecule (fig. 11.5) contains four polypeptide chains that make up the protein globin, and each chain is joined to **heme,** a complex iron-containing structure. It is the iron that forms a loose association with oxygen and in this way carries it in the blood.

Red Cells (Erythrocytes)

The red cells are small biconcave, disk-shaped cells without nuclei. They pass through several developmental stages during which time they lose the nucleus and acquire hemoglobin.

Red cells are continuously manufactured in the red bone marrow of the skull, ribs, vertebrae, and ends of the long bones (fig. 11.6). Normally there are between 4 and 6 million red cells per cubic millimeter of blood. Evidence indicates that the oxygen tension of arterial blood serves to regulate red cell formation and, consequently, also hemoglobin production. At high altitudes, where the oxygen tension is low, the red cell count increases. It is believed that low oxygen tension in the blood causes the kidneys to produce a substance called **renal erythropoietic factor (REF).** This joins with liver globulin to produce a combination that stimulates the red bone marrow to produce more red cells (fig. 11.7).

Red cells live only about 120 days and are destroyed chiefly in the liver and spleen, where they are engulfed by large phagocytic cells. When red cells are broken down, the hemoglobin is released. The iron is recovered and returned to the red bone marrow for reuse. The heme portion of the molecule undergoes chemical degradation and is excreted by the liver in the bile as bile pigments. The bile pigments are primarily responsible for the color of feces.

Anemia   When there is an insufficient number of red cells or the cells do not have enough hemoglobin, the individual suffers from **anemia** and has a tired, rundown feeling.[1]

In iron-deficiency anemia, the hemoglobin count is low. It may be that the diet does not contain enough iron. Certain foods, such as spinach, raisins,

[1]Sickle-cell anemia and Cooley's anemia are discussed in chapter 21.

Human Anatomy and Physiology

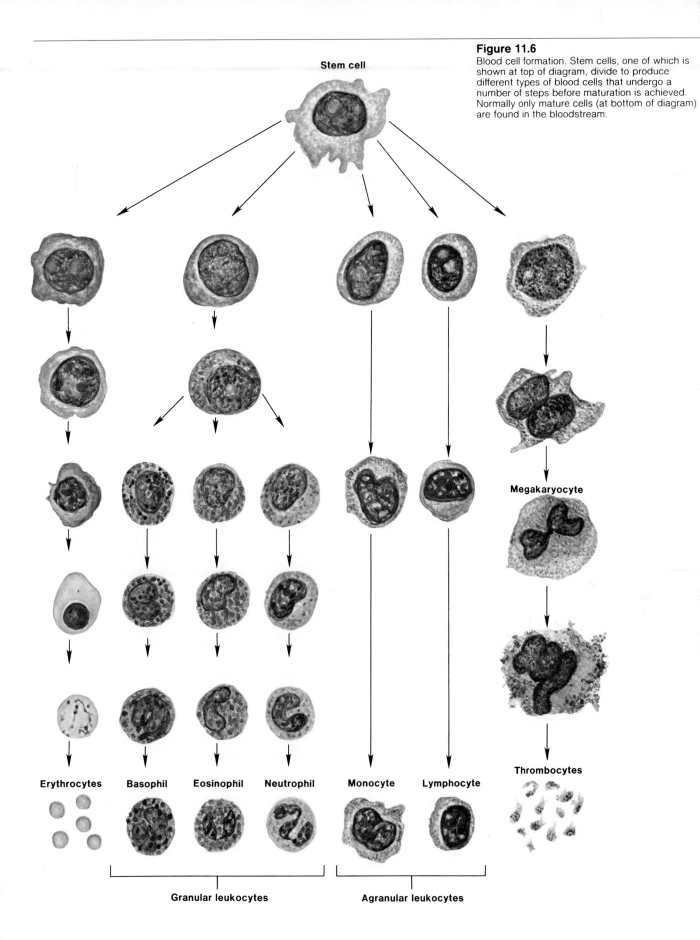

Stem cell

Figure 11.6
Blood cell formation. Stem cells, one of which is shown at top of diagram, divide to produce different types of blood cells that undergo a number of steps before maturation is achieved. Normally only mature cells (at bottom of diagram) are found in the bloodstream.

Megakaryocyte

Thrombocytes

Erythrocytes   **Basophil**   **Eosinophil**   **Neutrophil**   **Monocyte**   **Lymphocyte**

Granular leukocytes   **Agranular leukocytes**

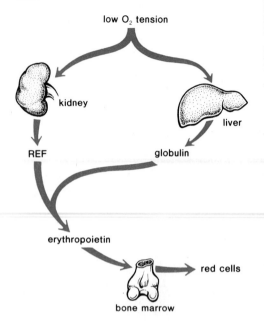

and liver, are rich in iron and the inclusion of these in the diet can help prevent
this type of anemia.

In another type of anemia, called **pernicious anemia,** the digestive tract
is unable to absorb enough vitamin B_{12}. This vitamin is essential to the proper
formation of red cells and without it, immature red cells tend to accumulate
in the bone marrow in large quantities. A special diet and administration of
vitamin B_{12} is an effective treatment for pernicious anemia.

Transport of Carbon Dioxide

Red cells that have given up oxygen to tissue fluid are now ready to take part
in the transport of carbon dioxide. Reduced hemoglobin will combine with
carbon dioxide to form carbaminohemoglobin:

$$Hb + CO_2 \underset{lungs}{\overset{tissues}{\rightleftharpoons}} HbCO_2$$

However, such a combination with hemoglobin actually represents only a small
portion of the carbon dioxide in the blood. Most of the carbon dioxide is transported as the **bicarbonate ion,** HCO_3^-. This ion is formed after carbon dioxide
has combined with water. Carbon dioxide combined with water forms **carbonic acid;** this dissociates (breaks down) to a hydrogen ion and a bicarbonate
ion:

$$CO_2 + H_2O \underset{lungs}{\overset{tissues}{\rightleftharpoons}} H_2CO_3 \underset{lungs}{\overset{tissues}{\rightleftharpoons}} H^+ + HCO_3^-$$

There is an enzyme within red cells, called **carbonic anhydrase,** that
speeds up this reaction. The released hydrogen ions, which could drastically
change the pH, are absorbed by the globin portions of hemoglobin, and the
bicarbonate ions diffuse out of the red cells to be carried in the plasma. Reduced hemoglobin, which combines with a hydrogen ion, may be symbolized
as HHb. The latter plays a vital role in maintaining the pH of the blood.

Once systemic venous blood has reached the lungs, the reaction just described takes place in the reverse: the bicarbonate ion joins with a hydrogen
ion to form carbonic acid and this splits into carbon dioxide and water. The
carbon dioxide diffuses out of the blood into the lungs for expiration. Now
hemoglobin is ready again to transport oxygen. Table 11.3 summarizes the
structure and function of hemoglobin.

Table 11.3	Hemoglobin
Heme	**Globin**
Nonprotein	Protein
Contains iron	
Carries oxygen	Carries carbon dioxide; acts as a buffer; absorbs H^+
Becomes bile pigments	May be reused

Capillary Exchange within the Tissues

Arterial Side

When arterial blood enters the tissue capillaries (fig. 11.8) it is bright red
because the red cells are carrying oxygen. It is also rich in nutrients that are
dissolved in the plasma. At this end of the capillary, blood pressure (40 mm
Hg) is higher than the osmotic pressure of the blood (15 mm Hg). The blood
pressure, you will recall, is created by the pumping of the heart. The osmotic
pressure is caused by the presence of salts and also, in particular, by the plasma
proteins that are too large to pass through the pores (fig. 11.9) of the capillary
wall. Since the blood pressure is higher than the osmotic pressure, fluid, together with oxygen and nutrients (glucose and amino acids), will exit from
the capillary. This is a **filtration** process because large substances, such as red
cells and plasma proteins, remain behind, while small substances such as water
and nutrient molecules leave the capillaries. *Tissue fluid,* created by this process, consists of all the components of plasma except the proteins.

Figure 11.8
Diagram of a capillary illustrating the exchanges that take place and the forces that aid the process. At the arterial end of a capillary, the blood pressure is higher than the osmotic pressure and therefore water, oxygen, and glucose tend to leave the bloodstream. At the venous end of a capillary, the osmotic pressure is higher than the blood pressure and, therefore, water, ammonia, and carbon dixoide tend to enter the bloodstream. Notice that the red cells and plasma proteins are too large to exit from a capillary.

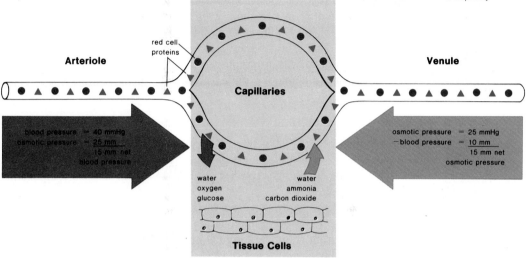

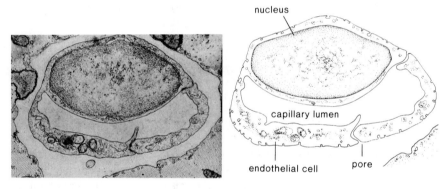

Figure 11.9
Electron micrograph of a cross section of a capillary in heart muscle and a drawing of this same cross section. The wall of a capillary consists of a single layer of flattened endothelial cells. The electron micrograph (and the drawing) shows parts of two endothelial cells: the nucleus and part of the cytoplasm of one cell, and a small part of the cytoplasm of another cell. Notice the pore that occurs where the two cells meet. Small molecules can pass through these pores.

Midsection

Along the length of the capillary, molecules will follow their concentration gradient as diffusion occurs. Diffusion, you will recall, is the movement of molecules from an area of greater concentration to an area of lesser concentration. The area of greater concentration for nutrients is always the blood, because after these molecules have passed into the tissue fluid they are taken up and metabolized by the tissue cells. The cells use glucose and oxygen in the process of cellular respiration, and they use amino acids for protein synthesis. Following cellular respiration, the cells give off caron dioxide and water. Whenever the cells break down amino acids, they remove the amino group, which is released as ammonia. Carbon dioxide and ammonia, being waste products of metabolism, leave the cell by diffusion. Since tissue fluid is always the area of greater concentration for these waste materials, they diffuse into the capillary.

Figure 11.10

Scanning electron micrograph showing an erythrocyte caught in the fibrin threads of a clot. Fibrin threads form from activated fibrinogen, a normal component of blood plasma.

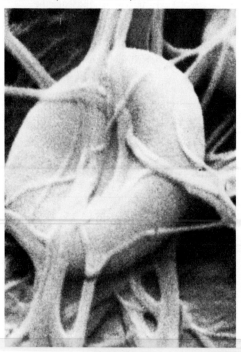

Figure 11.11

When blood clots, serum is squeezed out as a solid plug is formed. In a blood vessel this plug helps prevent further blood loss.

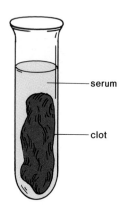

- serum
- clot

Table 11.4	Body fluids
Name	**Composition**
Blood	Formed elements and plasma
Plasma	Liquid portion of blood
Serum	Plasma minus fibrinogen
Tissue fluid	Plasma minus proteins
Lymph	Tissue fluid within lymph vessels

Venous Side

At the venous end of the capillary, blood pressure is much reduced (10 mm Hg), as can be verified by reviewing figure 10.16 in the previous chapter. However, there is no reduction in osmotic pressure (25 mm Hg), which tends to force fluid into the capillary. Therefore, fluid now enters the capillary. When it does, it brings with it additional amounts of waste molecules (carbon dioxide and ammonia). As the blood leaves the capillaries, it is deep purple in color because the red cells contain reduced hemoglobin. Carbon dioxide is carried as the bicarbonate ion, and this, along with ammonia, is dissolved in the plasma.

This system of retrieving fluid by means of osmotic pressure is not completely effective. There is always some fluid that is left and not picked up at the venous end. This excess tissue fluid enters the lymph vessels. *Lymph* is tissue fluid contained within lymph vessels. Lymph is returned to systemic venous blood when the major lymph vessels enter subclavian veins (p. 217).

Blood Clotting

When an injury occurs to a blood vessel, **clotting,** or coagulation, of the blood takes place. This is obviously a protective mechanism to prevent excessive blood loss. As such, blood clotting is another mechanism by which blood components maintain homeostasis.

Portions of the blood that have been identified as necessary to clotting are (1) platelets, (2) prothrombin, a globulin protein, and (3) fibrinogen. **Platelets** result from fragmentation of certain large cells, called megakaryocytes, in the red bone marrow (fig. 11.6). They are produced at a rate of 200 billion a day and the bloodstream possesses more than a trillion (fig. 11.10). **Fibrinogen** and **prothrombin** are manufactured and deposited in the blood by the liver. Vitamin K is necessary to the production of prothrombin, and if by chance this vitamin is missing from the diet, hemorrhagic disorders develop.

The steps necessary for blood clotting are quite complex but may be summarized in this simplified manner:

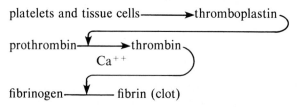

When a blood vessel is damaged, platelets clump at the site of the puncture and partially seal the leak. They and the injured tissues release an enzyme called **thromboplastin** that converts prothrombin to **thrombin.** Thrombin, in turn, acts as an enzyme that severs two short amino acid chains from each fibrinogen molecule. These activated fragments then join end to end, forming long threads of **fibrin.** Fibrin threads wind around the platelet plug in the damaged area of the blood vessel and provide the framework for the clot. A blood clot is red because red cells are also trapped within the fibrin threads (fig. 11.10). Although the red cells have no function in clotting, it is their presence that makes a clot appear red.

If blood is placed in a test tube, blood clotting can be prevented by adding citrate or any other substance that combines with calcium. This is because calcium (Ca^{++}) ions are required for the blood clotting reactions to occur. Also since blood clotting is an enzymatic process, clotting takes place at a faster rate if blood is warmed than if it is kept cool. If the blood is allowed to clot, a yellowish fluid comes to lie above the clotted material (fig. 11.11). This fluid is called **serum,** and it contains all the components of plasma except fibrinogen. Since we have now used a number of different terms to refer to portions of the blood, table 11.4 reviews these terms for you.

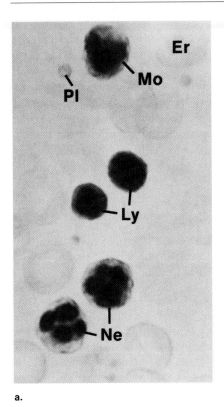

a.

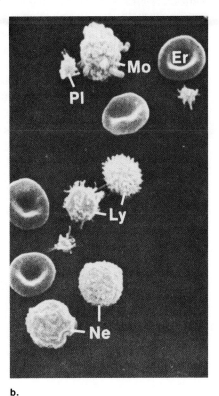

b.

Figure 11.12

White cells (leukocytes) as they appear under the a light microscope (a.) and the scanning electron microscope (b.). The neutrophils *(Ne)* are at least twice as common as the lymphocytes *(Ly)*. The monocytes *(Mo)* occur infrequently. Notice the multilobed nucleus in the neutrophil and the round nucleus in the other two types. The other cells, *Er* (erythrocyte) and *Pl* (platelet) are other types of formed elements.

From: TISSUES AND ORGANS: A TEXT-ATLAS OF SCANNING ELECTRON MICROSCOPY by R. G. Kessel and R. Kardon. W. H. Freeman and Company. © 1979.

A fibrin clot is only a temporary way to repair the blood vessel. An enzyme called **plasmin** destroys the fibrin network and restores the fluidity of plasma. This is a protective measure because a blood clot can act as a thrombus or an embolus. In either case it interferes with circulation and may even cause the death of tissues in the area (p. 220).

Infection Fighting

The body defends itself against parasites, such as bacteria and viruses, in several ways. The so-called first line of defense is the outer covering (skin and mucous membranes), which resists invasion by parasites. The second line of defense is dependent on two components of blood: white cells and gamma globulins. Infection fighting is the third of the three ways in which the blood components contribute to homeostasis.

White Cells (Leukocytes)

White cells (fig. 11.12) may be distinguished from red cells in that they are usually larger; have a nucleus; and without staining, would appear to be white in color. With staining, white cells characteristically appear a bluish shade in color. White cells are less numerous than red cells, with only 7,000 to 8,000 cells per cubic millimeter.

Table 11.5 lists the different types of white cells and figure 11.6 diagrams their maturation. On the basis of structure, it is possible to divide white cells into the **granulocytes** and the **agranulocytes**. The granulocytes have granules in the cytoplasm and a many-lobed nucleus joined by nuclear threads; therefore, they are called **polymorphonuclear**. Granulocytes are formed in the red bone marrow and perhaps are derived from the same type parent cell as the red cells. The agranulocytes do not have granules and have a circular, or indented, nucleus. They are produced in lymphoid tissue found in the bone

Table 11.5	White cells (leukocytes)	
	Granulocytes (polymorphonuclear)	
	Size	Granules stain
Neutrophils	9–12μm	Lavender
Eosinophils	9–12μm	Red
Basophils	9–12μm	Deep blue
	Agranulocytes	
	Size	Type of nucleus
Monocytes	12–20μm	Indented
Lymphocytes	8–10μm	Large

marrow and in the spleen, lymph nodes, and tonsils. **Leukemia** is a form of cancer characterized by an uncontrolled production of leukocytes in either the red bone marrow or in lymphoid tissue. In both types of leukemia, the white cells are numerous but nonfunctional; therefore the victim has a lowered resistance to infections.

Infection fighting by white cells is primarily dependent on the neutrophils, which comprise 60 to 70 percent of all leukocytes, and the lymphocytes, which make up 25 to 30 percent of the leukocytes. Neutrophils are **phagocytic;** they destroy bacteria and viruses by traveling to the site of invasion and engulfing the foe. Lymphocytes secrete gamma globulins[2] called immunoglobulins, or **antibodies,** that combine with foreign substances to inactivate them. Neutrophils and lymphocytes may be contrasted in the following manner:

Neutrophils	Lymphocytes
Granules in cytoplasm	No granules in cytoplasm
Polymorphonuclear	Mononuclear
Produced in bone marrow	Produced in lymphoid tissue
Phagocytic	Make antibodies

Like the neutrophils, monocytes and eosinophils are phagocytic.

Inflammatory Reaction

After bacteria and viruses have invaded the body, they destroy cells either by producing poisonous chemicals called **toxins** or by attacking the cells directly. Damaged tissues and basophils, too, release histamine and other substances that dilate blood vessels and increase capillary wall permeability. As fluid escapes the capillaries, swelling results; now the site of invasion is a region of **inflammation.**

Neutrophils and monocytes enter the inflamed area by squeezing through pores within the capillary walls. They move in an amoeboid manner; part of the cell protrudes and attaches to a stationary object and the remainder of the cell then pulls itself forward. When a neutrophil reaches a foreign substance and phagocytizes it (fig. 11.13), an intracellular vacuole is formed. Now the engulfed material is destroyed or neutralized by hydrolytic enzymes when the vacuole combines with a lysosome. (The granules of a neutrophil are in fact lysosomes.) Some neutrophils die and these, along with dead tissue, cells, bacteria, and living white cells, form **pus,** a thick yellowish fluid. The presence of pus indicates that the body is trying to overcome the infection.

Once monocytes have arrived on the scene, they swell five to ten times their original size and become **macrophages,** large phagocytic cells that are able to devour a hundred invaders and still survive. Macrophages often leave the bloodstream and take up residence in different tissues of the body; for example, the lungs, liver, and connective tissue contain active macrophages. Here they not only engulf bacteria but also act as scavengers that devour old blood cells, bits of dead tissue, and other debris.

Many infections bring about an explosive increase in the number of leukocytes, largely because the inflamed tissues liberate a substance that passes, by way of the blood, to the bone marrow, where it stimulates the production and release of white cells, usually neutrophils. Some illnesses, however, cause an increase in other types of white cells. For example, the characteristic finding in the viral disease **mononucleosis** is a great number of atypical lymphocytes that are larger than mature lymphocytes and stain more darkly. This condition takes its name from the fact that lymphocytes are *mononuclear.*

[2]The gamma globulins get their name from the fact that it was observed that if globulins underwent electrophoresis (were put in an electrical field), they separated into three major components called alpha globulin, beta globulin, and gamma globulin. Almost all circulating antibodies were found in the gamma globulin fraction and, as a result, this term is used for circulating antibodies.

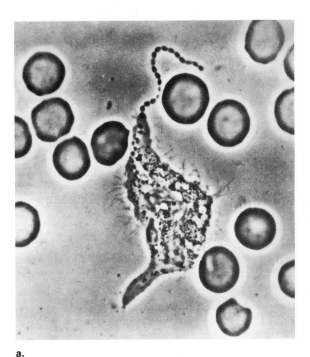

a.

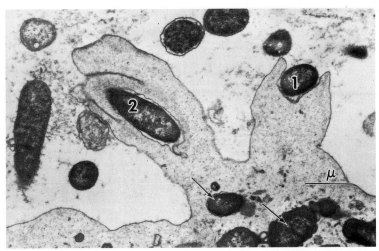

b.

Figure 11.13
Phagocytosis by neutrophils. *a.* Neutrophil ingesting a chain of streptococcal bacteria. *b.* Electron micrograph of a neutrophil ingesting bacteria (1 and 2) by means of pseudopodia. Subsequently, bacteria are seen within neutrophil *(arrows)*. When an infection occurs, the bone marrow sends out neutrophils by the millions. Many of these die in defending the body.

Antibodies

Parasites and their toxins cause lymphocytes to produce antibodies. Each lymphocyte produces one type of antibody that is specific for one type of antigen. **Antigens** are most often proteins and polysaccharides located, for example, in the outer covering of a parasite or present in its toxin. Antibodies combine with their antigens (fig. 11.14) in such a way that the antigens are rendered harmless. Sometimes the antibodies cause precipitation of the antigen, or agglutination (clumping) of the antigen, or simply prepare it for phagocytosis. In any case, it is well to keep in mind that the antigen is the foreigner, and the antibody is the substance prepared by the body. The antigen-antibody reaction may be symbolized as follows:

antigen   +   antibody ⟶ inactive complex
(foreign   (globulin
substance)   protein)

The **antigen-antibody reaction** is a type of lock-and-key reaction in which the two molecules fit together as do a lock and key. This seems surprising at first because it has been shown that all antibodies have the same overall shape. Even so, each type of antibody has a variable region, a unique sequence of amino acids that results in a receptor site that is capable of combining with one type antigen. In other words, this particular sequence of amino acids shapes a site where the antibody fits the antigen.

Immunity

An individual is actively immune when the body is forever capable of producing antibodies that can react to a specific disease-causing antigen. The blood in these individuals always contains lymphocytes that are capable of producing these antibodies. Exposure to an antigen, either naturally or by way of a vaccine, can cause active immunity to develop. Chapter 12 is about immunity and explores the topic in detail.

Figure 11.14
An electron micrograph that shows antibodies *(light areas)* attached to viruses *(dark areas)*. Each antibody will combine with only one type of antigen, usually a foreign protein such as those that occur in the coat of a virus.

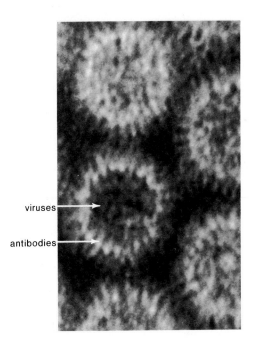

viruses

antibodies

Figure 11.15
Agglutination. *a.* In this blood sample, the red cells did not agglutinate. *b.* In this blood sample, the red cells did agglutinate because a specific antibody was present to react with an antigen on the red cells. It is always necessary to check for possible agglutination between the recipient's and donor's blood before blood is given to a patient.

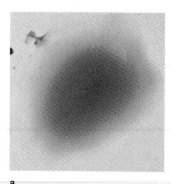

a.

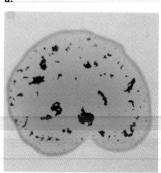

b.

Table 11.6 Blood groups

Type	Antigen	Antibody	% U.S. black	% U.S. Caucasian
A	A	b	25	41
B	B	a	20	7
AB	A,B	none	4	2
O	none	a,b	51	50

Blood Typing

ABO Grouping

After the preceding discussion, it should not be surprising to learn that the red cells of one individual may be antigenic to another individual. Two antigens that may be present on the red cells have been designated as **A** and **B**. As table 11.6 shows, an individual may have one of these antigens (i.e., type A or type B), or both (type AB), or neither (type O). Thus, blood type is dependent on which antigens are present on the red cells. As you can see, type O blood is most common in the United States.[3]

Within the plasma of an individual, there are antibodies to the antigens that are *not* present on that individual's red cells. Thus, for example, type A blood has antibody b. Type AB blood has no antibodies because both antigens are on the red cells.

In order to determine what type of blood a person may receive by transfusion, it is necessary to test whether *the donor's red blood cells are compatible with the recipient's plasma.* If they are not compatible, **agglutination,** or clumping, (fig. 11.15) of the donor's cells will occur, and this can be quite dangerous since it may stop circulation. Agglutination must not be confused with clotting, which refers to the change of fibrinogen to fibrin threads. Agglutination is simply a clumping of cells.

To illustrate incompatibility, let us consider the possibility of giving type A blood to type B:

donor A cells × recipient B plasma
antigen A   antibodies a

This combination will obviously result in agglutination, and therefore these two types of blood are considered incompatible. However, theoretically it would be possible to give type O blood to any recipient:

donor O cells × all recipient plasmas
no antigen   antibodies, a, b, ab, or none

Since no combination will result in agglutination, type O blood has been called the universal donor. On the other hand, type AB blood has been called the universal recipient because type AB can theoretically receive blood from anyone:

all donor cells   ×   recipient AB plasma
antigens A, B, AB, or O   no antibodies

In practice, determination of the ABO blood type alone is not sufficient to indicate which people are compatible donors and recipients because there are many more antigens than the ones just considered. Therefore, before blood is given to another person, a cross-match is done in which the donor's cells are

[3]Inheritance of blood type is discussed on page 453.

Human Anatomy and Physiology

● Rh antigen   ▽ Rh antibody

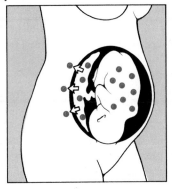 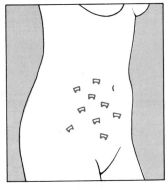 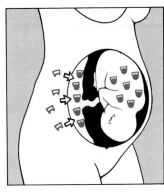

a. during pregnancy   **b.** before delivery   **c.** months and years later   **d.** subsequent pregnancy

first mixed with the recipient's plasma. Further, the recipient's red cells are mixed with the donor's plasma to test for any possible agglutination. Only if both of these combinations show no agglutination is blood of one person given to another.

Rh System

Another important antigen in matching blood types is the **Rh factor.** Persons with this particular antigen on the red cells are Rh positive; those without it are Rh negative. (Only 15 percent of Caucasians are negative.) Rh negative individuals do not normally make antibodies to the Rh factor but they will make them when exposed to the Rh factor. It is possible to extract these antibodies and use them for blood type testing. When Rh positive blood is mixed with Rh antibodies, agglutination occurs.

The Rh factor is particularly important during pregnancy (fig. 11.16). If the mother is Rh negative and the father is Rh positive, the child may be Rh positive. The Rh positive red cells begin leaking across into the mother's circulatory system as placental tissues normally break down before and at birth. This causes the mother to produce Rh antibodies. If the mother becomes pregnant with another Rh positive baby, Rh antibodies (but not antibodies a and b discussed earlier) may cross the placenta and cause destruction of the child's red cells. This is called **erythroblastosis.**

This problem has been solved by giving Rh negative women, an Rh immune globulin injection called RHOgam just after the birth of any Rh positive child. This injection contains Rh antibodies that attack the baby's red cells before these cells can stimulate the mother to produce her own antibodies.

Testing Blood Type

The standard test to determine blood type consists of putting a drop of anti-A (containing antibodies a), anti-B (containing antibodies b), and anti-Rh on a slide. A drop of the person's blood is added to each of these. If agglutination occurs, the person has this antigen on the red cells. Figure 11.17 illustrates the proper interpretation for several possible results.

Figure 11.17
A diagrammatic representation of ABO and Rh blood typing reactions. Plasma antibody types used in the tests are indicated at the top of the columns. When agglutination occurs, the individual has the antigen on the red cells and the blood type is then known.

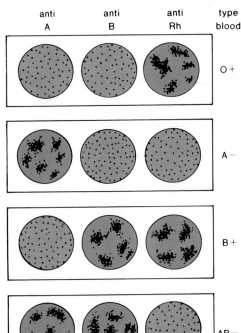

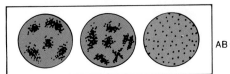

Summary

There are three types of extracellular body fluids: blood, tissue fluid, and lymph. When blood passes through a capillary, tissue fluid is formed. Lymph is tissue fluid contained with lymph vessels.

Blood, which is composed of two parts—formed elements and plasma— has three functions: transport to and from cells, clotting, and infection fighting. All of these can be related to its primary function of maintaining a constant internal environment, or homeostasis.

The transport of nutrients and wastes in the blood presents no problems because they are dissolved in the plasma. However, the transport of oxygen relies on a respiratory pigment, hemoglobin, that is contained within the red cells. There are 5 million red cells per cubic millimeter of blood; they are small ($8\mu m$), biconcave disks that lack a nucleus and are made in the bone marrow. The production of red cells is controlled by oxygen tension in the environment. When oxygen tension is low, more red cells are formed.

The end result of transport—capillary exchange in the tissues—is facilitated by blood pressure and osmotic pressure. At the arterial end of a capillary, blood pressure is greater than osmotic pressure; therefore, water leaves the capillary along with oxygen and nutrients that diffuse from the capillary. At the venous end, osmotic pressure created by the presence of proteins exceeds blood pressure, causing water to enter the capillary. Carbon dioxide and nitrogenous waste enter the capillary by diffusion.

Blood clotting requires a series of enzymatic reactions involving blood platelets, prothrombin, and fibrinogen. In the final reaction, fibrinogen becomes fibrin threads entrapping cells. The fluid that escapes from a clot is called serum and consists of plasma minus fibrinogen.

White cells and gamma globulin proteins are required in the process of fighting infections. The two most important of the white cells are the phagocytic neutrophils, which have granules in the cytoplasm and a polymorphonucleus (many-lobed nucleus), and the antibody-producing lymphocytes, which do not have granules and have a spherical nucleus. Neutrophils are produced in the bone marrow, while lymphocytes are produced in lymphoid tissue. The neutrophils are mobile and travel by amoeboid movement to the site of invasion. They squeeze out of the capillaries and begin to engulf the foreign bacteria and viruses they find there. Some die, and these, along with dead tissue cells, form pus. Any foreign protein is recognized by the body as an antigen, or a substance that can stimulate the production of antibodies, which are proteins that can combine with an antigen because their two shapes fit together.

Red blood cells of an individual are not necessarily received without difficulty by another individual. For example, antigens A and B are on the red cells. In the plasma there are two possible antibodies: a or b. If the corresponding antigen and antibody are put together, clumping, or agglutination, occurs; in this way, the blood type of an individual may be determined in the laboratory. After determination of the blood type, it is theoretically possible to decide who can give blood to whom. For this, it is necessary to consider the donor's antigens and the recipient's antibodies.

Another important antigen is the Rh antigen. This particular antigen is important during pregnancy because an Rh negative mother may form antibodies to the Rh antigen after the birth of a child who is Rh positive. These antibodies can cross the placenta to destroy the red cells of any subsequent Rh positive child.

Human Anatomy and Physiology

Study Questions

1. Define blood, plasma, tissue fluid, lymph, serum. (p. 236)
2. Name three functions of blood and tell how they are related to the maintenance of homeostasis. (p. 228)
3. State the major components of plasma. Name the plasma proteins and tell their common function as well as their specific functions. (p. 229)
4. Give the equation for the oxygenation of reduced hemoglobin. Where does this reaction occur? Where does the reverse reaction occur? (p. 229)
5. Discuss the life cycle of red blood cells. (p. 232) Compare the structure and function of heme to globin. (p. 234)
6. Give an equation that indicates how CO_2 is carried in the blood. Indicate the direction of the reaction in the tissues and in the lungs. (p. 234) In what way does hemoglobin aid the process of transporting CO_2? (p. 234)
7. Draw a diagram of a capillary, illustrating the exchanges that occur in the tissues. What forces operate to facilitate exchange of molecules across the capillary wall? (pp. 234–36)
8. Name the steps that take place when blood clots. Which substances are present in the blood at all times and which appear during the clotting process? (p. 236)
9. Name and discuss two ways that blood fights infection. Associate each of these with a particular type of white blood cell. (p. 238)
10. What are the four ABO blood types in humans? (p. 240) What formula can be used to determine which types can give blood to each other? (p. 240)
11. Problems can arise during childbearing if the mother is which Rh type and the father is which Rh type? Explain why this is so. (p. 247)

Selected Key Terms

tissue fluid (tish'u floo'id)
formed element (form'd el'ĕ-ment)
plasma (plaz'mah)
hemoglobin (he''mo-glo'bin)
anemia (ah-ne'me-ah)
bicarbonate ion (bi-kar'bo-nāt i'on)
platelet (plāt'let)
thromboplastin (throm''bo-plas'tin)
thrombin (throm'bin)
fibrin threads (fi'brin thredz)
granulocytes (gran'u-lo-sīts)

agranulocytes (ah-gran'u-lo-sīts'')
polymorphonuclear
  (pol''e-mor''fo-nu'kle-ar)
phagocytosis (fag''o-si-to'sis)
antibody (an'tĭ-bod''e)
neutrophil (nu'tro-fil)
lymphocytes (lim'fo-sīts)
macrophage (mak'ro-fāj)
agglutination (ah-gloo''ti-na'shun)
Rh factor (ar'āch fak'tor)

Further Readings

Berne, R. M., and Levy, M. N. 1981. *Cardiovascular physiology.* 4th ed. St. Louis: C. V. Mosby.
Doolittle, R. F. 1981. Fibrinogen and fibrin. *Scientific American* 245(6):126.
Lerner, R. A., and Dixon, F. J. 1973. The human lymphocyte as an experimental animal. *Scientific American* 228(6):12.
Perutz, M. F. 1964. The hemoglobin molecule. *Scientific American* 211(5):64.
Zucker, M. B. 1961. Blood platelets. *Scientific American* 204(2):58.
———. 1980. The functioning of blood platelets. *Scientific American* 242(6):86.

12

Immunity

Chapter concepts

1 The immune system consists of lymphocytes and the structures that produce them and within which they reside.

2 There are two types of immune responses. One type of lymphocyte, the B cell, is responsible for antibody-mediated immunity and another type, the T cell, is responsible for cell-mediated immunity.

3 Active immunity, the ability of the body to produce specific antibodies, can be promoted by immunization.

4 Passive immunity, acquired when antibodies are received from an outside source, has gained importance due to proposed methods of producing specific human antibodies in the laboratory.

5 While the immune system preserves our existence, it is also responsible for certain undesirable effects, such as tissue rejection, allergies, and autoimmune diseases.

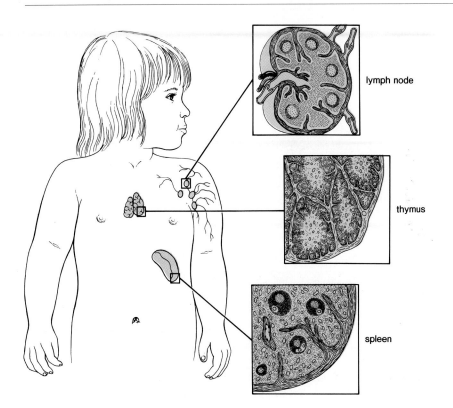

Figure 12.1
The immune system includes all those organs in which lymphocytes are produced and found, such as lymph nodes, thymus, and spleen. The thymus is important to the maturation of T cells, a particular type of lymphocyte. The lymphocytes within lymph nodes help purify lymph and those within the spleen help purify the blood.

lymph node

thymus

spleen

The immune system (fig. 12.1) protects us from disease. It consists of about a trillion **lymphocytes,** the second most common white cell (p. 238); the spleen; lymph vessels and lymph nodes, where lymphocytes are found; and the bone marrow and thymus, where lymphocytes are manufactured and processed. Lymphocytes carry cell membrane **receptors** that are capable of recognizing **epitopes,** small portions of antigens. **Antigens** are foreign proteins (in some instances, polysaccharides) that may be introduced into an animal by viruses and bacteria, plant pollen, or any cell or cell product not normally a part of the body. For example, incompatible red blood cells are antigenic to the recipient, as was discussed in chapter 11.

A lymphocyte's receptors are capable of recognizing just one particular epitope with which they combine in a lock-and-key manner. It is estimated that during our lifetime we encounter a million different antigens and therefore we need the same number of different lymphocytes to protect ourselves against antigens. How does the body generate such a great variety of lymphocytes, each with unique receptors? Lymphocytes are derived from bone marrow stem cells (fig. 12.2), and diversification occurs during the maturation process so that in the end each has a specific gene that codes for only one type of receptor. If by chance a lymphocyte develops that has receptors for the body's own proteins, it is suppressed and develops no further. Thus, it is said that the immune system *learns to tell self from nonself.* This is why lymphocytes do not normally carry receptors for the tissues of the organism that produced them.

Figure 12.2
Diversification of lymphocytes. A lymphoid stem
cell in bone marrow divides to produce a number of
cells that diversify during maturation so that each
one comes to have just one type of receptor in the
cell membrane. Each type of receptor is capable of
combining with only one type of antigen.

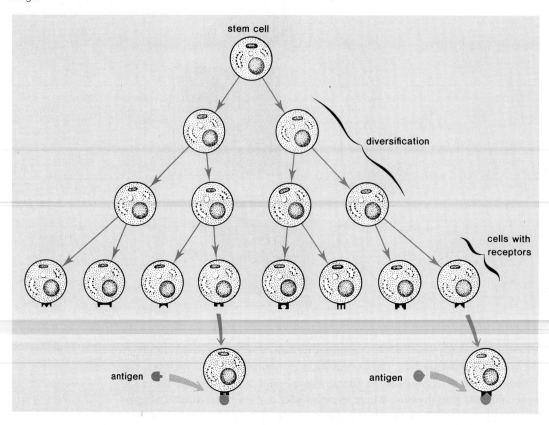

B Cells versus T Cells

Lymphocytes are of two types: **T (thymus-dependent) cells** and **B (bone-marrow-dependent) cells.** Both of these are derived from lymphoid bone marrow stem cells. However, the stem cells that produce T cells have passed through the thymus, a flat, pinkish gray, two-lobed organ that lies high in the chest, before they finally come to reside in lymphoid tissue. The stem cells that produce B cells have not passed through the thymus (fig. 12.3).

T cells and B cells have different functions (table 12.1). In general, T cells are responsible for **cell-mediated immunity,** which has two primary aspects: (1) the T cell itself attacks the foreign antigen, and (2) the antigen is usually a part of another cell. T cells will attack cells bearing antigens foreign to the individual, even in the tissues.

B cells are responsible for **antibody-mediated immunity,** which also has two primary aspects: (1) a B cell produces **antibodies,** proteins that are capable of combining with and inactivating antigens, which in some cases are toxins, or poisons, released by bacteria, and (2) the antibodies are released into the bloodstream or lymph.

Figure 12.3
Both B and T cells are derived from bone marrow
stem cells; however, the stem cells that produce T
cells have passed through the thymus, whereas
those that produce B cells have not. When a B cell
is stimulated, it divides to produce plasma cells
and memory cells. When a T cell is stimulated, it
divides to produce T cells that release
lymphokines.

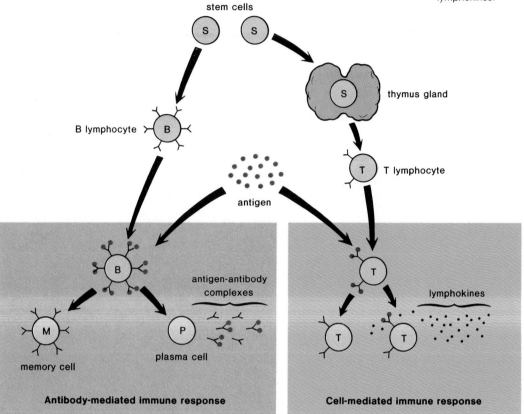

stem cells

B lymphocyte

thymus gland

T lymphocyte

antigen

antigen-antibody
complexes

lymphokines

memory cell

plasma cell

Antibody-mediated immune response

Cell-mediated immune response

Table 12.1 Some properties of T cells and B cells

Property	T cells (cell-mediated immunity)	B cells (antibody-mediated immunity)
Antigen-binding receptors on cell surface	Specific receptors	Specific receptors
Response to binding of antigen	Enlarge, multiply, liberate lymphokines	Enlarge, multiply to produce plasma cells that secrete antibodies
Cytotoxic activity	Antigen-stimulated T cells kill antigen-bearing target cells on contact	None
Function in antibody production	Stimulate antibody production by B cells	Synthesize and liberate antibodies
Effect on macrophages	Stimulate phagocytic activity of macrophages	None

Figure 12.4

Clonal selection theory. The antigen finds or selects
the B cell that clones, producing, by the fifth day,
many mature plasma cells, which actively secrete
antibodies and memory cells that retain the ability
to secrete these antibodies at a future time.

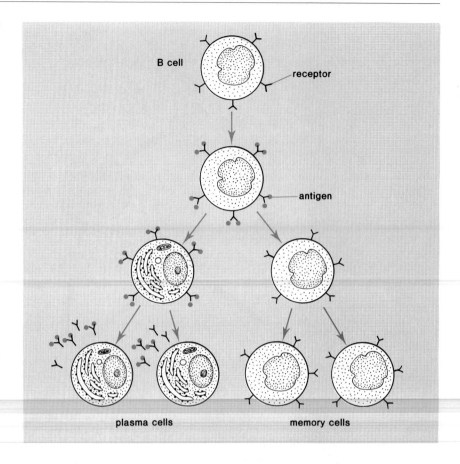

Figure 12.5

Antigen-antibody reaction. *a.* An antibody contains
two heavy *(long)* amino acid chains and two light
(short) amino acid chains arranged to give two
variable regions where a particular antigen is
capable of binding with the antibody in a lock-and-
key manner. *b.* Quite often the antigen-antibody
reaction produces complexes of antigens
combined with antibodies.

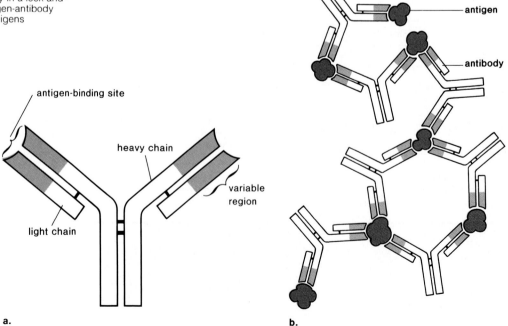

Human Anatomy and Physiology

Antibody-mediated Immunity

The antibodies produced by a B cell have a structure identical to the receptors projecting from the cell's membrane. B cells will not produce antibodies until they have come into contact with an antigen that combines with their receptors in a lock-and-key manner. Even then, they will only produce antibodies when stimulated to do so by T cells. So-called *helper T cells* encourage antigen-triggered B cells to transform into antigen-producing cells, and *suppressor T cells* prevent them from doing so.

When stimulated to do so, a B cell divides and produces a number of **plasma cells** (fig. 12.4), which are enlarged, active, antibody-producing lymphocytes. All the plasma cells derived from one parent lymphocyte are called **clones,** and they all produce the same type antibody. Notice that the presence of the antigen determines which lymphocyte will be stimulated to produce antibodies. This is called the **clonal selection theory** because the antigen has selected which B cells will produce a clone of plasma cells. When antibody production is high enough, suppressor T cells prevent further antibody production; the antigen disappears from the system and development of plasma cells ceases.

Some members of the clone do not participate in the current antibody production; instead they remain in the bloodstream as **memory cells** (fig. 12.4), forever capable of producing the antibody specific to a particular antigen. The individual is said to be **actively immune** because a certain number of antibodies are always present in the system and also because memory cells can produce more plasma cells if the same antigen invades the system again.

Antigen-Antibody Complexes

An antibody is a Y-shaped molecule having two long "heavy" chains and two short "light" chains of amino acids. At the end of all four chains of every antibody is a tiny segment, called the variable portion, that binds to an antigen in a lock-and-key manner. In figure 12.5, the variable region of each antibody is shown in color.

The antigen-antibody reaction can take several forms as indicated in table 12.2. Many times, the antigen-antibody complex, sometimes called the **immune complex,** marks the antigen for destruction by other forces. For example, the complex may be engulfed by neutrophils or macrophages or it may activate a portion of blood serum called complement. **Complement** refers to nine different proteins that become enzymatic when activated. Among various functions, complement has one that is not duplicated by other members of the immune system. Complement enzymes can break down cell membranes, allowing water and salts to enter a damaged cell until it bursts (fig. 12.6).

Table 12.2 Reactions of antibodies

Types of antibodies	Reactions
Antitoxins	Neutralize toxins of infective agents
Agglutinins	Cause clumping of certain infective agents
Opsonins	Make certain infective agents more susceptible to work of phagocytes
Lysins	Dissolve certain infective agents
Precipitins	Bring about a precipitation of flocculation of extracts of infective agents

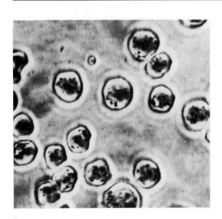
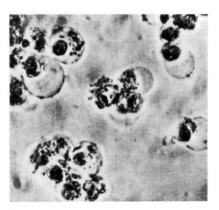

a.   b.

Figure 12.6
Photomicrographs of tumor cells. *a.* Normal appearance. *b.* After treatment with antibody and complement, the cells swell and burst. Complement is a series of about nine proteins that are converted to enzymes when activated by immune complexes.

Table 12.3 Infectious diseases caused by viruses

Respiratory Tract	Nervous System
Common colds	Encephalitis
*Flu	*Polio
Viral pneumonia	*Rabies

Skin Reactions	Liver
*Measles	*Yellow fever
*German measles	Infectious hepatitis
Chicken pox	
*Smallpox	**Other**
Warts	*Mumps
	Herpes
	Cancer

*Vaccines available. Yellow fever, rabies, flu, and smallpox vaccines are given if the situation requires them. Others are routinely given.

Table 12.4 Infectious diseases caused by bacteria

Respiratory Tract	Nervous System
Strep throat (sometimes causing rheumatic and scarlet fever)	*Tetanus
Pneumonia	Botulism
*Whooping cough	Meningitis
*Diphtheria	
*Tuberculosis	**Digestive Tract**
	Food poisoning (salmonella, botulism, and staph)
Skin Reactions	*Typhoid fever
Staph (pimples and boils)	*Cholera
*Gas gangrene (wound infections)	
	Venereal Diseases
	Gonorrhea
	Syphilis

*Vaccines are available. Tuberculosis vaccine is not used in this country. Typhoid fever, cholera, and gas gangrene vaccines are given if the situation requires it. Others are routinely given.

Figure 12.7
Immunization responses. The primary response after the first injection of a vaccine is minimal, but the secondary response after the second injection shows a dramatic rise in the amount of antibody present in serum.

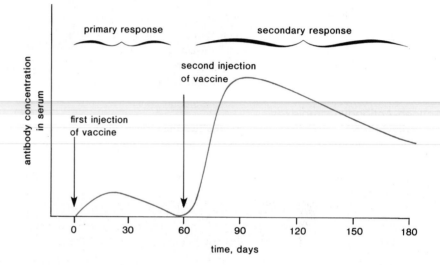

Active Immunity

Active immunity permanently protects an individual against disease organisms, such as most of the viral diseases listed in table 12.3 and many of the bacteral diseases listed in table 12.4. Immunity can occur after an individual has been infected with a disease-causing virus or bacterium. In many instances today, however, it is possible to be medically immunized against a disease. **Vaccines** are most often prepared from attenuated bacteria and viruses. Attenuation means that the bacteria and viruses have been treated in the laboratory so that they are no longer virulent (able to cause disease). New methods of preparing vaccines are becoming available. Recombinant DNA technique has been used to produce a vaccine for hoof-and-mouth disease, a serious illness among domesticated cattle, and investigators are also experimenting with producing synthetic vaccines. A synthetic vaccine is a sequence of amino acids, or epitope, assembled in the laboratory. The synthetic epitope has the ability to produce immunity to the disease-causing agent. These new methods allow the preparation of pure vaccine that cannot cause any untoward effects.

Human Anatomy and Physiology

After a vaccine is injected, it is possible to determine the amount of antibody present in the bloodstream—this is called the **antibody titer.** After the first injection, a primary response occurs. There is a period of several days during which no antibodies are present; then there is a slow rise in the titer, which is followed by a gradual decline (fig. 12.7). After a second injection, a secondary response occurs. The titer rises rapidly to a level much greater than before. A second injection is often called the **"booster shot"** since it boosts the antibody titer to a high level. The antibody titer is now high enough to prevent disease symptoms even if the individual is exposed to the disease. Thereafter the individual is immune to that particular disease.

Primary and secondary responses occur whenever an individual is exposed twice to the same disease-causing agent. The difference in the responses may be related to the number of plasma and memory cells. Upon the second exposure, these cells are already present and antibodies can be rapidly produced.

Immunization can be used in instances other than those that protect humans from future illnesses. For example, the reading on page 252 tells of a novel way that immunization is being used to grow larger sheep.

Passive Immunity

Passive immunity occurs when an individual is given antibodies to combat a disease. Since these antibodies are not produced by the individual's lymphocytes, passive immunity is short-lived. For example, newborn infants possess passive immunity because antibodies have crossed the placenta from their mother's blood. These antibodies soon disappear however, so that at about six months of age, infants become more susceptible to infections.

Even though passive immunity is not lasting, it is sometimes used to prevent illness in a patient who has been unexpectedly exposed to an infectious disease. Usually the person receives an injection of a serum containing antibodies. This may have been taken from donors who have recovered from the illness. In other instances, horses have been immunized and serum taken from them to provide the needed antibodies. Horses are used to produce antibodies against diphtheria, botulism, and tetanus. Occasionally a patient who receives these antibodies becomes ill because the serum contains proteins that the individual's immune system recognizes as foreign. This is called serum sickness.

Monoclonal Antibodies

The benefits of producing pure human antibodies in unlimited quantities are obvious. Methods for producing such antibodies are now being devised (fig. 12.8). In one method, lymphocytes are removed from the body and exposed in vitro (in laboratory glassware) to a particular antigen. Then they are fused with a cancer cell because cancerous cells, unlike normal cells, will divide an unlimited number of times. The fused cells are called **hybridomas;** *hybrid* because they are a fusion of two different cells and *oma* because one of the cells is a cancer cell.

The antibodies produced by hybridoma cells are called **monoclonal antibodies** because they are all the same type and because they are produced by cells derived from the same parent cell. As the reading on page 254 suggests, the hope is that one day monoclonal antibodies will be available in sufficient number to help patients fight infectious diseases and cancer. In the meantime, they are currently being used to allow quick and certain diagnosis of various illnesses, including some forms of cancer, prior to conventional therapy.

Figure 12.8
One possible method for producing human monoclonal antibodies. *a.* Blood sample is taken from patient. *b.* Inactive lymphocytes from sample are exposed to antigen. *c.* Activated lymphocytes are fused with cancer cells. *d.* Resulting hybridomas divide repeatedly, giving many cells that produce monoclonal antibodies.

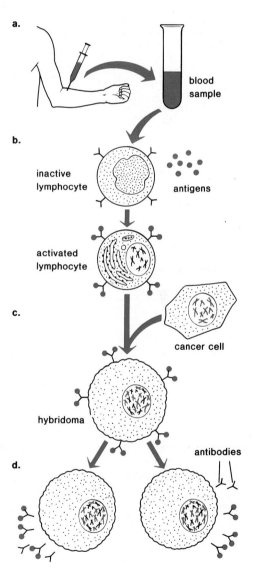

The New Shepherds
of St. Kilda

Grazing in a field at Langford, near Bristol, England, are 12 lambs—six sets of twins. They appear to be normal, contented members of the St. Kilda breed, except that one of each pair is about 15 percent taller than its twin. This came about not through chance, differences in nutrition, or some feat of futuristic breeding. Instead, Langford's extraordinary sheep were the products of a novel attempt to immunize the body against one of its own hormones— in this case, the hormone that slows growth. But similar methods of immunization are being used to probe and regulate several other physiological processes. Those in charge of the Langford experiments, Stuart Spencer and Diane Williamson, are animal physiologists with the Agricultural Research Council's Meat Research Institute. They knew the hazards associated with natural or synthetic hormones, known as steroids, that are used to make animals fatten quickly. Two years ago one class of synthetic steroids, illegally injected into calves, found its way into canned baby foods in Italy. As a result, there are now moves to ban all steroids throughout Europe. Though many veterinarians believe the hormones can be administered safely, residues in the meat of treated animals poses a potential threat to consumers. Currently, more than a quarter of meat animals in the United Kingdom and about half of all U.S. cattle receive steroids.

Spencer and Williamson wondered whether they could duplicate the effect of steroids— which make livestock fatten more quickly and thus boost farmers' profits—by interfering with the normal physiological controls over growth. As a target, they selected somatostatin—a hormone produced by the hypothalamus that acts as a brake on the release of other hormones that make animals grow. If somatostatin's effects could be neutralized, perhaps the animals would put on weight more rapidly.

One strategy was to trick the immune system into making antibodies—like those produced in response to infectious bacteria— that would specifically combine with somatostatin and demobilize it, giving free rein to the growth promoting hormones. But animals do not usually generate antibodies against their own constituents. So the Langford researchers combined sheep somatostatin with a protein taken from human blood plasma. This substance, they theorized, would behave as "foreign" if given to sheep, thereby stimulating the production of antibodies.

When Spencer and Williamson injected six St. Kilda lambs with the sheep-human complex, beginning when the lambs were three weeks old and repeating every two weeks until slaughter at the age of five months, the results were dramatic. The immunized animals grew almost twice as quickly as their respective twins. Though all 12 lambs enjoyed equal access to pasture and feed, the six test animals virtually doubled in weight. As the researchers had hoped, antibodies produced in response to the injected, modified somatostatin were also neutralizing the sheep's natural hormone.

Spencer and Williamson's strategy was not only successful but more effective than anticipated. They knew that the chief role of somatostatin is to regulate release of another class of liver-synthesized hormones that stimulate cells to divide, thus making tissues grow. But they discovered another benefit: As their lambs put on weight, they noticed that not only were the animals' soft tissues increasing abnormally rapidly, their bones were growing longer. In other words, the sheep were likely to have bigger frames and a greater quantity of muscle protein, which consumers recognize as meat.

Although still in its early days, the potential advantages of this kind of immunization over the use of steroids are becoming clear. Not

only does the technique eliminate the possibility of harmful residues, it promises to promote greater growth—more than 70 percent extra weight as compared to a maximum of 20 percent with steroids—because of its effect on bones. And the method might prove simpler and cheaper than using steroids, depending on current investigations into the number and frequency of injections needed for maximum benefit. It may be possible to achieve the same or better results with less frequent doses of vaccine.

More study, however, is required to determine the relationship between growth acceleration and the rate at which somatostatin antibodies increase in the bloodstream. Though no ill effects have come to light in the Langford work, the researchers are watching carefully for any such signs. It is remotely possible that the antibodies might cross-react with something else in the body. This happens, for example, when antibodies against streptococcus bacteria—the well-known culprit in strep throat, among other things—attack heart tissue and cause rheumatic fever.

These concerns are being addressed in a parallel project at the Dutch Research Institute for Animal Husbandry at Zeist in the Netherlands. There Jan Garssen, in collaboration with Stuart Spencer, is developing the technique for breeds of sheep other than St. Kilda (chosen for the Langford experiments because they have a high rate of twinning). Tests with Dutch moor sheep have already established perceptible gains in weight and height as early as a few weeks following the first immunization. Moreover, independent researchers have examined animals before and after slaughter and found no adverse effects on the quality of their carcasses. The meat was as good as usual, with no excess of fat over protein.

Encouraged by these early results, Meat Research Institute scientists are starting to apply their method in breeds of sheep raised commercially in Britain. In addition, they have begun experiments with Large White pigs and found that they too grow more quickly when immunized. They plan to extend the project to cattle next. The future could hold a revolution in animal husbandry, with benefits including not only increased growth but parallel effects such as enhanced milk and wool production.

Immunization can be used in numerous ways—even to make sheep grow fatter.

Antibodies for Sale

It all began as an experiment in basic immunology in a cluttered laboratory in England. Now people talk of a new era in medicine and a billion dollar market. Only seven years after the discovery of a method to produce large amounts of a specific antibody, the first fruits of the technique are appearing on the market. They are medical diagnostic tests. . . .

"It is easier for us and for the FDA to evaluate in a clinical situation markers and [targets] already well understood," says Cole Owen of Hybritech, Inc., in La Jolla, Calif., one of the companies specializing in monoclonal antibodies. In some cases the monoclonal antibody tests, compared with conventional tests, offer greater sensitivity, specificity and speed and can avoid the use of costly instruments or radioactive isotopes.

An example of greater sensitivity is the several pregnancy tests using monoclonal antibodies. These tests are being sold to physicians and hospitals. Like the tests that depend on conventional mixtures of antibodies, the new tests detect increased levels of the hormone human chorionic gonadotropin (HCG). Monoclonal Antibodies, Inc., another specialty company, in Palo Alto, Calif., claims its pregnancy test is 3 to 20 times more sensitive than conventional urine tests. Therefore it can detect pregnancy earlier—10 days after conception and before the first missed menstrual period. . . .

In addition, the time savings that can arise from monoclonal antibody tests are critical in treating some serious infectious diseases like meningitis. Becton Dickinson and Co. of Paramus, N.J., is introducing a diagnostic test for *Neisseria meningitidis* group B that can detect the bacterium in body specimens of cerebrospinal fluid, blood or urine without the necessity of having the bacteria grown in laboratory culture. In the previous methods specimens had to be tended under special conditions for up to two days before the bacteria would be present in high enough concentration for detection. In contrast, the new test gives results in 15 minutes.

Although there may be large earnings for companies in producing improved versions of widely used tests, the most exciting prospects for monoclonal antibody applications are the forays into the frontiers of medical practice. There are already on the market tests for microorganisms for which previously there were no practical and reliable diagnostic methods. There are also new tests to determine the body's immune system status and to detect cancers.

Chlamydia is an example of a widespread disease for which there has been no simple and inexpensive diagnostic test. It is considered the most common sexually transmissible disease, with about 10 million cases occurring annually in the United States. Chlamydia can be effectively treated with antibiotic, but if untreated the bacterial infection can cause such serious complications as pelvic inflammatory disease, which can result in infertility. In a newborn of an infected mother, chlamydia can cause eye and respiratory infections. . . .

Perhaps the most alluring possibilities for monoclonal antibody tests are in diagnosing and monitoring cancers. One test already on the market detects prostate cancer, which had been difficult to diagnose. Both Abbott Laboratories in Chicago and Hybritech have developed monoclonal antibodies to prostatic acid phosphatase, usually called PAP. This enzyme is present at high levels in men with the disease.

Monoclonal antibodies are also being used to assess patients' immune systems. Both Hybritech and Ortho Diagnostic of Raritan, N.J., have produced monoclonal antibodies that bind to specific types of white blood cells and allow them to be counted. "These can be used to evaluate disease states including cancer," says James A. Murray of Johnson and Johnson Co., the parent company of Ortho. They may also be important in diagnosing immune system diseases, such as leukemia and acquired immune deficiency syndrome (AIDS).

Related to diagnostics, but more difficult to develop, are monoclonal antibodies' use as medical imaging agents. Imaging is the localization of a disease-related substance in the human body. It can be considered an intermediate between diagnostics and therapeutics. As in diagnostics, the antibody seeks out and binds to a target substance, for instance a microorganism, injured tissue or cancer cell. An attached tag, most likely a radioactive material, allows clinicians to locate the antibody.

Cancer is the target of most of the imaging projects. Hybritech is working on monoclonal antibodies specific for each of three types of solid tumor—breast, lung and liver. "The use of a radioactive label to identify a tumor will help the surgeon to define where the tumor is and the extent of surgery necessary," says Owen. "It will also confirm that the tumor is not anywhere else—that is, whether it is localized or metastasized." . . .

It may be a short step from imaging with monoclonal antibodies to using them in therapy. If an antibody will seek out a tumor and bind to it, that same antibody may reduce the tumor size. However, if antibodies alone will not correct a medical condition, scientists predict that chemicals attached to the antibodies will make them better able to attack mircroorganisms or malignancies. . . .

Another therapeutic use of monoclonal antibodies is in suppressing in a very specific way the immune response of a patient receiving an organ or tissue transplant. With antibodies to specific white blood cells the patient may be unable to reject the transplant, but still have the immune system functioning to fight infections. Monoclonal antibodies against T cells also have been used successfully in a bone marrow transplant to prevent graft-versus-host disease.

Finally, there is promise that monoclonal antibodies will specifically counter human cancer. Clinical trials are underway for a variety of cancers including leukemias and colorectal carcinoma.

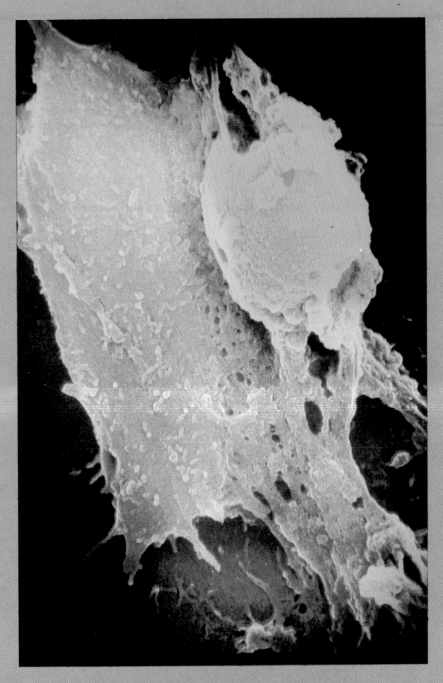

In a sense, it is ironic that this antibody production technique, which starts with a cancer cell, is supplying new diagnostic, imaging and therapeutic tools to fight malignancy diseases. The technique thus tames a cancer in the laboratory and employs its prolific growth to supply previously unobtainable amounts of the natural agents for fighting disease.

Spherical antibody-producing cell fuses with flatter cancerous cell to give a source of monoclonal antibodies.

Figure 12.9

Scanning electron micrograph of three T cells attached to a macrophage. Presumably, at this time, the macrophage presents concentrated antigens to the T cells so that they are stimulated to become "angry killers."

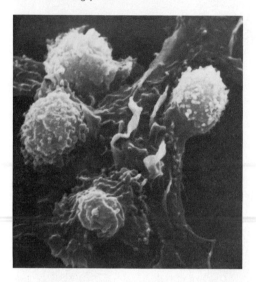

Figure 12.10

Cell-mediated immunity. *a.* Scanning electron micrograph showing a T cell *(left)* attacking a cancer cell to its right. *b.* Death of cancer cell is indicated by the blebs, or deep folds, that have appeared on its surface membrane.

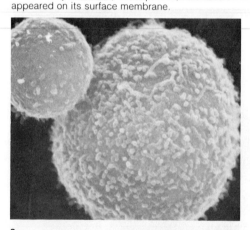

a.

b.

Cell-mediated Immunity

T cells (fig. 12.3) are responsible for **cell-mediated immunity,** the characteristics of which are listed in table 12.1. T cells become especially activated after they are stimulated by macrophages that concentrate antigens and then "present" them to T cells (fig. 12.9). After being stimulated, T cells enlarge, divide, and become "killer cells." These killer cells exhibit cytotoxicity, meaning that they destroy on contact cells that bear the antigen (fig. 12.10).

T cells specialize in providing resistance against antigens that have invaded the tissues of the body. They attack cells that display foreign antigens on their cell surface, often including the cells of a transplanted organ or a donated skin graft. T cells are believed to be the lymphocytes ordinarily responsible for preventing the development of cancer. As long as they can recognize newly developed cancer cells, a cancerous growth cannot begin. Therefore when cancer is present, it most likely signifies a failure of the immune system.

Activated T cells liberate **lymphokines** and/or **interferon,** potent chemicals that regulate the activity of other white cells. One type of lymphokine draws macrophages, neutrophils, and eosinophils toward T cells. After they have arrived, another signals them to remain in battle. Under the influence of lymphokine, macrophages become "angry killers," or cells with increased ability to phagocytize and destroy antigens.

Interferon

Interferon is a group of small proteins that are produced and released by T cells and by most other cells after they have been invaded by viruses. When viruses infect cells, they subvert the cell's synthetic apparatus and cause the cells to make more virus particles. When infected cells are synthesizing components of new virus particles, they also synthesize interferon, which they release into their environment. Interferon subsequently protects other cells because it interferes with viral replication in these cells. However, when interferon is received by a neighboring cell, it attaches to *species specific* receptors. Therefore, human interferon is effective only in humans.

Investigations are under way to see which infectious diseases respond best to interferon treatment and whether it is also effective against certain forms of cancer. In the past, supplies of interferon had to be extracted from white cells and the amount available was so scarce that treatment cost $20,000 to $30,000 per patient. Now, however, interferon is being commercially produced by the recombinant DNA technique (p. 491) and the cost of treatment has been reduced to $300 per patient. Soon it may be possible to determine which types of patients will benefit best from interferon therapy.

Immunological Side Effects and Illnesses

The immune system protects us from disease because it can tell *self from nonself.* Sometimes, however, the immune system is *under*protective as when an individual develops cancer or is *over*protective when an individual has allergies.

Allergies

Allergies are caused by an overactive immune system that forms antibodies to substances that are not usually recognized as being foreign substances. Unfortunately, allergies are usually accompanied by coldlike symptoms or, even at times, severe systemic reactions such as a sudden low blood pressure, termed shock.

Human Anatomy and Physiology

Among the five varieties of antibodies—immunoglobulin A (IgA), IgD, IgE, IgG, and IgM—it is IgE that causes allergies. IgE antibodies are found in the bloodstream but they, unlike the other types of antibodies, also reside in the membrane of *mast cells* found in the tissues. Some investigators contend that mast cells are basophils that have left the bloodstream and have taken up residence in the tissues. In any case, when the allergen attaches to the IgE receptors on mast cells, these cells release histamine, leukotriene, and other substances. Histamine causes increased secretion of mucus, and leukotriene causes constriction of the airways, resulting in the characteristic wheezing and labored breathing of someone with asthma (fig. 12.11). On occasion, basophils and other white cells release these chemicals into the bloodstream. Then increased capillary permeability can lead to fluid loss and shock.

Allergy shots sometimes prevent the occurrence of allergic symptoms. Injections of the allergen cause the body to build up high quantities of IgG antibodies and these combine with allergens received from the environment before they have a chance to reach the IgE antibodies located in the membrane of mast cells.

Tissue Rejection

Certain organs such as skin, the heart, and the kidney could be easily transplanted from one person to another if the body did not attempt to reject them. It is obvious that the transplanted organ is foreign to the individual, and for this reason the immune system reacts to it. At first T cells appear on the scene and then later, antibodies bring about a disintegration of the foreign tissue.

Organ rejection can be controlled in two ways: careful selection of the organ to be transplanted and the administration of **immunosuppressive drugs.** In the first instance, it is best if the organ is made up of cells having the same type glycoproteins as those on the cells of the prospective recipient. The glycoproteins act as antigens known as the **major histocompatibility complex,** and it is these that the T cell recognizes as foreign to the recipient. In regard to immunosuppression, there is now available a drug called Cyclosporine (fig. 12.12), which suppresses cell-mediated immunity only as long as it is administered. This means that after the body has adjusted to the new organ, the drug can be withdrawn and immunity will return. This is desirable because a person is more apt to catch an infectious disease when immunity is suppressed.

Figure 12.11
A patient takes a deep breath and exhales as hard as possible into a spirometer. The device measures the volume of air expelled in one second. An increase in volume after the administration of an antihistamine suggests that the patient has asthma.

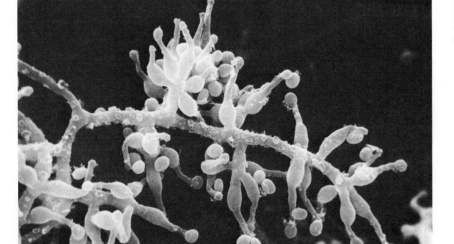

Figure 12.12
Scanning electron micrograph of the fungus that makes the immunosuppressive drug *cyclosporine*. It was only by chance that investigators discovered that one of the chemicals produced by the fungus was immunosuppressive.

Figure 12.13
Aging in mice. The possible importance of the immune system to the aging process is illustrated by the fact that the mouse in *a.* was injected with donor lymph node cells, whereas the mouse in *b.* was not injected.

a.

b.

Even if all the problems involved in transplantation were solved, there would still be the difficulty of finding enough organs. It is estimated that over 125,000 donated hearts could be utilized each year. Obviously, this number of natural organs is not likely to be available; therefore, it is hoped that one day artificial organs will be available.

Autoimmune Diseases

Certain human illnesses are believed to be due to the production of antibodies that act against an individual's own tissues. Some forms of hemolytic anemia in which the red cells are prematurely destroyed have been shown to be due to autoantibodies. In myasthenia gravis, autoantibodies attack the neuromuscular junctions so that the muscles do not obey nervous stimuli. Muscular weakness results. In multiple sclerosis (MS), antibodies attack the myelin sheath of nerve fibers, causing various neuromuscular disorders. A person with systemic lupus erythematosus (SLE) forms various antibodies to different constituents of the body, including the DNA of the cell nucleus. The disease sometimes results in death, usually due to kidney damage.

Autoimmune diseases are an area of active research. It could be that certain cells in these individuals really do carry foreign antigens due to a previous viral infection. Or it could be that the suppressor T cells are failing to prevent B cells from producing antibodies against the body's own tissues. Thus far, investigators have not settled on a preferred treatment for autoimmune diseases. However, the drug Cyclosporine or the administration of monoclonal antibodies may prove to be effective in these patients, and clinical trials are under way.

AIDS

The illness AIDS (acquired immune deficiency syndrome) is characterized by a weakened immune system. Persons who have this disease often die from a rare form of skin cancer or a rare form of pneumonia. AIDS, which is usually sexually transmitted, particularly among homosexual men, is discussed in more detail in the reading on page 408.

Researchers observed that AIDS could also be transmitted by way of blood-to-blood contact, for example, dirty needles or a transfusion, and that patients are extremely vulnerable to infections because they lack the normal number of T cells. This encouraged them to speculate that AIDS is caused by a viral infection, and it was recently announced that an HTLV (human T cell leukemia virus) is most likely the causative agent of AIDS. This virus attacks and destroys T cells leaving the body susceptible to infections.

Aging

The thymus (fig. 12.1) is large in relation to the rest of the body during fetal development and childhood; however, it stops growing by puberty and then begins to atrophy and get progressively smaller. This has led some researchers to suggest that aging may be associated with a general decline in the immune system. In an experiment that substantiates this possibility, mice injected with lymph node cells do not age as rapidly as those that are not injected (fig. 12.13).

Summary

The immune system consists of lymphocytes and associated structures. Lymphocytes have the ability to tell self from nonself. Each one carries receptors that combine only with one type of foreign protein and usually not with one of the body's own proteins.

There are two types of lymphocytes, B cells and T cells. Both types are derived from lymphoid bone marrow stem cells, but only those that produce T cells have passed through the thymus. B cells are responsible for antibody-mediated immunity, and T cells are responsible for cell-mediated immunity.

In addition, T cells regulate the function of B cells; there are helper T cells that stimulate B cells and suppressor T cells that suppress B cells. When an antigen combines with receptors on T and B cells, helper T cells stimulate the B cells to divide and develop into a number of plasma cells. This is called the clonal selection theory because the antigen has chosen the B cell that clones and produces antibodies. Antibodies have exactly the same structure as the receptors on the B cell. Antibodies combine with antigens, and the complex may be taken up by macrophages or it may activate the complement portion of plasma. Once there are sufficient antibodies, suppressor T cells cause antibody production to slow down.

Some of the cells that arise after B cell stimulation are memory cells, cells that are forever capable of producing antibodies and more plasma cells to combat a particular antigen in the future. Once the body has the capability of producing adequate numbers of antibodies to prevent illness, the person is said to be actively immune. Today it is often possible by means of vaccines to immunize individuals medically against various bacterial and viral diseases.

Passive immunity is also possible. In this case the individual is provided with preformed antibodies. These antibodies can be extracted from the serum of another individual or from a horse that has been immunized previously. Scientists are now able, however, to produce monoclonal antibodies in the laboratory. There is great hope that this technique will allow the production of antibodies to help fight cancer.

T cells are responsible for cell-mediated immunity. When an antigen combines with a T cell receptor, it is stimulated especially by macrophages to divide and produce more T cells, each one of which is a killer cell that attacks foreign cells in the tissues. T cells release lymphokines that stimulate other white cells, including macrophages, to assist in the battle. Most likely, T cells are responsible for destroying cancerous cells before they are able to develop.

Sometimes the process of immunity has unwanted effects. Allergies are due to an overactive immune system that forms antibodies to substances not normally considered foreign. The allergic response releases chemicals such as histamine and leukotriene that cause the unpleasant symptoms associated with allergies. Tissues and organs used for transplant operations contain cells with glycoproteins foreign to the recipient and so they are attacked. Autoimmune diseases occur when the body begins to produce antibodies against its own tissues. Perhaps the best known of the autoimmune diseases is MS (multiple sclerosis).

Study Questions

1. What is the function of the immune system? (p. 245) Describe the relationship of lymphocytes and antigens. (p. 245)
2. What is meant by the statement that the immune system "learns to tell self from nonself"? (p. 245)
3. State five differences between antibody-mediated and cell-mediated immunity. (pp. 246–47)
4. What is the clonal selection theory? (pp. 248, 249)
5. What normally happens to immune complexes? (p. 249)
6. Relate active immunity to the presence of plasma cells and memory cells. (p. 249)
7. How is active immunity achieved without accompanying illness? (pp. 250–51)
8. What is passive immunity and how might it be achieved today? (p. 255)
9. What is interferon, how can it be produced, and how might it be used? (p. 256)
10. Discuss tissue rejection, autoimmune diseases, cancer, and aging as they relate to the immune system. (pp. 257–58)

Selected Key Terms

lymphocytes (lim′fo-sīts)

receptor (re-sep′tor)

epitope (ep′i-tōp)

antigen (an′tĭ-jen)

thymus (thi′mus)

T cells (te selz)

B cells (be selz)

cell mediated immunity (sel me′de-āt″ed ĭ-mu′nĭ-te)

antibody mediated immunity (an′tĭ-bod″e me′de-āt″ed ĭ-mu′nĭ-te)

plasma cell (plaz′mah sel)

clonal selection (klōn′al sĕ-lek′shun)

memory cell (mem′o-re sel)

complement (kom″plĕ-ment)

monoclonal antibodies (mon″o-klōn′al an′ti-bod″ēz)

lymphokines (lim′fo-kīnz)

interferon (in″ter-fēr′on)

Further Readings

Buisseret, P. D. 1982. Allergy. *Scientific American* 247(2):86.

Burke, D. C. 1977. The status of interferon. *Scientific American* 236(4):42.

Collier, R. J., and Kaplan, D. A. 1984. Immunotoxins. *Scientific American* 251(i):56.

Edmundson, D. C., and Edmundson, A. B. 1977. The antibody combining site. *Scientific American* 236(1):50.

Immunology. 1976. Readings from *Scientific American.* San Francisco: W. H. Freeman.

Koffler, D. 1980. Systemic lupus erythematosus. *Scientific American* 243(1):52.

Leder, P. 1982. The genetics of antibody diversity. *Scientific American* 246(5):102.

Lerner, R. A. 1983. Synthetic vaccines. *Scientific American* 248(2):66.

Melnick, J. L. et al. 1977. Viral hepatitis. *Scientific American* 237(1):44.

Old, L. J. 1977. Cancer immunology. *Scientific American* 236(5):62.

Pestka, S. 1983. The purification and manufacture of human interferons. *Scientific American* 249(2):36.

Raff, M. C. 1976. Cell-surface immunology. *Scientific American* 234(5):30.

Rose, N. R. 1981. Autoimmune diseases. *Scientific American* 244(2):80.

13

Respiration

Chapter concepts

1 The respiratory tract of humans is designed in such a way that air is filtered, warmed, and saturated with water before gas exchange takes place across a very extensive moist surface.

2 Breathing is a required process for human life because it brings to the blood the oxygen needed by the cells for cellular respiration, and it rids the body of carbon dioxide, a by-product of cellular respiration.

3 The respiratory pigment hemoglobin has chemical and physical characteristics that promote its combination with oxygen in the lungs and its release of oxygen in the tissues. It also aids in the transport of carbon dioxide from the tissues to the lungs, largely by its ability to buffer.

4 The respiratory tract is especially subject to disease because it serves as an entrance for infectious agents. Polluted air contributes to cause for two major lung disorders—emphysema and cancer.

B reathing is more eminently necessary than eating. While it is possible to stop eating altogether for several days, it is not possible to remain alive for longer than several minutes without breathing. Breathing supplies the body with the oxygen needed for cellular respiration, as indicated in the following equation.[1]

$$38 \text{ ADP} + 38 \textcircled{P} \longrightarrow 38 \text{ ATP}$$

$$C_6H_{12}O_6 + 6 O_2 \longrightarrow 6 H_2O + 6 CO_2$$

The equation indicates that the body requires oxygen to convert the energy within glucose to phosphate-bond energy. Thus, the more energy expended, the greater the need for oxygen. The minimum amount of oxygen a person consumes at complete rest, without eating previously, is related to the basal metabolic rate (p. 195). The average young adult male utilizes about 250 milliliters of oxygen per minute in a basal, or restful, state. Exercise and digestion of food raise the need for oxygen. The average amount of oxygen needed with mild exercise is 500 milliliters of oxygen per minute. The equation for cellular respiration also indicates that cells produce carbon dioxide. This metabolic end product must be eliminated from the body by the breathing process.

Altogether, respiration may be used to refer to the complete process of getting oxygen to body cells for cellular respiration and the reverse process of ridding the body of carbon dioxide given off by cells. Thus, respiration may be said to include the following:

1. **Breathing:** entrance and exit of air into and out of the lungs
2. **External respiration:** exchange of gases ($O_2 + CO_2$) between air and blood
3. **Internal respiration:** exchange of gases between blood and tissue fluid
4. **Cellular respiration:** production of ATP in cells. In this chapter, we are studying the first three portions of the respiratory process. Cellular respiration is discussed at length in chapter 5.

Breathing

The normal breathing rate is about 14 to 20 times per minute. Breathing consists of taking air in, **inspiration,** and forcing air out, **expiration.** The composition of inspired and expired air is given in table 13.1.

Comparing inspired air with expired air, we see that the expired air contains less oxygen and more carbon dioxide than inspired air. This means that the body does indeed take in oxygen and give off carbon dioxide.

Passage of Air

During inspiration and expiration, air is conducted toward or away from the lungs by a series of cavities, tubes, and openings, which are listed in order in table 13.2 and illustrated in figure 13.1.

As air moves in along the air passages, it is filtered, warmed, and moistened. The filtering process is accomplished by coarse hairs and cilia in the region of the nostrils and by cilia alone in the rest of the nose and windpipe. In the nose, the hairs and cilia act as a screening device. In the trachea, cilia beat upward, carrying mucus, dust, and occasional bits of food that "went the wrong way" into the pharynx where the accumulation may be swallowed or expectorated. The air is warmed by heat given off by the blood vessels lying

[1]The body also requires oxygen for the respiration of fats and amino acids in addition to glucose.

Table 13.1 Composition of inspired and expired air

Component of air	Inspired air (%/vol)	Expired air (%/vol)
N_2	79.00	79.60
O_2	20.96	16.02
CO_2	0.04	4.38

Table 13.2 Path of air

Structure	Function
Nasal cavities	Filters, warms, and moistens
Nasopharynx	Passage of air from nose to throat
Pharynx (throat)	Connection to surrounding regions
Glottis	Passage of air
Larynx (voice box)	Sound production
Trachea (windpipe)	Passage of air to thoracic cavity
Bronchi	Passage of air to each lung
Bronchioles	Passage of air to each alveolus
Alveoli	Air sacs for gas exchange

Human Anatomy and Physiology

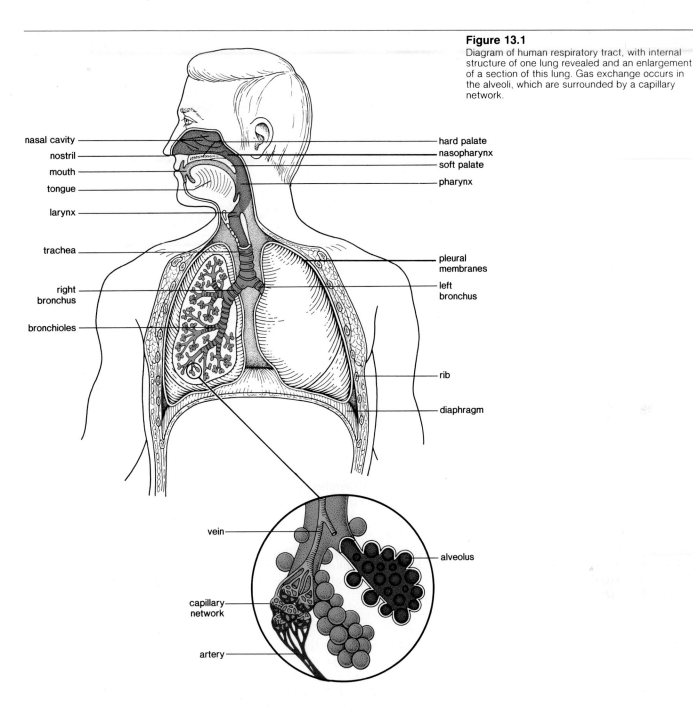

nasal cavity

nostril

mouth

tongue

larynx

trachea

right
bronchus

bronchioles

hard palate

nasopharynx

soft palate

pharynx

pleural
membranes

left
bronchus

rib

diaphragm

vein

capillary
network

artery

alveolus

close to the surface of the lining of the air passages, and it is moistened by the wet surface of these passages. By the time the inspired air reaches the lower end of the trachea, it is about 99.5 percent saturated with water.

On the other hand, as air moves out during expiration, it becomes progressively cooler and loses its moisture. As the gas cools, it deposits its moisture on the lining of the windpipe and nose, and the nose may even drip as a result of this condensation. But the air still retains so much moisture that upon expiration on a cold day it condenses and forms a small cloud.

Each portion of the air passage also has its own unique structure and function, as described in the sections that follow.

Figure 13.2
The odor receptors, termed olfactory epithelium, are located high up in the recesses of each nasal cavity. When they are stimulated, a stimulus is conducted to the brain by way of the olfactory nerve.

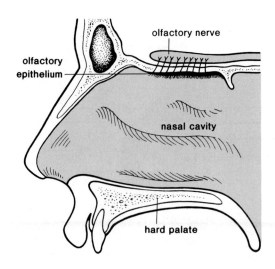

Figure 13.3
The tear (lacrimal) ducts drain into the nasal cavities. Thus, crying causes a "runny nose."

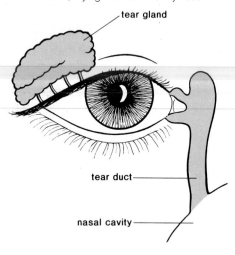

Figure 13.4
Pathway of air. When we breathe, the glottis is open, and when we swallow, the epiglottis covers the glottis.

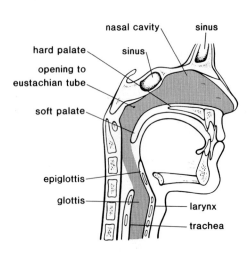

Nose

The **nose** contains two nasal cavities, narrow canals with convoluted lateral walls that are separated from one another by a septum. Up in the narrow recesses of the nasal cavities are special ciliated cells (fig. 13.2) that act as odor receptors. Nerves lead from these cells and go to the brain where the impulses are interpreted as smell.

The nasal cavities have a number of openings. The tears produced by tear (lacrimal) glands drain into the nasal cavities by way of tear ducts (fig. 13.3). Thus crying produces a runny nose. The nasal cavities open into the cranial sinuses (fig. 13.4), air-filled spaces in the skull, and empty into the nasopharynx, a chamber just beyond the soft palate. The **eustachian tubes** lead into the nasopharynx from the middle ears.

Pharynx

The air and food channels cross in the **pharynx** (figs. 9.5 and 13.4) so that the trachea (windpipe) lies in front of the esophagus, which normally opens only during the process of swallowing food. Just below the pharynx lies the larynx, or voice box.

Larynx

The **larynx** may be imagined as a triangular box whose apex, the Adam's apple, is located at the front of the neck. At the top of the larynx is a variable-sized opening called the **glottis.** When food is being swallowed, the glottis is covered by a flap of tissue called the **epiglottis** so that no food passes into the larynx. If, by chance, food or some other substance does gain entrance to the larynx, reflex coughing usually occurs to expel the substance.

At the edges of the glottis, embedded in mucous membrane, are elastic ligaments called the **vocal cords** (fig. 13.5). These cords, which stretch from the back to the front of the larynx just at the sides of the glottis, vibrate when air is expelled past them through the glottis. Vibration of the vocal cords produces sound. The high or low pitch of the voice depends upon the length, thickness, and degree of elasticity of the vocal cords and the tension at which they are held. The loudness or intensity of the voice depends upon the amplitude of the vibrations or the degree to which vocal cords vibrate.

At the time of puberty, the growth of the larynx and the vocal cords is much more rapid and accentuated in the male than in the female, causing the male to have a more prominent Adam's apple and a deeper voice. The voice "breaks" in the young male due to his inability to control the longer vocal cords.

Human Anatomy and Physiology

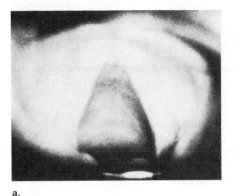

a.

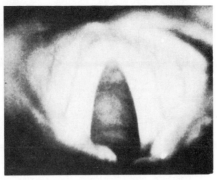

b.

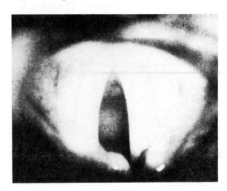

c.

Figure 13.5
Vocal cords. The vocal cords are seen from above at the edge of the glottis, the variable opening. When air is expelled from the larynx, the cords vibrate, producing the voice.

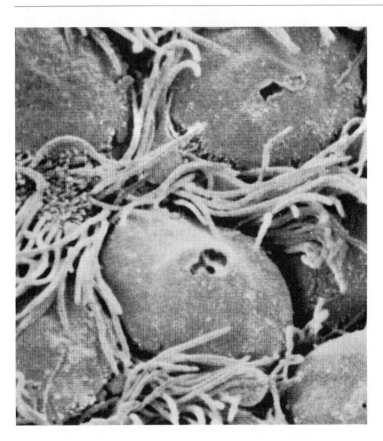

Figure 13.6
Scanning electron micrograph of ciliated cells that line the bronchi. These cilia beat toward the trachea and help keep impurities from entering the lungs. Openings on the surface of some cells through which mucus may have been discharged are visible.

Trachea

The larynx is continuous with the **trachea,** which is held open by cartilaginous rings. Ciliated mucous membrane (fig. 13.6) also lines the trachea, and normally these cilia keep the windpipe free of debris. Smoking is known to destroy the cilia, and consequently the soot in cigarette smoke collects in the lungs. Smoking will be discussed more fully at the end of this chapter.

If the trachea is blocked because of illness or accidental swallowing of a foreign object, it is possible to insert a tube by way of an incision made in the trachea; this tube acts as an artificial air intake and exhaust duct. The operation is called a **tracheostomy.**

Figure 13.7

An alveolus. The shape of an alveolus and the extent of the surrounding capillary network is revealed by this scanning electron micrograph of a cast that was prepared by injecting the surrounding capillary network with a resin, after which all tissues were removed.

From: TISSUES AND ORGANS: A TEXT-ATLAS OF SCANNING ELECTRON MICROSCOPY by R. G. Kessel and R. Kardon. W. H. Freeman and Company, © 1979.

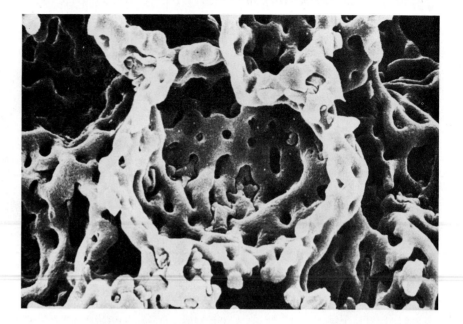

Bronchi

The trachea divides into two **bronchi** that enter the right and left lungs and branch into a great number of smaller passages called the **bronchioles.** The two bronchi resemble the trachea in structure, but as the bronchial tubes divide and subdivide, their walls become thinner and the small rings of cartilage do not occur. Each bronchiole terminates in an elongated space that is enclosed by a multitude of air pockets, or sacs, called **alveoli** (fig. 13.1), which make up the lungs.

Lungs

Within the lungs each alveolar sac is only one layer of squamous epithelium surrounded by blood capillaries. Gas exchange occurs between the air in the alveoli and the blood in the capillaries (fig. 13.7).

A film of lipoprotein that lines the alveoli of mammalian lungs lowers the surface tension and prevents them from closing up. Some newborn babies, especially premature infants, lack this film, resulting in lung collapse. This condition, called infant respiratory distress syndrome, often results in death.

There may be as many as 700,000,000 alveoli in the human lung. This is the equivalent of 75 square yards, or 100 times the surface of the skin! Because of their many air spaces, the lungs are very light; normally, a piece of lung tissue dropped in a glass of water will float.

Externally, the lungs are cone-shaped organs that lie on either side of the heart in the thoracic cavity. The branches of the pulmonary artery accompany the bronchial tubes and form a mass of capillaries around the alveoli.[2] The four pulmonary veins collect blood from these capillaries and empty into the left atrium of the heart. Figure 13.8 shows the relationship of the pulmonary vessels to the trachea and bronchial tubes. The lungs are enclosed by the pleural membranes, one of which adheres closely to the walls of the chest and diaphragm, while the other is fused to the lungs. The two pleural layers are very close to one another, being separated only by a slight amount of fluid

[2]These capillaries are used for gas exchange and do not furnish the lungs with oxygen. The bronchial artery supplies the lungs with their oxygen needs.

Human Anatomy and Physiology

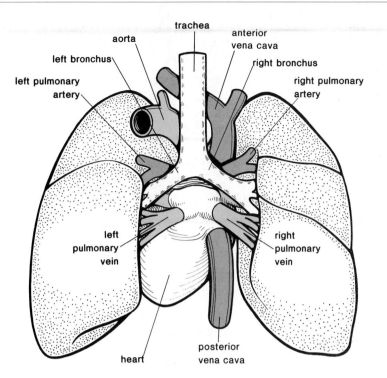

(fig. 13.9). This is important because there is little resistance to the lungs being pulled open when the chest expands. If a puncture occurs through the chest walls, causing air to enter between the two layers of the pleura, the lungs collapse. Sometimes this is done purposely to give a lung time to heal itself. If the amount of air entering is small, it will eventually be absorbed and the lung will reinflate.

Mechanism of Breathing

In order to understand the manner in which air is drawn into and expelled out of the lungs, it is necessary to remember first that there is a continuous column of air from the pharynx to the alveoli of the lungs; that is, the air passages are always open.

Secondly, we may note that the lungs lie within the sealed-off chest (thoracic) cavity. The **ribs,** which are hinged to the vertebral column at the back and to the sternum (breastbone) at the front, along with the muscles that lie between them, make up the top and sides of the chest cavity. The **diaphragm,** a dome-shaped horizontal muscle, forms the floor of the chest cavity. The lungs themselves are enclosed by the pleural membranes, which have an intrapleural pressure less than atmospheric pressure. With these facts in mind, let us now consider inspiration and expiration.

Inspiration (Breathing In)

It can be shown that carbon dioxide and hydrogen ions are the primary stimuli that cause us to breathe. When the concentration of CO_2 and/or H^+ reach a certain level in the blood, the breathing center in the medulla oblongata—the stem portion of the brain—is stimulated. This center is not affected by low oxygen levels, but there are chemoreceptors in the carotid bodies (located in the carotid arteries) and in the aortic bodies (located in the aorta) that do respond to low blood oxygen in addition to carbon dioxide and hydrogen ion concentration.

Figure 13.9
Pleural membranes. These membranes completely surround the lungs. Humans breathe by negative pressure and this requires that the intrapleural pressure be less than atmospheric pressure. Thus, when the thoracic cavity increases in size, the lungs expand and the air comes rushing in.

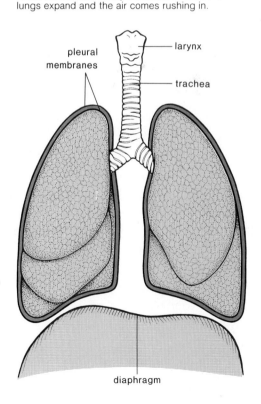

Figure 13.10

Nervous control of breathing. During inspiration, the respiratory center stimulates the rib (intercostal) muscles and the diaphragm to contract by way of the efferent (phrenic) nerve. Nerve impulses from the expanded lungs by way of the afferent (vagus) nerve then inhibit the respiratory center. Lack of stimulation causes the rib muscles and diaphragm to relax and expiration follows.

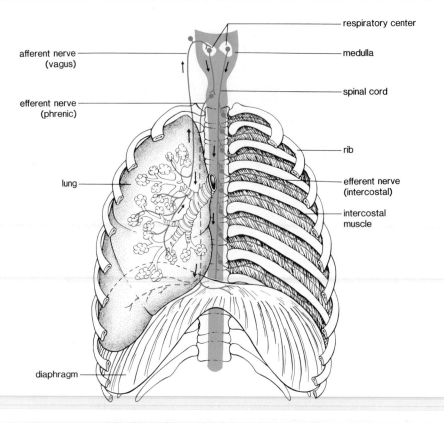

respiratory center

medulla

afferent nerve (vagus)

efferent nerve (phrenic)

spinal cord

rib

efferent nerve (intercostal)

intercostal muscle

lung

diaphragm

Figure 13.11

Breathing. *a.* During inspiration, the rib cage moves upward and outward, while the diaphragm moves down. Now the lungs expand and the air comes rushing in. *b.* During expiration, the rib cage moves downward and inward and the diaphragm moves up. Now the lungs recoil and the air is pushed outward.

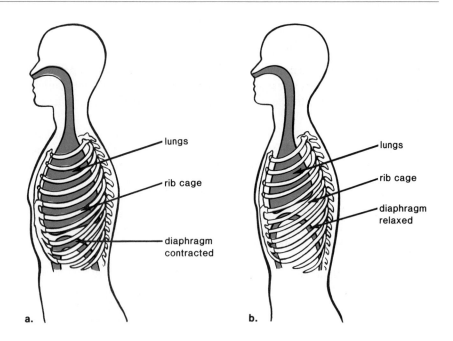

lungs

rib cage

diaphragm contracted

lungs

rib cage

diaphragm relaxed

a.

b.

Human Anatomy and Physiology

As diagrammed in figure 13.10, when the breathing center is stimulated, a nerve impulse goes out by way of nerves to the diaphragm and rib cage. In its relaxed state, the diaphragm is dome shaped, but upon stimulation it contracts and lowers. When the rib muscles contract, the rib cage moves upward and outward. Both of these contractions serve to increase the size of the chest cavity. As the chest cavity increases in size, the lungs expand. Consequently, air pressure within the enlarged alveoli lowers and is immediately rebalanced by air rushing in through the nose or mouth.

Inspiration (fig. 13.11) is the active phase of breathing. It is during this time that the diaphragm contracts, the rib muscles contract, and the lungs are pulled open, with the result that air comes rushing in. Note that the air comes in because the lungs have already opened up; the air does not force the lungs open. This is why it is sometimes said that *humans breathe by negative pressure.* It is the creation of a partial vacuum that sucks air into the lungs.

Expiration (Breathing Out)

When the lungs are expanded, the stretching of the alveoli stimulates special receptors in the alveolar walls, and these receptors initiate nerve impulses from the inflated lungs to the breathing center. When the impulses arrive at the medulla oblongata, the center is inhibited and stops sending signals to the diaphragm and the rib cage. The diaphragm relaxes and resumes its dome shape (fig. 13.11). The abdominal organs press up against the diaphragm. The rib cage moves down and inward. The elastic lungs recoil and air is pushed outward.

Table 13.3 summarizes the events causing inspiration and expiration. It is clear that while inspiration is an active phase of breathing, normally expiration is passive since the breathing muscles automatically relax following contraction. But it is possible, in deeper and more rapid breathing, for both phases to be active because there is another set of rib muscles whose contraction can forcibly cause the chest to move downward and inward. Also, when the abdominal wall muscles are contracted, there is an increase in pressure that helps expel air.

The breathing center may stimulate deeper and more rapid breathing or this may be done voluntarily. It is possible for a person to deliberately breathe more rapidly or more slowly, or to hold the breath for a short time. However, it is impossible to commit suicide by holding your breath; eventually, carbon dioxide buildup in the blood forces the resumption of breathing.

Consequences

There are certain physiological consequences that result from the manner in which we breathe. First of all, we may note that breathing is initiated and continues due to the presence of carbon dioxide in the blood. Therefore, when necessary, it is better to give a person oxygen gas containing carbon dioxide rather than pure oxygen alone. The mixture of gases stimulates the resumption of breathing, whereas pure oxygen does not.

Also, readers who have followed this discussion carefully will have no difficulty in realizing that the respiratory tract contains a certain amount of dead space, or space that contains air not used in gas exchange. If we diagram the tract as illustrated in figure 13.12, we see that the dead space can be considered to extend from the top *(A)* of the tube (pharynx) to a certain level within the lungs *(B)*. This fresh air does not immediately reach the respiratory

Table 13.3	Breathing process
Inspiration	**Expiration**
Medulla sends stimulatory message to diaphragm and rib muscles.	Stretch receptors in lungs send inhibitory message to medulla.
Diaphragm contracts and flattens.	Diaphragm relaxes and resumes dome position.
Rib cage moves up and out.	Rib cage moves down and in.
Lungs expand.	Lungs recoil.
Negative pressure in lungs.	Positive pressure in lungs.
Air is pulled in.	Air is forced out.

Figure 13.12
Distribution of air in lungs. The air between *A* and *B* does not immediately reach the alveoli; therefore this is called dead space. The air below *C* represents the amount of residual air that has not left the lungs. Only the air between *B* and *C* brings with it additional oxygen for respiration.

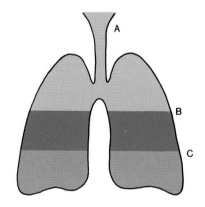

surface of the alveoli. Then to complicate matters even further, there is a certain amount of air that never leaves the lungs. This is called residual air and is indicated below *C* in the diagram. This means that the air indicated between *B* and *C* is the portion that immediately supplies additional oxygen. Normally we inhale about 500 cubic centimeters of air; of this amount, about 350 cubic centimeters immediately reaches the alveoli. With deep breathing we can inhale a maximum of 1,650 cubic centimeters. Even so, it may be readily seen that humans cannot breathe through a very long tube. Any device that increases the amount of dead space beyond maximal inhaling capacity spells death to the individual because the air inhaled would never reach the alveoli.

External and Internal Respiration

External Respiration

The term *external respiration* refers to the exchange of gases between the air in the alveoli and the blood within the pulmonary capillaries (fig. 13.1). The wall of an alveolus consists of a thin, single layer of cells, and the wall of a blood capillary also consists of such a layer. Since neither wall offers resistance

Human Anatomy and Physiology

a result of a breakdown of the cellular pump supply that maintains the balance of sodium, potassium, and water, cerebral edema can cause permanent brain damage. The brain tissue swells as water accumulates, resulting in severe headaches, vomiting, loss of coordination, coma and death. Mountaineers stricken with cerebral edema also report hallucinations: The survivors of an expedition at 22,000 feet in the Andes were convinced they saw bulldozers and palm trees on the summit and that tourists were stealing their supplies.

Whatever the underlying cause, altitude sickness usually can be avoided if visitors to the mountains take enough time for their ascent. Restricting a day's climb to less than 1,000 feet will protect most people from serious illness. Of the drugs prescribed to help prevent altitude sickness, only Diamox, an agent that helps overcome the effects of overbreathing, seems to reduce the symptoms of acute mountain sickness. It has little effect on the more serious illnesses. Once acute mountain sickness strikes, the most effective remedy is to drink large quantities of liquids, take aspirin for headaches, and stay mildly active. Because the pulmonary and cerebral edemas are more serious, the victim should start down immediately.

to the passage of gases, *diffusion* is believed to govern the exchange of oxygen and carbon dioxide between alveolar air and the blood. Active cellular absorption and secretion do not appear to play a role. Rather, the direction in which the gases move is determined by the pressure or tension gradients between blood and inspired air.

Atmospheric air contains little carbon dioxide, but blood flowing into the lung capillaries is almost saturated with the gas. Therefore, *carbon dioxide diffuses out of the blood into the alveolus.* The pressure pattern is the reverse for oxygen. Blood coming into the pulmonary capillaries is oxygen poor and alveolar air is oxygen rich; therefore, *oxygen diffuses into the capillary.* Breathing at high altitudes is less effective than at low altitudes because the air pressure is less, making the concentration of oxygen (and other gases) lower than normal; therefore less oxygen diffuses into the blood. The body responds in various ways, as described in the reading on page 270. Breathing problems do not occur in airplanes because the cabin is pressurized to maintain an appropriate pressure. Emergency oxygen is available in case the pressure should, for one reason or another, be reduced.

Figure 13.13

Diagram illustrating external and internal respiration. During external respiration in the lungs, CO_2 leaves the blood and O_2 enters the blood. During internal respiration in the tissues, O_2 leaves the blood and CO_2 enters the blood.

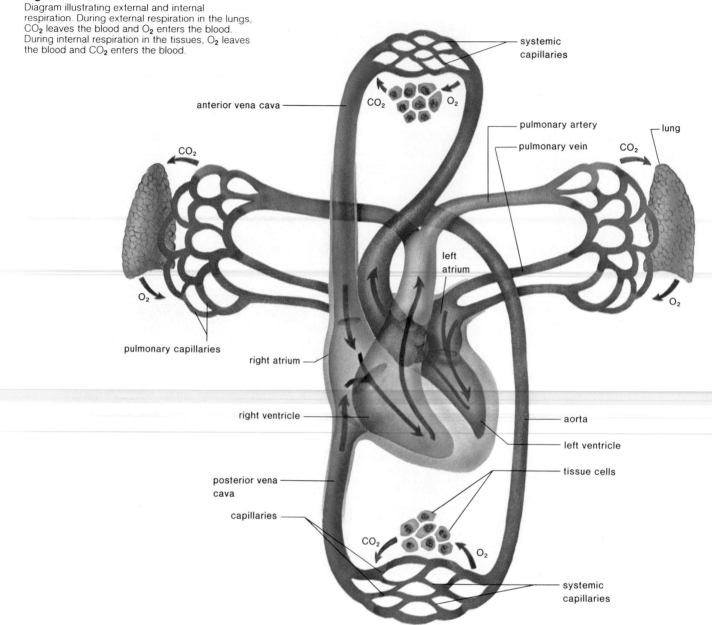

systemic capillaries

anterior vena cava

pulmonary artery

pulmonary vein

lung

CO_2

CO_2

O_2

O_2

left atrium

pulmonary capillaries

right atrium

right ventricle

aorta

left ventricle

tissue cells

posterior vena cava

capillaries

CO_2

O_2

systemic capillaries

As blood enters the pulmonary capillaries (fig. 13.13), most of the carbon dioxide is being carried as the bicarbonate ion, HCO_3^-. As the little free carbon dioxide remaining begins to diffuse out, the following reaction is driven to the right:

$$H^+ + HCO_3^- \rightarrow H_2CO_3 \rightarrow H_2O + CO_2$$

The enzyme carbonic anhydrase (p. 234), present in red cells, speeds up the reaction. As the reaction proceeds, hemoglobin gives up the hydrogen ions it has been carrying; HHb becomes Hb.

Now hemoglobin more readily takes up oxygen and the following reaction takes place:

$$Hb + O_2 \longrightarrow HbO_2$$

It is a remarkable fact that at the oxygen tension in inspired air (150 mm Hg), hemoglobin is about 100 percent saturated. As you can see in figure 13.14,

Human Anatomy and Physiology

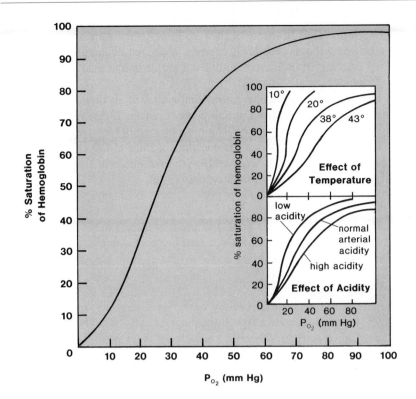

Figure 13.14
Properties of hemoglobin as revealed by the hemoglobin dissociation curve. The large curve shows the percentage of saturation of hemoglobin at 38° and normal arterial blood acidity. As the partial pressure of oxygen (PO_2) decreases, hemoglobin gives up its oxygen and this effect is also promoted by the higher temperature and higher acidity of the tissues.

while hemoglobin takes up oxygen in increasing amounts as the oxygen tension increases, the curve begins to level off at about 100 mm Hg. This means that hemoglobin easily retains oxygen at the oxygen tension in the lungs and tends to release it at the oxygen tension in the tissues. Further, another remarkable fact about hemoglobin is that it more readily takes up oxygen in the neutral pH and cool temperature of the lungs. On the other hand, it will give up oxygen more readily at the more acid pH and warmer temperature of the tissues.[3]

Internal Respiration

As blood enters the systemic capillaries (fig. 13.13), oxygen leaves hemoglobin and diffuses out into tissue fluid:

$$HbO_2 \longrightarrow Hb + O_2$$

Diffusion of oxygen out of the blood and into the tissues occurs because the oxygen tension in tissue fluid is low due to the fact that the cells are continuously using it up in cellular respiration. On the other hand, carbon dioxide tension is high because carbon dioxide is continuously being produced by the cells. Thus, *carbon dioxide will diffuse into the blood.*

A small amount of carbon dioxide is carried by the formation of carbamino hemoglobin (p. 234). But most carbon dioxide combines with water to form carbonic acid, which dissociates to $H^+ + HCO_3^-$. The enzyme carbonic anhydrase present in red cells speeds up the reaction:

$$CO_2 + H_2O \longrightarrow H_2CO_3 \longrightarrow H^+ + HCO_3^-$$

The globin portion of hemoglobin combines with the excess hydrogen ions produced by the reaction, and Hb becomes HHb. In this way, the pH of the blood remains fairly constant. The bicarbonate ion, HCO_3^-, diffuses out of the red cells to be carried in the plasma.

[3]	pH	Temperature
Lungs	7.40	37°C (98.6°F)
Body	7.38	38°C (100.4°F)

Figure 13.15
A sneeze. Atomization of droplets into the air during sneezing illustrates that germs are easily transferred from one person to another by this process.

We have seen that the full length of the respiratory tract is lined with a warm, wet mucosa lining that is constantly exposed to environmental air. The quality of this air, determined by the pollutants and germs it contains, can affect the health of the individual.

Germs frequently spread from one individual to another by way of the respiratory tract. Droplets from one single sneeze (fig. 13.15) may be loaded with billions of bacteria or viruses. The mucous membranes are protected by the production of mucus and by the constant beating of the cilia; but if the number of infective agents is large and/or the resistance of the individual is reduced, an upper respiratory infection may result.

Upper Respiratory Infections

An upper respiratory infection (URI) is one that affects only the nasal cavities, throat, trachea, bronchi, and associated organs.[4] Some of the most common infections are discussed here.

Common Cold

A cold is a viral infection that usually begins as a scratchy sore throat, followed by a water mucus discharge from the nasal cavities. There is rarely a fever and symptoms are usually mild, requiring little or no medication. While colds have a short duration, immunity is also brief. Since there are estimated to be over 150 cold-causing viruses, it is very difficult to isolate oneself in order to avoid infection.

Influenza

"Flu" is also a viral infection; but while it begins as an upper respiratory infection, it spreads to other parts of the body, causing aches and pains in the joints. There is usually a fever and the illness lasts for a longer length of time than a cold. Immunity is possible, but only the vaccine developed for the particular virus prevalent that season can usually be successful in protecting the individual during a current flu epidemic. Since flu viruses constantly mutate, there can be no buildup in immunity and a new viral illness rapidly spreads from person to person and from place to place. Pandemics, in which a newly mutated flu virus spreads about the world, have occurred on occasion.

Bronchitis

Viral infections can spread from the nasal cavities to the sinuses (sinusitis), to the middle ears (otitis media), to the larynx (laryngitis), and to the bronchi (bronchitis). *Acute* bronchitis is usually caused by a secondary bacterial infection of the bronchi, resulting in a heavy mucoid discharge with much coughing. Acute bronchitis usually responds to antibiotic therapy.

Chronic bronchitis is not necessarily due to infection. It is often caused by a constant irritation of the lining of the bronchi that undergoes degenerative changes, with the loss of cilia preventing the normal cleansing action. There is frequent coughing and the individual is more susceptible to upper respiratory infections. Chronic bronchitis most often affects cigarette smokers.

[4]Allergies, including asthma, are discussed on pages 256–57.

Strep Throat

This is a very severe throat infection caused by the bacterium *Streptococcus*. Swallowing is very difficult and there is a fever. Strep throat should be treated promptly because it may lead to complications, such as rheumatic fever in which the heart valves may be permanently affected.

Lung Disorders

Pneumonia and tuberculosis, two infections of the lungs, formerly caused a large percentage of deaths in the United States. Now they are controlled by antibiotics. The other two illnesses discussed in the following, emphysema and lung cancer, are not due to infections; in most instances they are due to cigarette smoking.

Pneumonia

Both viruses and bacteria can infect the lungs but a bacterial infection, usually caused by the bacteria *Pneumococcus,* is the most serious. In lobar pneumonia, the infection is localized in specific lobes of the lungs and these become inoperative as they fill with mucus and pus. The more lobes involved, the more serious the infection.

Tuberculosis

Tuberculosis is caused by the tubercle bacillus. It is possible to tell if a person has ever been exposed to tuberculosis by use of a skin test in which a highly diluted extract of the bacilli is injected into the skin of the patient. A person who has never been in contact with the bacillus will show no reaction, while one who has developed immunity to the organism will show an area of inflammation that peaks in about 48 hours. If these bacilli do invade the lung tissue, the cells build a protective capsule about the foreigners to isolate them from the rest of the body. This tiny capsule is called a **tubercle.** If the resistance of the body is high, the imprisoned organisms may die, but if the resistance is low the organisms may eventually be liberated. If a chest X ray detects the presence of tubercles, the individual is put on appropriate drug therapy to ensure the localization of the disease and eventual destruction of any live bacterial organisms.

Emphysema

Emphysema refers to the destruction of lung tissue, with accompanying ballooning or inflation of the lungs due to trapped air. The trouble stems from the destruction and collapsing of the bronchioles. When this occurs, the alveoli are cut off from renewed oxygen supply and the air within them is trapped. The trapped air very often causes *rupturing of the alveolar walls* together with a fibrous thickening of the walls of the small blood vessels in the vicinity. In any case, the victim is breathless and may have a cough. Since the surface area for gas exchange is reduced, not enough oxygen reaches the heart and brain. Even so, the heart works furiously to force more blood through the lungs and this may lead to a heart condition. Lack of oxygen for the brain may make the person feel depressed, sluggish, and irritable.

Since emphysema often develops in persons who smoke, it has been possible to follow the development of the disease by doing autopsies on smokers (fig. 13.16).

Figure 13.16
Development of emphysema. *a.* Normal alveoli. *b.* Rupture of some walls of alveoli. *c.* Absence of normal alveolar tissue. Emphysema causes a ballooning of the chest because of trapped air.

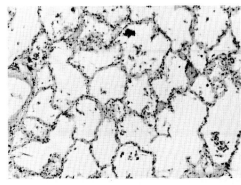

a.

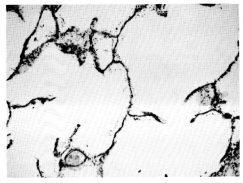

b.

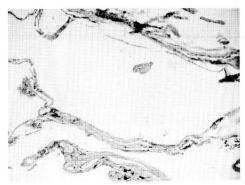

c.

Figure 13.17
Stages in the development of bronchial cancer.
a. Cilia have disappeared and epithelial cells have
increased in number. *b.* Precancerous cells
continue to invade other tissues layers.

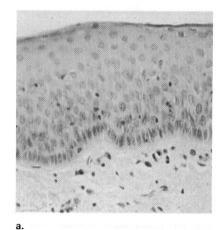 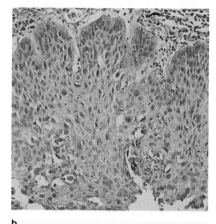

a.   b.

Figure 13.18
Normal versus cancerous lungs. *a.* Normal lungs
appear light in color. *b.* Lungs of heavy smoker.
Notice how black the lungs are except where
cancerous tumors have formed.

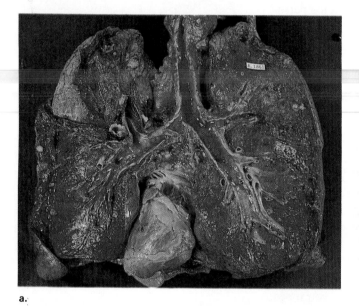 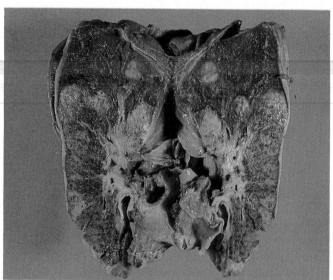

a.   b.

Lung Cancer

Autopsies on smokers have also revealed the progressive steps by which cancer
of the lung develops (fig. 13.17). The first event appears to be a *thickening of
the cells* that line the bronchi. (Callusing occurs whenever cells are exposed
to irritants.) Then there is a *loss of cilia* so that it is impossible to prevent
dust and dirt from settling in the lungs. Following this, cells with atypical
nuclei appear in the thickened lining. A disordered collection of cells with
atypical nuclei may be considered to be cancer *in situ* (at one location). The
final step occurs when some cells break loose and penetrate the other tissues,
a process called *metastasis*. This is true cancer (fig. 13.18). The tumor may
grow until the bronchus is blocked, cutting off the supply of air to that lung.
The lung then collapses, and the secretions trapped in the lung spaces become

Human Anatomy and Physiology

Table 13.4 Risks of smoking compared to benefits of quitting

Risks of smoking	Benefits of quitting
Shortened life expectancy. 25–year-old 2-pack a day smokers have life expectancy 8.3 years shorter than non-smoking contemporaries. Other smoking levels: proportional risk.	**Reduced risk of premature death** cumulatively. After 10–15 years, ex-smokers' risk approaches that of those who've never smoked.
Lung cancer. Smoking cigarettes "major cause in both men and women."	Gradual decrease in risk. **After 10–15 years, risk approaches that of those who never smoked.**
Larynx cancer. In all smokers (including pipe and cigar) it's 2.9 to 17.7 times that of nonsmokers.	**Gradual reduction of risk** after smoking cessation. **Reaches normal after 10 years.**
Mouth cancer. Cigarette smokers have 3 to 10 times as many oral cancers as nonsmokers. Pipes, cigars, chewing tobacco also major risk factors. Alcohol seems synergistic carcinogen with smoking.	Reducing or eliminating smoking/drinking reduces risk in first few years; **risk drops to level of nonsmokers in 10–15 years.**
Cancer of esophagus. Cigarettes, pipes and cigars increase risk of dying of esophageal cancer about 2 to 9 times. Synergistic relationship between smoking and alcohol.	Since risks are dose related, reducing or eliminating smoking/drinking **should have risk-reducing effect.**
Cancer of bladder. Cigarette smokers have 7 to 10 times risk of bladder cancer as nonsmokers. Also synergistic with certain exposed occupations: dye-stuffs, etc.	Risk decreases gradually to that of nonsmokers over 7 years.
Cancer of pancreas. Cigarette smokers have 2 to 5 times risk of dying of pancreatic cancer as nonsmokers.	Since there is evidence of dose-related risk, reducing or eliminating smoking should have risk-reducing effect.
Coronary heart disease. Cigarette smoking is major factor; responsible for 120,000 excess U.S. deaths from coronary heart disease (CHD) each year.	**Sharply decreases risk after one year.** After 10 years ex-smokers' risk is same as that of those who never smoked.
Chronic bronchitis and pulmonary emphysema. Cigarette smokers have 4–25 times risk of death from these diseases as nonsmokers. Damage seen in lungs of even young smokers.	**Cough and sputum disappear** during first few weeks. **Lung function may improve** and rate of deterioration slow down.
Stillbirth and low birthweight. Smoking mothers have more stillbirths and babies of low birthweight—more vulnerable to disease and death.	Women who stop smoking before 4th month of pregnancy **eliminate risk of stillbirth and low birthweight** caused by smoking.
Children of smoking mothers smaller, under-developed physically and socially, seven years after birth.	Since children of nonsmoking mothers are bigger and more advanced socially, inference is that **not smoking during pregnancy might avoid such underdeveloped children.**
Peptic ulcer. Cigarette smokers get more peptic ulcers and die more often of them; cure is more difficult in smokers.	Ex-smokers get ulcers but these are **more likely to heal rapidly and completely** than those of smokers.
Allergy and impairment of immune system.	Since these are direct, immediate effects of smoking, they are obviously **avoidable by not smoking.**
Alters pharmacologic effects of many medicines, diagnostic tests and greatly increases risk of thrombosis with oral contraceptives.	**Majority of blood components elevated by smoking return to normal after cessation.** Nonsmokers on Pill have much lower risks of thrombosis.

infected, with a resulting pneumonia or the formation of a lung abscess. The only treatment that offers a possibility of cure, before secondary growths have had time to form, is to remove the lung completely. This operation is called *pneumonectomy*.

Statistical studies have shown that smoking cigarettes may be associated with other illnesses, including other types of cancer. Table 13.4 lists the risks of smoking and benefits of quitting that have been established by appropriate investigations and studies. If a person does stop smoking, the body tissues, if not already cancerous, return to normal.

Summary

Respiration has been divided into the following components: breathing, external and internal respiration, and cellular respiration.

During the process of breathing, air enters and exits the lungs by way of the respiratory tract, which consists of the nose (which also smells the air), the nasopharynx, the pharynx, the larynx (which also contains the vocal cords), the trachea, the bronchi, and the bronchioles. The right and left lungs located on either side of the heart, are covered by pleural membranes. The chest cavity is bounded by the rib cage and by the diaphragm. The bronchi, along with the pulmonary arteries and veins, penetrate the lungs. Thereafter, the bronchi divide into the bronchioles, which enter the alveoli, air sacs surrounded by extensive pulmonary capillaries.

Inspiration begins when the breathing center in the medulla oblongata, stimulated by carbon dioxide in the blood, sends excitatory nerve impulses to the diaphragm and rib cage. As they contract, the diaphragm lowers and the rib cage moves up and out; the elastic lungs expand, creating a partial vacuum that causes air to rush in. Nerves within the expanded lungs send inhibitory impulses to the breathing center, stopping its excitatory messages to the breathing muscles. As they relax, the diaphragm resumes its dome shape and the rib cage retracts, pushing air out of the lungs during expiration.

External respiration occurs when gases are exchanged between the alveoli and the surrounding capillaries. Blood laden with the bicarbonate ion, enters the capillaries where the following reaction, speeded up by the enzyme carbonic anhydrase, takes place:

$$H^+ + HCO_3^- \rightarrow H_2CO_3 \rightarrow H_2O + CO_2$$

As carbon dioxide diffuses out of the blood into the alveoli, oxygen diffuses in to be taken up by hemoglobin:

$$Hb + O_2 \longrightarrow HbO_2$$

Internal respiration takes place in the tissues where hemoglobin gives up its oxygen:

$$HbO_2 \longrightarrow Hb + O_2$$

As oxygen diffuses out of the blood, carbon dioxide diffuses into the blood and combines with water:

$$CO_2 + H_2O \rightarrow H_2CO_3 \rightarrow H^+ + HCO_3^-$$

The resulting hydrogen ion is taken up by the globin portion of hemoglobin, and the bicarbonate ion is carried in the plasma.

There are a number of illnesses associated with the respiratory tract. Infections such as colds and flu are known to everyone. In addition, pneumonia and tuberculosis are lung infections of a serious nature. Both may be cured by drugs and rest. Two illnesses that have been attributed to breathing polluted air are emphysema and lung cancer. Both of these conditions may develop from smoking cigarettes. Emphysema refers to the inflating of the lungs by trapped air. This occurs when the bronchioles are destroyed and collapse. Cancer develops when cells with atypical nuclei appear in the lining of the bronchi.

Study Questions

1. What are the four parts of respiration? In which of these is oxygen actually used up and carbon dioxide produced? (p. 262)
2. List the parts of the respiratory tract. (p. 262) What are the special functions of the nasal cavity, larynx, and alveoli? (pp. 264, 266)
3. What are the steps in inspiration and expiration? How is breathing controlled? (pp. 267–69)
4. Why can't we breathe through a very long tube? (pp. 269–70)
5. What physical process is believed to explain gas exchange? (p. 271)
6. What two equations are needed to explain external respiration? (p. 272)
7. How is hemoglobin remarkably suited to its job? (p. 273)
8. What two equations are needed to explain internal respiration? (p. 273)
9. Name some infections of the respiratory tract. (pp. 274–75)
10. What is emphysema and how does it affect one's health? (p. 275)
11. By what steps is cancer believed to develop in the person who smokes? (p. 276)

Selected Key Terms

inspiration (in″spĭ-ra′shun)
expiration (eks″pĭ-ra′shun)
eustachian tube (u-sta′ke-an tūb)
pharynx (far′ingks)
larynx (lar′ingks)
glottis (glot′is)
epiglottis (ep″ĭ-glot′is)
trachea (tra′ke-ah)
tracheostomy (tra″ke-os′to-me)
bronchi (brong′ki)
bronchiole (brong′kē-ōl)
alveoli (al-ve′o-lī)
diaphragm (di′ah-fram)
emphysema (em′fĭ-se′mah)
hemoglobin (he″mo-glo′bin)
bicarbonate ion (bi-kar′bo-nāt i′on)
carbonic anhydrase (kar-bon′ik-an-hi′drās)
carbaminohemoglobin (kar-bam″ĭ-no-he″mo-glo′bin)

Further Readings

Avery, M. E. et al. 1973. Lung of the newborn infant. *Scientific American* 228(4):74.

Comroe, J. H., Jr. 1966. The lung. *Scientific American* 214(2):56.

Fenn, W. O. 1960. The mechanism of breathing. *Scientific American* 202(6):138.

Hammond, E. C. 1962. The effects of smoking. *Scientific American* 207(9):39.

Slonim, N. B., and Hamilton, L. H. 1981. *Respiratory physiology.* 4th ed. St. Louis: C. V. Mosby.

14

Excretion

Chapter concepts

1 Excretion rids the body of unwanted substances, particularly end products of metabolism.

2 Several organs assist in the excretion process, but the kidneys, which are a part of the urinary system, are the primary organs of excretion.

3 The formation of urine by the more than one million nephrons present in each kidney serves not only to rid the body of nitrogenous wastes but also to regulate the water content, the salt levels, and the pH of the blood.

4 The kidneys, whose malfunction brings illness and may cause death, are important organs of homeostasis.

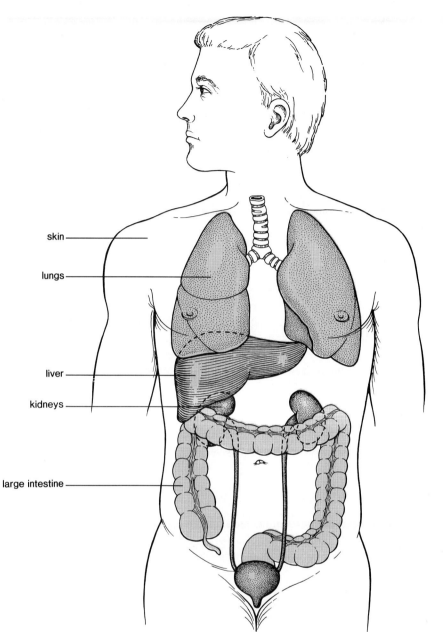

skin

lungs

liver

kidneys

large intestine

The composition of blood serving the tissues remains relatively constant due to both the continual addition of substances needed by cells and the continual removal of substances not needed by cells. In previous chapters, we have discussed how the digestive tract and lungs add nutrients and oxygen to the blood. In this chapter, we will discuss how the organs of excretion (fig. 14.1) remove substances from the blood and thereby help maintain homeostasis.

Table 14.1 Some metabolic end products

Name	End product of	Primarily excreted by
Nitrogenous Wastes		
Ammonia	Amino acid metabolism	Kidneys
Urea	Ammonia metabolism	Kidneys
Uric acid	Nucleotide metabolism	Kidneys
Creatinine	Creatine phosphate metabolism	Kidneys
Other		
Bile pigments	Hemoglobin metabolism	Liver
Carbon dioxide	Cellular respiration	Lungs

Excretory Substances

Excretion rids the body of metabolic wastes. Among these are the toxic products listed in table 14.1. In addition, salts and water are constantly being excreted.

Nitrogenous End Products

Several of the end products excreted by humans are related to nitrogen metabolism since amino acids, nucleotides, and creatine all contain nitrogen.

Ammonia (NH_3) arises from the deamination, or removal, of amino groups from amino acids. Ammonia is extremely toxic to the body, and only animals living in fresh water, who continually flush out their bodies with excess water, excrete ammonia. In our bodies, ammonia is converted to urea by the liver.

Urea is produced in the liver by a complicated series of reactions called the urea cycle. In this cycle, carrier molecules take up carbon dioxide and two molecules of ammonia to finally release urea:

$$H_2N - \overset{\overset{\displaystyle O}{\|}}{C} - NH_2$$

Uric acid occurs when nucleotides are metabolically broken down. Uric acid is relatively insoluble and if present in excess, will precipitate out of the plasma. Crystals of uric acid sometimes collect in the joints, producing a painful ailment called gout.

Creatinine is an end product of muscle metabolism. It results when creatine phosphate, a molecule that serves as a reservoir of high energy phosphate, breaks down.

Other Excretory Substances

Other excretory substances are bile pigments, carbon dioxide, ions (salts), and water.

Bile Pigments

Bile pigments are derived from the heme portion of hemoglobin and are incorporated into bile within the liver. The liver produces bile, which is stored in the gallbladder before passing into the small intestine. If for any reason the bile duct is blocked, bile spills out into the blood, producing a discoloration of the skin called jaundice (p. 190).

Carbon Dioxide

The lungs are the major organs of carbon dioxide excretion, although the kidneys are also important. The kidneys excrete bicarbonate ions, the form in which carbon dioxide is carried in the blood.

Ions

Ions (salts) are excreted, not because they are end products of metabolism but because their proper concentration in the blood is so important to the pH, osmotic pressure, and **electrolyte balance** of the blood.[1] The balance of potassium (K^+) and sodium (Na^+) is important to nerve conduction. The level of calcium (Ca^{++}) in the blood affects muscle contraction; iron (Fe^{++}) takes part in hemoglobin metabolism; magnesium (Mg^{++}) helps many enzymes function properly.

Water

Water is an end product of metabolism; it is also taken into the body when food and liquids are consumed. The amount of fluid in the blood helps determine blood pressure. Treatment of hypertension sometimes includes the administration of a diurectic drug that increases the excretion of sodium and water by the kidneys.

Organs of Excretion

The kidneys are the primary excretory organs, but there are other organs that also function in excretion, such as those discussed in the discussion that follows.

Skin

The sweat glands in the skin (fig. 8.12) excrete perspiration, a solution of water, salt, and some urea. The sweat glands are made up of a coiled tubule portion in the dermis and a narrow, straight duct that exits from the epidermis. Although perspiration is an excretion, we perspire not so much to rid the body of waste but to cool off the body. The body cools because heat is lost as perspiration evaporates. Thus, sweating keeps the body temperature within normal range during muscular exercise or when the outside temperature rises. In times of renal failure, more urea than usual may be excreted by the sweat glands, to the extent that a so-called urea frost is observed on the skin.

Liver

The liver excretes bile pigments, which are incorporated into bile, before it passes into the small intestine. The yellow pigment found in urine, called urochrome, is also derived from the breakdown of heme, but this pigment is deposited in the blood and therefore is excreted by the kidneys.

Lungs

The process of expiration (breathing out) not only removes carbon dioxide from the body, it also results in the loss of water. The air we exhale contains moisture, as demonstrated by blowing onto a cool mirror.

---

[1] Electrolytes are ions that can conduct electricity when in solution.

Table 14.2	Composition of urine
Water	95%
Solids	5%
Organic wastes (per 1500 ml of urine)	
urea	30 g
creatinine	1–2 g
ammonia	1–2 g
uric acid	1 g
Ions (Salts)	25 g

Positive	*Negative*
sodium	chlorides
potassium	sulfates
magnesium	phosphates
calcium	

Table 14.3	Urinary system
Organ	**Function**
Kidneys	Produce urine
Ureters	Transport urine
Bladder	Storage of urine
Urethra	Elimination of urine

Intestine

Certain salts, such as those of iron and calcium, are excreted directly into the cavity of the intestine by the epithelial cells lining it. These salts leave the body in the feces.

At this point, it might be helpful to remember that the term *defecation,* and not *excretion,* is used to refer to the elimination of feces from the body. Substances that are excreted are those that are waste products of metabolism. Undigested food and bacteria, which make up feces, have never been a part of the functioning of the body, but salts that are passed into the gut are excretory substances because they were once metabolites in the body.

Kidneys

The kidneys excrete urine, which contains a combination of the end products of metabolism (table 14.2). The kidneys are a part of the urinary system.

Urinary System

The urinary system includes the structures illustrated in figure 14.2 and listed in table 14.3. The organs are listed in order according to the path of urine.

The **kidneys** are reddish-brown organs about 4 inches long, 2 inches wide, and 1 inch thick, which lie on either side of the midline against the dorsal body wall to which they are anchored by connective tissue. The renal artery,

Figure 14.2
The urinary system. Urine is only found within the kidneys, ureters, bladder, and urethra.

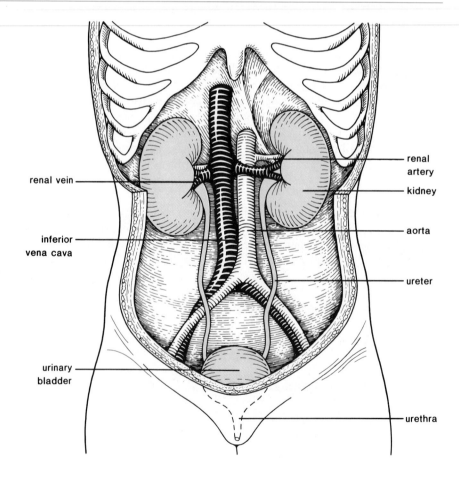

Human Anatomy and Physiology

renal vein, nerves, and ureters join the kidney on the concave side toward the midline.

The **ureters** are muscular tubes that convey the urine toward the bladder by peristaltic contractions. Urine enters the bladder in jets that occur at the rate of one to five per minute.

The **urinary bladder,** which can hold up to 600 milliliters of urine, is a hollow muscular organ that gradually expands as urine enters. Figure 14.3a and figure 14.3b show the bladder wall in its relaxed distended conditions. In the male, the bladder lies ventral to the rectum, seminal vesicles, and ductus deferens. In the female, it is ventral to the uterus and upper vagina.

The **urethra,** which extends from the urinary bladder to an external opening, differs in length in females and males. In females, the urethra lies ventral to the vagina and is only about 1½ inches long. The short length of the female urethra facilitates bacterial invasion and explains why females are more prone to urethral infections. In males, the urethra averages 6 inches when the penis is relaxed. As the urethra leaves the bladder, it is encircled by the prostate gland. In older men, enlargement of the prostate gland may prevent urination, a condition that can usually be corrected.

Notice that there is no connection between the genital (reproductive) and urinary systems in females, but there is a connection in males. When urinating, the urethra in the male carries urine, and during sexual orgasm the urethra transports semen. This double function does not alter the path of urine, and it is important to realize that urine is found only in those structures listed in table 14.3.

Urination

When the bladder fills with urine, stretch receptors send nerve impulses to the spinal cord; nerve impulses leaving the cord then cause the bladder to contract and the sphincters to relax so that urination may take place. In older children and adults, it is possible for the brain to control this reflex, delaying urination to a suitable time.

Figure 14.3
Photomicrograph of bladder wall. *a.* Relaxed. *b.* Distended. Because the bladder can distend, the intrabladder pressure does not increase greatly until the bladder fills beyond 200 to 400 milliliters. The bladder can hold up to a liter (quart) of urine.

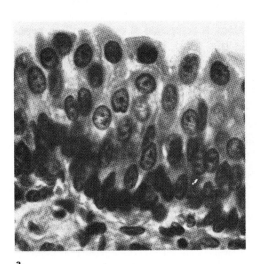

a.

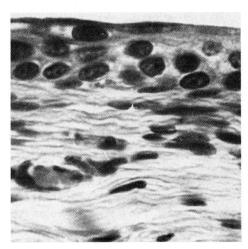

b.

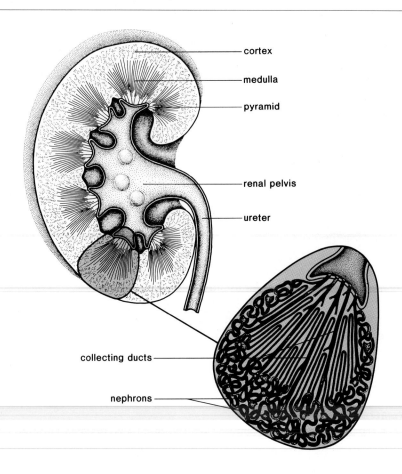

cortex

medulla

pyramid

renal pelvis

ureter

collecting ducts

nephrons

Kidneys

When a kidney is sliced longitudinally, it is shown to be composed of three macroscopic parts (fig. 14.4): (1) an outer granulated layer called the cortex, which dips down in between (2) a radially striated, or lined, layer called the **medulla** and (3) an inner space, or cavity, called the **pelvis,** where the urine collects before entering the ureters. Occasionally **kidney stones** form in the pelvis. They are formed from fairly insoluble substances such as uric acid and calcium salts that precipitate out of the urine instead of remaining in solution. The stones either pass naturally or they may be surgically removed.

Nephrons Microscopically, the kidney is composed of over 1 million nephrons, sometimes called renal tubules (fig. 14.5). Each nephron is made up of several parts. The blind end of the tubule is pushed in on itself to form a cuplike structure called **Bowman's capsule,** within which there is a capillary tuft called the **glomerulus.** The outer layer of Bowman's capsule is composed of squamous epithelial cells; the inner layer is composed of specialized cells that allow easy passage of molecules. Next, there is a **proximal** (meaning near the Bowman's capsule) **convoluted tubule** in which the cells are cuboidal, with many mitochondria and an inner brush border. Then the cells become flat, the tube narrows, and makes a U-turn to form the portion of the tubule called the **loop of Henle.** This leads to the **distal** (far from Bowman's capsule) **convoluted tubule,** where the cells are cuboidal, again with mitochondria, but no brush border. The distal convoluted tubule enters the **collecting duct.**

Figure 14.6 indicates the position of a single nephron within the kidney. Bowman's capsules and convoluted tubules lie within the cortex and account for the granular appearance of the cortex. Loops of Henle and collecting ducts lie within the triangular-shaped pyramids of the medulla. Since these are longitudinal structures, they account for the striped appearance of the pyramids.

Human Anatomy and Physiology

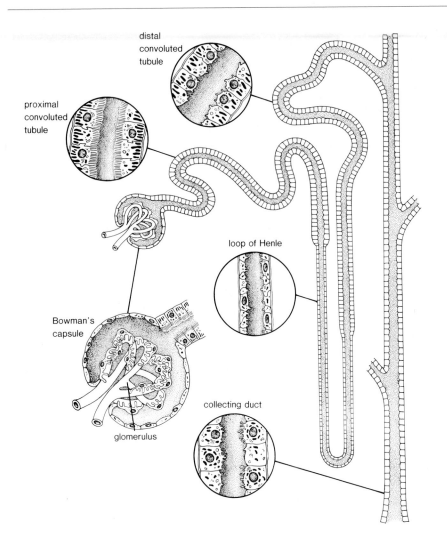

Figure 14.5
Nephron gross and microscopic anatomy. A nephron is made up of Bowman's capsule, proximal convoluted tubule, loop of Henle, distal convoluted tubule, and collecting duct. Bowman's capsule contains a capillary tuft called the glomerulus. The blowups show the tissue types at these different locations.

distal convoluted tubule

proximal convoluted tubule

loop of Henle

Bowman's capsule

glomerulus

collecting duct

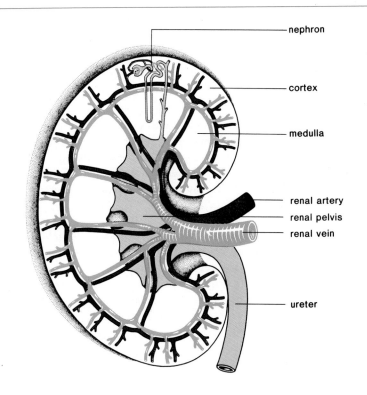

Figure 14.6
Each kidney receives a renal artery that divides into arterioles within the kidney. Venules leaving the kidney join to form the renal vein. This drawing also shows how one nephron is placed in the kidney. Parts of each nephron are in the cortex and other parts are in the medulla (see figure 14.8).

nephron

cortex

medulla

renal artery
renal pelvis
renal vein

ureter

Figure 14.7

Diagram of nephron showing steps in urine formation: filtration, reabsorption, and tubular excretion. Note also that water enters the tissues at the loop of Henle and collecting duct.

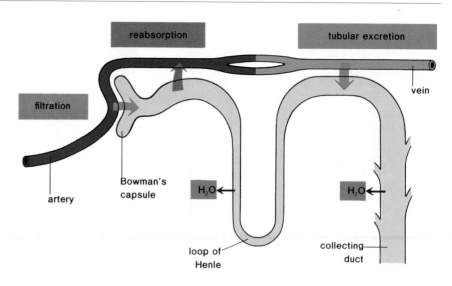

Table 14.4	Nephron	
Name of part	**Location in kidney**	**Function**
Bowman's capsule	Cortex	Forms filtrate
Proximal convoluted tubule	Cortex	Selective reabsorption
Loop of Henle	Medulla	Extrusion of sodium and reabsorption of water
Distal convoluted tubule	Cortex	Tubular excretion
Collecting duct	Medulla	Reabsorption of water

Urine Formation (Simplified)

Urine formation can be studied in brief (fig. 14.7) or in detail, on page 289. Table 14.4 lists the parts of a nephron, their location within the kidney, and their contribution to urine formation.

Overview of Urine Formation

Each nephron has its own blood supply, including two capillary regions; the glomerulus is a capillary tuft inside Bowman's capsule, and the **peritubular capillary network** surrounds the rest of the nephron (fig. 14.8). Urine formation requires the movement of molecules between these capillaries and the nephron.

1. *Pressure filtration* occurs at Bowman's capsule. During pressure filtration, water, nutrient molecules, and waste molecules move from the glomerulus to the inside of Bowman's capsule. The blood has been *filtered* because large molecules, such as protein molecules, remain within the blood while small molecules, such as glucose and urea, leave the blood to enter the tubule. The substances that are filtered are called the *glomerular filtrate*.
2. *Selective reabsorption* occurs primarily at the proximal convoluted tubule. During selective reabsorption, nutrient and salt molecules are actively reabsorbed from the proximal convoluted tubule into the peritubular capillary, and water follows passively.
3. *Tubular excretion* occurs primarily at the distal convoluted tubule. During tubular excretion, certain substances, penicillin and histamine, for example, are actively secreted from the peritubular capillary into the fluid of the tubule. Also, as mentioned on page 295, hydrogen ions and ammonia are secreted as required. This is an important way in which the kidneys contribute to homeostasis.

Human Anatomy and Physiology

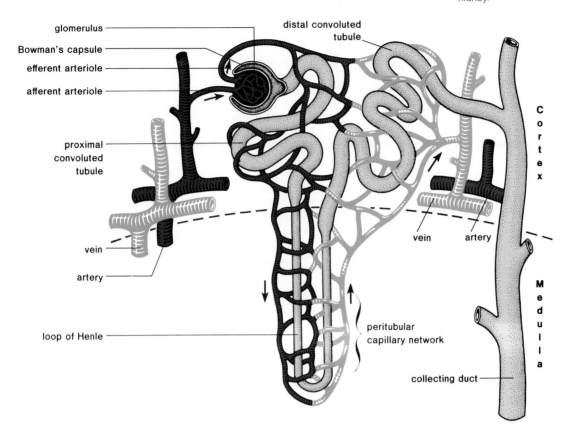

Table 14.5 Circulation within a nephron*

Name of structure	Comment
Afferent arteriole	Brings arteriolar blood toward Bowman's capsule
Glomerulus	A capillary tuft enveloped by Bowman's capsule
Efferent arteriole	Takes arteriolar blood away from Bowman's capsule
Peritubular capillary network	Capillary bed that envelopes the rest of the tubule
Venule	Takes venous blood away from the tubule

*Compare to figure 14.8

Concentrated Urine

Humans excrete a urine that contains only waste molecules dissolved in a minimum amount of water. This concentrated urine results because water is reabsorbed not only at the proximal convoluted tubule but along the entire length of the renal tubule, particularly at the loop of Henle and the collecting duct. Regulation of blood volume by excretion of water is discussed in more detail on page 294.

Urine Formation (in Detail)

In order to fully appreciate urine formation, it is necessary to understand the blood supply of the nephron. Table 14.5 lists the blood vessels associated with each part of the nephron. Locate these blood vessels in figure 14.8 and trace the path of blood about a nephron.

Blood Supply

After the renal artery leaves the aorta to enter the kidney, it branches into numerous smaller arteries. These small arteries branch off into tiny arterioles, one for each nephron. Each arteriole, called an **afferent arteriole,** divides to form the glomerulus (fig. 14.9), which is surrounded by Bowman's capsule. When the capillaries rejoin, they form another arteriole called the **efferent arteriole.** (This is unusual since there is usually an arteriole on one side of a capillary bed and a venule on the other side—here there is an arteriole on both sides.) The efferent arteriole soon divides into the peritubular capillary network, which serves the needs of the renal tubule. This capillary network supplies the tubule cells with nutrients and oxygen and leads to a venule that joins with the venules from other nephrons to form the renal vein, a vessel that enters the inferior vena cava.

Bowman's Capsule

Whole blood, of course, enters the afferent arteriole and the glomerulus. Under the influence of glomerular blood pressure, which is usually about 60 mm Hg, small molecules move from the glomerular capillaries to the inside of Bowman's capsule across the thin walls of each. Since large molecules (e.g., proteins) and formed elements are unable to pass through, the process is a filtration called **pressure filtration,** and a filtrate of glomerular blood is said to form within Bowman's capsule. In addition to water as its main component, the **glomerular filtrate** contains dissolved substances in approximately the same concentration as plasma. The proteins and formed elements that are too large to be a part of the filtrate leave the glomerulus by way of the efferent arteriole.

In effect, then, blood that enters the glomerulus is divided into two portions: the filterable components and the nonfilterable components.

Filterable Blood Components	Nonfilterable Blood Components
water	formed elements (blood cells
nitrogenous wastes	and platelets)
nutrients	proteins
salts (ions)	

Figure 14.9
A scanning electron micrograph of a section of kidney cortex showing a glomerulus (the outer layer of Bowman's capsule has been removed). The holes surrounding the glomerulus are cross sections of tubules.

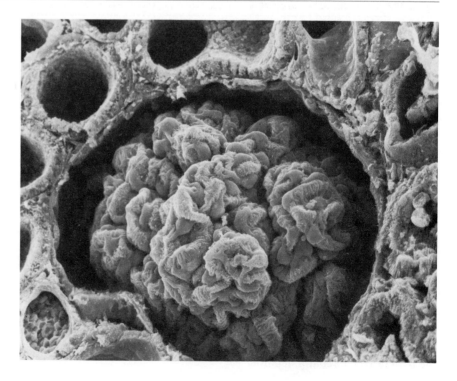

Human Anatomy and Physiology

The filterable components pass from the capillary tuft to the inside of Bowman's capsule, and the nonfilterable components stay within the vascular system surrounding the tubule.

The blood pressure within the glomerulus is higher than in other capillary beds because the renal artery is very short and, most important, the glomerulus lies between two arterioles whose circumference can be regulated to increase blood pressure. Also, if necessary, a special region of the afferent arteriole called the **juxtaglomerular apparatus,** or polar cushion (fig. 14.10), can release **renin,** a substance that brings about high blood pressure. While renin is normally released only when a higher blood pressure is needed for filtration, it seems to be continually released by persons with kidney disease and accounts for the hypertension that is usually present in these patients.

A consideration of the preceding filterable substances leads one to conclude that if the glomerular filtrate were the same as urine, the body would continually lose nutrients, water, and salts. Obviously, death from dehydration, starvation, and low blood pressure would quickly follow. Thus, we can assume that the composition of the filtrate must be altered as this fluid passes within the remainder of the tubule.

Proximal Convoluted Tubule

Both passive and active reabsorption of molecules from the tubule to the blood occur as the filtrate moves along the proximal convoluted tubule.

Passive reabsorption involves particularly the movement of water molecules from the area of greater concentration in the filtrate to the area of lesser concentration in the blood. Two factors aid the process. The nonfilterable proteins remain in the blood, where they exert an osmotic pressure that pulls water back into the bloodstream. Following the active reabsorption of sodium (Na^+), which is discussed later, chlorine (Cl^-) follows passively because, being a negative ion, it is attracted to the positive charge of sodium. Water follows passively because the reabsorption of sodium increases the osmotic pressure of the blood.

Active reabsorption involves active transport and this accounts for the large energy needs of the kidney. In figure 14.11, two membranes are shown; the first of these exists between the tubule cavity and the tubule cell, and the second lies between the tubule cell and the blood. A molecule, glucose, for example, diffuses passively into the tubule cell but is then actively transported from the cell into the blood, requiring the use of a carrier molecule. Reabsorption by active transport is selective since only molecules recognized by carrier molecules move across the membrane. This accounts for the homeostatic ability of the kidney to reabsorb only molecules needed by the body.

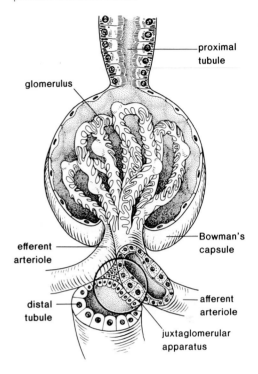

Figure 14.10
Drawing of glomerulus and adjacent distal convoluted tubule. The juxtaglomerular apparatus *(circled)* is sensitive to the fluid pressure within the distal convoluted tubule and releases renin if this pressure falls below normal.

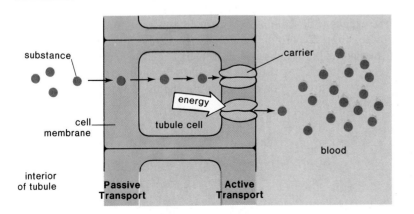

Figure 14.11
Nutrient molecules and sodium are actively reabsorbed from a kidney tubule in the manner illustrated. These molecules move passively into the tubule cell but then are actively transported out of the tubule cell into the blood. Active transport requires the participation of carrier molecules.

Figure 14.12

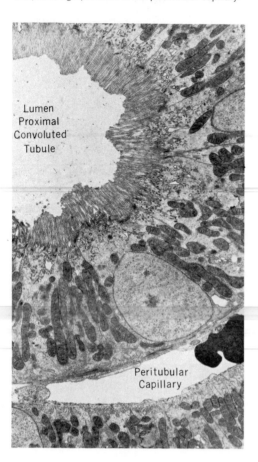

Electron micrograph of cells that line the lumen (inside) of proximal convoluted tubule where selective reabsorption takes place. The cells have a brush border composed of microvilli that greatly increase the surface area exposed to the lumen. Each cell has many mitochondria that supply the energy needed for active transport. A red blood cell *(lower right)* is seen in the peritubular capillary.

Lumen
Proximal
Convoluted
Tubule

Peritubular
Capillary

Threshold Level

Reabsorption occurs until the threshold level of a substance is obtained. Thereafter the substance will appear in the urine. For example, the threshold level for glucose is 15 grams glucose per 100 milliliters of blood. After this amount is reabsorbed, any excess present in the filtrate will appear in the urine. This is a common occurrence in diabetes mellitus (sugar diabetes, p. 384) because the liver fails to store glucose as glycogen. The real difficulty is that the pancreas is failing to produce insulin, a hormone that promotes glucose uptake in all cells.

In contrast to the high threshold level of glucose, urea has a very low threshold level that is quickly reached, so that nearly all urea remains in the urine.

Tubule Cells

The structure of the cells that line the proximal convoluted tubule is anatomically adapted for absorption (fig. 14.12). Within the tubule, these cells are covered by numerous microvilli, about one micron in length, that increase the surface area for reabsorption. In addition, the cells contain numerous mitochondria that produce the energy necessary for active transport.

Tubular Fluid

We have seen that the filtrate that enters the proximal convoluted tubule is divided into two portions—the components that are reabsorbed and the components that are not reabsorbed.

Reabsorbed Filtrate Components	Nonreabsorbed Filtrate Components
most water	some water
nutrients	wastes
required salts (ions; e.g., Na⁺, Cl⁻)	excess salts (ions)

The substances that are not reabsorbed become the tubular fluid that enters the loop of Henle.

Loop of Henle

Animals, including humans, whose nephrons contain a loop of Henle excrete a hypertonic urine (i.e., a larger amount of metabolic wastes per volume than blood). The loop of Henle is made up of a **descending** (going down) and an **ascending** (going up) **limb.** The ascending limb actively extrudes sodium into the tissue of the medulla, and thus this tissue is always hypertonic to the fluid within the descending limb. Water therefore passively diffuses out of the descending limb to be carried away by the blood. Sodium, however, passively diffuses into the descending limb because this is the area of lesser concentration of sodium. The arrows in figure 14.13 show the movement of these molecules.

As the exchange of sodium occurs, the fluid within the descending limb becomes increasingly hypertonic (concentrated). But as the fluid moves up within the ascending limb, sodium is actively extruded, as mentioned, and the fluid is hypotonic (dilute) by the time it reaches the distal convoluted tubule. (Water does not diffuse into or out of the ascending limb because it is impermeable to water.)

Human Anatomy and Physiology

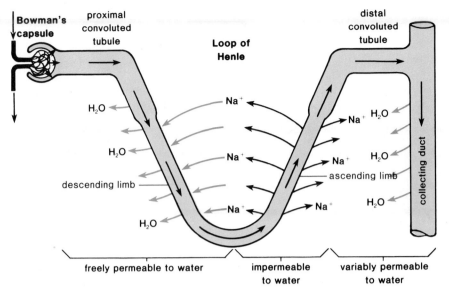

Figure 14.13
Diagram illustrating countercurrent exchange about a nephron. The ascending limb of the loop of Henle extrudes sodium and this creates the osmotic pressure that pulls water out of the descending limb and the collecting duct. Note that the ascending limb is impermeable to water. The permeability of the collecting duct is under hormonal control (see table 14.6).

By noting the direction of sodium in figure 14.13, it can be seen that there is a circulation of sodium between the ascending limb to the tissues to the descending limb, which turns to become the ascending limb. This is called a **countercurrent exchange** because as the fluid moves in the opposite direction within the two limbs there is an exchange of sodium between them.

Distal Convoluted Tubule

The distal convoluted tubule continues the work of the proximal convoluted tubule in that sodium and water are both reabsorbed. As before, sodium is actively reabsorbed into the blood capillary, and thereafter water follows passively.

In this region of the tubule also, substances may be added to the urine by a process called **tubular excretion,** or augmentation. The cells that line this portion of the tubule have numerous mitochondria. Tubular excretion is an active process just like selective reabsorption, but the molecules are moving in the opposite direction. Histamine and pencillin are actively excreted as are hydrogen and ammonia ions, as discussed in the following.

Collecting Duct

The fluid that enters the collecting duct is isotonic to the cells of the cortex. This means that to this point the net effect of reabsorption of water and sodium has been to produce a fluid in which the proportion of water to sodium is the same as in most tissues. However, in the medulla there is an increasing concentration of sodium along the length of the collecting duct, due to the extrusion of sodium by the ascending limb of the loop of Henle. Therefore water diffuses out of the collecting duct, and the urine within the collecting duct becomes hypertonic. Urine (table 14.1) now passes from the collecting duct to the pelvis of the kidney.

Figure 14.14

Presence or absence of loop of Henle in groups of animals. The internal fluids of marine protochordates (p. 616) are isotonic to sea water and there is no loop of Henle. Freshwater bony fishes and amphibians do not need to conserve water and there is no loop of Henle. Reptiles are terrestrial and conserve water by excreting a solid waste, uric acid, rather than by reabsorbing water. Only mammals are solely dependent on excreting a hypertonic urine to conserve water. They have a well-developed loop of Henle.

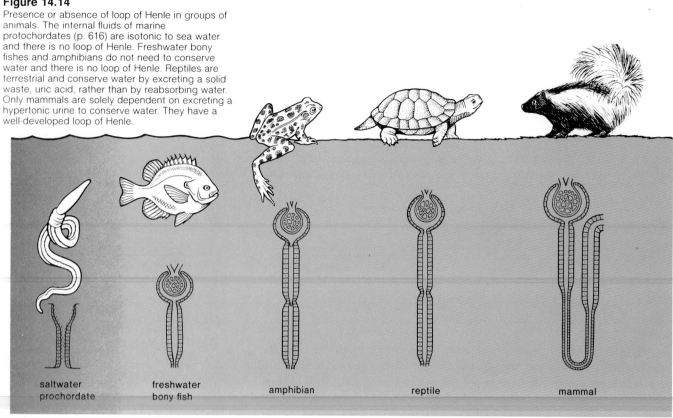

| saltwater protochordate | freshwater bony fish | amphibian | reptile | mammal |

In conclusion, then, human beings along with other mammals excrete a concentrated, or hypertonic, urine, not because water fails to enter the nephron but because the water that enters is reabsorbed. The consistent presence of a loop of Henle in all nephrons makes it possible for efficient reabsorption to occur. In lower vertebrates (fig. 14.14) up to and including the reptiles, the nephrons lack a loop of Henle and the kidney is incapable of producing a hypertonic urine. In birds, some nephrons do have a loop of Henle, but the arrangement is not as efficient as in mammals. Reptiles and birds, unlike mammals, conserve water by excreting uric acid, a solid nitrogenous waste. This is possible because the urinary system empties in the cloaca, a common repository for both the urinary and digestive system.

Regulatory Functions of the Kidneys

Blood Volume

Reabsorption of water is under the control of a hormone called **ADH (antidiuretic hormone),** which is released by the posterior lobe of the pituitary gland (p. 374). ADH increases the permeability of the distal convoluted tubule and collecting duct so that more water can be reabsorbed. In order to understand the function of this hormone, consider its name. *Diuresis* means increased excretion of urine, and *antidiuresis* means suppression of urinary excretion. When ADH is present, more water is reabsorbed and a decreased amount of urine results. This hormone is secreted according to whether blood volume needs to be increased or decreased. When water is reabsorbed at the collecting duct, blood volume increases, and when water is not reabsorbed, blood volume decreases.

Human Anatomy and Physiology

In practical terms (table 14.6), if an individual does not drink much water on a certain day, the posterior lobe of the pituitary releases ADH; more water is reabsorbed; blood volume is maintained at a normal level; and, consequently, there is less urine. On the other hand, if an individual drinks a large amount of water and does not perspire much, the posterior lobe of the pituitary does not release ADH; more water is excreted; blood volume is maintained at a normal level; and a greater amount of urine is formed.

Drinking alcohol causes diuresis because it inhibits the secretion of ADH. Even so, there is evidence to suggest that beer drinking causes diuresis simply because of the increased fluid intake. Drugs called diuretics, are often prescribed for high blood pressure. The drugs cause increased urinary excretion and thus reduce blood volume and blood pressure. Concomitantly, any edema present is also reduced.

Adjustment of pH

The kidneys aid in maintaining a constant pH of the blood, and the whole nephron takes part in this process. Figure 14.15 indicates that the excretion of hydrogen ions and ammonia ($H^+ + NH_3 + NH_4$), together with the reabsorption of bicarbonate ions and sodium, is adjusted in order to keep the pH within normal limits. If the blood is acid, hydrogen ions are excreted in combination with ammonia, while bicarbonate ions and sodium are reabsorbed. This will restore alkalinity because the bicarbonate ion promotes the formation of hydroxyl ions, which then are balanced by the sodium ions:

$$Na^+HCO_3^- + HOH \longrightarrow H_2CO_3 + Na^+OH^-$$

If the blood is alkaline, fewer hydrogen ions are excreted and fewer sodium and bicarbonate ions are reabsorbed.

The reabsorption and/or excretion of ions (salts) by the kidneys illustrates their homeostatic ability to maintain not only the pH of the blood but also the osmolarity of the blood. Osmolarity increases as salts are reabsorbed. Reabsorption of ions, such as K^+ and Mg^{++}, also maintains the proper electrolyte balance of the blood, as discussed on page 283.

Problems with Kidney Function

Because of the great importance of the kidney for maintenance of body fluid homeostasis, renal failure is a life-threatening event. As the reading on page 296 points out, there are many types of illnesses that bring on progressive renal disease and renal failure. Researchers believe that after kidney disease has begun, a rich protein diet contributes to high blood pressure in the glomerulus, a condition that finally results in failure of the kidneys.

Infections of the urinary tract themselves are a fairly common occurrence. If the infection is localized in the urethra, it is called **urethritis.** If it invades the bladder, it is called **cystitis.** And finally, if the kidneys are affected, it is called **nephritis.** Glomerular damage sometimes leads to blockage of the glomeruli so that no fluid moves into the tubules. Or it can cause the glomeruli to become more permeable than usual. This is detected when a **urinalysis** is done. If the glomeruli are too permeable, albumin, white cells, or even red cells may appear in the urine.

When kidney damage is so extensive that more than two-thirds of the nephrons are incapacitated, waste substances accumulate in the blood. This condition is called **uremia** because urea begins to accumulate in the blood. While the presence of nitrogenous wastes can cause serious damage, the retention of water and salts is of even more concern. The latter causes edema, fluid accumulation in the body tissues. Imbalance in ionic composition of body fluids can even lead to loss of consciousness and heart failure.

Table 14.6	Antidiuretic hormone	
Increase in ADH	Increased reabsorption of water	Less urine
Decrease in ADH	Decreased reabsorption of water	More urine

Figure 14.15

The kidneys maintain homeostasis, in part, by adjusting the ionic composition of the blood. In order to maintain the pH fairly constant, the excretion of H^+ and reabsorption of Na^+ and HCO_3^- is adjusted as needed. If the blood is acid, hydrogen ions are excreted in conjunction with ammonia. If the blood is alkaline, reabsorption of sodium bicarbonate is minimized.

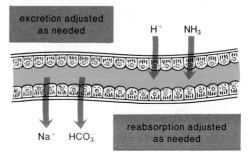

Excretion

295

Comprehending Kidney Disease

For many people whose kidneys are damaged by disease or physiologic disorder, the prognosis is grim: a long deterioration of kidney performance ending in complete renal failure, at which point either dialysis or kidney transplantation is required to sustain life. Even if the illness that caused the initial injury abates short of total kidney destruction, the steady deterioration of the kidneys may well continue.

Why so many kinds of kidney injury should evoke the same pattern of nearly inexorable decline—known clinically as "progressive renal disease"—has been a mystery for decades. Now, a team of medical researchers has produced what is in effect a "unified field theory" of chronic renal failure, incorporating diverse experimental and clinical evidence. Barry M. Brenner, Timothy W. Meyer and Thomas H. Hostetter, all of Harvard Medical School and Brigham and Women's Hospital in Boston [have] summarized their theory . . .

The new theory proposes that all forms of progressive renal disease, whatever their initial causes, share a single major mode of injury: the chronic increase of blood pressure in the kidney's glomerular capillaries. The glomeruli are the tiny cup-shaped membranes through which fluid from the bloodstream filters into the urinary tract. When blood pressure in the glomerular capillaries is elevated, the membranes are forced to filter larger volumes of fluid. According to the theory, this increased filtration rate in turn leads to a disabling sclerosis, or hardening, of the glomeruli. And as more glomeruli are disabled, the theory asserts, a greater burden of blood flow (and with it, pressure) is shifted to the units still functioning, and they, too, succumb to sclerosis.

Such a cycle of destruction could be set in motion by any trauma that inactivates enough glomeruli to cause a harmful elevation of blood pressure in those that survive. Hence, the theory might account for the wide range of maladies—from streptococcal infection to systemic lupus erythematosus—that can precipitate progressive renal disease.

While the majority of these maladies may act indirectly to raise intrarenal blood pressure by causing initial kidney destruction, other factors may work directly to dilate glomerular blood vessels and thus increase blood flow. One of the theory's most striking

Kidney Replacements

Kidney Transplant

Patients with renal failure can sometimes undergo kidney transplant operations during which they receive a functioning kidney from a donor. As with all organ transplants, there is the possibility of organ rejection, so that receiving a kidney from a close relative has the highest chance of success. The current one-year survival rate is 97 percent if the kidney is received from a relative and 90 percent if it is received from a nonrelative. Recently, investigators have discovered that the drug Cyclosporine (fig. 12.12) is most helpful in preventing organ rejection while at the same time allowing the patient to fight infections. Others are hopeful that monoclonal antibodies that react against killer T cells will be available soon.

Human Anatomy and Physiology

contentions is that a normal component of the human diet—protein—does exactly this. Animal studies conducted by several groups indicate that high-protein diets can lead to sustained increases in intrarenal blood pressure.

This hypothesis, if confirmed, could have a widespread impact on current medical practice. Brenner and other researchers have found that reducing protein intake can slow the progression of renal disease and prolong life in rats with injured kidneys. And four teams of medical researchers in Europe are currently testing this treatment in human patients. The researchers reported on their work at a meeting of the Third International Congress on Nutrition and Metabolism in Renal Disease held in September in Marseilles, France, and so far the results are encouraging: instituting protein restriction early in renal disease appears to slow the advance of kidney deterioration. Remarks Brenner, "The hope here is that we have a form of prevention, so kidney disease doesn't get to the point where you require dialysis. . . ."

While the new theory rests on a large body of data from many studies, some of its principal arguments are still hypothetical.

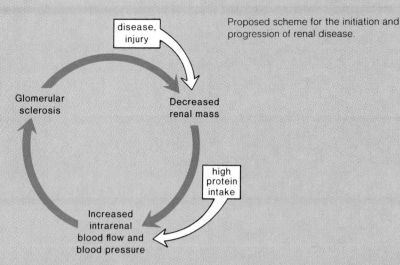

Proposed scheme for the initiation and progression of renal disease.

The exact relationship between the loss of kidney tissue and the increase of blood pressure in remaining tissue, for instance, is not well understood. And scientists are uncertain about the link between intrarenal blood pressure and glomerular sclerosis. "It's a major theory—an important new approach," comments Nancy Cummings of the National Institute of Arthritis, Diabetes, and Digestive and Kidney Diseases, "but many of the ideas still need to be tested." She adds that "much of the theory is based on work that was done with animals, not humans."

Dialysis

If a satisfactory donor cannot be found for a kidney transplant, which is frequently the case, the patient may undergo dialysis treatments, utilizing either a kidney machine or continuous ambulatory peritoneal (abdominal) dialysis, CAPD. **Dialysis** is defined as the diffusion of dissolved molecules through a semipermeable membrane. These molecules will, of course, move across a membrane from the area of greater concentration to one of lesser concentration.

In the case of the kidney machine (fig. 14.16), the patient's blood is passed through a semipermeable membranous tube that is in contact with a balanced salt (dialysis) solution. Substances more concentrated in the blood diffuse into the dialysis solution, also called the dialysate. Conversely, substances more concentrated in the dialysate diffuse into the blood. Accordingly, the artificial

Figure 14.16

Diagram of an artificial kidney. As the patient's blood circulates through dialysis tubing, it is exposed to a solution. Salts enter the blood from the solution and wastes exit from the blood into the solution because of a preestablished concentration gradient. In this way the blood is not only cleansed, the pH can also be adjusted.

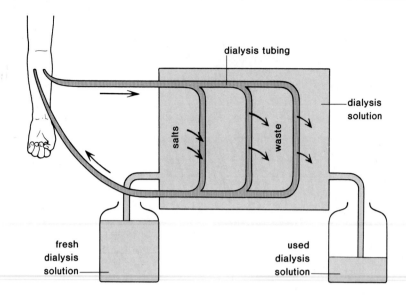

Figure 14.17

CAPD (continuous ambulatory peritoneal dialysis). *a*. Dialysate fluid is introduced into the abdominal cavity by way of a plastic bag. *b*. After bag is securely placed at waist, patient can move freely about. *c*. After four to eight hours, the old fluid is removed before the procedure is repeated.

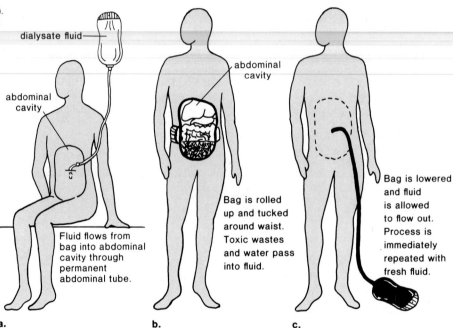

kidney can be utilized either to extract substances from the blood, including waste products or toxic chemicals and drugs, or to add substances to the blood, for example, bicarbonate ions if the blood is acid. In the course of a six-hour dialysis, from 50 to 250 grams of urea can be removed from a patient, which greatly exceeds the urea clearance of normal kidneys. Therefore, a patient need undergo treatment only about twice a week.

In the case of CAPD, a fresh amount of dialysate is introduced directly into the abdominal cavity from a bag attached to a permanently implanted plastic tube. Waste and water molecules pass into the fluid from the surrounding organs before the fluid is collected four or eight hours later (fig. 14.17).

Human Anatomy and Physiology

Summary

Excretory substances fall into at least two categories: end products of metabolism and substances in excess, such as water and ions (salts). The end products of metabolism are, for the most part, nitrogenous wastes, such as ammonia, urea, uric acid, and creatinine, all of which are excreted primarily by the kidneys. Other end products of metabolism, such as bile pigments and carbon dioxide, are excreted by the liver and lungs, respectively.

The kidneys are part of the urinary system, and they produce urine. Urine is made up primarily of nitrogenous end products and ions (salts) in a small amount of water. Urine passes to the ureters, which take it to the bladder where it may be stored for a time. From there it leaves the body by way of the urethra.

Macroscopically, the kidney is made up of three parts: a pelvis, medulla, and cortex. Microscopically, it is made up of over 1 million renal tubules, or nephrons. These tubules account for the kidney's macroscopic anatomy and they have several parts: Bowman's capsule, proximal convoluted tubule, loop of Henle, distal convoluted tubule, and the collecting duct. The loop of Henle and collecting duct are in the medulla, while the other portions are in the cortex. Each renal tubule has its own blood supply: the afferent arteriole approaches Bowman's capsule and divides to become a capillary tuft called the glomerulus, which is enclosed by the capsule. The efferent arteriole leaves the capsule and immediately branches into a capillary bed, which is in close contact with all other parts of the tubule.

During the first step of urine formation, termed pressure filtration, small components of plasma pass into Bowman's capsule from the glomerulus, due to blood pressure. This filtrate of blood contains water, nutrients, nitrogenous wastes, and ions (salts). During the second step, termed selective reabsorption, nutrients and sodium are actively reabsorbed from the proximal convoluted tubule back into the blood. During the third step in urine formation, termed tubular excretion, a few types of substances are actively secreted into the distal convoluted tubule.

Water is reabsorbed along the length of the tubule, but especially from the loop of Henle and collecting duct. Sodium is actively extruded from the ascending limb of the loop of Henle and this creates an osmotic pressure that draws water from the descending limb and collecting duct. Humans excrete a hypertonic urine; that is, one in which the concentration of nitrogenous waste is high. ADH, a hormone produced by the posterior pituitary, controls the reabsorption of water.

The whole tubule participates in maintaining the pH of the blood by regulating the pH of urine. In practice, hydrogen ions are excreted, and sodium and bicarbonate ions are reabsorbed to maintain the pH.

Various types of problems can lead to kidney failure, which necessitates that the person must either receive a kidney from a donor or undergo dialysis treatments by means of the kidney machine or CAPD. In the former, waste products are removed from the blood as it passes through a tube surrounded by a fluid and in the latter, a fluid is introduced into the abdomen where it collects waste products.

Study Questions

1. Name four nitrogenous end products and explain how each is formed in the body. (pp. 282–83)
2. Name several excretory organs and the substances they excrete. (pp. 283–84)
3. What is the composition of urine? (p. 284)
4. Give the path of urine. (p. 284)

5. Name the parts of the kidney tubule, or nephron. (p. 296)
6. Trace the path of blood about the tubule. (p. 290)
7. Describe how urine is made by telling what happens at each part of the tubule. (pp. 288, 290–91)
8. Explain these terms: pressure filtration, active reabsorption, and countercurrent exchange. (pp. 290–93)
9. How does the nephron regulate the pH of the blood? (p. 295)
10. Explain how the artificial kidney machine and CAPD work. (pp. 297–98)

Selected Key Terms

ammonia (ah-mo′ne-ah)
urea (u-re′ah)
uric acid (u′rik as′id)
creatinine (kre-at′ĭ-nin)
ureter (u-re′ter)
urinary bladder (u′rĭ-ner″e blad′der)
urethra (u-re′thrah)
medulla (mĕ-dul′ah)
pelvis (pel′vis)
nephron (nef′ron)
Bowman's capsule (bo′manz kap′sūl)
glomerulus (glo-mer′u-lus)
proximal convoluted tubule (prok′sĭ-mal kon′vo-lūt-ed tu′būl)
distal convoluted tubule (dis′tal kon′vo-lūt-ed tu′būl)
collecting duct (kŏ-lekt′ing dukt)
peritubular capillary (per″ĭ-tu′bu-lar kap′ĭ-lar″e)
glomerular filtrate (glo-mer′u-lar fil′trāt)
countercurrent exchange (kown″ter-kur′ent eks-chānj)
antidiuretic hormone (an″tĭ-di″u-ret′ik hōr′mōn)
urinalysis (u″rĭ-nal′ĭ-sis)

Further Readings

Langley, L. L. 1965. *Homeostasis*. New York: Reinhold Publishing.
Merrill, J. P. 1961. The artificial kidney. *Scientific American* 205(1):56.
Metabolic regulation. In *From cell to organisms*. 1967. Readings from *Scientific American*. San Francisco: W. H. Freeman.
Pitts, R. F. 1974. *Physiology of the kidney and body fluids*. 3d ed. Chicago: Year Book Medical Publishers.
Smith, H. W. 1953. The kidney. *Scientific American* 188(1):40.

Nervous System

Chapter concepts

1 The nervous system is made up of neurons that are specialized to carry nerve impulses. A nerve impulse is an electrochemical change that takes place when sodium ions (Na^+) move from the outside to the inside of a neuron and potassium ions (K^+) move from the inside to the outside.

2 The nervous system consists of the central and peripheral nervous systems. The two systems are joined when a reflex occurs: nerve impulses initiated by a sense organ are interpreted by the central nervous system, which then directs a proper muscular or glandular reaction.

3 The central nervous system, made up of the spinal cord and brain, is highly organized. Consciousness is a function only of the cerebrum, which is most highly developed in humans.

4 Transmission between neurons is accomplished by means of chemicals called neurotransmitter substances. Mood-altering drugs affect the transmission of these neurotransmitters.

Figure 15.1

Overall organization of the nervous system in human beings. The central nervous system is at the top of the diagram and the peripheral nervous system is below. These portions of the nervous system take their names from their locations in the body.

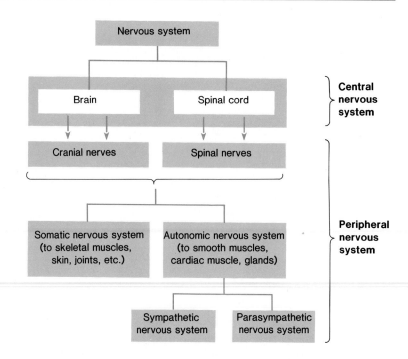

Figure 15.2

Location of central nervous system (brain and cord) and peripheral nervous system (nerves). The central nervous system (CNS) lies in the center of the body and the peripheral nervous system (PNS) lies to either side.

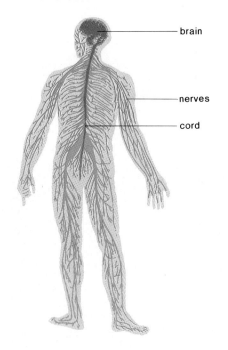

The nervous system tells us that we exist and, along with the muscles, accounts for our distinctly animal characteristics of mobility and quick reaction to environmental stimuli. It is the one system we associate most clearly with what we take to be our very essence or being. Yet the nervous system is composed simply of nerve cells called **neurons,** which are specialized to carry a nerve impulse.

As figure 15.1 indicates, the nervous system has two major divisions: the central and peripheral nervous systems. The **central nervous system** includes the brain and spinal cord (nerve cord), which lie in the midline of the body where the brain is protected by the skull and the spinal cord is protected by the vertebrae. The **peripheral nervous system,** which is further divided into the somatic division and the autonomic division includes all the cranial and spinal nerves. These nerves project out from the central nervous system; thus the name peripheral nervous system. Figure 15.2 illustrates what is meant by the central nervous system and the peripheral nervous system. The division is arbitrary; the two systems work together and are connected to one another.

Neurons

Structure

All neurons (fig. 15.3) have three parts: dendrite(s), cell body, and axon. A **dendrite** conducts nerve impulses (message) toward the cell body, and an **axon** conducts nerve impulses away from the cell body. There are three types of neurons: sensory, motor, and connector. A **sensory neuron** takes a message from a sense organ to the central nervous system and has a long dendrite and short axon, while a **motor neuron** takes a message away from the central nervous system to a muscle fiber or gland and has short dendrites and a long axon. Because motor neurons cause muscle fibers and glands to react, they are said to **innervate** these structures. Sometimes a sensory neuron is referred to as the **afferent neuron,** and the motor neuron is called the **efferent neuron.** These words,

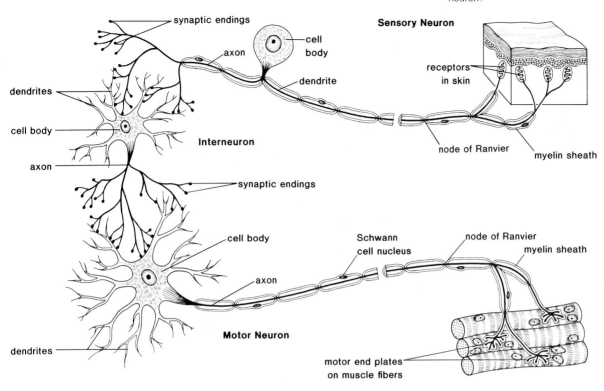

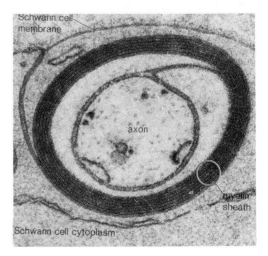

Figure 15.4
Electron micrograph of a cross section of the myelin sheath that encloses the long fibers of neurons. The myelin sheath is composed of Schwann cell membrane and has a white glistening appearance in the body.

Table 15.1 Neurons

Neuron	Structure	Function
Sensory (afferent)	Long dendrites, short axon	Carry nerve impulses (message) from periphery to CNS*
Motor (efferent)	Short dendrites, long axon	Carry nerve impulses (message) from CNS to periphery
Interneuron	Short dendrites, long or short axon	Carry nerve impulses (message) within CNS

*CNS = central nervous system.

which are derived from Latin, mean running to and running away from, respectively. Obviously, they refer to the relationship of these neurons to the central nervous system.

An **interneuron** (also called association neuron, or connector neuron) is always found completely within the central nervous system and conveys messages between parts of the system. An interneuron can have short dendrites and a long axon or short dendrites and a short axon. Table 15.1 summarizes the three types of neurons that are also illustrated in figure 15.3.

The dendrites and axons of neurons are sometimes called **fibers** or processes. Most long fibers, whether dendrites or axons, are covered by a **myelin sheath** (fig. 15.4), formed by tightly packed spirals of the cell membrane of Schwann cells. This sheath, which gives nerves their white appearance, is interrupted by intervals called the nodes of Ranvier.

Figure 15.5
The squid axons shown here produce rapid
muscular contraction so that the squid can move
quickly. These neurons are so large that a
microelectrode can be inserted in the axon to study
the nature of the nerve impulse and also inserted in
the cell bodies within ganglia to study the nature of
transmission between neurons.

Nerve Impulse

A neuron is specialized to conduct nerve impulses. The nature of a nerve impulse has been studied by using giant axons in the squid (fig. 15.5) and a type of voltmeter called an oscilloscope, which shows a trace or pattern indicating changes in voltage as time elapses (fig. 15.6). Voltage is a measurement of the potential difference between two points. When a potential difference exists, we can say that a plus and a minus pole exist; therefore, an oscilloscope indicates the existence of polarity and records polarity changes.

Resting Potential

In the experimental setup shown in figure 15.7a, an oscilloscope is wired to two electrodes, one of which is an internal recording electrode that records from inside a giant axon of the squid. The axon, being a process that extends from a cell body, is essentially a membranous tube filled with cytoplasm, or in this case, axoplasm. When the axon is not conducting an impulse, the oscilloscope records a potential difference across the membrane equal to -60 millivolts (mV). This is the **resting potential** because the axon is not conducting an impulse.

Such polarity is not unexpected because it is known that there is a difference in ion distribution on either side of the membrane. As figure 15.7b shows, there is a concentration of sodium ions (Na^+) outside the axon and a concentration of potassium ions (K^+) inside the axoplasm. Also, there are large organic negative ions in the axoplasm, which cause the resting fiber to be negative inside. These organic ions are held inside due to the selectively permeable nature of the axomembrane. The distribution of sodium and potassium is maintained by a form of active transport called the sodium/potassium pump (p. 74), which requires energy and is believed to function whenever the neuron is not conducting an impulse.

Action Potential

If the axon is stimulated to conduct a nerve impulse by an electric shock, by a sudden difference in pH, or by a pinch, a trace appears on the oscilloscope screen. This pattern caused by rapid polarity changes, called the **action potential,** has an upswing and a downswing (fig. 15.7c and 15.7d).

The Upswing (from -60 mV to $+40$ mV) Sophisticated experiments indicate that as the action potential goes to $+40$ mV, sodium ions are rapidly

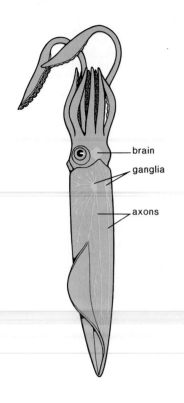

brain
ganglia
axons

Figure 15.6
Scientist working at an oscilloscope, the electrical
recording device that measures changes in voltage
wherever an electrode is placed on or inserted in a
neuron.

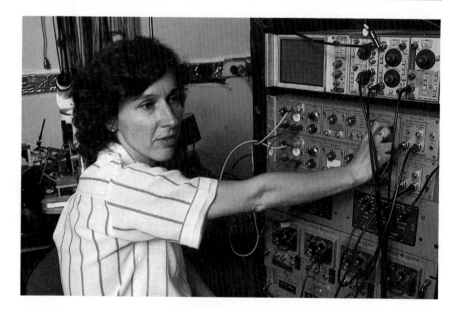

Human Anatomy and Physiology

moving to the inside of the axon. Evidence shows that stimulation of the axon has caused **sodium gates,** or channels, to open, allowing sodium to flow into the axon. This sudden permeability of the membrane causes the oscilloscope to record a **depolarization:** the inside of the fiber goes from negative to positive as sodium ions enter.

The Downswing (from +40 mV to −60 mV)   It is now known that the restoration of the resting potential (or the return to −60 mV) is caused by the exit of potassium ions from the axoplasm. The membrane has suddenly become permeable to potassium because the **potassium gates,** or channels, have opened. The oscilloscope records a **repolarization** as the inside of the axon returns to negative again.

Table 15.2 summarizes the events that occur during transmission of a nerve impulse.

Table 15.2   Summary of nerve impulse

Resting potential	Action potential or nerve impulse
−60 mV (inside negative). Sodium/ potassium pump at work. Large organic ions cause negativity inside.	a. −60 mV to +40 mV Sodium gates are open and sodium moves to inside. Inside becomes positive compared to outside. b. +40 mV to −60 mV Potassium gates are open and potassium moves to outside. Inside returns to negative again.

Figure 15.7
The resting and action potential. *a.* Resting potential. The oscilloscope reads a resting potential of −60 millivolts due to (*b.*) the presence of large negative organic ions inside the axoplasm. Note also the unequal distribution of Na⁺ and K⁺ across the membrane. *c.* Action potential. The action potential is a change in polarity that may be explained by (*d.*) first the movement of Na⁺ to the inside and second, by the movement of K⁺ to the outside of the axon.

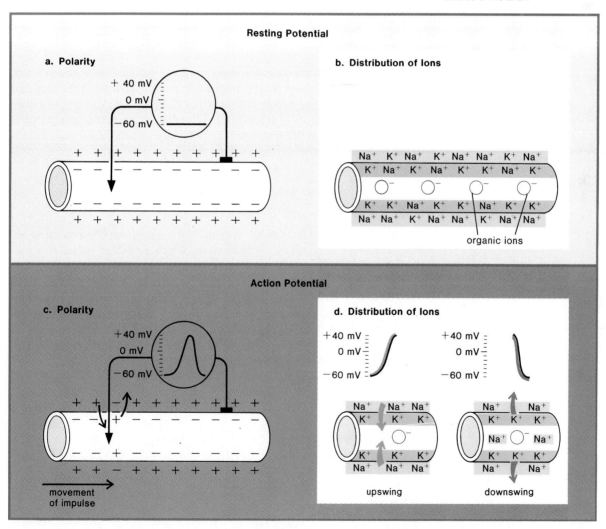

Figure 15.8
Longitudinal section of a vertebrate axon illustrates
the manner by which the nerve impulse travels
down a long nerve fiber. The speed of the impulse
is due to the fact that it jumps from one node of
Ranvier to the next.

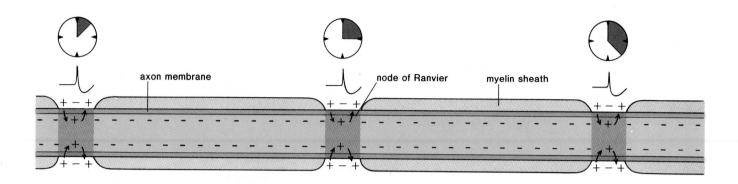

Recovery Phase   A fiber can conduct a volley of nerve impulses because only a small number of ions are exchanged with each impulse. When the fiber rests, however, there is a recovery phase during which the sodium/potassium pump returns the sodium to the outside and the potassium to the inside.

Conduction

Although the oscilloscope records from only one area of the axon, the nerve impulse actually travels along the length of an axon. In nonmyelinated fibers, the action potential at one point generates the action potential at the very next point so that the nerve impulse is propagated along the entire length of a fiber. Conduction is hundreds of times faster (200 meters per second compared to 0.5 meter per second) in myelinated fibers because depolarization occurs only at the nodes of Ranvier (fig. 15.8). Thus the action potential jumps from node to node rather than traveling from point to point.

Transmission across a Synapse

The mechanism by which an action potential passes from one neuron to another is not the same as the mechanism by which an action potential is conducted along a neuron. Each axon branches into many fine terminal branches, each of which is tipped by a small swelling, or terminal knob (fig. 15.9). Each knob lies very close to the dendrite (or cell body) of another neuron. This region is called a **synapse** and the knob is called a **synaptic ending.** The membrane of the knob is called the **presynaptic membrane,** and the membrane of the next neuron just beyond the knob is called the **postsynaptic membrane.** The small gap between is the **synaptic cleft.**

Figure 15.9
Synapse anatomy. *a.* Synapses occur where axon synaptic endings of one neuron lie near the dendrites, or cell bodies, of the next neuron. *b.* Diagram of a synapse based on electron microscopy studies shows that there is a space called the synaptic cleft between the two neurons. *c.* Transmission across a synapse occurs when a neurotransmitter substance released by vesicles at the presynaptic membrane diffuses across the synaptic cleft to the postsynaptic membrane where it changes the polarization of the membrane.

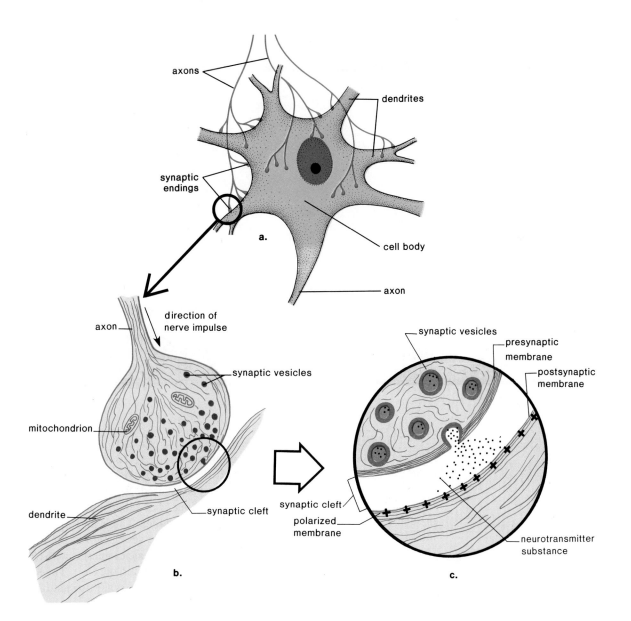

axons

dendrites

synaptic endings

cell body

a.

axon

direction of nerve impulse

axon

synaptic vesicles

mitochondrion

dendrite

synaptic cleft

b.

synaptic vesicles

presynaptic membrane

postsynaptic membrane

synaptic cleft

polarized membrane

neurotransmitter substance

c.

Figure 15.10
Electron micrograph of a synapse. Many mitochondria are seen in the upper portion, and many vesicles are seen in the lower portion of the cross section of a synaptic ending. The fuzzy dark areas along the cleft are believed to be the sites of transmitter release.

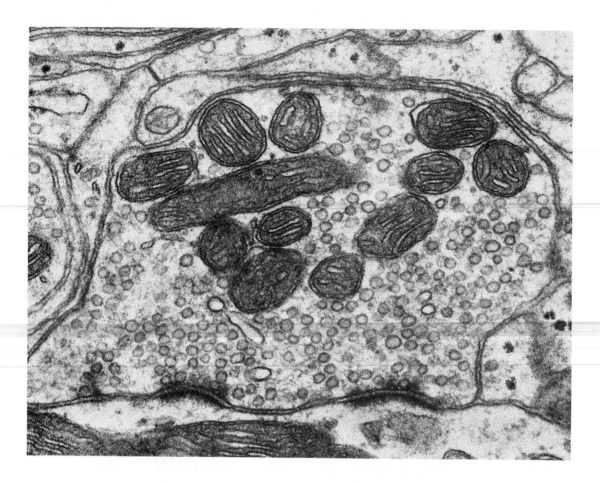

Transmission of nerve impulses across a synaptic cleft is carried out by chemicals called **neurotransmitter substances.** Figure 15.10 is an electron micrograph that shows synaptic vesicles at the end of an axon; the transmitter substance is stored in these vesicles before it is released. When nerve impulses traveling along an axon reach a synaptic ending, they modify the membrane in such a way that calcium ions flow into the ending. These ions appear to interact with contractile proteins to pull the synaptic vesicles up to the inner surface of the presynaptic membrane. When the vesicles merge with this membrane, a neurotransmitter substance is discharged into the cleft. The neurotransmitter molecules diffuse across the cleft to the postsynaptic membrane where they bind with **receptor sites** in a lock-and-key manner. This reception alters the potential of the postsynaptic membrane in either an excitatory or inhibitory direction; an excitatory neurotransmitter makes the potential less negative, whereas an inhibitory neurotransmitter makes the potential more negative. If sufficient excitatory neurotransmitter substance is received, the neuron fires (initiates nerve impulses.)

Human Anatomy and Physiology

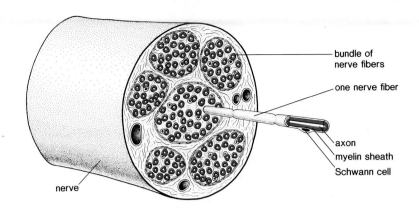

Figure 15.11
Diagram of a cross section of a nerve, with one axon extended to show that each fiber is enclosed by a myelin sheath. Because nerves contain so many fibers it has been difficult to successfully rejoin them after they are severed in an accident. Scientists have now found that if they hold well-cut pieces together, and then surround them by a solution that resembles cytoplasm, the nerve will repair itself and be functional.

bundle of nerve fibers

one nerve fiber

axon

myelin sheath

Schwann cell

nerve

Transmitter Substances

Acetylcholine (Ach) and **noradrenalin** (NA) are well-known excitatory transmitters active in both the peripheral and central nervous systems. Examples of inhibitory substances, so far discovered only in the central nervous system, are given on page 318.

Once a transmitter substance has been released into a synaptic cleft, it has only a short time to act. In some synapses, the cleft contains enzymes that rapidly inactivate the neurotransmitter. For example, the enzyme **acetylcholinesterase** (AchE), or simply cholinesterase, breaks down acetylcholine. In other synapses, the synaptic ending rapidly absorbs the transmitter substance, possibly for repackaging in synaptic vesicles or for chemical breakdown. The enzyme monoamine oxidase breaks down noradrenalin after it is absorbed. The short existence of neurotransmitters in the synapse prevents continuous stimulation (or inhibition) of postsynaptic membranes.

Summation and Integration

A neuron is on the receiving end of many synapses (fig. 15.9a). Whether a neuron fires or not depends on the summary effect of all the excitatory and/or inhibitory neurotransmitters received. If the amount of excitatory neurotransmitters received is sufficient to overcome the amount of inhibitory neurotransmitters received, the neuron fires. If the amount of excitatory neurotransmitters received is not sufficient, only **local excitation** occurs. Thus synapses are regions where a "summing up" occurs and, therefore, are also regions of **integration,** where the nervous system can fine tune its response to the environment. The structure and function of synapses allows them to carry on this very important activity.

One-way Propagation

Transmission across a synapse is one-way because only the ends of axons have synaptic vesicles that are able to release neurotransmitter substances to affect the potential of the next neuron. Also, neurons obey the **all-or-none law,** meaning that a neuron either fires maximally or it does not fire at all. A nerve does not obey the all-or-none law, because a nerve contains many fibers (fig. 15.11), any number of which may be carrying nerve impulses. Thus a nerve may have degrees of performance.

Table 15.3   Nerves*

Type of nerve	Consists of	Function
Sensory nerves	Long dendrites only of sensory neurons	Carry message from receptors to central nervous system (CNS)
Motor nerves	Long axons only of motor neurons	Carry message from CNS to effectors
Mixed nerves	Both long dendrites of sensory neurons and long axons of motor neurons	Carry message in dendrites to CNS and away from CNS in axons

*Compare to table 15.1.

Figure 15.12

Underside of the brain showing the origins of the cranial nerves. Many cranial nerves are either sensory or motor nerves that serve the sense organs and muscles of the face and neck.

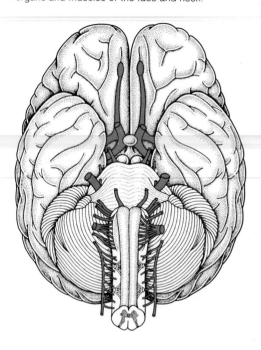

Peripheral Nervous System

Nerves

The **peripheral nervous system** (PNS) consists of nerves that contain only long dendrites and/or long axons. This arises because neuron cell bodies are found only in the brain, spinal cord, and ganglia. **Ganglia** are collections of cell bodies within the PNS.

There are three types of nerves (table 15.3). **Sensory nerves** contain only the long dendrites of sensory neurons; **motor neurons** contain only the long axons of motor neurons; while **mixed nerves** contain both the long dendrites of sensory neurons and the long axons of motor neurons. Each nerve fiber within a nerve is surrounded by a white myelin sheath (fig. 15.11) and therefore nerves have a white, shiny, glistening appearance.

Cranial Nerves

Humans have 12 pairs of **cranial nerves** attached to the brain (fig. 15.12). Some of these are sensory, some are motor, and others are mixed. Notice that although the brain is a part of the central nervous system (CNS), the cranial nerves are a part of the peripheral nervous system (PNS). All cranial nerves, except the vagus, are concerned with the head, neck, and face regions of the body; while the vagus nerve has many branches to serve the internal organs.

Spinal Nerves

Each spinal nerve emerges from the cord (fig. 15.13) by two short branches, or *roots,* which lie within the vertebral column. The **dorsal root** can be identified by the presence of an enlargement called the *dorsal root ganglion.* This ganglion contains the cell bodies of the sensory neurons whose dendrites conduct impulses toward the cord. The ventral root of each spinal nerve contains axons of motor neurons that conduct impulses away from the cord. These two roots join just before the spinal nerve leaves the vertebral column. Therefore all spinal nerves are mixed nerves that contain many sensory dendrites and motor axons.

Human beings have 31 pairs of spinal nerves, which give evidence that humans are segmented animals, especially since the spinal nerves serve the particular region of the body where they are located.

Somatic Nervous System

The **somatic nervous system** includes all those nerves that serve the musculoskeletal system and the exterior sense organs, including those in the skin. Exterior sense organs are **receptors** that receive environmental stimuli and

Human Anatomy and Physiology

Figure 15.16
Structure and function of the autonomic nervous system. The sympathetic fibers arise from the thoracic and lumbar portion of the cord; the parasympathetic fibers arise from the brain and sacral portion of the cord. Each system innervates the same organs but have contrary effects. For example, the sympathetic system speeds up and the parasympathetic system slows down the beat of the heart.

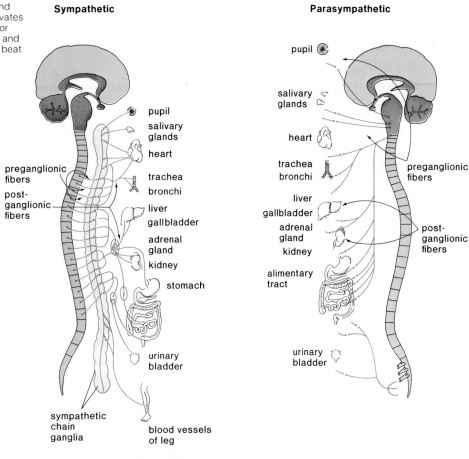

Sympathetic

- preganglionic fibers
- postganglionic fibers
- pupil
- salivary glands
- heart
- trachea
- bronchi
- liver
- gallbladder
- adrenal gland
- kidney
- stomach
- urinary bladder
- sympathetic chain ganglia
- blood vessels of leg

Parasympathetic

- pupil
- salivary glands
- heart
- trachea bronchi
- liver gallbladder
- adrenal gland
- kidney
- alimentary tract
- preganglionic fibers
- postganglionic fibers
- urinary bladder

Table 15.5 Sympathetic versus parasympathetic system

Sympathetic	Parasympathetic
Fight or flight	Normal activity
Noradrenalin is neurotransmitter	Acetylcholine is neurotransmitter
Postganglionic fiber is longer than preganglionic	Preganglionic fiber is longer than postganglionic
Preganglionic fiber arises from middle portion of cord	Preganglionic fiber arises from brain and lower portion of cord

Parasympathetic System

The vagus nerve and fibers that arise from the bottom portion of the cord form the **parasympathetic nervous system.** Therefore this system is often referred to as the *craniosacral portion* of the autonomic nervous system. In the parasympathetic nervous system, the preganglionic fiber is long and the postganglionic fiber is short because the ganglia lie near or within the organ (fig. 15.15). The parasympathetic system promotes all those internal responses we associate with a relaxed state; for example, it causes the pupil of the eye to contract, promotes digestion of food, and retards the heartbeat. The neurotransmitter utilized by the parasympathetic system is acetylcholine.

Figure 15.16 contrasts the sympathetic and parasympathetic systems, and table 15.5 lists all the differences we have noted between these two systems.

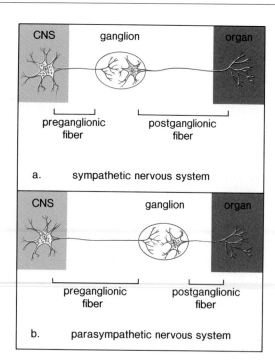

Figure 15.15
Location of ganglia in the sympathetic and parasympathetic nervous systems. *a.* In the sympathetic nervous system, each ganglion lies close to the spinal cord (CNS) and therefore the preganglionic fiber is short and the postganglionic fiber is long. *b.* In the parasympathetic nervous system, each ganglion lies close to the organ being innervated and therefore the preganglionic fiber is long and the postganglionic fiber is short.

Various other reactions usually accompany a reflex response; the person may look in the direction of the object, jump back, and utter appropriate exclamations. This whole series of responses is explained by the fact that the sensory neuron stimulates several connector neurons, which take impulses to all parts of the central nervous system, including the cerebrum, which, in turn, makes the person conscious of the stimulus and his or her reaction to it.

Autonomic Nervous System

The autonomic nervous system, a part of the PNS, is made up of motor neurons that control the internal organs automatically and usually without the need for conscious intervention. There are two divisions to the autonomic nervous system: the sympathetic and parasympathetic systems. Both of these (1) function automatically and usually subconsciously in an involuntary manner; (2) innervate all internal organs; and (3) utilize two motor neurons and one ganglion for each impulse. The first neuron has a cell body within the central nervous system and a **preganglionic axon.** The second neuron has a cell body within the ganglion and a **postganglionic axon.**

Sympathetic System

The preganglionic fibers of the **sympathetic nervous system** arise from the middle or *thoracic-lumbar portion* of the cord and almost immediately terminate in ganglia that lie near the cord. Thus, in this system the preganglionic is short, while the postganglionic fiber that makes contact with the organs is long (fig. 15.15).

The sympathetic nervous system is especially important during emergency situations and is associated with "fight or flight." For example, it inhibits the digestive tract but dilates the pupil, accelerates the heartbeat, and increases the breathing rate. It is not surprising, then, that the neurotransmitter released by the postganglionic axon is noradrenalin, a chemical close in structure to adrenalin, a well-known heart stimulant.

Figure 15.14

Diagram of a reflex arc, the functional unit of the nervous system. Trace the path of a reflex by following the black arrows. Name the three types of neurons that are required for a simple reflex, such as the rapid response to touching a hot object with the hand.

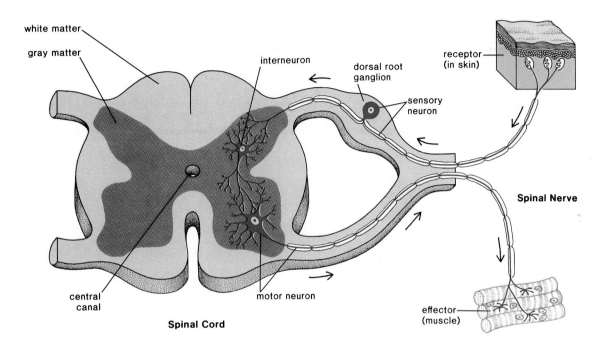

Table 15.4	Path of a simple reflex
1. Receptor (formulates message)*	Generates nerve impulses.
2. Sensory neuron (takes message to central nervous system)	Impulses move along dendrite (spinal nerve)†—and proceed to cell body (dorsal root ganglia) and then go from cell body to axon (gray matter of cord)
3. Interneuron (passes message to motor neuron)	Impulses picked up by dendrites and pass through cell body to axon (completely within gray matter)
4. Motor neuron (takes message away from central nervous system)	Impulses travel through short dendrites and cell body (gray matter of cord) to axon (spinal nerve)
5. Effector (receives message)	Receives nerve impulses and reacts: glands secrete and muscles contract.

*Phrases within parentheses state overall function.
†Words within parentheses indicate location of structure.

involve the brain. Figure 15.14 illustrates the path of the second type of reflex action. Whenever a person touches a very hot object, a receptor in the skin generates nerve impulses that move along the dendrite of a sensory neuron toward the cell body and central nervous system. The cell body of a sensory neuron is located in the dorsal root ganglion just outside the cord. From the cell body, the impulses travel along the axon of the sensory neuron and enter the cord; there they may pass to many interneurons, one of which lies completely within the gray matter and connects with a motor neuron. The short dendrites and cell body of the motor neuron are in the ventral region (horn) of the gray matter, and the axon leaves the cord by way of the ventral root. The nerve impulses travel along the axon to muscle fibers that then contract so that the hand is withdrawn from the hot object. (See table 15.4 for a listing of these events.)

Human Anatomy and Physiology

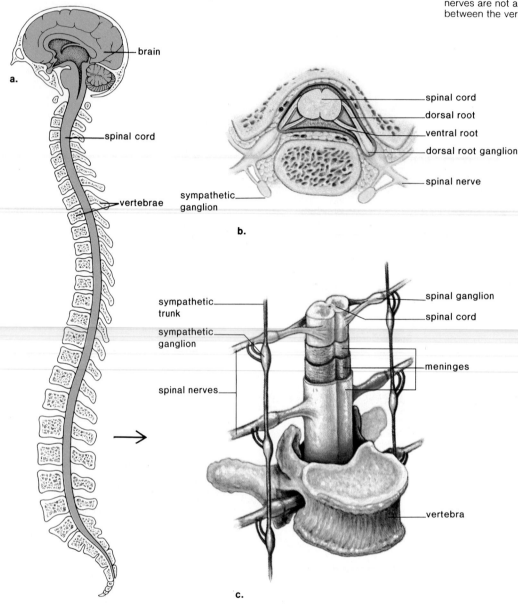

then initiate nerve impulses. Muscle fibers are **effectors** that bring about a reaction to the stimulus. Receptors are studied in chapter 17 and muscle effectors are studied in chapter 16.

Reflex Action

Reflexes are automatic, involuntary responses to changes occurring inside or outside the body. In the somatic nervous system, outside stimuli often initiate a reflex action. Some reflexes, such as blinking the eye, involve the brain, while others, such as withdrawing the hand from a hot object, do not necessarily

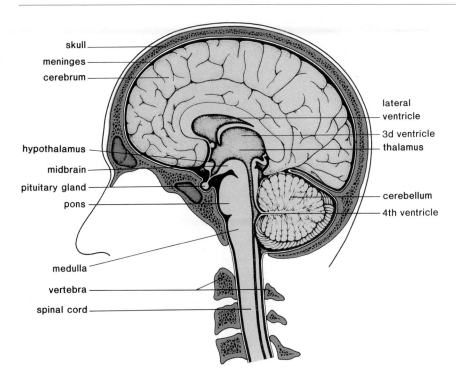

Central Nervous System

The central nervous system (CNS) consists of the spinal cord and brain. As figures 15.13 and 15.17 illustrate, the CNS is protected by bone: the brain is enclosed within the skull and the spinal cord is surrounded by vertebrae. Also, both the brain and spinal cord are wrapped in three protective membranes known as **meninges;** spinal meningitis is a well-known infection of these coverings. The spaces between the meninges are filled with **cerebrospinal fluid,** which cushions and protects the CNS. A small amount of this fluid is sometimes withdrawn for laboratory testing when a spinal tap is done (fig. 15.18). Cerebrospinal fluid is also contained within the **central canal** of the spinal cord and the **ventricles** of the brain. The latter are interconnecting spaces that produce and serve as a pathway for cerebrospinal fluid.

Spinal Cord
The cord (fig. 15.13 and fig. 15.14) contains (1) the central canal filled with cerebrospinal fluid, (2) gray matter containing cell bodies and short fibers, and (3) white matter containing long fibers of interneurons that run together in bundles called **tracts.** These tracts connect the cord to the brain.

The dorsal and ventral sides of the spinal cord are specialized to handle sensory and motor information, respectively. Sensory information from the spinal nerves enters the spinal cord through the dorsal roots, while motor information from the spinal cord is sent to spinal nerves through the ventral roots. In the gray matter, dorsal cells function primarily in receiving sensory information, while ventral cells are primarily motor. Within the white matter of the spinal cord, ascending tracts of dorsal axons take information to the brain, while descending tracts in the ventral part of the cord primarily carry information down from the brain.

Brain
The brain of the human, like that of all vertebrates, can be divided into three major sections: the hindbrain, the midbrain, and the forebrain. The **midbrain** is largely a relay station that has decreased in importance in humans, while the **forebrain** has increased markedly in size and importance.

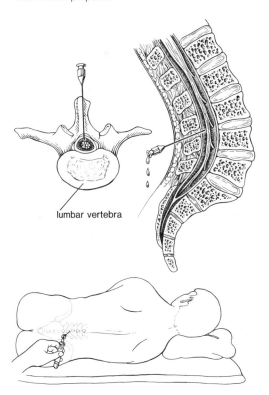

lumbar vertebra

Figure 15.19
The reticular formation is a part of the ascending
reticular activating system. The sensory information
received by the reticular formation is sorted out by
the thalamus before nerve impulses are sent to
other parts of the brain.

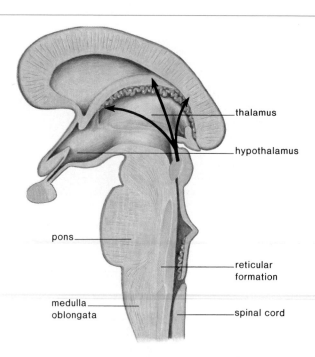

thalamus

hypothalamus

pons

reticular
formation

medulla
oblongata

spinal cord

Hindbrain

In the **hindbrain**, the **medulla oblongata** (fig. 15.17) contains centers for heartbeat, breathing, and vasoconstriction (blood pressure), and also reflex centers for vomiting, coughing, sneezing, hiccoughing, and swallowing. Since the medulla is located in the lower portion of the brain stem, it functions as a pathway for sensory and motor impulses between the brain and the spinal cord.

The **cerebellum,** a bilobed structure that resembles a butterfly, is also located in the hindbrain. The cerebellum, the second largest portion of the brain, functions in muscle coordination, integrating impulses received from higher centers to ensure that all the skeletal muscles work together to produce smooth and graceful motions. The cerebellum is also responsible for maintaining normal muscle tone and transmitting impulses that maintain posture. It receives information from the inner ear indicating the position of the body and thereafter sends impulses to those muscles whose contraction maintains or restores balance.

Forebrain

The forebrain includes the hypothalamus, thalamus, and the cerebrum (fig. 15.17). The **hypothalamus** is concerned with homeostasis, or the constancy of the internal environment, and contains centers for hunger, sleep, thirst, body temperature, water balance, and blood pressure. The hypothalamus controls the pituitary gland and thereby serves as a link between the nervous and endocrine systems.

The **thalamus** is the last portion of the brain for sensory input before the cerebrum. It serves as a central relay station for sensory impulses traveling upward from other parts of the cord and brain to the cerebrum. It receives all sensory impulses (except those associated with the sense of smell) and channels them to appropriate regions of the cortex for interpretation.

The thalamus has connections to various parts of the brain by way of the *diffuse thalamic projection system,* an extension of the *reticular formation* (fig. 15.19), a complex network of cell bodies and fibers that extends from the medulla to the thalamus. Together they form the **ARAS,** the ascending

Human Anatomy and Physiology

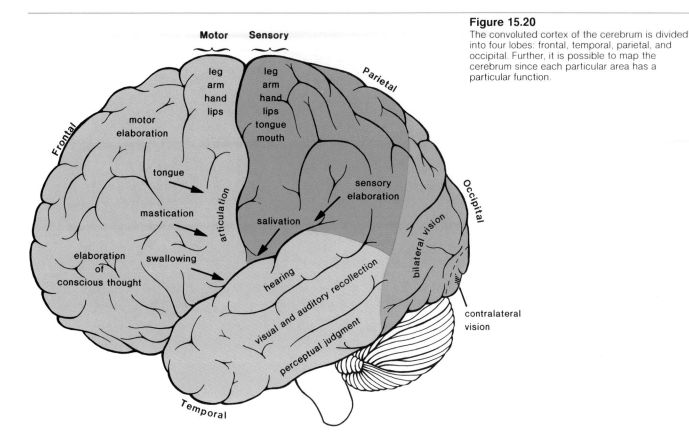

Motor   Sensory

leg
arm
hand
lips

motor
elaboration

tongue

mastication

elaboration
of
conscious thought

swallowing

articulation

Frontal

leg
arm
hand
lips
tongue
mouth

Parietal

sensory
elaboration

salivation

hearing

visual and auditory recollection

perceptual judgment

bilateral vision

Occipital

contralateral
vision

Temporal

Figure 15.20
The convoluted cortex of the cerebrum is divided into four lobes: frontal, temporal, parietal, and occipital. Further, it is possible to map the cerebrum since each particular area has a particular function.

reticular activating system, which is believed to sort out incoming stimuli, passing on only those that require immediate attention. For this reason, the thalamus is sometimes called the gatekeeper to the cerebrum because it alerts the cerebrum to only certain sensory input. Thus, it hopefully allows you to concentrate on your homework while the television is on.

Epilepsy is caused by a disturbance of the normal communication between the ARAS and the cortex. Petit mal seizures, which may be only a momentary numbness or a tingling sensation, seem to be due to the inability of the ARAS to transmit or the cerebrum to receive signals. In a grand mal seizure, the cerebrum becomes extremely excited. There is a reverberation of signals within the ARAS and cerebrum that continues until they become so fatigued that the signals cease. In the meantime, the individual loses consciousness, even while convulsions are occurring. Following an attack, the brain is so fatigued the person must sleep for a while.

The **cerebrum,** which is the only area of the brain responsbile for consciousness, is the largest portion of the brain in humans. The outer layer of the cerebrum, called the **cortex,** is gray in color and contains cell bodies and short fibers. The cerebrum is divided into halves known as the right and left **cerebral hemispheres.** Each half contains four types of lobes: **frontal, parietal, temporal,** and **occipital** (fig. 15.20). The cerebrum can be mapped according to the particular functions of each of the lobes (table 15.6). Association areas are believed to be areas for intellect, artistic and creative ability, learning, and memory. Sensory areas receive nerve impulses from the sense organs and produce what we call sensations. The particular sensation produced is the prerogative of the area of the brain that is stimulated, since the nerve impulse itself always has the same nature (described previously.) Motor areas of the cerebrum initiate nerve impulses that control muscle fibers. A momentary lack

Table 15.6 Functions of the cerebral lobes

Lobe	Functions
Frontal lobes	Motor areas control movements of voluntary skeletal muscles. Association areas carry on higher intellectual processes such as those required for concentration, planning, complex problem solving, and judging the consequences of behavior.
Parietal lobes	Sensory areas are responsible for the sensations of temperature, touch, pressure, and pain from the skin. Association areas function in the understanding of speech and in using words to express thoughts and feelings.
Temporal lobes	Sensory areas are responsible for hearing and smelling. Association areas are used in the interpretation of sensory experiences and in the memory of visual scenes, music, and other complex sensory patterns.
Occipital lobes	Sensory areas are responsible for vision. Association areas function in combining visual images with other sensory experiences.

Figure 15.21
The extrapyramidal and limbic systems. The
extrapyramidal region, which includes portions of
cerebrum, cerebellum, and pons, controls body
movement and posture. The limbic system, which
includes portions of the cerebrum, thalamus, and
hypothalamus, is concerned mainly with emotion
and memory.

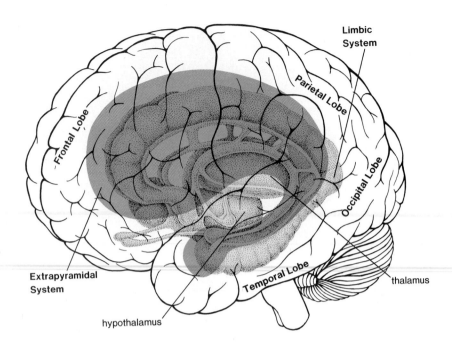

of oxygen during birth can damage the motor areas of the cerebral cortex so
that the individual develops the symptoms of cerebral palsy, a condition characterized by a spastic weakness of the arms and legs.

There has been a great deal of testing to determine whether the right
and left half of the cerebrum serve different functions. These studies have
tended to suggest that the left half of the brain is the verbal (word) half and
the right half of the brain is the visual (spatial relation) half. However, there
have also been results to indicate that such a strict dichotomy does not exist
between the two halves.

Extrapyramidal and Limbic Systems

Just beneath the cerebral cortex are masses of white matter that belong to the
ascending and descending tracts. Some of the latter, called the *extrapyramidal system* (fig. 15.21), pass into the *basal nuclei,* several masses of gray
matter that lie deep within each hemisphere of the cerebrum. The basal nuclei
are a part of the **limbic system,** an interior loop that connects portions of the
frontal lobes, temporal lobes, thalamus, and hypothalamus. Stimulation of different areas of the limbic system causes the subject to experience rage, pain,
pleasure, or sorrow. By causing pleasant or unpleasant feelings about experiences, the limbic system apparently guides the individual into behavior that
is likely to increase the chance of survival.

Recently there has been a great deal of investigation regarding the extrapyramidal and the limbic system (fig. 15.21). The synapses in this area
utilize both excitatory and inhibitory neurotransmitters. The excitatory transmitters include acetylcholine (Ach) and noradrenalin (NA). The inhibitory
transmitters include gamma aminobutyric acid (GABA), serotonin, and dopamine. It has been discovered that several neurological illnesses are due to
an imbalance in these neurotransmitters. Parkinson's disease and Huntington's chorea result from malfunctions of the extrapyramidal system. Parkinson's disease is a condition characterized by a wide-eyed, unblinking expression
and an involuntary tremor of the fingers and thumbs, muscular rigidity, and

a shuffling gait. All these symptoms are due to dopamine deficiencies. Huntington's chorea is characterized by a progressive deterioration of the individual's nervous system that eventually leads to constant thrashing and writhing movements until insanity precedes death. The problem in this case is believed to be GABA malfunctions. Most recently it has been discovered that Alzheimer's disease, a severe form of senility with marked memory loss that is found in 5 to 10 percent of all people over age 65, seems to be due to acetylcholine deficiencies. Alzheimer's disease is termed a disorder of the limbic system because the limbic system is concerned not only with emotion but also memory.

Treatment of individuals with brain disorders has thus far been directed toward restoring the proper balance of neurotransmitter substances. However, it now appears that it might be possible to implant new cells to replace the deficient cells in brains. The reading on page 320 describes the progress that has been made in this area so far.

Learning and Memory

Learning requires memory, but just what permits memory to occur is not yet known. Investigators have been working with invertebrates such as slugs and snails because their nervous systems are very simple and yet they can be taught to perform a particular behavior. In order to determine how learning takes place, it has been possible to insert electrodes into individual cells to alter or record the electrochemical responses of these cells (fig. 15.22). This work suggests that learning results from changes in nerve transmission at existing synapses and not from the creation of new synapses.

Thus far, learning requiring only short-term memory has been studied in this manner. A good example of **short-term memory** in humans is the ability to recall a telephone number long enough to dial the number. **Long-term memory,** such as the ability to recall the events of the day, has been shown to require (1) use of neurotransmitters found within the limbic system; (2) the hippocampus, a part of the temporal lobe in the limbic system; and (3) protein synthesis. Drugs that affect any one of these three elements can impair memory. Even so, memories appear to be stored throughout the association areas; when stimulated by an experimenter, no particular region is richer in memories than another.

Brain Waves

The electrical activity of the brain can be recorded in the form of an electroencephalogram (EEG). Electrodes are taped to different parts of the scalp, and an instrument called the electroencephalograph records the so-called brain waves (fig. 15.23).

When the subject is awake, two types of waves are usual: *alpha waves,* with a frequency of about 6 to 13 per second and a potential of about 45 microvolts, predominate when the eyes are closed, and *beta waves,* with higher frequencies but lower voltage, appear when the eyes are open.

During an eight-hour sleep there are usually five times when the brain waves become slower and larger than alpha waves. During each of these times, there are irregular flurries as the eyes move back and forth rapidly. When subjects are awakened during the latter, called **REM** (rapid eye movement) **sleep,** they always report that they were dreaming. The significance of REM sleep is still being debated, but some studies indicate that REM sleep is needed for memory to occur.

The EEG is a good diagnostic tool; for example, an irregular pattern can signify epilepsy or a brain tumor. A flat EEG signifies lack of electrical activity of the brain, or brain death, and thus it may be used to determine the precise time of death.

Figure 15.22
Individual nerve cells in a snail, *Hermissenda*, are being stimulated by microelectrodes, simulating the signals that scientists had previously recorded when a snail learns to avoid light. When this snail is freed, it automatically avoids the light and does not need to be taught like other snails. To teach snails to avoid light, they are placed on a table that rotates every time they venture toward light.

Figure 15.23
Encephalograms are recordings of the electrical activity of the brain. The alpha waves, which appear when the subject is awake with eyes closed, are the most common. Second most common are the beta waves recorded when the subject is awake with eyes open. Sleep has various stages, as indicated.

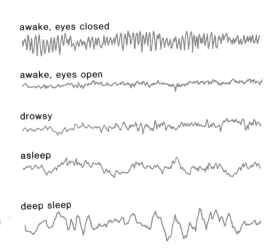

awake, eyes closed

awake, eyes open

drowsy

asleep

deep sleep

Brain Healing

Paraplegics may one day be cured by a transplant of brain cells.

Brain damage is forever, or so doctors once thought, but that longstanding medical axiom is now being proved wrong. In laboratories across the U.S. and Europe, researchers are finding that by creating the right chemical environment, and in some cases implanting new cells in the brain, damaged nervous systems can be coaxed to regenerate. Even more encouraging is the discovery, so far shown only in animals, that cellular regrowth can restore lost mental functions, and, in addition, improve memory and learning.

The latest achievement in this promising field is the work of Dr. Donald Stein and three colleagues at Clark University in Worcester, Mass. As reported in last week's issue of *Science*, the group attempted to restore mental functioning in 21 rats whose brains had been damaged by removal of large sections of the frontal cortex. This section of the brain is involved in the learning of complex spatial relationships. Typically, rats

sustaining such a severe injury would take 18 days or more to master a maze that required them to alternate right and left turns in the correct order to get a drink of water. "The rat has got to remember what he did the last time and then do the opposite," Stein explains. Normal rats can learn the task in just 2½ days.

Before attempting to repair the brain damage, Stein's team waited a week to allow for the natural accumulation of healing proteins called nerve growth factors. Then they implanted a pinhead-size lump of tissue that had been taken from the frontal cortex of normal rat embryos. The researchers used fetal cells because they are rich in growth factors and adapt easily to a new environment. Result of the operation: the brain-damaged rats were able to learn the maze in just 8½ days. While this is still slower than normal, says Stein, "the transplant was clearly producing some degree of functional recovery." Stein later found that new connections had grown

---

Figure 15.24

Drug action at synapses. *a.* Drug stimulates release of neurotransmitter. *b.* Drug blocks release of neurotransmitter. *c.* Drug combines with neurotransmitter preventing its breakdown. *d.* Drug mimics neurotransmitter. *e.* Drug blocks receptor so that neurotransmitter cannot be received.

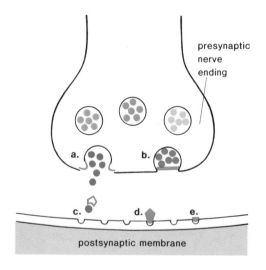

Drug Action

There is a wide variety of neurological drugs that are used to alter the mood and/or emotional state. Nevertheless, two general principles of drug action have been discovered: (1) mood-altering drugs particularly affect the ARAS (p. 316) and limbic system, and (2) they either promote or decrease the action of a particular neurotransmitter. There are a number of different ways in which drugs can influence transmission of neurotransmitters and some of these are described in figure 15.24. It is clear, as outlined in table 15.7, that stimulants can either enhance the action of an excitatory transmitter or block the action of an inhibitory transmitter. Also, depressants can either enhance the action of an inhibitory transmitter or block the action of an excitatory transmitter.

Domestic Drugs

People may not realize that they daily imbibe drugs that act on the CNS. For example, caffeine in coffee and theophylline in tea are stimulants that block the action of adenosine, a chemical that inhibits the release of neurotransmitters. Nicotine, the addicting substance in cigarettes, enhances the action of Ach (acetylcholine).

Poisons

Some poisons, such as strychnine, block inhibitory synapses in the spinal cord and brain stem. Other poisons, such as insecticides and nerve gas, inactivate

between the transplanted tissue and the rest of the brain.

According to Stein, the immediate lesson of his group's experiment is "that there is much more capacity for response to brain injury than previously thought." The same conclusion has been reached by researchers who have regenerated nerve fibers in other parts of animals' brains as well as in their spinal cords. At Saint Elizabeths Hospital in Washington, D.C., for instance, Neuroscientist William Freed has treated rats with fetal cell implants to relieve symptoms resembling Parkinson's disease in humans. The implanted cells are capable of producing dopamine, a vital brain chemical lacking in the afflicted rats and in Parkinson's patients. Such techniques used with humans, some researchers believe, may lead to a cure for Parkinson's disease within five to ten years. Eventually, it may also become possible to repair the spinal cords of paraplegics and regenerate parts of the extensively damaged brains of

patients with Alzheimer's disease, Huntington's chorea, multiple sclerosis and other degenerative disorders.

Before such steps can be attempted, however, certain ethical questions surrounding the use of tissue from human fetuses must be resolved. The sensitivity of this issue may be dramatized this week, when a funding bill for the National Institutes of Health, which finances many such experiments, is scheduled to come before the House of Representatives. California Republican William Dannemeyer plans to offer an amendment that would drastically increase the already stringent restrictions on the use of human fetuses for research purposes. A similar measure was passed by the House last year [1982] but defeated in the Senate. Congressman Henry Waxman, who opposes the amendment, argues that "fetal research saves lives, prevents or cures chronic diseases and makes pregnancy safer. As a result of such work, reductions in

infant mortality and treatments for diabetes, as well as for brain disorders, are on the horizon."

Researchers, hoping to avoid controversy, are looking for alternatives to fetal tissues. In the case of Parkinson's disease, says Freed, it may be possible to transplant dopamine-secreting cells taken from the patient's own adrenal gland. Other approaches were discussed at a conference on fetal cell research last month in Brookline, Mass. Among them: the possibility of altering monkey fetal cells for use in humans. Ultimately, as researchers become able to identify the chemicals that give fetal cells their regenerative powers, they may find ways to synthesize these substances or to develop cell cultures that produce them in the lab. Unlocking these secrets "is the best hope we have for those who have lost brain cells because of a stroke, an injury or a degenerative disease," says Vernon Mark, director of neurosurgery at Boston City Hospital. "Right now," he adds, "it's the only game in town."

AchE (acetylcholinesterase). The result is the same—increased excitability, convulsions, and death may occur. There are also poisons that have exactly the opposite effect. Botulism toxin and curare decrease the amount of Ach in the synaptic cleft. Death from paralysis follows since, as discussed in the next chapter, Ach causes muscle contraction.

Depressants

Table 15.8 lists some mood-altering drugs, many of which have legitimate medical uses when used under a doctor's care but which are often taken in excess and are therefore called "drugs of abuse." **Sedatives,** which include barbituates, depress all nervous functions, acting first on the cortex and then on the rest of the brain, depending on the dose taken. They induce a sleeplike EEG. **Tranquilizers,** such as Librium and Valium, and more recently, **alcohol** have been shown to enhance the action of the inhibitory transmitter GABA. Dependency develops when the body begins to produce less GABA. Consumption of a large amount of alcohol within a short period of time can cause death because of its depressing effect on brain functions. A habitual use of excess alcohol can cause damage to several areas of the brain, particularly the hippocampus, which results in memory impairment. The breakdown of alcohol in the liver leads to cirrhosis of the liver (p. 191) and other physiological side effects, eliminating alcohol as a drug of choice in treating anxiety.

Table 15.7	Drug action	
Drug action	**Neurotransmitter**	**Result**
Blocks	Excitatory	Depression
Enhances	Excitatory	Stimulation
Blocks	Inhibitory	Stimulation
Enhances	Inhibitory	Depression

Table 15.8 Mood-altering drugs of abuse

	Drugs	Often prescribed brand names	Medical uses	Potential physical dependence	Potential psychological dependence
Narcotics	Opium	Dover's Powder, Paregoric	Analgesic, antidiarrheal	High	High
	Morphine	Morphine	Analgesic	High	High
	Codeine	Codeine	Analgesic, antitussive	Moderate	Moderate
	Heroin	None	None	High	High
	Meperidine (pethidine)	Demerol, Pethadol	Analgesic	High	High
	Methadone	Dolophine, Methadone, Methadose	Analgesic, heroin substitute	High	High
	Other narcotics	Dilaudid, Leritine, Numorphan, Percodan	Analgesic, antidiarrheal, antitussive	High	High
Depressants	Chloral hydrate	Noctec, Somnos	Hypnotic	Moderate	Moderate
	Barbiturates	Amytal, Butisol, Nembutal, Phenobarbital, Seconal, Tuinal	Anesthetic, anti-convulsant, sedation, sleep	High	High
	Glutethimide	Doriden	Sedation, sleep	High	High
	Methaqualone	Optimil, Parest, Quaalude, Somnafac, Sopor	Sedation, sleep	High	High
	Tranquilizers	Equanil, Librium, Miltown, Serax, Tranxene, Valium	Anti-anxiety, muscle relaxant, sedation	Moderate	Moderate
	Other depressants	Clonopin, Dalmane, Dormate, Noludar, Placydil, Valmid	Anti-anxiety, sedation, sleep	Possible	Possible
Stimulants	Cocaine†	Cocaine	Local anesthetic	Possible	High
	Amphetamines	Benzedrine, Biphetamine, Desoxyn, Dexedrine	Hyperkinesis, narcolepsy, weight control	Possible	High
	Phenmetrazine	Preludin	Weight control	Possible	High
	Methylphenidate	Ritalin	Hyperkinesis	Possible	High
	Other stimulants	Bacarate, Cylert, Didrex, Ionamin, Plegine, Pondimin, Pre-sate, Sanorex, Voranil	Weight control	Possible	Possible
Hallucinogens	LSD	None	None	None	Degree unknown
	Mescaline	None	None	None	Degree unknown
	Psilocybin-psilocyn	None	None	None	Degree unknown
	MDA	None	None	None	Degree unknown
	PCP‡	Sernylan	Veterinary anesthetic	None	Degree unknown
	Other hallucinogens	None	None	None	Degree unknown
Cannabis	Marijuana Hashish Hashish oil	None	Glaucoma	Degree unknown	Moderate

From: *Drugs of Abuse* Produced by the Affairs in cooperation with the Office of Public Science and Technology.
†Designated a narcotic under the Controlled Substances Act.
‡Designated a depressant under the Controlled Substances Act.

Human Anatomy and Physiology

Tolerance	Duration of effects (in hours)	Usual methods of administration	Possible effects	Effects of overdose	Withdrawal syndrome
Yes	3 to 6	Oral, smoked	Euphoria, drowsiness, respiratory depression, constricted pupils, nausea	Slow and shallow breathing, clammy skin, convulsions, coma, possible death	Watery eyes, runny nose, yawning, loss of appetite, irritability, tremors, panic, chills and sweating, cramps, nausea
Yes	3 to 6	Injected, smoked			
Yes	3 to 6	Oral, injected			
Yes	3 to 6	Injected, sniffed			
Yes	3 to 6	Oral, injected			
Yes	12 to 24	Oral, injected			
Yes	3 to 6	Oral, injected			
Probable	5 to 8	Oral			
Yes	1 to 16	Oral, injected	Slurred speech, disorientation, drunken behavior without odor of alcohol	Shallow respiration, cold and clammy skin, dilated pupils, weak and rapid pulse, coma, possible death	Anxiety, insomnia, tremors, delirium, convulsions, possible death
Yes	4 to 8	Oral			
Yes	4 to 8	Oral			
Yes	4 to 8	Oral			
Yes	4 to 8	Oral			
Yes	2	Injected, sniffed	Increased alertness, excitation, euphoria, dilated pupils, increased pulse rate and blood pressure, insomnia, loss of appetite	Agitation, increase in body temperature, hallucinations, convulsions, possible death	Apathy, long periods of sleep irritability, depression, disorientation
Yes	2 to 4	Oral, injected			
Yes	2 to 4	Oral			
Yes	2 to 4	Oral			
Yes	2 to 4	Oral			
Yes	Variable	Oral	Illusions and hallucinations (with exception of MDA); poor perception of time and distance	Longer, more intense "trip" episodes, psychosis, possible death	Withdrawal syndrome not reported
Yes	Variable	Oral, injected			
Yes	Variable	Oral			
Yes	Variable	Oral, injected, sniffed			
Yes	Variable	Oral, injected, smoked			
Yes	Variable	Oral, injected, sniffed			
Yes	2 to 4	Oral, smoked	Euphoria, relaxed inhibitions, increased appetite, disoriented behavior	Fatigue, paranoia, possible psychosis	Insomnia, hyperactivity, and decreased appetite reported in a limited number of individuals

Stimulants

Amphetamines have a structure similar to the excitatory transmitter NA (noradrenalin) and are believed to promote the synaptic release of dopamine and NA, thereby increasing the amount received by the postsynaptic membrane. Unpleasant side effects of amphetamines, including hallucinations, may be due to stimulant-induced insomnia. Cocaine blocks the uptake of NA and thereby increases the length of time it is present in the synaptic cleft, and this may account for some of its psychological effects. One study found, however, that cocaine users who took the drug under controlled laboratory conditions could not distinguish its effects from other drugs or even from a placebo. These investigators concluded that environmental factors may play a role in the sensation of euphoria produced by the drug.

There are two types of **antidepressants.** One type, represented by Elavil, is believed, like cocaine, to prevent the reabsorption of NA and the other excitatory neurotransmitters. Another type, represented by Parnate, inhibit the enzyme monoamine oxidase, which breaks down NA. In any case, both types of antidepressants increase the amount of NA in the synaptic cleft and thereby relieve depression. Recently it has been found that both types of antidepressants also block receptors for the neurotransmitter histamine.

Antipsychotics

There is a delicate balance of neurotransmitters at the cerebral synapses, and it logically follows that mental illness might be caused by an imbalance. (This does not mean that all mental illnesses are caused by neurotransmitter imbalance.) Drugs that can restore the normal balance alleviate the symptoms of mental illness. For example, it is possible that Lithium is effective in treating manic-depressive symptoms because it blocks the release of NA from the presynaptic membrane and thereby controls the manic (euphoric) phase. When the manic phrase is prevented from developing, the depression phase does not follow. Drugs, such as Thorazine, relieve the symptoms of schizophrenia, apparently because they bind to dopamine receptors, interfering with its normal action.

Hallucinogens

LSD (lysergic acid diethylamide) and mescaline, a chemical derived from the peyote plant, affect the action of serotonin on certain ARAS cells involved in vision and emotion. This explains their ability to produce hallucinations. "Bad trips" may be due to a concomitant (simultaneous) effect on dopamine.

The active ingredient in marijuana (tetrahydrocannabinol, or THC) causes hallucinations only when taken in large doses. In low doses, THC is like a mild sedative, acting as a hypnotic drug whose effect resembles the effect of alcohol and tranquilizers. The mode of action of marijuana is not yet fully understood although it does impair short-term memory and slows learning.

Narcotics

Opium and heroin bind to receptors meant for the body's own natural opiates, the endorphins and enkephalins. The natural opiates are believed to alleviate pain by preventing the release of a neurotransmitter, termed **substance P,** from certain sensory neurons in the region of the spinal cord (fig. 15.25). When substance P is released, pain is felt, and when substance P is not released, pain

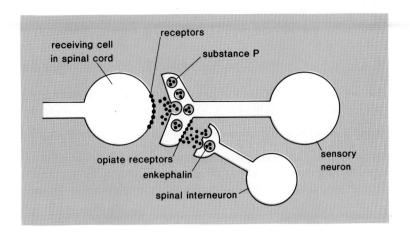

Figure 15.25
Suggested mode of action of the natural opiates, enkephalins. The reception of an enkephalin by a sensory neuron may inhibit the release of substance P, a neurotransmitter that ultimately promotes the sensation of pain.

is not felt. Some investigators have also shown that *acupuncture* may relieve pain by causing a release of natural opiates in the central nervous system. Evidence also indicates that there are opiate receptors in neurons that travel from the spinal cord to the limbic system, and that stimulation of these can cause a feeling of pleasure. This explains why opium and heroin not only kill pain but also produce a feeling of tranquility.

Endorphin and enkephalin research has offered an explanation for **drug addiction.** Narcotic intake causes the CNS to stop producing its own natural opiates. Therefore the addict must take more and more of the drug in order to get the same effect. Furthermore, when the intake of morphine or heroin is stopped, stimulatory neurotransmitters are no longer opposed, causing withdrawal symptoms. As the body gradually begins to produce its own supply of opiates again, the individual returns to a normal state.

Tolerance, which is discussed in the following, is another reason why an addict must take ever-increasing doses of a drug.

Drug Tolerance

Even when a drug does not gradually replace the natural substances produced by the body, an individual may find that ever-increasing consumption is needed to achieve the same effect. This may be related to drug breakdown. The liver produces enzymes to detoxify and break down drugs, preparing them for excretion. As more drugs, including alcohol, are consumed, more and more of these enzymes are produced. This explains, for example, the ability of a heavy drinker to consume more alcohol than others.

Drug Interaction

It is dangerous to take concomitantly two or more kinds of drugs that have the same effect on the CNS. At any one time, the liver contains only a certain quantity of detoxification enzymes. Therefore, if two drugs are taken in the usual dosage at the same time, the liver requires twice as long to detoxify them. In the meantime, the drugs cause a compound effect on the central nervous system. The consequences may be dire. For example, if alcohol and barbituates are taken together, the total depressive effect on the CNS can lead to coma and death.

Summary

The anatomical unit of the nervous system is the neuron, of which there are three types: sensory, motor, and interneuron. Each of these is made up of a cell body, an axon, and dendrites. Cell bodies are found first and foremost in the central nervous system, but they are also found in ganglia. Axons and dendrites make up nerves (table 15.3) that project from the brain and cord.

The nerve impulse (table 15.2) is the same in all neurons. It simply consists of a change in permeability of the membrane so that sodium ions move to the inside of a neuron and potassium ions move to the outside. This movement of ions produces an electrochemical charge that can be recorded in terms of millivolts by an oscilloscope. Depolarization occurs with the initiation of the action potential; repolarization restores the resting potential. Transmission of the nerve impulse from one neuron to another takes place when synaptic vesicles found at the ends of axons release a neurotransmitter substance that diffuses across a synapse to the postsynaptic membrane of the next neuron. Reception of excitatory neurotransmitters increases the likelihood that the next neuron will transmit a nerve impulse, and reception of inhibitory neurotransmitters decreases the likelihood that it will fire.

A simple reflex action requires the use of neurons that make up a reflex arc (table 15.4). A sensory neuron conducts the nerve impulse from a receptor to an interneuron, which in turn transmits the impulse to a motor neuron, which conducts it to an effector. Reflexes are automatic and some do not require involvement of the brain.

The peripheral nervous system contains cranial and spinal nerves, which contain the long fibers of sensory and/or motor neurons. Nerves have a white appearance because the long fibers are surrounded by a myelin sheath. The autonomic nervous system is a part of the peripheral nervous system and contains two parts: the sympathetic system, which is often associated with those reactions that occur during times of stress, and the parasympathetic system, which is often associated with those activities that occur during times of relaxation. Besides contrary physiological effects, these two systems can also be contrasted anatomically (table 15.5).

The central nervous system consists of the spinal cord and brain. The gray matter of the cord contains cell bodies; the white matter contains tracts that consist of the long axons of interneurons. These run from all parts of the cord, even up to the cerebrum. The brain integrates all nervous system activity and commands all voluntary activities. In the hindbrain, the medulla oblongata has centers for visceral functions and the cerebellum coordinates muscle contractions. In the forebrain, the hypothalamus, in particular, controls homeostasis; the thalamus specializes in sense reception; and the cerebrum gives us consciousness. The cerebrum can be mapped, and each lobe seems to have particular functions (table 15.6).

Neurological drugs, although quite varied, have been found to affect the ARAS and limbic system by either promoting or preventing the action of neurotransmitters. Stimulatory drugs enhance excitatory neurotransmitters or decrease the activity of inhibitory neurotransmitters. Depressants have the opposite effect. Drug addiction and tolerance can now be explained in biological terms.

Study Questions

1. What are the two main divisions of the nervous system? How are these divisions subdivided? (p. 302)
2. What are the three types of neurons? How are they similar and how are they different? (p. 303)
3. What does *resting potential* mean and how is it brought about? (p. 304) Describe the two parts of an action potential and the change that may be associated with each part. (pp. 304–5)
4. What is the sodium/potassium pump and when is it active? (p. 306)
5. What is a neurotransmitter substance; where is it stored; how does it function; and how is it destroyed? (p. 308) Name two well-known neurotransmitters. (p. 309) What is *summation?* (p. 309)
6. What are the three types of nerves and how are they anatomically different? (p. 310) Functionally different? (p. 310) Distinguish between cranial and spinal nerves. (p. 310)
7. Trace the path of a reflex action after discussing the structure and function of the spinal cord and spinal nerve. (p. 311)
8. What is the autonomic nervous system and what are its two major divisions? (p. 313) Give several similarities and differences between these divisions. (pp. 313–14)
9. Name the major parts of the brain and give a function for each. (pp. 315–17)
10. Describe the EEG and discuss its importance. (p. 319)
11. Name the various categories of drugs that affect the CNS. How does each type of drug affect transmission across the synapse? (pp. 320–25)

Selected Key Terms

neuron (nu′ron)
dendrite (den′drīt)
axon (ak′son)
myelin (mi′ĕ-lin)
resting potential (rest′ing po-ten′shal)
action potential (ak′shun po-ten′shal)
synapse (sin′aps)
neurotransmitter (nu″ro-trans-mit′er)
ganglion (gang′gle-on)
cranial nerve (kra′ne-al nerv)
reflex (re′fleks)
sympathetic nervous system (sim″pah-thet′ik ner′vus sis′tem)
parasympathetic nervous system (par″ah-sim″pah-thet′ik ner′vus sis′tem)
meninges (mĕ-nin′jēz)
medulla oblongata (mĕ-dul′ah ob″long-gah′tah)
cerebellum (ser″ĕ-bel′um)
hypothalamus (hi″po-thal′ah-mus)
thalamus (thal′ah-mus)
cerebrum (ser′ĕ-brum)
limbic system (lim′bik sis′tem)

Further Readings

Alkon, D. L. 1983. Learning in a marine snail. *Scientific American* 249(1):70.

Bloom, F. E. 1981. Neuropeptides. *Scientific American* 245(4):148.

Jacobson, M., and Hunt, R. K. 1973. The origins of nerve cell specificity. *Scientific American* 228(2):26.

Julien, R. M. 1981. *A primer of drug action.* 3d ed. San Francisco: W. H. Freeman.

Katz, B. 1961. How cells communicate. *Scientific American* 205(3):209.

Llinas, R. R. 1982. Calcium in synaptic transmission. *Scientific American* 247(4):56.

Menaker, M. 1972. Nonvisual reception. *Scientific American* 226(3):12.

Morell, P., and Nortin, W. T. 1980. Myelin. *Scientific American* 242(5):88.

Morrison, A. R. 1983. A window on the sleeping brain. *Scientific American* 248(4):94.

Nathanson, J. S., and Greengard, P. 1977. Second messengers in the brain. *Scientific American* 237(2):108.

Ray, O. S. 1983. *Drugs, society and human behavior.* 3d ed. St. Louis: C. V. Mosby.

Rubenstein, E. 1980. Diseases caused by impaired communication among cells. *Scientific American* 242(3):102.

Scientific American 1979. 241(3). Entire issue is devoted to nervous system.

Snyder, S. H. 1977. Opiate receptors and internal opiates. *Scientific American* 236(3):44.

Stent, G. S. 1972. Cellular communication. *Scientific American* 227(3):42.

Thompson, R. F., ed. 1976. *Progress in psychobiology: Readings from* Scientific American. San Francisco: W. H. Freeman.

Van Dyke, C., and Byck, R. 1982. Cocaine. *Scientific American* 246(3):128.

Wurtman, R. J. 1982. Nutrients that modify brain function. *Scientific American* 246(4):50.

Wurtz, R. H., et al. 1982. Brain mechanisms of visual attention. *Scientific American* 246(6):124.

Musculoskeletal System

Chapter concepts

1 The skeleton, which contributes greatly to our general appearance, has various functions and is divided into the axial and appendicular skeletons.

2 Macroscopically, skeletal muscles work in antagonistic pairs and exhibit certain physiological characteristics related to the fact that muscles are composed of muscle fibers.

3 Microscopically, muscle fiber contraction is dependent on actin and myosin filaments and a ready supply of Ca^{++} and ATP.

4 Nerve fibers cause muscle cells to contract and the sequence of events leading up to contraction has been studied in detail.

Table 16.1	Bones of skeleton
Part	**Bones**
Axial skeleton	Skull
	Vertebral column
	Sternum
	Ribs
Appendicular skeleton	
Pectoral girdle	Clavicle
	Scapula
Arm	Humerus
	Ulna
	Radius
Hand	Carpals
	Metacarpals
	Phalanges
Pelvic girdle	Innominate
Leg	Femur
	Tibia
	Fibula
Foot	Tarsals
	Metatarsals
	Phalanges

Muscles and bones working together allow humans to perform the many mechanical tasks of their daily lives. Body weight and appearance are largely accounted for by these organs (fig. 16.1), whose structure suits their functions as discussed in the following.

Skeleton

Functions
The skeleton, notably the large heavy bones of the legs, supports the body against the pull of gravity. The skeleton also protects soft body parts. For example, the skull forms a protective encasement for the brain, as does the ribcage for the heart and lungs. Flat bones, such as those of the skull, ribs, and breastbone, produce red blood cells in both adults and children. All bones are a storage area for inorganic calcium and phosphorus salts. Bones also provide sites for muscle attachment. The long bones, particularly those of the legs and arms, permit flexible body movement.

Structure
The skeleton may be divided into two parts: the **axial skeleton** and the **appendicular skeleton** (table 16.1).

Axial Skeleton
The **skull,** or cranium, is composed of many bones fitted tightly together in adults. In newborns, certain bones are not completely formed and instead are joined by membranous regions called **fontanels,** all of which usually close by the age of 16 months. The bones of the skull contain the **sinuses,** air spaces lined by mucous membrane. Two of these, called the mastoid sinuses, drain into the middle ear. Mastoiditis, a condition that can lead to deafness, is an inflammation of these sinuses. Whereas the skull protects the brain, the several bones of the face join together to support and protect the special sense organs and to form the jawbones.

The **vertebral column** extends from the skull to the pelvis and forms a dorsal backbone that protects the spinal cord (fig. 15.13). Normally, the vertebral column has four curvatures that provide more resiliency and strength than a straight column could. It is composed of many parts, called **vertebrae,** that are held together by bony facets, muscles, and strong ligaments. The vertebrae are named according to their location in the body (fig. 16.2).

There are disks between the vertebrae that act as a kind of padding. They prevent the vertebrae from grinding against one another and absorb shock caused by movements such as running, jumping, and even walking. Unfortunately, these disks become weakened with age and can slip, or even rupture. This causes pain when the damaged disk presses up against the spinal cord and/or spinal nerves. The body may heal itself, or else the disk can be removed surgically. If the latter occurs, the vertebrae can be fused together but this will limit the flexibility of the body. The presence of the disks allows motion between the vertebrae so that we can bend forward, backward, and side to side.

The vertebral column, directly or indirectly, serves as an anchor for all the other bones of the skeleton (fig. 16.1). All the 12 pairs of **ribs** connect directly to the thoracic vertebrae in the back, and all but two pairs connect either directly or indirectly via shafts of cartilage to the **sternum** (breastbone) in the front. The lower two pairs of ribs are called "floating ribs" because they do not attach to the sternum.

Human Anatomy and Physiology

Figure 16.1

Major bones and muscles of the human body. The axial skeleton, composed of the skull, vertebral column, sternum, and ribs, lies in the midline; the rest of the bones belong to the appendicular skeleton.

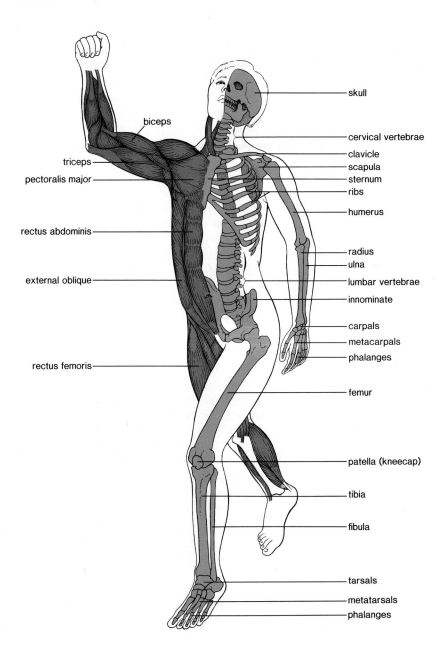

Figure 16.2

The vertebral column. The vertebrae are named according to their location in the column, which is flexible due to the presence of disks between the vertebrae. Note the presence of the coccyx, the vestigial "tailbone."

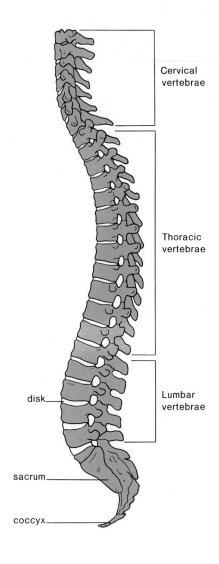

Appendicular Skeleton

The appendicular skeleton consists of the bones within the pectoral and pelvic girdle and the attached appendages. The pectoral (shoulder) girdle and appendages (arms and hands) are specialized for flexibility, whereas the pelvic girdle (hip bones) and appendages (legs and feet) are specialized for strength.

Figure 16.3
The bones of the pectoral girdle, arm, and hand. The humerus becomes the "funny bone" of the elbow.

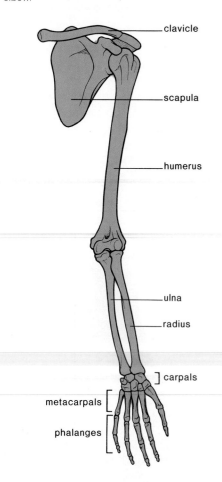

clavicle

scapula

humerus

ulna

radius

carpals

metacarpals

phalanges

The components of the **pectoral girdle** (fig. 16.3) are loosely linked together by ligaments rather than firm joints. Each **clavicle** (collar bone) connects with the sternum in front and the **scapula** (shoulder blade) behind, but the scapula is freely movable and held in place by muscles. This allows it to follow freely the movements of the arm. The single long bone in the upper arm (fig. 16.3), the **humerus**, has a smoothly rounded head that fits into a socket of the scapula. The socket, however, is very shallow and much smaller than the head. While this means that the arm can move in almost any direction, there is little stability. Therefore this is the joint that is most apt to dislocate. The opposite end of the humerus meets the two bones of the lower arm, the **ulna** and the **radius**, at the elbow. (The prominent bone in the elbow is the topmost part of the ulna.) When the arm is held so that the palm is turned frontward, the radius and ulna are about parallel to one another. When the arm is turned so that the palm is next to the body, the radius crosses in front of the ulna, a feature that contributes to the easy twisting motion of the lower arm.

The many bones of the hand increase its flexibility. The wrist has eight **carpal** bones that look like small pebbles. Then there are five **metacarpal** bones that fan out to form a framework for the palm. The metacarpal bone that leads to the thumb is placed in such a way that the thumb can reach out and touch the other digits. (**Digits** is a term that refers to either fingers and toes.) Beyond the metacarpals are the **phalanges,** the bones of the fingers and the thumb. The phalanges of the hand are long, slender, and lightweight.

The **pelvic girdle** (fig. 16.4) consists of two heavy, large **innominate** (hip bones). The innominate bones are anchored to the sacrum and together these bones form a hollow cavity, the pelvis. The weight of the body is transmitted through the pelvis to the legs and then onto the ground. The largest bone in the body is the **femur,** or thigh bone. While it is a strong bone, it is doubtful that the femurs of a fairy-tale giant could support the increase in weight. If a giant were 10 times taller than an ordinary human being, he would also be about 10 times wider and thicker, making him weigh about one thousand times as much. This amount of weight would break even giant-size femurs.

In the lower leg the larger of the two bones, the **tibia** (fig. 16.4), has a ridge we call the shin. Both of the bones of the lower leg have a prominence that contributes to the ankle—the tibia on the inside of the ankle and the **fibula** on the outside of the ankle. Although there are seven **tarsal** bones in the ankle, only one receives the weight and passes it on to the heel and the ball of the foot. If one wears high-heel shoes, though, the weight is thrown even further toward the front of the foot. The **metatarsal** bones form the arches of the foot. There is a longitudinal arch from the heel to the toes and a transverse arch across the foot. These provide a stable, springy base for the body. If the tissues that bind the metatarsals together become weakened, flat feet are apt to result. The bones of the toes are called **phalanges,** just like those of the fingers, but in the foot the phalanges are stout and extremely sturdy.

Joints
Bones are joined together at the joints, which are often classified according to the amount of movement they allow. Some bones, such as those that make up the skull, are sutured together and are **immovable.** Other joints are **slightly movable,** such as the joints between the vertebrae. The vertebrae are separated

by disks, described earlier, that increase their flexibility. Similarly, the two hip bones are slightly movable where they are ventrally joined by cartilage. Owing to hormonal changes, this joint becomes more flexible during late pregnancy and this allows the pelvis to expand during childbirth. Most joints are **freely movable synovial joints** in which the two bones are separated by a cavity. **Ligaments** composed of fibrous connective tissue bind the two bones to one another, holding them in place as they form a capsule. In a double-jointed individual, the ligaments are unusually loose. The joint capsule is lined by synovial membrane that produces **synovial fluid,** a lubricant for the joint. The knee is an example of a synovial joint (fig. 16.5). In the knee, as in other freely movable joints, the bones are capped by cartilage, but in the knee there are also crescent-shaped pieces of cartilage between the bones, called **menisci.** These give added stability, helping to support the weight placed on the knee joint. Unfortunately, as discussed in the reading on page 334, athletes often suffer injury of the menisci, known as torn cartilage. The knee joint also contains 13 fluid-filled sacs called **bursae** that ease friction between tendons and ligaments and between tendons and bones. Infection of bursae is called bursitis. Tennis elbow is a form of bursitis.

There are different types of movable joints. The knee and elbow joints are *hinge joints* because, like a hinged door, they largely permit movement in one direction only. More movable are the *ball-and-socket joints;* for example, the ball of the femur fits into a socket on the hip bone. Ball-and-socket joints allow movement in all planes and even allow a rotational movement.

Synovial joints are subject to **arthritis.** In *rheumatoid arthritis,* the synovial membrane becomes inflamed and grows thicker. Degenerative changes take place that make the joint almost immovable and painful to use. There is evidence that these effects are brought on by an autoimmune reaction. In old-age arthritis, or *osteoarthritis,* the cartilage at the ends of the bones disintegrates so that the two bones become rough and irregular. This type of arthritis is apt to affect the joints that have received the greatest use over the years.

Figure 16.4
The bones of the pelvic girdle, leg, and foot. The femur is our strongest bone and withstands a pressure of 1,200 pounds per cubic inch when we walk.

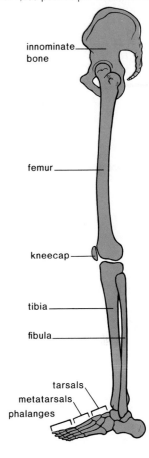

Figure 16.5
The knee joint, an example of a freely movable synovial joint. Notice that there is a cavity between the bones that is encased by ligaments and lined by synovial membrane. The kneecap protects the joint.

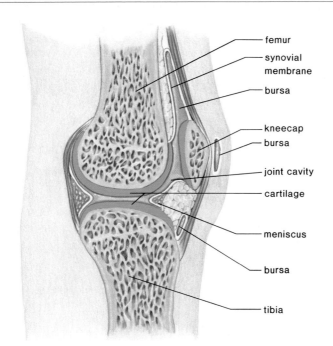

The Knee

Every year surgeons remove some 52,000 pieces of cartilage from knees that were previously propelling someone down the gridiron. "Football isn't just a contact sport, it's a collision sport," says Edward Percy, an orthopedic surgeon and head of the University of Arizona's Sports Medicine program. Even so, less than half the knee injuries in football are caused by a direct blow to the knee. Most arise during the tangling of bodies that occurs in tackling or blocking. Twenty percent come about with no contact at all . . . .

The knee's vulnerability to sports injuries is not the result of some intrinsic flaw in its design. In fact, the knee is a nice compromise of mobility and stability. It can bend 150 degrees, swing from side to side, and twist on itself. And it can absorb the force, equal to nearly seven times an athlete's weight, that occurs as a tightend makes the final cut of a down-and-out pass pattern.

The knee is the meeting place of the two major leg bones: the thighbone, called the femur, and the shinbone, called the tibia. The bottom end of the thighbone is shaped like a baby's behind and coated with cartilage. The top of the lower leg bone has two shallow scoops hollowed out of it and a ridge along the middle. The femur rides along this ridge like a cowboy in a saddle. Wedges of cartilage line the top of the tibia, acting as shock absorbers and keeping the femur from rocking too much from side to side. . . .

Cartilage can be torn if it's twisted too tight. This can occur as a result of a collision or fall, but it can also happen if an athlete simply combines a lot of pressure with a twist. For example, a tennis player may lunge to return a volley, bending her knee to the inside, as if she were knock-kneed. This puts pressure on the cartilage between the outer parts of the femur and tibia. The action is not unlike that of a mortar and pestle: The cartilage is ground between the two bones.

Because cartilage is the only tissue in the human body that has no blood supply of its own, it often cannot get enough nourishment to repair itself after it is damaged. Once it's torn, it stays torn. Most doctors recommend removing all or at least the damaged part of it.

Five years ago, this type of operation meant a stay in the hospital, five weeks on crutches, weeks of therapy, and a scar along the inside or outside of the knee that impressed friends. A new instrument has become widely used, however, that shortens both recovery time and scars. Called an arthroscope, it enables some patients to be out of the hospital in as little as a day, spend only a few days on crutches, and begin running after two weeks of therapy.

Essentially a tiny lens and light mounted on a pencil-thin shaft (and usually hooked up to a television camera), the arthroscope allows a surgeon to peer into the knee through a small slit in the skin. It also enables a surgeon to use similarly shaped knives, scissors, and other instruments to remove cartilage while leaving only a tiny scar. . . .

There is one injury to the knee, however, that can end a career if not immediately repaired: tearing a ligament. The knee would be nothing but a balancing act if not for the tough, fibrous ligaments that grip the bones like leather straps on a hinge and hold it steady. "The knee has minimal bony stability," says Reider. "If you cut away the ligaments, the joint falls apart."

There are five ligaments around the knee and two more buried deep within the joint. Most knee sprains involve the ligaments at the inside of the knee because collisions usually occur at the outside of the leg. If the blow is hard enough, the ligament will tear.

Ligaments are like taffy; once stretched, they stay stretched. "If the knee were meat, the ligaments would be gristle," says Reider. "It's very tough but only stretches six percent of its length before it breaks." A partially torn ligament will generally heal itself, but a completely torn ligament must be sewn back together. If the break is in the middle, that can be a difficult

task. "Ligaments consist of many strands," says Percy. "It's like trying to sew two hairbrushes together."

The sooner the severed ligament is repaired the better. The fluid in the knee will break down tissue, and if the injury is not repaired within 10 days, says Tab Blackburn, a physical therapist and trainer at the Rehabilitation Services of Columbus, Inc., in Georgia, "the fluids in the knee will turn the ligament to mush."

The front-most ligament, which runs from the front of the shinbone to the back of the thighbone, is particularly prone to injury. Doctors are not quite sure why, but this ligament, called the anterior cruciate is stretched or torn in nearly 70 percent of all serious knee injuries. Unfortunately, the ligament also has a poor blood supply and usually does not heal even if it is sewn back together.

Some athletes can continue to play without this ligament. It depends on the job the knee has to do. "A football player who has built up the muscles and other ligaments around his knee might not have any difficulty coming back," says Blackburn, "But a wiry gymnast with a lot of flexibility in her knees may have trouble." If a knee remains unstable, one solution is to jerry-rig a knee with new "ligaments" fashioned out of pieces of other tissue. A piece of tendon from the thigh, for example, is sometimes cut out and attached to the bone around the knee with staples. Percy and his colleagues are experimenting with using collagen, a fibrous substance extracted from cattle hide and woven into a band to help bind the bones together.

Knee specialists blame overtraining for the majority of knee problems afflicting amateur athletes, particularly those who are just beginning an exercise program. An estimated 40 percent of all women runners, for example, have knee problems severe enough to require a doctor within the first three months of training. One problem—actually a hodgepodge of muscle, tendon, and bone

Playing football often results in knee injuries.

irritation sometimes called runner's knee—affects nearly a third of the estimated 15 million joggers in the United States. During running each foot hits the ground 1,500 times a mile. Unlike walking, where both feet are on the ground 30 percent of the time, each knee must alternately bear the full load of the body hitting the pavement during running.

Figure 16.6
Anatomy of a long bone. A long bone is encased
by fibrous membrane except where it is covered by
articular cartilage at the ends. The central shaft is
composed of compact bone, but the ends are
spongy bone, which can contain red marrow. A
central medullary cavity contains yellow marrow.

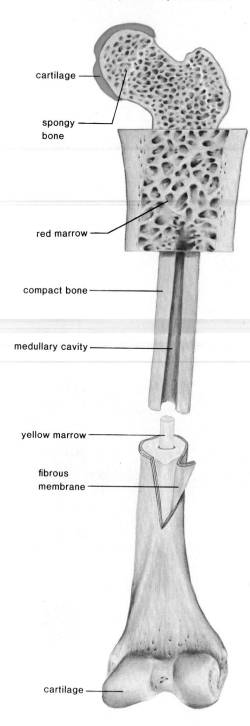

cartilage

spongy
bone

red marrow

compact bone

medullary cavity

yellow marrow

fibrous
membrane

cartilage

Long Bones

A long bone, such as the femur, illustrates principles of bone anatomy. When the bone is split open as in figure 16.6, the longitudinal section shows that it is not solid but has a cavity bounded at the sides by compact bone and at the ends by spongy bone. Beyond the spongy bone there is a thin shell of compact bone and finally a layer of cartilage.

Compact bone, as discussed previously (p. 169), contains bone cells in tiny chambers called *lacunae,* arranged in concentric circles around *Haversian canals,* which contain blood vessels and nerves. The lacunae are separated by a matrix that contains protein fibers of collagen and mineral deposits, primarily calcium and phosphorus salts.

Spongy bone contains numerous bony bars and plates separated by irregular spaces. Although lighter than compact bone, spongy bone is still designed for strength. Just as braces are used for support in buildings, the solid portions of spongy bone follow lines of stress. The spaces in spongy bone are often filled with **red marrow,** a specialized tissue that produces blood cells. The cavity of a long bone usually contains **yellow marrow,** which is a fat-storage tissue.

Growth and Development

Most of the bones of the skeleton are cartilaginous during prenatal development. Later, the cartilage is converted to bone by bone-forming cells known as **osteoblasts.** At first, there is only a primary ossification center at the middle of a long bone but later, secondary centers form at the ends of the bones. There remains a *cartilaginous disk* between the primary ossification center and each secondary center. The cartilage cells within the disk continue to divide and as they do so, the bone increases in length. Eventually, though, the disks disappear and the bone stops growing as the individual attains adult height.

In the adult, bone is continually being broken down and then built up again. Bone-absorbing cells, called **osteoclasts,** are derived from cells carried in the blood stream. As they break down bone, they remove worn cells and deposit calcium in the blood. Apparently after a period of about three weeks they disappear. The destruction caused by the work of osteoclasts is repaired by osteoblasts. As they form new bone, they take calcium from the blood. Eventually some of these cells get caught in the matrix they secrete and are converted to osteocytes, the cells found within Haversian systems, page 169.

Because of continual renewal, the thickness of bones can change, according to the amount of physical use or due to a change in certain hormone balances (see chapter 18). In most adults, the bones become weaker due to a loss of mineral content. Strange as it may seem, adults seem to require more calcium in the diet than do children in order to promote the work of osteoblasts. Many older women, in particular, suffer from **osteoporosis,** a condition in which weak and thin bones cause aches and pains and tend to fracture easily. Tendency toward osteoporosis is caused by lack of exercise and too little calcium in the diet.

Skeletal Muscles

Muscles are effectors, which enable the organism to respond to a stimulus (p. 311). Skeletal muscles are attached to the skeleton, and their contraction

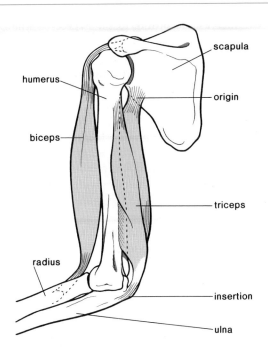

accounts for voluntary movements. Involuntary muscles, both smooth and cardiac, were discussed previously on page 170. Here we will divide our discussion of skeletal muscle into macroscopic anatomy and physiology, and microscopic anatomy and physiology.

Macroscopic Anatomy and Physiology

Muscles are typically attached to bone by **tendons** made of fibrous connective tissue. Tendons most often attach muscles to the far side of a joint so that the muscle extends across the joint (fig. 16.7). When the central portion of the muscle, called the belly, contracts, one bone remains fairly stationary and the other one moves. The **origin** of the muscle is on the stationary bone, and the **insertion** of the muscle is on the bone that moves.

When a muscle contracts, it shortens. Thus muscles can only pull; they cannot push. Because of this, muscles work in **antagonistic pairs.** The biceps and triceps are a pair of muscles that move the lower arm up and down (fig. 16.7). When the biceps contracts, the lower arm bends; and when the triceps contracts, the lower arm straightens.

In the laboratory, it is possible to study the contraction of individual whole muscles. Customarily, a calf muscle is removed from a frog and mounted in an apparatus, called a **physiograph,** figure 16.8. The muscle is stimulated electrically, and when it contracts, it pulls on a lever. The lever's movement is recorded, and the resulting pattern is called a **myogram.**

All-or-None Response

A *single* muscle fiber either responds to a stimulus and contracts or it does not. At first, the stimulus may be so weak that no contraction occurs but as soon as the strength of the stimulus reaches the **threshold stimulus,** the muscle fiber will contract completely. Thus a muscle fiber obeys the **all-or-none law.**

Figure 16.8
A physiograph. This apparatus can be used to record a myogram, a visual representation of the contraction of a muscle that has been dissected from an animal.

Figure 16.9

Physiology of muscle contraction. *a.* Simple muscle twitch is composed of three periods: latent, contraction, and relaxation. *b.* Summation and tetanic contraction. When a muscle is not allowed to relax completely between stimuli, the contractions increase in size and then the muscle remains contracted until it fatigues.

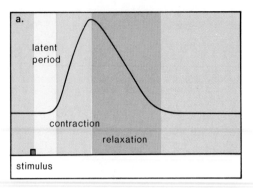

a.

latent
period

contraction

relaxation

stimulus

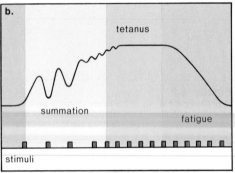

b.

tetanus

summation

fatigue

stimuli

Figure 16.10

Muscle spindle structure and function. After associated muscle fibers contract, spindle fibers aid in the coordination of muscular contraction by sending sensory impulses to the central nervous system, which then directs only certain other muscle fibers to contract.

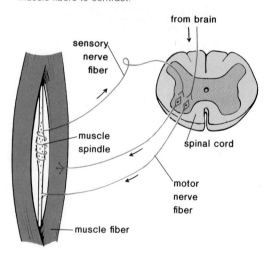

from brain

sensory
nerve
fiber

muscle
spindle

spinal cord

motor
nerve
fiber

muscle fiber

Contrary to that of an individual fiber, the strength of contraction of a whole muscle can increase according to the degree of stimulus beyond the threshold stimulus. A whole muscle contains many fibers, and the degree of contraction is dependent on the total number of fibers contracting. The **maximum stimulus** is the one beyond which the degree of contraction does not increase.

Muscle Twitch

If a muscle is placed in a physiograph (fig. 16.8) and given a maximum stimulus, it will contract and then relax. This action—a single contraction that lasts only a fraction of a second—is called a muscle **twitch.** Figure 16.9a is a myogram of a twitch, which is customarily divided into a **latent period,** or the period of time between stimulation and initiation of contraction; the period of **contraction;** and the period of **relaxation.**

If a muscle is exposed to two maximum stimuli in quick succession, it will respond to the first but not to the second stimulus. This is because it takes an instant following a contraction for the muscle fibers to recover in order to respond to the next stimulus. The very brief moment, following stimulation, during which a muscle remains unresponsive is called the **refractory period.**

Summation and Tetanus

If a muscle is given a rapid series of threshold stimuli, it may respond to the next stimulus without relaxing completely. In this way, muscle tension **summates** until maximal sustained **tetanic contraction** is achieved (fig. 16.9b). The myograms no longer show individual contractions; rather, they are completely fused and blended into a straight line. Tetanic contraction continues until the muscle fatigues, due to depletion of energy reserves. **Fatigue** is apparent when a muscle relaxes even though stimulation is continued.

Tetanic contractions occur whenever skeletal muscles are being actively used. Ordinarily, however, only a portion of any particular muscle is involved—while some fibers are contracting others are relaxing. Because of this, intact muscles rarely fatigue completely.

Muscle Tone

Intact skeletal muscles also have **tone,** a condition in which some fibers are always contracted. Muscle tone is particularly important in maintaining posture. If the muscles of the neck, trunk, and legs suddenly become relaxed, the body collapses.

The maintenance of tone requires the use of special sense receptors called **muscle spindles** (fig. 16.10). A muscle spindle consists of a bundle of modified muscle fibers with sensory nerve fibers wrapped around a short, specialized region somewhere near the middle of their length. A spindle contracts along with muscle fibers, but thereafter it sends stimuli to the CNS that enable it to regulate muscle contraction so that tone is maintained.

Adenosine Triphosphate (ATP)

A plentiful supply of ATP is required for muscle contraction. There are three ways in which ATP is made available, as described in figure 16.11 and in the following.

Human Anatomy and Physiology

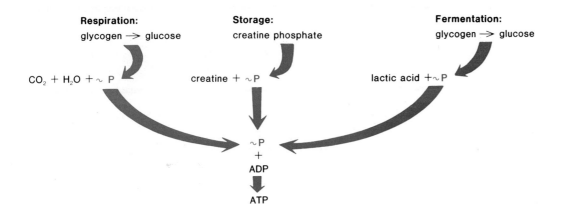

1. Muscle fibers are generously supplied with mitochondria (fig. 2.14) within which, by means of aerobic cellular respiration, ATP is formed.
2. Muscle fibers contain **creatine phosphate** (phosphocreatine), which is used as a storage supply of high-energy phosphate. Creatine phosphate does not participate directly in muscle contraction. Instead, it is used to regenerate ATP by the following reaction:

   creatine $\sim$ P + ADP $\longrightarrow$ ATP + creatine

3. When all of the creatine phosphate has been depleted, a muscle fiber can still produce ATP by means of anaerobic respiration (p. 111). Anaerobic respiration occurs, of course, when the fibers are not being supplied with oxygen quickly enough to make aerobic respiration possible. This occurs during times of strenuous exercise. In practice, anaerobic respiration can supply ATP for only a short time because lactic acid buildup produces muscular aching and fatigue.

Oxygen Debt

We all have had the experience of having to continue deep breathing following strenuous exercise. This continued intake of oxygen is required to complete the metabolism of lactic acid that has accumulated during exercise and represents an oxygen debt that the body must pay to rid itself of lactic acid. The lactic acid is transported to the liver, where one-fifth of it is completely broken down to carbon dioxide and water by means of the Krebs cycle and respiratory chain. The ATP gained by this respiration is then used to convert four-fifths of the lactic acid back to glucose. Figure 16.12 indicates how oxygen is used after vigorous exercise to pay off oxygen debt.

Figure 16.12
Lactic acid, which accumulates during strenuous muscular action, is converted back to pyruvic acid once exercise is over. One-fifth of the pyruvic acid is catabolized to produce energy so that the other four-fifths of the pyruvic acid may be converted to glucose by anabolism.

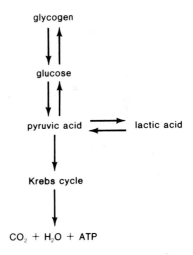

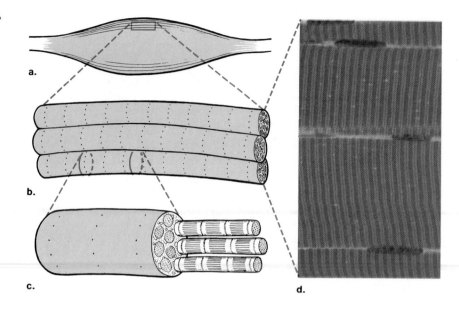

a.

b.

c.

d.

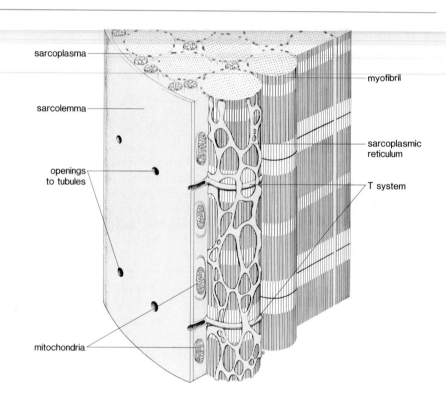

sarcoplasma

sarcolemma

openings to tubules

mitochondria

myofibril

sarcoplasmic reticulum

T system

Table 16.2	Muscle cells
Component	**Term**
Cell membrane	Sarcolemma
Cytoplasm	Sarcoplasm
Endoplasmic reticulum	Sarcoplasmic reticulum

Microscopic Anatomy and Physiology

A whole skeletal muscle is composed of muscle fibers (fig. 16.13). Each fiber is a cell and contains the usual cellular components, but special terminology has been assigned to certain ones, as indicated in table 16.2. Also, a muscle fiber has some unique anatomical characteristics. For one thing, the **sarcolemma,** or cell membrane, forms tubules that penetrate or dip down into the cell so that they come into contact but do not fuse with expanded portions of modified ER, which is termed the **sarcoplasmic reticulum** in muscle cells. The tubules comprise the so-called *T* (for *transverse*) *system* (fig. 16.14). The ex-

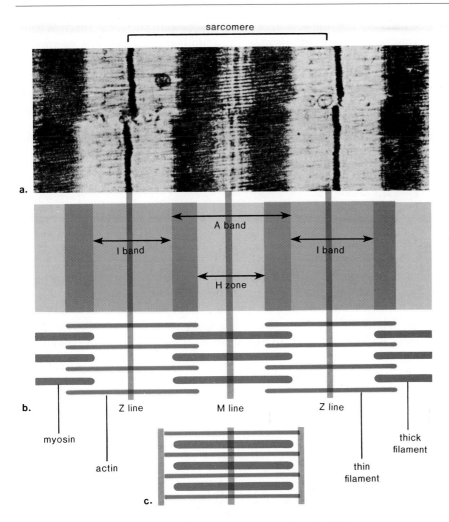

sarcomere

A band

I band

I band

H zone

Z line

M line

Z line

myosin

actin

thick filament

thin filament

a.

b.

c.

Figure 16.15
a. Electron micrograph of a sarcomere showing the typical striations of skeletal muscle. *b.* The striations contain various bands and dark lines. The I band contains the Z line and thin filaments. The A band contains both thin and thick filaments except at the center where the H zone has only thick filaments anchored by the M line. *c.* Notice that the I band has decreased in size and H zone has disappeared in the contracted sarcomere because the thin filaments have moved to the center.

panded portions of the sarcoplasmic reticulum, called **calcium-storage sacs,** contain calcium, Ca^{++}, an element that is essential to muscle contraction.

The sarcoplasmic reticulum encases hundreds, and even sometimes thousands, of cell inclusions called **myofibrils,** which are the contractile portions of the fibers. Myofibrils are cylindrical in shape and run the length of the fiber. The light microscope shows that the myofibrils have light and dark bands called striations (figs. 16.13c and 16.13d). It is the banding pattern of myofibrils that cause skeletal muscle to be striated. Electron micrographs (figs. 16.14 and 16.15a) have revealed that there are even areas of light and dark within the bands themselves. These areas can be studied in relation to a unit of a myofibril called a sarcomere.

Sarcomere Structure and Function

A sarcomere extends between two dark lines called the *Z lines* (fig. 16.15b). The *I band* is a light region that takes in the sides of two sarcomeres. Therefore an I band includes a Z line. The center dark region of a sarcomere is called the *A band*. The A band is interrupted by a light center, the *H zone*. A fine, dark stripe called the *M line* cuts through the H zone.

These bands and zones relate to the placement of filaments within each sarcomere. A sarcomere contains two types of filaments, **thin filaments** and **thick filaments** (fig. 16.15b). The thin filaments are attached to a Z line, and the thick filaments are anchored by an M line. The I band is light because it contains only thin filaments; the A band is dark because it contains both thin and thick filaments, except at the center where, in the lighter H zone, only thick filaments are found.

Table 16.3	Contractile elements
Component	**Definition**
Myofibril	Muscle cell contractile subunit
Sarcomere	Functional unit of myofibril
Myosin	Thick filament
Actin	Thin filament

Table 16.4	Muscle contraction
Name	**Function**
Actin filaments	Slide past myosin causing contraction
Ca^{++}	Needed for actin to bind to myosin
Myosin filaments	a. Enzyme that splits ATP b. Pulls actin by means of cross bridges
ATP	Supplies energy for bonding between actin and myosin = actomyosin

Sarcomere contraction is dependent on two proteins, **actin** and **myosin,** that make up the thin and thick filaments respectively (table 16.3). When a sarcomere contracts (fig. 16.15c), the actin filaments slide past the myosin filaments and approach one another. This causes the I band to become shorter and the H zone to almost or completely disappear. The movement of actin in relation to myosin is called the **sliding filament theory** of muscle contraction. During the sliding process, the sarcomere shortens even though the filaments themselves remain the same length (fig. 16.16).

The overall formula for muscle contraction can be represented as follows:

$$actin + myosin \xrightarrow[Ca^{++}]{ATP \quad ADP + \textcircled{P}} actomyosin$$

The participants in this reaction have the functions listed in table 16.4. Even though the actin filaments slide past the myosin filaments, it is the latter that does the work. In the presence of Ca^{++}, portions of the myosin filaments, called **cross bridges** (fig. 16.16b), reach out and attach to the actin filaments, pulling them along. Following attachment, ATP is broken down as detachment occurs. Myosin brings about ATP breakdown and therefore it is not only a structural protein, it is also an ATPase enzyme. The cross bridges attach and detach some 50 to 100 times as the thin filaments are pulled to the center of a sarcomere. If, by chance, more ATP molecules are not available, detachment cannot occur. This explains *rigor mortis,* permanent muscle contraction after death.

Muscle Fiber Contraction and Relaxation

Nerves innervate muscles, and nerve impulses cause muscles to contract. A motor axon within a nerve sends branches to several muscle fibers, which collectively are termed a *motor unit.* Each branch becomes unmyelinated and has terminal knobs that contain synaptic vesicles filled with the neuromuscular transmitter acetylcholine (Ach). A terminal end knob lies in close proximity to the sarcolemma of the muscle fiber. This region, called a neuromuscular junction (fig. 16.17) has the same components as a synapse: a

Figure 16.16
Sliding filament theory. *a.* Relaxed sarcomere. *b.* Contracted sarcomere. Note that during contraction, the I band and H zone decrease in size. This indicates that the thin filaments slide past the thick filaments. Even so, the thick filaments do the work by pulling the thin filaments by means of cross bridges.

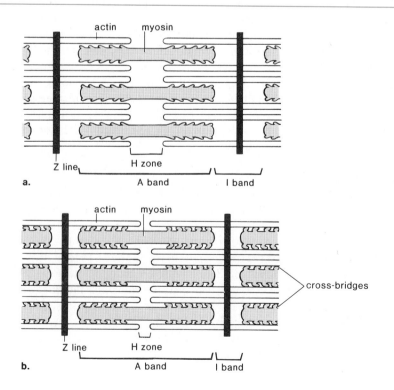

Human Anatomy and Physiology

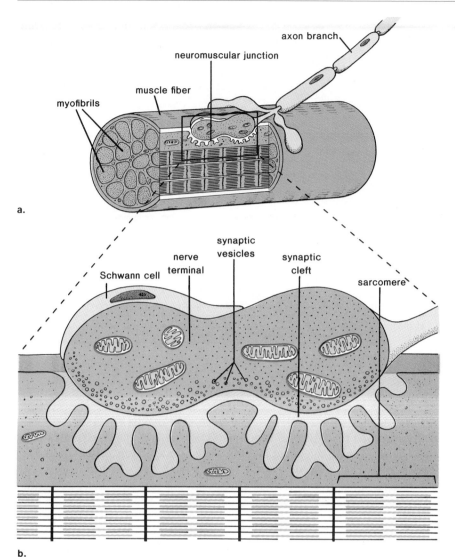

axon branch

neuromuscular junction

muscle fiber

myofibrils

a.

synaptic vesicles

nerve terminal

synaptic cleft

Schwann cell

sarcomere

b.

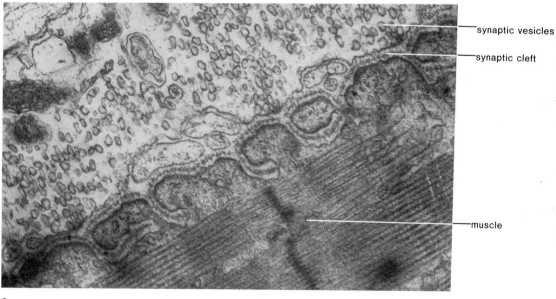

synaptic vesicles

synaptic cleft

muscle

c.

Figure 16.17
Anatomy of a neuromuscular junction. *a.* A neuromuscular junction occurs where a synaptic ending of an axon branch comes in close proximity to a muscle fiber. *b.* The synaptic ending contains synaptic vesicles filled with Ach. When these vesicles fuse with the presynaptic membrane, Ach diffuses across the synaptic cleft to initiate a muscle action potential. *c.* Actual electron micrograph.

Figure 16.18

Detailed structure and function of sarcomere contraction. After calcium, Ca++ is released from its storage sacs, it combines with troponin, a protein that occurs periodically along tropomyosin threads. This causes the tropomyosin threads to shift their position so that cross-bridge binding sites are revealed along the actin filaments. The myosin filament extends its globular heads, forming cross bridges that bind to these sites. The breakdown of ATP by myosin causes the cross bridges to detach and reattach farther along the actin. In this way, the actin filaments are pulled along past the myosin filaments.

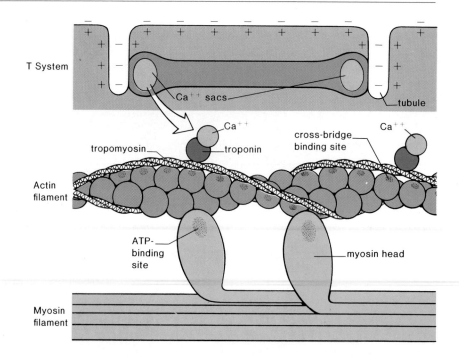

presynaptic membrane, a synaptic cleft, and a postsynaptic membrane. Only in this case, the postsynaptic membrane is a portion of the sarcolemma of a muscle fiber. The sarcolemma, just like a neural membrane, is polarized; the inside is negatively charged and the outside is positively charged.

Nerve impulses cause synaptic vesicles to merge with the presynaptic membrane and release Ach into the synaptic cleft. When Ach reaches the sarcolemma, it is depolarized. The result is a **muscle action potential** that spreads over the sarcolemma and down the T system to where calcium ions are stored in calcium-storage sacs of the sarcoplasmic reticulum. When the action potential reaches a sac, Ca++ ions are released and diffuse into the sarcoplasm, after which they attach to the thin filaments. A thin filament is a twisted double strand of globular actin molecules. Associated with the actin filaments are two other proteins: *tropomyosin* forms threads that twist about the actin filaments, while *troponin* is located at intervals along the tropomyosin. The calcium combines with the troponin and this causes the tropomyosin threads to shift in position so that myosin binding sites are exposed (fig. 16.18).

The thick myosin filament is a bundle of myosin molecules, each having a globular head. These heads form the cross bridges that attach to the binding sites on actin. Also, as they break down ATP, detachment and reattachment to a site farther along occurs. In this way actin filaments are pulled past the myosin filament.

Contraction discontinues when nerve impulses no longer stimulate the muscle fiber. With the cessation of a muscle action potential, Ca++ ions are pumped back into their storage sacs by active transport. Relaxation now occurs.

Effect of Contraction on Size of Muscle

Forceful muscular activity causes muscles to increase in size. This increase, called **hypertrophy,** occurs only if the muscle contracts to at least 75 percent of its maximum tension. However, only a few minutes of forceful exercise a day are required for hypertrophy to occur. The number of muscle fibers do not increase, instead the size of each fiber increases. The fibers show a gain in metabolic potential as well as in the number of myofibrils. This means that the muscle can work longer before it gets tired. Some athletes take steroids, either testosterone or related chemicals, to promote muscle growth. A possible means by which these hormones act is discussed on page 371.

When muscles are not used or used for only very weak contractions, they decrease in size, or **atrophy.** Atrophy can occur when a limb is placed in a cast or when the nerve serving a muscle is damaged. If nerve stimulation is not restored, the muscle fibers will gradually be replaced by fat and fibrous tissue. Unfortunately, atrophy causes the fibers to progressively shorten, leaving body parts in contorted positions.

Summary

The skeleton aids movement of the body while it also supports and protects the body. Bones serve as deposits for inorganic salts, and some bones are sites for blood-cell production. The skeleton is divided into two parts: (1) the axial skeleton, which is made up of the skull, ribs, and vertebrae; and (2) the appendicular skeleton, which is composed of the appendages and their girdles. A shoulder girdle consists of a clavicle and scapula. The arm contains the humerus, radius, and ulna. The wrist contains the carpals; the palm, the metacarpals; and the fingers, the phalanges.

Joints are regions where bones are joined together. Some joints are immovable, some are slightly movable, and others are freely movable. In the latter, the bones are separated by a cavity filled with synovial fluid and joined together by ligaments.

A long bone has a cavity bounded by spongy bone and compact bone. The spongy bone contains red marrow where blood cells are produced. The cavity itself is the location for yellow marrow. Growth of a long bone continues as long as the end cartilaginous disks are still present.

Muscles are attached to bones, usually on the far side of a joint, by means of tendons. Since muscles only pull and do not push, they work in antagonistic pairs. Studies of whole muscle contraction illustrate that a whole muscle has degrees of contraction dependent on how many muscle fibers contract at one time. Also, a myogram of a muscle twitch shows that it consists of a latent period, a period of contraction, and a period of relaxation. Similarly it is possible to observe summation and tetanic contractions that occur when a muscle does not relax between stimuli. Intact skeletal muscles exhibit tone, a partial contraction, at all times. Muscle spindles communicate with the CNS, allowing it to coordinate the contraction of fibers within a muscle.

A constant supply of ATP is required for muscle contraction. There are three ways by which ATP is made available: aerobic cellular respiration, transfer of a high-energy P from creatine phosphate to ADP, and fermentation. Following vigorous exercise, we continue to breathe deeply because oxygen is needed to metabolize lactic acid that has accumulated due to fermentation.

Muscle fibers are cells that contain the usual components of cells. In addition, they have some unique features. The cell membrane, called the sarcolemma, dips into the cell, forming the tubules of the T system that lie next to Ca^{++}-containing portions of the sarcoplasmic reticulum.

A longitudinal section of a muscle fiber shows that myofibrils are made up of functional units called sarcomeres. Within a sarcomere are the thin filaments of actin and the thick filaments of myosin, whose arrangement accounts for the striations of muscle fibers. The light regions at the sides of a sarcomere contain actin only; the central dark region contains myosin and actin filaments. When a sarcomere contracts, the actin filaments slide past the myosin filaments. Nevertheless it is myosin that does the work. Myosin extends cross bridges that attach to and pull the actin filaments along, as they first attach and then detach. ATP breakdown by myosin is necessary for detachment to occur.

Nerves innervate muscles, and nerve impulses cause them to contract. An axon gives off unmyelinated branches that terminate in knobs embedded in a muscle fiber. Here synaptic vesicles release Ach into the synaptic cleft of the neuromuscular junction. When the sarcolemma receives the Ach, a muscle action potential moves down the T system to the calcium-storage sacs. When calcium is released, it enters the sarcoplasm and attaches to troponin, a protein located on tropomyosin, a molecule that forms threads twisted about the actin filament. This causes the tropomyosin to shift its position so that myosin binding sites are exposed. Myosin crossbridges now bind to actin and the sarcomere contracts.

When nerve impulses and muscle action potentials cease, the calcium is actively transported back into the storage sacs. Muscle relaxation now occurs.

Study Questions

1. Distinguish between the axial and appendicular skeletons. (pp. 330–31)
2. List the bones that form the pectoral and the pelvic girdles. (p. 332)
3. Describe the anatomy of a freely movable joint; of a long bone. (p. 333)
4. Describe how muscles are attached to bones. (p. 337) Why do muscles act in antagonistic pairs? (p. 337)
5. The study of whole muscle physiology often includes observing both threshold and maximal stimulus, muscle twitch, summation, and tetanic contraction. Describe the significance of each of these. (p. 338)
6. How is the tone of a muscle maintained and how do muscle spindles contribute to the maintenance of tone? (p. 338)
7. Cite three ways in which ATP is made available for muscle contraction. (pp. 338–39)
8. What is oxygen debt and how is it repaid? (p. 339)
9. Discuss the microscopic anatomy of a muscle fiber and the structure of a sarcomere. What is the sliding filament theory? (pp. 340–42)
10. Give the function of each participant in the following reaction:

$$\text{actin} + \text{myosin} \xrightarrow[Ca^{++}]{\text{ATP} \longrightarrow \text{ADP} + \textcircled{P}} \text{actomyosin}$$

(p. 342)
11. What causes a muscle action potential? How does the muscle action potential bring about sarcomere and muscle fiber contraction? (pp. 342–44)

Selected Key Terms

vertebra (ver′tĕ-brah)
clavicle (klav′ĭ-k′l)
scapula (skap′u-lah)
humerus (hu′mer-us)
ulna (ul′nah)
radius (ra′de-us)
innominate (ĭ-nom′ĭ-nāt)
femur (fe′mur)
tibia (tib′e-ah)
fibula (fib′u-lah)
ligament (lig′ah-ment)
synovial fluid (sĭ-no′ve-al floo′id)
meniscus (mĕ-nis′kus)
compact bone (kom′pakt bōn)
spongy bone (spun′je bōn)
tendon (ten′don)
tone (tōn)
creatine phosphate (kre′ah-tin fos′fāt)
actin (ak′tin)
myosin (mi′o-sin)

Further Readings

Cohen, C. 1975. The protein switch of muscle contraction. *Scientific American* 233(5):36.

Hoyle, G. 1970. How is muscle turned on and off? *Scientific American* 222(4):84.

Huxley, H. E. 1962. The mechanism of muscular contraction. *Scientific American* 207(6):18.

Lester, H. A. 1977. The response to acetylcholine. *Scientific American* 236(2):106.

Margaria, R. 1972. The sources of muscular energy. *Scientific American* 226(5):83.

Merton, P. A. 1972. How we control the contraction of our muscles. *Scientific American* 226(5):30.

Murray, J. M., and Weber, A. 1974. The cooperative action of muscle proteins. *Scientific American* 230(2):58.

Porter, K. R., and Franzini-Armstrong, C. 1965. The sarcoplasmic reticulum. *Scientific American* 212(3):73.

17

Senses

Chapter concepts

1 Sense organs are sensitive to environmental stimuli and are therefore termed receptors. Each receptor responds to one type of stimulus, but they all initiate nerve impulses.

2 The sensation realized is the prerogative of the region of the cerebrum receiving nerve impulses.

3 There are sense receptors that respond to mechanical stimuli, chemical stimuli, and radiant energy. Our knowledge of the outside world is dependent on these stimuli.

Sense organs receive external and internal stimuli; therefore they are called **receptors.** Each type of receptor is sensitive to only one type of stimulus. Table 17.1 lists the receptors discussed in this chapter and the stimulus to which each reacts.

Receptors are the first component of the reflex arc that was described in chapter 15. When a receptor is stimulated, it generates nerve impulses that are transmitted to the spinal cord and/or brain; but only if the impulses reach the cerebrum are we conscious of a sensation. The sensory portion of the cerebrum can be mapped according to the parts of the body and the type of sensation realized at different loci (fig. 15.20).

General Receptors

Microscopic receptors (table 17.1) are present in the skin, visceral organs, and in the muscles and joints. They are all specialized nerve endings for the detection of touch, pressure, pain, temperature (hot and cold), and proprioception. Proprioception refers to the sense of knowing the position of the limbs; for example, if you close your eyes and move your arm about slowly, you still have a sense of where your arm is located.

Skin

The skin (fig. 17.1) contains receptors for touch, pressure, pain, and temperature. It is a mosaic of these tiny receptors, as you can determine by passing a metal probe slowly over the skin. At certain points there will be a feeling of pressure; at others, a feeling of hot or cold (depending on the temperature of the probe). Certain parts of the skin contain more receptors for a particular sensation; for example, the fingertips have an abundance of touch receptors.

A simple experiment suggests that temperature receptors are sensitive to the flow of heat. Fill three bowls with water—one cold, one warm, and one hot. Put your left hand in the cold water and your right in the hot water for a few moments. Your hands will adjust, or adapt, to these temperatures so that when you put both hands in warm water, each hand will indicate a different temperature of the water. Therefore, it seems that when the outside

Table 17.1	Receptors	
Receptors	**Sense**	**Stimulus**
General		
Temperature	Hot-cold	Heat flow*
Touch	Touch	Mechanical displacement of tissue†
Pressure	Pressure	Mechanical displacement of tissue†
Pain	Pain	Tissue damage‡
Proprioceptors	Limb placement	Mechanical displacement†
Special		
Eye	Sight	Light*
Ear	Hearing	Sound waves†
	Balance	Mechanical displacement†
Taste buds	Taste	Chemicals‡
Olfactory cells	Smell	Chemicals‡

*Radioreceptors
†Mechanoreceptors
‡Chemoreceptors

Figure 17.1
Receptors in human skin. Free nerve endings are
pain receptors; Pacinian corpuscles are pressure
receptors; Merkel's disks and Meissner's
corpuscles are touch receptors, as are the nerve
endings surrounding the hair follicle.

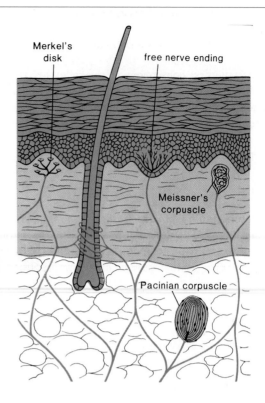

temperature is higher than the temperature to which we have adjusted, we
detect a sensation of warm or hot as heat flows into the skin. When the outside
temperature is low enough that heat flows out from the skin, we detect cool-
ness or cold.

Other skin receptors beside those for temperature also demonstrate adaptation. **Adaptation** occurs when the receptor becomes so accustomed to
stimuli that it stops generating impulses even though the stimulus is still present. The touch receptors are of this type. They can quickly adapt to the clothing we put on so that we are not constantly aware of the feel of the clothes
against our skin.

The receptors of the skin can be used to illustrate that sensation actually
occurs in the brain and not in the sense organ itself. If the nerve fiber from
the sense organ is cut, there is no sensation. Also, since a nerve impulse is
always the same electrochemical change, the particular sensation realized does
not have to do with the nerve impulse. It is the brain that is responsible for
the type of sensation felt and the localization of the sensation. For example,
if we connected a pain receptor in the foot to a nerve normally receiving impulses from a heat receptor in the hand, and then proceeded to stick the pain
receptor in the foot, the subject would report the feeling of warmth in the
hand. The brain indicates the sensation and the localization. This realization
is mildly disturbing, because it makes us aware of how dependent we are on
the anatomical wholeness of the body in order to be properly aware of our
surroundings.

Viscera

The internal organs have receptors that aid the maintenance of homeostatis.
We have already mentioned that stretch receptors in the lungs respond to lung

Human Anatomy and Physiology

expansion; the aortic and carotid bodies are sensitive to pH and oxygen levels of the blood, and the osmoreceptors in the hypothalamus detect blood osmolarity. We are unaware of the functioning of these receptors because the impulses they generate never reach the conscious levels of the brain.

The viscera have only one receptor that corresponds to those found in the skin. We are sensitive to pain arising from internal disorders. Sometimes this pain is felt as pain from the skin. This is called **referred pain** (fig. 17.2). It has been shown that the different internal organs have a more or less definite relationship to certain areas of the skin. Pain arising from the intestine is located in the skin of the back, groin, and abdomen; pain from the heart is felt in the left shoulder and arm. An explanation for this is that nerve stimuli from the pain receptors of the internal organs travels to the spinal cord where it makes contact with neurons also receiving messages from the skin. The brain interprets this as pain in the skin.

Acupuncture is a technique founded on the principle that needles inserted in the skin can relieve internal pain and promote healing. The exact mechanism by which acupuncture may be helpful is still under investigation. One hypothesis relies on the "gate-control theory" of pain, which states that transmission-of-pain signals from the body to the spinal cord and brain can be transmitted or blocked, depending on physiological conditions. Acupuncture needles may produce stimuli that block the transmission of pain sensations from internal organs. Or it could be that acupuncture causes the level of the natural opiates, endorphins and enkephalins, in the central nervous system to rise so that pain is not felt (fig. 15.25).

Muscles and Joints

The sense of position and movement of limbs (i.e., proprioception) is dependent upon receptors termed **proprioceptors.** Muscle spindles discussed earlier in chapter 14 are sometimes considered to be proprioceptors. Stretching of associated muscle fibers causes muscle spindles to increase the rate at which they fire and for this reason, they are sometimes called **stretch receptors.** The *knee jerk* is a common example of the manner in which muscle spindles act as stretch receptors (fig. 17.3). When the legs are crossed at the knee and the tendon at the knee is tapped, both the tendon and a muscle in the thigh are stretched. Stimulated by the stretching, muscle spindles transmit impulses to the spinal cord and thereafter the thigh muscle contracts. This causes the lower leg to jerk upward in a kicking motion.

There are proprioceptors located in the joints and associated ligaments and tendons that respond to stretching, pressure, and pain. Nerve endings from these receptors are integrated with those received from other types of receptors so that the person knows the position of body parts.

Special Senses

The special senses include the chemoreceptors for taste and smell, the light receptors for sight, and the mechanoreceptors for hearing and balance.

Chemoreceptors

Taste and smell are called the *chemical senses* because these receptors are sensitive to certain chemical substances in the food we eat and the air we breathe.

Taste buds are located on the tongue. Many lie along the walls of the papillae (figs. 17.4 and 17.5), the small elevations visible to the naked eye. Isolated ones are also present on the palate, pharynx, and epiglottis.

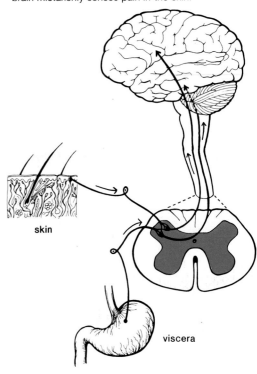

Figure 17.2
Referred pain. The receptors in the abdominal organs send impulses to a connector neuron that also receives impulses from the skin. Thereafter the brain mistakenly senses pain in the skin.

skin

viscera

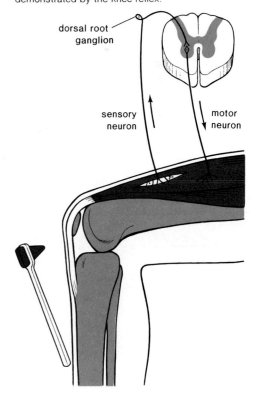

Figure 17.3
Proprioceptors in muscles and joints. Muscle spindle fibers are stretch receptors as can be demonstrated by the knee reflex.

dorsal root ganglion

sensory neuron

motor neuron

Figure 17.4

Taste buds. *a.* Elevations, called papillae, indicate the presence of taste buds. *b.* Drawing of a taste bud shows the various cells that make up a taste bud. Sensory cells in a taste bud end in microvilli that have receptors for chemicals in food. When the chemicals combine with the receptors, nerve impulses are generated.

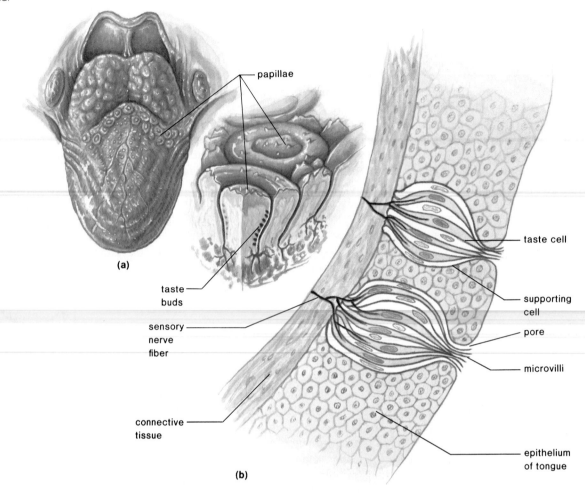

papillae

(a)

taste buds

sensory nerve fiber

connective tissue

(b)

taste cell

supporting cell

pore

microvilli

epithelium of tongue

Figure 17.5

Scanning electron micrograph of papillae of tongue. Cells of one papilla have been removed to reveal the presence of three taste buds.

From: TISSUES AND ORGANS: A TEXT-ATLAS OF SCANNING ELECTRON MICROSCOPY by R. G. Kessel and R. Kardon. W. H. Freeman and Company, © 1979.

Taste buds (fig. 17.4b) are pockets of cells that extend through the tongue epithelium and open at a taste pore. Within the oval pocket are supporting cells and a number of elongated cells that end in microvilli. These cells, which have associated nerve fibers, are sensitive to chemicals dissolved in the pore. Nerve impulses are most probably generated when the chemicals bind to receptor sites found on the microvilli.

The sense of taste has been shown to be genetically inherited, and foods taste differently to various people. This might very well account for the fact that some persons dislike a food that is preferred by others.

It is believed that there are four types of tastes (bitter, sour, salty, sweet) and that taste buds for each are concentrated on the tongue in particular regions. Sweet receptors are most plentiful near the tip of the tongue. Sour receptors occur primarily along the margins of the tongue. Salt receptors are most common on the tip and the upper front portion of the tongue. And bitter receptors are located toward the back of the tongue.

Human Anatomy and Physiology

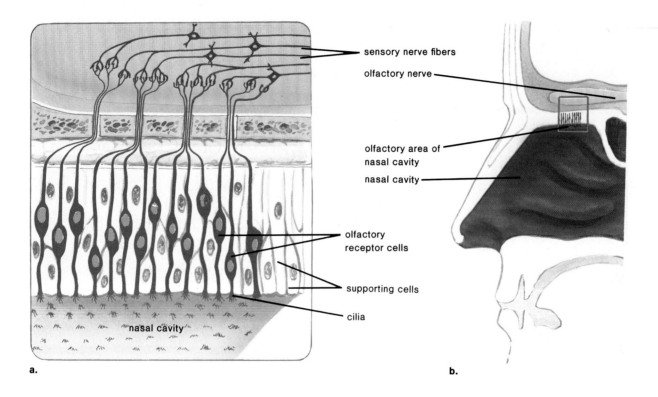

sensory nerve fibers

olfactory nerve

olfactory area of nasal cavity

nasal cavity

olfactory receptor cells

supporting cells

cilia

nasal cavity

a.   b.

The **olfactory cells** (fig. 17.6) are located high in the roof of the nasal cavity. These cells, which are specialized endings of the fibers that make up the olfactory nerve, lie among supporting epithelial cells. Each cell ends in a tuft of six to eight cilia that probably bear receptor sites for various chemicals. Research, resulting in the stereochemical theory of smell, suggests that different types of smell are related to the various shapes of molecules rather than to the atoms that make up the molecules. When chemicals combine with the receptor sites, nerve impulses are generated. The olfactory receptors, like the touch and temperature receptors, also adapt to outside stimuli. In other words, we can become accustomed to a smell and no longer take notice of it.

The sense of taste and the sense of smell supplement each other, creating a combined effect when interpreted by the cerebral cortex. For example, when we have a cold, we think that our food has lost its taste when actually we have lost the ability to sense its smell. This may work in the reverse also. When we smell something, some of the molecules move from the nose down into the mouth region and stimulate the taste buds there. Thus, part of what we refer to as smell may actually be taste. The reading on page 354 illustrates other ways in which taste and smell affect our everyday life.

Enough of Atkins, Stillman and Scarsdale. It may be that the Boring Diet works as well—or better. Scientists have long reasoned that if good taste and smell can increase appetite, terrible taste and odor, or one flavor eaten over and over, should be boring enough to decrease it. Last week, at an international conference on "The Determination of Behavior by Chemical Stimuli," a pair of biologists reported findings suggesting that any tedious diet helps weight loss. If it were possible to eat one food all the time, according to Israeli Nutritional Biochemist Michael Naim, all but the genetically obese would quickly shed pounds.

At Jerusalem's Hebrew University, Naim and French Neurobiologist Jacques Le Magnen spoke to a gathering sponsored by the European Chemoreception Research Organization, joining some two dozen other scholars who reported on such topics as the sniffing power of infants, the sex life of guinea pigs and a three-nation T shirt-smelling study.

Le Magnen's work shows that when their diet is monotonous, rats eat only what they need, but overeat and become fat when fed a different flavor every 30 minutes. At the conference, he reported that the taste and smell of the food encouraged hunger and obesity by causing a reflexive increase in insulin. Le Magnen also reported evidence, in both rats and humans, that each new tasty food produces a conditioned insulin release. In other words, even if a varied meal and a one-food meal are equal in size and good taste, the varied meal may prove more fattening because it increases appetite. According to Naim, the best way for humans to lose weight may be to adopt a varied and revolting diet: food that is too sweet one day, too salty the next, or alternately bitter and sour.

Hebrew University Physiologist Jacob Steiner told the scientists that all tastes and smells accelerate the heart rates of adolescents. A sweet taste speeds the heart by 2% or 3%, bitter and sour tastes race the pulse 17% to

Photoreceptor—the Eye

The eye (fig. 17.7), an elongated sphere about 1 inch in diameter, has three layers or coats. The outer **sclera** is a white fibrous layer except for the transparent cornea, the window of the eye. The middle thin, dark brown layer, the **choroid,** contains many blood vessels and absorbs stray light rays. Toward the front, the choroid thickens and forms a ring-shaped structure, the ciliary body, containing the **ciliary muscle,** which controls the shape of the lens for near and far vision. Finally the choroid becomes a thin, circular, muscular diaphragm, the **iris,** which regulates the size of the pupil. The **lens,** attached to the ciliary body by ligaments, divides the cavity of the eye into two chambers. A viscous and gelatinous material, the **vitreous humor,** fills the large cavity behind the lens. The chamber between the cornea and the lens is filled with an alkaline, watery solution secreted by the ciliary body and called the **aqueous humor.**

Glaucoma   A small amount of aqueous humor is continually produced each day. Normally it leaves the anterior chamber by way of tiny ducts that are located where the iris meets the cornea. If these drainage ducts are blocked, pressure rises and presses on capillaries that feed nerve fibers located in the retina. With the passage of time, some of these fibers die and the result is partial or total blindness.

20% faster. Steiner, in long-term test studies of infants, discovered that first reactions to smells are inborn, not acquired. Newborns react positively to pleasant odors and screw up their faces in response to unpleasant ones, even before they have tasted any food at all.

Psychobiologist Gary Beauchamp of Philadelphia reported that the odor of female guinea pig urine is such a powerful stimulus to the male that it loses interest in mating if its sense of smell is impaired. When Beauchamp removed the male's vomeronasal organ, which relays odor information to the brain, sexual activity declined. In the wild, after the removal of the vomeronasal organ, even guinea pigs near their mates sometimes cannot find them.

The T shirt tests, conducted by West German Biologists Margret Schleidt and Barbara Hold, showed that men and women blindfolded could identify perspiration odors of their mates. Schleidt tested 75 couples in West Germany, Italy and Japan, asking them to wear cotton T shirts to bed for a week and avoid using perfumes or deodorants. In all three sets of tests, results were the same: subjects were generally able to sniff out the shirts worn by their mates, and both men and women considered male odors more unpleasant than female odors. But when women selected the shirts they thought belonged to their husbands, only the Japanese women labeled the odors more unpleasant than their own. Why should women of Japan be harsher on their husbands? Offering an unscientific but provocative hypothesis, Schleidt replied: "One reason may be that in Japan women's marriages are usually arranged."

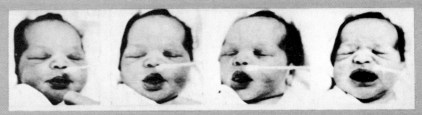

Baby, before first feeding, reacts to smell of (1) banana, (2) fish, (3) butter, (4) rotten eggs.

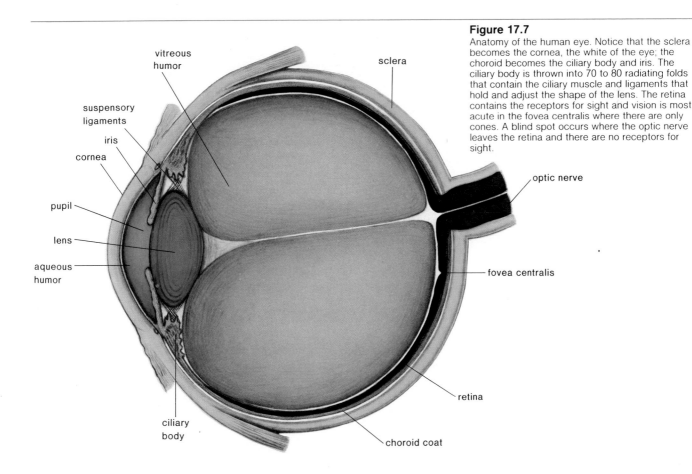

Figure 17.7
Anatomy of the human eye. Notice that the sclera becomes the cornea, the white of the eye; the choroid becomes the ciliary body and iris. The ciliary body is thrown into 70 to 80 radiating folds that contain the ciliary muscle and ligaments that hold and adjust the shape of the lens. The retina contains the receptors for sight and vision is most acute in the fovea centralis where there are only cones. A blind spot occurs where the optic nerve leaves the retina and there are no receptors for sight.

Figure 17.8
Structure of retina. Rods and cones are located toward the back of the retina, followed by the bipolar cells and the ganglionic cells whose fibers become the optic nerve.

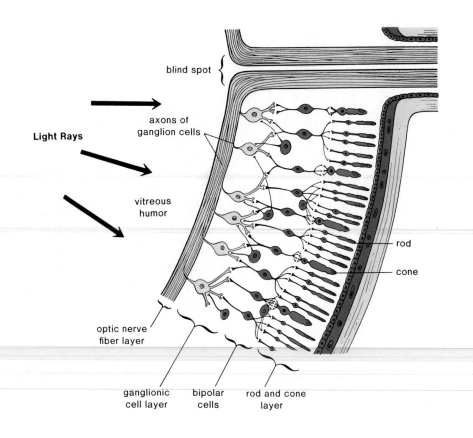

blind spot

Light Rays

axons of
ganglion cells

vitreous
humor

rod

cone

optic nerve
fiber layer

ganglionic
cell layer

bipolar
cells

rod and cone
layer

Table 17.2 Name and function of parts of the eye

Part	Function
Lens	Refraction and focusing
Iris	Regulates light entrance
Pupil	Hole in iris
Choroid	Absorbs stray light
Sclera	Protection
Cornea	Refraction of light
Humors	Refraction of light
Ciliary body	Holds lens in place
Retina	Contains receptors
Rods	Black-and-white vision
Cones	Color vision
Optic nerve	Transmits impulse
Fovea	Region of cones in retina
Ciliary muscle	Accommodation

Retina
The inner layer of the eye, the **retina,** has three layers of cells (fig. 17.8). The layer closest to the choroid contains the sense receptors for sight, the **rods** and **cones** (fig. 17.9); the middle layer contains bipolar cells; and the innermost layer contains ganglion cells whose fibers become the **optic nerve.** Only the rods and cones contain light-sensitive pigments and thus light must penetrate to the back of the retina before nerve impulses are generated. Nerve impulses initiated by the rods and cones are passed to the bipolar cells, which in turn pass them to the ganglion cells whose fibers pass in front of the retina, forming the optic nerve, which turns to pierce the layers of the eye. The presence of these three layers of nerve cells permits a certain amount of integration before nerve impulses are sent to the brain. There are no rods and cones where the optic nerve passes through the retina; therefore, this is a **blind spot** where vision is impossible.

The retina contains a very special region called the **fovea centralis,** an oval yellowish area with a depression where there are only cone cells. Vision is most acute in the fovea centralis. Table 17.2 lists the parts of the eye and their functions.

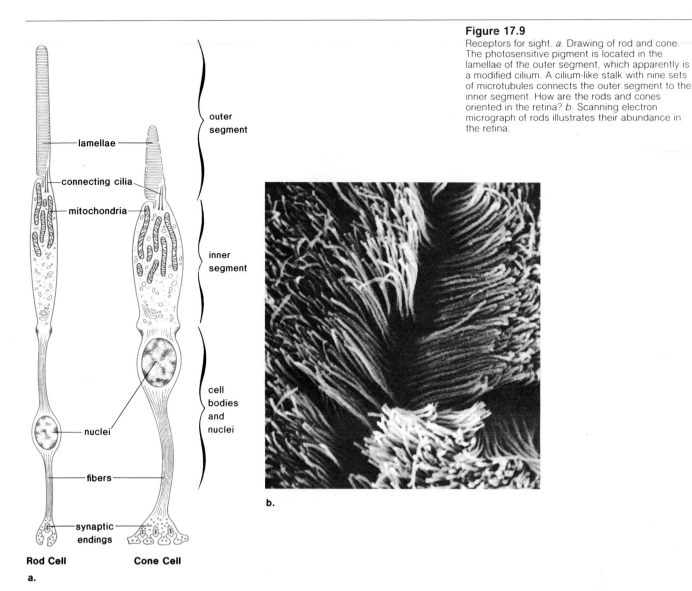

outer segment

lamellae

connecting cilia

mitochondria

inner segment

nuclei

cell bodies and nuclei

fibers

synaptic endings

Rod Cell   **Cone Cell**

a.

b.

Physiology

Focusing   When we look at an object, light rays are focused on the retina (fig. 17.10). In this way, an **image** of the object appears on the retina. The image on the retina occurs when the rods and cones in a particular region are excited. Obviously, the image is much smaller than the object. In order to produce this small image, light rays must be bent (refracted) and brought to a focus. They are bent as they pass through the cornea. Further bending occurs as the rays pass through the lens and humors.

Accommodation   Light rays are reflected from an object in all directions. If the eye is distant from an object, only nearly parallel rays enter the eye and the cornea alone is needed for focusing. But if the eye is close to the object, many of the rays are at sharp angles to one another and additional focusing is required. The lens provides this additional focusing power. While the lens

Figure 17.10
Focusing. Light rays from each point on an object are bent by the cornea in such a way that they are directed to a single point after emerging from the lens. By this process an inverted image of the object forms on the retina.

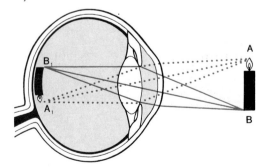

Figure 17.11

Accommodation. *a.* When the eye focuses on a far object, the lens is flat because the ciliary muscle is relaxed and the suspensory ligament is taut.
b. When the eye focuses on a near object, the lens rounds up because the ciliary muscle contracts, causing the suspensory ligament to relax.

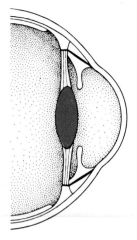

**a. Normal
Distant Focus**

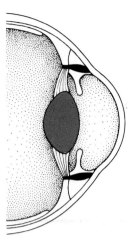

b. Near Focus

Table 17.3	Accommodation	
Object	**Ciliary muscle**	**Lens**
Near object	Ciliary muscle contracts, ligaments relax	Lens becomes round
Far object	Ciliary muscle relaxes, ligaments under tension	Lens is flattened

remains flat when we view distant objects, it rounds up when we view close objects. When rounded, the lens provides the additional refraction required to bring the diverging light rays to a sharp focus on the retina (fig. 17.11). The shape of the lens is controlled by the ciliary muscle within the ciliary body. When we view a distant object, the ciliary muscle is relaxed, causing the ligaments attached to the ciliary body to be under tension; therefore the lens remains relatively flat. When we view a close object, the ciliary muscle contracts, releasing the tension on the ligaments and, therefore, the lens rounds up due to its natural elasticity (table 17.3). Since close work requires contraction of the ciliary muscle, it very often causes "eye strain."

With aging, the lens loses some of its elasticity and is unable to accommodate in order to bring close objects into focus. This usually necessitates the wearing of glasses, as is discussed on page 361. The lens is also subject to **cataracts;** it can become cloudy and opaque and unable to transmit rays of light. Special cells within the interior of the lens contain proteins called crystallin. Recent research suggests that cataracts are brought on when these proteins become oxidized, causing their three-dimensional shape to change. If so, researchers believe that they may eventually be able to find ways to restore the normal configuration of crystallin so that cataracts can be treated medically instead of surgically.

Inverted Image   The image on the retina is upside down, and it is thought that perhaps by experience this image is righted in the brain. In one experiment, scientists wore glasses that inverted the field. At first, they had difficulty adjusting to the placement of the objects, but then they soon became accustomed to their inverted world. Experiments such as this suggest that if we see the world upside down, the brain learns to see it right side up.

Human Anatomy and Physiology

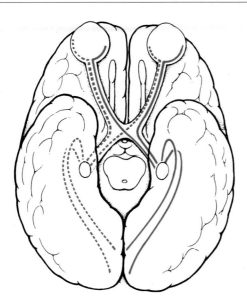

Figure 17.12
Optic chiasma. Solid lines indicate that nerve impulses from the right half of each retina go to the right visual cortex and the dotted lines indicate that the nerve impulses from the left half of each retina go to the left visual cortex. This is due to a crossing over that occurs in the optic chiasma.

Stereoscopic Vision   We can see well with either eye alone, but the two eyes functioning together provide us with stereoscopic vision. Normally, the two eyes are directed by the eye muscles toward the same object and therefore the object is focused on corresponding points of the two retinas. But each eye sends to the brain its own information about the placement of the object because each forms an image from a slightly different angle. These data are pooled to produce depth perception by a two-step process. First, because the optic nerves cross at the optic chiasma (fig. 17.12), one-half of the brain receives information from both eyes about the same part of an object. Later, the two halves of the brain communicate to arrive at a complete three-dimensional interpretation of the whole object.

Biochemistry   In *dim light,* the pupils enlarge so that more rays of light can enter the eyes. As the rays of light enter, they strike the rods and cones, but only the 160 million rods located in the periphery, or sides, of the eyes are sensitive enough to be stimulated by this faint light. The rods do not detect fine detail or color, so at night, for example, all objects appear to be blurred and have a shade of gray. Rods do detect even the slightest motion, however, because of their abundance (fig. 17.9b) and position in the eyes.

The rods contain **rhodopsin,** a pigment called visual purple. When light strikes rhodopsin it breaks down into a protein, **opsin,** and the pigment portion of the molecule, **retinal.** This leads to a depolarization of the rod cells and a release of transmitter substance from the rod cells so that nerve impulses are generated. When the eye is exposed to a flash of light, the stimulus generated lasts one-tenth of a second. This is why we continue to see an image if we close our eyes immediately after looking at an object. It also allows us to see motion if still frames are presented at a rapid rate, as in "movies."

The more rhodopsin present in the rods, the more sensitive are our eyes to dim light. Therefore, during the time required for adaptation (adjustment) to dim light when we find it difficult to see, rhodopsin is being formed in the rods.

As figure 17.13 shows, retinal eventually breaks down to vitamin A. Vitamin A is abundant in carrots, thus the notion that we all should eat carrots for good vision is not without foundation. Actually, most of the vitamin A is transformed back to retinal, which then combines in the dark with opsin to again form rhodopsin.

Figure 17.13
Biochemistry of vision. Light causes the chemical rhodopsin to break down into its component parts, retinal and opsin. This initiates the nerve impulse. In the dark, the pigment retinal, for which vitamin A is a precursor, and the protein opsin recombine to form rhodopsin.

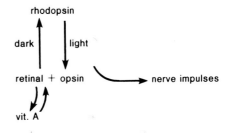

Figure 17.14
Scanning electron micrograph of rods and cones.
The cones in the foreground are responsible for
color vision and the rods in the background are
responsible for night vision.

In *bright light* the pupils get smaller, so that less light enters the eyes. The cones (fig. 17.14), located primarily in the fovea, are active and detect the fine detail and color of an object. In order to perceive depth, as well as to see color, we turn our eyes so that reflected light from the object strikes the fovea.

To understand color vision, it is necessary to know something about light. Light is radiant energy, which, like other forms of radiant energy, travels through the air as waves. The different kinds of radiant energy may be arranged in a series—those with the shortest wavelengths at one end and those with the longest wavelengths at the other. This series is called the electromagnetic spectrum (p. 121). Visible white light comprises only a small portion of the spectrum and can itself be divided into wavelengths of different sizes and colors. When all the wavelengths of the visible spectrum enter the eye at the same time in nearly equal quantities, we see the light as colorless or white. This is why sunlight or the light from an electric lamp appears colorless. When white light is passed through a prism, it is broken up into its constituent colors (p. 121). Raindrops in the air may act as prisms. Sunlight passing through them is separated into its component wavelengths, and a rainbow occurs.

An object has color when it absorbs some wavelengths but not others. A black object absorbs all wavelengths, a white object absorbs none, and a red object absorbs all colors but red. Red wavelengths, for example, are reflected from red apples to our eyes, and therefore we see them as red. But researchers have found that the clarity of the color depends on contrast: apples will appear more red when they are on a green tree. This can be related to the *opponent-process theory of color vision.* Color vision has been shown to depend on three kinds of cones that contain pigments sensitive to either blue, green, or red light. The nerve impulses generated from one type cone not only stimulate certain cells in the visual cortex of the brain, they also inhibit the reception of impulses from other type cones. For example, when we see red, certain cells in the brain are prohibited from receiving impulses from green cones. Similarly, impulses sent through blue cones tend to oppose the combination of signals sent by red and green cones—which together produce yellow. This process assists integration and enables the brain to tell the location of various colors in the environment.

Complete **color blindness** is extremely rare. In most instances a particular type of cone is lacking or deficient in number. The lack of red or green cones is the most common, affecting about 5 percent of the American population. If the eye lacks red cones, the green colors become accentuated, and vice versa.

Glasses

The majority of people can see what is designated as a size "20" letter 20 feet away and thus are said to have 20/20 vision. Persons who can see close objects but cannot see the letters from this distance are said to be nearsighted. Nearsighted persons can see near better than they can see far. These individuals often have an elongated eyeball and when they attempt to look at a far object, the image is brought to focus in front of the retina (fig. 17.15). They can see near because they can adjust the lens to allow the image to focus on the retina; but to see far, these people must wear concave lenses that diverge the light rays so that the image can be focused on the retina.

Persons who can easily see the optometrist's chart but cannot see close objects well are farsighted; these individuals can see far away better than they can see near. They often have a shortened eyeball, and when they try to see near, the image is focused behind the retina. When the object is far away, the lens can compensate for the short eyeball, but when the object is close, these persons must wear a convex lens to increase the bending of light rays so that the image will be focused on the retina.

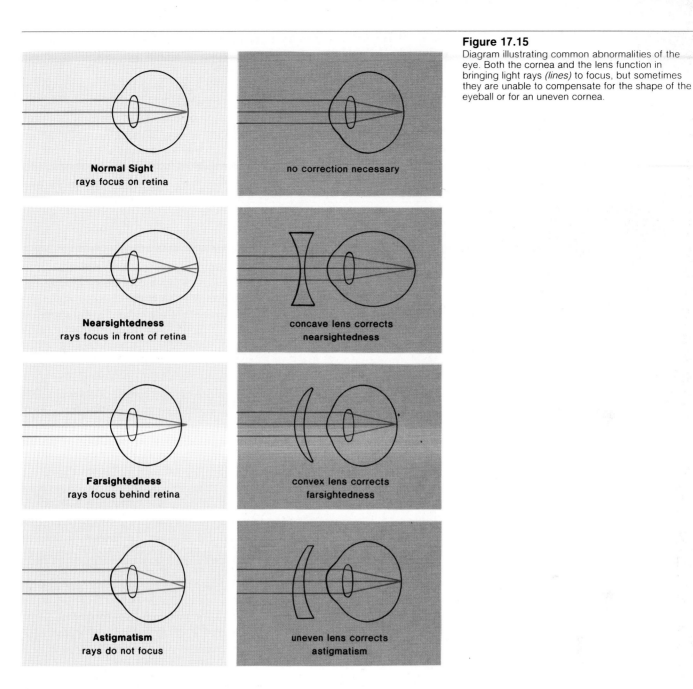

Figure 17.15
Diagram illustrating common abnormalities of the eye. Both the cornea and the lens function in bringing light rays *(lines)* to focus, but sometimes they are unable to compensate for the shape of the eyeball or for an uneven cornea.

Normal Sight
rays focus on retina

no correction necessary

Nearsightedness
rays focus in front of retina

concave lens corrects nearsightedness

Farsightedness
rays focus behind retina

convex lens corrects farsightedness

Astigmatism
rays do not focus

uneven lens corrects astigmatism

Table 17.4   Common abnormalities of the eye

Name	Effect	Fault	Result	Correction
Nearsighted	Can't see far, can see near	Long eyeball	Image focused in front of retina	Concave lens
Farsighted	Can't see near, can see far	Short eyeball	Image focused behind retina	Convex lens
Astigmatism	Can't focus	Irregular eyeball	Image not focused	Irregular lens

When the cornea or lens is uneven, the image is fuzzy because the light rays cannot be focused evenly on the retina. This fault can be corrected by an unevenly ground lens to compensate for the uneven cornea. Table 17.4 summarizes these conditions and their corrections.

Table 17.5   The ear

	Outer ear	Middle ear	Inner ear	
			Cochlea	Sacs plus semicircular canals
Function	Directs sound waves to tympanic membrane	Picks up and amplifies sound waves	Hearing	Maintains equilibrium
Anatomy	Pinna Auditory canal	Tympanic membrane Ossicles	Contains organ of Corti Auditory nerve starts here	Saccule and utricle Semicircular canals
Media	Air	Air (eustachian tube)	Fluid	Fluid

Path of vibration: Sound waves—vibration of tympanic membrane—vibration of hammer, anvil, and stirrup—vibration of oval window—fluid pressure waves of fluids in canals of inner ear lead to stimulation of hair cells—bulging of round window.

Bifocals   As mentioned earlier, with normal aging, the lens loses some of its ability to change shape in order to focus on close objects. Since nearsighted individuals still have difficulty seeing objects clearly in the distance, they must wear bifocals, which means that the upper part of the lens is for distant vision and the remainder for near vision.

Mechanoreceptor—the Ear

The ear accomplishes two sensory functions: balance and hearing. The sense cells for both of these are located in the inner ear and consist of hair cells with cilia that respond to mechanical stimulation. Each hair cell has from 30 to 150 extensions that are called cilia despite the fact that they contain tightly packed filaments rather than microtubules.[1] When the cilia of any particular hair cell are displaced in a certain direction, the cell generates nerve impulses that are sent along a cranial nerve to the brain.

Anatomy

Table 17.5 lists the parts of the ear, and figure 17.16 is a drawing of the ear. The ear has three divisions: outer, middle, and inner. The **outer ear** consists of the **pinna** (external flap) and **auditory canal.** The opening of the auditory canal is lined with fine hairs and sweat glands. In the upper wall are modified sweat glands that secrete earwax, a substance that helps guard the ear against the entrance of foreign materials such as air pollutants.

The **middle ear** begins at the **tympanic membrane** (eardrum) and ends at a bony wall in which are found two small openings covered by membranes. These openings are called the **oval** and **round windows.** The posterior wall of the middle ear leads to many air spaces within the **mastoid process.**

Three small bones are found between the tympanic membrane and the oval window. Collectively called the **ossicles,** individually they are the **hammer** (malleus), **anvil** (incus), and **stirrup** (stapes) (fig. 17.16) because their shapes resemble these objects. The hammer adheres to the tympanic membrane, while the stirrup touches the oval window.

The eustachian tubes extend from the middle ear to the nasopharynx and permit equalization of air pressure. Chewing gum, yawning, and swallowing in elevators and airplanes helps move air through the eustachian tubes upon ascent and descent.

[1]There is usually one extension that does have the structure of a true cilium, but apparently it need not be present for the hair cell to function properly.

Human Anatomy and Physiology

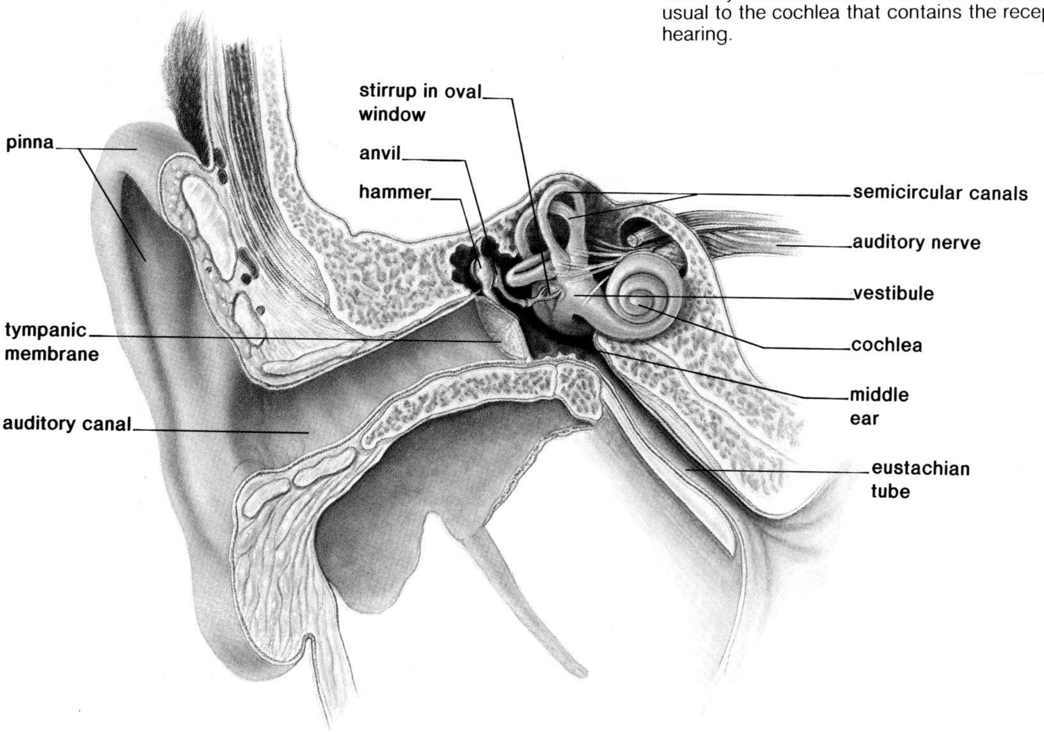

Figure 17.16
Anatomy of the human ear. In the middle ear, the hammer, anvil, and stirrup amplify sound waves. Otosclerosis is a condition in which the stirrup becomes attached to the inner ear and unable to carry out its normal function. It can be replaced by a plastic piston and thereafter the individual hears normally because sound waves are transmitted as usual to the cochlea that contains the receptors for hearing.

Whereas the outer ear and middle ear contain air, the inner ear is filled with fluid. The **inner ear** (fig. 17.17a), anatomically speaking, has three areas: the first two, called the vestibule and semicircular canals, are concerned with balance; and the third, the cochlea, is concerned with hearing.

The **semicircular canals** are arranged so that there is one in each dimension of space. The base of each canal, called the **ampulla,** is slightly enlarged. Within the ampullae are little hair cells (fig. 17.17b) whose cilia are inserted into a gelatinous medium.

The **vestibule** is a chamber that lies between the semicircular canals and the cochlea. It contains two small sacs called the **utricle** and **saccule.** Within both of these are little hair cells whose cilia protrude into a gelatinous substance. Resting on this substance are calcium carbonate granules, or **otoliths** (fig. 17.17c).

The **cochlea** (fig. 17.17a) resembles the shell of a snail because it spirals. Within the tubular cochlea are three canals: the vestibular, the **cochlear canal,** and the tympanic canal. Along the length of the basilar membrane, which forms the lower wall of the cochlear canal, are little hair cells whose cilia come into contact with another membrane called the **tectorial membrane.** The hair cells plus the tectorial membrane are called the **organ of Corti** (fig. 17.17d). When this organ sends nerve impulses to the cerebral cortex, it is interpreted as sound.

Figure 17.17

The inner ear in detail. *a.* The entire inner ear. The regions that are drawn in detail to the side and below are indicated in black. *b.* Ampullae of the semicircular canals contains hair cells with cilia that bend as the fluid moves in the canals. *c.* Utricle and saccule contain hair cells with cilia that bend when the body moves horizontally (utricle) or vertically (saccule). *d.* Organ of Corti contains hair cells with cilia that touch the tectorial membrane. When the basilar membrane moves upward the cilia are bent.

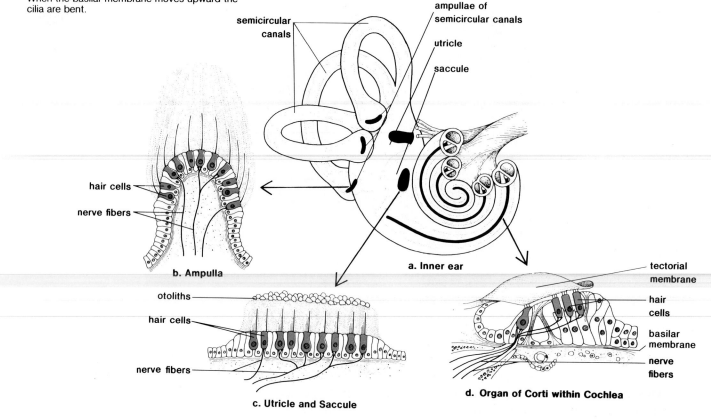

b. Ampulla

c. Utricle and Saccule

a. Inner ear

d. Organ of Corti within Cochlea

Figure 17.18

Diagram of cochlea unwound. Refer to figures 17.16 and 17.17 for the placement of the cochlea. The arrows represent the pressure waves that move from the oval window to the round window. These cause the basilar membrane to vibrate and the cilia of at least a portion of the 15,000 hair cells to bend against the tectorial membrane. The resulting nerve impulses result in hearing.

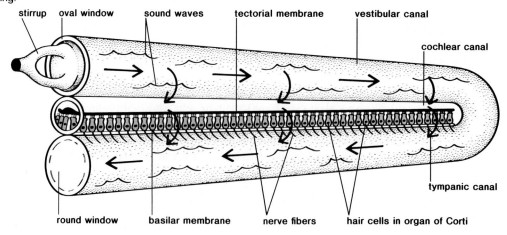

Human Anatomy and Physiology

Physiology

Balance (Equilibrium)   The sense of balance has been divided into two senses: *static equilibrium,* referring to knowledge of movement in one plane, either verticle or horizontal, and *dynamic equilibrium,* referring to knowledge of angular and/or rotational movement.

When the body is still, the otoliths in the utricle and saccule rest on the hair cells. When the body moves horizontally, the granules in the utricle are displaced, and when the body moves vertically, the granules in the saccule are displaced. Displacement causes the cilia to bend slightly so that the cell generates nerve impulses that travel by way of a cranial nerve to the brain.

When the body is moving about, the fluid within the semicircular canals moves back and forth. This causes bending of the cilia attached to hair cells within the ampullae, and they initiate nerve impulses that travel to the brain. Consistent movement of the fluid in the semicircular canals causes one form of motion sickness.

Hearing   The process of hearing begins when sound waves enter the auditory canal. Just as ripples travel across the surface of a pond, sound travels by the successive vibrations of molecules. The pitch of a sound is determined by the frequency of the vibrations, or cycles. The human ear is able to hear sounds from a range of about 20 to 20,000 cycles per second (cps) if the loudness is sufficient. For example, most of the noises made by our internal organs are inaudible because the intensity is insufficient. Otherwise we would always hear the beating of our heart and the rumbling of our stomach. Loudness depends upon the amplitude of the sound waves. It is customary to use the decibel to compare the loudness of sounds. A sound of zero decibels (db) is the faintest audible sound, and a sound of 140 db is loud enough to be painful. Rock music may be as loud as 130 db and a jet plane at takeoff is estimated at 150 db.

Ordinarily, sound waves do not carry much energy, but when a large number of waves strike the eardrum, it moves back and forth (vibrates) ever so slightly. The hammer then takes the pressure from the inner surface of the eardrum and passes it by way of the anvil to the stirrup in such a way that the pressure is multiplied about 20 times as it moves from the eardrum to the stirrup. The stirrup strikes the oval window, causing it to vibrate and in this way the pressure is passed to the fluid within the inner ear.

If the cochlea is unwound, as shown in figure 17.18, the vestibular canal is seen to connect with the tympanic canal; thus, as the figure indicates, pressure waves move from one canal to the other toward the round window, a membrane that can bulge to absorb the pressure. As a result of the movement of the fluid within the cochlea, the basilar membrane moves up and down, and the cilia of the hair cells rub against the tectorial membrane. This bending of the cilia initiates nerve impulses that pass by way of the auditory nerve to the brain, where the impulses are interpreted as a sound.

The organ of Corti is narrow at its base but widens as it approaches the tip of the cochlear canal. Each part of the organ is sensitive to different wave frequencies, or pitch. Near the apex, the organ of Corti responds to low pitches such as a tuba, and near the base it responds to higher pitches such as a bell or whistle. The neurons from each region along the length of the cochlea lead to slightly different areas in the brain. The pitch sensation we experience depends upon which of these areas of the brain is stimulated. Volume is a function of the amplitude of sound waves. Loud noises cause the fluid of the cochlea to oscillate to a greater degree, and this, in turn, causes the basilar membrane to move up and down to a great extent. The resulting increased stimulation is interpreted by the brain as loudness. It is believed that tone is an interpretation of the brain based on the distribution of hair cells stimulated.

Figure 17.19
Damage to organ of Corti due to loud noise. *a.* Normal organ of Corti. *b.* Organ of Corti after 24-hour exposure to a noise level typical of rock music. Note scars where cilia have been worn away.

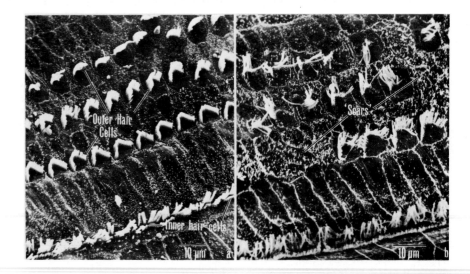

Deafness   There are two major types of deafness: *conduction* and *nerve deafness*. Conduction deafness can be due to a congenital defect, as those that occur when a pregnant woman contracts German measles during the first trimester of pregnancy. (For this reason every female should be sure to be immunized against rubella before the childbearing years.) Or, conduction deafness can be due to infections that have caused the ossicles to fuse together, restricting the ability to magnify sound waves. As was mentioned in chapter 13, respiratory infections can spread to the ear by way of the eustachian tubes; therefore, every cold and ear infection should be taken seriously.

Nerve deafness most often occurs when cilia on the sense receptors within the cochlea have worn away (fig. 17.19). Since this may happen with normal aging, old people are more likely to have trouble hearing; however, nerve deafness also occurs when young people listen to loud music amplified to 130 decibels. Because the usual types of hearing aids are not helpful for nerve deafness, it is wise to avoid subjecting the ears to any type of continuous loud noise. Costly cochlear implants that directly stimulate the auditory nerve are available. Those who have these electronic devices report that the speech they hear is like that of a robot.

Summary

All receptors are the first part of a reflex arc. Each type is sensitive to a particular kind of stimulus in the external or internal environment. When stimulation occurs, receptors initiate nerve impulses that are transmitted to the spinal cord and/or brain. Only when nerve impulses reach the cerebrum are we conscious of sensation. The cerebrum can be mapped according to the type of sensation felt and the localization of sensation.

The sense organs can be divided into the general senses and the special senses. The skin contains receptors for touch, pressure, temperature, and pain. The viscera contain pain receptors sparsely distributed. Because the impulses from these receptors pass to the same neurons in the spinal cord as do those

Further Readings

Amoore, J. E. et al. 1964. The stereochemical theory of odor. *Scientific American* 210(2):28.

Favreau, O. E., and Corballis, M. C. 1976. Negative aftereffects in visual perception. *Scientific American* 235(6):42.

Glickstein, M., and Gibson, A. R. 1976. Visual cells in the pons of the brain. *Scientific American* 235(5):90.

Hudspeth, A. J. 1983. The hair cells of the inner ear. *Scientific American* 248(1):54.

Johansson, G. 1975. Visual motion perception. *Scientific American* 232(6):76.

Land, E. H. 1977. The retinex theory of color vision. *Scientific American* 237(6):108.

MacNichol, E. F., Jr. 1964. Three-pigment color vision. *Scientific American* 211(6):48.

Neisser, U. 1968. The processes of vision. *Scientific American* 219(3):204.

Pettigrew, J. D. 1972. The neurophysiology of binocular vision. *Scientific American* 227(2):13.

Regan, et al. 1979. The visual perception of motion in depth. *Scientific American* 241(1):136.

Ross, J. 1976. The resources of binocular perception. *Scientific American* 234(3):80.

Rushton, W. A. H. 1975. Visual pigments and color blindness. *Scientific American* 232(3):64.

van Heyningen, R. 1975. What happens to the human lens in cataract? *Scientific American* 233(6):70.

Wolfe, J. M. 1983. Hidden visual processes. *Scientific American* 248(2):94.

Young, R. W. 1970. Visual cells. *Scientific American* 223(4):80.

18

Hormones

Chapter concepts

1 The hormonal glands usually are ductless glands that secrete directly into the bloodstream.

2 Hormonal secretions coordinate the biochemical functioning of the body by acting on target organs.

3 In general, hormonal glands are controlled by a negative feedback mechanism.

4 Malfunctioning of hormonal glands can bring about a dramatic change in appearance and can cause early death.

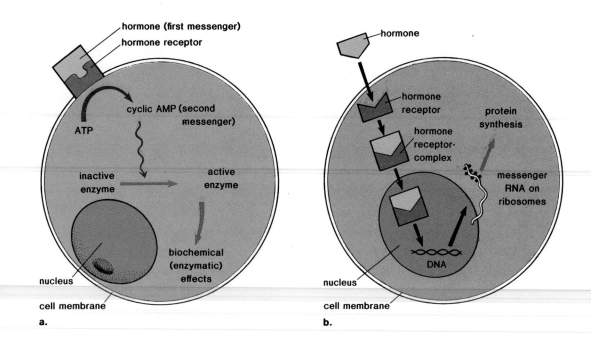

A long with the nervous system, hormones coordinate the functioning of body parts. Their presence or absence affects our metabolism, our appearance, and our behavior. It is now known that hormones are chemical regulators of cellular activity.

Hormones are organic substances that fall into two basic categories: (1) amino acids, polypeptides, or proteins and (2) steroid hormones. Steroids are complex rings of carbon and hydrogen atoms (fig. 1.32). The difference between steroids is due to the atoms attached to these rings.

When hormones of the first type are received by a cell, they bind to specific receptor sites (fig. 18.1a) in the membrane. This hormone-receptor complex activates an enzyme that produces **cAMP** (cyclic adenosine monophosphate), a compound made from ATP, that contains only one phosphate group attached to adenine at two locations. The cAMP now activates the enzymes of the cell to carry out their normal functions.

Unlike peptide hormones, steroid hormones pass through the cell membrane with no difficulty because they are relatively small and lipid soluble. There are receptor molecules present in the cytoplasm (fig. 18.1b). After the hormone has combined with the receptor, the hormone-receptor complex moves into the nucleus where it binds with chromatin at a location that promotes activation of a particular gene. Protein synthesis follows. In this manner, steroid hormones lead to protein synthesis.

Figure 18.2

Origination of hormones. *a.* The substances we call hormones today may have at one time been chemical messengers between like cells. Chemical messengers are sent between like tissue cells (1), including nerve cells (2). The chemical messengers that are sent between nerve cells include neurotransmitter substances. *b.* In vertebrates, hormones are produced by specialized glands and are distributed to target organs by way of the circulatory system. Tissue cells (1), including nerve cells (2), produce hormones. For example, the hypothalamus produces hormones that control the anterior pituitary.

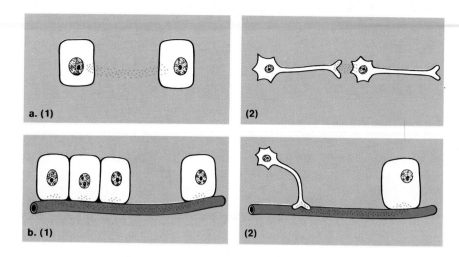

Figure 18.3

Anatomical location of major endocrine glands in the body. The hypothalamus controls the pituitary, which in turn controls the hormonal secretions of the thyroid, adrenal cortex, and sex organs. Both sets of sex organs are shown; ordinarily an individual has only one set of these.

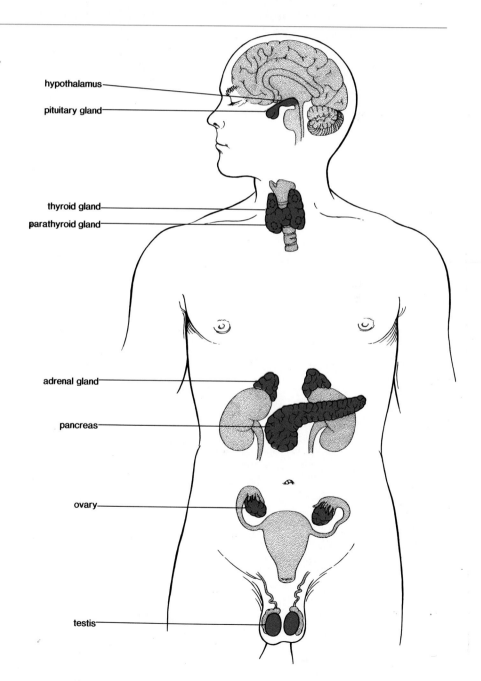

Table 18.1 The principal endocrine glands and their hormones

Gland	Hormones	Chief functions	Disorders too much/too little
Anterior pituitary	Thyroid-stimulating (TSH, thyrotropin)	Stimulates thyroid	*See* thyroid
	Adrenocorticotropic (ACTH)	Stimulates adrenal cortex	*See* adrenal cortex
	Gonadotropic	Stimulates gonads	*See* gonads
	Follicle-stimulating (FSH)	Egg and sperm	
	Leuteinizing (LH)	Sex hormones	
	Lactogenic (LTH, prolactin)	Milk production	
	Growth (GH, somatotropic)	Growth	Giant, acromegaly/midget
Posterior pituitary	Antidiuretic (ADH, vasopressin)	Water retention by kidneys	Diverse*/diabetes insipidus
	Oxytocin	Uterine contraction	
Thyroid	Thyroxin	Increases metabolic rate (cellular respiration)	Exophthalmic goiter/simple goiter myxedema, cretinism
	Calcitonin	Plasma level of calcium	Tetany/weak bones
Parathyroid	Parathormone (PTH)	Plasma levels of calcium and phosphorus	Weak bones/tetany
Adrenal cortex	Glucocorticoids (cortisol)	Gluconeogenesis	
	Mineralocorticoids (aldosterone)	Sodium retention; potassium excretion by kidneys	Cushing's syndrome/Addison's disease
	Sex hormones	Sex characteristics	
Adrenal medulla	Adrenalin (norepinephrine)	Fight or flight	
Pancreas	Insulin	Lowers blood sugar	Shock/diabetes mellitus
	Glucagon	Raises blood sugar	
Testes	Androgens (testosterone)	Secondary male characteristics	Diverse/eunuch
Ovary	Estrogen (by follicle)	Secondary female characteristics	Diverse/masculinization
	Progesterone (by corpus luteum)		

*The word *diverse* in this chart means that the symptoms have not been described as a syndrome in the medical literature.

Formerly it was believed that only certain glands in the human body produced hormones and that each type could be received only by certain other cells, called the target cells or organs. Investigators have been surprised of late to find that various tissues in the body produce hormones that possibly are for use only within the organ itself. For example, we normally associate insulin production only with the pancreas. However, it has been found that brain cells produce a type of insulin and that brain cells have insulin receptors, even though pancreatic insulin normally has little access to the brain. It would seem that the insulin made by brain cells is acting as a chemical messenger between brain cells. Perhaps even the chemicals we term hormones today were originally simply chemical messengers between like cells in the first multicellular organisms to evolve. Then with the evolution of complex animals, hormones began to be produced by specialized glands that secreted them into the bloodstream (fig. 18.2). In reference to vertebrates, *a hormone is usually defined as a chemical messenger sent by one part of the body to another part by way of the blood stream.* The glands (fig. 18.3) that produce hormones are called **endocrine glands** because they secrete their products internally, placing them directly in the blood. Since these glands do not have ducts for the transport of their secretions, they are sometimes called **ductless glands.** The parts of the body that notably respond to a particular hormone can still be called its **target organ(s)** although we should realize that many other types of cells besides these may have receptors.

Table 18.1 lists the major endocrine glands in humans, the hormones produced by each, and the associated disorders that occur when there is an abnormal level of the hormones—either too much or too little. The adrenal cortex and sex glands produce steroid hormones; the other glands produce hormones that are either amino acids, polypeptides, or proteins.

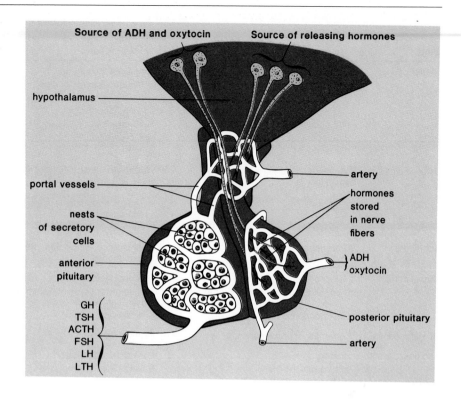

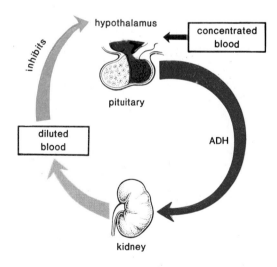

Pituitary Gland

The pituitary gland, which has two portions called the **anterior pituitary** and
the **posterior pituitary,** is a small gland about 1 centimeter in diameter that
lies at the base of the brain. The posterior pituitary is connected by means of
a stalk to the hypothalamus (fig. 18.4), that portion of the brain concerned
with homeostasis (p. 175). The anterior pituitary has no direct anatomical
connection with the posterior pituitary, but there is a portal system composed
of tiny blood vessels that connects it to the hypothalamus.

The hypothalamus controls both the anterior and posterior pituitary. In
fact, it produces the hormones released by the posterior pituitary. These hormones are made in the hypothalamus and pass along nerves that end in the
posterior pituitary. The hypothalamus controls the anterior pituitary by producing **hypothalamic-releasing hormones,** which are transported to the anterior pituitary by the blood within the portal system that connects the two organs.
Each type of releasing hormone causes the anterior pituitary to either secrete
or to stop secreting a specific hormone.

Posterior Pituitary

The posterior pituitary releases antidiuretic hormone (ADH), sometimes called
vasopressin, and oxytocin.

ADH, as discussed in chapter 14, promotes the reabsorption of water
from the loop of Henle and the collecting duct, portions of the kidney tubules.
It is believed that the hypothalamus contains osmoreceptors, or cells that are
sensitive to the amount of water in the blood. When these cells detect that the
blood lacks sufficient water, ADH is produced by hypothalamic neurons and
is transported by their fibers to the posterior pituitary, where it is released (fig.
18.4). As the blood becomes more dilute, the hormone ceases to be produced
and released. Figure 18.5 illustrates how the level of this hormone is controlled

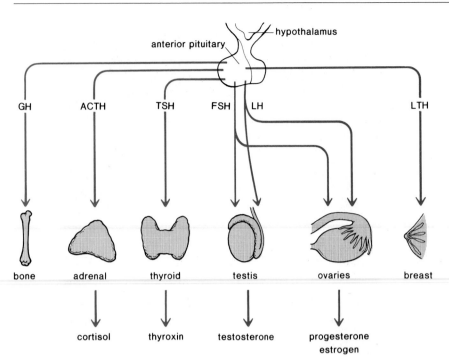

by a circular pattern in which the effect of the hormone (dilute blood) acts to shut down the production and release of the hormone. This is an example of control by **negative feedback.** Negative feedback mechanisms regulate the activities of most hormonal glands.

Inability to reproduce ADH causes **diabetes insipidus** (watery urine) in which a person produces copious amounts of urine with a resultant loss of salts from the blood. The condition can be corrected by the administration of ADH.

Oxytocin, the other hormone released by the posterior pituitary, causes the uterus to contract and may be used to artificially induce labor. It also stimulates the release of milk from the breast when a baby is nursing.

Anterior Pituitary

The anterior pituitary produces at least six different hormones (figs. 18.4 and 18.6). The production of each anterior pituitary hormone is controlled by a hypothalamic releasing hormone. Two of the hormones produced by the anterior pituitary have a direct effect on the body. These are growth hormone and lactogenic hormone.

Growth hormone (GH), or somatotropin, affects the phenotype dramatically since it determines the size of the individual. If little or no growth hormone is secreted by the anterior pituitary during childhood, the person will become a midget, of perfect proportions but quite small in stature. If too much growth hormone is secreted, the person will become a giant (fig. 18.7). Giants usually have poor health, primarily because growth hormone has a secondary effect on blood sugar level, promoting an illness called diabetes (sugar) mellitus, which is discussed in the following.

Growth hormone promotes cell division, protein synthesis, and bone growth. It stimulates the transport of amino acids into cells and increases the activity of ribosomes, both of which are essential to protein synthesis. In bones, it promotes growth of the cartilaginous plates and causes osteoblasts to convert cartilage to bone (p. 336). Evidence suggests that the effects on cartilage and bone may actually be due to hormones called somatomedins, released by the liver. Growth hormone causes the liver to release somatomedins.

Figure 18.7
Size determination. The difference in size between a giant and a midget can be explained by a difference in production of growth hormone by the anterior pituitary.

Figure 18.8
Acromegaly is caused by the production of growth hormone in the adult. It is characterized by an enlargement of the bones in the face and fingers of an adult. Compare the normal-size fingers (hand at left) to those of the patient.

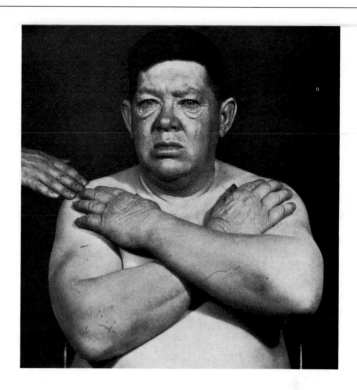

Figure 18.9
Control of hormone secretion. The level of thyroxin in the body is controlled in three ways, as shown: (a.) the level of TSH exerts feedback control over the hypothalamus; (b.) the level of thyroxin exerts feedback control over the anterior pituitary, and (c.) over the hypothalamus. In this way thyroxin controls its own secretion. Substitution of the appropriate terms would also allow this diagram to illustrate control of cortisol and sex hormone levels.

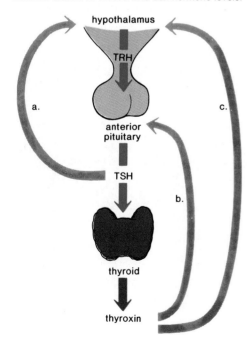

If the production of GH increases in an adult after full height has been obtained, only certain bones respond. These are the bones of the jaw, eyebrow ridges, nose, fingers, and toes. When these begin to grow, the person takes on a slightly grotesque look with huge fingers and toes, a condition called **acromegaly** (fig. 18.8).

Lactogenic hormone (LTH), also called prolactin, is produced in quantity only after childbirth. It causes the mammary glands in the breasts to develop and produce milk.

Master Gland
The anterior pituitary is called the master gland because it controls the secretion of other endocrine glands (fig. 18.6). As indicated in table 18.1, the anterior pituitary secretes the following hormones, which have an effect on other glands:

1. TSH, thyroid-stimulating hormone
2. ACTH, a hormone that stimulates the adrenal cortex
3. Gonadotropic hormones (FSH and LH) that stimulate the gonads, the testes in males and the ovaries in females

TSH causes the thyroid to produce thyroxin; ACTH causes the adrenal cortex to produce cortisol; and gonadotropic hormones cause the gonads to secrete sex hormones. Notice that it is now possible to indicate a three-tiered relationship between the hypothalamus, pituitary, and other endocrine glands. The hypothalamus produces releasing hormones that control the anterior pituitary, and the anterior pituitary produces hormones that control the thyroid, adrenal cortex, and gonads. Figure 18.9 illustrates the feedback mechanism that controls the activity of these glands.

Human Anatomy and Physiology

Figure 18.10
Thyroid gland. The thyroid is composed of many follicles, lined by epithelial cells, that secrete a precursor to thyroxin. *ES* = epithelial cell surface; *Lu* = lumen of follicle; *CT* = connective tissue.
From: TISSUES AND ORGANS: A TEXT-ATLAS OF SCANNING ELECTRON MICROSCOPY by R. G. Kessel and R. Kardon. W. H. Freeman and Company, © 1979.

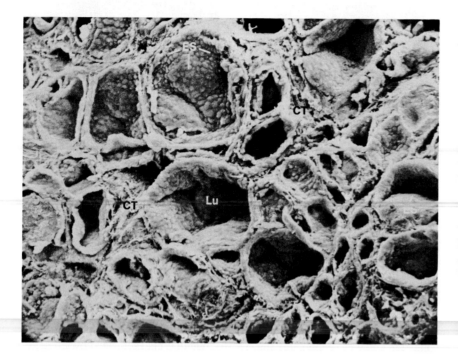

Figure 18.11
Simple goiter. An enlarged thyroid gland is often caused by a lack of iodine in the diet. Without iodine the thyroid is unable to produce thyroxin and continued anterior pituitary stimulation causes the gland to enlarge.

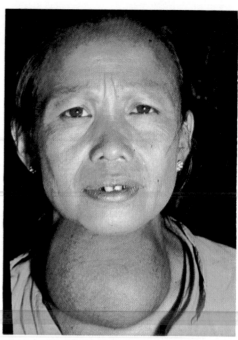

Thyroid Gland

The thyroid gland (fig. 18.3) is located in the neck and is attached to the trachea just below the larynx. Internally (fig. 18.10), the gland is composed of a large number of follicles filled with thyroglobulin, the storage form of thyroxin. The production of both of these requires iodine. Iodine is actively transported into the thyroid gland, where the concentration may become as much as 25 times that of the blood. If iodine is lacking in the diet, the thyroid gland enlarges, producing a goiter (fig. 18.11). The cause of this enlargement becomes clear if we refer to figure 18.9. When there is a low level of thyroxin in the blood, a condition called hypothyroidism, the anterior pituitary is stimulated to produce TSH. TSH causes the thyroid to increase in size so that enough thyroxin usually is produced. In this case, enlargement continues because enough thyroxin is never produced. An enlarged thyroid that produces some thyroxin is called a simple goiter.

Activity and Disorders
Thyroxin increases the metabolic rate. It does not have a target organ; instead, it stimulates most of the cells of the body to metabolize at a faster rate. The number of respiratory enzymes in the cell increases, as does oxygen uptake.

If the thyroid fails to develop properly, a condition called **cretinism** results. Cretins (fig. 18.12) are short, stocky persons who have had extreme hypothyroidism since childhood and/or infancy. Thyroid therapy can initiate growth, but unless treatment is begun within the first two months, mental

Figure 18.12
Cretinism. Cretins are individuals who have suffered from thyroxin insufficiency since birth or early childhood. Skeletal growth is usually inhibited to a greater extent than soft tissue growth; therefore, the child appears short and stocky. Sometimes the tongue becomes so large that it obstructs swallowing and breathing.

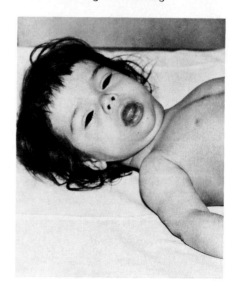

Figure 18.13
Myxedema is caused by thyroid
insufficiency in the older adult. An unusual
type of edema leads to swelling of the
face and bagginess under the eyes.

Figure 18.14
Exophthalmic goiter. Protruding eyes occur
when an active thyroid gland enlarges.

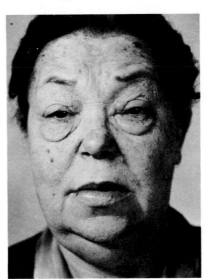

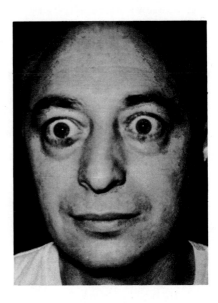

Figure 18.15
Adrenal gland. Each adrenal gland has two parts,
an outer cortex and an inner medulla, both of which
have specific functions. The outer area in the figure
is the cortex, and the inner area is the medulla.

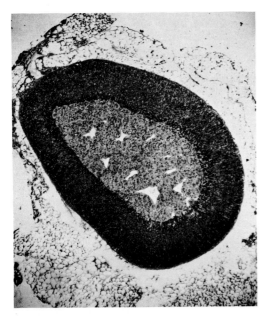

retardation results. The occurrence of hypothyroidism in adults produces the
condition known as **myxedema** (fig. 18.13), which is characterized by lethargy,
weight gain, loss of hair, slower pulse rate, decreased body temperature, and
thickness and puffiness of the skin. The administration of adequate doses of
the thyroid hormone restores normal function and appearance.

In the case of **hyperthyroidism** (too much thyroxin), the thyroid gland
is enlarged and overactive, causing a goiter to form and the eyes to protrude
for some unknown reason. This type of goiter is called **exophthalmic goiter**
(fig. 18.14). The patient usually becomes hyperactive, nervous, irritable, and
suffers from insomnia. Removal or destruction of a portion of the thyroid by
means of radioactive iodine is sometimes effective in curing the condition.

Calcitonin

In addition to thyroxin, the thyroid gland also produces the hormone calcitonin. This hormone helps regulate the calcium level in the blood and opposes
the action of parathyroid hormone. The interaction of these two hormones is
discussed on page 381.

Adrenal Glands

The adrenal glands, as their name implies (*ad* = near; *renal* = kidneys), lie
atop the kidneys (fig. 18.3). Each consists of an outer portion, called the cortex, and an inner portion, called the medulla (fig. 18.15). These portions, like
the anterior and posterior pituitary, have no connection with one another.

The hypothalamus exerts control over the activity of both portions of
the adrenal glands. The hypothalamus can initiate nerve impulses that travel
by way of the brain stem, nerve cord, and sympathetic nerve fibers (fig. 15.16)

to the adrenal medulla, which then secretes its hormones. The hypothalamus, by means of ACTH-releasing hormone, controls the anterior pituitary's secretion of ACTH, which in turn stimulates the adrenal cortex. *Stress* of all types, including both emotional and physical trauma, prompts the hypothalamus to stimulate the adrenal glands.

Adrenal Medulla

The adrenal medulla secretes adrenalin and noradrenalin. The postganglionic fibers of the sympathetic nervous system also secrete noradrenalin. In fact, the adrenal medulla is often considered to be an adjunct to the sympathetic nervous system.

Adrenalin and noradrenalin are involved in the body's immediate response to stress. They bring about all those effects that occur when an individual reacts to an emergency. Blood glucose level rises, the metabolic rate increases, as does breathing and the heart rate. The blood vessels in the intestine constrict, while those in the muscles dilate. Increased circulation to the muscles causes them to have more strength than usual. The individual has a wide-eyed look and is extremely alert. Adrenalin has such a profound effect on the heart that it is often injected directly into a heart that has stopped beating in an attempt to stimulate its contraction.

Adrenal Cortex

While the adrenal medulla may be removed with no ill effects, the adrenal cortex is absolutely necessary to life. The two major types of hormones made by the adrenal cortex are the glucocorticoids and the mineralocorticoids. It also secretes a small amount of male and female sex hormones. All of these hormones are steroids.

Glucocorticoids

Of the various glucocorticoids, the hormone responsible for the greatest amount of activity, is cortisol. The secretion of **cortisol** helps an individual recover from stress (fig. 18.16). Cortisol raises the level of amino acids in the blood, which, in turn, leads to an increased level of glucose when the liver converts these amino acids into glucose. It is said that the adrenal cortex brings about gluconeogenesis, or the production of glucose from nonglucose substances. Gluconeogenesis aids recovery because it allows an individual to maintain cellular respiration, especially in the brain, even when the body is not being supplied with dietary glucose. The amino acids not converted to glucose can be used for tissue repair should injury occur.

Cortisol also counteracts the inflammatory response (p. 238). During the inflammatory response, capillaries become more permeable and fluid leaks out, causing swelling in surrounding tissues. This causes the pain and swelling of joints that accompany arthritis and bursitis. The administration of cortisol aids these conditions because it reduces inflammation.

Mineralocorticoids

The secretion of mineralocorticoids, the most important of which is **aldosterone,** is not believed to be under the control of the anterior pituitary. These hormones regulate the level of sodium (Na^+) and potassium (K^+) in the blood, their primary target organ being the kidney where they promote renal reabsorption of sodium and renal excretion of potassium. The level of sodium in the blood seems to regulate the secretion of aldosterone (fig. 18.17).

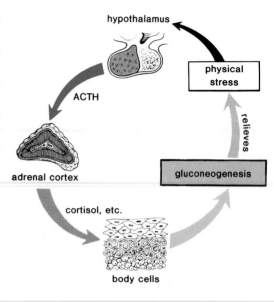

Figure 18.16
Reaction to stress. Stress causes the hypothalamus to produce a releasing hormone that stimulates the anterior pituitary to produce ACTH. This hormone causes the adrenal cortex to produce cortisol, a hormone that brings about gluconeogenesis, which is thought to relieve stress.

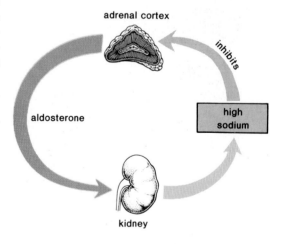

Figure 18.17
Regulation of aldosterone secretion. The adrenal cortex produces aldosterone, which acts on the kidneys to retain sodium. Once the sodium level in the blood is sufficient, the adrenal cortex no longer produces aldosterone.

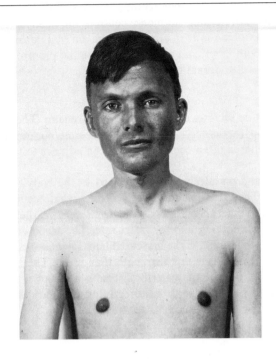

Figure 18.19
Adrenal cortex malfunction can lead to either Addison's disease or Cushing's syndrome.

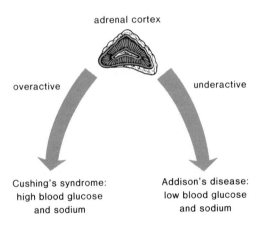

The concentration of sodium in the blood affects the reabsorption of water by the kidneys. If the level of sodium falls too low, too little water is reabsorbed, blood volume falls, and hypotension results. If the level of sodium rises too high, too much water is reabsorbed, blood volume increases, and hypertension results. Sodium and potassium levels are also critical for nerve conduction and muscle contraction; in fact, cardiac failure may result from too low a level of potassium.

Sex Hormones

The adrenal cortex produces a small amount of both male and female sex hormones. In males, the cortex is a source of female sex hormones and in females, it is a source of male hormones. The function of male sex hormones in the female is not known, nor is the function of female sex hormones in the male known. A tumor in the adrenal cortex can cause the production of a large amount of sex hormones, which can lead to feminization in males and masculinization in females.

Disorders

Addison's Disease   When there is a low level of adrenal cortex hormones in the body, the person begins to suffer from so-called **Addison's disease.** Because of the lack of cortisol, the patient is unable to maintain the glucose level of the blood, tissue repair is suppressed, and there is a high susceptibility to any kind of stress. Even a mild infection can cause death. Due to the lack of aldosterone, the blood sodium level is low and the person experiences low blood pressure along with acidosis and low pH. In addition, the patient has a peculiar bronzing of the skin (fig. 18.18).

Cushing's Syndrome   When there is a high level of adrenal cortex hormones in the body, the person suffers from **Cushing's syndrome** (fig. 18.19). Cortisol causes a tendency toward diabetes mellitus, a decrease in muscular protein, and an increase in subcutaneous fat. Because of these effects, the person usually develops thin arms and legs and an enlarged trunk. Due to the high level

of sodium in the blood, the patient has alkaline blood, hypertension, and edema of the face, which gives the face a moon shape. Masculinization may occur in women. The circus lady with a beard may be suffering from Cushing's syndrome.

Sex Organs

The sex organs are the testes in the male and the ovaries in the female. As will be discussed in detail in the following chapter, the testes produce the androgens, which are the male sex hormones, and the ovaries produce estrogen and progesterone, the female sex hormones. The hypothalamus and pituitary gland control the hormonal secretions of these organs in the same manner as described for the thyroid gland in figure 18.9.

The sex hormones control the secondary sex characteristics of the male and female (p. 396 and p. 402). Among other traits, males have greater muscle strength than do females. Among athletes, it is generally believed that the intake of so-called *anabolic steroids,* that is, the male sex hormone testosterone or synthetically related steroids, will cause greater muscle strength. The reading on page 382 discusses the pros and cons of taking these steroids, which are considered illegal by the International Olympic Committee. Any Olympic athlete whose urine tests positive for steroids at the time of an event is immediately disqualified from winning a medal.

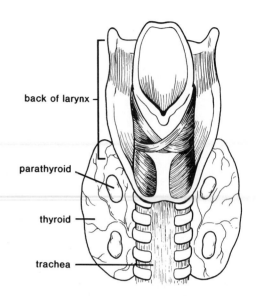

Figure 18.20
Parathyroid glands. These small glands are embedded in the posterior surface of the thyroid gland. Yet the parathyroids and thyroid glands have no anatomical nor physiological connection with one another.

back of larynx

parathyroid

thyroid

trachea

Parathyroid Glands

The parathyroid glands are embedded in the posterior surface of the thyroid gland, as shown in figure 18.20. Many years ago, these four small glands were sometimes removed by mistake during thyroid surgery. Under the influence of **parathyroid hormone** (PTH), also called parathormone, the calcium (Ca^{++}) level in the blood increases and the phosphate (PO_4^{-3}) level decreases. The hormone stimulates the absorption of calcium from the gut, the retention of calcium by the kidneys, and the demineralization of bone. In other words, PTH promotes the activity of osteoclasts (p. 336), the bone-resorbing cells. Although this also raises the level of phosphate in the blood, PTH acts on the kidneys to excrete phosphate in the urine. When a woman stops producing the female sex hormone estrogen following menopause, she is more likely to suffer from osteoporosis, characterized by a thinning of the bones. It is therefore reasoned that estrogen makes bones less sensitive to PTH. As discussed on page 336, a diet sufficient in calcium may help prevent osteoporosis.

If insufficient parathyroid hormone is produced, the level of calcium in the blood drops, resulting in **tetany.** In tetany, the body shakes from continuous muscle contraction. The effect is really brought about by increased excitability of the nerves, which fire spontaneously and without rest. Calcium plays an important role in both nervous conduction and muscle contraction.

The level of PTH secretion is controlled by a feedback mechanism involving calcium (fig. 18.21). When the calcium level rises, PTH secretion is inhibited, and when the calcium level lowers, PTH secretion is stimulated.

As mentioned previously, the thyroid secretes calcitonin, which also influences blood calcium level. Although calcitonin has the opposite effect of PTH, particularly on the bones, its action is not believed to be as significant. Still, the two hormones function together to regulate the level of calcium in the blood.

Figure 18.21
Regulation of parathyroid hormone secretion. A low blood level of calcium causes the parathyroids to secrete parathyroid hormone, which causes the kidneys and gut to retain calcium and osteoclasts to break down bone. The end result is an increased level of calcium in the blood. A high blood level of calcium inhibits hormonal secretion of parathyroid hormone.

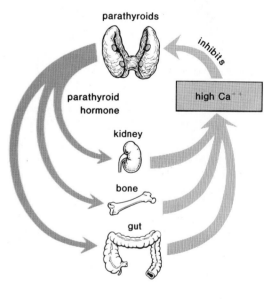

parathyroids

inhibits

parathyroid hormone

high Ca^{++}

kidney

bone

gut

The Toughest Test for Athletes

Lunging across the finish line or regaining balance after a hammer throw, an athlete competing in this summer's [1984] Los Angeles Olympic Games will usually know quickly if he has won a medal. What he may not know for a while is whether he can keep it. The reason: Olympic officials will be on the lookout for more than 300 drugs that athletes are forbidden to use. And, in what is the toughest action to date against drug users by any athletic body, the International Olympic Committee has instituted a testing system that seems almost certain to catch anyone who aims at getting a medal with the aid of a pill or a needle. Among the targets of the tests: amphetamines and, possibly the most dangerous drugs ever taken by athletes, anabolic steroids.

. . . Under the I.O.C.'s new testing system, a representative of the Los Angeles Olympic Organizing Committee will contact all three medal winners, as well as a fourth competitor selected at random, immediately after each event is completed. The escort will take the athletes to a doping control station, where two samples of urine will be taken from each person: one will be stored under strict security and the other will be analyzed in a $1.5 million laboratory operated for the I.O.C. by Don H. Catlin, chief of the University of California, Los Angeles, Medical School Division of Clinical Pharmacology.

If a test turns out to be positive, indicating drug use, the laboratory will notify the I.O.C.'s medical commission, which will tell the athlete and his team officials. A second analysis will then be carried out in the presence of observers from the I.O.C. and the athlete's team. If this is also positive, the athlete will be stripped of his medal. "If someone is using the banned drugs," says Catlin, "we'll find him."

The Olympic Committee added anabolic steroids to its list of banned substances in 1973. Since then, the drugs have become ever more widely used as men and women seek to push their bodies to still higher levels of attainment. In the U.S., where synthetic steroids were developed about half a century ago, their use is thought to extend from world-class athletes to high school football players. Soviet and East European trainers are widely believed to have been giving the drugs to their athletes since the 1950s. Nor will it be possible for athletes to escape detection simply by stopping use of steroids immediately before the Los Angeles Games: their presence in the body can be detected as much as six months after the drugs have been taken. Indeed, some doctors suspect that fear of detection may have contributed to the Soviet decision to boycott the Los Angeles Games.

Being a steroid user may cost an athlete far more than his or her Olympic medal: a growing body of medical evidence indicates that athletes who take steroids have experienced problems ranging from sterility to loss of libido, and the drug has been implicated in the deaths of young athletes from liver cancer and a type of kidney tumor. Steroid use has also been linked to heart disease. "Athletes who take steroids are playing with dynamite," says Robert Goldman, 29, a former wrestler and weight lifter who is now a research fellow in sports medicine at Chicago Osteopathic Medical Center and who has just published a book on steroid abuse, *Death in the Locker Room* (Icarus; $19.95). "Any jock who uses these drugs is taking chances not just with his health but with his life."

Anabolic steroids are essentially the male hormone testosterone and its synthetic derivatives. They were developed to alleviate strictly medical problems: correcting delayed puberty and preventing the withering of muscle tissue in people undergoing prolonged recovery from surgery, starvation or other traumas. Curiously, U.S. athletes were indirectly introduced to the drugs by Soviet athletes. In 1956 the late Dr. John Ziegler attended a world weight-lifting championship in Vienna and was told that the drugs were greatly improving the performance of the lifters from the Soviet Union. Ziegler, believing that U.S. athletes could also be helped by the drugs, worked with CIBA Pharmaceutical Co. to develop a steroid drug called Dianabol for use

by athletes. He quickly abandoned his research, however, when he saw that the drug was being abused. CIBA ceased production of Dianabol for the same reason, although the company continues to make these steroids for medical use. These drugs, doctors say, are being brought into the U.S. illegally from Europe and Mexico and are being used by athletes without a doctor's prescription. Said a disillusioned Ziegler, shortly before he died last year: "I wish I had never heard the word steroid."

The great majority of physicians say the drugs upset the body's natural hormonal balance, particularly that involving testosterone, which is present, though in different amounts, in both men and women. Normally, the hypothalamus, the part of the brain that regulates many of the body's functions, "tastes" the testosterone levels; if it finds them too low, it signals the pituitary gland to trigger increased production. When the hypothalamus finds the testosterone levels too high, as it does in the case of steroid abusers, it signals the pituitary to stop production. Problems can also arise in some cases after athletes stop taking the drugs and the hypothalamus fails to get the system started again.

The results can be traumatic. Many men experience atrophy, or shrinking, of the testicles, falling sperm counts, temporary infertility and a lessening of sexual desire; some men grow breasts, while others may develop enlargement of the prostate gland, a painful condition not usually found in men under 50. Women who take too many steroids can develop male sexual characteristics. Some grow hair on their chests and faces and lose hair from their heads; many experience abnormal enlargement of the clitoris. Some cease to ovulate and menstruate, sometimes permanently.

There are several other health risks. Steroids can cause the body to retain fluid, which results in rising blood pressure. This often tempts users to fight "steroid bloat" by taking large doses of diuretics. A postmortem on a young California weight lifter who had a fatal heart attack after using steroids within

the past year showed that by taking diuretics he had purged himself of electrolytes, chemicals that help regulate the heart. Convincing athletes of the dangers of steroids is far from easy. Earlier this year, Author Goldman asked this hypothetical question of 198 world-class athletes: would they take a pill that would guarantee them a gold medal even if they knew that it would kill them in five years? One hundred and three said that they would.

Figure 18.22

Islet of Langerhans. Photomicrograph of the pancreas shows the islets *(lighter area)*, which secretes insulin, and surrounding acinar tissue, which secretes digestive enzymes. In juvenile onset diabetes, the islets have lost the potential to produce insulin.

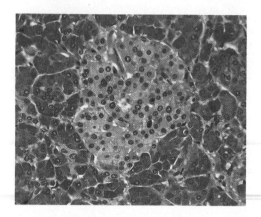

Figure 18.23

Regulation of insulin secretion. In response to a high blood sugar level, the pancreas secretes insulin, which promotes the uptake of glucose in body cells, muscles, and the liver. As a result of a low blood glucose level, the pancreas stops secreting insulin.

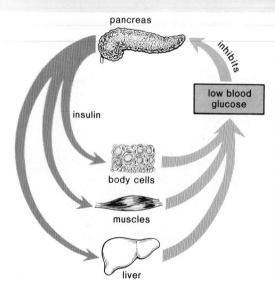

Pancreas

The pancreas is a long, soft organ that lies transversely in the abdomen (fig. 18.3) between the spleen and the duodenum of the small intestine. It is composed of two types of tissues (fig. 18.22); one of these produces and secretes the digestive juices that go by way of the pancreatic duct to the small intestine, and the other type, called the **islets of Langerhans,** produces and secretes the hormones insulin and glucagon directly into the blood. Insulin and glucagon are hormones that affect the blood glucose level in opposite directions, since insulin decreases the level and glucagon increases the level of glucose. Although the functions of glucagon do not seem to be as important as those of insulin, glucagon does function in times of stress, causing the liver to convert glycogen to glucose.

Insulin is secreted when there is a high level of glucose in the blood, which usually occurs just after eating. Once the glucose level returns to normal, insulin is not secreted, as illustrated in figure 18.23. Insulin is believed to cause all the cells of the body to take up glucose. When the liver and muscles take up glucose, they convert to glycogen any glucose not needed immediately. Therefore insulin promotes the storage of glucose as glycogen.

Diabetes Mellitus

A significantly large number of people suffer from **diabetes mellitus** (sugar diabetes). As much as 5 percent of the population may have diabetes, and the incidence is increasing yearly. One type of diabetes develops slowly and is usually seen in individuals after the age of 40. In this type of diabetes, termed **maturity-onset diabetes,** physiological symptoms are often mild and they can usually be controlled by diet alone. Curiously, there is enough insulin in the blood; therefore this type of diabetes is probably due to a defect in the molecular machinery of tissue cells. There is evidence that the cells have a decreased number of insulin receptors.

The second type of diabetes develops quickly and occurs mostly in individuals under the age of 20. In this type of diabetes, termed **juvenile-onset diabetes,** there is a deficiency of insulin in the blood. The symptoms of diabetes mellitus include the following:

Sugar in the urine
Frequent, copious urination
Abnormal thirst
Rapid loss of weight
Extreme hunger
General weakness
Drowsiness and fatigue
Itching of the genitals and skin
Visual disturbances, blurring
Skin disorders, such as boils, carbuncles, and infection[1]

Many of these symptoms develop because sugar is not being metabolized by the cells. The liver fails to store glucose as glycogen, and all the cells fail to utilize glucose as an energy source. This means that the blood glucose level rises very high after eating, causing glucose to be excreted in the urine. More water than usual is therefore excreted so that the diabetic is extremely thirsty.

Since carbohydrates are not being metabolized, the body turns to the breakdown of proteins and fat for energy. Unfortunately, the breakdown of

[1]Reprinted from *How to Live With Diabetes* by Henry Dolger, M.D., and Bernard Seeman. By permission of W. W. Norton & Company, Inc. Copyright © 1972, 1965, 1958 by Henry Dolger and Bernard Seeman.

Human Anatomy and Physiology

these molecules leads to the buildup of acids in the blood, with resulting acidosis and respiratory distress. It is the latter that can eventually cause coma and death of the diabetic. The symptoms that lead to coma (table 18.2) develop slowly.

The patient with juvenile-onset diabetes takes daily insulin injections. These injections control the diabetic symptoms but may still cause inconveniences since either an overdose of insulin or the absence of regular eating can bring on the symptoms of insulin shock, which are given in table 18.2. These symptoms appear because the blood sugar level has decreased below normal levels. Since the brain requires a constant supply of sugar, unconsciousness results. The cure is quite simple: an immediate source of sugar, such as a sugar cube or fruit juice, can counteract insulin shock immediately.

Prostaglandins

Prostaglandins (PG) are considered hormones even though they are lipid compounds, made from cell membrane fatty acids, and are sometimes active in the cells that produce them. Prostaglandins were first discovered in semen and were named after the prostate gland from which they were thought to originate. The prostaglandins in semen cause the uterus to contract during sexual intercourse, and this is believed to assist the movement of sperm as they traverse this organ. Because of their effect on the uterus, they are sometimes used to abort a fetus. Prostaglandins are also being considered for possible use as a safe, self-administered, once-a-month means of preventing pregnancy. As a contraceptive, prostaglandins cause menstruation to begin soon after administration. Prostaglandins are involved in many other aspects of reproduction and development.

Prostaglandins also function in nonreproductive organs. They are being considered as possible treatment for ulcers because they reduce gastric secretion; as treatment of hypertension because they lower blood pressure; and as prevention of thrombosis because they inhibit platelet aggregation.

Sometimes prostaglandins have contrary effects. For example, one type helps prevent blood clots while another helps blood clots to form. Also a large dose of PG may have an effect that is opposite to that of a small dose. Therefore, it has been very difficult to standardize PG therapy and, in most instances, prostaglandin drug therapy is still considered experimental.

It has been shown that aspirin prevents formation of prostaglandins, and this may account for aspirin's many therapeutic effects. In particular, it has been shown that prostaglandin biosynthesis occurs during the inflammatory response; since aspirin blocks prostaglandin synthesis, it helps prevent the swelling and pain of arthritis.

Summary

Hormones are chemical messengers produced by endocrine glands, or glands of internal secretion. Research into the mechanism of action of hormones suggests that there are two mechanisms, depending on whether the hormone is a peptide or steroid. Peptide hormones combine with receptor sites in the cell membrane. This leads to the formation of cAMP, which activates an enzyme within the cell. Steroid hormones combine with receptor molecules in the cytoplasm of the cell and the hormone-receptor complex moves into the nucleus to combine with and activate DNA to carry out protein synthesis.

The hypothalamus produces the hormones released by the posterior pituitary and also controls the anterior pituitary by means of releasing hormones. The posterior pituitary secretes two hormones: ADH, whose target

Table 18.2 Symptoms of insulin shock and diabetic coma

Insulin shock	Diabetic coma
Sudden onset	Slow, gradual onset
Perspiration, pale skin	Dry, hot skin
Dizziness	No dizziness
Palpitation	No palpitation
Hunger	No hunger
Normal urination	Excessive urination
Normal thirst	Excessive thirst
Shallow breathing	Deep, labored breathing
Normal breath odor	Fruity breath odor
Confusion, disorientation, strange behavior	Drowsiness and great lethargy leading to stupor
Urinary sugar absent or slight	Large amounts of urinary sugar
No acetone in urine	Acetone present in urine

organ is the kidney, and oxytocin. Diabetes insipidus occurs in persons who do not produce sufficient ADH, resulting in the production of large amounts of urine. Oxytocin causes the uterus to contract and is important during the time of childbirth.

The anterior pituitary secretes two hormones that have no noticeable effect on other endocrine glands. These are the lactogenic (LTH, prolactin) and growth hormone (GH). The former is secreted after childbirth to promote the development of the mammary glands (breasts). Growth hormone is especially important during childhood because it determines the size of the adult. Midgets and giants represent the extremes of too little or too much of this hormone. If this hormone is produced in excess in the adult, the condition called acromegaly develops.

In addition to these hormones, the anterior pituitary secretes hormones that control other endocrine glands: TSH stimulates the thyroid, ACTH stimulates the adrenal cortex; and gonadotropic stimulates the gonads.

The thyroid gland secretes thyroxin, which speeds up metabolism. Iodine is needed by the thyroid gland to make thyroxin; if iodine is missing from the diet, a simple goiter develops. An overactive thyroid results in an exophthalmic goiter, with an enlarged neck and protruding eyes. Cretins are individuals who lack sufficient thyroxin from birth; adults with hypothyroidism develop myxedema.

The adrenal cortex produces three types of hormones: glucocorticoids (e.g., cortisol), mineralocorticoids (e.g., aldosterone), and sex hormones. Glucocorticoids promote the conversion of amino acids into glucose, raising the level of glucose in the blood. They are important in preventing deterioration of the body due to stress. Mineralocorticoids stimulate the kidney tubules to reabsorb sodium and excrete potassium. Reabsorption of sodium is important in maintaining blood volume and blood pressure. Adrenal cortex sex hormones are important if the sex glands fail to secrete their normal amount of hormone. Two very important illnesses may result from adrenal cortex imbalance: Addison's disease occurs when the adrenal cortex fails to secrete its hormones, and Cushing's syndrome occurs when it oversecretes.

The parathyroids are embedded in the thyroid gland. They control the level of calcium and phosphate in the blood. Calcium level is maintained by promoting the absorption of calcium by the gut, reabsorption by the kidneys, and demineralization of bone. Too little parathyroid hormone results in tetany. These actions are opposed by calcitonin, which is produced by the thyroid.

The most common illness due to hormonal imbalance is diabetes mellitus (sugar diabetes). This condition occurs when the islets of Langerhans within the pancreas fail to produce insulin. Insulin promotes the uptake of glucose by the cells and the conversion of glucose to glycogen, thus lowering blood glucose levels. Without the production of insulin, the blood sugar level rises and some of it spills over into the urine. The real problem in diabetes mellitus, however, is acidosis, which may cause the death of the diabetic if therapy is not begun.

The secretion of hormones is controlled by negative feedback mechanism. For example, the level of the substances listed in the first column controls the secretion of the hormones in the second column.

Feedback by:	To control
Thyroxin	TSH
Calcium	Parathyroid hormone
Sodium	Aldosterone
Sugar	Insulin

Human Anatomy and Physiology

Study Questions

1. Hormones fall into what two groups, chemically speaking? (p. 371) Name some hormones in each group. (p. 373)
2. What mechanisms of action have been suggested to explain how hormones work? (p. 371)
3. How might vertebrate hormones have evolved? (p. 373) Define endocrine gland and target organ. (p. 373)
4. How does the hypothalamus control the posterior pituitary; the anterior pituitary? (p. 374)
5. Discuss two hormones secreted by the anterior pituitary that have an effect on the body proper rather than on other glands. (pp. 375–76) Why is the anterior pituitary called the master gland? (p. 376)
6. For each of the following endocrine glands name the hormone(s) secreted, the effect of the hormone(s), and the medical illnesses, if any, that result from too much or too little of each hormone: posterior pituitary, thyroid, parathyroids, adrenal cortex, adrenal medulla, pancreas. (p. 373)
7. Give the anatomical location of each of the endocrine glands listed in #6. (p. 372)
8. Draw a diagram to describe the action and control of ADH, thyroxin, glucocorticoids (e.g., cortisol), aldosterone, parathyroid hormone, and insulin. (pp. 374–87)

Selected Key Terms

endocrine (en'do-krin)

pituitary (pĭ-tu'ĭ-tār''e)

diabetes insipidus
(di''ah-be'tēz in-sip'ĭ-dus)

oxytocin (ok''se-to'sin)

lactogenic hormone
(lak''to-jen'ik hor'mōn)

cretinism (kre'tin-izm)

myxedema (mik''sĕ-de'mah)

exophthalmic goiter
(ek''sof-thal'mik goi'ter)

glucocorticoids
(gloo''ko-kôr'tĭ-koidz)

cortisol (kor'tĭ-sol)

mineralocorticoids
(min''er-al-o-kor'tĭ-koidz)

aldosterone (al''do-ster'ōn)

Addison's disease
(ad'ĭ-sonz dĭ-zēz')

Cushing's syndrome
(koosh'ingz sin'drōm)

parathyroid hormone
(par''ah-thi'roid hor'mōn)

tetany (tet'ah-ne)

Islets of Langerhans
(i'lets uv lahng'er-hanz)

insulin (in'su-lin)

diabetes mellitus
(di''ah-be'tēz mĕ-li'tus)

prostaglandins
(pros''tah-glan'dinz)

Further Readings

Gardiner, L. 1971. Deprivation dwarfism. *Scientific American* 224(1):76.

Gillie, R. B. 1971. Endemic goiter. *Scientific American* 224(6):92.

Guillemin, R., and Burgus, R. 1972. The hormones of the hypothalamus. *Scientific American* 227(5):24.

Levine, S. 1971. Stress and behavior. *Scientific American* 224(1):12.

Notkins, A. L. 1979. The causes of diabetes. *Scientific American* 241(5):62.

O'Malley, B. W., and Schrader, W. T. 1976. The receptors of steroid hormones. *Scientific American* 234(2):32.

Pastan, I. 1972. Cyclic AMP. *Scientific American* 227(2):97.

Pike, J. E. 1971. Prostaglandins. *Scientific American* 225(5):84.

Rassmussen, H., and Pechet, M. M. 1970. Calcitonin. *Scientific American* 223(4):42.

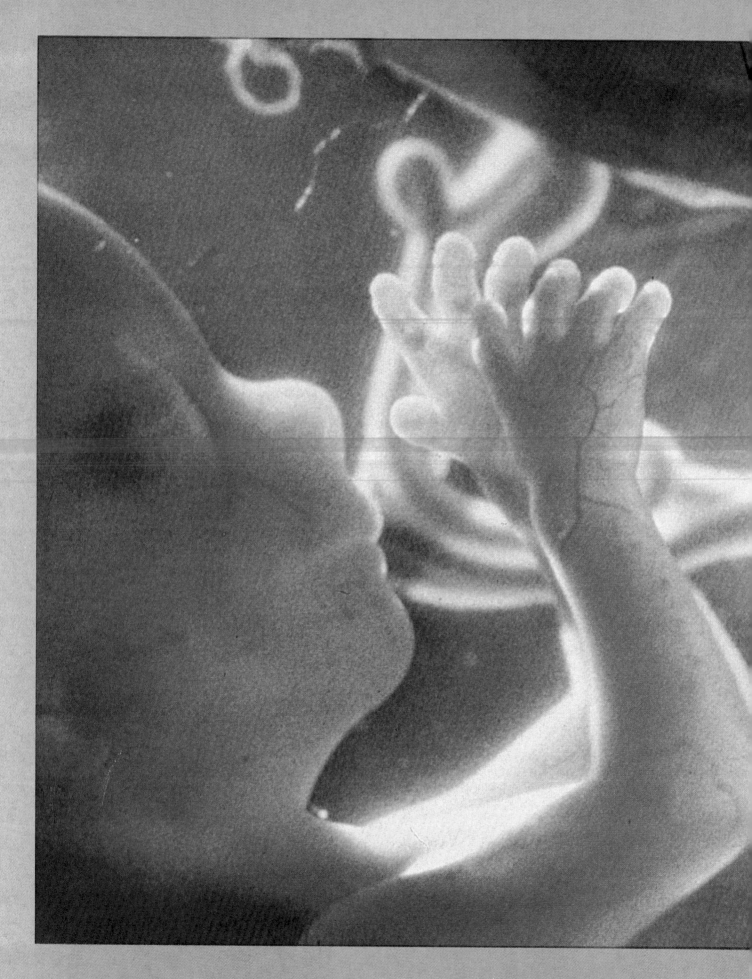

Part Four

Reproduction, Development, and Inheritance

The anatomy of the human male and female serve to bring the sperm to the egg, resulting in fertilization, followed by the gradual steps of development. Sexual reproduction results in a recombination of genes and therefore produces offspring that are, at the same time, both similar to and different from the parents. Knowing the genes carried by the parents sometimes makes it possible to predict certain features of the offspring. However, this simple relationship can be influenced by the interaction of genes during development and by the prenatal environment.

Biochemical knowledge of the makeup of the hereditary material and how it operates has forged a biological revolution. It is now possible to manipulate the genes, an advance that offers hope as a means of curing genetic diseases and cancer but, nevertheless, is viewed by some with misgivings.

19

Human Reproductive System

Chapter concepts

1 The male reproductive system is designed for the continuous production of a large number of sperm within a fluid medium.

2 The female reproductive system is designed for the monthly production of an egg and preparation of the uterus for possible implantation of the fertilized egg.

3 Hormones control the reproductive process and the sex characteristics of the individual.

4 Birth control measures vary in effectiveness from those that are very effective to those that are minimally effective.

5 There are new methods for treating infertility, including in vitro fertilization followed by artificial implantation.

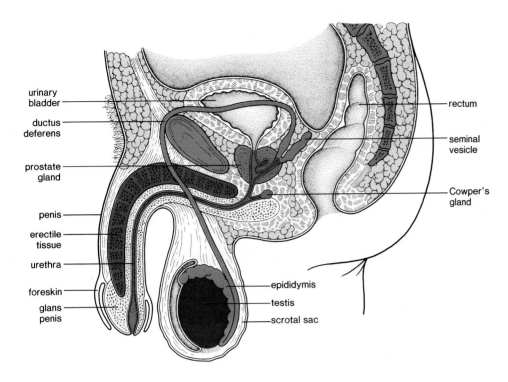

In advanced forms of sexual reproduction there are two types of gametes (sex cells) both of which contribute the same number of chromosomes to the new individual (fig. 4.3). The sperm are small and swim to the stationary egg, a much larger cell that contains food for the developing embryo. It seems reasonable that there should be a large number of sperm to ensure that a few will find the egg. In humans, the male continually produces sperm, which are temporarily stored before being released.

Male Reproductive System

Figure 19.1 shows the reproductive system of the male and table 19.1 lists the anatomical parts of this system.

Testes

The **testes** lie outside the abdominal cavity of the male within the **scrotal sacs.** The testes begin their development inside the abdominal cavity but descend into the scrotal sacs during the last two months of fetal development. If by chance the testes do not descend and the male is not treated or operated on to place the testes in the scrotum, sterility—the inability to produce offspring—usually follows. This is because the internal temperature of the body is too high to produce viable sperm.

Table 19.1 Male reproductive system

Organ	Function
Testes	Produce sperm and sex hormones
Epididymis	Maturation and some storage of sperm
Ductus deferens	Conducts and stores sperm
Seminal vesicles	Contribute to seminal fluid
Prostate gland	Contributes to seminal fluid
Urethra	Conducts sperm
Cowper's glands	Contribute to seminal fluid
Penis	Organ of copulation

Figure 19.2
Diagram of male reproductive system, front view. Left side shows location of cut for vasectomy; right side shows longitudinal cut of a testis. Enlargement shows microscopic anatomy of a testis, consisting of seminiferous tubules and interstitial cells.

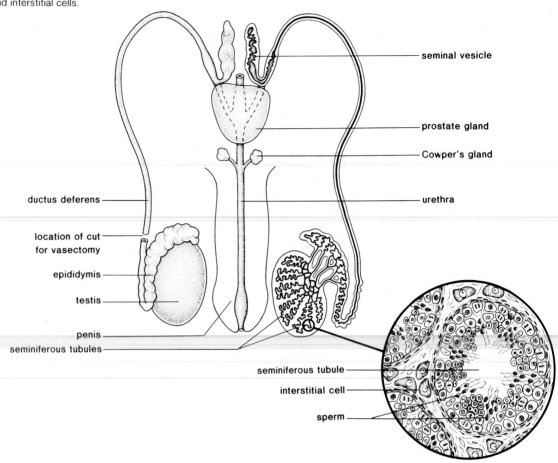

seminal vesicle

prostate gland

Cowper's gland

ductus deferens

urethra

location of cut for vasectomy

epididymis

testis

penis

seminiferous tubules

seminiferous tubule

interstitial cell

sperm

Seminiferous Tubules

Connective tissue forms the wall of each testis and divides it into lobules (fig. 19.2). Each lobule contains one to three tightly coiled **seminiferous tubules,** which have a combined length of approximately 750 feet. A microscopic cross section through a tubule shows that it is packed with cells undergoing spermatogenesis (fig. 4.20), during which time meiosis occurs and the chromosome number is reduced from 46 to 23.

Sperm The mature sperm, or spermatozoan (fig. 19.3), has three distinct parts: a head, a mid-piece, and a tail. The **tail** contains the 9 + 2 pattern of microtubules typical of cilia and flagella (fig. 2.19) and the **mid-piece** contains energy-producing mitochondria. The **head** contains the 23 chromosomes within a nucleus. The tip of the nucleus is covered by a cap called the **acrosome,** which is believed to contain enzymes needed for fertilization. The human egg is surrounded by several layers of cells and a mucoprotein substance. The acrosome enzymes are believed to aid the sperm in reaching the surface of the egg and allowing a single sperm to penetrate the egg. It is hypothesized that each acrosome contains such a minute amount of enzyme that it requires the action of many sperm to allow just one to actually penetrate the egg. This may explain why so many sperm are required for the process of fertilization. A normal human male usually produces several hundred million sperm per day, an adequate number for fertilization. Sperm are continually produced throughout a male's reproductive life.

Reproduction, Development, and Inheritance

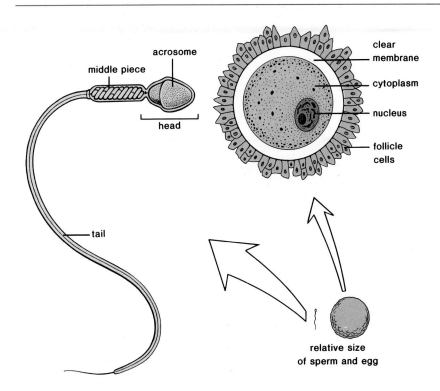

middle piece

acrosome

head

tail

clear membrane

cytoplasm

nucleus

follicle cells

relative size
of sperm and egg

Interstitial Cells

The male sex hormones, the androgens, are secreted by cells that lie between the seminiferous tubules and are therefore called **interstitial cells.** The most important of the androgens is testosterone, whose functions are discussed on page 396.

Genital Tract

Sperm are produced in the testes, but they mature in the **epididymis** (fig. 19.1), a tightly coiled tubule about 20 feet in length that lies just outside each testis. During the two-to-four day maturation period, the sperm develop their characteristic swimming ability. Also, it is possible that during this time defective sperm are removed from the epididymis. Each epididymis joins with a **ductus (vas) deferens,** which ascends through a canal called the **inguinal canal** and enters the abdomen where it curves around the bladder and empties into the urethra. Sperm are stored in the ductus deferens. They pass from each ductus into the urethra only when ejaculation (p. 395) is imminent.

Spermatic Cords   The testes are suspended in the scrotum by the **spermatic cords,** each of which consists of connective tissue and muscle fibers that enclose the ductus deferens, blood vessels, and nerves. The region of the inguinal canal, where the spermatic cord passes into the abdomen, remains a weak point in the abdominal wall. As such, it is frequently the site of hernias. A **hernia** is an opening or separation of some part of the abdominal wall through which a portion of an internal organ, usually the intestine, protrudes.

Seminal Fluid

At the time of ejaculation (p. 395), sperm leave the penis in a fluid called **seminal fluid.** This fluid is produced by three types of glands—the seminal vesicles, the prostate gland, and Cowper's glands. The **seminal vesicles** lie at the base of the bladder, and each has a duct that joins with a ductus deferens.

Figure 19.4
Erection contrasted to a flaccid penis and relaxed scrotum. The erectile tissue fills with blood when the penis becomes erect. Note that the penis lacks a foreskin, due to circumcision.

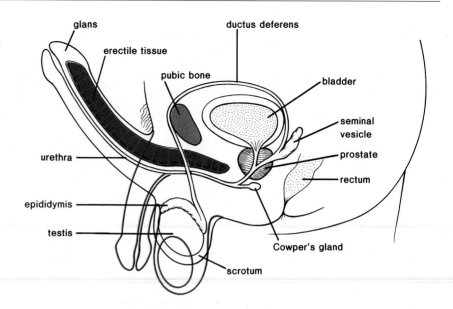

The **prostate gland** is a single doughnut-shaped gland that surrounds the upper portion of the urethra just below the bladder. In older men, the prostate may enlarge and cut off the urethra, making urination painful and difficult. This condition may be treated medically or surgically. **Cowper's glands** are pea-sized organs that lie posterior to the prostate on either side of the urethra.

Each component of seminal fluid seems to have a particular function. Sperm are more viable in a basic solution; seminal fluid, which is white and milky in appearance, has a slightly basic pH (about 7.5). Swimming sperm require energy, seminal fluid contains the sugar fructose, which presumably serves as an energy source. Seminal fluid also contains prostaglandins, chemicals that cause the uterus to contract. Some investigators now believe that uterine contraction is necessary to propel the sperm and that the sperm swim only when they are in the vicinity of the egg.

Penis

The penis has a long shaft and enlarged tip called the **glans penis.** In uncircumcised males, the glans penis is covered by a loose-fitting hood of skin. In circumcised males, the hood, called the **foreskin,** or prepuce, has been surgically removed. While circumcision is sometimes done for religious reasons, there are also medical reasons. Glands under the foreskin discharge a white, cheesy substance called **smegma.** If smegma accumulates under the foreskin, irritation and infection can result. Daily cleansing of the glans penis can prevent smegma accumulation and eliminate possible problems. However, it has also been suggested, but not proven, that there is less penile cancer in circumcised males and less cervical cancer in their mates. Even so, there are those who do not favor circumcision on the ground that the foreskin may have an important function that has not yet been determined.

The penis is the copulatory organ of males. When the male is sexually aroused, the penis becomes erect and ready for intercourse (fig. 19.4). **Erection** is achieved because blood sinuses within the erectile tissue of the penis become filled with blood. Parasympathetic impulses dilate the arteries of the penis while the veins are passively compressed so that blood flows into the erectile tissue under pressure. If the penis fails to become erect, the condition is called **impotency.** While it was formerly believed that almost all cases of impotency were due to psychological reasons, it has recently been reported that some cases may be due to hormonal imbalances. Treatment consists of finding the precise imbalance and restoring the proper level of testosterone.

Reproduction, Development, and Inheritance

Figure 19.5
Fertilization. *a.* Scanning electron micrographs of
an echinoderm egg surrounded by a large number
of sperm. *b.* Only one sperm penetrates the egg to
achieve fertilization.

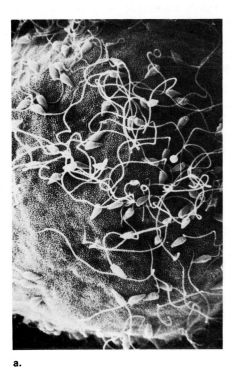

a.

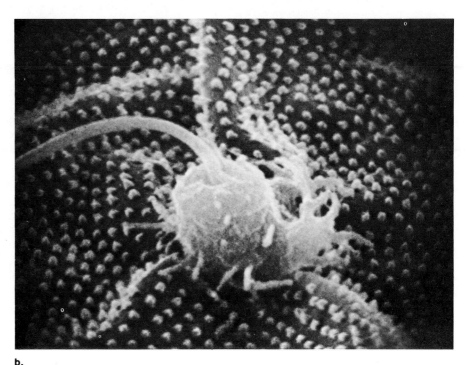

b.

Ejaculation

As sexual stimulation becomes intense, sperm enter the urethra from each
ductus deferens and the glands secrete seminal fluid. Sperm and seminal fluid
together are called **semen.** Once semen is in the urethra, rhythmical muscle
contractions cause it to be expelled from the penis in spurts. During ejaculation, a sphincter closes off the bladder so that no urine enters the urethra.
(Notice that the urethra carries either urine or semen at different times.)

The contractions that expel semen from the penis are a part of male
orgasm, the physiological and psychological sensations that occur at the climax of sexual stimulation. The psychological sensation of pleasure is centered
in the brain, while the physiological reactions involve the genital (reproductive) organs and associated muscles as well as the entire body. Marked muscle
tension is followed by contraction and relaxation.

Following ejaculation and/or loss of sexual arousal, the penis returns to
its normal flaccid state. After ejaculation, a male typically experiences a period of time, called the refractory period, during which stimulation does not
bring about an erection.

There may be in excess of 400 million sperm in 3.5 milliliters of semen
expelled during ejaculation. The sperm count can be much lower than this,
however, and fertilization (fig. 19.5) will still take place.

Hormonal Regulation in the Male

The hypothalamus has ultimate control of the testes' sexual functions because
it secretes a releasing hormone that stimulates the anterior pituitary to produce the gonadotropic hormones. Two gonadotropic hormones, **FSH** (follicle-
stimulating hormone) and **LH** (luteinizing hormone), are named for their
function in females but exist in both sexes, stimulating the appropriate gonads
in each. It is believed that FSH promotes spermatogenesis in the seminiferous

Figure 19.6
Hypothalamic-pituitary-gonad system as it
functions in the male. GnRH is a hypothalamic-
releasing hormone that stimulates the anterior
pituitary to secrete LH and FSH. These
gonadotropic hormones act on the testes. LH
promotes the production of testosterone, and FSH
promotes spermatogenesis. Negative feedback
controls the level of all hormones involved.

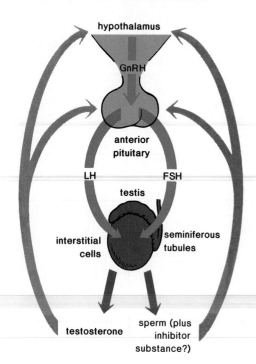

Table 19.2	Female reproductive system
Organ	**Function**
Ovaries	Produce egg and sex hormones
Fallopian tubes (oviducts)	Conduct egg
Uterus (womb)	Location of developing fetus
Cervix	Contains opening to uterus
Vagina	Organ of copulation and birth canal

tubules and that LH promotes the production of testosterone in the interstitial cells. Sometimes LH in males is given the name interstitial cell-stimulating hormone (ICSH).

The hormones mentioned are involved in a feedback process (fig. 19.6) that maintains the production of testosterone at a fairly constant level. For example, when the amount of testosterone in the blood rises to a certain level, it causes the hypothalamus to decrease its secretion of releasing hormone, which causes the anterior pituitary to decrease its secretion of LH. As the level of testosterone begins to fall, the hypothalamus increases secretion of the releasing hormone and the anterior pituitary increases its secretion of LH, and stimulation of the interstitial cells reoccurs. It should be emphasized that only minor fluctuations of testosterone level occur in the male and that the feedback mechanism in this case acts to maintain testosterone at a normal level. Feedback control of FSH secretion is still being investigated, and the possibility exists that the seminiferous tubules release a substance that inhibits FSH secretion.

Testosterone

The male sex hormone, testosterone, has many functions. It is essential for the normal development and functioning of the primary sex organs, those structures we have just been discussing. It is also necessary for the maturation of sperm, probably after diffusion from the interstitial cells into the seminiferous tubules.

Greatly increased testosterone secretion at the time of puberty stimulates the growth of the penis and testes. Testosterone also brings about and maintains the secondary sex characteristics in males that develop at the time of puberty. Testosterone causes growth of a beard, axillary (underarm) hair, and pubic hair. It prompts the larynx and vocal cords to enlarge, causing the voice to change. It is responsible for the greater muscle strength of males and this is the reason why some athletes take supplemental amounts of *anabolic steroids,* which are either testosterone or related chemicals. The pros and cons of taking anabolic steroids is discussed in the reading on page 382. Testosterone also causes oil and sweat glands in the skin to secrete; thus it is largely responsible for acne and body odor. Another side effect of testosterone activity is baldness. Genes for baldness are probably inherited by both sexes, but baldness is seen more often in males because of the presence of testosterone.

Testosterone is believed to be largely responsible for the sex drive and may even contribute to the supposed aggressiveness of males.

Female Reproductive System

Table 19.2 lists the anatomical parts of this system, and figure 19.7 shows the reproductive system of the female.

Ovaries

The ovaries (fig. 19.8) lie in shallow depressions, one on each side of the upper pelvic cavity. A longitudinal section through an ovary shows that it is made up of an outer cortex and an inner medulla. The cortex contains ovarian **follicles** at various stages of maturation. A female is born with a large number of follicles (400,000) in both ovaries, each containing a potential egg. In contrast to the male, the female produces no new gametes after she is born. Only a small number of eggs (about 400) ever mature because a female produces only one egg per month during her reproductive years. Since eggs are present at birth, they age as the woman ages. This is one possible reason why older women are more likely to produce children with genetic defects.

Reproduction, Development, and Inheritance

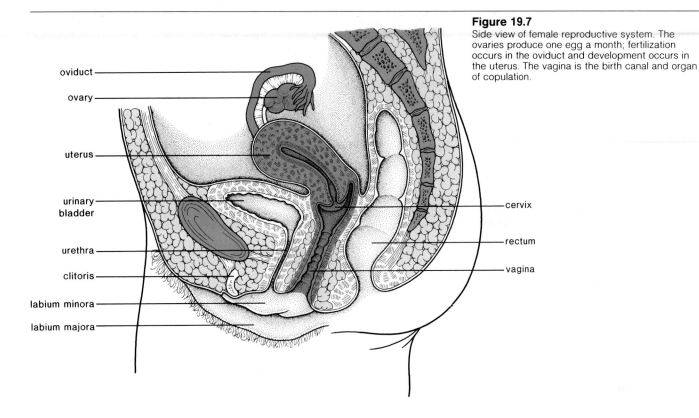

Figure 19.7
Side view of female reproductive system. The ovaries produce one egg a month; fertilization occurs in the oviduct and development occurs in the uterus. The vagina is the birth canal and organ of copulation.

oviduct

ovary

uterus

urinary bladder

urethra

clitoris

labium minora

labium majora

cervix

rectum

vagina

Figure 19.8
Reproductive organs of female, front view. Left side is surface view and right side is longitudinal view. The enlargement shows maturation of the follicle, release of egg, and resulting corpus luteum.

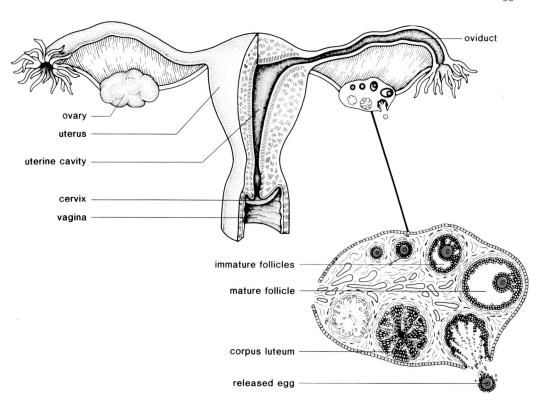

oviduct

ovary

uterus

uterine cavity

cervix

vagina

immature follicles

mature follicle

corpus luteum

released egg

Human Reproductive System

397

Figure 19.9
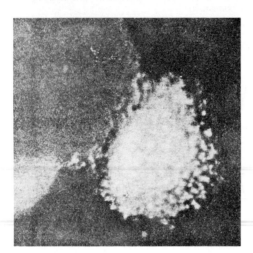
Ovulation. Human ovarian wall ruptures as egg is released. Egg is not visible because it is surrounded by follicular cells.

Once a month a follicle undergoes complete maturation, developing from a primary to a secondary to a mature **Graafian follicle.** A fluid-filled cavity within the Graafian follicle contains the developing egg, which will contain only 23 chromosomes once oogenesis is complete. When the egg is mature, **ovulation** occurs. The follicle balloons out on the surface of the ovary and bursts, releasing the egg, which is smaller than the period at the end of this sentence. However, the egg is surrounded by a mucoprotein substance and a few follicular cells (fig. 19.3). Once a follicle has lost its egg, it develops into a **corpus luteum,** a glandlike structure. If pregnancy does not occur, the corpus luteum begins to degenerate after about 10 days. If pregnancy does occur, the corpus luteum persists for three to six months.

The follicle and corpus luteum secrete the female sex hormones estrogen and progesterone, as discussed on page 401.

Genital Tract
The female genital tract includes the oviducts, uterus, and vagina.

Oviducts
The oviducts, also called uterine or fallopian tubes, extend from the uterus to the ovaries. The oviducts are not attached to the ovaries but instead have fingerlike projections called **fimbria** that sweep over the ovary at the time of ovulation. When the egg bursts (fig. 19.9) from the ovary during ovulation, it is usually swept up into an oviduct by the combined action of the fimbria and the beating of cilia that line the tubes.

Since the egg must traverse a small space before entering an oviduct, it is possible for the egg to get lost and instead enter the abdominal cavity. Such eggs usually disintegrate but in some rare cases have been fertilized in the abdominal cavity and have implanted themselves in the wall of an abdominal organ. Very rarely, such embryos have come to term, the child being delivered by surgery.

Once in the oviduct, the egg is propelled slowly by cilia movement and tubular muscular contraction toward the uterus. Fertilization usually occurs in an oviduct because the egg only lives approximately 6 to 24 hours. The developing zygote normally arrives at the uterus after several days and then embeds, or implants, itself in the uterine lining, which has been prepared to receive it. Occasionally, the zygote becomes embedded in the wall of an oviduct, where it begins to develop. Tubular pregnancies cannot succeed because the tubes are not anatomically capable of allowing full development to occur.

Uterus
The **uterus** is a thick-walled, muscular organ about the size and shape of an inverted pear. Normally it lies above and is tipped over the urinary bladder. The oviducts join the uterus anteriorly, while posteriorly, the cervix enters into the vagina nearly at a right angle. A small opening in the cervix leads to the vaginal canal. Development of the embryo takes place in the uterus. This organ, sometimes called the womb, is approximately 2 inches wide in its usual state but is capable of stretching to over 12 inches to accommodate the growing baby. The lining of the uterus, called the **endometrium,** participates in the formation of the placenta (p. 427), which supplies nutrients needed for embryonic and fetal development. The endometrium has two layers: a basal layer and an inner functional layer. In the nonpregnant female, the functional layer

of the endometrium varies in its thickness according to a monthly reproductive cycle, called the menstrual cycle (p. 401).

Cancer of the cervix is a common form of cancer in women. Early detection is possible by means of a Pap test, which requires that the doctor remove a few cells from the region of the cervix for microscopic examination. If the cells are cancerous, a hysterectomy may be recommended. A hysterectomy is the removal of the uterus. Removal of the ovaries in addition to the uterus, is termed an ovariohysterectomy. Since the vagina remains, the woman may still engage in sexual intercourse.

Vagina

The **vagina** is a tube that makes a 45-degree angle with the small of the back. The mucosal lining of the vagina lies in folds that extend as the fibromuscular wall stretches. This capacity to extend is especially important when the vagina serves as the birth canal, and it may also facilitate intercourse when the vagina receives the penis during copulation.

External Genitalia

The external genital organs of the female (fig. 19.10) are known collectively as the **vulva.** The vulva includes two large, hair-covered folds of skin called the **labia majora.** They extend backward from the **mons pubis,** a fatty prominence underlying the pubic hair. The **labia minora** are two small folds lying just inside the labia majora. They extend forward from the vaginal opening to encircle and form a foreskin for the **clitoris,** an organ that is homologous to the penis. Although quite small, the clitoris has a shaft of erectile tissue and is capped by a pea-shaped glans. The clitoris also has sense receptors that allow it to function as a sexually sensitive organ.

The **vestibule,** a cleft between the labia minora, contains the openings of the urethra and the vagina. The vagina may be partially closed by a ring of tissue called the hymen. The hymen is ordinarily ruptured by initial sexual intercourse; however, it can also be disrupted by other types of physical activities. If the hymen persists after sexual intercourse, it can be surgically ruptured.

Notice that the urinary and reproductive systems in the female are entirely separate. For example, the urethra carries only urine and the vagina serves only as the birth canal and the organ for sexual intercourse.

Orgasm

Sexual response in the female may be more subtle than in the male, but there are certain corollaries. The clitoris is believed to be an especially sensitive organ for initiating sexual sensations. It is possible for the clitoris to become ever so slightly erect as its erectile tissues become engorged with blood. But vasocongestion is more obvious in the labia minora, which expand and deepen in color. Erectile tissue within the vaginal wall also expands with blood, and the added pressure in these blood vessels causes small droplets of fluid to squeeze through the vessel walls and lubricate the vagina.

Release from muscle tension occurs in females, especially in the region of the vulva and vagina but also throughout the entire body. Increased uterine motility may assist the transport of sperm toward the oviducts. Since female organism is not signaled by ejaculation, there is a wide range in normality regarding sexual response.

Figure 19.10
External genitalia of female. At birth, the opening of the vagina is partially occluded by a membrane called the hymen. Physical activities and sexual intercourse disrupt the hymen.

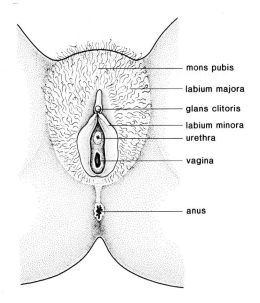

mons pubis
labium majora
glans clitoris
labium minora
urethra
vagina
anus

Figure 19.11

Hypothalamic-pituitary-gonad system (simplified) as it functions in the female. GnRH is a hypothalamic-releasing hormone that stimulates the anterior pituitary to secrete LH and FSH. These gonadotropic hormones act on the ovaries. FSH promotes the development of the follicle that later, under the influence of LH, becomes the corpus luteum. Negative feedback controls the level of all hormones involved.

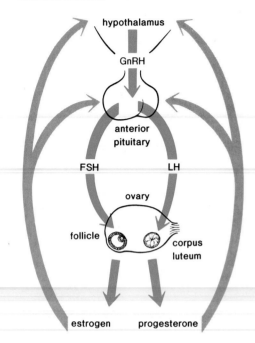

Table 19.3 Ovarian and menstrual cycles (simplified)

Ovarian cycle phases	Events	Menstrual cycle phases	Events
Follicular Days 1–13	FSH secretion by pituitary	Menstruation Days 1–5	Endometrium breaks down
	Follicle maturation and secretion of estrogen	Proliferation Days 6–13	Endometrium rebuilds
Ovulation Day 14*			
Luteal Days 15–28	LH secretion by pituitary	Secretory Days 15–28	Endometrium thickens and glands are secretory
	Corpus luteum formation and secretion of progesterone		

*Assuming a 28-day cycle.

Hormonal Regulation in the Female

Hormonal regulation in the female is quite complex, so we will begin with a simplified presentation and follow with a more in-depth presentation (p. 403) for those who wish to study the matter in more detail. The following glands and hormones are involved in hormonal regulation:

Hypothalamus: secretes GnRH gonadotropic-releasing hormone
Anterior pituitary: secretes *FSH* (follicle-stimulating hormone) and *LH* (luteinizing hormone), the gonadotropic hormones
Ovaries: secrete estrogen and progesterone, the female sex hormones

Hormonal Regulation (Simplified)

Ovarian Cycle

The gonadotropic and sex hormones are not present in constant amounts in the female and instead are secreted at different rates during a monthly ovarian cycle, which lasts an average of 28 days but may vary widely in specific individuals. For simplicity's sake it is convenient to emphasize that during the first half of a 28-day cycle (days 1 to 13, table 19.3) FSH from the anterior pituitary is promoting the development of a follicle in the ovary and that this follicle is secreting estrogen. As the blood estrogen level rises, it exerts feedback control over the anterior pituitary secretion of FSH so that this follicular phase comes to an end (fig. 19.11). The end of the follicular phase is marked by ovulation on the fourteenth day of the 28-day cycle. Similarly it may be emphasized that during the last half of the ovarian cycle (days 15-28, table 19.3) anterior pituitary production of LH is promoting the development of a corpus luteum, which is secreting progesterone. As the blood progesterone level rises, it exerts feedback control over anterior pituitary secretion of LH so that the corpus luteum begins to degenerate. As the luteal phase comes to an end, menstruation occurs.

Reproduction, Development, and Inheritance

Menstrual Cycle

The female sex hormones estrogen and progesterone have numerous functions, one of which is discussed here. The effect these hormones have on the endometrium of the uterus causes the uterus to undergo a cylical series of events known as the **menstrual cycle** (table 19.3). Cycles that last 28 days, are divided as follows:

During *days 1 to 5* there is a low level of female sex hormones in the body, causing the uterine lining to disintegrate and its blood vessels to rupture. A flow of blood, known as the **menses,** passes out of the vagina during a period of **menstruation,** also known as the menstrual period.

During *days 6 to 13* increased production of estrogen by an ovarian follicle causes the endometrium to thicken and become vascular and glandular. This is called the proliferation phase of the menstrual cycle.

Ovulation usually occurs on the fourteenth day of the 28-day cycle.

During *days 15 to 28* increased production of progesterone by the corpus luteum causes the endometrium to double in thickness and the uterine glands to become mature, producing a thick mucoid secretion. This is called the secretory phase of the menstrual cycle. The endometrium is now prepared to receive the developing zygote, but if pregnancy does not occur, the corpus luteum degenerates and the low level of sex hormones in the female body causes the uterine lining to break down. This is evident, due to the menstrual discharge that begins at this time. Even while menstruation is occurring, the anterior pituitary begins to increase its production of FSH and a new follicle begins maturation. Table 19.3 indicates how the ovarian cycle controls the menstrual cycle.

Pregnancy

If pregnancy occurs, menstruation does not occur. Instead, the developing zygote embeds itself in the endometrium lining several days following fertilization. This process, called **implantation,** is what causes the female to become **pregnant.** During implantation, an outer layer of cells surrounding the zygote produces a gonadotropic hormone (**HCG,** or **H**uman **C**horionic **G**onadotropic hormone) that prevents degeneration of the corpus luteum and instead causes it to secrete even larger quantities of progesterone. The corpus luteum may be maintained for as much as six months, even after the placenta is fully developed.

The **placenta** (fig. 20.14) originates from both maternal and fetal tissue and is the region of exchange of molecules between fetal and maternal blood although there is no mixing of the two types of blood. After its formation, the placenta continues production of HCG and begins production of progesterone and estrogen. The latter hormones have two effects: they shut down the anterior pituitary so that no new follicles mature, and they maintain the lining of the uterus so that the corpus luteum is not needed. There is no menstruation during the period of pregnancy.

Pregnancy Tests   Pregnancy tests, which are readily available in hospitals, clinics, and, now, even drug and grocery stores, are based on the fact that HCG is present in the blood and urine of a pregnant woman.

Figure 19.12
Anatomy of breast. The female breast contains lobules consisting of ducts and alveoli. The alveoli are lined by milk-producing cells in the lactating (milk-producing) breast.

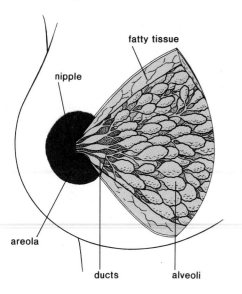

Before the advent of monoclonal antibodies, only a blood test requiring the use of radioactive material in the hospital was available to detect pregnancy before the first missed menstrual period. Now there is a monoclonal antibody (p. 254) test for the detection of pregnancy 10 days after conception. This test can be done on a urine sample in a doctor's office and the results are available within the hour.

The physical signs that might prompt a woman to have a pregnancy test are cessation of menstruation, increased frequency of urination, morning sickness, and increase in the size and fullness of the breasts, as well as darkening of the areolae.

Female Sex Hormones

The female sex hormones, estrogen and progesterone, have many effects on the body. In particular, estrogen secreted at the time of puberty stimulates the growth of the uterus and vagina. Estrogen is necessary for egg maturation and is largely responsible for the secondary sex characteristics in females. For example, it is responsible for the onset of the menstrual cycle, as well as female body hair and fat distribution. In general, females have a more rounded appearance than males because of a greater accumulation of fat beneath the skin. Also, the pelvic girdle enlarges in females so that the pelvic cavity has a larger relative size compared to males; this means that females have wider hips. Both estrogen and progesterone are also required for breast development.

Breasts A female breast contains 15 to 25 lobules (fig. 19.12), each with its own milk duct that begins at the nipple and divides into numerous other ducts that end in blind sacs called **alveoli.** In a nonlactating (nonmilk-producing) breast, the ducts far outnumber the alveoli because alveoli are made up of cells that can produce milk.

Milk is not produced during pregnancy. **Lactogenic hormone** (prolactin) is needed for lactation (milk production) to begin, and the production of this hormone is suppressed because of the feedback inhibition estrogen and progesterone have on the pituitary during pregnancy. It takes a couple of days after delivery for milk production to begin and, in the meantime, the breasts produce a watery, yellowish white fluid called **colostrum,** which differs from milk in that it contains more protein and less fat. The continued production of milk requires continued breastfeeding. When a breast is suckled, the nerve endings in the areola are stimulated and nerve impulses travel to the hypothalamus, which causes oxytocin to be released by the posterior pituitary. When this hormone arrives at the breasts, it causes contraction of the lobules so that milk flows into the ducts (called milk letdown).

Menopause

Menopause, the period in a woman's life during which the menstrual cycle ceases, is likely to occur between ages 45 and 55. The ovaries are no longer responsive to the gonadotropic hormones produced by the anterior pituitary, and the ovaries no longer secrete estrogen or progesterone. At the onset of menopause, the menstrual cycle becomes irregular, but as long as menstruation occurs it is still possible for a woman to conceive and become pregnant. Therefore, a woman is usually not considered to have completed menopause until there has been no menstruation for a year. The hormonal changes during menopause often produce physical symptoms, such as "hot flashes" that are caused by circulatory irregularities, dizziness, headaches, insomnia, sleepiness, and depression. Again, there is a great variation among women and any of these symptoms may be absent altogether.

Reproduction, Development, and Inheritance

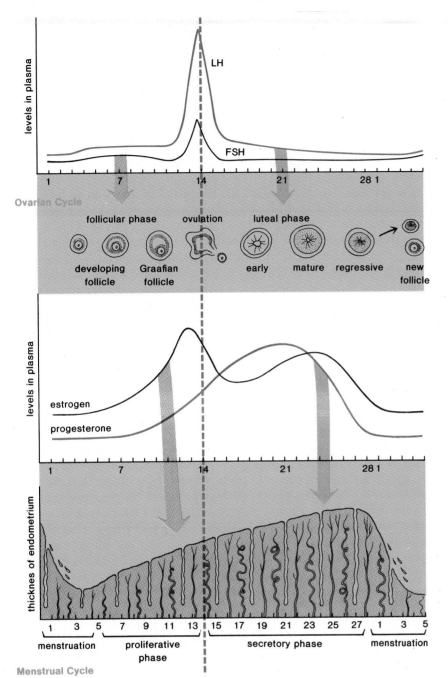

Women sometimes report an increased sex drive following menopause, and it has been suggested that this may be due to androgen production by the adrenal cortex.

Hormonal Regulation (in Detail)

Figure 19.13 shows the changes in blood concentration of all four hormones participating in the ovarian and menstrual cycles. Notice that all four of these hormones (FSH, LH, estrogen, and progesterone) are present during the entire 28 days of the cycle. Thus, in actuality, both FSH and LH *are* present during the follicular phase and both are needed for follicle development and

maturation of the egg. The follicle secretes primarily estrogen and a very minimal amount of progesterone. Similarly, both LH and FSH are present in decreased amounts during the luteal phase. LH may be primarily responsible for corpus luteum formation, but the corpus luteum secretes both progesterone and estrogen. The effect that these hormones have on the endometrium has already been stated. Estrogen stimulates growth of the endometrium and readies it for reception of progesterone, which causes it to thicken and become secretory.

Feedback Control   It has been frequently mentioned that a hormone can exert feedback inhibition. Therefore it comes as no surprise to find that as estrogen level increases during the first part of the follicular phase, FSH secretion begins to decrease. However, toward the end of the follicular phase there is a sharp increase in FSH and LH secretion at the point when the estrogen level is the highest. Thus it is believed that in this instance the high level of estrogen exerts *positive feedback on the hypothalamus,* causing it to secrete gonadotropic-releasing hormone, after which the pituitary momentarily produces an unusually large amount of FSH and LH. It is the surge of LH that is believed to promote ovulation. During the luteal phase, estrogen and progesterone bring about feedback inhibition as expected and the level of both LH and FSH declines steadily. Thus all four hormones eventually reach their lowest levels, causing menstruation to occur. It still is not known what causes the corpus luteum to degenerate if pregnancy does not occur. In some mammals there is evidence to suggest that prostaglandins (p. 385) cause degeneration, but this is not believed to be the case in humans.

Figure 19.14
Effectiveness (the percentage of women who are not expected to be pregnant within one year) of various birth control measures. Sterilization and the pill offer the best protection, while creams and so forth offer the least protection from the occurrence of pregnancy.

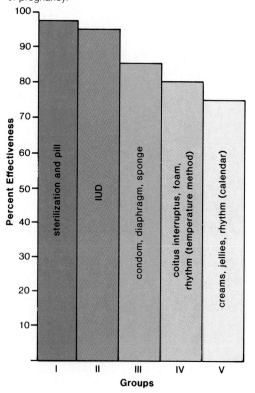

Birth Control

The use of birth control (contraceptive) methods decreases the probability of pregnancy. A common way to discuss pregnancy rate is to indicate the number of pregnancies expected per 100 women per year. For example, it is expected that 80 out of 100 young women, or 80 percent, who are regularly engaging in unprotected intercourse will be pregnant within a year. Another way to discuss birth control methods is to indicate their effectiveness, in which case the emphasis is placed on the number of women who will not get pregnant. For example, with the least effective method given in figure 19.14, we expect that 70 out of 100, or 70 percent, sexually active women will not get pregnant, while 30 women will get pregnant within a year.

Group I
Sterilization is a surgical procedure that renders the individual incapable of reproduction. Sterilization operations do not affect the secondary sex characteristics nor sexual performance.

In the male, a **vasectomy** consists of cutting and tying the ductus (vas) deferens on each side (fig. 19.2) so that the sperm are unable to reach the seminal fluid that is ejected at the time of orgasm. The sperm are then largely reabsorbed. Following this operation, which can be done in a doctor's office, the amount of ejaculate remains normal because sperm account for only about 1 percent of the volume of semen. Also, there is no effect on the secondary sex characteristics since testosterone continues to be produced by the testes.

In the female, **tubal ligation** consists of cutting or otherwise sealing the oviducts. Pregnancy rarely occurs because the passage of the egg through the oviducts has been blocked. Whereas major abdominal surgery was formerly required for a tubal ligation, today there are simpler procedures. Using a method called laparoscopy, which requires only two small incisions, the surgeon inserts a small, lighted telescope to view the oviducts and a small surgical

blade to sever them. An even newer method called hysteroscopy uses a telescope within the uterus to seal the tubes by means of an electric current.

While recently developed microsurgical methods allow either a ductus deferens or oviduct to be rejoined, it is still wise to view a vasectomy or tubal ligation as permanent. Even following successful resectioning, fertility is usually reduced by about 50 percent.

The **birth control pill** (fig. 19.15d) is usually a combination of estrogen and progesterone that is taken for 21 days of a 28-day cycle (beginning at the end of menstruation). The estrogen and progesterone in the pill effectively

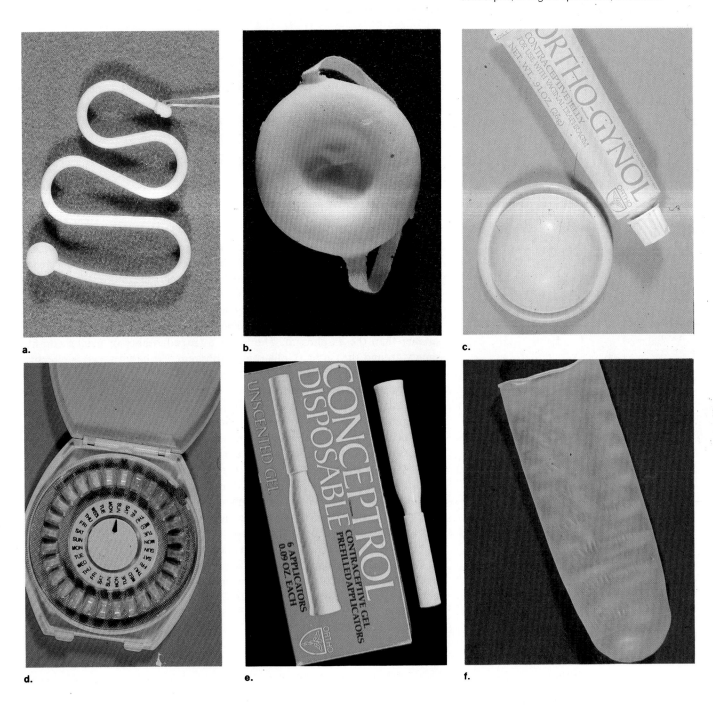

Figure 19.15
Various types of birth control methods: *a.* IUD; *b.* contraceptive sponge; *c.* diaphragm; *d.* birth control pills; *e.* vaginal spermicide; *f.* condom.

a.

b.

c.

d.

e.

f.

shut down the pituitary production of both FSH and LH so that no follicle begins to develop in the ovary; and since ovulation does not occur, pregnancy cannot take place. Both beneficial and adverse side effects have been linked to the pill. Women report relief of menstrual discomforts and acne. They also report several minor adverse side effects such as nausea and vomiting. Less common complaints are weight gain, headaches, and chloasma, areas of darkened skin on the face. One serious side effect of the pill is increased incidence of thromboembolism—almost exclusively in women who are over 35 and who smoke. An increased incidence of cervical and breast cancer, but a decreased incidence of uterine and ovarian cancer, has been observed among women who have taken the pill for several years. Since there are possible side effects, those taking the pill should always be seen regularly by a physician.

Group II

An intrauterine device (**IUD**) (fig. 19.15a) is a small piece of molded plastic that is inserted into the uterus by a physician. They are generally used in women who have given birth to at least one child. Most likely, IUDs prevent implantation by the embryo since there is often an inflammatory reaction where the device presses against the endometrium. The minor side effects of the IUD are expulsion, pain, irregular bleeding, or profuse menses. The major side effects of IUDs are rare and include pelvic infection and perforation of the uterus. The infection can usually be treated by antibiotics, but deaths have been known to occur.

Group III

The **diaphragm** (fig. 19.15c) is a soft rubber or plastic cup with a flexible rim that lodges behind the pubic bone and fits over the cervix. Each woman must be properly fitted by a physician, and the diaphragm must be inserted into the vagina two hours at most before sexual relations. It must also be used with a spermicidal jelly or cream and should be left in place for at least six hours after relations. If intercourse is repeated during this time, more jelly or cream should be inserted by means of a plastic insertion tube.

The **cervical cap,** a widely used contraceptive device popular in Europe, is currently being introduced in this country. The cervical cap is thicker and smaller than the diaphragm. The thimble-shaped rubber or plastic cup fits snugly around the cervix. Unlike the diaphragm, the cervical cap is effective even if left in place for several days.

A **condom** (fig. 19.15f) is a thin latex rubber sheath that fits over the erect penis. The ejaculate is trapped inside the sheath and thus does not enter the vagina. When used in conjunction with a spermicidal foam, cream, or jelly, the protection is better than with the condom alone. Today it is possible to purchase condoms that are already lubricated with a spermicide.

A **vaginal sponge** (fig. 19.15b) permeated with spermicide and shaped to fit the cervix is a new contraceptive recently made available to the general public after seven years of testing. Unlike the diaphragm and cervical cap, the sponge need not be fitted by a physician since one size fits everyone. It is effective immediately after placement in the vagina and remains effective for 24 hours. Like the other means of birth control in this category, the sponge is about 85 percent effective in preventing pregnancy. Both the condom and the sponge give some protection against sexually transmitted diseases such as those discussed in the reading on pages 408–9.

Reproduction, Development, and Inheritance

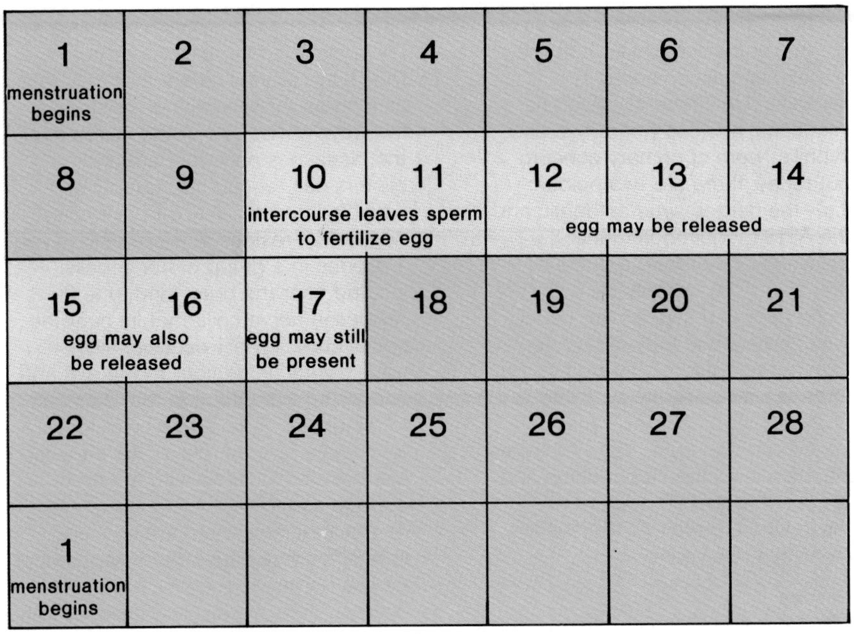

1 menstruation begins	2	3	4	5	6	7
8	9	10 intercourse leaves sperm to fertilize egg	11	12 egg may be released	13	14
15 egg may also be released	16	17 egg may still be present	18	19	20	21
22	23	24	25	26	27	28
1 menstruation begins						

Figure 19.16
Natural family planning. The calendar shows the "safe" and "unsafe" days for intercourse. This calendar is appropriate *only* for women with regular 28-day cycles. Few women have regular cycles month after month.

Group IV

It is possible for the male to withdraw the penis just before ejaculation so that the semen is deposited away from the vaginal area. This method of birth control, called **coitus interruptus,** has a relatively high failure rate because a few drops of seminal fluid may escape from the penis before ejaculation takes place. Even a small amount of semen can contain numerous sperm.

Spermicidal jellies, creams, and foams (fig. 19.15e) contain sperm-killing ingredients and may be inserted into the vagina with an applicator up to 30 minutes before each occurrence of intercourse. Foams are considered the most effective of this group of contraceptives. Used alone, these are not highly effective means of birth control for those who have frequent intercourse. They do offer some protection against sexually transmitted diseases.

Group V

Natural family planning, formerly called the **rhythm method** of birth control, is based on the realization that a woman ovulates only once a month and that the egg and sperm are viable for a limited number of days. If the woman has a consistent 28-day cycle, then the period of "safe" days can be determined, as in figure 19.16. This method of birth control is not very effective because the days of ovulation can vary from month to month, and the viability of the egg and sperm varies, perhaps monthly but certainly from person to person.

A more reliable way to practice natural family planning is to await the day of ovulation each month and then wait three more days before engaging in intercourse. The day of ovulation can be more accurately determined by noting the body temperature early each morning (body temperature rises at ovulation) or by taking the pH of the vagina each day (near the day of ovulation the vagina becomes more alkaline) or by noting the consistency of the mucus at the cervix (at ovulation the mucus is thicker and heavier). Physicians can instruct women how to do these procedures.

Sexually Transmitted Diseases

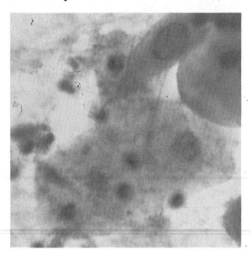

Cell infected with herpes virus.

There are about a dozen sexually transmitted diseases that are sometimes called venereal diseases. Until recently, bacterial infections, such as gonorrhea and syphilis, were of primary concern; but today, there are two epidemics—the herpes virus epidemic and the AIDS (acquired immune deficiency syndrome) epidemic—that all persons should be aware of. Gonorrhea and syphilis are usually curable because they most often respond to antibiotic treatment, but there are no currently available drugs to cure a herpes virus infection nor to cure AIDS. A herpes infection is a chronic condition and is not lethal, but 70 percent of those who contract AIDS usually die within two years.

Herpes

There are two types of herpes viruses that are involved in the current epidemic. *Herpes simplex virus 1* usually causes cold sores and fever blisters on the lips, and *herpes simplex virus 2* is the type that usually causes genital lesions. An estimated 20.5 million Americans now have genital herpes, and an estimated half-million additional cases are expected each year according to the Center for Disease Control in Atlanta. Some have even estimated that one out of every six sexually active individuals may be capable of spreading the disease to another.

Immediately after infection there are no symptoms, but once inside the virus begins to multiply rapidly. The individual may experience a tingling or itching sensation before blisters appear at the infected site within 2 to 20 days. Once the blisters rupture, they leave painful ulcers that may take as long as three weeks or as little as five days to heal. After the ulcers heal, the disease is only dormant and blisters can reoccur repeatedly at variable intervals. While the virus is dormant, it resides in nerve cells. Type 1 resides in a group of nerve cells located near the brain, and type 2 resides in nerve cells that lie near the spinal cord. Type 1 occasionally travels via a nerve fiber to the eye and causes an eye infection that can lead to blindness. Type 2 has been known to cause a form of meningitis and also has been associated with a form of cervical cancer.

Sunlight, sex, menstruation, and stress seem to cause the symptoms of genital herpes to reoccur. The ointment form of acyclovir relieves only initial symptoms, but the oral form, only recently developed, prevented outbreaks in about 70 percent of patients in a trial study.

In addition to dealing with the infection, people who have herpes often undergo psychological distress. Those who do not tell potential sexual partners risk the danger of passing it on, particularly if the disease is then active, and those who do tell potential sexual partners risk the danger of being rejected. Since the disease is most common among the promiscuous, some young people have changed their life-style to prevent infection. Although it is possible to learn to cope with an infection, those who already have herpes often feel that their entire lives will be affected.

Only a medically recognized method of birth control such as those discussed here should be used. Douching is of little value and position of intercourse will not prevent pregnancy at all. In fact, the proximate location of the penis (at the time of ejaculation) near but not in the vagina has been known to result in pregnancy.

Future Means of Birth Control

There are three areas in which birth control investigations have been directed. There is a need for long-lasting birth control methods, morning-after medication, and medication that is specifically for males.

Depo-Provera is an injectable contraceptive that is commercially available in many countries outside the United States. The injection contains crystals that gradually dissolve over a period of three months. The crystals contain a chemical related to progesterone and, like the pill, it suppresses ovulation.

Reproduction, Development, and Inheritance

AIDS

The cause of AIDS is not definitely known but doctors think a blood-borne cancer virus is to blame. The culprit appears to be a leukemia retrovirus that reduces the number of T cells needed to promote the production of antibodies. Retroviruses integrate their genes into that of the host cell and therefore they are difficult to detect, but as many as 50 percent of AIDS patients have tested positive for HTLV (human T cell leukemia virus).

Only members of certain high-risk groups usually contract AIDS. Presently, homosexual men make up 71 percent of all cases; 17 percent are intravenous drug abusers; 4 percent are Haitian immigrants; and 1 percent are hemophiliacs who use blood-clotting agents derived from blood plasma. Sexual partners and the children of these individuals are also considered a high-risk group. The common denominator of all groups is a weakened immune system. Many homosexual men have a history of venereal diseases and other infections that may have suppressed their immune system. Intravenous drug users often have had hepatitis; and children do not yet have a fully functioning immune system. Intimate contact, such as sexual contact, is still needed before one is exposed to the disease. Blood-to-blood contact will also spread AIDS and this alone most likely accounts for its presence among hemophiliacs.

AIDS is characterized by an immune system that is so weak that its victims are prey to all manner of viruses and bacteria. Patients often die due to a rare form of skin cancer or a rare form of pneumonia. Interferon injections have been effective against the skin cancer, but patients still contract and die from other diseases. Scientists have discovered that a T cell lymphokine, called interleukin-2, will stimulate white cells taken from AIDS patients and, therefore, this substance is now being tested as a possible treatment.

It is projected that there could be more than 20,000 cases of AIDS by 1985. But this may never materialize, especially because many homosexuals are changing their life-style. They are aware that the chances of contacting AIDS increase with promiscuity and therefore they are establishing monogamous relationships or remaining celibate.

Gonorrhea

Gonorrhea is the best-known bacterial disease that is transmitted sexually. In men, gonorrhea is usually readily detected because pus is discharged from the penis and burning sensations during urination develop within a few days after infection. But in women, the infection concentrates in the cervical canal and produces very mild, if any, symptoms. Women may unknowingly develop extended infections in the reproductive tract, resulting in infection and inflammation of the oviducts. This can lead to partial or complete blockage of the ducts, and thus sterility. Physicians have routinely treated gonorrhea with large doses of penicillin, but penicillin-resistant strains are becoming more common, and treatment is more complex than it once was.

Syphilis

Syphilis is a bacterial disease with an incubation period of about three weeks. After this amount of time, the first symptom—a hard, painful ulcer called a chancre—develops at the site of the infection. Even if the disease is left untreated, the chancre disappears after a short time, and the individual sees no more sign of disease for two to four months. At this time, the disease enters its secondary stage as a generalized skin rash, and infections of various organs sometimes develop. Then, the disease goes into a latent period that can last throughout life. But in some people, the disease can enter a phase that produces severe nervous system or circulatory system damage and even death. Syphilis also usually yields to penicillin treatment, but as in the case of gonorrhea, increasingly common antibiotic resistance is a growing problem.

Immunization

Immunization for these diseases is not yet possible. Resistance to antibiotic treatment has spurred interest in developing a vaccine for gonorrhea and syphilis. Several promising vaccines have been prepared and are presently being tested for herpes. Of 300 persons who received a herpes vaccine in England, only two developed the infection despite the fact that their sexual partners had the disease. Possibly a vaccine for herpes will be available to the general public by 1985. Researchers are hopeful that a vaccine for AIDS will also be developed in the near future.

The drug has not been approved for use in the United States because cancer developed in some test animals receiving the injections. More animal studies are now underway. An even more potent progesterone-like molecule is now being tested for implantation under the skin. The *implant* consists of narrow tubes that slowly release the drug over a period of five years.

Experimentation also goes forward on a *morning-after pill,* but as yet such a pill is not available to the general public. DES, a synthetic estrogen, which affects the uterine lining, making implantation difficult, is sometimes given following intercourse. Since large doses are required, causing nausea and vomiting, DES is usually given only for incest or rape. A synthetic steroid that has a high affinity for progesterone receptors is also under investigation as a post-fertilization means of birth control. This drug prevents progesterone from acting on the uterine lining and reduces the probability of implantation and/or the continuance of pregnancy. It is possible that prostaglandins that

cause contraction of the uterus and disintegration of the corpus luteum could be taken whenever menstruation is late. But since the prostaglandins have many side effects they are only given now under close supervision.

Various possibilities exist for a *"male pill."* Scientists have made analogs of gonadotropic-releasing hormone that interfere with the action of this hormone and prevent it from stimulating the pituitary. Experiments in both animal and human subjects suggest that one of these might possibly inhibit spermatogenesis in males (and ovulation in females) without affecting the secondary sex characteristics. It is believed that the seminiferous tubules produce a chemical that inhibits FSH production by the pituitary (p. 396). It is hoped that this chemical may someday be produced commercially and made available in pill form for males. Testosterone and/or related chemicals can be used to inhibit spermatogenesis in males, but there are usually feminizing side effects because the excess is changed to estrogen by the body.

Abortion

Whereas a miscarriage is the unexpected loss of an embryo or fetus, an **abortion** is the purposeful removal of an embryo or fetus from the womb. If an abortion is done within the first three months, hospitalization is not needed and recovery requires only a few hours. The most common technique of removing the embryo from the uterus involves dilation of the cervix and aspiration of the embryo and its membranes. As a suction machine removes the inner lining of the uterus, the patient experiences only a strange pulling sensation.

Delayed abortion, which is also available from four to six months, is more complicated and hospitalization is required. During the fourth month, dilation and sharp curretage can be utilized. In sharp curretage, the physician scrapes the uterine lining in order to remove the fetus. After the fourth month, the fetus must be expelled by way of the vagina. In the saline procedure, several ounces of fluid are removed from the amniotic sac around the fetus by a needle inserted in the abdomen and uterus. When this is replaced by a salt solution, expulsion of a nonliving or short-lived fetus follows. The patient will experience at least some labor pains just as if a normal birth were going to occur.

A study of the public health record indicates that there has been a decline in deaths due to abortions since they became legal in the 1970s. The present risk of dying from an abortion induced during the first 15 weeks of pregnancy is now considered to be one-seventh the risk of dying from pregnancy and childbirth.

Infertility

Sometimes couples do not need to prevent pregnancy; conception or fertilization does not occur despite frequent intercourse. The American Medical Association estimates that 15 percent of all couples in this country are unable to have any children and are therefore properly termed sterile; another 10 percent have fewer children than they wish and are therefore termed infertile.

Infertility can be due to a number of factors. There may be a congenital malformation of the reproductive tract, or the venereal disease gonorrhea (p. 409) may have caused an obstruction of the oviduct or ductus deferens. Sometimes these physical defects can be corrected surgically.

A hormonal imbalance, which can cause failure to ovulate in females and a low sperm count in males, can be treated medically. As a last resort, it is possible to give females a substance rich in FSH and LH that is extracted from the urine of postmenopausal women. This treatment causes multiple ovulations and sometimes multiple pregnancies, however. If the sperm count

Reproduction, Development, and Inheritance

in the male is low, it is possible to concentrate the sperm and use this to artificially inseminate the female. However, the emphasis of late has been on the quality of the sperm rather than the total number of sperm since men with very low sperm counts are known to father children naturally.

One area of concern is that radiation, chemical mutagens, and the use of psychedelic drugs can contribute to sterility, possibly by causing chromosomal defects.

In Vitro Fertilization

Assuming that the sperm and egg are normal, the new method of external fertilization followed by implantation of the zygote is a possibility for some couples. After appropriate hormonal treatment of the woman, laparoscopy (p. 404) allows the removal of eggs from Graafian follicles that have ballooned out of the ovary. Sperm from the males are placed in a solution that approximates the conditions of the female genital tract. When the eggs are introduced, fertilization occurs. The resultant zygotes begin development and after about two to four days they are inserted into the uterus of the woman, who is now in the secretory phase of her menstrual cycle. If an implantation is successful, development continues in the usual manner. Since the first successful birth by in vitro fertilization in 1978, in vitro fertilization clinics have opened in several countries, including the United States. Thus far, the number of birth defects has been no greater using this method than for normally conceived pregnancies. Despite the cost of about $3,000 and the low success rate of only about 20 percent per attempt, the demand is so great that at least 100 to 200 more clinics are expected in the United States during the next year or so.

Summary

In males, spermatogenesis occurs within the seminiferous tubules of the testes, which also produces testosterone within the interstitial cells. Sperm have a head, capped by an acrosome, where 23 chromosomes reside in the nucleus, a mitochondria-containing mid-piece, and a tail with a 9 + 2 pattern of microtubules. Sperm mature in the epididymis and are stored in the ductus deferens before entering the urethra just prior to ejaculation. The inguinal canal, which marks the area where the testes descended into the scrotal sac, is often a region where a hernia may develop.

The accessory glands (seminal vesicles, prostate gland, and Cowper's gland) produce seminal fluid. Semen, which contains sperm and seminal fluid, leaves the penis during ejaculation. The penis, which has a foreskin that may be removed by circumcision, must become erect in order to be placed in the vagina of a female. Erection occurs when blood sinuses within erectile tissue fill with blood. If sexual stimulation is sufficient, ejaculation follows an erection and this is an obvious sign of male orgasm. As many as 400 million sperm may be ejaculated. Although only one sperm fertilizes an egg, the others may assist the process. The acrosome contains enzymes that can digest away the cells and the material surrounding the egg.

Hormonal regulation in the male maintains testosterone at a fairly constant level. The hypothalamus produces a gonadotropic hormone that stimulates the anterior pituitary to produce FSH and LH, which are present in both sexes. In males, FSH promotes spermatogenesis and LH promotes testosterone production. Via a feedback mechanism, testosterone, in particular, inhibits the anterior pituitary production of LH. Evidence suggests that a chemical produced by the seminiferous tubules controls FSH secretion. Testosterone stimulates growth of the male genitals during puberty and is necessary for maturation of sperm and development of the secondary sex characteristics—those features that we associate with the male body aside from the genitals.

In females, oogenesis occurs within the ovaries where one follicle reaches maturity each month. This follicle balloons out of the ovary and bursts to release the egg. The ruptured follicle develops into a corpus luteum. The follicle and corpus luteum produce the female sex hormones estrogen and progesterone.

The egg must cross a small space to enter the oviducts, which conduct it toward the uterus. Fertilization takes place within the oviducts. If fertilization occurs, the zygote embeds itself in the uterine lining and the corpus luteum is maintained. If fertilization does not take place, the corpus luteum degenerates. The vagina, the copulatory organ in females, opens into the vestibule where the urethra also opens. The vestibule is bounded by the labia minora, which come together at the clitoris, a highly sensitive organ. Outside the labia minora are the labia majora. The entire region of the external genitalia is called the vulva. There is no ejaculation in the female and therefore orgasm is harder to detect and varies widely in normality.

Hormonal regulation in the female results in an ovarian cycle. For those who do not wish to study hormonal regulation in detail, it is simplest to emphasize that during the first half of the cycle, FSH from the anterior pituitary causes maturation of the follicle, which secretes estrogen. After ovulation, and during the second half of the cycle, LH from the anterior pituitary converts the follicle into the corpus luteum, which produces progesterone. Estrogen and progesterone regulate the menstrual cycle. At first, during the menstrual cycle, a low level of hormones causes the endometrium to break down as menstruation occurs (days 1 to 5). As estrogen begins to be produced by the follicle, the endometrium begins to rebuild (proliferation phase, days 6 to 13). Ovulation occurs on the fourteenth day of a 28-day cycle. As progesterone is produced by the corpus luteum, the endometrium thickens and becomes secretory (secretory phase, days 14 to 28).

Estrogen and, to some extent, progesterone affect the female genitals, promote development of the egg, and maintain the secondary sex characteristics. Lactogenic hormone causes the breasts to begin to secrete milk after delivery while another hormone, oxytocin, is responsible for milk letdown. When menopause occurs, FSH and LH are still produced by the anterior pituitary, but the ovaries are no longer able to respond.

Numerous birth control methods and devices are available for those who wish to prevent pregnancy. The methods that are 85 to 100 percent effective, are sterilization, the pill, the IUD, the sponge, and the diaphragm. A condom used with a spermicidal jelly or foam is also this effective. Methods that are more in the range of 70 to 85 percent effective are spermicidal foam and jelly alone, coitus interruptus, and natural family planning. The latter method, which is based on abstinence during the "fertile period," can lead to unexpected pregnancy because ovulation *can* occur at any time during the cycle.

Some couples are infertile. There may be a blockage in the oviducts that can be corrected surgically in some cases. Either sex may have a hormonal imbalance that can also be corrected. Also available to a few is in vitro fertilization, followed by implantation of the developing zygote.

Study Questions

1. Discuss the anatomy and physiology of the testes. (p. 391) Describe the structure of sperm. (p. 392)
2. Give the path of sperm. (p. 393)
3. What glands produce seminal fluid? (pp. 393–94)
4. Discuss the anatomy and physiology of the penis. (p. 394) Describe ejaculation. (p. 395)
5. Discuss hormonal regulation in the male. Name three functions for testosterone. (pp. 395–96)

Reproduction, Development, and Inheritance

6. Discuss the anatomy and physiology of the ovaries. (pp. 396–98) Describe ovulation. (p. 398)
7. Give the path of the egg. Where does fertilization and implantation occur? Name two functions for the vagina. (pp. 398–99)
8. Describe the external genitalia in females. (p. 399)
9. Compare male and female orgasm. (pp. 395, 399)
10. Discuss hormonal regulation in the female, either simplified and/or in detail. (pp. 400, 403) Give the events of the menstrual cycle and relate them to the ovarian cycle. (p. 401) In what way is menstruation prevented if pregnancy occurs? (p. 401)
11. Name four functions of the female sex hormones. (p. 402) Describe the anatomy and physiology of the breast. (p. 402)
12. Discuss the various means of birth control and their relative effectiveness. (pp. 404–10)

Selected Key Terms

testes

seminiferous tubules (sem″ĭ-nif′er-us tu′bulz)

interstitial cells (in″ter-stish′al selz)

epididymis (ep″ĭ-did′ĭ-mis)

ductus deferens (duk′tus def′er-enz)

seminal fluid (sem′ĭ-nal floo′id)

seminal vesicles (sem′ĭ-nal ves′ĭ-k′lz)

prostate gland (pros′tāt gland)

Cowper's glands (kow′perz glandz)

ovary (o′var-e)

follicles (fol′ĭ-klz)

ovulation (o″vu-la′shun)

corpus luteum (kor′pus lu′e-um)

uterus (u′ter-us)

endometrium (en″do-me′tre-um)

vagina (vah-ji′nah)

vulva (vul′vah)

menstrual cycle (men′stroo-al si′kl)

placenta (plah-sen′tah)

menopause (men′o-pawz)

Further Readings

Boston Women's Health Book Collective. 1983. *The new our bodies, ourselves.* New York: Simon and Schuster.

Demarest, R. J., and Sciarra, J. J. 1969. *Conception, birth and contraception: A visual presentation.* New York: Blakiston.

Epel, D. 1977. The program of fertilization. *Scientific American* 237(5):128.

Goldstein, B. 1976. *Human sexuality.* New York: McGraw-Hill.

Jaffe, F. S. 1973. Public policy on fertility control. *Scientific American* 229(1):17.

Katchadourian, H. 1977. *The biology of adolescence.* San Francisco: W. H. Freeman.

Mader, S. 1980. *Human reproductive biology.* Dubuque, Iowa: Wm. C. Brown Publishers.

Masters, W. H., and Johnson, V. E. 1966. *Human sexual response.* Boston: Little, Brown.

Steen, E. B., and Price, J. H. 1977. *Human sex and sexuality.* New York: Wiley.

U.S. Department of Health, Education, and Welfare, Bureau of Disease Prevention and Environmental Control. 1968. *Syphilis: A synopsis.* Public Health Service Publication No. 1660, January. Washington, D.C.: U.S. Government Printing Office.

20

Human
Development

1 The first stages of human development, which lead to the
establishment of the embryonic germ layers, are the same as those
of other animal embryos.

2 Induction, or the ability of one tissue to influence the development of
another, can explain differentiation, or specialization of parts, and can
account for the orderliness of development.

3 Human embryos have the same extraembryonic membranes as the
embryos of reptiles and birds but their function has been altered to
suit internal development.

4 It is possible to outline in a precise manner the steps in human
development from the fertilized egg to the birth of a child.

Figure 20.1
Amphioxus development. Since the egg has little
yolk, cleavage is complete (*first row*), gastrulation
occurs by invagination (*second row*), and the
coelom develops by outpocketing (*third row*).
Presumptive notochord induces the formation of
the neural tube.

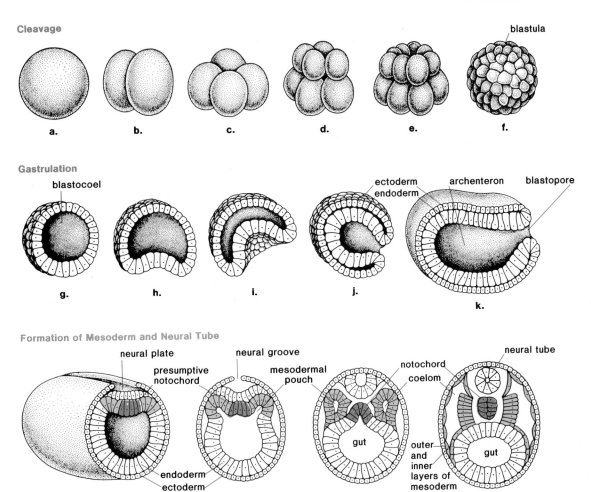

Cleavage

blastula

a.   b.   c.   d.   e.   f.

Gastrulation

blastocoel

ectoderm
endoderm

archenteron   blastopore

g.   h.   i.   j.   k.

Formation of Mesoderm and Neural Tube

neural plate   neural groove   neural tube

presumptive
notochord

mesodermal
pouch

notochord
coelom

gut

endoderm
ectoderm

outer
and
inner
layers of
mesoderm

gut

l.   m.   n.   o.

Development encompasses the time from **conception** (fertilization) to **birth** (parturition). In humans the **gestation period,** or length of pregnancy, is approximately nine months. It is customary to calculate the time of birth by adding 280 days to the start of the last menstruation because this date is usually known, whereas the day of fertilization is usually unknown. Because the time of birth is influenced by so many variables, only about 5 percent of babies actually arrive on the forecasted date.

Human development is very often divided into **embryonic development** (first two months) and **fetal development** (three through nine months). The embryonic period consists of early development, during which all the major organs form, and fetal development consists of a refinement of these structures. The fetus, not the embryo, is recognizable as a human being.

Much has been learned about human embryonic development by studying the early development of other animals whose eggs are more accessible, easier to see, and may be freely subjected to experimentation. As we discuss human embryonic development, we will have occasion to refer to the development of amphioxus (fig. 20.1), the frog, and chick in addition to humans. All these animals are chordates, animals that at some time in their life history

Table 20.1 Early developmental stages

Stage	Process	Result
Cleavage	Cell division without growth	Many-celled morula
Blastula	Morphogenesis and growth	Hollow ball of cells
Gastrula	Morphogenesis and growth	An embryo with three germ layers
Neurula	Differentiation by induction	Nervous system development

have an elastic supporting rod known as a **notochord.** In vertebrates, this rod is replaced by the vertebral column. All the animals mentioned except amphioxus are vertebrates.

Developmental Processes

All animal embryos develop by means of the following processes, which are also listed in table 20.1.

Cleavage   Immediately after fertilization, the zygote begins to divide so that at first there are 2, then 4, 8, 16, and 32 cells, and so forth. Since increase in size does not accompany these divisions, the embryo is at first no larger than the zygote was. Cell division during cleavage is mitotic, and each cell receives a full complement of chromosomes and genes.

Growth   Later cell division is accompanied by an increase in size of the daughter cells, and growth in the true sense of the term takes place.

Morphogenesis   Morphogenesis refers to the shaping of the embryo and is first evident when certain cells are seen to move, or migrate, in relation to other cells. By these movements, the embryo begins to assume various shapes.

Differentiation   When cells take on a specific structure and function, differentiation occurs. The first system to become visibly differentiated is the nervous system.

Early Developmental Stages

All embryos of higher animals go through the same early stages of development, as listed in table 20.1. Each stage can be identified by the events or the results of that stage.

Cleavage

This first stage is cell division without growth. It is best observed in an embryo such as amphioxus, which has little yolk, a rich nutrient material. (The yellow portion of a chick egg is the yolk.) Since amphioxus has little yolk, cell division is about equal and the cells are of a fairly uniform size (fig. 20.1a-f). Cleavage continues until there is a solid ball of cells called the **morula.**

Blastula

The second stage occurs when the cells of the morula more or less position themselves to create a space. In amphioxus a completely hollow ball, the **blastula,** results and the space within the ball is called the **blastocoel.** The human blastula, termed the **blastocyst,** consists of a hollow ball with a mass of cells—the **inner cell mass**—at one end. Figure 20.2 compares the appearance of a human embryo to that of amphioxus during the first stages of development.

Inner Cell Mass   Each cell within the inner cell mass has the genetic capability of becoming a complete individual. Sometimes during human development, the inner cell mass splits and two embryos start developing rather than one. These two embryos will be **identical twins** (fig. 20.3) because they have

Reproduction, Development, and Inheritance

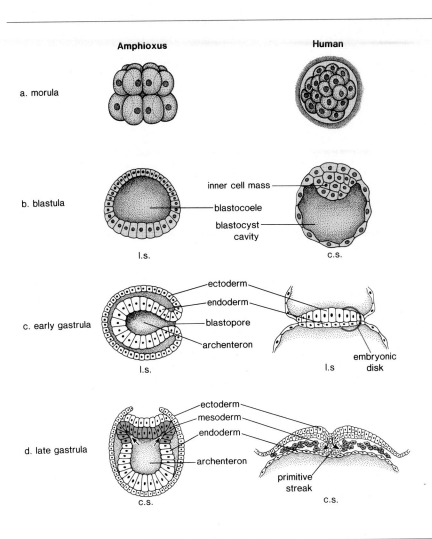

Amphioxus

Human

a. morula

b. blastula

inner cell mass
blastocoele
blastocyst cavity

l.s.

c.s.

c. early gastrula

ectoderm
endoderm
blastopore
archenteron

l.s.

embryonic disk

l.s

d. late gastrula

ectoderm
mesoderm
endoderm
archenteron

primitive streak

c.s.

c.s.

Figure 20.2
Comparison of amphioxus development and human development. *a.* Comparative morula stages. Cleavage is complete in both. *b.* Comparative blastula stages. The observed cavity is the blastocoele in amphioxus, but the blastocyst in humans. *c.* Comparative gastrula stages. The gastrula in amphioxus is spherical; the gastrula in humans is flattened. *d.* Comparative late gastrula stages. Outpocketing produces mesoderm in amphioxus, while invagination between ectoderm and endoderm produces mesoderm in humans.

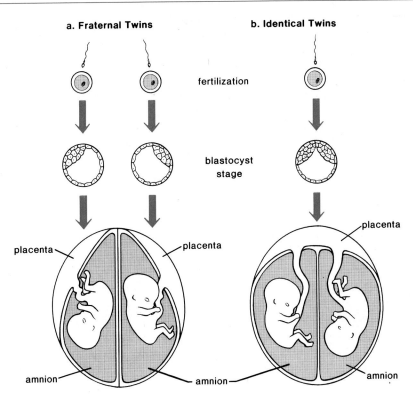

a. Fraternal Twins

b. Identical Twins

fertilization

blastocyst stage

placenta

placenta

placenta

amnion

amnion

amnion

Figure 20.3
Conception of fraternal versus identical twins.
a. Fraternal twins are formed when two eggs are released and fertilized. Fraternal twins receive a different genetic inheritance from both the mother and father. They can even have separate fathers.
b. Identical twins occur when an embryo breaks in two during an early stage of development. Identical twins have the exact same genetic inheritance from both the mother and father.

Closing in on Cloning

In the book and movie *The Boys from Brazil*, a demented Nazi doctor uses blood and tissue cells from Adolf Hitler to clone dozens of copies of the German dictator in the hope that at least one of them will seize power and conquer the world. Though the cloning of human beings is likely to be confined to fantasy for decades—perhaps forever—other kinds of cloning have long been possible. The Greek word *klon* means twig, and the simplest kind of vegetable cloning consists of cultivating cuttings from a plant. By the mid-1950s scientists had succeeded in cloning amphibians, producing frogs that were genetically identical to each other and carried the inherited characteristics of only a single parent. Most animal cloning has been done by transplanting nuclei into egg cells to produce an entire organism from a single cell. But the cloning of higher forms of life, like mammals, is hard to achieve. Mammal eggs are microscopic, ten to 20 times smaller in diameter than frogs' eggs, and vastly more difficult to manipulate. Consequently, the barriers to cloning laboratory mice had, until now, proved insurmountable. But [in 1981] . . . biologists had successfully done just that.

The work was carried out in Switzerland by Karl Illmensee of the University of Geneva and Peter Hoppe of the Jackson Laboratory in Bar Harbor, Me., both veteran researchers in cell biology. Their breakthrough was not in conception—since the procedures for cloning are familiar. It lay rather in the surgeon-like skill and persistence with which they used microscopic instruments to transplant nuclei from cell to cell.

First they scooped a mass of embryonic cells from the womb of a pregnant gray mouse. Using microscopes and a micropipette much finer than a human hair, they sucked out the cells' nuclei and, one by one, transplanted each into a recently fertilized egg extracted from another mouse. That mouse was black and functioned as a kind of genetic control. The researchers drew out the egg and sperm nuclei that were already in the black mouse's egg so that their genetic information could not influence the resulting clone. Next they cultured the cell in a solution of nutrients until it divided and grew into an early embryo, which was then inserted into the womb of a third mouse, this one white. The white mouse gave birth to a gray mouse, genetically identical to the original embryo.

In 363 tries, Illmensee and Hoppe managed to produce three such mice. The high failure rate was due mainly to the delicacy and complexity of the micromanipulative technique involved. In subsequent experiments, however, Illmensee and Hoppe had better luck. They generated several mice from a single embryo, all genetically identical to each other and thus true clones.

Almost every cell in an organism contains all the genetic information needed for reproducing the entire organism. But getting that

inherited exactly the same chromosomes. **Fraternal twins,** which arise when two different eggs are fertilized by two different sperm, do not have identical chromosomes. It has even been known to happen that these "twins" have different fathers.

Scientists have been able to demonstrate the genetic potential of inner cell mass cells within the laboratory as described in the reading on this page. Using mice embryos, scientists transplanted nuclei from these cells to newly enucleated eggs of a different mouse. The mice that developed had the genetic characteristics of the original embryo. Mice developing from the same inner cell mass are clones because they have exactly the same genes and are propagated by an asexual means. Cloning of human embryos has not as yet been achieved and cloning of an adult human, while frequently imagined, is considered to be very much in the future.

Reproduction, Development, and Inheritance

information to turn on or, as biologists put it, to "express" itself, is the main problem in animal cloning. It has been done now with immature cells. But as cells become differentiated, they seem to lose the ability to release genetic instructions for anything other than what they have become. A red blood cell can become only another red blood cell, for example. For that reason, Illmensee and Hoppe were only able to clone mice from embryonic cells that had not yet differentiated into cells for skin, bones, brains, eyes and other parts of the body. So far, there have been no undisputed reports of cloning from mature animal cells.

Nonetheless, the Illmensee-Hoppe mice, if they are produceable in large numbers, open many new avenues for research. The mice are, in fact, less important as clones than as vehicles for experiments in embryology, cell differentiation and immunology as well as in the study of birth defects and cancer. They enable medical researchers to introduce variables into otherwise genetically identical subjects, and then observe the results.

Some variation of the Illmensee-Hoppe technique may one day be used to clone prize bulls or even human beings. But other scientists question the ethics, as well as the scientific use, of trying to clone humans from undifferentiated cell masses. Whatever the original genetic imprint, the results would not be predictable, and mistakes would be stamped indelibly not on mice but on men.

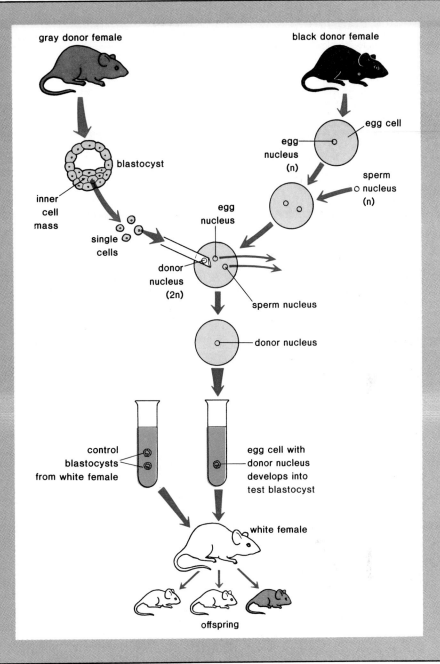

Gastrula

Gastrulation is evident in amphioxus when certain cells begin to push, or invaginate, into the blastocoel (fig. 20.1h-k). This creates a double layer of cells. The outer layer of cells is now called the **ectoderm** and the inner layer is called the **endoderm.** The space created by invagination will become the gut and is called either the **primitive gut** or the archenteron. The pore or hole created by the invagination is the **blastopore** and in amphioxus, as well as other vertebrates, this pore becomes the anus.

Gastrulation is not complete until a third middle layer of cells, the **mesoderm,** has been formed. In amphioxus this layer begins as outpocketings from the primitive gut; these outpocketings grow in size until they meet and fuse. In effect, then, two layers of mesoderm are formed and the space between them is called the coelom. A **coelom** is defined as a body cavity lined by mesoderm and within which the internal organs form.

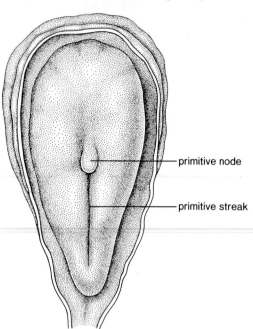

primitive node

primitive streak

Table 20.2   Organs developed from the three primary germ layers

Ectoderm	Mesoderm	Endoderm
Skin epidermis including hair, nails, and sweat glands	All muscles	Lining of digestive tract, trachea, bronchi, lungs, gallbladder, and urethra
Nervous system including brain, spinal cord, ganglia, nerves	Dermis of skin	Liver
	All connective tissue including bone, cartilage, and blood	Pancreas
Retina, lens, and cornea of eye	Blood vessels	Thyroid, parathyroid, and thymus glands
Inner ear	Kidneys	Urinary bladder
Lining of nose, mouth, and anus	Reproductive organs	
Teeth enamel		

Figure 20.2 compares human gastrulation to that of amphioxus. In humans a space called the amniotic cavity appears within the inner cell mass. The portion of the mass below this cavity is the embryonic disc, which elongates to form the primitive streak (fig. 20.4). Some of the upper cells within the primitive streak invaginate and spread out between the lower layer, now called endoderm, and the remaining cells of the upper layer, now called ectoderm. The invaginating cells are the mesoderm layer. Mesoderm later forms blocklike portions called **somites** in the posterior half of the embryo, and these become muscle tissue.

It is interesting to note that human development resembles chick development. In the chick there is a primitive streak rather than a spherical gastrula because the yolk does not participate in the early stages of development. The human egg contains very little yolk, yet early development resembles that of the chick. The evolutionary history of these two animals can provide an answer for this amazing resemblance. Both birds (e.g., chicks) and mammals (e.g., humans) are related to the reptiles, and this evolutionary relationship manifests itself in the manner in which development proceeds.

Germ Layers   Ectoderm, mesoderm, and endoderm are called the primary **germ layers** of the embryo, and no matter how gastrulation takes place the end result is the same: three germ layers are formed. It is possible to relate the development of future organs to these germ layers, as is done in table 20.2.

Neurula

This is the stage of development when **differentiation,** or specialization, of cells first becomes apparent. Differentiation cannot be explained by a parceling out of genes to the various cells since each and every cell of the animal's body receives a full complement of genes. Rather, genes must be controlled in such a way that only certain ones are active in certain cells. Recent investigative studies have suggested that the cytoplasm may contain either inhibitors or stimulators that combine with the chromosome and inactivate or activate certain genes. Support for the importance of cytoplasmic control of genes comes from a study of frog development. The frog's egg contains a special section of cytoplasm called the **gray crescent.** If an experimenter ties the fertilized egg so that each half has a portion of the gray crescent, both halves develop successfully into a complete embryo. If the experimenter ties an egg so that only one-half has the gray crescent, only that half develops successfully. The other half becomes a mass of nondifferentiated cells (fig. 20.5).

Early investigators referred to the gray crescent as the **primary organizer** for the embryo. An organizer is believed to be a group of cells that can

Reproduction, Development, and Inheritance

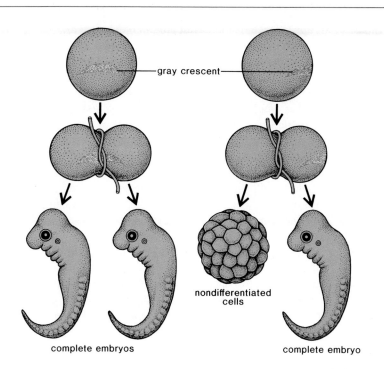

Figure 20.5
Gray crescent and development. If a frog's egg is divided so that each half receives an ample share of gray crescent, then each half develops into a complete embryo. On the other hand, if a frog's egg is divided so that only one-half receives the gray crescent, only that half develops into a complete embryo.

gray crescent

nondifferentiated cells

complete embryos

complete embryo

Figure 20.6
Induction. Presumptive notochord induces the formation of the neural tube. This series of drawings shows how the neural tube forms by closure of the neural folds and how the coelom forms by a splitting of mesoderm.

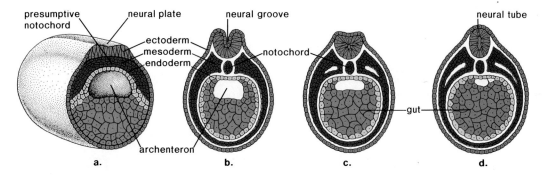

presumptive notochord   neural plate   neural groove   neural tube

ectoderm
mesoderm
endoderm

notochord

gut

archenteron

a.   b.   c.   d.

influence the development of other cells. The concept of an organizer was further developed when it was found that mesoderm tissue located at the upper lip of the blastopore brings about or induces the formation of the nervous system in animals.

Induction In vertebrate animals a central portion of the mesoderm is the future **notochord,** which will later be replaced by the vertebral column. The nervous system develops from ectoderm located just above the notochord. At first, a thickening of cells called the neural plate is seen along the dorsal surface of the embryo. Then neural ridges develop on either side of a neural groove that becomes the **neural tube** when the ridges fuse. Figure 20.6 shows cross sections of frog development, which illustrate the internal development of the

Figure 20.7

Human embryo at 21 days. The neural folds still need to close at the anterior and posterior of the embryo. The pericardial area contains the primitive heart, and the somites are the precursors of the muscles.

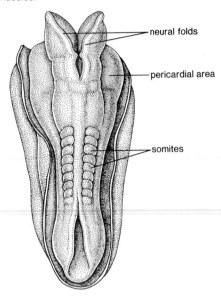

- neural folds
- pericardial area
- somites

nervous system. Figure 20.7 shows an intact human embryo, allowing you to see the external appearance of the developing nervous system.

Experiments have been performed with frogs to show that if the presumptive (potential) nervous system, lying just above the notochord, is cut out and transplanted to another region of the embryo, it will not form a neural tube. On the other hand, if the presumptive notochord is cut out and transplanted beneath what would be belly ectoderm, this ectoderm now differentiates into neural tissue (fig. 20.8). These experiments indicate that notochord mesoderm brings about the formation of the nervous system, and it is said that this mesoderm **induces** the formation of the neural tube.

The process of **induction** can explain the orderly development of the embryo. One tissue is induced by another and this, in turn, induces another tissue, and so forth, until development is complete. Thus, induction can account for the stepwise and timewise progression of development.

This theory of orderly development is supported by the fact that once the closure of the neural tube is complete, the presumptive forebrain induces the formation of the lens of the eyes. First, the sides of the forebrain bulge out and widen just beneath overlying ectoderm. This seems to trigger a thickening of these ectoderm cells. Then the bulge dips in to form a cuplike structure that will be the future eyeball, while the overlying thickened ectoderm grows into a ball of cells to form the lens of the eye.

Figure 20.8

Experiments proving importance of presumptive notochord. In experiment *A*, presumptive nervous system tissue does not complete its development when moved from its location above the notochord. On the other hand, in experiment *B*, presumptive notochord can cause even presumptive belly ectoderm to develop into a nervous system.

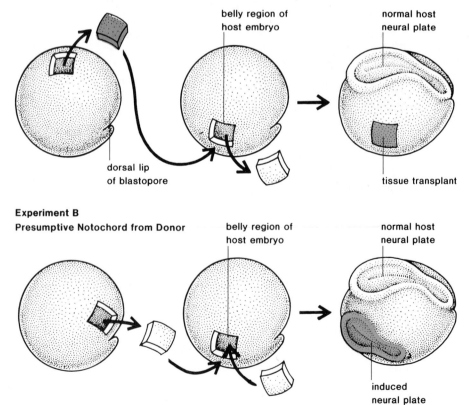

Experiment A
Presumptive Nervous System from Donor

belly region of host embryo

normal host neural plate

dorsal lip of blastopore

tissue transplant

Experiment B
Presumptive Notochord from Donor

belly region of host embryo

normal host neural plate

induced neural plate

Reproduction, Development, and Inheritance

Experimentation has suggested that RNA formation and diffusion may be responsible for the process of induction. For example, if ectoderm is placed in a solution that formerly contained notochord mesoderm, the ectoderm differentiates into nervous tissue. This indicates that the mesoderm must have left some chemicals behind in the solution and these chemicals carry out the inductive process. An analysis of such a solution, following mesoderm removal, shows that nucleic acids have been added to the solution by the mesoderm tissue. The nucleic acid most likely to appear is RNA.

Vertebrate Cross Section

With the formation of the nervous system, it is possible to show a generalized diagram (fig. 20.9) of a vertebrate embryo to illustrate placement of parts. Correlation of figure 20.9 with table 20.2 will help you relate the formation of vertebrate structures and organs to the three embryonic layers of cells: the ectoderm, mesoderm, and endoderm. Thus, the skin and nervous system develop from the ectoderm; muscles, skeleton, kidneys, circulatory system, and gonads develop from the mesoderm; and the digestive tract, lungs, liver, and pancreas develop from the endoderm.

The diagram illustrates that embryonic vertebrates have a notochord and a dorsal hollow nerve cord. Another characteristic is the presence of gill pouches or slits supported by gill arches (fig. 20.10). Obviously, only the lower

Figure 20.9
Typical vertebrate embryonic cross section. The nervous system and outer covering are derived from the ectoderm; somites and notochord are derived from the mesoderm, which also lines the coelom; the digestive tract is derived from the endoderm. The consistency of these derivations supports the germ layer theory.

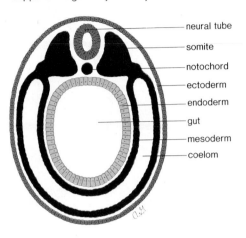

neural tube
somite
notochord
ectoderm
endoderm
gut
mesoderm
coelom

Figure 20.10
Human embryo at beginning of fifth week.
a. Scanning electron micrograph. *b.* Drawing. The embryo is curled so that the head touches the heart, two organs whose development is further along than the rest of the body. The organs of the gastrointestinal tract are forming. The presence of the tail is an evolutionary remnant; its bones will regress and become those of the coccyx. The arms and legs will develop from the bulges that are called limb buds.

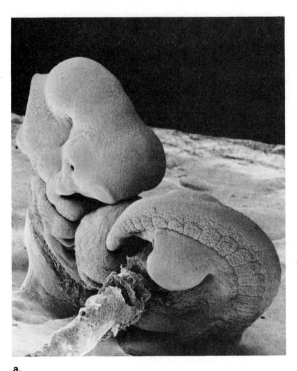

a.

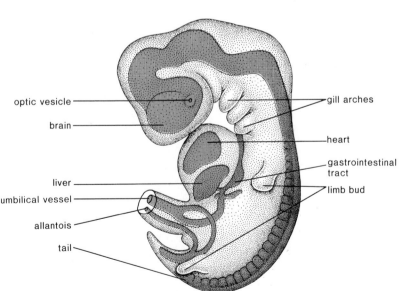

optic vesicle
brain
liver
umbilical vessel
allantois
tail

gill arches
heart
gastrointestinal tract
limb bud

b.

Figure 20.11

Extraembryonic membranes in bird and human embryos. In the human the chorion contributes to the fetal side of the placenta; the allantois and yolk sac are vestigial.

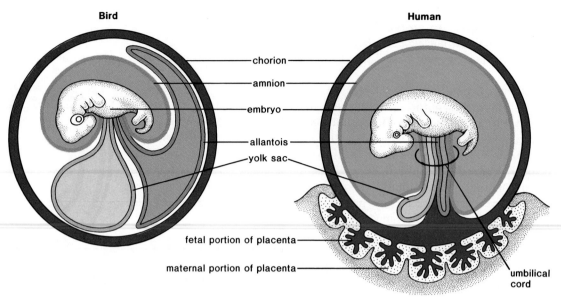

Bird   **Human**

chorion
amnion
embryo
allantois
yolk sac

fetal portion of placenta
maternal portion of placenta

umbilical cord

vertebrates (fishes and amphibian larvae) have actual use for gill slits as functioning structures. The fact that higher forms go through this embryonic stage of the lower forms indicates a relationship between them. The phrase **ontogeny** (development) **recapitulates** (repeats) **phylogeny** (evolutionary history) was coined some years ago as a dramatic way to suggest that all animals share the same embryonic stages. This theory has been modified today since embryos proceed only through those stages that are necessary to their later development. For example, in higher vertebrates actual gill slits never form; instead, the first gill pouch becomes the cavity of the middle ear and eustachian tube. The second pouch becomes the tonsils, while the third and fourth pouches become the thymus and parathyroids, respectively. The fifth pouch disappears. Thus, gill pouches develop because they are necessary to later development.

Extraembryonic Membranes

Before we consider human development chronologically, we must understand the placement of extraembryonic membranes. Extraembryonic membranes are best understood by considering their function in the chick. The formation of these membranes in reptiles first made development on land possible. If an embryo develops in the water, the water supplies oxygen for the embryo and takes away waste products. The surrounding water prevents drying out and provides a protective cushion.

On land, all these functions are performed by the extraembryonic membranes. Figure 20.11 shows the chick within its hard shell surrounded by the membranes. The **chorion** lies next to the shell and carries on gas exchange. The **yolk sac** surrounds the remaining yolk. The **allantois** collects nitrogenous waste and the **amnion** contains the amniotic fluid that bathes the developing embryo.

As figure 20.11 indicates, humans also have these extraembryonic membranes. The chorion develops into the fetal half of the placenta; the yolk sac

Figure 20.12
Fertilization and implantation of human embryo.
After the egg is fertilized, it begins cleavage as it
moves toward the uterus. At the time of
implantation, the embryo is in the gastrula stage of
development.

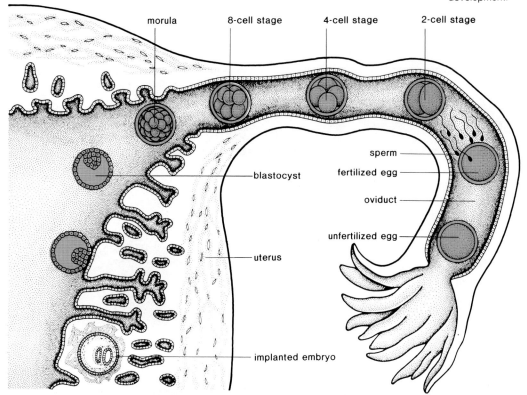

is present but lacks yolk and is largely nonfunctional; the allantoic blood vessels become the umbilical blood vessels; and the amnion contains fluid to cushion and protect the fetus. Thus, the function of the membranes has been modified to suit internal development, but their very presence indicates our relationship to birds and reptiles. It is interesting to note that all animals develop in water, either directly or within amniotic fluid.

Human Development

We are now prepared to consider development in a stepwise manner. Human embryonic development is complicated by the fact that the extraembryonic membranes develop very early. This may be related to the fact that the human egg has little yolk and the mother supplies the needs of the developing embryo by way of the placenta.

Embryonic Development

First Week

Days 1 to 4 Immediately after fertilization within the oviduct, the human zygote begins to undergo cleavage as it travels down the oviduct to the uterus (fig. 20.12).

Figure 20.13

Development of extraembryonic membranes during first month of human development. *a.* The blastocyst is surrounded by the trophoblast. *b.* As gastrulation occurs, the amniotic cavity appears. *c.* The chorion and yolk sac are now apparent *d.* With complete gastrulation, the body stalk is apparent. *e.g.* Embryo becomes more differentiated as the umbilical cord forms.

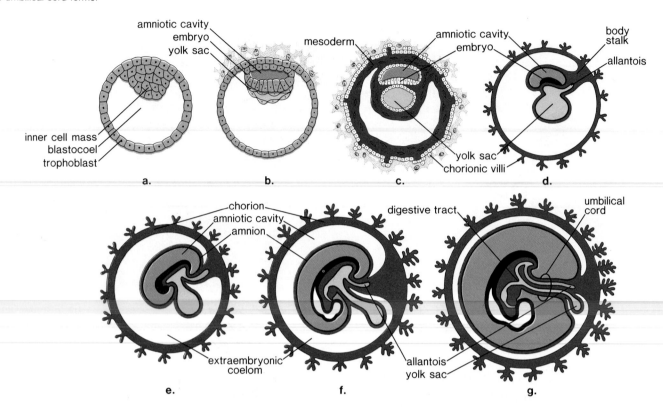

Days 5 to 6 The morula becomes the blastocyst (fig. 20.13a), which is the blastula stage in humans. The blastocyst consists of the inner cell mass and with a single layer of surrounding cells known as the **trophoblast.** The trophoblast signals the formation of the first of the extraembryonic membranes, since it will become the outer layer of cells of the chorion.

Day 7 At the end of the first week, the embryo begins the process of **implanting** itself in the wall of the uterus as the trophoblast secretes enzymes to digest away some of the tissue and blood vessels of the uterine wall (fig. 20.12). The embryo is now about the size of the period at the end of this sentence.

Second Week

Days 8 to 14 Implantation continues as the embryo undergoes gastrulation. The first stage of gastrulation is the formation of a layer of endodermal cells beneath the ectoderm. Above the embryo, an amnionic cavity is now visible as is a yolk sac cavity below (fig. 20.13b, c). Thus, two more extraembryonic membranes have made their appearance. The formation of the chorion is complete when a layer of mesodermal cells is seen lining the trophoblast. An outer mesodermal layer is added to both the amnion and yolk sac. Thus, by the end of the second week, three extraembryonic membranes are present. The embryo itself is referred to as an embryonic disk and there are signs of a primitive streak.

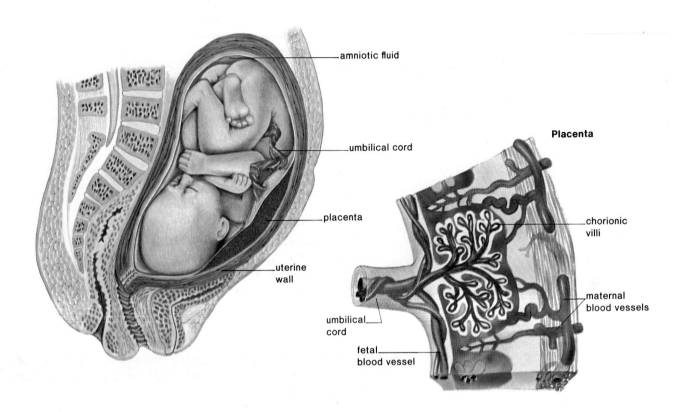

amniotic fluid

umbilical cord

placenta

uterine wall

Placenta

chorionic villi

maternal blood vessels

umbilical cord

fetal blood vessel

The embryo is completely implanted as the uterine wall closes over it. Treelike extensions of the chorion called chorionic villi project into the maternal tissues. Later, these disappear in all areas except the one where the **placenta** develops. By the tenth week the placenta (fig. 20.14) is fully formed. It has a fetal side contributed by the chorion and a maternal side consisting of uterine tissues. Carbon dioxide and other wastes move from the fetal to the maternal side, and nutrients and oxygen move from the maternal to the fetal side of the placenta. Notice in figure 20.14 how the chorionic villi are surrounded by maternal blood sinuses; yet the blood of the mother and fetus never mix since exchange always takes place across cell membranes.

Third Week

(Days 14 to 21) The second stage of gastrulation is complete as the mesoderm forms by invagination of cells at the primitive streak. A bridge of mesoderm called the body stalk connects the caudal end of the embryo with the chorion (fig. 20.13d). The vestigial allantois is contained within this stalk, and its blood vessels become the umbilical blood vessels. The umbilical cord that connects the developing embryo (⅛ inch or 3.2 mm. in size) to the placenta is fully formed when the head and tail lift up and the body stalk is moved toward the ventral side by a constriction (fig. 20.13e-g). The notochord is present and the neural tube is almost closed. The heart becomes conspicuous and a system of paired blood vessels is established. Somites are rapidly increasing in number, with 16 somites present at this time.

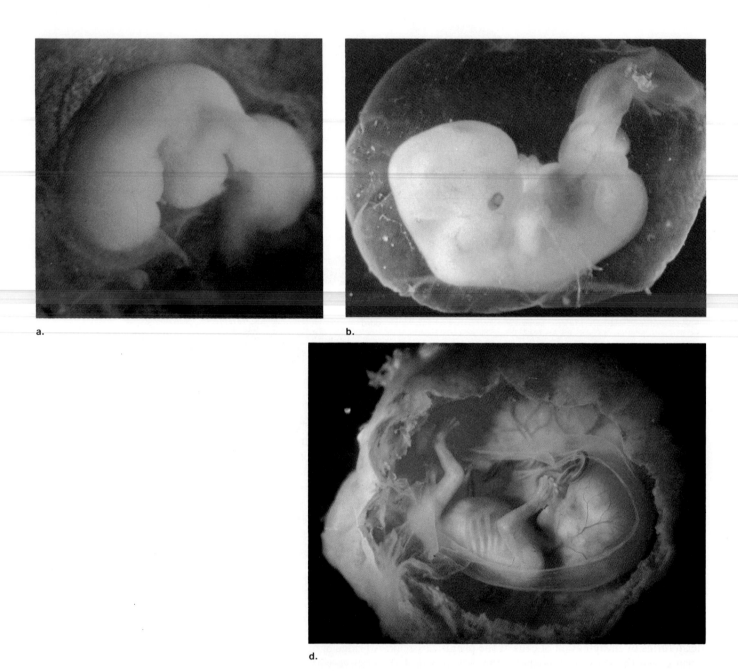

a.

b.

d.

Figure 20.15
Human development from the fourth to the sixteenth week. *a.* In the four-week-old embryo, the body is flexed and C shaped. *b.* At the end of six weeks, the head becomes disproportionately large. *c.* In the eight-week-old embryo, the nose is flat, the eyes are far apart, and the eyelids are fused. *d.* Surrounded by the extraembryonic membranes, this 12–week-old fetus appears to be sucking its thumb. *e.* At 16 weeks, the blood vessels are easily visible through the transparent skin.

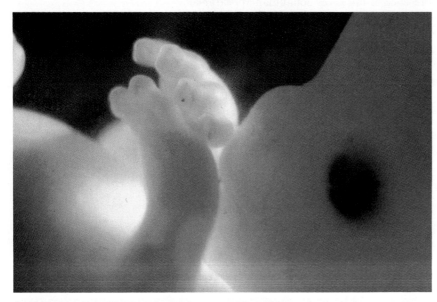

c.

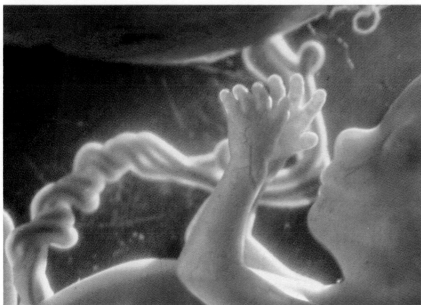

e.

Fourth Week (fig. 20.15)
3/16 inch (4.8 mm)
Heart pulsating and pumping blood
All somites (40) present
Eyes, ears, and nose forming
Digestive system forming
Limb buds begin to form
Body flexed; C shaped
Presence of tail gives nonhuman appearance

Sixth Week
9/16 inch (14.3 mm)
Head becomes disproportionately large
Face and neck forming
Limb buds developing; digits forming
Cartilaginous skeleton forming
Tail regressing

Eighth Week (Two Months)
1⅛ inches; about 1/30 ounce (28.6 mm; .95 gm)
Nose flat, eyes far apart, eyelids fused
Limbs beginning to take shape; digits well formed
Ossification beginning
All internal organs have formed
Recognizable as human

Fetal Development

Third Month
3 inches; 1 ounce (7.62 cm, 28.4 gm)
Head prominent
Eyes formed but lids still fused
External ears present; nose gains bridge
Tooth sockets and buds forming in jawbones
Nails forming
Ossification continuing
Heartbeat can be detected with special instruments
Sex can be determined by inspection

Fourth Month
6½ to 7 inches; 4 ounces (16.5 cm to 17.8 cm; 113.4 gm)
Eyes, ears, nose, and mouth have typical human appearance; eyebrows appear
Skin bright pink, transparent, and covered with fine, downlike hair
Bony skeleton now visible, body catching up with head size
Active muscles; movement may be felt as baby moves in womb

Fifth Month (fig. 20.16)
10 to 12 inches; ½ to 1 pound (25.4 cm to 30.5 cm; 226.8 gm to 453.6 gm or .227 kg to .454 kg)
Eyelids still completely fused; some hair may be present on head
Skin bright red and still covered by fine hair
Internal organs maturing; heartbeat can be heard without special instruments

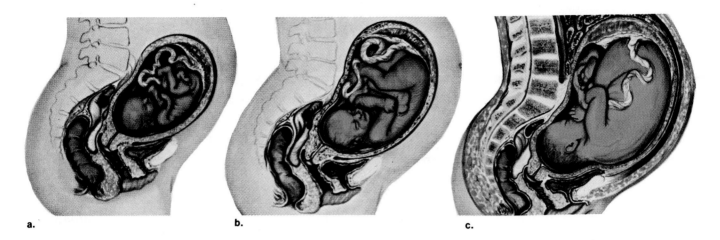

a. b. c.

Sixth Month

11 to 14 inches; 1¼ to 1½ pounds (27.9 cm to 35.6 cm; .57 kg to .68 kg)
Eyelids finally separated and eyelashes formed
Skin quite wrinkled and somewhat red; covered with heavy protective creamy coating
Nails now extend to end of digits

Seventh Month

14 to 17 inches; 3 pounds (35.6 cm to 43.2 cm; 1.36 kg)
Eyes are open
Skin still quite red and covered with wrinkles
Testes have descended into scrotal sacs
Premature baby at this stage has a slight chance for survival in nurseries staffed by skilled physicians and nurses

Eighth Month

16½ to 18 inches; 5 pounds (41.9 cm to 45.7 cm; 2.27 kg)
Subcutaneous fat deposition leads to weight gain
Bones of head soft and flexible
Growth and maturation of baby in last two months extremely valuable for survival

Ninth Month

Average baby weighs about seven pounds (3.14 kg) if a girl and seven and a half (3.4 kg) if a boy
Length about 20 inches (50.8 cm)
Skin still coated with creamy coating
Fine downy body hair has largely disappeared
Fingernails may protrude beyond ends of fingers
Size of soft spot between bones of skull varies considerably from one child to another, but generally will close within 12 to 18 months

Fetal Circulation

As figure 20.17 shows, the fetus has four features that are not present in adult circulation:

1. *Oval opening or foramen ovale:* an opening between the two atria. This opening is covered by a flap of tissue that acts as a valve.
2. *Arterial duct or ductus arteriosus:* a connection between the pulmonary artery and the aorta.
3. *Umbilical arteries and vein:* vessels that travel to and from the placenta, leaving waste and receiving nutrients.
4. *Venous duct or ductus venosus:* a connection between the umbilical vein and the vena cava.

All of these features may be related to the fact that the fetus does not use its lungs for gas exchange since it receives oxygen and nutrients from the mother's blood by way of the placenta, a structure formed by fetal tissue and uterine tissue.

If we trace the path of blood in the fetus, we may begin with the right atrium (fig. 20.17). From the right atrium, the blood may pass directly into the left atrium by way of the oval opening or it may pass through the atrioventricular valve into the right ventricle. From the right ventricle the blood goes into the pulmonary artery, but because of the arterial duct, most of the blood then passes into the aorta. Thus, by whatever route the blood takes, most of the blood will reach the aorta instead of the lungs.

Blood within the aorta travels to the various branches, including the iliac arteries that connect to the umbilical arteries leading to the placenta, where exchange between mother's blood and fetal blood takes place. It is interesting to note that the blood in the umbilical arteries, which travels to the placenta, is low in oxygen, but the blood in the umbilical vein, which travels from the placenta, is high in oxygen. The umbilical vein enters the venous duct, which passes directly through the liver. The venous duct then joins with the posterior vena cava, which goes to the right atrium again.

At birth, the oval opening usually closes. With the tying of the cord and the expansion of the lungs, blood enters the lungs in quantity. Return of this blood to the left side of the heart usually causes a flap to cover over the opening. However, the most common of all cardiac defects in the newborn is a persistence of the oval opening. Incomplete closure occurs in nearly one out of four individuals; but even so, passage of the blood from the right to the left atrium rarely occurs because either the opening is small or it closes when the atria contract. In a small number of cases, the passage of impure blood from the right to the left side of the heart is sufficient to cause a "blue baby." Such a condition may now be corrected by open-heart surgery.

The arterial duct closes because endothelial cells divide and block off the duct. Remains of the arterial duct and parts of the umbilical arteries and vein remain within the body but are transformed into connective tissue.

Birth

The uterus characteristically contracts throughout pregnancy. At first, light, often indiscernible contractions last about 20 to 30 seconds and occur every 15 to 20 minutes, but near the end of pregnancy they become stronger and more frequent so that the woman may falsely think that she is in labor. The onset of true labor is marked by uterine contractions that occur regularly every 15 to 20 minutes and last for 40 seconds or more. **Parturition,** which includes labor and expulsion of the fetus, is usually considered to have three stages.

Reproduction, Development, and Inheritance

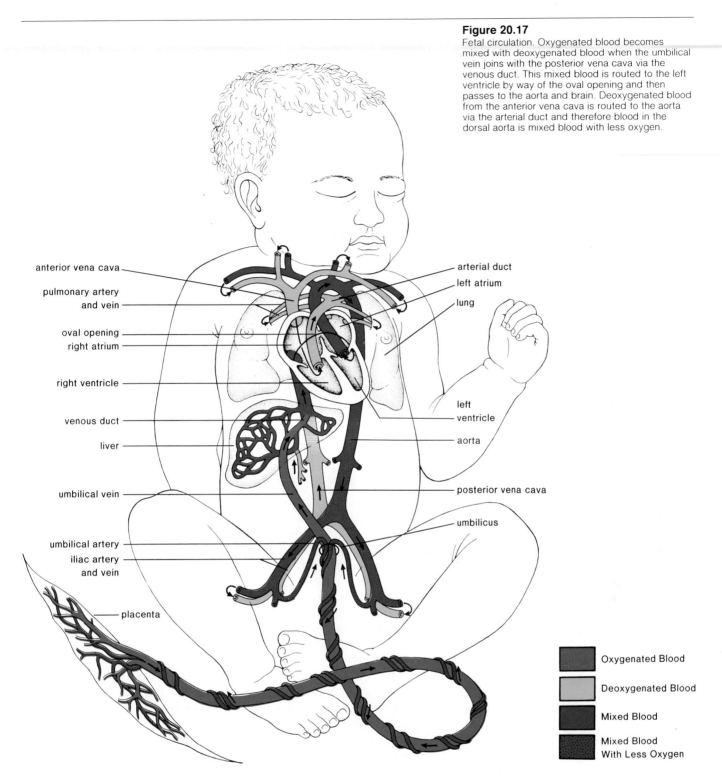

Figure 20.17
Fetal circulation. Oxygenated blood becomes mixed with deoxygenated blood when the umbilical vein joins with the posterior vena cava via the venous duct. This mixed blood is routed to the left ventricle by way of the oval opening and then passes to the aorta and brain. Deoxygenated blood from the anterior vena cava is routed to the aorta via the arterial duct and therefore blood in the dorsal aorta is mixed blood with less oxygen.

anterior vena cava

pulmonary artery and vein

oval opening

right atrium

right ventricle

venous duct

liver

umbilical vein

umbilical artery

iliac artery and vein

placenta

arterial duct

left atrium

lung

left ventricle

aorta

posterior vena cava

umbilicus

Oxygenated Blood

Deoxygenated Blood

Mixed Blood

Mixed Blood With Less Oxygen

During the *first stage*, the cervix dilates; during the *second*, the baby is born; and during the *third*, the afterbirth is expelled.

The events that cause parturition are still not entirely known but there is now evidence suggesting the involvement of prostaglandins. It may be too that the prostaglandins cause the release of oxytocin from the maternal posterior pituitary. Both prostaglandins and oxytocin do cause the uterus to contract and either can be given to induce parturition.

Stage 1

Prior to the first stage of parturition or concomitant with it, there may be a "bloody show" caused by the expulsion of a mucus plug from the cervical canal. This plug prevents bacteria and sperm from entering the uterus during pregnancy.

Uterine contractions during the first stage of labor occur in such a way that the cervical canal slowly disappears (fig. 20.18a-c) as the lower part of the uterus is pulled upward toward the baby's head. This process is called **effacement,** or "taking up the cervix." With further contractions, the baby's head acts as a wedge to assist cervical dilation. The baby's head usually has a diameter of about 4 inches and therefore the cervix has to dilate to this diameter in order to allow the head to pass through. If it has not occurred already, the amniotic membrane is apt to rupture now, releasing the amniotic fluid, which escapes out the vagina. The first stage of labor ends once the cervix is completely dilated.

Stage 2

During the second stage, the uterine contractions occur every one to two minutes and last about one minute each. They are accompanied by a desire to push or bear down. As the baby's head gradually descends into the vagina, the desire to push becomes greater. When the baby's head reaches the exterior, it turns so that the back of the head is uppermost (fig. 20.18d). Since the vagina may not expand enough to allow passage of the head without tearing, an **episiotomy** is often performed. This incision, which enlarges the opening, is stitched later and will heal more perfectly than a tear would. As soon as the head is delivered, the baby's shoulders rotate so that the baby faces either to the right or left. The physician may at this time hold the head and guide it downward while one shoulder and then the other emerges (fig. 20.18e-f). The rest of the baby follows easily.

Once the baby is breathing normally, the umbilical cord is cut and tied, severing the child from the placenta. The stump of the cord shrivels and leaves a scar, which is the navel.

Stage 3

The placenta, or **afterbirth,** is delivered during the third stage of labor (fig. 20.18g-h). About 15 minutes after delivery of the baby, uterine muscular contractions shrink the uterus and dislodge the placenta. The placenta is then expelled into the vagina. As soon as the placenta and its membranes are delivered, the third stage of labor is complete.

Prepared Childbirth

Some doctors and expectant couples feel that nervous system depressants may be harmful not only to the expectant mother but to the baby as well. This sentiment, together with a desire to enjoy and share the process of giving birth, has given impetus to the prepared childbirth movement. Usually couples who wish to practice prepared childbirth use the methods espoused by Dr. Fernand LaMaze and attend several teaching sessions in which they learn about the events of labor and delivery, the phenomenon of conditioned pain, and suggestions for behavior during labor and delivery.

It is believed that the woman may help prevent discomfort during labor by concentrating on mild, shallow breathing at the time of contractions. This breathing method prevents the diaphragm from exerting pressure on the abdominal organs and guarantees an adequate supply of oxygen for uterine contraction. When delivery begins and the woman feels a great need to push, her partner coaches her to use deep inhalation along with a controlled type of pushing at the time of each strong contraction. Advocates of the Lamaze

Reproduction, Development, and Inheritance

Figure 20.18
Process of birth. *a.* Dilation of cervix. *b.* Amnion
bursts. *c.* Fetus descends into pelvis. *d.* Head
appears. *e.* Rotation. *f.* Delivery of shoulders.
g-h. Expulsion of afterbirth.

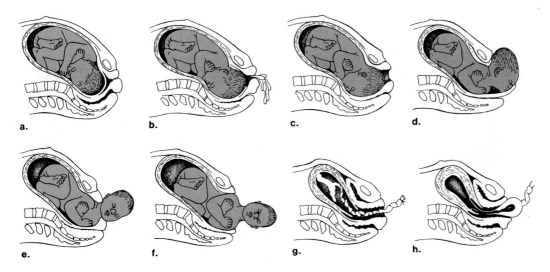

a.

b.

c.

d.

e.

f.

g.

h.

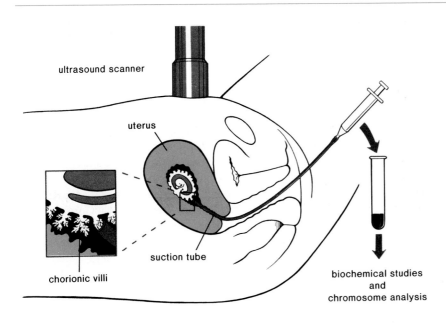

Figure 20.19
Chorionic villi sampling. In this procedure the
physician removes a sampling of embryonic cells
from the chorionic villi. They then can be tested for
genetic defects.

ultrasound scanner

uterus

chorionic villi

suction tube

biochemical studies
and
chromosome analysis

method of prepared childbirth feel that this active participation on the part
of the couple will not only help the woman overlook discomfort but will also
give the pair the pleasant reward of seeing the baby when it first appears.

Birth Defects

It is believed that at least 1 in 16 newborns has a birth defect, either minor
or serious, and the actual percentage may be even higher. Most likely only 20
percent of all birth defects are due to heredity. Those that are can sometimes
be detected before birth by subjecting embryonic and/or fetal cells to various
tests. A new method called **chorionic villi sampling** (fig. 20.19) allows physicians to collect embryonic cells as early as the fifth week. The doctor inserts

a long thin tube through the vagina into the uterus. With the help of ultrasound, which gives a picture of the uterine contents, the tube is placed between the lining of the uterus and the chorion. Then suction is used to remove a sampling of the chorionic villi cells. Chromosomal analysis and biochemical tests for about 200 different genetic defects can be done immediately on these cells.

Previous to the development of chorionic villi sampling, physicians had to wait until about the sixteenth week of pregnancy to perform **amniocentesis.** In amniocentesis, a long needle is passed through the abdominal wall to withdraw a small amount of amniotic fluid along with fetal cells. Since there are only a few cells in the amniotic fluid, testing must be delayed until cell culture has caused them to grow and multiply. Therefore it may be another two to four weeks until the prospective parents are told whether their child has a genetic defect.

Treatment of the fetus in the womb is a rapidly developing area of medical expertise. Biochemical defects can sometimes be treated by giving the mother appropriate medicines. For example, if a baby is unable to synthesize vitamin B_{12} and/or is unable to use biotin efficiently, the mother can take these substances in doses large enough to prevent any untoward effects. Structural defects can sometimes be corrected by surgery. For example, if the fetus has water on the brain or is unable to pass urine, tubes that temporarily allow the fluid to pass out into the amniotic fluid can be inserted even while the fetus is still in the womb. Physicians are hopeful that eventually all sorts of structural defects can be corrected by lifting the fetus from the womb just long enough for corrective surgery to be done.

Birth defects not due to heredity, called congenital defects, cannot be passed on to the next generation. These are often caused by microbes or substances that have crossed the placenta and altered normal development. The defect that results is dependent on the stage of development. For example, during the time the heart is developing, a foreign agent may cause a heart defect. Since most organs develop during the first two months, most defects occur during this time—when the woman may not yet realize she is pregnant.

It is recommended that all women capable of reproduction and who are sexually active take precautions to protect developing embryos. An Rh negative woman who has an Rh positive child should receive a RhoGam injection to prevent the production of Rh antibodies (p. 241), which can cause birth defects, including nervous system and heart defects during a subsequent pregnancy.

Also, women should be immunized for German measles before the childbearing years. German measle viruses can cause defects, such as deafness. Drugs such as aspirin, caffeine (present in coffee, tea, and cola), and alcohol should be severely limited. Mood-altering drugs (p. 320) most likely should not be taken at all. It is not unusual for babies of drug addicts and alcoholics to display withdrawal symptoms and to be mentally retarded. Most people are also aware of the fact that women taking the tranquilizer thalidomide produced children with deformed arms and legs (fig. 20.20).

If pregnancy is suspected, all medications and hormones, such as those in the birth control pill, should be discontinued. When the synthetic hormone DES was given to pregnant women to prevent miscarriage, their daughters showed an increased tendency toward cervical cancer. Other sex hormones can possibly cause abnormal fetal development, including abnormalities of the sex organs.

So-called fetotoxic chemicals should also be avoided. These include pesticides and many organic industrial chemicals. Cigarette smoke includes some of these very same chemicals so that babies born to smokers are often underweight and subject to convulsions.

X-ray diagnostic therapy should be avoided during pregnancy because X rays are mutagenic to a developing embryo or fetus (p. 493). Children born to women who have received X-ray treatment are apt to have birth defects and/or develop leukemia later on.

Now that physicians and lay people are aware of the various ways in which birth defects can be prevented, it is hoped that all types of birth defects, both genetic and congenital, will decrease dramatically.

Summary

In this chapter, we have seen that all animals, including human beings, develop similarly. The processes of development are cleavage, growth, morphogenesis, and differentiation. Cleavage results in many cells that become first a morula, then a blastula, and finally a gastrula. The result of gastrulation is the establishment of three germ layers: ectoderm, endoderm, and mesoderm. Later development can be related to these germ layers.

The control of differentiation, whereby tissues take on specific structures and function, may be accounted for by the process of induction. Good examples of this are the induction of the nervous system by the notochord and the induction of the lens by the forebrain.

All vertebrates at some time in their development portray a similar cross section that displays typical vertebrate embryonic characteristics: dorsal hollow nerve cord, notochord, and a coelom completely surrounded by mesoderm. Also, at some time in their embryonic history, vertebrates have gill pouches or slits. The fact that this occurs supports the statement that "ontogeny recapitulates phylogeny." This statement is not taken literally today since higher animals only pass through those stages that are helpful to their later development. Thus, although gill pouches develop in reptiles, birds, and mammals they are eventually modified for other functions.

The presence of extraembryonic membranes in reptiles made development on land possible. The chorion serves for gas exchange; the yolk sac provides nourishment; the allantois collects nitrogenous waste; and the amnion surrounds the embryo with fluid for protection. Humans also have these membranes, but their function has been modified for internal development: the chorion becomes the fetal part of the placenta; the yolk sac and allantois are largely nonfunctional; and the amnion again surrounds the embryo with fluid. The human embryo lies within the amniotic cavity and is connected to its lifeline, the placenta, by way of the umbilical cord. During the first two months, all major organs are formed and the embryo takes on a human appearance. After this, we refer to the developing new life as the fetus. Various features of the fetus are refined during the next several months, while the last few months serve largely to increase the size of the soon-to-be newborn. Certain changes must occur in fetal circulation as the baby is born. The oval opening and arterial duct close and most of the umbilical arteries and veins are cut off.

The process of birth requires three stages: dilation, delivery, and afterbirth. During the first stage, the cervix is increasing in size to permit the passage of the baby; during the second, the baby is forced out into the world by strong contractions of the uterus and abdominal wall. After the expulsion of the fetus, the membranes and placenta follow, as the afterbirth.

Birth defects are both genetic and congenital. Genetic defects can sometimes be detected by amniocentesis and fetoscopy. Many congenital defects can be prevented if certain precautions are taken, such as having a RhoGam injection when appropriate and receiving a German measles vaccination prior to pregnancy, and limiting the intake of drugs and refraining from smoking and X-ray diagnostic therapy during pregnancy.

Study Questions

1. List the processes of development for any organism. (p. 416)
2. List the early stages of development and describe what occurs during each stage. (pp. 416–20) Compare the appearance of amphioxus and the human embryo during these stages. (p. 417)
3. Explain why it is believed that the cytoplasm controls the genetic potential of cells. (p. 420) How do experiments with frogs support this belief? (pp. 420–21)
4. What are the three germ layers? What structures are associated with each germ layer? (p. 420)
5. What is induction and what experiments have been done to show that induction takes place? (pp. 421–23)
6. Draw a generalized cross section of a vertebrate embryo and label the parts. (p. 423)
7. What are the extraembryonic membranes for the chick? (p. 424) For the human? (p. 424) And what are their respective functions? (p. 424)
8. Describe in general the happenings during embryonic development of the human; during fetal development of the human. (pp. 425–31)
9. Trace the path of blood in the fetus from the umbilical vein to the aorta using two different routes. (p. 432)
10. Describe the three stages of parturition. (pp. 432–34)
11. Give several ways in which birth defects can be prevented. (pp. 435–37)

Selected Key Terms

cleavage (klēv'-ij)
growth (grōth)
morphogenesis (mor''fo-jen'ĭ-sis)
differentiation (dif''er-en''she-a'shun)
blastula (blas'tu-lah)
gastrula (gas'troo-lah)
ectoderm (ek'to-derm)
endoderm (en'do-derm)
mesoderm (mes'o-derm)
somites (so'mīts)
neurula (nu'roo-lah)
differentiation (dif''er-en''she-a'shun)
gray crescent (gra kres'ent)
induction (in-duk'shun)
notochord (no'to-kord)
chorion (ko're-on)
allantois (ah-lan'to-is)
amnion (am'ne-on)
placenta (plah-sen'tah)
afterbirth (af'ter-berth'')

Further Readings

Balinsky, B. J. 1981. *An introduction to embryology.* 5th ed. Philadelphia: W. B. Saunders.
Beaconsfield, et al. 1980. The placenta. *Scientific American* 243 (3):94.
Fuchs, F. 1980. Genetic amniocentesis. *Scientific American* 242 (6):47.
Rugh, R. et al. 1971. *From conception to birth: The drama of life's beginnings.* New York: Harper & Row.
Wessells, N. K., and Rutter, W. J. 1969. Phases in cell differentiation. *Scientific American* 220 (3):14.
Wolpert, L. 1978. Pattern formation in biological development. *Scientific American* 239 (4):154.

21

Chapter concepts

1 Genes, located on chromosomes, are passed from one generation to the next.

2 The Mendelian laws of genetics relate the genotype (inherited genes) to the phenotype (physical appearance).

3 The karyotype (chromosomal inheritance) may also be related to the phenotype. For example, persons receiving two X chromosomes are females and persons receiving an X and a Y chromosome are males.

Human Inheritance

Figure 21.1
Diagrammatic representation of a homologous pair
of chromosomes. The letters rR, Ss, and so forth
stand for alleles.

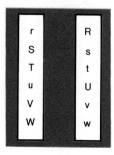

When a sperm fertilizes an egg, a new individual with the diploid number of chromosomes begins development. These chromosomes determine what the individual will be like; even if the egg develops in a surrogate mother, the individual will still resemble the original parents.

Today we say that the chromosomes located within the nuclei of cells contain the **genes.** By this we mean that it is possible to imagine that the chromosomes can be divided up into sections and that each section controls a particular trait of the individual. We will use the word **trait** to mean some aspect of the individual, such as height. In figure 21.1, the rectangles stand for a pair of homologous chromosomes and the letters stand for genes that control particular traits. Genes, like the letters in the rectangles, are in a particular sequence and are at particular spots, or loci, on the chromosomes. Alternate forms of a gene having the same position on a pair of chromosomes and affecting the same trait are called alleles. In our example, A is an **allele** of a,

Mendel's Results

Mendel's use of pea plants as his experimental material was a good choice because pea plants are easy to cultivate, have a short generation time, and can be self-pollinated or cross-pollinated at will. Mendel selected certain traits for study and before beginning his experiments made sure his parental (P_1 generation) plants bred true. He observed that when these plants self-pollinated, the offspring were like one another and like the parent plant. For example, a parent with yellow seeds always had offspring with yellow seeds; a plant with green seeds always had offspring with green seeds. Following that observation Mendel cross-pollinated the plants by dusting the pollen of plants with yellow seeds on the stigma of plants with green seeds whose own anthers had been removed, and vice versa. Either way the offspring (called F_1, or first filial, generation) resembled the parents with yellow seeds. These results caused Mendel to allow the F_1 plants to self-pollinate. Once he had obtained an F_2 generation, he observed the color of the peas produced. As table 21.A (p. 441) indicates, he counted over 8,000 plants and found an approximate 3:1 ratio (about three plants with yellow seeds for every plant with green seeds) in the F_2 generation.

Mendel realized that these results were explainable, assuming (a) there are two factors for every trait; (b) one of the factors can be dominant over the other that is recessive; (c) the factors separate when the gametes are formed. He assigned letters to these factors and displayed his results similar to this:

P_1 yellow $\times$ green
              YY   yy
F_1 all yellow
$F_1 \times F_1$ yellow $\times$ yellow
              Yy   Yy
F_2 3 yellow : 1 green

He believed that the F_2 plants with yellow seeds carried a dominant factor because his results could be related to the binomial equation, $a^2 + 2ab + b^2$, in this manner: $a^2 = YY$; $2ab = 2Yy$, and $b^2 = yy$. Thus the plants with yellow seeds would be YY or Yy, and there would be three plants with yellow seeds for every plant with green seeds.

As a test to determine if the F_1 generation was indeed Yy, Mendel back-crossed it with the recessive parent, yy. His results of 1:1 indicated that he had reasoned correctly. Today when a testcross is done, a suspected heterozygote is crossed with the recessive phenotype because this cross gives the best chance of producing the recessive phenotype.

Mendel performed a second series of experiments in which he crossed true-breeding plants that differed in two traits. For example, he crossed plants with yellow, round peas by plants with green, wrinkled peas. The F_1 generation always had both dominant characteristics and therefore he allowed the F_1 plants to self-pollinate. Among the F_2 generation he achieved an almost perfect ratio of 9:3:3:1 (table 21.A). For example, for every plant that had green, wrinkled

Reproduction, Development, and Inheritance

and vice versa. Also, *F* is an allele of *f*, and vice versa. *F* could never be an allele for *A* because *F* and *A* are different genes at different loci. Each allelic pair controls some particular trait of the individual, such as color of hair, type of fingers, length of nose.

Mendel's Laws

The first person to conduct a successful study of genetic or particulate inheritance was Gregor Mendel, a Catholic priest who grew peas in a small garden plot in 1860. Mendel knew nothing about cell structure, but his studies, described in the reading on this page, led him to conclude that inheritance is governed by **factors** that exist within the individual and are passed on to offspring. Mendel said *that every trait,* for example, height, *is controlled by two factors, or a pair of factors.* We now call these factors alleles. He also ob-

seeds he had approximately nine that had yellow, round seeds, and so forth. Mendel saw that these results were explainable if pairs of factors separate independently from one another when the gametes form, allowing all possible combinations of factors to occur in the gametes. This would mean that the probability of achieving any two factors together in the F_2 offspring was the product of their chances of occurring separately. Thus, since the chance of yellow peas was 3/4 (in a

one-trait cross) and the chance of round peas was 3/4 (in a one-trait cross), the chance of their occurring together was 9/16, and so forth.

Mendel achieved his success in genetics by studying large numbers of offspring, keeping careful records, and treating his data quantitatively. He showed that the application of mathematics to biology was extremely helpful in producing testable hypotheses.

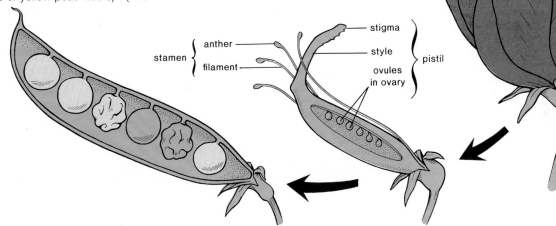

Pea flower anatomy. Self-pollination normally occurs within the flowers of a pea plant. In order to cross the dominant by the recessive, Mendel removed the anthers of the plant with the dominant trait and used pollen from the plant with the recessive trait to bring about cross-fertilization. He also performed the reverse cross in the same manner. The cross under discussion pertains to the shape and color of peas in a pod. All possible shapes and colors are illustrated.

Table 21.A   Mendel's results

Single-trait cross	F₁	F₂	Actual F₂ ratio
yellow × green	all yellow	6,022 yellow 2,001 green	3.01:1

Two-trait cross	F₁	F₂	Actual F₂ ratio
yellow, round × green, wrinkled	all round yellow	315 yellow, round 101 yellow, wrinkled 108 green, round 32 green, wrinkled	9.8:2.9:3.11:1.0

served that one of the factors controlling the same trait can be **dominant** over the other, which is **recessive.** The individual may show the dominant characteristic, for example, tallness[1], while the recessive factor for shortness, although present, is not expressed.

Mendel's experimental crosses made him realize that it was possible for a tall individual to pass on a factor for shortness. Therefore he concluded that while the individual has two factors for each trait, the gametes contained only one factor for each trait. This is often called Mendel's law of segregation:

Law of segregation: The factors separate when the gametes are formed, and only one factor of each pair is present in each gamete.

Inheritance of a Single Trait

Mendel suggested that letters be used to indicate factors so that **crosses** (gamete union resulting in offspring) might be more easily described. A capital letter indicates a dominant factor, and a lowercase letter indicates a recessive factor. The same procedure is used today, only the letters are now said to represent alleles. Also, Mendel's procedure and laws are applicable not only to peas but to all diploid individuals. Therefore, we will take as our example not peas but human beings. Figure 21.2 illustrates some differences between human beings that are known to be dominant or recessive. In doing a problem concerning hairline, the **key** would be represented as:

W = Widow's peak (dominant allele)

w = Continuous hairline (recessive allele)

The key simply tells us what letter of the alphabet to use for the gene in a particular problem and tells which allele is dominant, a capital letter signifying dominance.

Genotype and Phenotype

When we indicate the genes of a particular individual, two letters must be used for each trait mentioned. This is called the **genotype** of the individual. The genotype may be expressed not only by using letters but also by a short descriptive phrase, as table 21.1 shows. Thus the word **pure,** or **homozygous,** means that the two members of the allelic pair in the zygote (*zygo*) are the same (*homo*); genotype *WW* is called *homozygous dominant* and *ww* is called *homozygous recessive.* The word **heterozygous** means that the members of the allelic pair are different (*hetero*); only *Ww* is heterozygous. Another term, **hybrid,** is sometimes used in place of heterozygous and means that the genotype is not pure.

[1]Tallness is dominant in peas, but not in humans.

Table 21.1 Genotype versus phenotype

Genotype	Genotype	Phenotype
WW	Homozygous (pure) dominant	Widow's peak
Ww	Heterozygous (hybrid)	Widow's peak
ww	Homozygous (pure) recessive	Continuous hairline

Reproduction, Development, and Inheritance

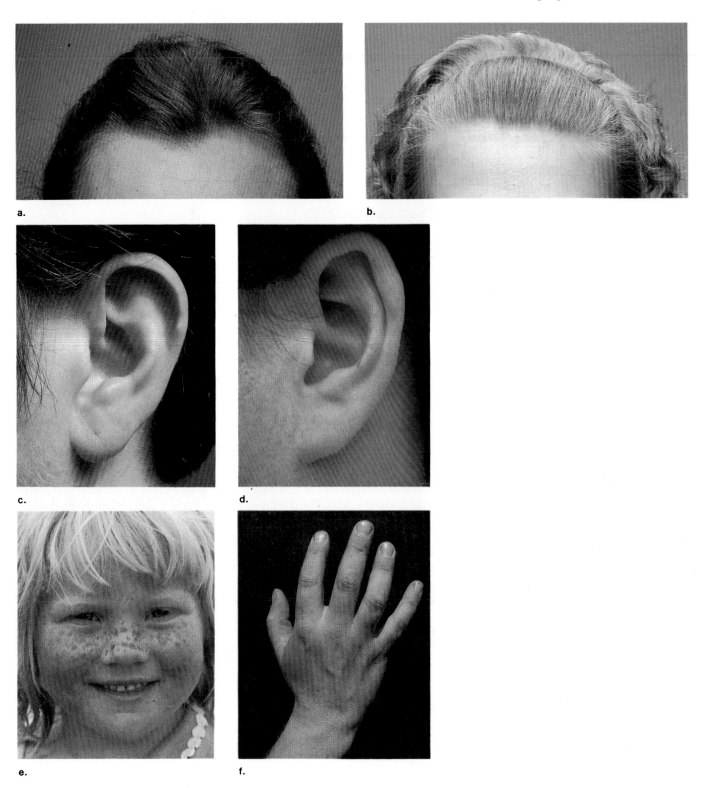

Figure 21.2
Common inherited characteristics in human beings.
Widow's peak (a.) is dominant over continuous
hairline (b.). Unattached earlobe (c.) is dominant
over attached earlobe (d.). Freckles (e.) are
dominant over no freckles. Short fingers (f.) are
dominant over long fingers.

a.

b.

c.

d.

e.

f.

As table 21.1 also indicates, the word **phenotype** refers to the physical characteristics of the individual. What the individual actually looks like is the phenotype. Notice that both homozygous dominant and heterozygous show the dominant phenotype.

Gamete Formation

Whereas the genotype has two alleles for each trait, the gametes have only one allele for each trait. This, of course is related to the process of meiosis. The alleles are present on a homologous pair of chromosomes and these chromosomes separate during meiosis (fig. 21.5). Thus the members of each allelic pair separate during meiosis, and there is only one allele for each trait in the gametes. When doing genetic problems it should be kept in mind that no two letters in a gamete may be the same. Thus Ww would represent a possible genotype and the gametes for this individual could contain either a W or w. Thus the possible gametes for this individual are W, w—the comma indicating two possible gametes.

Practice Problems 1

1. For each of the following genotypes, give all possible gametes.
   a. WW
   b. WWSs
   c. Tt
   d. Ttgg
   e. AaBb
2. For each of the following state whether it represents a genotype or a gamete.
   a. D
   b. Ll
   c. Pw
   d. LlGg

*Answers to problems are on page 469.

Crosses

It is now possible for us to consider a particular cross. If a homozygous man with a widow's peak (fig. 21.2a) marries a woman with a continuous hairline (fig. 21.2b), what kind of hairline will their children have?

In solving the problem, we must indicate the genotype of each parent by using letters, determine what the gametes are, and what the genotypes of the children are after reproduction. In the format that follows, P_1 stands for the parental generation, and the letters in this row are the genotypes of the parents. The second row shows that each parent has only one type of gamete in regard to hairline, and therefore all the children (F_1 = first filial generation) will have a similar genotype, that is, heterozygous. Heterozygotes show the dominant characteristic, and so all the children will have a widow's peak.

P_1	Widow's peak	×	Continuous hairline
	WW		ww
Gametes	W		w
F_1		Widow's peak	
		Ww	

Reproduction, Development, and Inheritance

These individuals are **monohybrids** because they are heterozygous for only one pair of alleles. If they marry someone else with the same genotype, what type of hairline will their children have?

P₁	Widow's peak	×	Widow's peak
	Ww		Ww
Gametes	W, w		W, w

In this problem, each parent has two possible types of **gametes**. In calculating F₁, it is assumed that either type of sperm has an equal chance to fertilize either type of egg. One way to assure that we have accounted for this is to use a **Punnett square** (fig. 21.3) in which all possible types of sperm are lined up vertically and all possible types of eggs are lined up horizontally (or vice versa), and every possible fertilization is considered.

When this is done, the results show a 3:1 phenotypic ratio; that is, three with widow's peak to one without. Such a ratio will actually be observed only if a large number of crosses of the same type take place and a large number of offspring result. Only then will all possible sperm have equal chance to fertilize all possible eggs. It is obvious that we do not routinely observe hundreds of offspring from a single type cross in humans, and so it is customary to merely state that each child has three chances out of four to have a widow's peak, or one chance out of four to have a continuous hairline. It is important to realize that **chance has no memory;** for example, if two heterozygous parents have already had three children with a widow's peak and are expecting a fourth child, this child still has three chances out of four to have a widow's peak and only one chance out of four of not having one. Each individual child has the same chances.

Probability

Another way to calculate the possible results of a cross is to realize that the chance, *or probability of receiving a particular combination of alleles, is simply the product of the individual probabilities.* In the cross just considered:

$Ww \times Ww$

the offspring have an equal chance of receiving W or w from each parent. Therefore:

Probability of W	$= \frac{1}{2}$
Probability of w	$= \frac{1}{2}$

and

Probability of WW	$= \frac{1}{2} \times \frac{1}{2}$	$=$	$\frac{1}{4}$
Probability of Ww	$= \frac{1}{2} \times \frac{1}{2}$	$=$	$\frac{1}{4}$
Probability of wW	$= \frac{1}{2} \times \frac{1}{2}$	$=$	$\frac{1}{4}$
$\frac{3}{4}$	$=$ Widow's peak		
Probability of ww	$= \frac{1}{2} \times \frac{1}{2}$	$=$	$\frac{1}{4}$
$\frac{1}{4}$	$=$ Continuous hairline		

Testcross

If a plant, animal, or person has the dominant phenotype, it is not possible to tell by inspection if the organism is homozygous dominant or heterozygous. However, if the plant or animal is crossed with a homozygous recessive, the

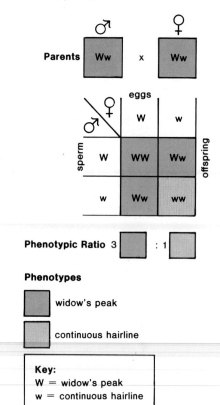

Figure 21.3
Monohybrid cross. In this cross, the parents are heterozygous for widow's peak. The chances of any child having a continuous hairline are one out of four, or 25 percent.

Key:
W = widow's peak
w = continuous hairline

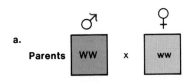

a.

Parents

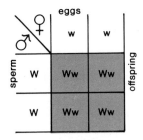

All with Widow's Peak

b.

Parents

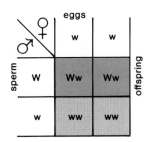

Ratio 1 : 1

Phenotypes

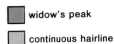

widow's peak

continuous hairline

Key:
W = widow's peak
w = continuous hairline

results may indicate what the original genotype was. For example, figure 21.4 shows the different results if a man with a widow's peak is homozygous dominant, or if he is heterozygous, and married to a woman with a continuous hairline. (She must be homozygous recessive or she would not have a continuous hairline.) In the first case, the man can only sire children with widow's peaks, and in the second the chances are 2:2 or 1:1 that a child will or will not have one. Thus, the cross of a possible heterozygote with an individual having the recessive phenotype gives the best chance of producing the recessive phenotype among the offspring. Therefore, this type cross is called the **testcross.**

Practice Problems 2
Using the information provided in figure 21.2, solve the following problems.

1. Both a man and a woman are heterozygous for freckles. What are the chances that their children will have freckles?
2. A woman is homozygous dominant for short fingers. Will any of her children have long fingers?
3. Both you and your sister or brother have attached earlobes, yet your parents have unattached ones. What are the genotypes of your parents?
4. A father has dimples, the mother does not have dimples; all the children have dimples. Dimples are dominant over no dimples. Give the probable genotype of all persons concerned.

*Answers to problems are on page 469.

Inheritance of Multitraits

Two Traits (Unlinked)
While it is possible to consider the inheritance of just one trait, actually each individual passes on to his or her offspring many genes for many traits. In order to arrive at a general understanding of multitrait inheritance, we will consider the inheritance of two traits. The same principles will apply to as many traits as we might wish to consider.

When Mendel performed two trait crosses, he formulated his second law:

Law of independent assortment: Pairs of factors separate independently of one another to form gametes, and therefore all possible combinations of factors may occur in the gametes.

Figure 21.5 illustrates that the law of segregation and the law of independent assortment hold because of the manner in which meiosis occurs. The law of segregation is dependent on the separation of members of homologous pairs of chromosomes; and the law of independent assortment is dependent on the random arrangement of homologous pairs with respect to one another during metaphase I prior to the separation process.

Crosses
When doing a two-trait cross, we realize that the genotypes of the parents require four letters because there is an allelic pair for each trait. Second, the gametes of the parents contain one letter of each kind in every possible combination, as predicted by Mendel's law of independent assortment. Finally, in order to produce the probable ratio of phenotypes among the offspring, all possible matings are presumed to occur.

Reproduction, Development, and Inheritance

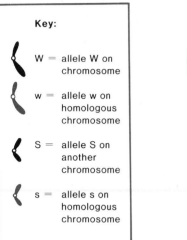

Key:

W = allele W on chromosome

w = allele w on homologous chromosome

S = allele S on another chromosome

s = allele s on homologous chromosome

Figure 21.5
Segregation and independent assortment during meiosis. Homologous chromosomes separate (segregate) during meiosis I. Separation by independent assortment results in all possible combinations of alleles. This shows that Mendel's law of segregation and law of independent assortment hold because of the manner in which meiosis occurs.

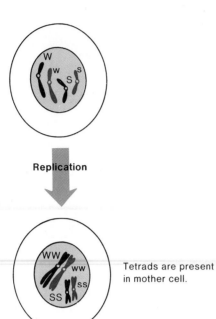

Replication

Tetrads are present in mother cell.

Homologous chromosomes separate

either   or

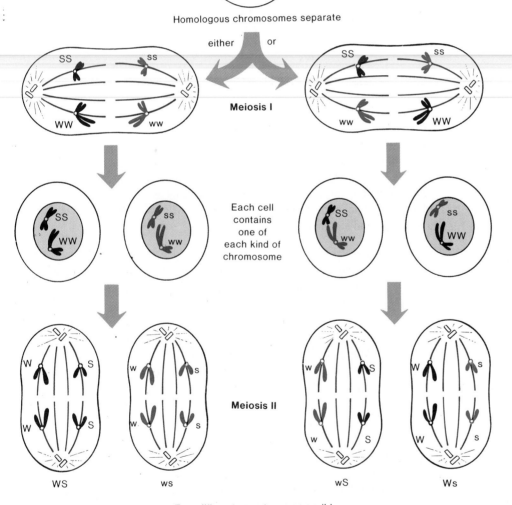

Meiosis I

Each cell contains one of each kind of chromosome

Meiosis II

WS   ws   wS   Ws

Four different gametes are possible.

Human Inheritance   447

To give an example, let us cross a person homozygous for widow's peak and short fingers with a person who has a continuous hairline and long fingers. The key for such a cross is:

W = Widow's peak S = Short fingers
w = Continuous hairline s = Long fingers

P_1 Widow's peak   Continuous hairline
          Short fingers   Long fingers
          $WWSS$ $wwss$

Gametes WS ws

F_1 Widow's peak
                         Short fingers
                         $WwSs$

In this particular cross, only one type of gamete is possible for each parent; therefore, all of the F_1 will have the same genotype ($WwSs$) and the same phenotype (widow's peak with short fingers). This genotype is called a *dihybrid* because the individual is heterozygous in two regards: hairline and fingers.

When a dihybrid reproduces with a dihybrid, each parent has four possible types of gametes:

$WwSs$ × $WwSs$
  ↓   ↓
Gametes: WS WS
           Ws Ws
           wS wS
           ws ws

The Punnett square (fig. 21.6) for such a cross shows the expected genotypes among 16 offspring if all possible sperm fertilize all possible eggs. An inspection of the various genotypes in the square shows that among the offspring, *nine* will have a widow's peak and short fingers, *three* will have a widow's peak and long fingers, *three* will have a continuous hairline and short fingers, and *one* will have a continuous hairline and long fingers. This is called a 9:3:3:1 phenotype ratio, and this ratio always results when a dihybrid is mated with a dihybrid and simple dominance is present.

Probability

We can use the previous ratio to predict the chances of each child receiving a certain phenotype. For example, the possibility of getting the two dominant phenotypes together is 9 out of 16 (9+3+3+1 = 16), and that of getting the two recessive phenotypes together is 1 out of 16.

We can also calculate the chance, or probability, of these various phenotypes occurring by knowing that the *probability of combinations of independent events is the product of the probabilities of each of the events.* Thus:

Probability of widow's peak = ¾
Probability of short fingers = ¾
Probability of continuous hairline = ¼
Probability of long fingers = ¼

Therefore

Probability of widow's peak and short fingers = ¾ × ¾ = 9/16
Probability of widow's peak and long fingers = ¾ × ¼ = 3/16
Probability of continuous hairline and short fingers = ¼ × ¾
    = 3/16
Probability of continuous hairline and long fingers = ¼ × ¼ = 1/16

Reproduction, Development, and Inheritance

Figure 21.6
Dihybrid cross. A dihybrid results when an individual homozygous dominant in two regards reproduces with an individual homozygous recessive in two regards. When a dihybrid reproduces with a dihybrid, there are four possible phenotypes among the offspring and the phenotypic ratio is 9:3:3:1 as indicated below.

P₁

Gametes

F₁

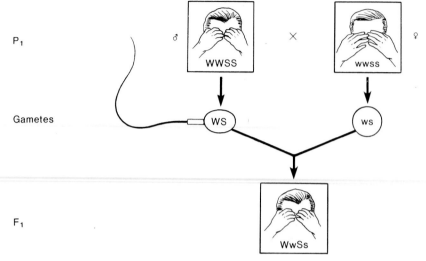

F₁ gametes

eggs

	WS	Ws	wS	ws
WS	WWSS	WWSs	WwSS	WwSs
Ws	WWSs	WWss	WwSs	Wwss
wS	WwSS	WwSs	wwSS	wwSs
ws	WwSs	Wwss	wwSs	wwss

sperm

offspring

Phenotypic Ratio   9 ▢ :3 ▢ :3 ▢ :1 ▢

Phenotypes

▢ widow's peak, short fingers

▢ widow's peak, long fingers

▢ continuous hairline, short fingers

▢ continuous hairline, long fingers

Key:
W = widow's peak
w = continuous hairline
S = short fingers
s = long fingers

Figure 21.7

Testcross. In this example it is impossible to tell by inspection if the male parent is homozygous dominant or if he is heterozygous for both traits. However, reproduction with a female who is recessive for both traits is likely to show which he is. If he is heterozygous, there is a 25 percent chance that the offsprings will show both recessive characteristics and a 50 percent chance that they will show one or the other of the recessive characteristics.

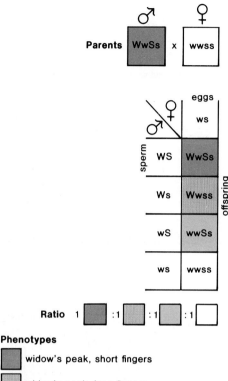

Ratio 1 ▦ :1 ▒ :1 ░ :1 □

Phenotypes

▦ widow's peak, short fingers

▒ widow's peak, long fingers

░ continuous hairline, short fingers

□ continuous hairline, long fingers

Key:
W = widow's peak
w = continuous hairline
S = short fingers
s = long fingers

Table 21.2	Phenotypic ratios of common crosses
Monohybrid × monohybrid	3:1 (dominant to recessive)
*Monohybrid × recessive	1:1 (dominant to recessive)
Dihybrid × dihybrid	9:3:3:1 (9 both dominant, 3 one dominant, 3 one dominant, 1 both recessive)
*Dihybrid × recessive	1:1:1:1 (all possible combinations in equal number)

*Called a backcross because it is as if the F_1 were mated back to the recessive parent. Also called a testcross because it can be used to test if the individual showing the dominant gene is homozygous or heterozygous. For a definition of all terms, see the Glossary.

Testcross

A plant or animal that shows the dominant traits can be tested for the dihybrid genotype by a mating with the recessive in both traits.

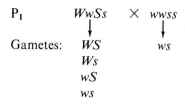

The Punnett square (fig. 21.7) shows that the resulting ratio is 1 widow's peak with short fingers : 1 widow's peak with long fingers : 1 continuous hairline with short fingers : 1 continuous hairline with long fingers, or 1:1:1:1.

Summary

Table 21.2 lists all of the crosses we have studied thus far, which show a frequently observed ratio. When these types of crosses are done, these ratios are observed.

Practice Problems 3

Using the information in figure 21.2, solve these problems.

1. What is the genotype of the offspring if a man homozygous recessive for type of earlobes and homozygous dominant for type of hairline is married to a woman who is homozygous dominant for earlobes and homozygous recessive for hairline.
2. If the offspring of this cross marries someone of the same genotype, then what are the chances that this couple will have a child with a continuous hairline and attached earlobes?
3. A person who has dimples and freckles marries someone who does not. This couple produces a child that does not have dimples nor freckles. What is the genotype of all persons concerned?

*Answers to problems are on page 469.

Beyond Mendel's Laws

While the study of Mendel's laws is helpful, we know today that they are an oversimplification. There are many exceptions, such as gene interactions, codominance, and gene linkage.

Gene Interactions

Contrary to Mendel's first law, and exemplified in the next section, more than one gene can control a single trait, but also one gene can affect various traits.

Reproduction, Development, and Inheritance

Figure 21.8
Inheritance of skin color. This white husband (aabb) and his mulatto wife (AaBb) had fraternal twins, one of which was white and one of which was mulatto.

Polygenic Inheritance

Two or more genes may affect the same trait in an additive fashion. When a black person has children by a white person, the children are mulatto, but two mulattoes can produce children who range in skin color from black to white. This can be explained if we assume that there are two genes that control skin color and that only capital letters in the example contribute equally to skin color.

Black = $AABB$
Dark = $AABb$ or $aABB$
Mulatto = $AaBb$ or $AAbb$ or $aaBB$
Light = $Aabb$ or $aabB$
White = $aabb$

Polygenic inheritance can cause the distribution of human traits according to a bell-shaped curve, with most individuals exhibiting the average phenotype. The more genes that control the trait, the more continuous the distribution will be. Just how many genes control skin color and height, another possible example of polygenic inheritance, is not known.

Epistasis

One gene can negatively affect the action of one or more other genes. An **epistatic gene** is one that masks the phenotypic expression of a gene at a different locus. For example, if a person inherits a pair of alleles that causes **albinism,** the inability to produce the pigment melanin, it does not matter if genes for eye color and hair color were also inherited. The latter genes cannot be expressed and the person will be an albino, lacking pigment in all parts of the body (fig. 21.9).

Pleiotropy

Pleiotropic genes are ones that have multiple effects. Therefore the individual exhibits a combination of traits known as a **syndrome;** yet all of these traits can be traced to one dominant allele or two recessive alleles. For example, a second reading for this chapter (p. 452) describes **Marfan's syndrome,** which is recognized by skeletal, eye, and cardiovascular defects. All of these are due to the inability to produce normal connective tissue. Abraham Lincoln is believed to have suffered from this disease caused by a dominant allele.

Figure 21.9
Albinism. Albinos are unable to produce the pigment melanin and therefore any genes received for coloring cannot be expressed. This is the rock singer Johnny Winter who is an albino.

Abe's Malady

To Dr. Harold Schwartz, the signs left little doubt. The seven-year-old boy visiting his Huntington Park, Calif., office in 1959 had Marfan's syndrome, a genetic disorder of the connective tissue that can cause heart and eye problems, affect skeletal growth and occasionally be fatal. A few months later, the boy's grandmother dropped in to inquire about his condition and revealed that her husband had died of Marfan's. The grandmother's married name was Lincoln.

Says Schwartz: "I call that my 'burning bush' moment. I had read Carl Sandburg's biography of Abraham Lincoln, which contains a great deal about Lincoln's physical characteristics." Suddenly everything connected. The Great Emancipator, Schwartz realized, was probably afflicted by Marfan's syndrome.

Since then, Schwartz, now 60, has traced the Lincoln Marfan gene back to 16th century England and now is more certain than ever about his theory. In the *Western Journal of Medicine*, he strongly suggests that had John Wilkes Booth not fired the fatal shot on April 14, 1865, Lincoln would have died within a year from complications of Marfan's syndrome—for which there is still no cure.

Schwartz points to the well-documented fact that Lincoln had disproportionately long arms, legs, hands and feet, even for a man of his height. While watching a regiment of Maine lumbermen during the Civil War, the President himself noted: "I don't believe that there is a man in that regiment with longer arms than mine." In 1907 a sculptor working with Lincoln casts observed that "the first phalanx of the middle finger is nearly half an inch longer than that of an ordinary hand." The President sometimes squinted with his left eye. All of these characteristics, according to Schwartz, are typical of Marfan's syndrome. In fact, Lincoln's "spider-like legs," a phrase used by one of the President's contemporaries, was the very simile used in 1896 by French Physician Bernard-Jean Antonin Marfan when he described the syndrome that was named for him.

Schwartz has also presented an ingenious bit of evidence that Lincoln had a specific cardiovascular problem also associated with Marfan's syndrome: imperfect closure of the valves of the aorta, the large artery that carries blood from the heart. The clue appeared in a picture of the President taken in 1863. Lincoln had his legs crossed, and in an otherwise sharp photo, the left foot—suspended in the air—is blurred. When viewing the print, Lincoln asked why the foot was fuzzy. A friend familiar with physiology suggested that the throbbing arteries in the leg might have caused some movement. Lincoln promptly crossed his legs and watched. "That's it!" he exclaimed. "Now that's very curious, isn't it?" Not to Schwartz. The Marfan-caused defect, he points out, results in "aortic regurgitation," which causes pulses of blood strong enough to shake the lower leg.

Schwartz has also found in the President's own words what he believes to be good evidence that before Lincoln was shot he was "in a state of early congestive heart failure"—brought on by his aortic condition. About seven weeks before Lincoln's assassination, for example, he told his friend Joshua Speed: "My feet and hands of late seem to be always cold, and I ought perhaps to be in bed." Though he was only 56 in 1865, Abe was also easily fatigued toward the end. "There is only one word that can express my condition," he said, "and that is 'flabbiness.' " Once, shortly before his death, he tried to get out of bed but fell back, too weak to rise. Only a day before Lincoln was shot, his wife Mary wrote of the President's "severe headache" and indisposition. Concludes Schwartz: the faulty aortic valves resulted in "a decompensating left ventricle which was the undiagnosed or concealed cause of the President's failing health."

Reproduction, Development, and Inheritance

Multiple Alleles

Three alleles for the same gene control the inheritance of A-B-O blood types. These alleles determine the presence or absence of antigens on the red blood cells. Therefore, *I* standing for immunogen (antigen) is used to signify the gene, and a superscript letter is used to signify the particular allele:

I^A = type A antigen on red cells
I^B = type B antigen on red cells
i^o = no antigens on the red cells

Each person has only two of the three possible alleles and both I^A and I^B are dominant over i^o. Therefore as table 21.3 shows, there are two possible genotypes for type A blood and two possible genotypes for type B blood. On the other hand, I^A and I^B are fully expressed in the presence of the other. Therefore if a person inherits one of each of these alleles, that person will have type AB blood. Type O blood can only result from the inheritance of two i^o alleles.

An examination of possible matings between different blood types sometimes produces surprising results; for example,

P_1 $I^A i^o \times I^B i^o$
F_1 $I^A I^B$, $i^o i^o$, $I^A i^o$, $I^B i^o$

Thus from this particular mating every possible phenotype (AB, O, A, B blood type) is possible.

Blood typing can sometimes aid in paternity suits. However, a blood test of a supposed father can only suggest that he *might* be the father, not that he definitely *is* the father. For example, it is possible, but not definite, that a man with blood type A (having genotype $I^A i^o$) is the father of a child with blood type O. On the other hand, a blood test can sometimes definitely prove that a man is not the father. For example, a man with blood type AB could not possibly be the father of a child with blood type O. Therefore, blood tests may legally be used only to exclude a man from possible paternity.

It might be noted here that the blood factor called Rh is inherited separately from A, B, AB, or O type blood. In each instance it is possible to be Rh^+ or Rh^-, meaning in the first case that an Rh factor is present on the red cells and in the second that an Rh factor is not present. It may be assumed that Rh is controlled by a single allelic pair in which simple dominance prevails: Rh^+ is dominant over Rh^-. Complications arise when an Rh^- woman marries an Rh^+ man and the child in the womb is Rh^+. With the birth of the first child of this phenotype, the mother may begin to build up antibodies to the factor and in later pregnancies these antibodies may cross the placenta to destroy the baby's blood cells, as was discussed in chapter 11.

Incomplete Dominance and Codominance

In incomplete dominance, neither member of an allelic pair is dominant over the other and the phenotype is intermediate between the two. For example, when a curly-haired white person reproduces with a straight-haired white person, their children will have wavy hair. And two wavy-haired individuals can produce all possible phenotypes: straight hair, curly hair, and wavy hair. To signify that neither allele is dominant, one allele is designated as *H* and the other as *H'*, as is done in figure 21.10.

In codominance, each member of an allelic pair is dominant and the phenotype exhibits both characteristics. For example, an individual with the genotype $I^A I^B$ has the blood type AB.

Table 21.3 Blood groups

Phenotype	Genotypes
A	$I^A I^A$, $I^A i^o$
B	$I^B I^B$, $I^B i^o$
AB	$I^A I^B$
O	$i^o i^o$

I = immunogen gene

Figure 21.10
Incomplete dominance. Among Caucasians, neither straight nor curly hair is dominant. When two wavy-haired individuals reproduce, the offspring has a 25 percent chance of having either straight or curly hair and a 50 percent chance of having wavy hair, the intermediate phenotype.

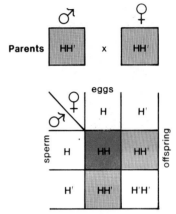

Results: 1 curly: 2 wavy: 1 straight

Genotypes   Phenotypes

HH = curly hair

HH' = wavy hair

H'H' = straight hair

Figure 21.11
Complete linkage (hypothetical). In this example,
the genes for dimples and type of fingers are
linked. Even though the parents are dihybrids, the
offspring show only two possible phenotypes.

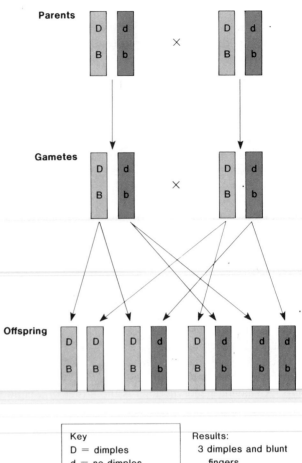

Parents

Gametes

Offspring

Key	Results:
D = dimples	3 dimples and blunt
d = no dimples	fingers
B = blunt fingers	1 no dimples and
b = pointed fingers	pointed fingers

Linkage

As illustrated in figure 21.1, a chromosome pair has a series of genes. Genes on the same chromosome are said to form a **linkage group.** Mendel's law of independent assortment cannot hold for linked genes since they tend to appear together in the same gamete. Thus, traits controlled by linked genes tend to be inherited together.

To take a hypothetical example, let us remember that dimples are dominant over no dimples and blunt fingers are dominant over pointed fingers. If a dihybrid were married to a dihybrid, you would expect four possible phenotypes among the offspring. But as figure 21.11 shows, only two phenotypes would appear if the genes were absolutely linked in the manner illustrated. When doing linkage problems, it is better to use the method illustrated in figure 21.11 rather than a Punnett square so that the genes can be shown on a single chromosome.

When a dihybrid is crossed with a recessive, you normally would expect all possible phenotypes among the offspring. If linkage is present, however, the number of possible phenotypes could possibly be reduced to two types. To take an actual example, it has been reported that A-B-O blood type gene and the gene for a very unusual dominant condition called nail-patella syndrome (NPS) are on the same chromosome. A person with NPS has fingernails and

Reproduction, Development, and Inheritance

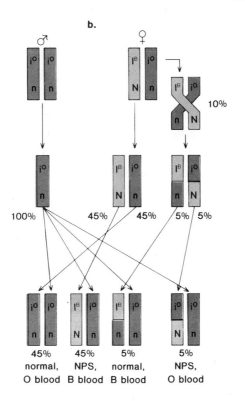

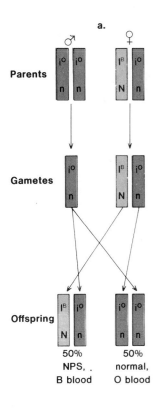

toenails that are reduced or absent and a kneecap (patella) that is small. In one family, the female parent had the genotype $I^B i^o$ for blood type and the genotype Nn for NPS; furthermore, it could be established that the allele I^B was on the same chromosome as N and that the allele i^o was on the same chromosome as n. Notice in figure 21.12a that if linkage holds, this individual would form only two possible gametes.

The male parent in this example had the recessive genotype for both traits and therefore could form only one type gamete carrying the recessive alleles of each gene as illustrated in figure 21.12a. Therefore, assuming linkage, the children of this couple should have only two possible phenotypes: blood type B with NPS and blood type O without NPS.

However, at least 10 percent of all offspring show a phenotype in which blood type B is found without NPS and blood type O is found with NPS. This indicates that crossing over occurred (fig. 21.12b).

Crossing Over

When tetrads form during meiosis, the inner chromatids may exchange portions by a process of breaking and then reassociating (fig. 4.17). The gametes that receive recombined chromosomes are called recombinant gametes. Recombinant gametes indicate that the linkage between two genes has been broken by crossing over. Figure 21.12 shows how recombinant gametes produced the unexpected phenotypes in our example.

In lower organisms, it has been possible to map chromosomes by studying the crossover frequency of linked genes. Genes distant from one another are more likely to be separated by crossing over than genes that are close together. Thus, the crossover frequency indicates the distance between two genes on a chromosome. Each percentage of crossing over is taken to mean a distance of one map unit. Using these frequencies, then, it is possible to indicate the order of the genes on the chromosome.

Human disorders, such as the one illustrated in figure 21.10, have helped to map human chromosomes, but laboratory investigations involving the chromosomes themselves have been more helpful. For example, human and mouse cells can be fused together in tissue cultures. As the cells grow and divide, some of the human chromosomes are lost and eventually the daughter cells contain only a few human chromosomes. Analysis of the proteins[2] made by the various man-mouse cells has enabled investigators to determine which genes are to be associated with which human chromosomes. Sometimes a man-mouse hybrid cell has formed in which the human chromosome is shortened and certain genes are missing. In this way, it has even been possible to determine the approximate location of the genes along the chromosomes.

Practice Problems 4

1. What is the genotype of a person with straight hair? Could this individual ever have a child with curly hair?
2. What is the darkest child that could result from a mating between a light individual and a white individual?
3. What is the lightest child that could result from a mating between two mulatto individuals?
4. From the following blood types, determine which baby belongs to which parents:

Mrs. Doe	Type A
Mr. Doe	Type A
Mrs. Jones	Type A
Mr. Jones	Type AB
Baby 1	Type O
Baby 2	Type B

5. Prove that a child does not have to have the blood type of either parent by indicating what blood types *might* be possible when a person with type A blood reproduces with a person with type B blood.
6. Imagine that ability to curl the tongue is dominant and that this characteristic is linked to a rare form of mental retardation, which is also dominant. The parents are both dihybrids, with the two dominant alleles on one chromosome and the two recessive alleles on the other. What phenotypic ratio is possible among the offspring if crossing over does not occur?

*Answers to problems are on page 469.

---

[2]Gene control of protein synthesis is discussed in the next chapter.

Reproduction, Development, and Inheritance

Table 21.4 Estimated incidence and prevalence of selected genetic diseases, U.S.A., 1976

Condition	Newly affected	Under age 20 with condition
Polydactyly (extra fingers and toes)	9,300	184,000
Cystic fibrosis	2,000	20,000
Hemophilia	1,200	12,400
Phenylketonuria	310	3,100
Sickle-cell anemia	1,200	16,000
Tay-Sachs disease	30	100
Thalassemia (Cooley's anemia)	70	1,000

Human Genetic Diseases

Birth defects can be environmentally induced, but many are genetic in origin. The incidence of some genetic diseases among newborns is given in table 21.4. It is well to keep in mind that these diseases are not contagious; they are inherited.

Dominant

We have already had occasion to mention Marfan's syndrome and nail-patella syndrome, which are both due to dominant alleles, but these are not common disorders. More common genetic diseases due to the inheritance of a single dominant allele are the following:

Achondroplasia: a form of dwarfism
Chronic simple glaucoma (some forms): a major cause of blindness if untreated
Huntington's chorea: progressive nervous system degeneration
Hypercholesterolemia: high blood cholesterol levels, propensity to heart disease
Polydactyly: extra fingers or toes[3]

If one parent is heterozygous for the characteristic and the other is recessive, each child has a 50:50 chance of receiving the characteristic or escaping it. Notice that if the characteristic is dominant, one parent must necessarily show it. Genetic counseling, then, is aided by this fact, except that some phenotypes do not appear until a person is grown. For example, **Huntington's chorea** (or Huntington's disease) does not appear until the thirties or early forties. There is a progressive deterioration of the individual's nervous system that eventually leads to constant thrashing and writhing movements until insanity precedes death. Recent studies suggest that Huntington's chorea is due to the inability to make use of the neurotransmitter substance GABA, in which case there is now hope for a cure.

People with Huntington's chorea seem to be more fertile than others. It is amazing that more than a thousand of the cases in the United States in the past century can be traced to one man born in 1831. Investigators are presently studying 100 victims in a single village in Venezuela where a total of 1,100 are at risk of developing Huntington's.

[3]National Foundation/March of Dimes.

Recessive

We have already had occasion to mention albinism (fig. 21.9), a recessively inherited disorder, but more common are the following, which are also controlled by one pair of alleles:

Cystic fibrosis: disorder affecting function of mucous and sweat glands
Galactosemia: inability to metabolize milk sugar
Phenylketonuria: essential liver enzyme deficiency
Thalassemia: blood disorder primarily affecting persons of Mediterranean ancestry
Tay-Sachs disease: fatal brain damage primarily affecting infants of East European Jewish ancestry[4]

In these cases both parents may appear to be normal but are carriers of the defective allele. A carrier is a person who does not show the trait but who can pass on the allele to an offspring. If both parents carry the allele, each child will run a 25 percent risk of manifesting the disease. Each child has a 25 percent risk of being homozygous normal but a 50:50 chance of receiving a single defective allele and being a carrier. Should these carriers marry other carriers, they run the same risk as the parents of transmitting the disease to the next generation.

Cystic fibrosis, characterized by abnormal mucus-secreting tissues, is now one of the most commonly inherited disorders among Caucasian children. At first the infant may have difficulty regaining the birth weight despite good appetite and vigor. A cough associated with a rapid respiratory rate but no fever indicates lung involvement. Large, frequent, and foul-smelling stools are due to abnormal pancreatic secretions. Whereas children previously died in infancy due to infections, they now often survive because of antibiotic therapy.

Phenylketonuria (PKU) is characterized by severe mental retardation due to an abnormal accumulation of the common amino acid phenylalanine within cells, including neurons. The disorder takes its name from the presence of a breakdown product, phenylketone, in the urine and blood. Newborn babies are routinely tested at the hospital and, if necessary, are placed on a diet low in phenylalanine. This diet improves the mental capabilities of affected individuals.

Tay-Sachs disease is caused by the inability to break down a certain type of fat molecule that accumulates around nerve cells until they are destroyed. Afflicted newborns appear normal and healthy at birth, but they do not develop normally. At first, they may learn to sit up and stand, but later they regress and become mentally retarded, blind, and paralyzed. Death usually occurs between ages three and four.

Codominant

Sickle-cell anemia occurs when the individual has the genotype $Hb^S Hb^S$. In these individuals the red blood cells are sickle shaped (fig. 21.13) because the abnormal hemoglobin molecule is less soluble than the normal hemoglobin, Hb^A. Sickle-shaped cells have a limited ability to transport oxygen. Inheritance involves codominance in the following manner: Individuals with the genotype $Hb^A Hb^A$ are normal; those with $Hb^S Hb^S$ have sickle-cell anemia; and those with $Hb^A Hb^S$ have **sickle-cell trait,** a condition in which the cells are sometimes sickle-shaped, as described in the paragraphs that follow. Two individuals with sickle-cell trait can produce children with all three phenotypes, as indicated in figure 21.14.

[4]National Foundation/March of Dimes.

Figure 21.13
Sickled cells. Individuals with sickle-cell anemia
have sickled red blood cells that tend to clump as
illustrated here.

Sickle-cell anemia is prevalent among members of the black race because the shape of the cells seems to give protection against the malaria parasite, which utilizes red cells during its asexual reproductive phase (p. 549). While infants with sickle-cell anemia often die, those with the trait are protected from malaria, especially during ages two to four. This means that in Africa these children survived and grew up to reproduce and pass on the allele to their offspring. As many as 60 percent of tribes in malaria-infected regions of Africa have the allele. In the United States, about 10 percent of the black population carry it.

The blood cells in persons with sickle-cell anemia cannot easily pass along small blood vessels. The sickle-shaped cells either break down or they clog blood vessels. Thus the individual suffers from poor circulation, anemia, and sometimes internal hemorrhaging. Jaundice, episodic pain of the abdomen and joints, poor resistance to infection, and damage to internal organs are all symptoms of sickle-cell anemia. Few patients live beyond age 40.

Persons with the sickle-cell trait do not usually have any difficulties unless they are exposed to air that is low in oxygen. At such times, the cells become sickle shaped, with accompanying disturbances in circulation.

Genetic Counseling

Now that persons are becoming aware that many illnesses are caused by faulty genes, more couples are seeking genetic counseling. The counselor studies the background of the couple and tries to determine if any immediate ancestor may have had a genetic disease. Then the counselor studies the couple themselves. As much as possible, laboratory tests are performed on all persons involved.

Figure 21.14
Inheritance of sickle-cell anemia. In this example, both parents have sickle-cell trait and are therefore carriers. Therefore, each child has a 25 percent chance of having sickle-cell anemia or of being perfectly normal and a 50 percent chance of having sickle-cell trait.

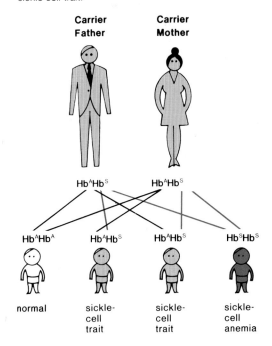

Figure 21.15
A pedigree chart indicates the phenotype of family members for several generations. The pattern of inheritance often reveals whether a trait is (a) dominant, (b) sex-linked recessive, or (c) recessive. Prove this by filling in the genotypes for each individual.

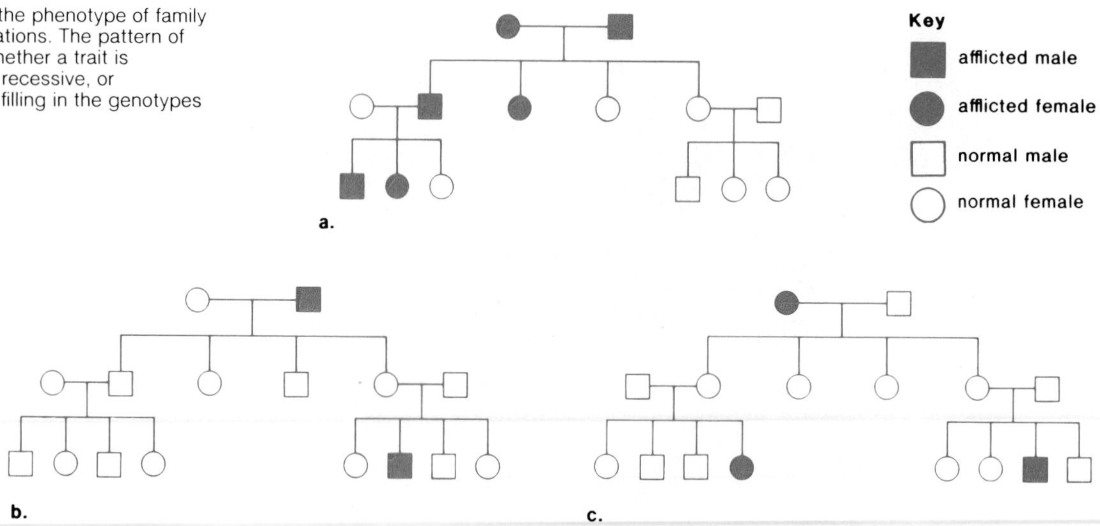

Key

■ afflicted male

● afflicted female

□ normal male

○ normal female

a.

b.

c.

Carrier tests are now available for a large number of potential genetic diseases. Blood tests can identify carriers of thalassemia (Cooley's anemia) and sickle-cell anemia. By measuring enzyme levels in blood or skin cells, carriers of enzyme defects can now be identified for some inborn metabolic errors, such as Tay-Sachs disease, in which a defect of a single enzyme interferes with a vital metabolic process. From this information, the counselor can sometimes predict the chances of the couple having a defective child.

Whenever the woman is pregnant, chorionic villi sampling can be done early and amniocentesis can be done later in the pregnancy. These procedures, which were discussed in the previous chapter (p. 435), allow the testing of embryonic and fetal cells, respectively, in order to determine if the child has a genetic disease. Most of the time the baby is normal, but should a defect be discovered the couple has an opportunity to elect to abort the pregnancy.

Pedigree Charts

Pedigree charts (fig. 21.15) are often constructed to show the inheritance of a certain condition within a family. Such charts are a great help in deciding whether a phenotype is controlled by a dominant or recessive allele. For example, if only one parent shows the trait and yet all or several of the children show it, then it is most probably dominant. Or if two individuals do not show the characteristic but their children do, then the characteristic must be determined by a recessive allele. On the other hand, if the characteristic appears primarily in males and passes from grandfather to grandson, then it must be controlled by a X-linked recessive allele (p. 464).

Chromosome Inheritance

Normal Inheritance

Occasionally, it is more useful to discuss the inheritance of chromosomes rather than the inheritance of a particular gene on a chromosome. As was discussed and illustrated in chapter 4 (p. 81), a karyotype may be prepared to display an individual's chromosomes. The autosomes are arranged by pairs and are numbered from 1 to 22. The sex chromosomes are the 23rd pair. Males have XY sex chromosomes and females have two X chromosomes in their karyotype.

Note that it is the father who determines the sex of the child (fig. 21.16) because the father forms two types of sperm—those containing X and those containing Y. Since the ratio of these is 50:50, the chance of bearing a male

Reproduction, Development, and Inheritance

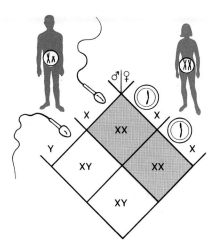

Figure 21.17
Down's syndrome. Common characteristics include a wide, rounded face and a fold of the upper eyelids that roughly resembles that of the Mongolian races. Mental retardation, along with an enlarged tongue, makes it difficult for persons with Down's syndrome to learn to speak coherently.

offspring is just as great as that of bearing a female. However, there are those who believe that the pH of the vagina can determine which type of sperm will fertilize the egg. An alkaline vagina at the time of ovulation seems to favor the Y-bearing sperm while the acidic vagina just before ovulation seems to favor the X-bearing sperm.

Abnormal Chromosome Inheritance

Sometimes individuals are born with either too many or too few chromosomes (table 21.5). It is possible also that even though there is the correct number of chromosomes, one chromosome may be defective in some way. Both autosomal and sex chromosome abnormalities occur.

Autosomal

The most common autosomal abnormality is seen in individuals with **Down's syndrome** (fig. 21.17). This syndrome is sometimes called mongolism because the eyes of the person seem to have an oriental-like fold, but this term is not

Table 21.5 Incidence of selected chromosomal abnormalities

Name	Frequency/ live births
Down's syndrome	1/800–1,000
Turner's syndrome	1/10,000
Superfemale	1/950
Klinefelter's syndrome	1/1,000
XYY	1/1,000

Figure 21.18

Nondisjunction during oogenesis. Nondisjunction can occur during meiosis I if the chromosome pairs fail to separate and during meiosis II if the chromatids fail to separate completely. In either case, the abnormal eggs carry an extra chromosome. Nondisjunction of both autosomes and sex chromosomes is possible. In males, nondisjunction of Y chromosomes during spermatogenesis can lead to an offspring that is an XYY male.

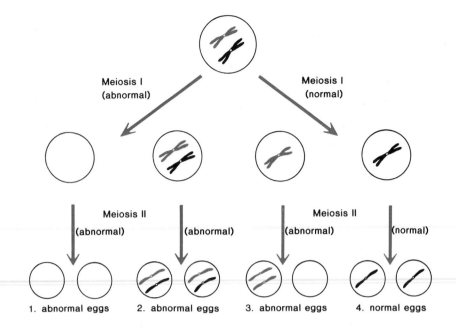

Figure 21.19

Karyotype of a male with Down's syndrome. Note that there are three number 21 chromosomes. Karyotypes can be done on embryonic or fetal cells collected by means of chorionic villi sampling or amniocentesis.

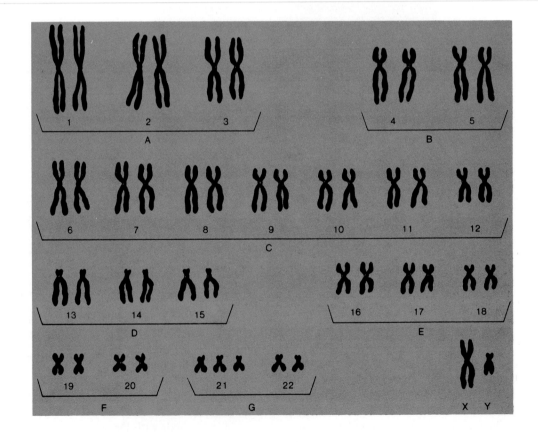

Reproduction, Development, and Inheritance

considered scientific. Other characteristics are short stature; stubby fingers; a wide gap between the first and second toes; a large, fissured tongue; a round head; a palm crease, the so-called simian line; and, unfortunately, mental retardation that can sometimes be severe.

Persons with Down's syndrome usually have three number 21 chromosomes because the egg had two number 21 chromosomes instead of one. In 23 percent of the cases studied, however, the sperm had the extra 21st chromosome. Either the chromosome pair, or the chromatids, failed to separate completely and instead went into the same daughter cell (fig. 21.18). Either of these occurrences is called **nondisjunction.** It would appear that nondisjunction is most apt to occur in the older female since children with Down's syndrome are usually born to women over age forty. If a woman wishes to know whether or not her unborn child is affected by Down's syndrome, she may elect to undergo chorionic villi testing or amniocentesis. Following this procedure, a karyotype (fig. 21.19) can reveal whether the child has Down's syndrome. If so, she may elect to continue or to abort the pregnancy.

Another chromosomal abnormality called **deletion** is responsible for a syndrome known as *cri du chat* (cat's cry). Affected individuals meow like a kitten when they cry, but more important perhaps is the fact that they tend to have a small head with malformations of the face and body and that mental defectiveness usually causes retarded development. Chromosomal analysis shows that a portion of chromosome number 5 is missing (deleted), while the other number 5 chromosome is normal.

Sexual

Due to nondisjunction of the sex chromosomes during oogenesis, an egg may be produced that has either no X chromosome or two X chromosomes. When the first of these is fertilized by an X-bearing sperm, a female with **Turner's syndrome** may be born (fig. 21.20b). These XO individuals have only one sex chromosome—an X; the O signifies the absence of the second sex chromosome. The ovaries never become functional, regressing to ridges of white streaks. Because of this, these females do not undergo puberty or menstruate, and there is a lack of breast development. Generally, these individuals have a stocky build, a webbed neck, and subnormal intelligence.

When an egg having two X chromosomes is fertilized by an X-bearing sperm, a **superfemale** having three X chromosomes results. While it might be supposed that the XXX female with 47 chromosomes would be especially feminine, this is not the case. Although there is a tendency toward mental retardation, most superfemales have no apparent physical abnormalities and many are fertile and have children with a normal chromosome count.

When an egg having two X chromosomes is fertilized by a Y-bearing sperm, a male with **Klinefelter's syndrome** results (fig. 21.20a). This individual is male in general appearance, but the testes are underdeveloped and the breasts may be enlarged. The limbs of these XXY males tend to be longer than average, body hair is sparse, and many are mentally defective.

XYY males also occur possibly due to nondisjunction during spermatogenesis. Afflicted males are usually taller than average, suffer from persistent acne, and tend to have barely normal intelligence. At one time it was suggested that these men were likely to be criminally aggressive, but it has been shown that the incidence of such behavior is quite low.

From an examination of these abnormal sex chromosome constituencies, it can be deduced that at least one X chromosome is required for human survival. There are no YO males. Also, the presence of a Y signifies a male regardless of the number of X chromosomes.

Figure 21.20
Sex chromosome inheritance in person's with Turner's or Klinefelter's syndrome is abnormal.

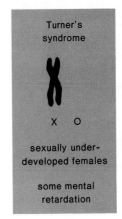
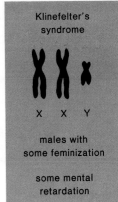

Sex-linked Inheritance

The genes which determine the development of the sexual organs are on the sex chromosomes. The fetus is unisexual and will automatically develop into a female unless a Y chromosome is present. There is evidence for a gene on the Y chromosome which directs the nondifferentiated gonad to produce testosterone and thereafter the fetus begins to develop into a male. Several other genes are necessary to complete the process of normal male sexual development. Genes are also required to complete normal female maturation.

Some genes on the sex chromosomes have nothing to do with sexual development and are instead concerned with other body traits. These genes are said to be **sex linked** because they are on the sex chromosomes. A few sex-linked genes are on the Y chromosome, but the most important ones discovered so far are on only the X chromosome.

X-linked Genes

X-linked genes have alleles on the X chromosome but determine body traits unrelated to sex. Since there are no alleles for these genes on the Y chromosome, any recessive allele present on the X chromosome in males will be expressed. In X-linked trait problems, the females are indicated by XX and the males by XY. An X-linked allele appears as a letter attached to the X chromosome. For example, in human beings color blindness is controlled by an X-linked recessive allele and therefore the key is:

X^C = normal vision
X^c = color blindness

The possible genotypes in both males and females are:

X^CX^C = a female with normal color vision
X^CX^c = a carrier female with normal vision
X^cX^c = a female who is color-blind
X^CY = a male with normal vision
X^cY = a male who is color-blind

Sometimes the first of these genotypes is called a completely normal female because a female with this genotype cannot pass on an allele for color blindness. The second genotype is a carrier female because although a female with this genotype appears normal, she is capable of passing on an allele for color blindness. Color-blind females are rare because they must receive the allele from both parents, but color-blind males are more common since they need only one recessive allele in order to be color-blind. The allele for color blindness had to have been inherited from their mother because it is on the X chromosome; males only inherit the Y chromosome from their father.

Cross

If a heterozygous woman is married to a man with normal vision, what are their chances of having a color-blind daughter? A color-blind son?

Parents: $X^CX^c \times X^CY$

Inspection indicates that all daughters will have normal vision because they will all receive the X^C combination from their father. The sons, however, have a 50:50 chance of being color-blind, depending on whether they receive the X^C or X^c from their mother. The inheritance of a Y from their father cannot offset this inheritance from their mother. Figure 21.21 illustrates the use of the Punnett square in doing sex-linked problems.

Figure 21.21
Cross involving X-linked genes. The male parent is normal, but the female parent is a carrier; an allele for color blindness is located on one of her chromosomes. Therefore, each son stands a 50 to 50 chance of being colorblind.

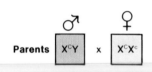

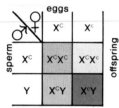

Results: female—all normal
males—1 normal: 1 colorblind

Genotypes	Phenotypes
X^CX^C	female, normal vision
X^CX^c	female, normal vision
X^cX^c	female, colorblind
X^CY	male, normal vision
X^cY	male, colorblind

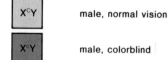

Key:
X^C = normal vision
X^c = colorblind

Reproduction, Development, and Inheritance

X-linked Traits
Some of the ways in which it is possible to recognize X-linked traits are:

1. More males than females are afflicted.
2. In order for a female to have the characteristic, her father must also have it. Her mother must have it or be a carrier.
3. The characteristic often skips a generation from the grandfather to the grandson (fig. 21.15b).
4. If a woman has the trait, all of her sons will have it.

Practice Problems 5
1. A woman is color-blind. What are the chances that her sons will be color-blind? If she is married to a man with normal vision, what are the chances that her daughters will be color-blind? Will be carriers?
2. Both parents are right-handed (R = right-handed, r = left-handed) and have normal vision. Their son is left-handed and color-blind. Give the genotype of all persons involved.
3. Both the husband and wife have normal vision. A woman has a color-blind daughter. What can you deduce about the girl's father?
4. Determine if the trait possessed by the darkened-in squares (males) and circles (females) following is dominant, recessive, or sex-linked recessive. Assume these traits are controlled by a single allelic pair.

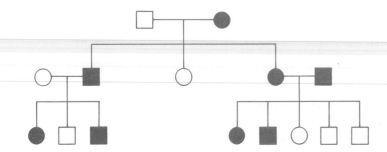

*Answers to problems are on page 469.

X-linked Genetic Diseases
Among 150 human disorders transmitted by a gene or genes on the X chromosome are:[5]

Agammaglobulinemia: lack of immunity to infections
Color blindness: inability to distinguish certain colors
Hemophilia: defect in blood-clotting mechanisms
Muscular dystrophy (some forms): progressive wasting of muscles
Spinal ataxia (some forms): spinal cord degeneration

The most common type of **hemophilia,** hemophilia A, is due to the absence, or minimal presence, of a particular clotting factor called Factor VIII. Hemophilia is called the bleeder's disease because the afflicted person's blood is unable to clot. While hemophiliacs do bleed externally after an injury, they also suffer from internal bleeding, particularly around joints. Years ago, a hemophiliac received blood plasma to stop the bleeding. Today, concentrated Factor VIII is available and may be self-injected to stop the bleeding.

[5]National Foundation/March of Dimes.

Figure 21.22
The boy pictured here lived in a germ-free bubble for 12 years because his body failed to produce antibodies to fight infections. It was hoped that a bone marrow transplant would give him the capability to produce antibodies so that he could leave his bubble home. When a marrow transplant from his sister failed, he was removed from his bubble for treatment, but he lived for a brief two weeks.

Muscular dystrophy, as the name implies, is characterized by the wasting away of muscles. The most common form, Duchenne type, is X linked; other types are not. Symptoms such as waddling gait, toe-walking, frequent falls, and difficulty in rising may appear as soon as the child starts to walk. Muscle weakness intensifies until the individual is confined to a wheelchair. Death usually occurs during the teenage years.

Persons with the genetic disease **agammaglobulinemia** are unable to manufacture antibodies and are therefore susceptible to repeated infections. Some are able to receive bone marrow transplants to effect a cure; others may have to live their lives in a protective sterile bubble or in a protective suit (fig. 21.22).

Summary

Two categories of genetics are considered: genic and chromosomal. A consideration of genic inheritance involves a study of how individual genes are inherited. Mendel concluded that there is a pair of factors (now called alleles) for every trait in all body cells and that one allele may be dominant over the other. According to his law of segregation, there is one allele for every trait in the sex cells; and according to his law of independent assortment, every possible combination of alleles (on separate chromosomes) may occur in the gametes.

Certain terminology and conventions are used to indicate the genotype and the gametes of the individuals. The same alphabetic letter is used for both the dominant and recessive alleles; a capital letter indicates the dominant and a small letter indicates the recessive. A homozygous dominant individual is indicated by two capital letters, and a homozygous recessive individual is indicated by two lowercase letters. The genotype of a heterozygous individual is indicated by a capital and a lowercase letter. Contrary to the individual, gametes have one letter of each type, either capital or lowercase as appropriate. All possible combinations of letters indicate all possible gametes (except if the genes are linked).

In doing an actual cross, it is assumed that all possible types of sperm fertilize all possible types of eggs. The results may be expressed as a probable phenotype ratio; it is also possible to state the chances of an offspring showing a particular phenotype. The results of some crosses may be determined by simple inspection, but certain others that commonly reoccur are given in table 21.2.

There are many exceptions to Mendel's laws and these include polygenic inheritance (skin color), epistasis (albinism), pleiotropy (syndromes), codominance (sickle-cell anemia), and multiple alleles (blood type).

Genes that appear on the same chromosomes are linked to one another and their alleles tend to go into the same gamete together. Linkage reduces the number of different genotypes and phenotypes among the offspring. However, recombinant gametes do occur by the process of crossing over and the frequency of crossing over can be used to map chromosomes.

Studies of human genetics have led to an understanding that many diseases of humans are genetically inherited and therefore it is wise to be aware that genetic counseling is available to couples who seek it. Among such diseases in the forefront today are Huntington's chorea, inherited as a dominant allele; Tay-Sachs, inherited as a recessive allele; and sickle-cell anemia, inherited as a codominant allelic pair.

A consideration of chromosomal inheritance is especially useful in regard to sex inheritance. A male inherits the chromosomes X and Y, and a female inherits two X chromosomes. Abnormal chromosomal inheritance can markedly affect the phenotype. An extra twenty-first chromosome causes

Down's syndrome; and abnormal sex chromosomal inheritance includes the genotypes XO (underdeveloped female), XXX (superfemale), XXY (underdeveloped male), and XYY (male). Some genes are sex linked, meaning that although they do not determine sex they are carried on the sex chromosomes. Most of the alleles for these genes are carried on the X chromosome and the Y is blank. It is possible to tell that genes are X-linked when many more males than females have the characteristic and/or it skips a generation, going from grandfather to grandson. For a daughter to show a recessive X-linked characteristic, her father must also show it and her mother must either show it or be a carrier. Several genetic diseases in humans are X-linked, including color blindness and hemophilia.

Study Questions

1. Explain why there is a pair of alleles for every trait except for sex-linked traits in males. (p. 447)
2. Relate Mendel's laws of inheritance to one-trait and two-trait problems. (pp. 442, 446)
3. What is the difference between genotype and phenotype? (p. 442)
4. What are the expected results from these crosses: heterozygous × heterozygous; heterozygous × recessive; dihybrid × dihybrid; dihybrid × double recessive? (p. 450)
5. What does the phrase "chance has no memory" mean? (p. 445)
6. Give six examples of exceptions to Mendel's laws. (pp. 450–56)
7. How would linkage (assuming dominant alleles on one chromosome and recessive on the other) affect the results of the last two crosses mentioned in question 4? (pp. 454–55)
8. Which parent determines the sex of the baby? (p. 460) What are the chances of having a boy or girl? (p. 460)
9. What is nondisjunction and how does it occur? (p. 463) What is the most common autosomal chromosomal abnormality? (p. 461) Name four sex chromosomal abnormalities. (p. 463)
10. What is sex linkage? (p. 463) Give all possible genotypes for an X-linked trait and discuss each. (p. 464)
11. How would a genetic counselor recognize an X-linked trait after studying the family history? (p. 465)

Additional Genetic Problems

1. A woman heterozygous for polydactyly (dominant) is married to a normal man. What are the chances that their children will have six fingers and toes? (p. 442)
2. John cannot curl his tongue (recessive) but both his parents can curl their tongues. Give the genotypes of all persons involved. (p. 442)
3. Parents who do not have Tay-Sachs (recessive) produce a child who has Tay-Sachs. What are the chances that each child born to this couple will have Tay-Sachs? (p. 445)
4. A man with widow's peak (dominant) who cannot curl his tongue (recessive) is married to a woman with a continuous hairline who can curl her tongue. They have a child who has a continuous hairline and cannot curl the tongue. Give the genotype of all persons involved. (p. 446)
5. Both Mr. and Mrs. Smith have freckles (dominant) and attached earlobes (recessive). Some of the children do not have freckles. What are the chances that the next child will have freckles and attached earlobes? (p. 446)

6. Mary has wavy hair (incomplete dominance) and marries a man with wavy hair. They have a child with straight hair. Give the genotype of all persons involved. (p. 453)

7. A man has type AB blood. What is his genotype? Could this man be the father of a child with type B blood? If so, what blood types could the child's mother have? (p. 453)

8. A woman with white skin has mulatto parents. If this woman married a light man, what is the darkest skin color possible for their children? the lightest? (p. 451)

9. What is the genotype of a man who is color-blind (X-linked recessive) and has a continuous hairline? If this man has children by a woman who is homozygous dominant for normal color vision and widow's peak, what will be the genotype and phenotype of the children? (pp. 453, 465)

10. Is the characteristic represented by the darkened individuals inherited as a dominant, recessive, or X-linked recessive? (p. 465)

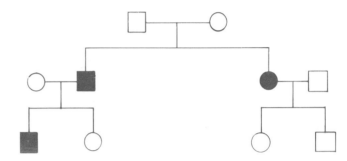

11. Fill in this pedigree chart to give the probable genotypes of the twins pictured in figure 21.8.

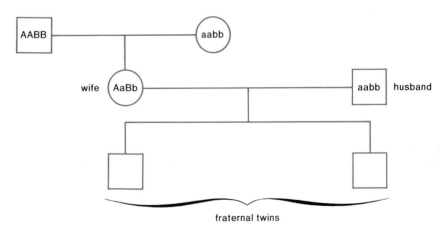

fraternal twins

Practice Problems 1
1. a. W b. WS, Ws, c. T, t d. Tg, tg
   e. AB, Ab, aB, ab
2. a. gamete b. genotype c. gamete
   d. genotype

Practice Problems 2
1. 75%
2. No
3. Heterozygous
4. DD × dd; Dd

Practice Problems 3
1. Dihybrid
2. 1/16
3. DdFf × ddff; ddff

Practice Problems 4
1. H'H', No
2. Light
3. White
4. Baby 1 = Doe;
   Baby 2 = Jones
5. AB, O, A, B
6. 3:1

Practice Problems 5
1. 100%, None, 100%
2. $RrX^CX^c \times RrX^CY$; rrX^cY
3. The husband is not father.
4. Dominant

Selected Key Terms

allele   (ah-lēl')
dominant allele   (dom'ĭ-nant ah-lēl')
recessive allele   (re-ses'iv ah-lēl')
segregation   (seg''re-ga'shun)
genotype   (je'no-tīp)
phenotype   (fe'no-tīp)
homozygous   (ho''mo-zi'gus)
heterozygous   (het''er-o-zi'gus)
monohybrid   (mon''o-hi'brid)
dihybrid   (di-hi'brid)
independent assortment   (in''de-pen'dent ah-sort'ment)
epistasis   (ĕ-pis'tah-sis)
pleiotropy   (pli-ot'ro-pe)
incomplete dominance   (in-kom-plēt' dom'ĭ-nans)
codominance   (ko-dom'ĭ-nans)
linkage   (lingk'ij)
crossing over   (kros'ing o'ver)
nondisjunction   (non''dis-jungk'shun)
deletion   (de-le'shun)
sex-linked genes   (seks-linkt' jēnz)

Further Readings

Baer, A. S. 1977. *The genetic perspective.* Philadelphia: W. B. Saunders.

Brady, R. O. 1973. Hereditary fat-metabolism diseases. *Scientific American* 229(2):88.

Cerami, A., and Peterson, C. M. 1975. Cyanate and sickle-cell disease. *Scientific American* 232(4):44.

Mader, S. S. 1980. *Human reproductive biology.* Dubuque, Iowa: Wm. C. Brown Publishers.

McKusick, V. A. 1969. *Human genetics.* 2d ed. Englewood Cliffs, N.J.: Prentice-Hall.

Ruddle, F. H., and Kucherlapati, R. S. 1974. Hybrid cells and human genes. *Scientific American* 231(1):36.

Volpe, E. P. 1979. *Man, nature, and society.* 2d ed. Dubuque, Iowa: Wm. C. Brown Publishers.

Winchester, A. M. 1979. *Human genetics.* 3rd ed. Columbus, Ohio: Charles E. Merrill.

22

Molecular Basis of Inheritance

Chapter concepts

1 DNA is the genetic material and therefore its structure and function constitute the molecular basis of inheritance.

2 DNA is able to replicate, mutate, and control the phenotype.

3 DNA controls the phenotype by controlling protein synthesis, a process that also requires the participation of RNA.

4 The manner in which gene action is regulated is an area of important research today.

5 Genetic engineering, especially recombinant DNA research, is being used to help treat genetic diseases.

6 Mutations, which account for the origin of genetic diseases and cancer, can be caused by environmental mutagens.

Figure 22.1
Life cycle of a T virus. A T virus is a complex virus with a head and a tail. Even so, it is composed of just a protein coat and inner core of DNA. Experimenters labeled the coat with ^{35}S and the DNA with ^{32}P and allowed the viruses to attack bacteria. Later, they found only ^{32}P in the cell and yet the cell produced many new viruses. From this they knew that DNA is the genetic material.

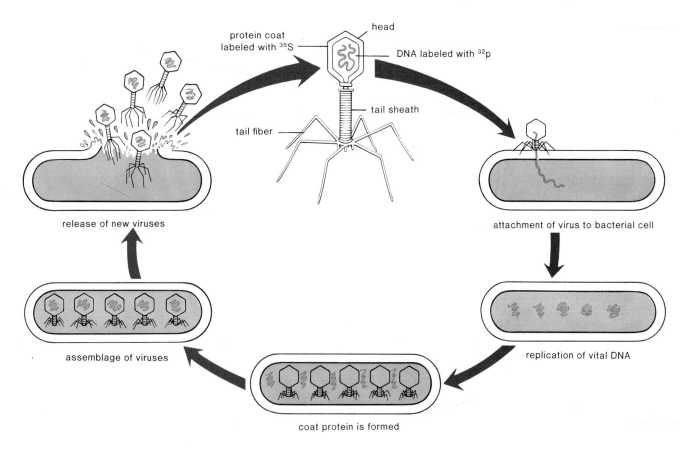

protein coat labeled with ^{35}S

head

DNA labeled with ^{32}p

tail sheath

tail fiber

release of new viruses

attachment of virus to bacterial cell

assemblage of viruses

replication of vital DNA

coat protein is formed

When the sperm fertilizes the egg, a new individual comes into being. Each of the gametes contributes genes that direct not only the development but the continued functioning of the individual from birth to death. In the previous chapter we saw how the inheritance of various genes affected the phenotype, but of what are the genes composed?

In the mid-1900s it was known that the genes are on the chromosomes and that the chromosomes contain both DNA and protein, but it was uncertain which of these was the genetic material. Scientists turned to experiments with viruses to determine which of these is the genetic material because they knew that viruses are tiny particles having just two parts: an outer coat of protein and an inner core of nucleic acid, most often DNA.

They chose to work with a virus called a T virus (the T simply means "type") that infects bacteria. They wished to determine which part of a T virus, the outer protein coat or the inner DNA core, enters a bacterium, taking over its machinery so that it produces more viruses. They began by culturing bacteria and viruses in radioactive sulfur and phosphorus until they had a batch with^{35}S, labeled protein coats and ^{32}P, labeled DNA.[1] Then they allowed these viruses to attack new bacteria (fig. 22.1) and determined whether the labeled

[1]In actuality, two different labeled batches were needed, but for simplicity's sake they are described as one batch here.

Figure 22.2

Nucleotides in DNA. Each nucleotide is composed of phosphate, the sugar deoxyribose, and a base.
a. The purine bases are adenine and guanine.
b. The pyrimidine bases are cytosine and thymine.

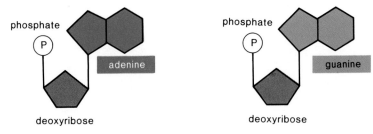

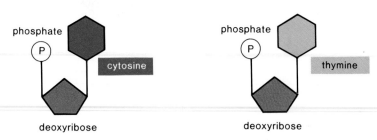

a. DNA nucleotides with purine bases

b. DNA nucleotides with pyrimidine bases

Figure 22.3

Structure of DNA. *a.* DNA is a double helix. The backbone of each strand is composed of phosphate and sugar molecules, and the bases project to the side. *b.* Complementary pairing between the bases in which a purine is always paired with a primidine. Specifically, (1) Thymine (T) is paired with Adenine (A) and (2) Guanine (G) is paired with Cytosine (C).

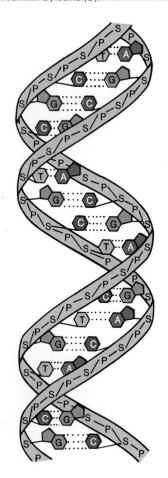

a.

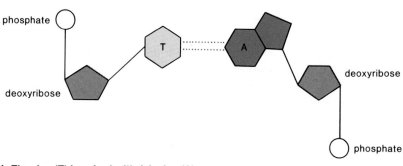

1. Thymine (T) is paired with Adenine (A).

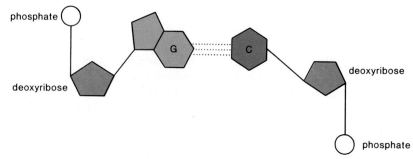

b. **2. Guanine (G) is paired with Cytosine (C).**

Reproduction, Development, and Inheritance

Figure 22.4
Overview of DNA structure. *a*. Double helix.
b. Ladder configuration. Notice that the uprights
are composed of sugar and phosphate molecules
and the rungs are complementary paired bases.
c. Nucleotide structure.

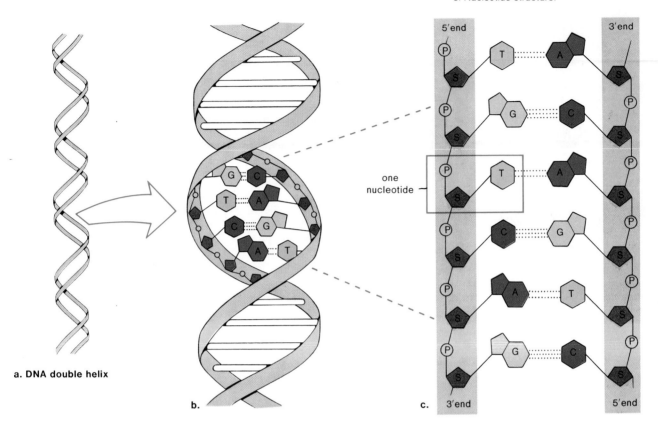

a. DNA double helix

b.

c.

protein or the labeled DNA entered the cell. They found that only DNA enters the cell to take over the metabolism of the cell so that more viral particles are made. In other words, DNA is the genetic material.

Structure of DNA

DNA is a nucleic acid that contains multiple copies of just four nucleotides (fig. 22.2). Notice that each nucleotide is a complex of three united subunits: phosphoric acid (phosphate), a pentose sugar (deoxyribose), and a nitrogen base. The bases can be either the **purines** adenine or guanine, which have a double ring, or the **pyrimidines** thymine or cytosine, which have a single ring. These structures are called bases because they have basic characteristics that raise the pH of a solution.

When nucleotides join together, they form a polymer, or **strand,** in which the backbone is made up of phosphate-sugar-phosphate-sugar —with the bases to one side of the backbone (fig. 1.34). DNA contains two such strands and therefore it is **double stranded.** The two strands of DNA twist about one another in the form of a **double helix** (fig. 22.3). The two strands are held together by hydrogen bonds between purine and pyrimidine bases. Thymine (T) is always paired with adenine (A), and guanine (G) is always paired with cytosine (C) (fig. 22.3b). This is called **complementary base pairing.**

If we unwind the DNA helix, it resembles a ladder (fig. 22.4). The sides of the ladder are made entirely of phosphate and sugar molecules, and the rungs of the ladder are made only of the complementary paired bases. The bases can be in any order but A is always paired with T and G is always paired

Solving the Puzzle

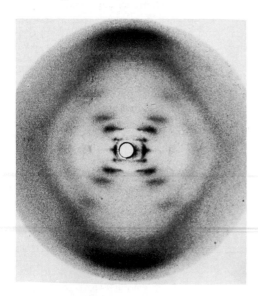

Figure 22.A X-ray diffraction photograph of DNA taken by Rosalind Franklin. The crossing pattern of dark spots in the center of the picture indicated that DNA was helical. The dark regions at the top and bottom of the photograph showed that the base pairs were stacked on top of one another.

In 1951, James Watson, an American biologist, began an internship at the University of Cambridge, England. There he met Francis Crick, an English physicist, who was interested in molecular structures. They began to try to build a model that would show the molecular structure of DNA, the known hereditary material. They knew that their model should explain the manner in which DNA can vary from species to species and even from individual to individual, and also that it should show how it is possible for DNA to replicate (make a copy of itself) so that copies of the genetic material can be passed on to daughter cells and from the parent to the offspring.

Bits and pieces of data were available to Watson and Crick, and they undertook to solve the puzzle by putting the pieces together. This is what they knew from research done by others:

1. DNA is a polymer of nucleotides, each one having a phosphate group, the sugar deoxyribose, and a nitrogenous base. There are four different nucleotides that differ according to the base: adenine (A) and guanine (G) are purines while cytosine (C) and thymine (T) are pyrimidines.

2. A chemist, Erwin Chargaff, had determined in the late 1940s that, regardless of the species under consideration, the number of purines in DNA always equals the number of pyrimidines and that the adenine content equals the thymine content, that is, [A] = [T], and the guanine content equals the cytosine content, that is, [G] = [C]. These findings came to be known as *Chargaff's rules*.

3. Rosalind Franklin and Maurice Wilkins, working at King's College, London, had just prepared an X-ray diffraction photograph (fig. 22.A) of DNA. It showed that DNA is a double helix of constant diameter and that the bases are regularly stacked above one another.

Using this data, Watson and Crick deduced that DNA has a twisted ladder-type structure: the sugar-phosphate molecules make up the sides of the ladder and the bases make up the rungs of the ladder. Further, they determined that if A was always hydrogen-bonded with T and G was always hydrogen-bonded with C (in keeping with Chargaff's rules), then the rungs would always have a constant width (as required by the X-ray photograph).

Watson and Crick built an actual model of DNA out of wire and tin (fig. 22.B). This double-helix model does indeed allow for differences in DNA structure between species because, while A must always pair with T and G must always pair with C, there is no set order in the sequencing of these pairs. Also, the model provides a means by which DNA can replicate, as Watson and Crick alluded to in their original paper; "It has not escaped our notice that the specific pairing we have postulated immediately suggests a possible copying mechanism for the genetic material."

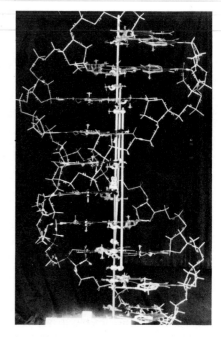

Figure 22.B A portion of the actual wire and tin model constructed by Watson and Crick.

with C, and vice versa. Therefore no matter what the order or the quantity of any particular base pair, the number of purine bases *always equals* the number of pyrimidine bases.

The structure of the DNA molecule was first determined by two young scientists, James Watson and Francis Crick. The data that was available to them and the way they used it to deduce DNA's structure is reviewed in the reading above.

Reproduction, Development, and Inheritance

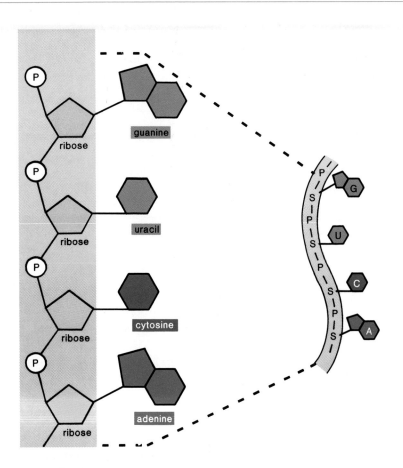

Structure of RNA

RNA is also a nucleic acid composed of multiple copies of nucleotides. However, the pentose sugar in RNA is ribose—not deoxyribose. Also, the pyrimidine thymine does not appear in RNA; it is replaced by the pyrimidine uracil. Therefore RNA contains the four bases, adenine (A), guanine (G), cytosine (C) and uracil (U). Finally RNA is **single stranded.** Figure 22.5 shows the structure of RNA and table 22.1 summarizes the differences between DNA and RNA structure.

Functions of DNA

Any hereditary material will have at least three functions. The hereditary material must be able to

1. replicate, make copies of itself, that may be passed on from cell to cell and from generation to generation;
2. control the activities of the cell; thereby producing the phenotypic characteristics of the individual and the species;
3. undergo **mutations**—permanent genetic changes passed on to the offspring—in order to account for the evolutionary history of life.

We now wish to explore the manner in which DNA carries out these functions.

Replication

The double-stranded structure of DNA lends itself to replication because each strand can serve as a template for the formation of a complementary strand. A **template** is most often a mold, used to produce a shape opposite to itself.

Table 22.1 DNA structure compared to RNA structure

	DNA	RNA
Sugar	Deoxyribose	Ribose
Bases	Adenine, guanine, thymine, cytosine	Adenine, guanine, uracil, cytosine
Strands	Double stranded with base pairing	Single stranded
Helix	Yes	No

Figure 22.6
DNA replication. Replication is called
semiconservative because each new double helix
is composed of an old parental strand and a new
daughter strand.

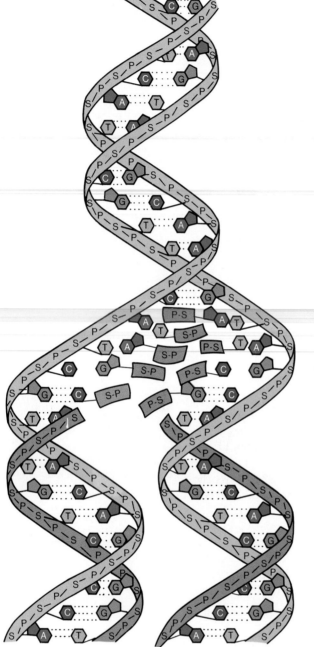

Region of parental DNA helix. (Both backbones are light).

Region of replication. Parental DNA is unzipped and new nucleotides are pairing with those in parental strands.

Region of completed replication. Each double helix is composed of an old parental strand (light) and a new daughter strand (dark). Notice that each double helix is exactly like the other one.

In this case, the word *template* is appropriate because each new strand of DNA has a sequence of bases complementary to the bases of the old strand of DNA.

Replication requires the following steps (fig. 22.6):

1. The two strands that make up DNA become "unzipped" (i.e., the weak hydrogen bonds between the paired bases are broken).
2. New complementary nucleotides, always present in the nucleus, move into place by the process of complementary base pairing.

Figure 22.7
Metabolic pathway by which phenylalanine is
converted to other metabolites. a. If the enzyme
that converts phenylalanine to tyrosine is defective,
tyrosine is converted to phenylpyruvic acid instead,
and the accumulation of this substance leads to
PKU (phenylketonuria). b. If the enzyme that
converts tyrosine to melanin is defective, albinism
results. c. If homogenistic acid can be metabolized,
alkaptonuria, meaning that the urine turns a dark
color, results.

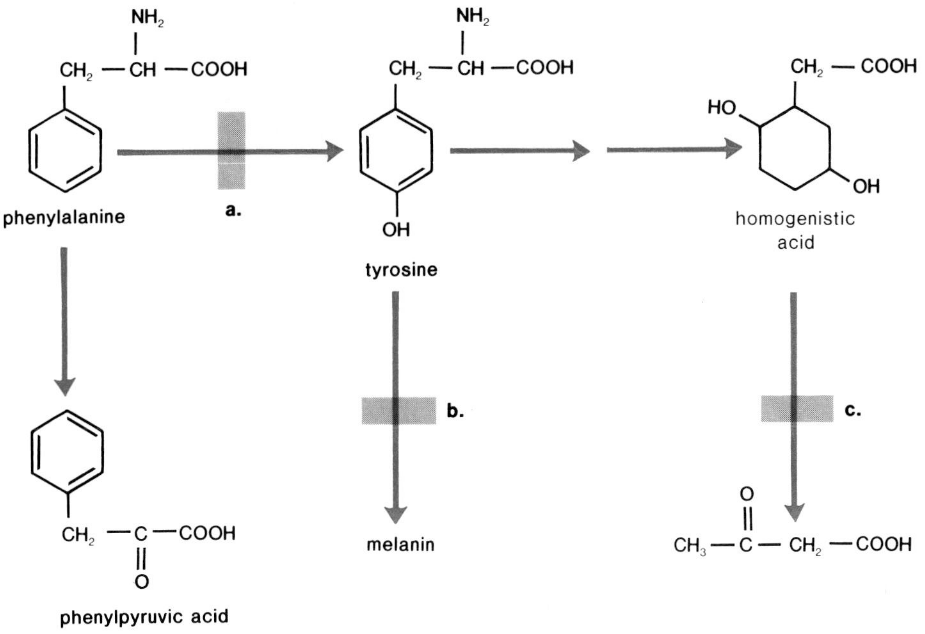

3. The adjacent nucleotides through their sugar-phosphate components
   become joined together along the newly forming chain.
4. When the process is finished, two complete DNA molecules are present,
   identical to each other and to the original molecule.

This replication process is described as **semiconservative** because each
double strand of DNA contains one old strand and one new strand. Although
DNA replication can be easily explained, it is in actuality an extremely complicated process involving many steps and enzymes. There are enzymes that
assist the unwinding process, that join together the nucleotides, and that assist
the rewinding process, just to mention a few. On occasion, errors are made
that cause a change in the DNA and, in this way, a mutation can arise.

Scientists were surprised to discover that the time required for replication of DNA in human cells is only about 20 times longer than in bacterial
cells even though a human chromosome is much longer than a bacterial one.
The explanation is that DNA replication in eukaryotic cells is initiated at
hundreds of sites, almost simultaneously, along each chromosome, whereas
bacterial chromosome replication proceeds from a single site.

Genes and Enzymes

Many novel experiments and years of research allowed scientists to conclude
that there is a relationship between genetic inheritance and the structure of
enzymes and other proteins. For example, the metabolic pathway outlined in
figure 22.7 was discovered in the early 1900s. In this pathway, three genetic
diseases are known. In the disease known as *phenylketonuria (PKU)*, phenylpyruvic acid accumulates in the body and spills over into the urine because

Figure 22.8

The first seven amino acids of the β chain of human hemoglobin in normal individuals and in individuals with sickle-cell anemia. Notice that in sickle-cell hemoglobin, valine has been substituted for glutamic acid. This is evidence that the sickle-cell gene, and thus any other gene, controls the sequence of amino acids in a polypeptide.

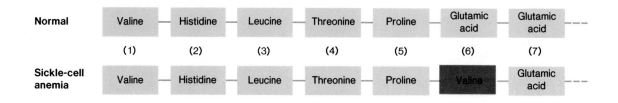

Normal	Valine	Histidine	Leucine	Threonine	Proline	Glutamic acid	Glutamic acid
	(1)	(2)	(3)	(4)	(5)	(6)	(7)
Sickle-cell anemia	Valine	Histidine	Leucine	Threonine	Proline	Valine	Glutamic acid

Figure 22.9

Eukaryotic cell structure. Ribosomal RNA is produced in the nucleolus. Chromatin contains DNA. Protein synthesis occurs at the ribosomes. Proteins that are synthesized at ribosomes attached to the endoplasmic reticulum are usually for export. Those that are synthesized at polysomes are usually for use inside the cell.

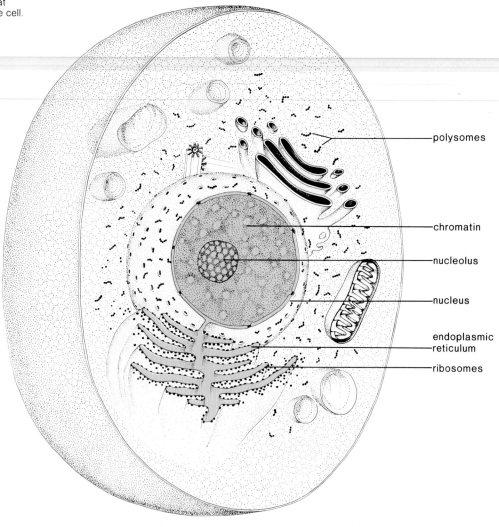

polysomes

chromatin

nucleolus

nucleus

endoplasmic reticulum

ribosomes

the enzyme needed to convert phenylalanine to tyrosine is missing. If the condition is not treated, the continued accumulation of phenylpyruvic acid can cause mental retardation. *Albinism* (fig. 21.9) results because tyrosine cannot be converted to melanin, the natural pigment in human skin. The genetic disease alkaptonuria results if the enzyme needed to metabolize homogentisic acid is missing. Because these diseases are inherited in a simple Mendelian fashion, and appeared to be controlled by recessive alleles, investigators concluded that there was a relationship between genes and enzymes.

Much later, it was discovered that hemoglobin in people with sickle-cell anemia differs from normal hemoglobin by a single amino acid (fig. 22.8). This suggested that a gene controls the sequence of amino acids in a protein. The protein, hemoglobin, actually contains two different types of polypeptide chains (designated α and β) (fig. 11.5) and only one of these types is affected in people with sickle-cell anemia. Thus it was pointed out that it might be more proper to state that a gene controls the sequence of amino acids in a polypeptide. Today, we define a gene as a section of a DNA molecule that determines the sequence of amino acids in a single polypeptide chain of a protein.

Protein Synthesis

The fact that DNA controls the production of proteins may at first seem surprising when we consider that genes are located in the nucleus of higher cells while proteins are synthesized at the ribosomes in the cytoplasm. However, while DNA is found only in the nucleus, RNA exists in both the nucleus and the cytoplasm (fig. 22.9).

Biochemical genetic research indicates that a type of RNA, called **messenger RNA (mRNA),** serves as a go-between for DNA in the nucleus and the ribosomes in the cytoplasm. This is possible because one strand of DNA can serve as a template for the production of a complementary strand of RNA, as well as a template for another strand of DNA. The RNA molecule contains a sequence of nucleotides that are complementary to those of a single gene. The mRNA moves from the nucleus to the ribosomes in the cytoplasm where it dictates the sequence of amino acids in a polypeptide. This concept is often called the central dogma of modern genetics and can be diagrammed as follows:

$$\text{DNA} \xrightarrow[\text{transcription}]{} \text{mRNA} \xrightarrow[\text{translation}]{} \text{protein}$$

The diagram indicates that the control of protein synthesis requires transcription and translation. During the process of **transcription,** complementary mRNA is formed in the nucleus, and during **translation,** its message is used in the cytoplasm to produce the correct order of amino acids in a polypeptide (fig. 22.10).

Code of Heredity

DNA provides mRNA with a message that directs the order of amino acids during protein synthesis, but what is the nature of the message? The message cannot be contained in the sugar-phosphate backbone because it is constant in every DNA molecule. However, the order of the bases in DNA and mRNA can and does change. Therefore, it must be the bases that contain the message. The order of the bases in DNA must code for the order of the amino acids in a polypeptide. Can four bases provide enough combinations to code for 20 amino acids? If the code were a doublet one (any two bases stand for one

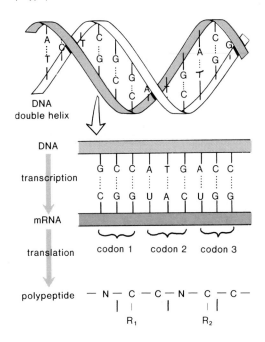

Figure 22.10

Transcription and translation in protein synthesis. Transcription occurs when DNA acts as a template for mRNA synthesis. Translation occurs when sequence of the codons found in mRNA determines the sequence of the codons in a polypeptide.

Table 22.2 Number of bases in code

Number of bases in code	Number of amino acids coded for
1	4
2	16
3	64

amino acid), it would not be possible to code for 20 amino acids (table 22.2). But if the code were a triplet code, then the four bases would be able to supply 64 different triplets, far more than are needed to code for 20 different amino acids. It should come as no surprise, then, to learn that the code is a triplet code.

To crack the code, artificial RNA was added to a medium containing bacterial ribosomes and a mixture of amino acids. Comparison of the bases in the RNA with the resulting polypeptide allowed investigators to decipher the code. Each three-letter unit of a messenger RNA is called a **codon.** All 64 codons have been determined (table 22.3). Sixty-one triplets correspond to a particular amino acid; the remaining three code for chain termination. The one codon that stands for the amino acid methionine also signals polypeptide initiation.

The DNA code is **degenerate**—many amino acids are designated by more than one codon. Codons specifying the same amino acid often differ only in the third nucleotide; this sometimes permits DNA to mutate without altering the amino acid sequence of the proteins encoded by the DNA. Therefore it is possible that degeneracy is a protective device that reduces the detrimental effect of changes in the DNA due to mutation.

The Genetic Code Is Universal   Research indicates that the code is essentially universal. The same codons stand for the same amino acids in all living things, including bacteria, plants, and animals. This illustrates the remarkable biochemical unity of living things and suggests that all living things have a common evolutionary ancestor.

Transcription

During transcription, the DNA code is passed to mRNA, and thus the code is "transcribed" or rewritten.

Messenger RNA

The process of transcription allows the formation of a mRNA that contains a sequence of bases complementary to DNA. A segment of the DNA helix unravels; complementary RNA nucleotides pair with DNA nucleotides of one of the strands. After an enzyme joins the nucleotides by way of their sugar-phosphate components, the resulting mRNA molecule carries a sequence of bases that are triplet codons complementary to the DNA triplet code (fig. 22.11).

The mRNA strand then passes from the cell nucleus into the cytoplasm, carrying the transcribed DNA code.

Table 22.3 Three-letter codons of messenger RNA, and the amino acids specified by the codons

Codons	Amino acid	Codons	Amino acid	Codons	Amino acid	Codons	Amino acid
AAU, AAC	Asparagine	CAU, CAC	Histidine	GAU, GAC	Aspartic acid	UAU, UAC	Tyrosine
AAA, AAG	Lysine	CAA, CAG	Glutamine	GAA, GAG	Glutamic acid	UAA, UAG	(Terminator)*
ACU, ACC, ACA, ACG	Threonine	CCU, CCC, CCA, CCG	Proline	GCU, GCC, GCA, GCG	Alanine	UCU, UCC, UCA, UCG	Serine
AGU, AGC	Serine	CGU, CGC, CGA, CGG	Arginine	GGU, GGC, GGA, GGG	Glycine	UGU, UGC	Cysteine
AGA, AGG	Arginine					UGA	(Terminator)*
						UGG	Tryptophan
AUU, AUC, AUA	Isoleucine	CUU, CUC, CUA, CUG	Leucine	GUU, GUC, GUA, GUG	Valine	UUU, UUC	Phenylalanine
AUG	Methionine					UUA, UUG	Leucine

*Terminating codons signal the end of the formation of a polypeptide chain.

Reproduction, Development, and Inheritance

mRNA

Translation

During translation, the sequence of codons in mRNA dictates the order of amino acids in a polypeptide. This is called translation because the sequence of bases in DNA is finally translated into a particular sequence of amino acids. Translation requires the involvement of several enzymes and two other types of RNA: ribosomal RNA (rRNA) and transfer RNA (tRNA).

Ribosomal RNA

Ribosomal RNA is sometimes called structural RNA because it was formerly believed that the ribosomes were like an inert workbench on which the amino acids were assembled. High-energy utilization by ribosomes, however, suggests that ribosomes probably play an important role in coordinating protein synthesis.

Ribosomes (fig. 2.9c) are composed of two subunits, each with characteristic RNA and protein molecules. The rRNA molecules are transcribed from DNA in the region of the nucleolus; the proteins are manufactured in the cytoplasm but then migrate to the nucleolus, where the ribosomal subunits are assembled before they migrate into the cytoplasm.

Transfer RNA

Located in the cytoplasm are small molecules of tRNA that transfer the amino acids from the cytoplasm to the ribosomes. Each molecule of tRNA attaches at one end to a particular amino acid. Attachment requires ATP energy and results in a high energy bond. Therefore this bond is indicated by a wavy line and the entire complex is designated by *tRNA ~ amino acid*. At the other end of each tRNA there is a specific **anticodon** complementary to an mRNA codon. (Each tRNA molecule is transcribed from DNA and then, due to intramolecular binding of complementary bases, the anticodon is exposed. Just as more than one codon can stand for a particular amino acid, so more than one anticodon or tRNA can signify the same amino acid.)

Complementary base pairing between codons and anticodons determines the order in which tRNA~ amino acid complexes come to a ribosome, and this in turn determines the final sequence of the amino acids in a polypeptide. The making of a protein is accomplished codon by codon.

The Process of Protein Synthesis

Protein synthesis requires three processes: initiation, elongation, and termination. During the initiation process, a ribosome becomes attached to a mRNA. Initiation always begins with a codon that stands for the amino acid methionine. First, the smaller ribosomal subunit binds to mRNA, and then the larger subunit joins to the smaller subunit, giving a complete ribosomal structure. **Elongation** occurs as the polypeptide chain grows in length. Figure 22.12 shows the process of elongation some time after initiation. A ribosome is large enough to accommodate two codons; at one codon, a tRNA is just about to leave and, at the other, a tRNA~amino acid complex has just arrived. The bond holding the amino acid chain to the previous tRNA molecule is enzymatically broken and the chain immediately becomes attached by a peptide bond to the newly arrived amino acid. The ribosome then moves laterally so that the next mRNA codon becomes available to receive the next tRNA~ amino acid complex. In

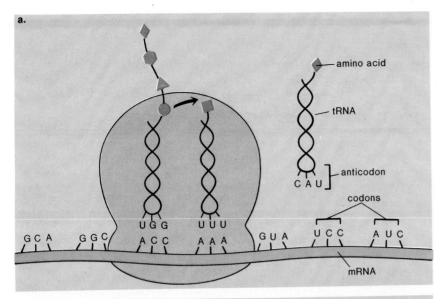

a.

amino acid

tRNA

anticodon

C A U

codons

U G G   U U U   U C C   A U C
A C C   A A A   G U A

mRNA

GCA   GGC

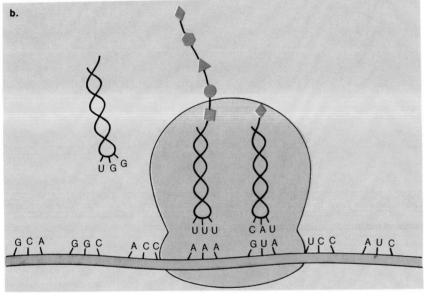

b.

U G G

U U U   C A U
A A A   G U A

GCA   GGC   ACC   UCC   AUC

this manner, the peptide chain grows and the primary structure of a protein comes about. The secondary and tertiary structures occur after termination, as the predetermined sequence of amino acids within the polypeptide chain interact with one another.

Termination of protein synthesis occurs at a specific nucleotide sequence on the mRNA where the last tRNA and completed polypeptide are liberated from the ribosomal complex. The ribosome dissociates into its two subunits and falls off the messenger. Several ribosomes, collectively called a **polysome,**

Figure 22.13

a. Polysome structure. Several ribosomes, called a polysome, move along a mRNA at a time. They function independently of each other so that several polypeptides can be made at the same time. *b.* Electron micrograph of several polysomes.

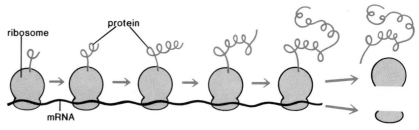

a.

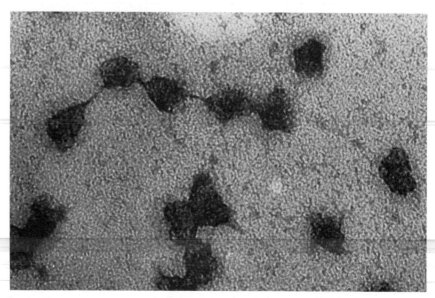

b.

Table 22.4   Steps in protein synthesis

Name of molecule	Special significance	Definition
DNA	Code	Sequence of bases in threes
mRNA	Codon	Complementary sequence of bases in threes
tRNA	Anticodon	Sequence of three bases complementary to codon
Amino acids	Building blocks	Transported to ribosomes by tRNAs
Protein	Enzyme	Amino acids joined in a predetermined order

may move along one mRNA at a time; therefore, several of the same type proteins may be synthesized at once (fig. 22.13). After the translation process is completed, the mRNA disintegrates.

Overview of Protein Synthesis

The following list, along with table 22.4, provides a brief summary of the steps involved in protein synthesis.

1. DNA, which always remains in the nucleus, contains a series of bases that serve as a *triplet code* (every three bases codes for an amino acid).
2. During transcription, one strand of DNA serves as a template for the formation of messenger RNA (mRNA), which contains *triplet codons* (sequences of three bases complementary to DNA code).
3. Messenger RNA goes into the cytoplasm and becomes associated with the *ribosomes,* which are composed of ribosomal RNA (rRNA) and proteins.

Reproduction, Development, and Inheritance

Figure 22.14

Summary of protein synthesis. Transcription occurs in the nucleus (color) and translation occurs in the cytoplasm (white). During translation, the ribosome moves along the mRNA. In the diagram, as the ribosome moves to the left, a tRNA bearing an amino acid will come to the ribosome. Thereafter the polypeptide chain will be passed to this tRNA ~ amino acid complex. Each time the ribosome moves a tRNA departs.

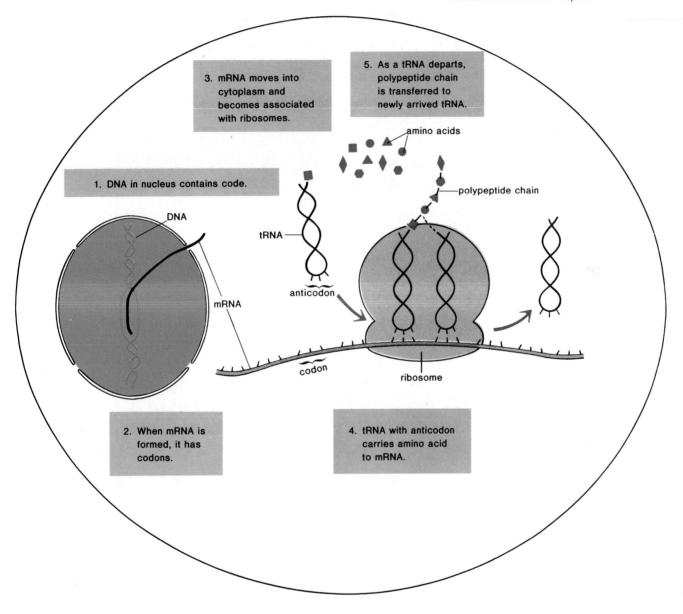

4. Transfer RNA (tRNA) molecules, each of which is bonded to a particular amino acid, have *anticodons* that pair complementarily to the codons in mRNA.

5. As the ribosome moves along mRNA, a newly arrived tRNA ~ amino acid complex receives the growing polypeptide chain from a tRNA molecule, which then leaves to pick up another amino acid. During translation, therefore, the linear sequence of codons determines the order in which the tRNA molecules arrive at the ribosomes and thus determines the *primary structure* of a protein (i.e., the order of its amino acids).

The transcription-translation process is illustrated in figure 22.14.

Figure 22.15

Prokaryote models for the regulation of protein synthesis. Notice that the first model explains how an operon can get turned on and the second model explains how an operon can get turned off.
a. Inducible model: (1.) A regulator gene codes for a repressor that can immediately bind to the operator preventing transcription from taking place. (2.) When an inducer combines with the repressor it is no longer capable of binding to the operator and therefore transcription takes place.
b. Repressible models (1.) The regulator gene codes for a repressor that is unable to bind to the operator and therefore transcription takes place. (2.) When a corepressor is combined with the repressor, it is now able to bind to the operator and therefore transcription cannot take place.

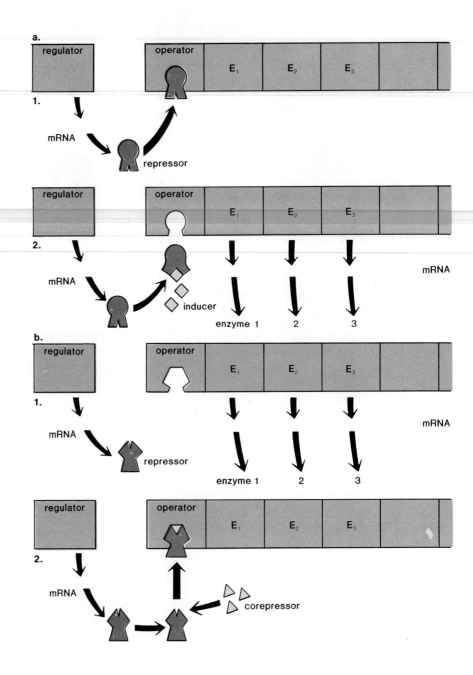

Reproduction, Development, and Inheritance

Regulatory Genes

Thus far, genes that provide coded information for the synthesis of proteins (polypeptides) have been discussed. These genes are called **structural genes** because they determine protein structure. Evidence indicates that another class of genes, called regulatory genes, exists. **Regulatory genes** regulate the activity of structural genes. For example, we previously mentioned that differentiation occurs during development of the human embryo and fetus. Differentiation is possible because although each cell receives a full complement of genes, only certain ones are active in each cell type. First, however, we will consider regulation in prokaryotes before we consider regulation in eukaryotes.

Prokaryotes

Research with the bacterium *E. coli* has resulted in at least two models that explain the regulation of gene transcription and, therefore, protein synthesis.

Prokaryotic regulatory models have the following components (fig. 22.15):

1. An **operon** is a group of genes, called structural genes, that code for enzymes active in a particular metabolic pathway, such as enzymes 1, 2, 3 in this pathway:

$$A \xrightarrow{\ \ 1\ \ } B \xrightarrow{\ \ 2\ \ } C \xrightarrow{\ \ 3\ \ } D$$

2. An **operator** is a segment of DNA that acts as an on/off switch for transcription of the operon.
3. A **regulatory gene** is a gene that codes for a protein that either
   a. immediately combines with the operator preventing transcription or
   b. must first join with a metabolite before it combines with the operator. The two models (fig. 22.15) are dependent on whether (a) or (b) controls transcription of a particular operon.

In figure 22.15a, the operon is normally inactive because the regulatory gene codes for a protein, called a **repressor,** that combines with the operator, preventing transcription. The operon becomes active when the repressor joins with an inducer molecule, and the complex is unable to bind with the operator. The inducer, so named because it induces protein synthesis, is a metabolite in a metabolic pathway. For example, *A* in the pathway above could be an inducer. In this **inducible operon model,** then, the first metabolite indicates the need for particular enzymes.

In figure 22.15b, the operon is normally active because the regulatory gene codes for an inactive repressor that must join with the corepressor before the complex combines with the operator. A **corepressor,** so named because it prevents protein synthesis, is a metabolite in a metabolic pathway. For example, *D* in the pathway above could be a corepressor. In this **repressible operon model,** the presence of the end product indicates that particular enzymes are no longer needed.

Notice that the inducible model accounts for the fact that some structural genes are normally inactive and the repressible model accounts for the fact that some structural genes are normally active. Thus some genes could normally be turned on, while others could normally be turned off.

Evidence suggests that these models are not the only means by which regulation of protein synthesis can occur. Much interest of late has centered on control elements that can move into and out of various locations on the chromosome.

Table 22.5 Participants in regulatory models

Participants	Action
Operon	Genes that code for enzymes in a metabolic pathway
Operator	An on/off switch for transcription of the operon
Regulatory gene	A gene that codes for a repressor
Inducer	A metabolite that inactivates a repressor
Corepressor	A metabolite that activates a repressor

Figure 22.16

Eukaryotic chromosome. The chromosomes of higher organisms contain segments that do not code for polypeptides. Repetitive DNA is noncoding and consists of sequences of base pairs repeated many times, one after another. Structural genes themselves are interrupted by intervening sequences, or introns, which do not code for polypeptides. During transcription, the entire gene, including these segments, are copied into mRNA. Before the mRNA leaves the nucleus, these segments are spliced out. Structural genes are flanked at one end by regulatory genes that control whether transcription takes place or not. Codes at the ends of the structural gene show where to start and stop transcription.

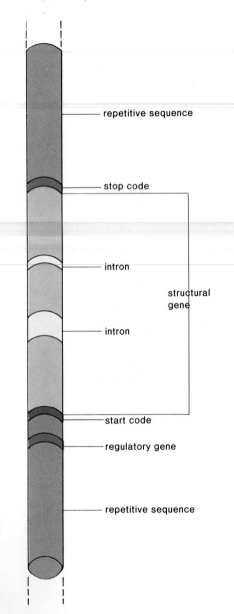

repetitive sequence

stop code

intron

structural gene

intron

start code

regulatory gene

repetitive sequence

Eukaryotes

Eukaryotic chromosomes, unlike the single circular prokaryotic chromosome, contains not only DNA but also proteins.

Proteins

The eukaryotic chromosome contains proteins called **histones** and many other types of proteins that are grouped together as **nonhistones.** While it was believed at first that the histones might have a regulatory function, this does not seem possible because all types of cells have similar histone proteins. It is possible, however, that the nonhistone proteins might eventually be found to have a regulatory function.

DNA

Both plants and animals seem to have more DNA than they need. The single chromosome of a prokaryote can code for about 2 thousand proteins and this is judged to be a reasonable number for these small cells. In mammals there is enough DNA to code for 3 million proteins and this does not seem reasonable, even for these more complex organisms.

There are currently two questions being asked about the "extra" DNA present in eukaryotic cells. Where is the excess DNA located on the chromosome, and does it have a function? First of all, there are most likely duplicate copies of many genes. Certainly in polyploid plants, the same gene is present several times over. In animals there is evidence that duplication followed by gene mutation has contributed to the complexity of higher organisms. For example, this process could account for the many forms of globin: the globin found in fetal hemoglobin, in muscle myoglobin, and even the two types of globin (α and β) found in adult hemoglobin are different. Multiple copies of the histone-generating genes have been found in various organisms; for example, *Drosophila* has 110 copies of different genes that code for histones.

Even accounting for duplicate genes, there is still a lot of DNA in eukaryotes that apparently does not code for functional proteins. Evidence is building that these sections have regulatory functions. At least 20 percent of the DNA seems to be sections called **repetitive DNA.** By this is meant that the same short sequence of nucleotides is repeated over and over again. These sections are normally found between functional genes (fig. 22.16). Their function is unknown, but it is speculated that they could be control elements that can move from one part of the DNA to another. Perhaps when present near a gene, they turn it on. **Movable elements,** sometimes called "jumping genes," have been found in both prokaryotes and eukaryotes. When movable elements happen to enter a portion of DNA that codes for a functional protein, they cause a mutation that is reversed when they move out again.

The most surprising finding of late has been that eukaryotic genes are normally interrupted by sections of DNA that are not part of the structural gene although they may have a control function. Therefore, eukaryotic genes are sometimes referred to as "split genes." Figure 22.17 shows an electron micrograph of the gene for ovalbumin, a primary component of the white part of a chicken's egg. This gene has been matched up (hybridized) with its mRNA taken from the cytoplasm. As shown in figure 22.17b, the loops occur because there are portions of DNA that have no complementary sequence in the mRNA. Those portions are called **introns** because they are **intra**gene segments. The portions of DNA that do match up with the mRNA are called **exons** because they are **ex**pressed. In other words, the exons are the actual gene coding for the protein, in this case ovalbumin.

As opposed to prokaryotic genes, eukaryotic genes are transcribed in a nucleus. When DNA is transcribed, the mRNA contains bases that are complementary to both the exons and the introns. But before the mRNA exits

Reproduction, Development, and Inheritance

Figure 22.17

Split gene. The presence of split genes in eukaryotic cells is demonstrated by allowing mRNA taken from the cytoplasm to hybridize (match up) with DNA taken from the nucleus. *a.* Actual electron micrograph shows that there are segments of the DNA that loop out from the hybrid because they have no complementary sequences in the mRNA. *b.* An interpretative drawing shows the mRNA and the DNA. Those portions of the DNA that have no counterpart in the mRNA have been assigned a letter from A to G, and those portions that do have a counterpart in the mRNA are numbered from 1 to 7. Only the latter are translated into protein.

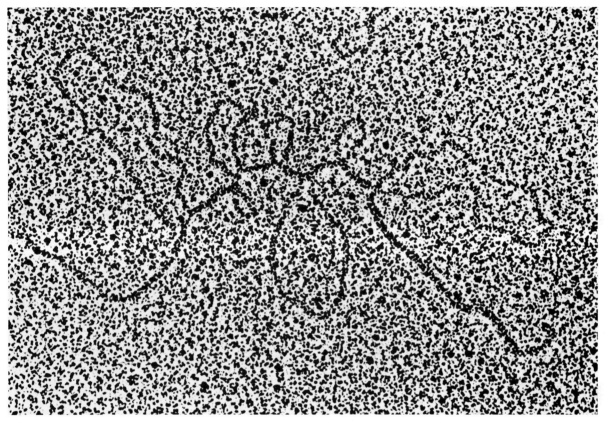

a.

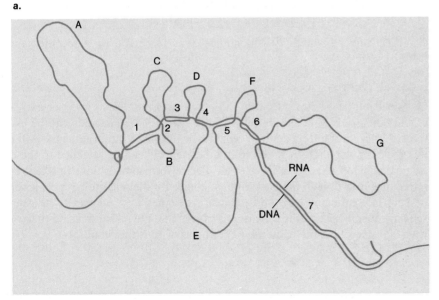

b.

Figure 22.18
Recombinant DNA experimentation. *a.* Plasmid DNA is removed from bacterium. *b.* Foreign DNA is incorporated into the plasmid. *c.* Plasmid is reintroduced into the bacterium. *d.* The incorporated gene is cloned when the bacterium reproduces and, if all goes well, will also direct protein synthesis.

a.

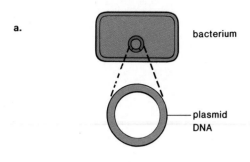

bacterium

plasmid DNA

b.

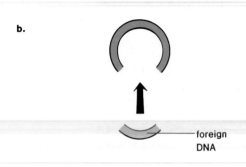

foreign DNA

c.

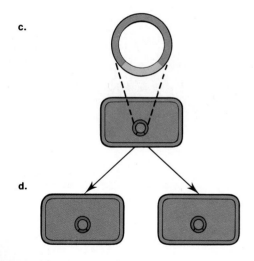

d.

from the nucleus, the nucleotides complementary to the introns are removed. Is this simply a signal that the mRNA is ready to exit from the nucleus? Or do these excised portions of mRNA have a regulatory role? Perhaps they are complementary to other sections of the DNA and when they pair with these sections they indicate that certain other genes should also be transcribed. Is this the manner in which the genes "talk to one another"? These are some of the questions that are being asked about the function of introns. Others are interested in the possibility that introns can facilitate the occurrence of useful mutations as opposed to harmful mutations. It has been shown that the exons sometimes are rearranged to give a different gene than previously. In other words, perhaps eukaryotic genes are composed of modules that can be put together in various ways, just as sectional furniture can be arranged differently. Each rearrangement would be a mutation. The techniques to be discussed in the following section have been useful in helping geneticists answer these questions.

Knowledge about regulatory genes is extremely important because genes can best be manipulated when we know how to turn them on and off. Also, mutations of regulatory genes probably account for some genetic disease and/or the development of cancer. It is also possible that the mutation of a single regulatory gene could initiate evolution toward speciation since many structural genes would thereby be affected.

Genetic Engineering

Genetic engineering refers to certain kinds of laboratory procedures and research involving genes, including human genes. Both in vitro and in vivo experiments are commonplace in genetic research. *In vitro* means that cell parts and molecules are studied in a test tube, while *in vivo* means that living organisms are used in the experiments. The following procedures are common in experiments involving both prokaryotic DNA and eukaryotic DNA, including human DNA.

a. Isolation of a gene; that is, removal of a particular portion of DNA from a cell
b. Determination of the sequence of nucleotides in a gene
c. Manufacture of a gene; that is, joining nucleotides together in the proper sequence
d. Placement of an isolated or synthetic gene in another cell where it undergoes replication and directs protein synthesis

The procedure by which (*d*) is done has come to be known as recombinant DNA research.

Recombinant DNA (Plasmid) Experiments

Certain bacteria, such as *E. coli* and *A. tumefaciens,* have rings of extrachromosomal DNA called **plasmids** (fig. 22.18). These plasmids can be extracted from the bacteria and then enzymatically sliced into fragments. After this has been done, DNA taken from other sources can be attached to these fragments before the plasmid re-forms. The re-formed plasmids are often reintroduced into *E. coli,* the bacterium of choice for making multiple copies of a gene. Recombinant DNA procedures are sometimes referred to as **gene splicing** techniques and the gene that is spliced into the plasmid is said to have been "*cloned*" once the bacteria make copies of it as they multiply.

Reproduction, Development, and Inheritance

Applications
More and more applications have been found for genetic engineering procedures and we will discuss a few of these.

Protein Products
The recombinant DNA procedure has been used to cause *E. coli* to produce various proteins, including the human hormones insulin and growth hormone. Drug companies are hopeful that they will soon be able to make these hormones available to the public for the treatment of diabetes and dwarfism. But the first commerically available product is a vaccine for an animal ailment called foot-and-mouth disease. This disease, which causes many cattle to die each year, is caused by a virus. *E. coli* has been engineered by the procedure just described to produce a protein from the virus coat, and this protein can be used as a vaccine to immunize cattle against the disease.

Cloned plasmids can be removed from *E. coli* and introduced into other types of cells. In this way, yeast cells have been engineered to make human interferon (p. 256), a chemical useful in immunity and cancer research. Since yeast is used in the production of beer, technology is already available for mass production of these cells. Perhaps, then, interferon will soon be available in bulk.

Alteration of Genetic Inheritance
Plasmids which originally were derived from *A. tumefaciens* have been introduced into plant cells because *A. tumefaciens* normally infects plant cells. In this way, it has been possible to transfer a foreign gene into plant cells. Further, these cells can be treated so that they give rise to entire plants. The hope is that one day it may be possible to provide plants with the necessary genes to resist pests and/or to fix aerial nitrogen, reducing the need for pesticide and fertilizer use.

Mammalian cells will not take up plasmids, but cloned genes have been injected into egg cells or treated with crystals of calcium phosphate, after which they will enter mammalian cells. In the former instance it has been possible to show that a second generation of mice still carried the injected gene.

Diagnostic Tests
Cloned DNA fragments are now being used in medical diagnosis of infectious and genetic diseases. In the former case, these fragments were originally removed from infectious organisms and in the latter case, they were removed from the cells of individuals who had a particular genetic disease. The fragments are called **DNA probes** because they search out and bind to complementary sequences in cells. The occurrence of binding is often detected by radioactive or fluorescent techniques.

Some researchers believe that usefulness of DNA probes will continue to expand and will greatly increase the ability of the medical profession to not only diagnose infectious diseases rapidly and accurately but also to characterize the genetic inheritance of individuals including the unborn. Such characterization can be used to prevent the effects of genetic disease and facilitate transplant surgery, for example.

Table 22.6 Chromosomal mutations

Type	Description	
Normal	Prior to any structure change	a b c d. e f g h i a b c d e f g h i
Translocation	Exchange of chromosome pieces between nonhomologous pairs	a b g f e d c h i a b c d e f g h i
Deletion	Loss of a piece of chromosome	a b c d e h i a b c d. e f g h i
Duplication	More than one copy of the same gene is present	a b. c c d. e f g h i a b c d. e f g h i
Inversion	Portion of chromosome breaks loose and rejoins with the ends reversed	a b c d. e h g f i a b c d e f g h i

Table 22.7 Gene mutations

Base change		Result
Normal	TAC'GGC'ATG'TCA	
Deletion	ACG'GCA'TGT'CA	Polypeptide completely altered
Addition	ATA'CGG'CAT'GTC'A	Polypeptide completely altered
Substitution	TAG'GGC'ATG'TCA	Change in only one amino acid

Mutations

Changes in the DNA code that are passed on to subsequent cell generations are called mutations. Mutations may arise spontaneously (due to no known cause) or they may be environmentally induced, as discussed in the following. Mutations are of great concern today because they promote genetic diseases and cancer.

Various **chromosomal mutations** are described in table 22.6. One instance of translocation sometimes causes Down's syndrome. This condition occurs when a tiny piece of chromosome number 21 becomes attached to another chromosome, such as numbers 13, 14, 15, or 22. The inheritance of this chromosome along with two normal number 21 chromosomes means that the person will have Down's syndrome.

Gene mutations (table 22.7) involve one or more nucleotide changes in a single gene. If a single nucleotide base is added to or deleted from a segment of DNA, the entire code can be so altered that a nonfunctional polypeptide results. A single base substitution only causes one amino acid to change in a polypeptide. Still, this can have a dramatic effect on the phenotype. As discussed on page 478, sickle-cell hemoglobin contains a polypeptide that differs

from normal hemoglobin only by the substitution of the amino acid valine for the amino acid glutamic acid. Gene mutations due to a single nucleotide change can occur, due to errors in the replication of DNA. Gene mutations can also occur during meiosis whenever crossing over takes place within genes, rather than between genes as is usual. Recently, as we have discussed, investigators have reported that mutations are often due to a rearrangement of genes within and between chromosomes.

Environmental Mutagens

An environmental factor that increases the chances of a mutation is a **mutagen.** When a mutagen leads to an increase in the incidence of cancer, it is called a **carcinogen.** There are two broad categories of mutagens: chemical mutagens and radiation. Suspected chemical mutagens range from food additives and hallucinogenic drugs to manufacturing chemicals and pesticides. Scientists have even discovered that natural food contains plant chemicals that act as mutagens and some that act as antimutagens. Suspected mutagens are often tested in bacteria, fruit flies, and finally, mice. If cancer develops in the mice, then the government either allows the product to be sold with an appropriate warning label or else it bans the sale of the product entirely.

All nonvisible short wavelengths of radiation are believed to be mutagenic, but the greater the amount, the greater the risk. Organisms are exposed to radiation from natural and man-made sources. Short electromagnetic waves, such as X rays and gamma rays, and shortwave radiation from radioactive elements, such as iodine 131 and plutonium 210, have the ability to penetrate tissues and cause both somatic and germinal mutations. Longer electromagnetic waves, such as ultraviolet waves and microwaves, do not cause germinal mutations although they can cause skin cancer and/or burns, as well as other superficial somatic mutations in susceptible individuals.

While many people are concerned about radiation from the nuclear power industry, medical diagnostic radiation actually accounts for at least 90 percent of human exposure to man-made radiation. Everyone should be aware of this potential danger and have X rays only when necessary.

Germinal Mutations

Germinal mutations occur during the maturation of gametes. Germinal mutations are recognized when the offspring exhibits a genetic disease. It is estimated, for example, that the gene mutation that causes achondroplastic dwarfism, characterized by disproportionate shortening of the arms and legs, occurs 10 to 70 times per 1 million gametes. A famous instance of a germinal mutation has been traced to Queen Victoria of England (1819–1901). She most likely received a mutant allele for hemophilia from one of her parents. She, in turn, passed the mutant allele along so that one of her sons had hemophilia and two of her daughters were carriers. Through marriage, the daughters spread the allele to the royal houses of Europe. Her great-grandson, Alexis, who was the only heir to the Russian throne just prior to the Russian Revolution, had hemophilia.

Somatic Mutations

Somatic mutations are mutations that affect the individual's body cells. We are most familiar with somatic mutations that accompany cancer. Ordinarily, cells grow in an orderly fashion and cease growing when they contact one another. But cancer cells are characterized by uncontrolled growth and they fail to stay in the organ where they arose (fig. 22.19). The outer membrane of cancer cells carries different markers and displays special tumor antigens.

Figure 22.19
Normal versus cancerous growth. *a.* Normal growth and repair. *b.* During the development of cancer, abnormal cells proliferate in a manner similar to that shown. The cells are no longer held in check by cell-to-cell contact and begin to pile up.

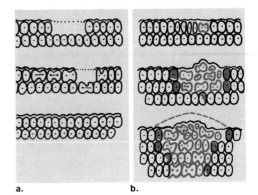

a.   b.

Figure 22.20

Transformation of normal cells into cancerous cells by means of a virus infection. *a.* The normal cell contains one oncogene but is not cancerous. The virus also contains an oncogene, which it passes to the cell. Now the cell is transformed and becomes cancerous because it produces a protein that causes the cell to become abnormal. *b.* The normal cell contains an oncogene but is not cancerous. The virus contains an enhancer that it passes to the cell. Now the cell is transformed and becomes cancerous because it produces a protein that causes the cell to become abnormal.

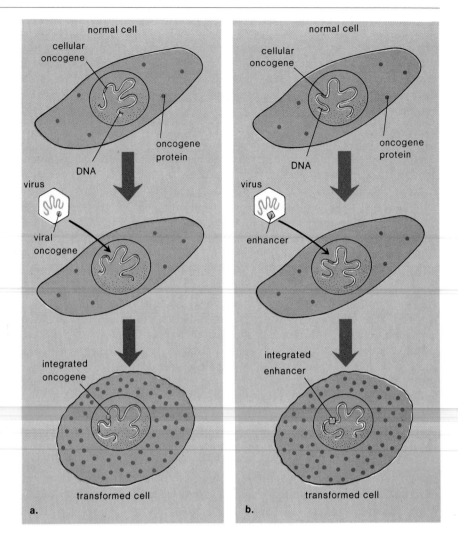

Long before now it was believed that cancer begins with a change in the DNA, but the exact nature of the change was unknown. Recently, investigators have been able to determine that cells contain genes, called proto-oncogenes (*proto* = before; *onco* = tumor) that can become *oncogenes,* cancer-causing genes. These genes are not alien to the cell, they are normal essential genes that have undergone a mutation. For example, by using recombinant DNA techniques, investigators have shown that an oncogene that causes both lung cancer and bladder cancer differs from a normal gene by a change in only one nucleotide.

The reading on page 496 tells how scientists working in different research areas are now able to explain their results in a similar manner. For example, a cancer-causing virus may have picked up an oncogene when it reproduced in a previous host. After that it is capable of passing the oncogene to a new host (fig. 22.20a). Or even if it lacks an oncogene, a virus may introduce a control element that causes a cell to become cancerous (fig. 22.20b). Investigators have discovered that regions of DNA, called **enhancers,** increase the activity of any nearby gene and that such elements are often carried by viruses. It seems that a cell turns cancerous either when a normally inactive gene is transcribed or when a gene begins to be overactive. In any case, there is a change in the regulation of the gene. Most likely, hormones can in some instances help cause cancer because they, too, influence the activity of genes (fig. 18.1b).

Reproduction, Development, and Inheritance

Figure 22.21
Summary of the development of cancer. A virus can pass an oncogene to a cell. A normal gene, called a proto-oncogene, can become an oncogene due to a mutation caused by a chemical or radiation. The oncogene either expresses itself to a greater degree than normal or else expresses itself inappropriately. Thereafter the cell becomes cancerous. Cancer cells are usually destroyed by the immune system and the individual only develops cancer when the immune system fails to perform this function.

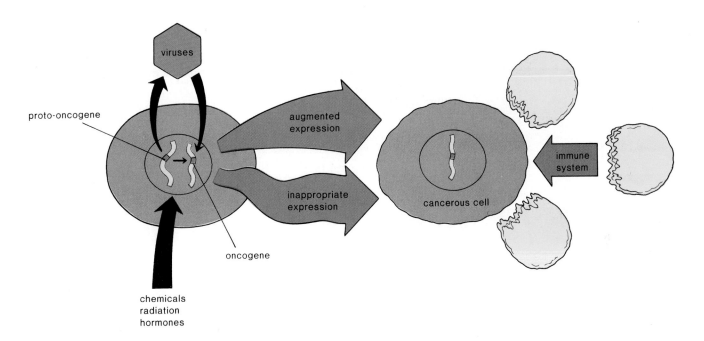

One would not expect a single oncogene to transform a perfectly normal cell. Therefore it is not surprising that biologists believe that cancer is a multi-step process. For example, it may be that a cell becomes cancerous only after it contains at least two oncogenes that cooperate with one another. Then, too, we should keep in mind that a strong immune system usually attacks and destroys cancerous cells. Cancer develops when the immune system fails in this capacity. Cancer may be more common in older individuals because, with age, the immune system is not as active as it once was. Figure 22.21 gives a summary of the ways in which cancer is now believed to come about.

What exactly do oncogenes code for? Evidence is gathering that oncogenes do code for metabolically active proteins that affect growth and organization of tissues. This would explain why cancerous cells exhibit uncontrolled, disorganized growth. During the time that cancer remains at the site of origin (**in situ**), it is usually curable; but eventually cancer cells may invade underlying tissues, become detached, and be carried by the lymphatic and circulatory systems to other parts of the body where new cancer growth may begin. The process by which cancer spreads to other parts of the body is called **metastasis;** when this occurs, the cancer may spread throughout the body and is then usually considered incurable by current methods of treatment. This is the rationale for early detection and treatment of cancer.

The American Cancer Society has publicized seven danger signals for cancer; these are listed in table 22.8. Any person who notices one of these signals should consult a physician promptly.

Table 22.8   Danger signals for cancer

1. Unusual bleeding or discharge
2. A lump or thickening in the breast or elsewhere
3. A sore that does not heal
4. Change in bowel or bladder habits
5. Persistent hoarseness or cough
6. Persistent indigestion or difficulty in swallowing
7. Change in a wart or mole

Advances in the War on Cancer

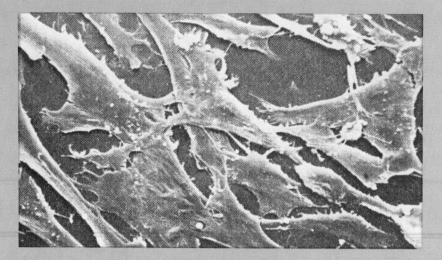

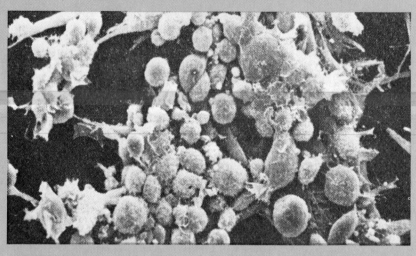

Figure 22.C *Top*. Normal fibroblasts are flat and extended. *Bottom*. After being infected with Rouse sarcoma virus, the cells become round and cluster together in piles. It is possible that the virus carried an oncogene.

What causes nice, healthy, law-abiding cells to go berserk, proliferate wildly and thus produce the phenomenon called cancer? The answer, scientists have long suspected, lies in the genetic material of the cells. Somehow genes, composed of the molecule deoxyribonucleic acid (DNA), have been made abnormal, perhaps by such environmental factors as cigarette smoke and radiation.

In recent months, scientists in three fields have made remarkable progress in documenting how these genetic changes take place. Cell geneticists, studying chromosomes, the discrete units into which genes are packaged, have begun to associate specific abnormalities— an extra chromosome, a missing piece of a chromosome—with specific types of cancer. Virologists have clarified the role that viruses can play in altering DNA. And molecular biologists have used the new tools of genetic engineering to pinpoint precisely which gene, out of the tens of thousands present in every human cell, is responsible for causing a tumor. . . .

The first breaks in cancer genetics came from the field of virology. Scientists have known since 1908 that a virus could cause malignant tumors in chickens. Over the decades, it was found that viruses could cause tumors in mice, cats, cows and a menagerie of other species. But not until 1980 did anyone identify a virus that causes cancer in human cells. Dr. Robert Gallo of the National Cancer Institute in Bethesda, Md., isolated a virus that can transform

Summary

Experiments with a T virus indicated that DNA is the genetic material. DNA has a structure like a twisted ladder: sugar-phosphate backbones make up the sides of the ladder; hydrogen-bonded bases make up the rungs of the ladder. The base A is always paired with the base T, and the base C is always paired with the base G. This so-called complementary base pairing helps explain replication. During replication, DNA becomes "unzipped," and a complementary strand forms opposite to each original strand. This is called semiconservative replication.

DNA controls the phenotype because it controls protein synthesis. DNA and several forms of RNA participate in protein synthesis. RNA differs from DNA in several respects (table 22.1). DNA, which always stays within the

Reproduction, Development, and Inheritance

normal human white blood cells into the malignant type found in a rare cancer called T-cell leukemia. The same virus was found last year to be responsible for a relatively high rate of both T-cell leukemia and a form of lymphoma (cancer of the lymphatic system) in Japan. Says Gallo: "There is strong evidence that this virus will be important for a number of human cancers."

Researchers are also learning just how a cancer virus can alter normal cell DNA. Some cancer viruses contain cancer-causing genes, or oncogenes. When these genes are isolated and then transferred into healthy cell cultures in the laboratory, they create malignant cells. In the past decade, more than 15 oncogenes have been found in cancer viruses.

Yet many cancer viruses do not appear to contain oncogenes. These viruses seem to cause cancer in a less direct way, perhaps by altering genes that are already in the normal cell. Investigating this possibility, Microbiologists J. Michael Bishop and Harold Varmus of the University of California School of Medicine at San Francisco made an astonishing discovery: genes almost identical to the cancer-causing genes in viruses can be found in the normal cells of all manner of creatures, from fruit flies to humans. The supposition is that these harmless genes can easily be turned into the dangerous genes they so closely resemble. In this sense, says Bishop, "cancer may be part of the genetic dowry of every living cell. . . ."

Competing laboratories are racing to identify the genes capable of inducing cancer in human cell cultures. About a dozen such genes have already been isolated from leukemic cells and from tumors of the lung, bladder, colon and breast. Many of these genes are nearly identical to oncogenes isolated years earlier from cancer viruses. Moreover, certain tumors (colon and lung carcinomas, for example) were found to contain the same oncogene. This suggests that perhaps several dozen genes are responsible for producing the 100 or more known forms of cancer.

Since all of the work with human cancer genes has been carried out in laboratory dishes, the role played by oncogenes in causing tumors in the body itself remains unproved. Evidence is, however, mounting. Researchers have known for two decades that certain forms of cancer are associated with certain visible changes in the 23 pairs of chromosomes found in human cells. Recent studies suggest, for example, that in some lung cancers a piece is often missing from chromosome No. 3. Better-documented changes occur in certain leukemias and lymphomas. In one form of chronic leukemia, a piece of chromosome 22 changes place with a piece of chromosome 9. In most patients with a cancer known as Burkitt's lymphoma, a piece of chromosome 8 has changed places with a piece of No. 14.

How or why these "translocations" might cause

cancer has never been known, but the first real insight was announced at the Chicago conference. Researchers representing teams at Harvard and at Philadelphia's Wistar Institute reported that they had analyzed the piece of chromosome 8 involved in Burkitt's lymphoma. Both found that it contained a gene virtually identical to a cancer-causing gene isolated years earlier from a virus. The oncogene is located at the precise point where the fragment of chromosome 8 broke off. Thus the first link has been made between a known oncogene and an easily detectable change in the chromosomes of cancer patients.

Before these new discoveries can be applied, more must be learned about the precise function of oncogenes in the cell. Every gene in nature carries a code, or recipe, for creating a specific protein. If a single gene is responsible for producing cancer, it must be a result of the protein that the gene creates. Researchers are just beginning to figure out what kinds of proteins oncogenes make. Once they do, scientists may be able to develop tests to detect tiny traces of these proteins; such tests could allow the diagnosis of cancer at a far earlier stage than is now possible. Identifying the proteins made by cancer genes will enable biologists to create antibodies that specifically attack cancer cells and, perhaps, to block the creation of those proteins, thus stopping the cancer process itself.

nucleus, contains a triplet code: a series of three bases codes for one particular amino acid. During transcription, messenger RNA is made complementary to one of the DNA strands. It then contains codons and moves to the cytoplasm and becomes associated with the ribosomes. During translation, transfer RNA molecules, attached to their own particular amino acid, travel to the mRNA, and through complementary base pairing, the tRNAs and thus the amino acids in a polypeptide chain become sequenced in a predetermined order.

Bacterial research has permitted the development of two models for the control of gene transcription. In one model, a regulatory gene codes for a repressor that binds with an operator gene in such a way that transcription of an operon is normally impossible. When this repressor joins with an inducer, however, the complex cannot bind with the operator; therefore operon transcription begins. In the other model, the regulatory gene codes for a repressor

that must join with a corepressor before the complex can bind with the operator to prevent transcription. Therefore, in this model, transcription is normally taking place.

Recent research has shown that a large portion of the eukaryotic chromosome has no known function. Some sections, called movable elements, may move from place to place regulating transcription. The genes themselves are split; the portions that code for a polypeptide are called the exons, and the portions between are called the introns. After mRNA is formed, the sequences corresponding to the introns are excised.

Recombinant DNA techniques allow eukaryotic DNA to be placed in the bacterium *E. coli,* where it usually replicates and functions normally. If so, this cloned DNA can be removed and put into other types of cells.

Two types of mutations are recognized: chromosomal mutations are large enough to be detected by the light microscope, but gene mutations may be as small as one nucleotide change. If a nucleotide is added or deleted, the entire code changes and a nonfunctional polypeptide can result. A base pair substitution is not expected to have a dire effect, but there are exceptions. Radiation and certain chemicals are now known to be mutagens and carcinogens. Germinal mutations often lead to nonfunctional enzymes. Somatic mutations often accompany cancer. It now appears that viruses sometimes introduce oncogenes and/or enhancers that transform a cell into a cancerous cell.

Study Questions

1. Describe the experiment that designated DNA rather than protein as the genetic material. (pp. 471–72)
2. Describe DNA and RNA structure. (pp. 473–75)
3. Explain how DNA replicates. (pp. 475–77)
4. Various genetic diseases indicate that DNA controls the formation of proteins. Name and discuss some of these diseases. (pp. 477–79)
5. If the code is TTA′TGC′TCC′TAA, what are the codons and what is the sequence of amino acids? (p. 480)
6. List the five steps involved in protein synthesis. (pp. 484–85)
7. Define operon, operator, and regulatory gene. Describe the two prokaryote models for control of gene transcription. (p. 487)
8. With reference to figure 22.16, discuss the recent findings in regard to eukaryotic chromosomes. (pp. 488–90)
9. You are a scientist who has decided to "clone a gene." Tell precisely how you would proceed. (p. 490)
10. Name four types of chromosomal mutations and give examples. (p. 492)
11. Show how a deletion or addition of a nucleotide can affect the code given in number 5 and the resulting polypeptide. (p. 492)
12. What are the two main categories of mutagens? What source of radiation poses the greatest threat to human health? (p. 493)
13. With reference to figure 22.21, explain the current findings regarding the development of cancer. (p. 495)

Reproduction, Development, and Inheritance

Selected Key Terms

purines (pu'rinz)
pyrimidines (pi-rim'ĭ-dinz)
strand (strand)
complementary base pairing
    (kom''pli-men'tă-re bās pār'ing)
helix (he'liks)
replication (re''plĭ-kā'shun)
mutation (mu-ta'shun)
template (tem'plāt)
transcription (trans-krip'shun)
translation (trans-la'shun)
codon (ko'don)
degenerate (de-jen'er-āt)
anticodon (an''tĭ-ko'don)
operon (op'er-on)
histone (his'tōn)
intron (in'tron)
exon (eks'on)

Further Readings

Anderson, W. F., and Diacumakos, E. G. 1981. Genetic engineering in mammalian cells. *Scientific American* 245(1):106.

Bishop, J. M. 1982. Oncogenes. *Scientific American* 246(3):80.

Brady, R. O. 1973. Hereditary fat-metabolism diseases. *Scientific American* 229(2):88.

Chambon, P. 1981. Split genes. *Scientific American* 244(5):60.

Chilton, M. 1983. A vector for introducing new genes into plants. *Scientific American* 248(6):50.

Darnell, J. E. 1983. The processing of RNA. *Scientific American* 249(4):90.

Dickerson, R. E. 1983. The DNA helix and how it is read. *Scientific American* 249(6):94.

Fedoroff, N. V. 1984. Transposable genetic elements in maize. *Scientific American* 250(6):84.

Freifelder, D., ed. 1978. *Recombinant DNA: Readings from Scientific American.* San Francisco: W. H. Freeman.

Gilbert, W., and Villa-Komaroff, L. 1980. Useful proteins from recombinant bacteria. *Scientific American* 242(4):74.

Hunter, T. 1984. The proteins of oncogenes. *Scientific American* 251(2):70.

Maniatis, R., and Ptashe, M. 1976. A DNA operator-repressor system. *Scientific American* 234(1):64.

Nicolson, G. L. 1979. Cancer metastasis. *Scientific American* 240(3):66.

Nomura, M. 1984. The control of ribosome synthesis. *Scientific American* 250(1): 102.

Ptashe, M., et al. 1982. A genetic switch in a bacterial virus. *Scientific American* 247(5):128.

Rich, A., and Kim, S. H. 1978. The three-dimensional structure of transfer RNA. *Scientific American* 238(1):52.

Weinberg, R. A. 1983. A molecular basis of cancer. *Scientific American* 249(5): 126.

Part Five

Evolution and Diversity

E volution depends on the retention of genetic changes that have been tested by the environment. This process termed natural selection results in adaptation to both the abiotic and biotic environment.

A gradual increase in chemical complexity produced the first cell(s) and this (these) evolved into all the forms of life we see about us. Taxonomists try to classify living things according to their evolutionary relationship; therefore, when we study taxonomy we are also studying evolutionary history. This text recognizes five kingdoms: Monera (bacteria and blue-green bacteria), Protista (protozoans and unicellular algae), Fungi (molds and mushrooms), Plants (multicellular algae and terrestrial plants), and Animals.

Humans are primates, animals adapted to living in trees. They share a common ancestor with apes, some of whom still live in trees. The first manlike ancestor may have left the trees when grasslands replaced trees in Africa. Walking erect could have evolved in association with this change of habitat. Later, tool use and intelligence evolved together, in association with a newly acquired hunting way of life, which had a profound effect on the behavior of humans.

Culture, which began with tool use, soon also included art, science, religion, and so forth. Unfortunately, twentieth-century culture tends to make humans unaware of their natural place in the biosphere.

23

Evolution

Chapter concepts

1 Life evolved from the first cell(s) into all the forms of life now present and extinct.

2 The fossil record, comparative anatomy, embryology, and biochemistry all provide evidences of evolution.

3 Evolution, defined as a change in frequency of genes in the gene pool of a population, results in adaptation to the environment as a result of natural selection.

4 Natural selection occurs when the better adapted members of a population reproduce to a greater degree than the less well adapted members.

5 New species come about when a population is at first geographically isolated and then later reproductively isolated from other similar populations.

6 Adaptation to a variety of environments explains the diversity of life. Inability to adapt to a changing environment explains extinction.

E volution is the process that explains the history and diversity of life. Data from various fields of biology give us evidence that evolution produced the myriad of organisms that are now present on earth.

Evidences for Evolution

Fossil Record

Our knowledge of the history of life, which is depicted in figure 23.1, is based primarily on the fossil record. **Fossils** are the remains or evidence of some organism that lived long ago. Most fossils are formed when an organism is buried in mud or sand before the hard mineralized parts have decayed. A fossil may be the remains of this part, or it may be the impression or mold that the part made in the rock developing about it. Sometimes, fossils are formed by a replacement of the original organic material by a durable mineral, such as silica.

The fossil record clearly indicates that the major groups of organisms appeared on earth in a sequential manner. The reasonable assumption is that the older groups gave rise to the younger groups. Supporting this supposition, fossils have been found that seem to be intermediate between major groups. A well-known example of such a transitional form is *Archeopteryx* (fig. 23.2), whose fossil remains suggest that it was a flying reptile except that it clearly had the feathers and the beak of a bird. A living example of the same principle is *Peripatus* (fig. 23.3), a two-inch-long animal that looks like a caterpillar and has characteristics of both the annelids (segmented worms) and arthropods (insects, crustaceans). These two animals may not be those of the precise species that respectively gave rise to birds or arthropods, but they do indicate a relationship between reptiles and birds and a relationship between annelids and arthropods.

Comparative Anatomy

A comparative study of the anatomy of groups of organisms has shown that each has a **unity of plan.** For example, the reproductive organs of all flowering plants are basically similar and all vertebrate animals have essentially the same type skeleton. Unity of plans allow organisms to be classified into various groups. Organisms most similar to one another are placed in the same **species,** similar species are placed in a **genus,** similar genera in a family; thus, we proceed from **family** to **order** to **class** to **phylum** (animals) or **division** (plants) to **kingdom.** The classification of any particular organism indicates to what kingdom, phylum, class, order, family, genus, and species the organism belongs. According to the **binomial system** of naming organisms, each organism is given a two-part name, which consists of the genus and species to which it belongs. Thus, for example, a human is *Homo sapiens* and the domesticated cat is *Felis domestica.* **Taxonomy** is the branch of biology that is concerned with classification, and biologists who specialize in classifying organisms are called taxonomists.[1]

Table 23.1	The classification of modern humans
Kingdom	Animalia (animals)
Phylum	Chordata (chordates)
Class	Mammalia (mammals)
Order	Primates (primates)
Suborder	Anthropoidea (anthropoids)
Superfamily	Hominoidea (hominoids)
Family	Hominidae (hominids)
Genus	*Homo* (humans)
Species	*sapiens* (modern humans)

[1]The classification system given on page A–3 is utilized in this text.

Figure 23.1

The fossil record provides a history of life. The record indicates that primitive, single-celled organisms resembling modern bacteria, or blue-green bacteria, were the first life forms to appear. In time, more complex cellular forms appeared, followed by multicellular plants and animals. After that, there was an increase in complexity.

Era	Period	Epoch	Years from Start of Period to Present	
Cenozoic	Quaternary	Recent	10,000	
		Pleistocene (Ice Age)	3 million	
		Tertiary	63 million	
Mesozoic "Age of Reptiles"	Cretaceous		135 million	
	Jurassic		181 million	
	Triassic		230 million	
Paleozoic	Permian		280 million	
	Carboniferous		345 million	
	Devonian		405 million	
	Silurian		425 million	
	Ordovician		500 million	
	Cambrian		600 million	
Proterozoic			1.5 billion	
			2.5 billion	
Archeozoic			4.5 billion	

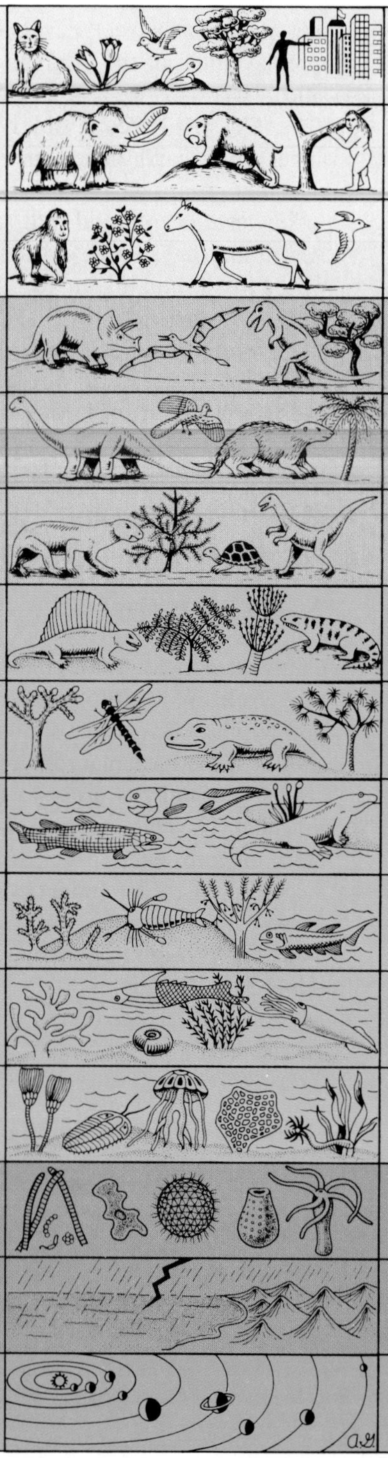

Evolution and Diversity

Plant Life	Animal Life
Increase in the number of herbaceous plants	Age of human civilization
Extinction of many species of plants	Great mammals such as woolly mammoth and saber-toothed tiger became extinct First human social life
Dominance of land by angiosperms	Dominance of land by mammals, birds, insects Mammalian radiation First humans
Angiosperms prevalent, gymnosperms decline Trees resembling modern-day maples, oaks, and palms flourish	Dinosaurs reach peak, then become extinct Second great radiation of insects First primates
Gymnosperms such as cycads and conifers still prevalent	Dinosaurs large, specialized, more abundant First mammals appear First birds appear
Dominance of land by gymnosperms and ferns Decline of club mosses and horsetails	First dinosaurs appear Mammallike reptiles evolve
Gymnosperm and angiosperm(?) evolve	Expansion of reptiles Decline of amphibians
Age of great coal forests including club mosses, horsetails, and ferns	"Age of Amphibians" First great radiation of insects First reptiles appear
Expansion of land plants; first forests of club mosses, horsetails, and ferns	"Age of Fishes" First land vertebrates, the amphibians, appear
First vascular plants, modern groups of algae and fungi	First air-breathing land animals, such as land scorpion, appear Rise of fishes
Invasion of land by plants(?)	Diverse marine invertebrates, coral and nautaloid common First vertebrates appear as fish
Marine algae common	Diverse primitive marine invertebrates, trilobites common Animals with skeletons appear
Multicellular acoelomate and coelomate animals evolve Eukaryotic protists and fungi evolve	
Prokaryotes abundant	
Anaerobic and photosynthetic bacteria evolve Formation of earth and rest of solar system	

Figure 23.2
Archaeopteryx a. This fossil had features common to both reptiles and birds. Note the indication of feathers, a birdlike feature, and the long bony tail, a reptilian feature. *b.* An artist's conception of *Archaeopteryx.*

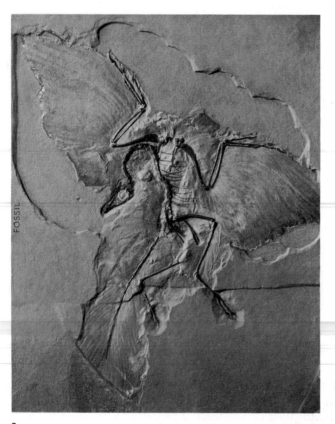

a.

b.

Figure 23.3
Peripatus. This animal has features common to both annelids and arthropods. It is obviously a segmented animal; its excretory, reproductive, and nervous systems are similar to those of the annelids, while its circulatory and respiratory systems are similar to those of the arthropods.

Evolution and Diversity

A *unity of plan* is explainable by descent from the same **common ancestor.** Species that share a recent common ancestor will share a large number of the same genes and thus will be quite similar to each other and to this ancestor. Species that share a more distant common ancestor will have fewer genes in common and will be less similar to each other and to this ancestor because differences arise as organisms continue on their own evolutionary pathways. This principle allows biologists to construct **evolutionary trees,** diagrams that tell how various organisms are believed to be related to one another. All evolutionary trees have a branchlike pattern (fig. 23.4), indicating that evolution does not proceed in a single steplike manner; rather, evolution proceeds by way of common ancestors that often give rise to two different groups of organisms. For example, reptiles are believed to have produced both birds and mammals.

Even after related oganisms have become adapted to different ways of life, they may continue to show similarities of structure. For example, the forelimb of all vertebrates contains the same fundamental bone structure (fig. 23.5) despite their specific specializations. Similarities in structure that have arisen through descent from a common ancestor are called **homologous structures.** Homologous structures indicate that organisms are related. Sometimes two groups of organisms have structures that function similarly but are constructed differently. In contrast to homologous structures, **analogous structures,** such as an insect wing and bird wing, have similar functions but differ in their anatomy and thus we know they evolved independently from one another.

Figure 23.4
The evolutionary tree pattern used in this text. Note that the black areas indicate common ancestors. For example, among animals, both birds and mammals trace their ancestry back to reptiles. Among plants, the two major groups of flowering plants are derived from an earlier common ancestor.

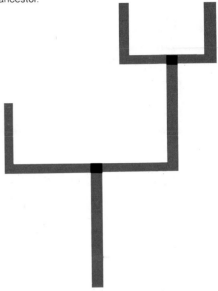

Figure 23.5
Homologous structures. The bones are coded so that you may note the similarity in the bones of the forelimbs of these vertebrates. This similarity is to be expected since all vertebrates trace their ancestry back to a common ancestor.

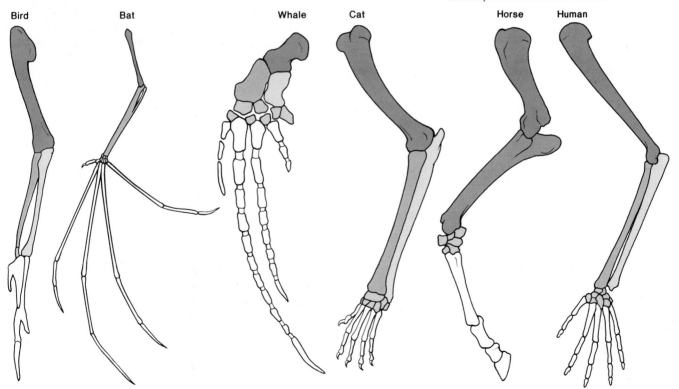

Bird   Bat   Whale   Cat   Horse   Human

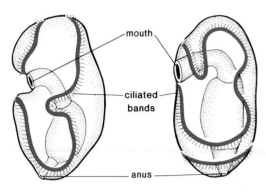

mouth

ciliated bands

anus

Comparative Embryology

Some groups of organisms share the same type of embryonic stages. For example, as shown in figure 23.6, the larva of certain lower chordates (the phylum containing vertebrates) is strikingly similar to that of certain echinoderms (e.g., starfish). As we would expect, the embryonic stages of all vertebrates are also similar. Thus, during development a human embryo at one point has gill pouches, even though it will never breathe by means of gills as do fishes, and a rudimentary tail, even though it will never have a long tail as do some four-legged vertebrates. In this way, embryological observations indicate evolutionary relationships.

Vestigial Structures

An organism may have structures that are underdeveloped and seemingly useless and yet they are fully developed and functional in related organisms. These types of structures are called **vestigial.** Figure 23.7 illustrates numerous vestigial structures in humans. The presence of these structures is understandable when we realize that related organisms share genes in common. Presumably, genes that code for vestigial structures are in the process of slowly being phased out although there are various reasons, to be discussed in this chapter, why this may not happen as soon as expected. For example, it has been found that chickens still retain genes for the development of tooth dentine, even though modern chickens do not have teeth.

Comparative Biochemistry

Almost all living organisms use the same basic biochemical molecules, including DNA, ATP, and many identical or nearly identical enzymes. It would seem that these molecules evolved very early in the evolution of life and have been passed on ever since.

Analyses of amino acid sequences in certain proteins like hemoglobin and cytochrome C have been done in various animals in order to determine how distantly related they are. The rationale is that the number of differences will reflect how long ago the two species shared a common ancestor. Also, analyses of DNA nucleotide differences of the genome (all the genes) have been done for the same purpose. Figure 23.8 shows the results of one such study. Investigators have been gratified to find that evolutionary trees based on biochemical data are quite similar to those based on anatomical data. Whenever the same conclusions are drawn from independent data, they substantiate scientific theory, in this case organic evolution, even more than usual.

Biogeography

In chapter 32, we will have an opportunity to study the geographic distribution of plants and animals. It is observed that similar but geographically separate environments have different plants and animals that are similarly adapted. For example, figure 32.8 compares North American animals with African animals that are adapted to living in a grassland environment. First, you will notice that although each group of animals could live in the other's biogeographic region, they do not. Why? Because geographic separation made it impossible for a common ancestor to produce descendants for both regions. On the other hand, notice that the same type of adaptations are seen in both groups of animals. This phenomenon, called **convergent evolution,** supports the belief that the evolutionary process causes organisms to be adapted to their environments.

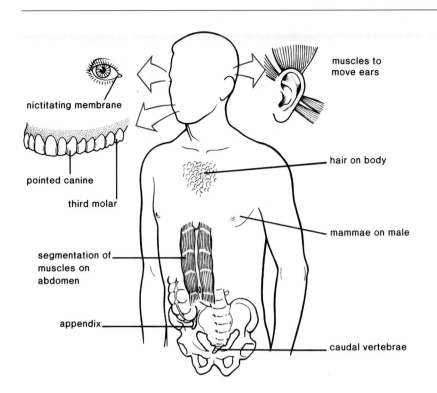

Figure 23.7
Vestigial structures. Human beings have various vestigial structures such as those shown. These show our relationship to animals in which these structures are fully developed and functional.

nictitating membrane

muscles to move ears

pointed canine

third molar

hair on body

mammae on male

segmentation of muscles on abdomen

appendix

caudal vertebrae

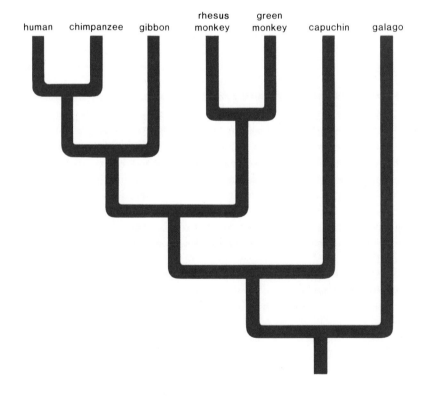

Figure 23.8
Evolutionary tree of primate species based on a biochemical study of their genomes. The length of the branches indicates the approximate number of nucleotide pair differences that were found between groups.

human chimpanzee gibbon rhesus monkey green monkey capuchin galago

a. No evolution

First generation

Genotype	PP	Pp	pp
Frequency	0.64	0.32	0.04

Gametes

P=0.8   p=0.2

	0.8P	0.2p
0.8P	PP 0.64	Pp 0.16
0.2p	pP 0.16	pp 0.04

Second generation

Genotype	PP	Pp	pp
Frequency	0.64	0.32	0.04

b. Evolution

First generation

Genotype	PP	Pp	pp
Frequency	0.64	0.32	0.04

Gametes

p=0.83   p=0.17

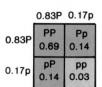

	0.83P	0.17p
0.83P	PP 0.69	Pp 0.14
0.17p	pP 0.14	pp 0.03

Second generation

Genotype	PP	Pp	pp
Frequency	0.69	0.28	0.03

Microevolution

In order to study the mechanism of evolution, it is best to consider the evolution of populations rather than the evolution of major groups of organisms. Since this is evolution in miniature, it is called microevolution.

For the sake of discussion, a **population** will be defined as a group of interbreeding individuals living in a particular area. The various alleles and their frequencies constitute the genetic makeup of the population, which is very often referred to as the **gene pool.** If there were nothing to upset the equilibrium, we would expect the gene pool to remain constant generation after generation. For example, suppose it is known that one-fourth of all persons in a human population are homozygous dominant for widow's peak, one-half are heterozygous, and one-fourth are homozygous recessive for continuous hairline. What will be the ratio of genotypes in the next generation?

Using the key given in the previous chapter, W = widow's peak and w = continuous hairline, we can describe the population in this manner:

$$¼ \ WW + ½ \ Ww + ¼ \ ww$$

Necessarily, the homozygous dominant individuals will produce one-fourth of all the gametes of the population, and these gametes will all carry the dominant allele, W; the heterozygotes will produce one-half of all the gametes, but one-fourth will be W and one-fourth will be w; the homozygous recessive will produce one-fourth of all the gametes and they will be w. Therefore, in summary, one-half of the gametes will be W and one-half will be w.

Assuming that all possible gametes have an equal chance to combine with one another, then, as the Punnett square shows, the next generation will have exactly the same ratio of genotypes as the previous generation.

	½ W	½ w
½ W	¼ WW	¼ Ww
½ w	¼ Ww	¼ ww

Results:

$$¼ \ WW + ½ \ Ww + ¼ \ ww$$

To take another example, let's suppose that 64 percent of the population is homozygous dominant and does not have PKU; 32 percent are heterozygous; and 4 percent are homozygous recessive and have PKU. Figure 23.9 shows that the genetic makeup of the next population will be exactly the same as the parental generation. The same results would be obtained no matter what alleles we consider and no matter how many generations were included. This means that (a) dominant alleles do not tend to take the place of recessive alleles and that recessive alleles do not tend to disappear and (b) sexual reproduction in and of itself cannot bring about a change in the allele frequency of the population.

Hardy-Weinberg Law

The gene pool may also be described by means of the quadratic equation:

$$p^2 + 2pq + q^2 = 1.00$$

In this case p represents the frequency of the dominant allele and q represents the frequency of the recessive allele. Therefore:

p^2 = homozygous dominant individuals
q^2 = homozygous recessive individuals
$2pq$ = heterozygous individuals

The real value of this mathematical approach to population genetics is that by observation or inspection it is possible to determine the percentage of individuals who are recessive, and from this it is possible to calculate the frequencies of the alleles and genotypes. (Usually these frequencies are given in decimals rather than fractions because the frequencies of the two alleles are not always easily converted to a fraction.) Using this method:

$$
\begin{aligned}
\text{if } q^2 &= 0.25, & q &= 0.50 \\
p^2 &= 0.25, & p &= \underline{0.50} \\
2pq &= \underline{0.50} & & 1.00 \\
& \ \ \ 1.00
\end{aligned}
$$

Notice that $p + q$ (frequencies of the two alleles) must equal 1.00 and $q^2 + p^2 + 2pq$ (frequencies of the various genotypes) must also equal 1.00.

To take another example, suppose by inspection we determine that 1 percent of the population has dimples. Therefore, 99 percent of the population does not have dimples. Of these, how many are homozygous dominant? How many are heterozygous?

To answer these questions, first convert 1 percent to a decimal. Then we know that $q^2 = 0.01$ and that therefore $q = 0.1$. Since $p + q = 1.0$, then we know that $p = 0.9$ and that therefore p^2 (frequency of the population that is homozygous dominant) $= 0.81$. To determine the frequency of the heterozygote, we simply realize that thus far we have accounted for only 0.82 of the population, and that therefore $0.18 = $ heterozygous. Or if you prefer, calculate that $2pq = 0.18$. In summary we have found that

Homozygous recessive	= 0.01	= 1 percent do have dimples
Homozygous dominant	= 0.81⎫	
Heterozygous	= 0.18⎭	= 99 percent do not have dimples

Practice Problems

1. A student places 600 fruit flies with the genotype *Ll* and 400 with the genotype *ll* in a culture bottle. Assuming that evolution does not occur, what will be the genotype frequencies in the next generation and each generation thereafter?
2. Four percent of the members of a population of pea plants are short. What is the frequency of the recessive allele and the dominant allele? What are the genotype frequencies in this population?
3. Twenty-one percent of a population is homozygous dominant, 49 percent are heterozygous, and 29 percent are recessive. What percentage of the next generation is predicted to be recessive?

Answers to problems are on page 525.

Theoretically, it would be possible for the gene pool of a population to remain constant generation after generation. In other words, sexual recombination, in and of itself, cannot alter gene frequencies in large populations. But, in fact, the gene pool rarely if ever remains constant and the Hardy-Weinberg law recognizes this when it states: The gene pool stays constant only if (1) the population is large and mating is random, (2) no mutations occur, (3) there is no gene flow, and (4) there is no natural selection. *When the gene pool does not stay constant, then evolution has occurred.*

Figure 23.10

Variations among individuals of a population. It is easy for us to note that humans vary one from the other. However, the same is true for populations of any organism.

Table 23.2 Mechanism of evolution

Produce variation	Reduce variation
Mutations	Genetic drift
Gene flow	Natural selection
Recombination	

Figure 23.11

Gene interactions. This diagram illustrates that one gene can affect many characteristics of the individual (pleiotropy) and one characteristic can be controlled by several genes (polygeny).

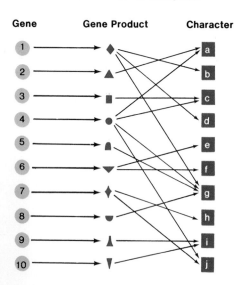

Evolutionary Process

The evolutionary process depends on individual variations (fig. 23.10). It is readily apparent that the members of a human population vary, but just as humans differ one from the other, so do members of other populations. The daisies on the hill and the earthworms in your backyard are not genotypically or phenotypically the same. Metabolic, structural, and behavioral differences exist between them. The evolutionary process (table 23.2) requires both the (1) production of genotype and phenotype variations and (2) the reduction of these variations.

Production of Variations

Genetic variations in the gene pool of sexually reproducing diploid organisms have three sources: *mutations, gene flow,* and *recombinations.*

Mutations are the raw material for evolution and the ultimate source of all variations found in natural populations. Many times, observed mutations, such as those that cause human genetic diseases, seem to harm rather than benefit an individual. This may be because members of a population are so adapted (suited) to an environment that only nonbeneficial changes are apparent. Nevertheless, recessive nonobserved mutations may be occurring that could be beneficial should the environment change.

In organisms that lack sexual reproduction, such as bacteria, genotype variability is dependent entirely on mutations. This is sufficient because the short generation time of these organisms allows new mutations to be immediately tested by the environment.

Gene flow, which occurs when individuals immigrate and emigrate between populations, brings new genes into the pool of each population. The sharing of genes can cause the two populations to become similarly adapted but can also keep each one from becoming very closely adapted to a local environment.

Recombination of genes occurs during meiosis and fertilization. During meiosis, crossing over and independent assortment produces unlike gametes. During fertilization, gamete union also brings about a genotype unlike those of the parents.

Recombination is an important source of variability in sexually reproducing organisms. It allows different combinations of genes to be tested by the environment. After all, it is the entire genotype, represented by the combination of genes, not an individual gene, that determines whether the individual is suited to the environment. The phrase **unity of the genotype** means that the genotype should be viewed not as a composite of individual genes but as a cohesive whole. Only then is it possible to take into account gene interactions (fig. 23.11), such as pleiotropy (one gene can affect several different characteristics), polygeny (one characteristic can be controlled by several genes) and regulatory genes, which modify the action of other genes.

Reduction in Variations

Genetic drift and *natural selection* both act in such a way that variations in a population are sorted out and reduced. But only natural selection consistently results in adaptation. Adaptations may be structural (land animals breathe by means of lungs), physiological (desert animals make do with metabolic water), or behavioral (some animals forage at night and others forage in the daytime).

Genetic drift, as diagrammed in figure 23.12a, is a reduction in gene pool variation that occurs purely due to chance. It should be visualized as a chance drifting toward certain genes so that others are eliminated. In the diagram, "chance sampling" means that only a few individuals among all the various phenotypes available produce offspring.

Figure 23.12
Genetic drift versus adaptation. In both cases,
mutation, gene flow, and recombination are sources
of genetic variation. The various genotypes develop
into various phenotypes. *a*. A chance sampling of
these phenotypes can lead to genetic drift.
b. Selection of adapted phenotypes can lead to
adaptation to the environment.

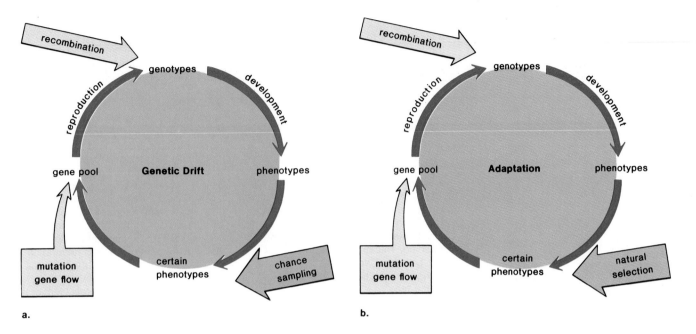

a.

b.

Genetic drift operates in both large and small populations, but it is significant only in small populations. Imagine that by chance a small number of individuals, representing a fraction of the gene pool, found a colony and then, during gamete formation, only certain of their genes are passed on to the next generation. As a result, a severe reduction in genetic variation compared to the original population has taken place and genetic drift has occurred. This combination of circumstances has been historically observed and is called the *founder principle*. For example, an investigation of a small religious group, called the Dunkers, showed that the blood type was 60 percent blood group A. Since the frequency of this blood group is 40 percent in the United States and 45 percent in West Germany (the country from which the Dunkers are derived), the high occurrence of blood group A among the Dunkers can only be explained by drift.

Genetic drift is *not* expected to produce adaptation to the environment because phenotypes are not selected for reproduction; rather, chance alone determines who will reproduce. We can imagine that a natural disaster, for example, would severely reduce a large population so that only a few individuals would remain to reproduce. Only those genes that happened to be passed on to the next generation would then be available in the gene pool. As gene pool variation decreases, the possibility of fixation of a few genes and thus genetic drift increases.

Natural selection, as diagrammed in figure 23.12b, is the process by which populations become better adapted to their environment. Individuals that are more suited to the environment are the very ones that are more likely to survive and reproduce. In this way adaptation of the species to the environment tends to increase with each generation. Adaptations to various environments explains the diversity of life.

Darwin and the Theory of Natural Selection

Charles Darwin was born in England in 1809. All his life he loved nature. As a young man, he was an ardent naturalist, walking the fields and woodlands of the countryside, studying plants and animals and observing rock formations. At school, he preferred to associate himself with professors of biology and geology even though his father wanted him to be either a physician or a clergyman.

In 1831, Darwin agreed to serve as ship's naturalist on the H.M.S. Beagle, a ship commissioned to spend five years on a survey mission around the world because he saw this as an opportunity to study organisms in other parts of the world. While aboard ship, he read a book by Charles Lyell called *Principles of Geology*. Lyell gave evidence that the earth was not static and unchanging; rather, it was subject to continuous cycles of erosion and uplifting. This made Darwin wonder if the biological world is also dynamic rather than static as people believed at the time.

During the voyage, Darwin observed that similar environments about the globe contained different types of organisms. He wondered if this wasn't possibly because they were geographically separated. He also found and studied various fossils that made him think that organisms do change over time. In the Galápagos Islands off the coast of South America he found 13 species of finches whose adaptations could be explained if one assumed they had descended from a common ancestor. These and other observations encouraged Darwin to believe that each species becomes adapted or suited to its environment with time.

When Darwin arrived home, he spent the next 20 years gathering data to support a principle of organic evolution. His most significant contribution to this principle was his theory of *natural selection,* which explains how a species becomes adapted to its environment. Before formulating the theory, he read an essay on human population growth written by Thomas Malthus. Malthus observed that although the human population has a tendency to increase in size, factors like fire, war, plague, and famine serve to keep the population more or less in check. If it were not for these, the human population would increase explosively. Darwin could see that all populations of organisms have the same potential. He calculated that a single pair of elephants could have 19 million descendants in 750 years.

Other organisms have even greater reproductive potential than this pair of elephants; yet usually the number of each type organism remains about the same. This is because most offspring are eaten by predators, succumb to disease, or are unable to obtain adequate amounts of food or a place to live. Darwin decided there is a constant *struggle for existence,* and only a few survive to reproduce. The ones that do survive, and contribute to the evolutionary future of the species are, by and large, the better adapted individuals. This so-called "survival of the fittest" causes the next generation to be better adapted than the previous generation.

Darwin's theory of natural selection was nonteleological. Organisms do not strive to adapt themselves to the environment. Rather, the environment acts on them to select those individuals that are best adapted. These are the ones that have been "naturally selected" to pass on their characteristics to the next generation. In order to emphasize the nonteleological nature of Darwin's theory, it is often contrasted to Lamarck's theory of *acquired characteristics.* Lamarck was also a 19th-century naturalist. Lamarck thought that as the species strove toward perfection, new characteristics were acquired in response to a need imposed by the environment. For example, the Lamarckian explanation for the long neck of the giraffe is that elongation occurred because the ancestors of the giraffe needed to reach into the trees to feed on high-growing vegetation (fig. 23.A). This repeated stretching of the neck over time caused the neck to vary and become longer. This acquired characteristic was passed on to the next generation and in this way the species became adapted to the environment. Lamarck's theory is teleological because, according to him, the species does shape its own future.

Evolution and Diversity

a. Lamarck's Theory

Early giraffes probably had short necks which they stretched to reach food.

Their offspring had longer necks which they stretched to reach food.

Eventually the continued stretching of the neck resulted in today's giraffe.

Existing data do not support this theory.

b. Darwin's Theory

Early giraffes probably had necks of various lengths.

Competition and natural selection led to survival of the longer-necked giraffes and their offspring.

Eventually only long-necked giraffes survived the competition.

Existing data support this theory.

Darwin's theory of natural selection, along with numerous observations and examples, appeared in his book *The Origin of Species,* which was published in 1859. The principles of genetics, which were developed soon after, support Darwin's theory of natural selection rather than Lamarck's theory of acquired characteristics. Lamarck's theory suggests that the phenotype controls the genotype, and modern genetics has shown that the genotype controls the phenotype. Changes in the genotype can bring about variations in the phenotype. Now these variations are "tested" by the environment so that a larger proportion of the better adapted individuals have offspring than do those that are not as well adapted. Whereas Darwin emphasized the importance of survival, modern evolutionists emphasize the importance of reproductive success.

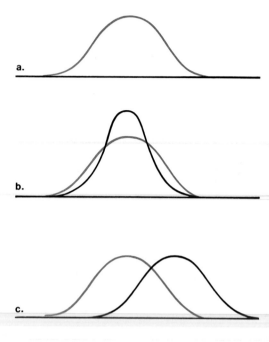

a.

b.

c.

Table 23.3 Maintenance of variation	
Genetic causes	**Ecological factors**
Diploidy	Opposing ecological pressures
Heterozygote superiority (heterosis)	Geographic variations
Pleiotropy	Changing environment

Natural selection operates by means of both biotic and abiotic factors that influence survival and reproduction. Biotic factors involve other organisms in the environment and abiotic factors involve physical conditions. So, for example, an organism that is better able to escape a predator would be selected, that is, would be more likely to have offspring. Similarly, an organism better able to withstand the climate would be selected and again would be more likely to have offspring. "Survival of the fittest," a phrase often used to refer to natural selection, means, according to modern thinking, that a better adapted genotype has a greater *probability* of producing offspring than a less adapted one.

Figure 23.9b illustrates that natural selection brings about a change in frequency of genes in the gene pool. This occurs when better adapted individuals increase their contribution to the gene pool and less well adapted individuals decrease their contribution. As this process continues generation after generation, the gene pool naturally tends toward stabilization.

Figure 23.13 indicates that natural selection has two common effects: (*a*) to stabilize variations and (*b*) to direct variations. **Stabilizing selection** tends to eliminate atypical phenotypes and enhances adaptation of the population to current environmental circumstances, but **directional selection** selects an extreme phenotype better adapted to a new environmental circumstance. Directional selection occurs during a time when the environment is changing rapidly or when members of a population are adapting to a new environmental situation.

Maintenance of Variations

Even though stabilization of the gene pool is an expected result of the evolutionary process, genetic and phenotypic variation may still be retained for reasons such as those listed in table 23.3.

Diploidy helps maintain variation because recessive genes may remain hidden in the gene pool, serving as a potential source of future phenotypic variations. Pleiotropic genes may produce variations aside from those that adapt the organism to a current environment. Such phenotypic variations may be neutral or harmful to the organism.

The environment may actually promote the maintenance of two distinctly different phenotypes due to opposing selection pressures. In Africa, persons with normal red blood cells carry oxygen more efficiently but are more susceptible to malaria, while those with sickle-shaped cells escape malaria but die from sickle-cell anemia. Clearly the heterozygote is best adapted, but in order to perpetuate it, both homozygotes must be maintained. Cases such as this are called *balanced polymorphism* (*poly* = many; *morphism* = shape).

The environment itself may vary and therefore may call for different adaptations at different times, such as seasonal changes within certain environments. Members of a population may become generally adapted, never specializing for any particular season, or they may become *polymorphic*, having a different phenotype for each season. Polymorphic adaptation is demonstrated by the arctic hare, which has a white coat in the winter and a brown coat in the summer.

Examples of Natural Selection

While natural selection usually takes many hundreds or even thousands of years to produce a noticeable change in the phenotype, there are a few examples of rapid adaption. Also, humans carry out rapid artificial selection in order to produce a certain phenotype.

Industrial Melanism

Before the industrial revolution in England, collectors of a moth called the peppered moth (fig. 23.14) noted that most moths were light colored, although occasionally a dark-colored moth was captured. Several decades after the industrial revolution, however, the black moths made up 99 percent of the moth

Evolution and Diversity

population in polluted areas. An explanation for this rapid change can be found in natural selection. The color of the moths, dark or light, is caused by their genetic makeup; black is a mutation that occurs with some regularity. Moths rest on the trunks of trees during the day; if they are seen there by predatory birds, they are eaten. As long as the trees in the environment are light in color, the light-colored moths live to reproduce. But once the trees turn black due to industrial pollutants, natural selection enables the dark moths to avoid being eaten and to survive and reproduce; thus, the black phenotype becomes the more frequent one in the population. This explanation has been supported by experiments in which both dark- and light-colored moths were released into industrial and nonindustrial areas. In the industrial areas, the light moths suffered more attrition; in the nonindustrial areas, the dark moths did not survive. This shows that the phenotype most adapted to the environment is the one that is preserved in nature. Industrial melanism has also been noted in the United States; around major cities the insects have taken on a darker color than in the nonpolluted countryside.

Antibiotic Adaptation

Since the introduction of antibiotics, it has been noted that various strains of bacteria have become resistant to them, thus encouraging continued research to find new and different antibiotics. Experiments have shown that bacterial resistance is also an example of natural selection. In one such experiment, depicted in figure 23.15, bacteria are grown on growth medium with and without the antibiotic streptomycin. Only a few bacteria survive on the streptomycin, but these few can grow abundantly when plated on either normal or streptomycin medium. These results favor this explanation: the few bacteria

Figure 23.15
Natural selection among bacteria exposed to antibiotics. *a.* The circle shows that the bacteria are capable of growing on normal medium. *b.* Dots show that only certain bacteria can grow on streptomycin medium. *c.* Those from (*b.*) can still grow on normal medium but can also grow abundantly on streptomycin medium. It is obvious that the adaptive mutation occurred prior to growth on streptomycin.

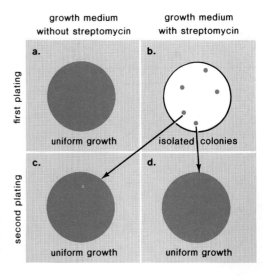

Figure 23.16
Comparison of an old variety of wheat (in background) with a new highly productive dwarf wheat (foreground). The dwarf wheat was produced under the direction of Norman Borlaug by crossing wheat varieties from around the world. Japanese varieties gave these plants their short stature, which prevents them from toppling over even though they produce excessive amounts of grain. Mexican varieties gave them their ability to grow under various conditions.

Figure 23.16
Comparison of an old variety of wheat (in background) with a new highly productive dwarf wheat (foreground). The dwarf wheat was produced under the direction of Norman Borlaug by crossing wheat varieties from around the world. Japanese varieties gave these plants their short stature, which prevents them from toppling over even though they produce excessive amounts of grain. Mexican varieties gave them their ability to grow under various conditions.

that can grow on streptomycin possess a gene that mutated before they were exposed to the antibiotic and that enables them to grow in what would otherwise be a hostile environment. These bacteria now give rise to a progeny that are adapted to the new environment.

DDT Adaptation

Similar to bacterial resistance to antibiotics, insects have become resistant to DDT. This example, too, can be explained by the fact that the few insects capable of surviving in an environment of DDT pass on this resistance to the next generation. Since the parents were resistant, the next generation is completely resistant.

Artificial Selection

Human beings have long acted as a selective force to bring about desired phenotypes in domestic plants and animals. Cattle, poultry, hogs, sheep, and goats have all been bred to have the characteristics that humans find most beneficial, as have house pets. High-yield crops have also been produced by employing a method that relies on **hybrid vigor.** Two inbred parental strains of little value can sometimes give a hybrid that has many desirable traits, including high yield, disease resistance, and proper growth habits. Hybrid seeds are used in the planting of wheat (fig. 23.16), rice, and rye, and these have been important in the world's effort to feed the growing masses of human beings.

Origin of Species

For our present discussion, we are defining a *species* as a group of interbreeding populations that share a gene pool that is reproductively isolated from other species. The populations that belong to the same species are spread over a certain geographic range. If the geographic range of a species is large, more phenotypic variations between populations are apt to be seen than if the geographic range is small. For example, differences in the song sparrows pictured in figure 23.17 can be accounted for in that this species, *Melospiza melodia,* has a range that extends across the United States from the east to the west coast. Not only do the birds differ anatomically, they also sing a slightly different song. However, as long as there is gene flow, or the movement of genes

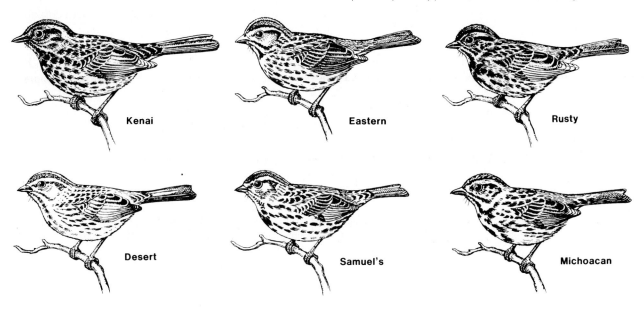

| Kenai | Eastern | Rusty |
| Desert | Samuel's | Michoacan |

Table 23.4   Reproductive isolating mechanisms

Isolating mechanisms	Example
Premating	
Habitat	Species at same locale occupy different habitats
Temporal	Species mate at different seasons or different times of day
Behavioral	In animals, courtship behavior differs or they respond to different songs, calls, pheromones, or other signals
Mechanical	Genitalia unsuitable to one another
Postmating	
Gametic mortality	Sperm cannot reach or fertilize egg
Zygote mortality	Hybrid dies before maturity
Hybrid sterility	Hybrid survives but is sterile and cannot reproduce
F_2 fitness	Hybrid is fertile but F_2 hybrid has lower fitness

from one population to another as a consequence of the immigration of individuals from one population to another, it is possible to regard these sparrows
as belonging to one species. Sometimes, if interbreeding between populations
is possible but rarely occurs, taxonomists assign subspecies or race designa-
tions to populations by giving them a third name in addition to their normal
binomial name.

The populations of the same species exchange genes, but different species do not exchange genes. Reproductive isolation of the gene pools of similar
species is accomplished by such mechanisms as those listed in table 23.4. **Premating isolating mechanisms** are those that prevent intercourse from ever taking place, and **postmating isolating mechanisms** are those that prevent hybrid
offspring from developing or breeding if reproduction is attempted. In evolutionary terms, a hybrid is an offspring of individuals who belong to populations that do not normally reproduce with one another.

It is important to realize at this point that there are two ways to prevent
members of two species from reproducing with one another. Some species are
prevented from reproducing simply because they are geographically isolated
from one another. These species are said to be **allopatric species.** Other species
do not reproduce with one another even though they are present in the same
locale. These species are said to be **sympatric species.**

Figure 23.18

Allopatric speciation. *a.* A hypothetical species has many populations, each one represented by a single rabbit. *b.* Geographic isolation occurs, and gene flow between all populations is not possible. *c.* Divergent evolution occurs because the populations represented by white rabbits and those represented by black rabbits are exposed to different selective pressures. *d.* When geographic isolation has ended, the populations represented by black rabbits are not able to reproduce successfully with those represented by white rabbits. This shows we are now dealing with two different species.

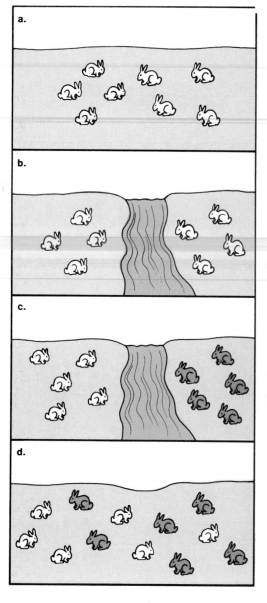

Speciation

One species can give rise to two different species and when this happens, speciation has occurred. There are two criteria by which we can recognize that speciation has occurred. The two criteria are, of course, the presence of structural differences and the failure to reproduce successfully in the same natural setting. It is now generally accepted that in most instances *speciation is a two-step process: geographic isolation is later followed by reproductive isolation.*

The two-step process for speciation is illustrated in figure 23.18. Initially, all populations of the species live in the same environment and share a common gene pool. In our example, this is exemplified by giving all populations, each represented by a rabbit, the same color. Then let us suppose that a canal is dug to divert water from a nearby source and that the canal separates our populations of rabbits into two groups so that they are now geographically isolated (fig. 23.18b). Therefore, the first step of speciation has occurred and for the sake of clarity, we will refer to the two sets of populations as subspecies (represented in our example by coloring the two sets of populations different colors (fig. 23.18c).

Once geographic isolation has occurred, the subspecies begin to undergo divergent evolution. There are three reasons for this. (1) First, the subspecies have different gene pools. This is obvious when we consider that the original populations had slight differences. Since gene flow between certain populations is now prevented, the genes for these structural differences have a greater chance of being passed on to the next generation. (2) Each new gene pool will

a.

b.

c.

Evolution and Diversity

now separately be subject to the normal increase in variation caused by mutation and recombination of genes during gamete formation. (3) The subspecies have different environments and therefore experience different selective pressures. Consequently, natural selection will cause the subspecies to diverge further genetically and phenotypically from one another. Given enough time, this divergence will eventually result in reproductive isolation so that even if the geographic barrier is removed, the two population systems cannot successfully reproduce with one another. In our example, we can imagine that the canal was so destructive to the environment that it was decided that the land should be restored as it was previously (fig. 23.18d). Now, however, gene flow between the two population sets does not occur and the second step of speciation has occurred; what was formerly one species has become two species.

Adaptive Radiation

One of the best examples of speciation is provided by the finches on the Galápagos Islands, which are very often called Darwin's finches because Darwin (p. 514) first realized their significance as an example of how evolution works. The Galápagos Islands (fig. 23.19) located 600 miles west of Ecuador, South America, are volcanic but do have forest regions at higher elevations. The 13 species of finches (fig. 23.20), placed in three genera, are believed to be descended from mainland finches that migrated to one of the islands some years ago. Thus, Darwin's finches are an example of *adaptive radiation,* or the proliferation of a species by adaptation to different ways of life. We can imagine

Figure 23.19
Location of the Galápagos Islands. These islands are close enough to South America that finches could have come there from the mainland. Once they arrived, the original population on one island is presumed to have spread out to the other islands where varying selective pressures would have caused divergent evolution to occur. With time, the various populations of finches were unable to reproduce with one another.

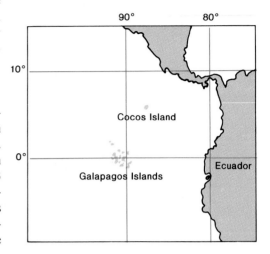

Figure 23.20
The ancestral form from which the Galápagos finches are descended most likely had a strong beak capable of crushing seeds. Each of the present-day thirteen species of finches is adapted to a particular way of life. *a.* The small tree finch has a delicate bill and eats aphids and small berries. *b.* The large tree finch grinds fruit and insects with a parrotlike bill. *c.* The small ground finch has a pointed bill and eats tiny seeds and ticks picked from iguanas. *d.* The large ground finch has a conical bill that enables it to eat big, hard seeds. *e.* The cactus finch has a long bill that probes for nectar in cactus flowers. *f.* The woodpecker finch has a stout, straight bill that chisels through tree bark to uncover insects, but because it lacks a woodpecker's long tongue it uses a tool—usually a cactus spine or a small twig—to ferret insects out.

e.

d.

f.

Figure 23.21
Adaptive radiation among reptiles and mammals.
Both reptiles and mammals evolved into the types
of forms shown. However, the mammals did not
begin their adaptive radiation until the reptiles had
declined. Thus each group shows similar
adaptations because they were subjected to the
same selective pressures.

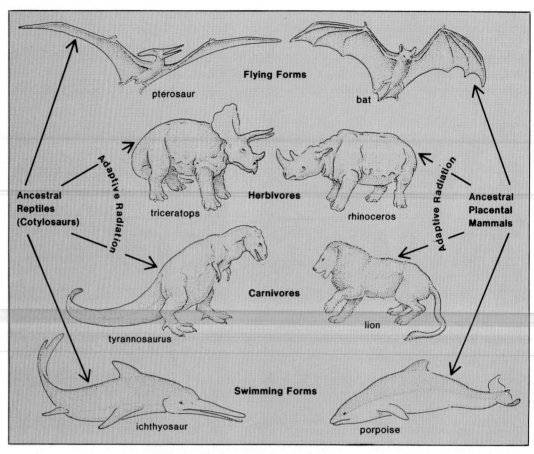

that after the original population of a single island increased, some individuals dispersed to other islands. The islands are ecologically different enough to have promoted divergent feeding habits. This is apparent because the birds, although physically resembling each other in many respects, have different bills, each of which is adapted to a particular food-gathering method. There are seed-eating ground finches with bills appropriate to cracking small, medium, or large size seeds; insect-eating tree finches, also with different size bills; and a warbler-type finch with a bill adapted to nectar gathering. Among the tree finches there is a woodpecker type, which lacks the long tongue of a true woodpecker but who makes up for this by using a cactus spine or twig to ferret out insects.

Most likely as the populations on the various islands began to assume their own evolutionary pathways, they began to develop postmating isolating mechanisms. When secondary contact occurred between the original and the derived populations, premating mechanisms would subsequently have been selected for.

Higher Taxonomic Categories

Adaptive radiation is observed not only among species but among higher taxonomic categories also. Both the reptiles and mammals underwent large-scale adaptive radiation as they became adapted to different ways of life, as illustrated in figure 23.21. Because the mammals began their radiation after many

Figure 23.22
Sizes of vertebrate groups during past time
periods. Width of geometric shapes indicates
abundance of species, and dashed lines indicate
unknown ancestral forms. Amphibians underwent
adaptative radiation before reptiles but by now
they have both declined. Adaptative radiation of
birds and mammals extends until the present.

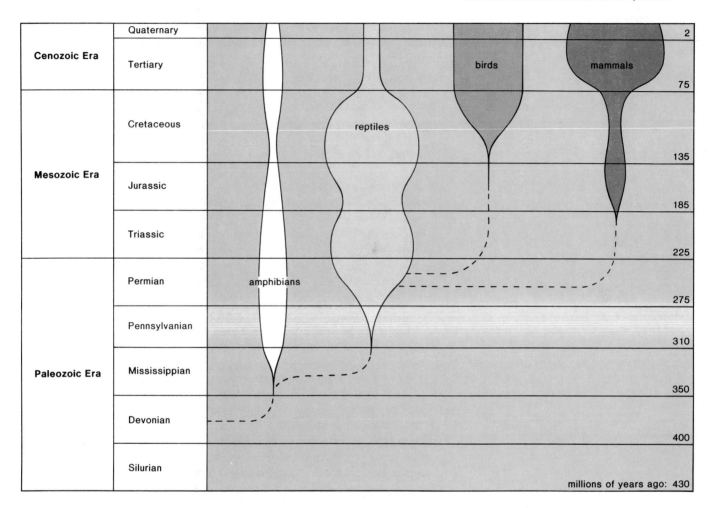

reptiles had become extinct (fig. 23.22), they were exposed to the same selection pressures and the resulting adaptations were similar. This is also an example of *convergent evolution*. Therefore figure 23.21 can be used to illustrate
both adaptive radiation and convergent evolution.

Extinction

Extinction can help explain the origin of higher taxonomic categories. As related species become extinct, wider and wider gaps are observed until it becomes obvious that the ones remaining should be placed in entirely different
taxonomic categories.

Summary

Evolution explains the history and diversity of life. Evidences for evolution
can be taken from the fossil record, comparative anatomy, embryology, and
biochemistry. Also, vestigial structures and biogeography support the occurrence of evolution. Taxonomists classify organisms according to their evolutionary relationships, first into species, then into genera, orders, classes, phyla

(animals) or divisions (plants), and finally kingdoms. Each organism is given a binomial name composed of its genus and species.

Although evolution can be studied on a grand scale called macroevolution, the process of evolution is better understood by studying microevolution because actually evolution is believed to take place by means of a gradual change in the frequency of genes in the gene pool of a population. Such change continuing for a vast period of time (millions of years) will produce large changes. The Hardy-Weinberg law, which is a fundamental statement of population genetics, says that the frequency of different alleles in a population remains constant if (1) the population is large and mating is random, (2) no mutations occur, (3) there is no gene flow, and (4) there is no natural selection. These conditions are rarely if ever met. The Hardy-Weinberg law shows that dominance of a gene does not lead to its increase and recessiveness of a gene does not lead to its disappearance.

Evolution is a two-step process requiring the production of genotypic variations and a sorting out of these variations. Genetic variation in a population of sexually reproducing diploid organisms has three sources: mutation (both chromosomal and gene), gene flow (emigration or immigration), and recombination (at the time of meiosis or fertilization). Although it is possible and sometimes useful to consider the inheritance of individual genes, it is the entire genotype that produces the phenotype. Sexual recombination can produce variation because new combinations of genes come together by the process of meiosis and random union of gametes.

Genetic drift and natural selection act in such a way that gene pool variations are reduced, but only natural selection consistently results in adaptation. Genetic drift occurs when by chance a few genotypes contribute inordinately to the next generation's gene pool. Genetic drift is particularly effective when a small population breaks away from a large population. This is called the founder principle.

Natural selection is brought about by the increased probability of survival and reproduction of the better adapted members of a population. The phrase "survival of the fittest" simply means that better adapted individuals are more likely to produce offspring than poorly adapted individuals. Adaptation to particular environments accounts for the diversity of life. It has been possible to witness the process of natural selection in several recent instances: the adaptation of moths to polluted areas, the adaptation of bacteria to modern drugs, and the adaptation of insects to pesticides.

Evolution has two possible primary effects: (1) to stabilize population variations and (2) to direct population variations. Eventually, stabilization always occurs, except that genetic variations are maintained because of genetic and ecological reasons (table 23.3).

Speciation is the origin of species and this usually requires geographic isolation followed by reproductive isolation. One frequently cited example of speciation is the evolution of several species of finches on the Galápagos Islands. This is also an example of adaptive radiation into unfilled niches, as was the evolution of many types of reptiles and mammals on a grander scale. The adaptive radiation of the latter two groups is also an example of convergent evolution because each group evolved to fill similar niches.

The extinction of intermediate species can help explain the wide separation we now observe between major groups of animals.

Evolution and Diversity

Study Questions

1. Show that the fossil record, comparative anatomy, vestigial structures, comparative embryology, and comparative biochemistry all give evidence that evolution has occurred. (pp. 503–8)
2. What is an evolutionary tree and how are evolutionary trees constructed? (p. 507)
3. What is the Hardy-Weinberg law? (pp. 510–11) If genotype *gg* is found in 16 percent of a population, what is the frequency of the *g* allele? The *G* allele? What proportion of the next generation will be *gg* if the law holds? (p. 510)
4. Name and describe the sources of gene pool variations in a population made up of diploid sexually reproducing individuals. (p. 512)
5. What factors prevent the Hardy-Weinberg law from operating? (pp. 511–12)
6. Discuss the concept of the "unity of the genotype." (p. 512)
7. Name and contrast two processes that reduce gene pool variations. (pp. 512–16)
8. Name several reasons for the maintenance of variation in a gene pool. (p. 516)
9. Give three modern examples of natural selection. (pp. 516–18)
10. Define a species. How do new species originate? (pp. 518–21)
11. When are adaptive radiation and convergent evolution apt to take place? (pp. 521–22)

*Answers to Practice Problems

1. 60 percent *Ll* and 40 percent *ll*   2. recessive allele = 0.2 and dominant allele = 0.8; homozygous dominant = 0.64; homozygous recessive = 0.04; heterozygous = 0.32   3. 29 percent

Selected Key Terms

fossils (fos''lz)
binomial system (bi-no′me-al sis′tem)
taxonomy (tak-son′o-me)
common ancestor (kŏ′mun an′ses-tor)
homologous (ho-mol′o-gus)
vestigial (ves-tij′e-al)
biogeography (bi''o-je-og′rah-fe)
population (pop''u-la′shun)
gene pool (jēn pool)
gene flow (jēn flo)
recombination (re''kom-bĭ-na′shun)
evolution (ev''o-lu′shun)
variations (va''re-a′shunz)
genetic drift (jĕ-net′ik drift)
natural selection (nat′u-ral sĕ-lek′shun)
isolating mechanism (i′so-lāt-ing mek′ah-nizm)
speciation (spe''se-a′shun)
adaptive radiation (ah-dap′tiv ra''de-a′shun)
convergent evolution (ken-ver′jent ev''o-lu′shun)
extinction (eks-ting′shun)

Further Readings

Cavalli-Sforza, L. L. 1969. "Genetic drift" in an Italian population. *Scientific American* 221(2):30.

————. 1974. The genetics of human populations. *Scientific American* 231(3):80.

Dickerson, R. E. 1972. The structure and history of an ancient protein. *Scientific American* 226(4):58.

Dodson, E. O., and Dodson, P. 1976. *Evolution: Process and product.* 2d ed. New York: D. Van Nostrand.

Ehrlich, P. R., et al. 1974. *The process of evolution.* 2d ed. New York: McGraw-Hill.

Grant, V. 1977. *Organismic evolution.* San Francisco: W. H. Freeman Co.

Laporte, L. F., ed. 1978. *Evolution and the fossil record: Readings from* Scientific American. San Francisco: W. H. Freeman.

Lerner, I. M., and Libby, W. J. 1976. *Heredity, evolution and society.* 2d ed. San Francisco: W. H. Freeman.

Mayr, E. 1970. *Populations, species and evolution.* Cambridge, Mass.: Belknap Press.

Scientific American. 1978. 239(3), Entire issue devoted to evolution.

Stebbins, G. L. 1977. *Processes of organic evolution.* 2d ed. Englewood Cliffs, N.J.: Prentice-Hall.

Volpe, E. P. 1977. *Understanding evolution.* Dubuque, Iowa: Wm. C. Brown Publishers.

Wallace, B., and Srb, A. M. 1978. *Adaptation.* Englewood Cliffs, N.J.: Prentice-Hall.

24

Origin of Life

Chapter concepts

1 It is now possible to suggest that the first cell or cells arose by a slow process of chemical evolution.

2 It is generally accepted that the first cell was a heterotroph that utilized anaerobic respiration.

3 Autotrophic nutrition made life on land possible by releasing free oxygen into the atmosphere.

4 Although it is believed that at one time life did originate by chemical evolution, this process does not occur today. Today, life comes only from life.

Figure 24.1

A model for the origin of life. *a.* When the earth was formed, atoms sorted themselves out according to weight. *b.* The primitive atmosphere contained the gases hydrogen, methane, ammonia, and water vapor; as the latter cooled, some gases were washed into the ocean. *c.* The availability of energy, represented here by ultraviolet rays and volcanic eruption, allowed gases to form simple organic molecules that (*d.*) reacted to form macromolecules in the ocean (*e.*). After autotrophs arose, aerobic respiration became possible.

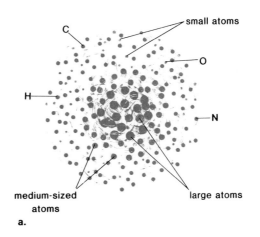

a.

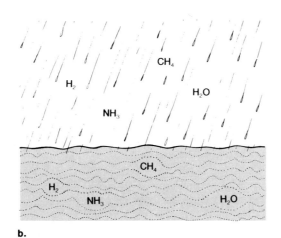

b.

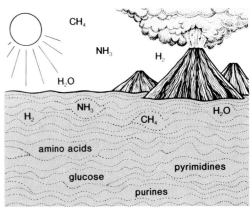

c.

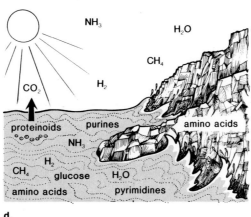

d.

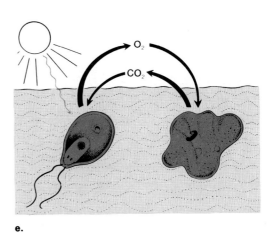

e.

oday we don't believe that life arises spontaneously from nonlife, and we say that "life comes only from life." But if this is so, how did the first form of life come about? First, we could assume that the first form of life was very simple; for example, a single cell (or cells). As soon as this cell could grow, reproduce, and mutate, it could be said to be alive. Since it was the very first living thing, it had to come from nonliving chemicals. In fact, it is possible that slow progression of chemicals from the simple to the complex finally resulted in a live cell. In other words, a **chemical evolution** produced the first form of life. Table 24.1 lists the steps that could have occurred to produce life. The evidence for these steps is based on our knowledge of the primitive earth and on experiments that have been performed in the laboratory.

Chemical Evolution

When the earth was first formed about 5 billion years ago, it was a glowing mass of free atoms (fig. 24.1), which sorted themselves out according to weight. The heavy ones, such as iron and nickel, sank toward the center of the earth; the lighter atoms, such as silicon and aluminum, formed the middle shell; and the very lightest atoms, hydrogen, nitrogen, oxygen, and carbon, may have collected on the outside. The temperature was so hot that atoms could not permanently bind together; whenever bonds formed, they were quickly broken.

As the earth cooled, the heavy atoms tended to liquefy and solidify, but the intense heat at the center prevented complete solidification, and even today the earth contains a hot, thickly flowing molten core. In the middle shell, the lighter atoms congealed and formed the outer surface of the earth, the so-called crust. Cooling may also have allowed the first atmosphere to form.

Primitive Atmosphere

There are two current hypotheses about the origin of the primitive atmosphere. One hypothesis suggests that the gases of the primitive atmosphere came about when cooling allowed the lightest of the atoms to react with one another. Since hydrogen was the most abundant of these atoms, it combined with itself and with carbon, oxygen, and nitrogen to form hydrogen gas (H_2), methane gas (CH_4), water vapor (H_2O), and ammonia vapor (NH_3). An abundance of hydrogen atoms would have caused the first atmosphere to be a highly reducing atmosphere (p. 25).

The second hypothesis is also compatible with the steps listed in table 24.1. This hypothesis suggests that the gases of the primitive atmosphere were released from volcanic eruptions. Further, it is believed that while the atmosphere may have been mildly reducing, it probably also contained carbon dioxide (CO_2) and nitrogen gas (N_2).

Both hypotheses support the contention that the first atmosphere contained little or no free oxygen. Therefore, the first atmosphere is believed to have lacked O_2.

Simple Organic Molecules

Water, present at first as vapor in the atmosphere, formed dense, thick clouds, but cooling eventually caused the vapor to condense to liquid, and rain began to fall. This rain was in such quantity that it produced the oceans of the world. The gases, dissolved in the rain, were carried down into the newly forming oceans. The remaining steps shown in table 24.1 took place in the sea, where life arose.

Table 24.1   Origin of the first cell

Primitive earth

↓ **Cooling**

Gases

↓ **Energy capture**

Small molecules: amino acids, glucose, nucleotides

↓ **Polymerization**

Macromolecules: proteins, carbohydrates

↓ **Anaerobic respiration**

Protocell: heterotrophic fermenter

↓ **Nucleic acid formation**

Cell: reproduction

↓ **Evolution**

Autotrophic cell: gives off oxygen

↓ **Aerobic respiration**

Animal-like heterotroph

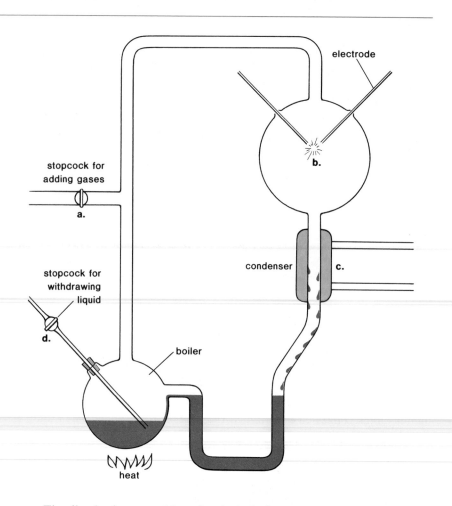

The dissolved gases, although relatively inert, are believed to have reacted together to form simple organic compounds when they were exposed to the strong outside **energy sources** present on the primitive earth. These energy sources included heat from volcanoes and meteorites, radioactivity from the earth's crust, powerful electric discharges in lightning, and solar radiation, especially ultraviolet radiation. In a classic experiment (fig. 24.2), Stanley Miller showed that an atmosphere containing methane and ammonia could have produced organic molecules. These gases were dissolved in water and circulated in a closed container past an electric spark. After a week's run, he analyzed the contents of the reaction mixture and found, among other organic compounds, amino acids and nucleotides. Other investigators have achieved the same results by utilizing carbon monoxide and nitrogen gas dissolved in water.

These experiments indicate that the primitive gases not only could have but probably did react with one another to produce simple organic compounds that accumulated in the ancient seas. Neither oxidation (there was no free oxygen in the ancient atmosphere) nor decay (there were no bacteria) would have destroyed these molecules, and they would have accumulated in the oceans for hundreds of millions of years. With the accumulation of these simple organic compounds, the oceans became a thick, hot **organic soup,** containing a variety of organic molecules.

Macromolecules

These organic molecules combined to form still larger molecules and macromolecules. Perhaps this came about by a chance combination of molecules in the ocean, or perhaps some of the smaller molecules were washed ashore where dry heat would have encouraged polymerization.

Evolution and Diversity

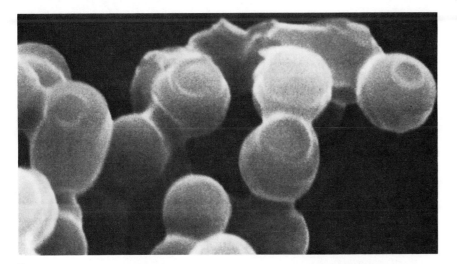

a.

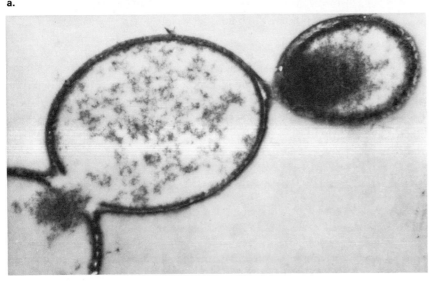

b.

Figure 24.3
Microspheres. *a.* Scanning electron micrograph of the exterior of proteinoid microspheres, which may be similar to the structure of the protocells. *b.* Electromicrograph of the interior of protein microspheres.

Sidney Fox of the University of Miami supports the idea that amino acid polymers[1] were the first macromolecules to form. In various experiments, he has shown that amino acids will combine in a preferred order when exposed to dry heat. Presumably, amino acids collected in shallow puddles along the shore, and the sun caused polymerization to occur as drying took place.

Other investigators believe that nucleic acid may have formed first in exactly the same manner. These scientists point out that since DNA is the genetic material, it is logical that it formed first.

Biological Evolution

Protocells

Fox has shown that when amino acid polymers are exposed to water they form **proteinoid microspheres** (fig. 24.3), which have many properties similar to today's cells (table 24.2). Such microspheres may have been cell precursors, called **protocells.** He feels that protocells could have then evolved to contain the macromolecules characteristic of true cells. He calls this a **cell-first hypothesis,** meaning that the protocell came fist—before true proteins and nucleic acids.

[1]Fox recently suggests that these polymers should not be called proteins because they do not have all the characteristics of true cellular proteins.

Table 24.2 Proteinoid microparticles possess many properties similar to contemporary cells

Stability (to standing, centrifugation, sectioning)
Microscopic size
Variability in shape but uniformity in size
Numerousness
Stainability
Ultrastructure (electron microscope)
Double-layered boundary
Selective passage of molecules through boundary
Catalytic activities
Patterns of association
Propagation by "budding" and fission
Growth by accretion
Motility
Propensity to form junctions and to communicate

Figure 24.4
Coacervates are polymer-rich colloidal droplets. It is
theorized by some that coacervates could have
been protocells.

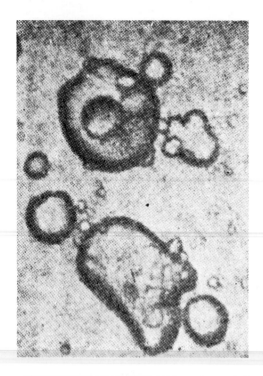

Other investigators believe that protocell formation may have required
the presence of various macromolecules aside from amino acid polymers. Some
researchers support the theoretical work of Oparin, who wrote a definitive book
on the topic in 1938. Oparin pointed out that, under appropriate conditions
of temperature, ionic composition, and pH, concentrated mixtures of macromolecules tend to give rise to complex units called **coacervate** droplets (fig.
24.4). Coacervate droplets have a tendency to adsorb and incorporate various
substances from the surrounding solution. Eventually, a semipermeable-type
boundary may form about the droplet.

Heterotroph Hypothesis   We might now ask, How did the protocell carry on
nutrition and respiration? Nutrition would have been no problem because the
protocell lived in the "organic soup," which contained simple organic molecules that would have been its food. Thus, the protocell would have been a
heterotroph, an organism that takes in preformed food. Notice, too, that heterotrophs are believed to have preceded autotrophs, organisms that make their
own organic food.

In regard to respiration, the protocell must have carried on **anaerobic
respiration,** or **fermentation,** which does not require free oxygen:

$$glucose \longrightarrow alcohol \text{ and/or acids} + CO_2 + energy$$

The protocell would have been capable of growing since it carried on
nutrition and respiration; but without genes, it could not have mutated and
evolved. However, we can theoretically assume that the protocell may have
incorporated nucleotides or nucleic acids that eventually formed genes. With
this development the protocell would have become a **cell** capable of passing
on genetic traits by means of cell division.

True Cells

The first cells would have been anaerobic heterotrophs just like the protocells.
There are anaerobic bacteria living today that may resemble very closely the
first true cells. Other evolutionary lines, however, produced the autotrophs.

Autotrophs   The oldest microfossils (microscopic fossils) are believed to be
those of autotrophs. They are the remains of giant colonies of bacteria found
in a rock dated 3.5 billion years ago. The rock was found during a geological
expedition in Western Australia (fig. 24.5).

When bacteria photosynthesize, they do not give off oxygen. Therefore
further evolution was required before there were autotrophs similar to today's
algae. These autotrophs would have possessed chlorophyll *a* and would have
photosynthesized and made their own food by using carbon dioxide and water:

$$\begin{array}{ccc} carbon\ dioxide & energy & glucose \\ + & \xrightarrow{\hspace{3cm}} & + \\ water & from\ the\ sun & oxygen \end{array}$$

The evolution of autotrophs was critical because the preformed organic
molecules in the ocean no doubt would have been running out. The newly
evolved autotrophs would have provided a much needed continual source of
food. These autotrophs also made aerobic respiration possible because they
put oxygen gas into the atmosphere.

Aerobic Respiration   The presence of oxygen in the atmosphere permitted
aerobic respiration:

$$oxygen + glucose \longrightarrow carbon\ dioxide + water + energy$$

Figure 24.5
Field researchers examine rocks near North Pole,
Australia, for evidence of the earth's earliest forms
of life.

Evolution and Diversity

It also permitted these plants and animals to invade the land because oxygen in the upper atmosphere forms ozone (O_3), which filters out the ultraviolet rays of the sun. Before the formation of this so-called **ozone shield,** the amount of radiation would have destroyed land-dwelling organisms. There is concern today that the ozone shield may be in danger of breaking down due to air pollutants, particularly nitric oxides and chlorine, that are capable of reacting with ozone. Nitric oxides are given off by jet planes, and chlorine is released from an aerosol propellant called freon. Although freon is no longer used in this country, it is still used in other countries around the world. It has been suggested that a 10 percent annual increase in the use of aerosols could reduce the ozone layer by 10 percent in 20 years and by 40 percent by the year 2014.

Origin of Life Today

We have shown that life could have come into existence by means of a chemical evolution. However, we do not believe that this same chemical evolution is occurring today for the following reasons:

1. Appropriate energy sources, particularly ultraviolet radiation, are unavailable. The ozone layer now acts as a shield to prevent these rays from reaching the earth in quantity.
2. Whereas the first atmosphere was believed to be a reducing one, which promoted the buildup of organic molecules, today's atmosphere is an oxidizing one, which tends to break down organic molecules.
3. Living organisms, already present, would use any newly formed organic molecules for food.

Summary

Evidence now exists to suggest the manner in which the first form of life, the single cell, arose. The earth began as a mass of hot, glowing individual atoms. The primitive atmosphere may have formed as the lightest of these atoms cooled and joined together to form hydrogen gas, methane, water vapor, and ammonia vapor. Another hypothesis proposes that volcanic eruptions produced a primitive atmosphere much like today's atmosphere, except that it lacked free oxygen. In any case, further cooling caused water vapor to turn to rain, and the quantity of this rain was great enough to produce the oceans in which the gases were dissolved. Here in the ocean, as Miller and more recent investigators have shown, gases could have reacted with one another under the influence of an outside energy source, such as lightning or ultraviolet radiation. Small organic molecules resulting from this process later polymerized to produce large organic macromolecules similar to proteins or nucleic acids. Fox has proposed a cell-first hypothesis, which is dependent on the observation that amino acid polymers can form spheres resembling cell-like bodies called proteinoids. Proteinoids have many features in common with cells. The first cell-like structure, called the protocells, would have carried on anaerobic respiration and heterotrophic nutrition. Once the protocells acquired genes they became true cells, which allowed other types of cells to evolve. Eventually autotrophs arose that not only supplied food for themselves and other organisms but also released oxygen into the atmosphere. This oxygen allowed the evolution of eukaryotic plants and animals and formed an ozone shield that permits these organisms to live on the land. The great variety of living things that evolved from the first cell(s) are classified by this text into five kingdoms.

Today, life is believed to come only from life because an appropriate energy source is lacking to cause the formation of organic molecules, and if they did form, they would be oxidized by oxygen or eaten by pre-existing life.

Study Questions

1. What was the primitive earth like when it first formed? (p. 529)
2. What were the lightest atoms that may have been present outside the new planet? (p. 529) What gases may have formed from these atoms? (p. 529)
3. Under what conditions was it possible for gases to react with one another to produce small organic molecules? (p. 530)
4. Describe a type of experiment that shows that organic molecules can form from primitive gases. (p. 530)
5. What is the cell-first hypothesis? (p. 531)
6. What type of respiration and nutrition did the protocell have? (p. 532) The protocell became a true cell when it could do what? (p. 532) How could this have come about? (p. 532)
7. How did the evolution of autotrophs change the primitive atmosphere? (p. 532)
8. Why doesn't life arise by chemical evolution today? (p. 533)

Selected Key Terms

chemical evolution (kem'ĭ-kal ev''o-lu'shun)
primitive atmosphere (prim'ĭ-tiv at'mos-fēr)
reducing atmosphere (re-dūs'ing at'mos-fēr)
energy source (en'er-je sors)
organic soup (or-gan'ik sōōp)
polymerization (pol''ĭ-mer''ĭ-za'shun)
proteinoid (pro'te-in-oid)
protocell (pro'to-sel)
coacervate (ko-as'er-vāt)
anaerobic (an-a''er-ōb'ik)
heterotroph hypothesis (het'er-o-trof hi-poth'ē-sis)
cell first hypothesis (sel ferst hi-poth'ē-sis)
biological evolution (bi-o-loj'ē-kal ev''o-lu'shun)
autotroph (aw'to-trōf)
aerobic (a''er-ōb'ik)
oxidizing atmosphere (ok'sĭ-dīz-ing at'mos-fēr)
ozone shield (o'zōn shēld)

Further Readings

Day, W. 1979. *Genesis on planet earth.* East Lansing, Mich.: House of Talos.
Dickerson, R. E. 1978. Chemical evolution and the origin of life. *Scientific American* 239(3):70.
Folsom, C. E. 1979. *The origin of life.* San Francisco: W. H. Freeman.
Groves, D. I., et al. 1981. An early habitat of life. *Scientific American* 245(4):64.
Oparin, A. I. 1968. *Genesis and evolutionary development of life.* New York: Academic Press.
Schopf, J. W. 1978. The evolution of the earliest cells. *Scientific American* 239(3):110.
Vidal, G. 1983. The oldest eukaryotic cells. *Scientific American* 250(2):48.

25

Chapter concepts

1 Viruses are noncellular; whether they should be considered living organisms is questionable.

2 The monerans are prokaryotes, while the protistans and fungi are eukaryotes.

3 The kingdom Monera includes bacteria and cyanobacteria, which despite their small size are important organisms.

4 The kingdom Protista contains unicellular organisms that may resemble the first eukaryotic cells.

5 The kingdom Fungi contains the most complex organisms to rely on saprophytic nutrition.

Viruses and Kingdoms Monera, Protista, and Fungi

Figure 25.1
Electron micrographs of viruses. *a.* The tobacco mosaic virus is helical. The drawing shows how the nucleic acid core circles within a coat composed of individual units. *b.* An adenovirus is polyhedral. The insert shows how this viral coat is also composed of individual units. *c.* T viruses are complex because they are composed of two major parts, a head (*upper*) and tail fibers (*below*).

a.

It is our aim to discuss living organisms from the simple to the complex and from the primitive (earliest evolved) to the most advanced (latest evolved). The designation *primitive* means only that these organisms have changed less with time (fig. 23.1) than have the advanced organisms. It is also well to remember that no living group of organisms is the direct ancestor of another living group of organisms, although it is possible for two living groups to have shared a common ancestor. The great variety of living organisms within any particular group is the result of adaptive radiation from a common ancestor, which may at times be determined from the fossil record.

It is curious that we must begin our discussion with viruses when they are not even included in the classification table found in the Appendix (p. A–3). We begin with viruses only because they are on the borderline between living and nonliving things.

Evolution and Diversity

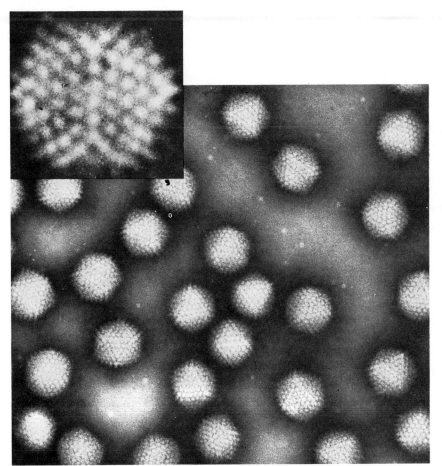

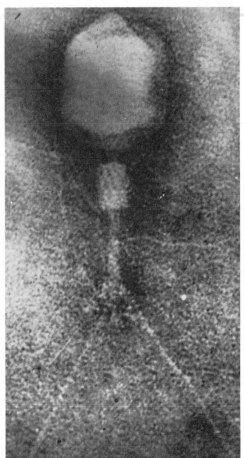

b.

c.

Viruses

Viruses are considered noncellular because they do not have a cellular type of organization. They are tiny particles (5–200 nm) composed of just two parts: *an outer coat of protein and an inner core of nucleic acid.* Although viruses cannot be seen with the light microscope, the electron microscope has permitted a study of their structure. Typically they are either helical, like the tobacco mosaic virus, or polyhedral, like the adenovirus, or a combination of both, like the T viruses (fig. 25.1). Regardless, the coat, which is made up of repeating protein subunits, surrounds the nucleic acid (either DNA or RNA).

Viruses are capable of reproduction, but only within living cells; therefore they are called **obligate parasites.** In the laboratory, active animal viruses

Figure 25.2

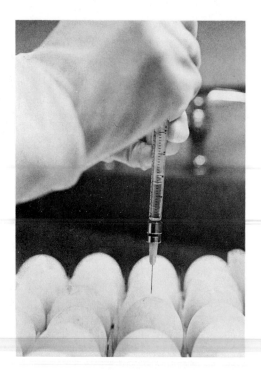

are maintained by injecting them into live chick embryos (fig. 25.2). Outside living cells, viruses are nonliving and can be stored just as chemicals are stored. Therefore, it is proper to ask if viruses should be considered alive.

Most viruses are extremely specific. Certain ones attack only plants; others attack only animals, either birds or mammals; and the viruses called **bacteriophages** attack only bacteria. Thus, viruses are said to have a specific host range. An interesting theory about the origin of viruses suggests that they began as pieces of DNA or RNA from the host they attack. The human disease-causing viruses (table 12.3) even prefer specific tissue types.

Reproduction

Two types of life cycles, termed the lytic and the lysogenic cycle (fig. 25.3) have been observed.

Lytic Cycle

When a T-even virus (meaning that it is designated by an even rather than odd number) happens to collide with an *E. coli* cell, it attaches to specific receptors by means of its protein tail fibers. An enzyme digests away part of the bacterial cell wall and the viral DNA enters the bacterial cell by way of the tail. Once inside, the viral DNA brings about disintegration of host DNA and takes over the operation of the cell. Viral DNA replication, utilizing the nucleotides within the host cell, produces many copies of viral DNA. Transcription occurs and mRNA molecules utilize host ribosomes to bring about the production of multiple copies of coat protein. Viral DNA and coat protein are assembled to produce about 100 viral particles. Lysozyme, synthesized under virus direction, disrupts the cell wall and the particles are released.

This is the **lytic cycle** because the host cell is lysed (broken open) to allow dispersal of the particles.

Lysogenic Cycle

Some bacteriophages do not immediately undergo a lytic life cycle. Instead, the viral DNA becomes integrated into the bacterial DNA. In this stage it is called a **prophage.** The prophage is replicated along with the host DNA and all subsequent cells, called **lysogenic cells,** carry a copy of the prophage. Certain environmental factors such as ultraviolet radiation can induce the lytic cycle. The prophage leaves the bacterial chromosome; replication of viral DNA, production of coat protein, assemblage, and lysis follow.

RNA Viruses

Some viruses have RNA genes, that is, only RNA is enclosed within coat protein. How does such a virus bring about replication and transcription in order to undergo the lytic cycle?

In most of these viruses, the single strand of RNA present serves as a template for the production of double-stranded RNA. This unique molecule then serves as a template for the replication of multiple copies of the original genetic material and for the transcription of mRNA molecules. Special enzymes, termed *RNA replicase and transcriptase,* are coded for by the RNA genes.

Other RNA viruses have an enzyme called *reverse transcriptase* that carries out RNA---→DNA transcription. This DNA is used as a template

Evolution and Diversity

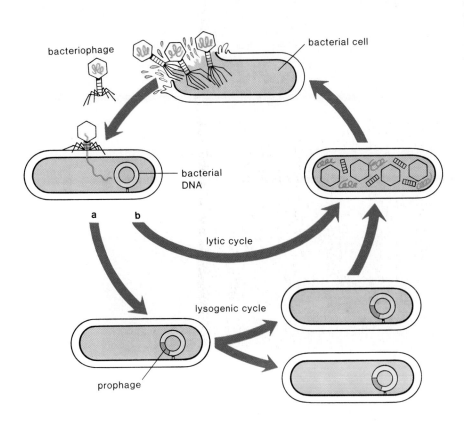

Figure 25.3
Lytic versus lysogenic life cycle. In the lytic cycle, a bacteriophage reproduces within the host bacterial cell and then the cell is lysed (broken open), allowing the viral particles to escape. In the lysogenic cycle, the DNA of a bacteriophage integrates into the host DNA and becomes a prophage. Thereafter, the prophage is replicated along with the bacterial chromosome and passed to all the daughter cells. When and if the viral DNA leaves the chromosomes, the lysogenic cycle can be followed by the lytic cycle.

bacteriophage

bacterial cell

bacterial DNA

a   b

lytic cycle

lysogenic cycle

prophage

to produce double-stranded DNA. Up until the discovery of these so-called **retroviruses** in the late 1960s, it was never thought that RNA could be a template for the production of DNA. But these viruses are able to bring this about.

The double-stranded DNA produced in retroviruses can instigate the lytic cycle or it can become integrated into the host DNA.

Importance of Viruses

Viruses are best known for causing infectious diseases in animals and plants. In plants, infectious diseases can only be controlled by destroying those plants that show symptoms of disease. In animals, especially humans, they are controlled by administering vaccines and, only recently, by the administration of antiviral drugs, as discussed in the reading on page 544. Some well-studied viral diseases in humans are flu, mumps, measles, polio, rabies, and infectious hepatitis.

Evidence is gathering that certain viruses may also at least contribute to the development of cancer by bringing oncogenes (p. 494) into the host cell. These viruses are ones that have a tendency to undergo the lysogenic cycle. Since the resulting proviruses are passed on to offspring, this would explain why some kinds of cancer tend to run in families.

Figure 25.4

Scanning electron micrographs of bacteria.
a. Spherical-shaped bacteria. *b.* Rod-shaped
bacteria. *c.* Spiral-shaped bacteria with flagella
used for locomotion. See Figure 2.20 for a
generalized drawing of a bacteria.

(*a.* and *c.*) Dr. R. G. Kessel and Dr. C. Y. Shih, From
SCANNING ELECTRON MICROSCOPY, Springer-Verlag,
Berlin, Heidelberg, New York, 1976.

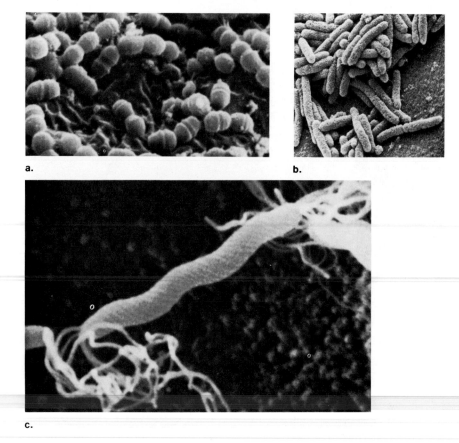

a.   b.

c.

Figure 25.5

Some bacteria move about by means of flagella,
which are seen here on either side of the organism.
Each flagellum is attached to a basal structure
having a hook within a shaft. It is thought that the
inner ring portion of the shaft rotates somewhat like
an electric motor, and this twists the rigid flagellum,
moving the bacterium at speeds up to about 80–90
meters per second.

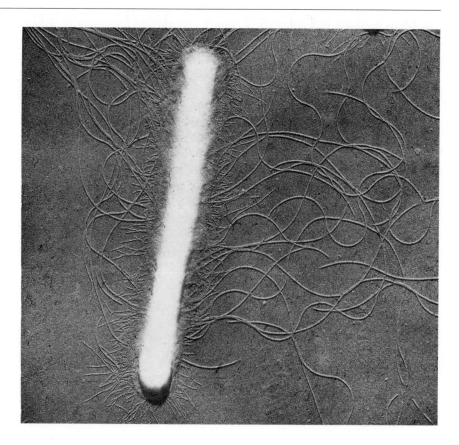

Evolution and Diversity

Kingdom Monera

Kingdom Monera[1] contains only prokaryotes, the first types of cells to evolve. The prokaryotes include bacteria and cyanobacteria. Cyanobacteria were formerly called blue-green algae. The term **algae** is now used merely to mean an aquatic photosynthesizing organism, and as we shall see, different types of algae are classified in the kingdoms Monera, Protista, and Plantae.

A prokaryote has a cell wall that contains unique amino sugars cross-linked by peptide chains. Its composition is entirely different from the cell wall of any eukaryotic cell. Some bacteria are also surrounded by a polysaccharide or polypeptide capsule that enhances their virulence (ability to cause disease). Cyanobacteria also have a sheath of gelatinous material outside their cell walls.

Prokaryotic cells do not have the cytoplasmic organelles found in eukaryotic cells, except ribosomes (table 2.3). They do have DNA, but it is not contained within a nuclear envelope; therefore they are said to lack a nucleus. They have respiratory enzymes, but no mitochondria; and if they possess chlorophyll, it is found within thylakoids, but there are no chloroplasts.

Asexual reproduction is the norm for prokaryotes. Bacteria are able to exchange portions of DNA, but similar occurrences have not been observed in cyanobacteria.

Bacteria

Bacterial cells are small but are generally larger than viruses; they range in size from 1 to 10 μm (micrometers) in length and from 0.2 to 0.3 μm in width. Since they are microscopic, it is not always obvious that they are abundant in the air, water, soil, and on most objects. It has even been suggested that the combined weight of all bacteria would exceed that of any other type organism on earth.

There are many different types of bacteria, but we will consider only the eubacteria.

Structure

Eubacteria occur in three basic shapes (fig. 25.4): **rod** (bacillus), **round,** or spherical (coccus), and **spiral** (a curved shape called a spirillum). Some bacteria can locomote by means of **flagella** (fig. 25.5), which are composed only of a protein called flagellin, and some can adhere to surfaces by means of **pili,** projections also composed of protein.

Reproduction

Bacteria reproduce asexually by **binary fission.** First, the single chromosome duplicates and then the two chromosomes move apart into separate areas; then the cell membrane grows inward and partitions the cell into two daughter cells, each of which now has its own chromosome (fig. 25.6).

Sexual exchange has been observed between two bacteria when the so-called male passes DNA to the female by way of a pilus called a conjugation pilus. Recombination of the genetic material occurs within the female, which then divides. Bacterial cells can also pick up fragments of DNA released into the medium from dead cells. This is called **transformation** because the cell receiving the DNA is transformed into a cell with the genotype and phenotype of the absorbed DNA. Bacteriophages provide a third means by which it is possible to recombine bacterial DNA. In this process, called **transduction,** the phages carry portions of bacterial DNA from one cell to another.

When faced with unfavorable environmental conditions, some bacteria can form **endospores.** During spore formation, the cell shrinks, rounds up within the former cell membrane, and secretes a new, thicker wall inside the old one

[1]The classification system utilized in this text is given on page A–3.

Figure 25.6
Reproduction in bacteria. Above, the single chromosome is seen to be attached to the cell membrane where it is replicating. As the cell membrane lengthens, the two chromosomes separate. Once fission has taken place, each bacterium has its own chromosome.

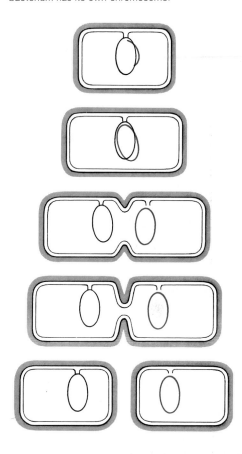

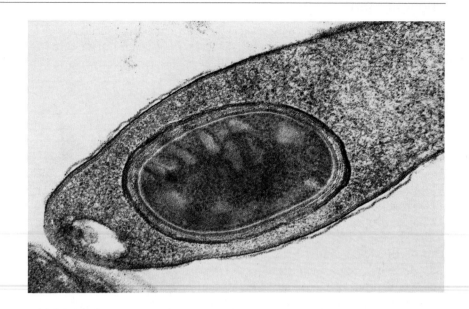

Figure 25.7
Spore formation. This bacterium contains a spore, the dark oval at the lower end of the cell. A spore protects the organism's DNA from exposure to environmental conditions that could destroy it.

(fig. 25.7). Endospores are amazingly resistant to extreme temperatures, drying out, and harsh chemicals, including acids and bases. When conditions are again suitable for growth, the spore absorbs water, breaks out of the inner shell, and becomes a typical bacterial cell.

Metabolism

Some living bacteria, like the first cell (p. 532), are obligate anaerobes and are unable to live in the presence of oxygen. A few serious illnesses, such as botulism, gas gangrene, and tetanus, are caused by anaerobic bacteria. Some bacteria, called **facultative anaerobes,** are indifferent to oxygen and can survive whether or not it is present. Most bacteria, however, are aerobic and, like animals, require a constant supply of oxygen to carry out complete cellular respiration.

Every type of nutrition is found among bacteria except holozoism (eating of whole food). A few bacteria are autotropic, being either chemosynthetic or photosynthetic. The **chemosynthetic bacteria** oxidize inorganic compounds to obtain the necessary energy to produce their own food. Among the inorganic compounds oxidized by specific bacteria are ammonia, nitrite, sulfur, hydrogen, and ferrous iron. **Photosynthetic bacteria** contain chlorophyll and can obtain energy from the sun to produce their own food. The bacterial pigment is not the same as that in cyanobacteria, and bacteria do not release oxygen because instead of using water as a hydrogen source, they use either molecular hydrogen, hydrogen sulfide, or organic compounds. None of these gives off oxygen when broken down.

Most bacteria are heterotrophic **saprophytes,** which send out digestive enzymes into the environment to break down large molecules into small molecules that can be absorbed across the cell membrane.

Importance

Within ecosystems, bacteria are decomposers (p. 713) and when they break down dead organic matter, inorganic nutrients are returned to producers. As decomposers, they also play a role in the carbon cycle, preventing the buildup of organic matter and releasing carbon as carbon dioxide. In the nitrogen cycle,

Evolution and Diversity

Figure 25.8
Sterilization by autoclaving. The sterilizing room in a hospital contains many autoclaves in which steam under pressure kills bacterial cells and endospores. Sterilization permits surgical procedures to be done with reduced fear of subsequent infection.

nitrogen-fixing bacteria in the soil or in the nodules of legumes (fig. 31.9) provide a usable source of nitrogen for producers. Also, the nitrifying bacteria oxidize ammonia, an excretion product of animals, to nitrate, a nutrient for plants.

Bacteria may be free living or symbiotic, forming mutualistic, commensalistic, or parasitic relationships (p. 703). The nitrogen-fixing bacteria in the nodules of legumes are mutualistic, as are the bacteria, mainly *E. coli,* within our own intestinal tract. We provide the bacteria with a home and they provide us with certain vitamins. Commensalistic bacteria reside on our skin where they usually cause no difficulty. Due to their presence, parasitic bacteria and fungi may have difficulty establishing residence.

The parasitic bacteria cause numerous diseases, such as those listed in table 12.4. General cleanliness is the first step toward preventing the spread of these diseases. Disinfectants and antiseptics also help reduce the number of infectious bacteria. Typhoid is kept in check by adding chlorine to water; diphtheria, typhoid, and tuberculosis are also partially controlled by the pasteurization (heating to 145° F. for 30 minutes) of milk. Sterilization, a process that kills all living things, even endospores, is used whenever all bacteria and viruses must be killed. Sterilization can be achieved by use of an autoclave (fig. 25.8), a container that admits steam under pressure.

Canned foods have been sterilized to kill all bacteria that might be present. Several other methods are also possible to preserve food not only from disease-causing bacteria but from general spoilage. The pH of acidic foods is too low and dried foods lack enough moisture to be subject to bacterial contamination. Food that has a high content of salt and/or sugar are osmotically unfavorable and do not need special treatment. Refrigeration keeps the bacteria count of susceptible foods low and freezing prevents growth of bacteria completely.

Antibiotics and Antiviral Drugs

Pencillium chrysogenum

An antibiotic is a chemical that selectively kills bacteria when it is taken into the body as a medicine. There has been a dramatic reduction in the number of deaths due to pneumonia, tuberculosis, and other infections since 1900 and this can in part be attributed to the increasing use of antibiotic therapy.

Most antibiotics are produced naturally by soil microorganisms. Penicillin is made by the fungus *Penicillium;* streptomycin, tetracycline, and erythromycine are all produced by a bacterium, *Streptomyces.* Sulfa, an analog of a bacterial growth factor, can be produced in the laboratory.

Antibiotics are metabolic inhibitors specific for bacterial enzymes. This means that they poison bacterial enzymes without harming host enzymes. Penicillin blocks the synthesis of the bacterial cell wall, streptomycine, tetracycline, and erythromycine block protein synthesis, and sulfa prevents the production of a coenzyme.

There are problems associated with antibiotic therapy. Some patients are allergic to antibiotics and their reaction to them may even be fatal. Antibiotics not only kill off disease-causing bacteria, they also reduce the number of beneficial bacteria in the intestinal tract. The latter may have held in check a pathogen that now is free to multiply and invade the body. The use of antibiotics sometimes prevents natural immunity from occurring, leading to the necessity for recurring antibiotic therapy. Most important, perhaps, is the growing resistance of certain strains of bacteria. While penicillin used to be 100 percent effective against hospital strains of *Staphylococci aureus,* today it is far less effective. Tetracycline and penicillin, long used to cure gonorrhea, now have a failure rate of more than 20 percent against certain strains of *Gonococcus.* Most physicians believe that antibiotics should only be administered when absolutely necessary and some believe that if this is not done then resistant strains of bacteria will completely replace present strains and antibiotic therapy will no longer be effective at all. They are very much opposed to the current practice of adding antibiotics to livestock feed in order to make animals grow fatter because resistant bacteria are easily transferred from animals to humans.

The development of antiviral drugs has lagged far behind the development of antibiotics. Viruses lack most enzymes and instead utilize the metabolic machinery of the host cell. Rarely has it been possible to find a drug that successfully interferes with viral reproduction without also interfering with host metabolism. One such drug, however, called Vidarabine was approved in 1978 for treatment of viral encephalitis, an infection of the nervous system, and another called Acyclovir (ACV) seems to be helpful in treating genital herpes.

Bacteria have long been used by humans to commercially produce various products. Chemicals, such as ethyl alcohol, acetic acid, butyl alcohol, and acetones, are produced by bacteria. Bacterial action is involved in the production of butter, cheese, sauerkraut, rubber, cotton, silk, coffee, and cocoa. By means of gene splicing, bacteria are now used to produce human insulin and interferon, as well as other types of proteins (p. 256). Even antibiotics, as discussed in the reading on this page, are produced by bacteria.

Cyanobacteria (Blue-greens)

Cyanobacteria (fig. 25.9) are sometimes called "blue-greens" because they contain a blue pigment in addition to the green pigment chlorophyll. Actually, however, many have additional pigments and may appear black, brown, yellow, or red.

Blue-greens may be **unicellular, filamentous,** or **colonial.** The filaments and colonies are not considered multicellular because each cell is independent of the others. Blue-greens lack any visible means of locomotion although some oscillate (sway back and forth).

Figure 25.9
Anabena, a filamentous cyanobacterium.
a. Photomicrograph. Note the heterocysts
(enlarged cells) where nitrogen fixation occurs.
b. Electron micrograph. Chlorophyll-containing
membranes line the entire cell. The large round
white body is reserved food. See figure 2.20 for a
generalized drawing of a cyanobacterium.

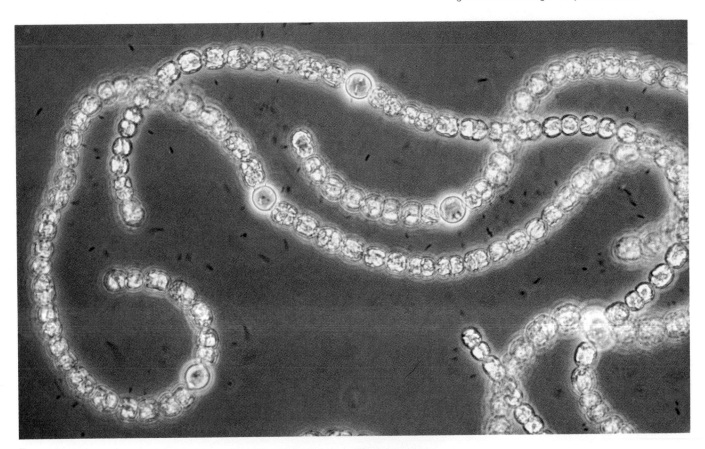

a.

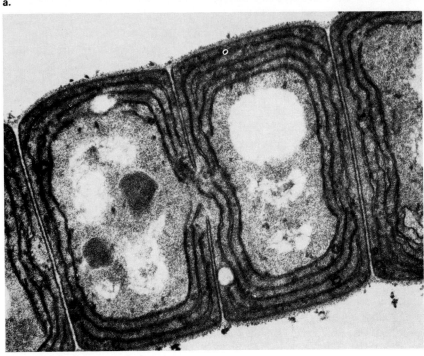

b.

Figure 25.10

Lichen anatomy. Lichens are made up of two types of organisms, an alga and a fungus. In this drawing, the algal cells are represented by circles and the fungus is represented by the filaments. Biologists are still debating whether this is a completely mutualistic relationship or whether the fungus is at least somewhat parasitic on the alga.

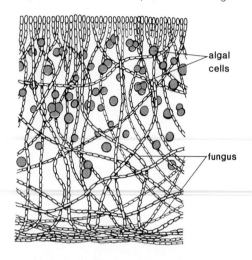

algal cells

fungus

Table 25.1	Types of protozoans	
Protozoans	**Locomotor organelles**	**Example**
Amoeboid	Pseudopodia	*Amoeba*
Ciliate	Cilia	*Paramecium*
Flagellate	Flagella	*Trypanosoma*
Sporozoa	No locomotor organelle	*Plasmodium*

Blue-greens are common in fresh water, soil, and moist surfaces. They are also found in habitats devoid of other types of life, such as hot springs. With the fungi, they form lichens that can grow on rocks. A lichen (figs. 25.10 and 25.22) is a mutualistic relationship in which the blue-green provides organic nutrients to the fungus, while the latter possibly protects and furnishes inorganic nutrients to its partner. Some blue-greens have a special advantage because they can fix aerial nitrogen and therefore their nutrient requirements are minimal. Lichens help transform rocks to soil so that other forms of life may follow.

In fresh water, cyanobacteria are sometimes responsible for the bloom associated with cultural eutrophication (p. 733). The occasional red color of bodies of fresh water and rivers is due to a species of blue-green that contains a red pigment.

Blue-greens are believed to have evolved from photosynthetic bacteria, although they probably diverged early in the history of prokaryotes. Comparison of their methods of photosynthesis reveals that blue-greens are more complex.

Bacteria
Photosystem I only
Do not give off oxygen
Unique type of chlorophyll

Blue-greens
Photosystems I and II
Do give off oxygen
Type of chlorophyll found in plants

Kingdom Protista

The protists are unicellular eukaryotes. They have all the organelles with which we are familiar: a nucleus, mitochondria, endoplasmic reticulum, and Golgi apparatus, for example.

We will include in the kingdom Protista[2] unicellular (or colonial) organisms whose relationships to either plants or animals has not clearly been established. (Unicellular organisms whose relationship is clearer are included as members of these other kingdoms.) While unicellular organisms must cope with the environment without benefit of tissues and organ systems, they are not simple; their complexity resides in the organization of their cells.

Protozoans

Protozoans are small (2 μm to 1,000 μm), usually colorless, unicellular organisms that lack a cell wall. They tend to have special structures for food gathering and locomotion; excretion and respiration are carried out across the cell membrane. Their animallike characteristics include the ability to react quickly to outside stimuli; they generally move toward food and better environmental conditions or away from obstacles and unfavorable environments. Although sexual exchange is sometimes observed, reproduction occurs by cell division. Protozoans are classified according to their type of locomotor organelle (table 25.1).

Amoeboids

An amoeba, such as *Amoeba proteus* (fig. 25.11), is a small mass of cytoplasm without any definite shape. It moves about and feeds by means of cytoplasmic extensions called pseudopodia, or false feet. A pseudopod forms when the cytoplasm streams forward in a particular direction.

The organelles within an amoeba include food, or digestive vacuoles, and contractile vacuoles. *Food vacuoles* are characteristic of holozoic protozoans and are formed within an amoeba when a morsel of food is surrounded by pseudopodia. This is a form of phagocytosis, which produces a vacuole that later becomes a digestive vacuole. **Contractile vacuoles** first collect excess water

[2]The classification system utilized in this text is given on page A–3.

Evolution and Diversity

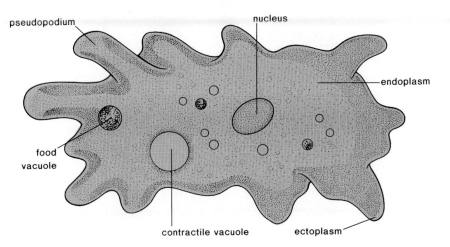

pseudopodium

nucleus

endoplasm

food vacuole

contractile vacuole

ectoplasm

Figure 25.11
Amoeba anatomy. Cytoplasmic streaming occurs to form pseudopodia. It has been observed that the gel-like ectoplasm must become more fluid before streaming of the sol-like endoplasm is possible. Specialized vacuoles carry on digestive and excretory functions. At one time it was thought that the contractile vacuoles had the power to contract, but most likely they are simply squeezed by the cytoplasm.

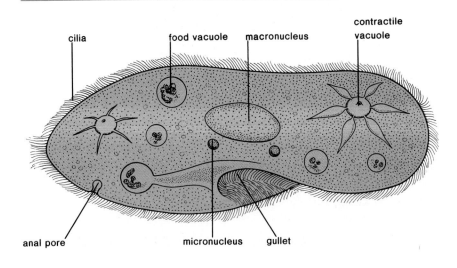

cilia

food vacuole

macronucleus

contractile vacuole

anal pore

micronucleus

gullet

Figure 25.12
Paramecium anatomy. This single cell is the most complex of the protozoans. The macronucleus controls the vegetative cell; the micronuclei only serve to allow sexual exchange of DNA. Food moves from the gullet to a vacuole to the anal pore. An extensive network of fibers (not shown) keeps the cilia beating in one direction to propel the organism.

from the cytoplasm and then appear to "contract," releasing the water through a temporary opening in the cell membrane. Contractile vacuoles are most often seen in freshwater protozoans.

Two forms of marine amoeba have shells. **Foraminifera** have a chalky, sometimes many-chambered shell; these organisms were so numerous at one time that their remains built the White Cliffs of Dover. **Radiolaria** secrete a beautiful skeleton of silica that becomes the bottom ooze in deeper parts of the ocean. One form of amoeba, *Entamoeba histolytica*, causes amoebic dysentery in humans.

Ciliates

The ciliates, such as those in the genus *Paramecium* (fig. 25.12), are the more complex of the protozoans. Hundreds of cilia project through tiny holes in the outer covering, or pellicle. Lying in the ectoplasm just beneath the pellicle are numerous oval capsules that contain **trichocysts.** The contents of the trichocysts may be discharged as long threads, which are used for defense. When a *Paramecium* feeds, food is swept down a gullet to the cell mouth, below which food vacuoles form. Following digestion, the soluble nutrients are absorbed by the cytoplasm and the indigestible residue is eliminated at the anal pore.

Figure 25.13
Conjugation. During this time of close contact the paramecia exchange nuclear material.

Dr. R. G. Kessel and Dr. C. Y. Shih, From SCANNING ELECTRON MICROSCOPY, Springer-Verlag, Berlin, Heidelberg, New York, 1976.

Figure 25.14
Trypanosome infection. *a.* A stained blood smear from a patient suffering with African sleeping sickness showing trypanosomes among the blood cells. *b.* Structure of trypanosome as revealed by the electron microscope. The flexible pellicle allows indulating movement of the cell.

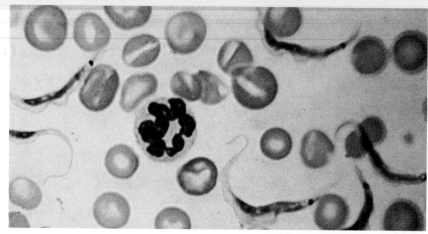

a.

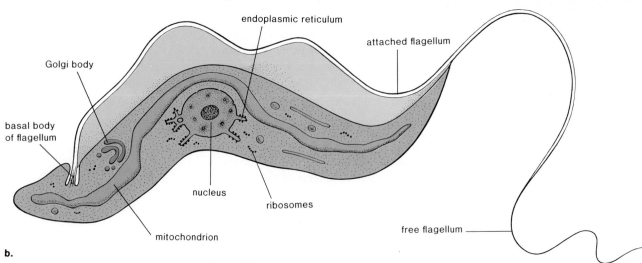

b.

Evolution and Diversity

Ciliates have two types of nuclei: a large macronucleus and one or more small micronuclei. The macronucleus controls the normal metabolism of the cell, while the micronuclei are concerned with reproduction. Following meiotic division of the micronuclei, two *Paramecia* may exchange these in a sexual process called **conjugation** (fig. 25.13).

Flagellates

Protozoans that move by means of flagella are sometimes called zooflagellates to distinguish them from unicellular algae that have flagellates.

Many zooflagellates enter symbiotic relationships (p. 703). *Trichonympha collaris* lives in the gut of termites and enzymatically converts wood to soluble carbohydrates easily digested by the insect. The trypanosomes (fig. 25.14) cause African sleeping sickness and are transmitted to vertebrates by the tsetse fly. The tsetse fly, which becomes infected when it takes a blood meal from a diseased animal, passes on the disease when it feeds on another victim. The white cells in an infected animal accumulate around the blood vessels leading to the brain and cut off the circulation. The lethargy characteristic of the disease is caused by an inadequate supply of oxygen for normal brain alertness.

Sporozoa

The sporozoa are nonmotile parasites with a complicated life cycle that always involves the formation of infective spores. The most important human parasite among the sporozoa is *Plasmodium vivax* (fig. 25.15), the causative agent of

Figure 25.15

Life cycle of *Plasmodium vivax,* a protozoan that causes one type of malaria. Asexual reproduction in humans: The bite of an infected *Anopheles* mosquito releases minute, spindle-shaped sporozoites that pass from the blood into the liver where asexual spores are produced. These invade and reproduce inside the red blood cells. Periodic chills and fever are caused by toxins that pour into the bloodstream when the spores are released. After a time, some of these spores become cells capable of developing into gametes if taken up by another mosquito. Sexual reproduction in mosquito: In the host stomach, the gametes mate and a wormlike zygote works its way through the stomach wall to encyst in the outer layer. Many divisions produce sporozoites that find their way to the salivary glands of the mosquito, and the cycle repeats.

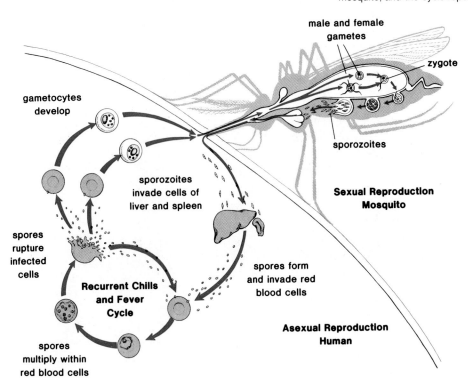

Figure 25.16

Euglena anatomy. Euglena is typical of those protozoans that have both animal-like and plantlike characteristics. Movement is by means of a very long flagellum. A photoreceptor allows *Euglena* to seek the light where photosynthesis occurs in the numerous chloroplasts.

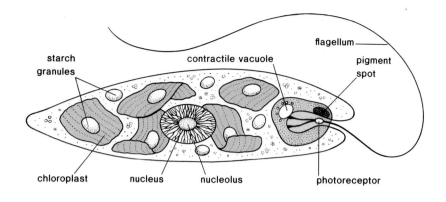

starch granules

contractile vacuole

flagellum

pigment spot

chloroplast

nucleus

nucleolus

photoreceptor

Figure 25.17

Scanning electron micrograph of a dinoflagellate. The cell can be either naked or covered by a cellulose envelope, sometimes divided into plates like this. There are two flagella, one circles the cell, while the other extends posteriorly.

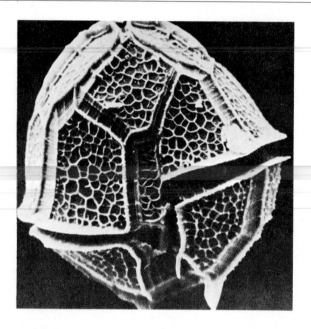

Figure 25.18

Diatoms. These unicellular algae have a unique golden-brown pigment in addition to chlorophyll, but actually may be various colors as shown here. These beautiful patterns are markings on their silica-embedded walls.

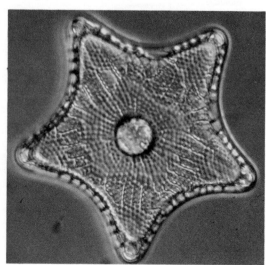

Evolution and Diversity

one type of malaria. When a human being is bitten by an infected female *Anopheles* mosquito, the parasite eventually invades the blood cells. The chills and fever of malaria occur when the infected cells burst and release toxic substances into the blood.

The eradication of malaria has centered on the destruction of the mosquito, since without this host the disease cannot be transmitted from one human being to another. However the use of pesticides has caused the development of resistant strains of mosquitos. Therefore it is hoped that a vaccine will one day be available.

Unicellular Algae
The unicellular algae placed in the kingdom Protista all have chloroplasts and carry out photosynthesis in a manner similar to plants. They do not represent forms that are believed to be direct ancestors of plants, however.

Euglenoids
Organisms belonging to the genus *Euglena* (fig. 25.16) are freshwater organisms having both animal-like and plantlike characteristics. Their animal-like characteristics include motility and a flexible body wall. They move by means of flagellae, typically having one much longer than another, that project anteriorly. Because they are bounded by a flexible pellicle instead of a rigid cell wall, they can assume different shapes as the underlying cytoplasm undulates or contracts.

Their plantlike characteristics include the possession of chloroplasts. A light-sensitive swelling at the base of one flagellum is shaded by a pigment spot and this allows *Euglena* to judge the direction of the light. *Euglena* moves toward light so that photosynthesis can take place.

Dinoflagellates
Dinoflagellates (fig. 25.17) have two external grooves or furrows, each containing a single flagellum. One furrow is transverse and completely encircles the cell; the other is longitudinal and extends along one side only. The beating of these flagella causes the organism to spin like a top. The cell wall, when present, is frequently divided into polygonal plates of cellulose closely joined together. At times there are so many of these organisms that they cause a "red tide." The toxins they give off cause widespread fish kill and can cause a paralysis in humans if they eat shellfish that have fed on the dinoflagellates.

Usually the dinoflagellates are an important source of food for small animals in the ocean and they live as symbiotes within the bodies of some invertebrates. For example, corals usually contain large numbers of these organisms and this allows them to grow much faster than otherwise.

Diatoms
Diatoms (fig. 25.18) are golden brown in color due to an accessory pigment in their chloroplasts that masks the color of chlorophyll. Diatoms have a cell wall composed of two halves, or valves, of unequal size; the larger fits over the smaller like the lid of a box. The cell wall has an outer layer of silica, a common ingredient of glass. The valves are covered with a great variety of striations and markings that form beautiful patterns when observed under the microscope.

Usually diatom reproduction is by cell division. Each daughter cell receives a valve and manufactures a new valve to fit inside it. One daughter cell is always slightly smaller than the parental cell and eventually, after several divisions, gets much smaller than normal. Then the diatoms carry out sexual reproduction, which allows them to return to their original maximum size (fig. 25.19).

Diatoms are the most numerous of all unicellular organisms in the oceans. As such, they serve as an important source of food for other organisms. Their

Figure 25.19
Diatom life cycle. Reproduction by cell division causes the descendants to become ever smaller in size. Eventually, sexual reproduction restores the original size.

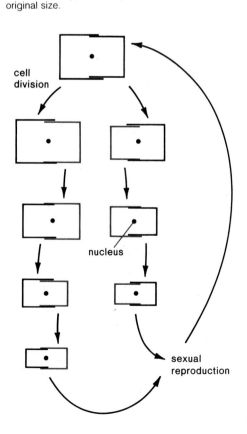

Figure 25.20
Life cycle of black bread mold *Rhizopus* including
both asexual (lower left) and sexual cycles. Notice
that the sexual cycle requires two mating strains.

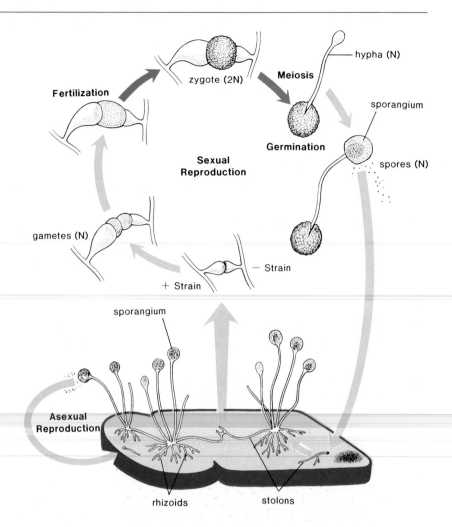

remains, called diatomaceous earth, accumulate in the ocean floor and are
mined and used commercially in making insulation, as an abrasive in tooth-
paste and silver polish, and in water-filtering systems.

Kingdom Fungi

Fungi, being saprophytic like bacteria, are also decomposers. Since they are
the most complex of the saprophytic organisms, they are placed in a separate
kingdom in the classification system used by this text.

The body of all fungi, except unicellular yeast, is made up of filaments
called hyphae. A **hypha** consists of an elongated cylinder, containing a mass
of cytoplasm and hundreds of haploid nuclei, which may or may not be separated by cross walls. A collection of hyphae is called a **mycelium.** Most fungi
reproduce both asexually and sexually. They are adapted to reproduction on
land and produce windblown spores during both these aspects of their life cycle
(fig. 25.20). Classification emphasizes mode of sexual reproduction.

Black Bread Molds

Black bread molds belonging to the genus *Rhizopus* are often used as an example of this group. These molds exist as a whitish or grayish haploid mycelium on bread or fruit (fig. 25.20). During asexual reproduction, some hyphae
grow upright and bear a spherical **sporangium** within which thousands of spores
are formed.

During sexual reproduction, hyphae of different mating strains (usually
referred to as plus and minus) approach each other. Upon contact, swellings

a.

b.

c.

d.

of these hyphae are partitioned off by formation of cross walls. These portions, which function as gametes, join and the nuclei fuse to form a diploid zygote nucleus. The zygote wall then thickens and blackens. The resulting zygospore remains dormant for several months. At germination, the nucleus of the zygospore undergoes meiosis to produce a short hypha, which immediately forms a sporangium, with the subsequent release of windblown spores.

Sac Fungi

There are many different types of sac fungi (fig. 25.21), most of which produce asexual spores, called **conidia.** During sexual reproduction, sac fungi form spores called ascospores within saclike cells called **asci.** In most species, the asci are supported within **fruiting bodies,** a collection of specialized hyphae.

a. b.

In cup fungi, the largest group of sac fungi, the fruiting body takes the shape of a cup.

Yeasts are sac fungi that do not form fruiting bodies. In fact, yeasts are different from all other fungi in that they are unicellular and most often reproduce asexually by budding. Yeasts, as you know, carry out fermentation as follows:

$$glucose \longrightarrow carbon\ dioxide + alcohol$$

In baking, the carbon dioxide from this reaction makes bread rise. In the production of wines and beers, it is the alcohol that is desired, and the carbon dioxide is allowed to escape into the air.

Blue-green molds, notably *Penicillium,* are sac fungi. This mold grows on many different organic substances, such as bread, fabrics, leather, and wood. It is purposefully used by humans to provide the characteristic flavor of Camembert and Roquefort cheese; more important, it produces the antibiotic penicillin. Most penicillin today, however, is synthetically produced. Another mold, the red bread mold *Neurospora,* was used in the experiments that helped decipher the function of genes.

Unfortunately, sac fungi also cause many diseases of plants; chestnut tree blight and Dutch elm disease are caused by sac fungi, as are powdery mildews, apple scab, and ergot. Ergot, a disease of cultivated cereals, contains LSD; persons who eat infected grain are likely to experience an LSD trip.

The fungus portion of lichens (fig. 25.22) is usually a sac fungus, while the alga may be cyanobacteria or green algae. Lichens can live on bare rock or poor soil and are able to survive great extremes of heat, cold, and dryness in all regions of the world. Reindeer moss is a lichen that is an important food source for arctic animals.

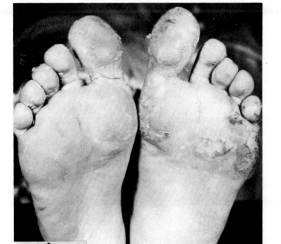

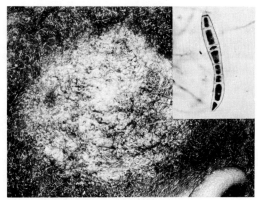

Figure 25.23
Human diseases caused by fungi. *a.* Athlete's foot
can be caused by *Trichophyton mentagrophytes,*
which reproduces by means of spores (shown in
insert). *b.* Ringworm can be caused by
Microsporum audouini, which also reproduces by
means of spores (insert).

a.

b.

Club Fungi

Among the club fungi (fig. 25.21), asexual reproduction may also be accomplished by formation of conidia spores. As a result of sexual reproduction, members of this group form club-shaped structures called **basidia,** often within fruiting bodies. The mushroom, puffball, and bracket (shelf) fungi are club fungi; the visible portions of these are actually fruiting bodies, and the mycelia lie beneath the surface. On the underside of a mushroom cap, the basidia project from the gills. Within each basidium, a diploid nucleus undergoes meiosis to produce windblown spores.

Many mushrooms are considered gourmet delicacies and have been prized since the time of the Roman Empire. The best-known edible mushroom is *Agaricus campestris,* one of the few gilled mushrooms that can be cultivated commercially. About 65,000 tons of mushrooms are produced annually in this country. Unfortunately, some mushrooms (particularly members of the genus *Amanita*) are extremely poisonous and have killed unwary mushroom hunters.

Rusts and smuts are parasitic club fungi that attack grains, resulting in great economic loss and necessitating expensive control measures. They do not have a conspicuous fruiting body and consist of vegetative hyphae, together with spores of various kinds. On the other hand, the mycelia of club fungi that lie beneath the soil often form beneficial symbiotic relationships with plants, notably pine trees. These so-called "fungus roots" help the trees garner nutrients from the soil as described in the reading on page 556.

Fungi Imperfecti

Fungi imperfecti cannot be assigned to a definite group because the sexual portion of the life cycle has not been observed. **Ringworm** and **athlete's foot** (fig. 25.23) belong to this group, as does *Candida albicans,* which causes moniliasis, a fairly common vaginal infection in females who take the birth control pill.

Fungus Roots (Mycorrhizae)

A symbiotic relationship between fungal hyphae and the roots of plants has been discovered. The relationship is so close that it is hard to distinguish the hyphae from the root proper. The hyphae penetrate the root and apparently help the plant absorb inorganic nutrients from the soil. In exchange, the plant most likely supplies the fungi with organic nutrients.

When the association is present, plants can survive and thrive in poor soils. If the association is absent, the plants require excess nutrients in order to achieve the same productivity. Is it possible that farmers who use pesticides will automatically need to apply more fertilizer because the fungi of the soil have been destroyed?

One-year-old loblolly pine seedings with the normally low levels of naturally occurring mycorrhizae.

One-year-old loblolly seedlings with abundant *Pisolithus tinctorius* mycorrhizae following artificial inoculation.

Evolution and Diversity

Figure 25.24
Plasmodium of slime mold. At this stage in its life cycle, the slime mold is a naked mass of protoplasm that creeps along over rocks or vegetation.

Slime Mold

A slime mold (fig. 25.24) has a naked mass of multinucleated protoplasm that moves about on decaying leaves much like a giant amoeba. At certain times, this so-called **plasmodium** constricts and forms fruiting bodies within which spores are produced. When these spores are released, they may, depending on the slime mold, germinate to produce either amoeboid cells or flagellated cells. In both cases, the cells act like gametes and join to form a zygote, which develops into the plasmodium again.

The slime molds have characteristics in common with amoeboid protistans, and it is difficult to assign them clearly to any particular group, but the presence of a fruiting body shows a relationship to fungi.

Summary

Viruses are noncellular obligate parasites made up of a coat of protein and a nucleic acid core. The life cycle of the bacteriophage T virus shows that only viral DNA need enter a cell for reproduction to occur. Following entrance, host DNA disintegrates as viral DNA replicates; then coat protein formation precedes assemblage of complete viruses that are released.

The kingdom Monera includes the bacteria and cyanobacteria (blue-greens), which are prokaryotic, meaning that their cells lack the organelles found in eukaryotic cells. Bacteria, which commonly have either a rod or spherical or spiral shape, display a wide range of metabolic activities. Most are aerobic, but some are facultative anaerobes or even obligate anaerobes. Although most bacteria are saprophytic, all types of nutrition are found except holozoism (eating of whole food). Reproduction is by binary fission, but sexual exchange does occasionally take place. Some bacteria form endospores that can survive the harshest of treatment except sterilization. Blue-greens, which are actually different colors, photosynthesize in a manner similar to plants. Some can fix aerial nitrogen and therefore have very minimal nutrient requirements. An overabundant supply of nutrients causes them to produce a bloom of ecological fame. Blue-greens, together with fungi, form lichens, important soil formers due to their action on rocks.

The kingdom Protista contains protozoans and unicellular eukaryotic algae whose relationship to plants is not well established. Protozoans are animal-like in that they are heterotrophic and motile. They are classified according to the type of locomotor organelle employed (table 25.1). The protistan unicellular algae include the euglenoids, dinoflagellates, and diatoms. They all have chloroplasts and carry on photosynthesis in the same manner as plants.

Fungi are saprophytic heterotrophic eukaryotes composed of hyphae filaments that form a mycelium. Along with heterotrophic bacteria, they are organisms of decay. The fungi produce spores during both sexual and asexual reproduction. The major groups of fungi are distinguishable on the basis of their mechanism of sexual reproduction. The black-bread molds produce spores in sporangia; the sac fungi produce spores in saclike cells; and the club fungi produce spores in club-shaped structures. The sac and club fungi typically have fruiting bodies.

Study Questions

1. Compare and contrast the viral lytic and lysogenic cycles. (pp. 538–39)
2. Describe the prokaryote characteristics shared by bacteria and blue-greens. (p. 541)
3. What are the three shapes of bacteria? (p. 541) How do bacteria reproduce? (p. 541) What are endospores? (p. 541)
4. Discuss the importance of bacteria and blue-greens. (pp. 542–43)
5. What two major groups of organisms are included in the kingdom Protista? (pp. 546, 551)
6. Give an example for each type of protozoan studied. Describe the anatomy of those that are free living and the life cycle of the parasitic ones. (pp. 546–51)
7. Discuss the significance of the euglenoids, dinoflagellates, and diatoms. (pp. 551–52)
8. Describe the structure of black bread mold and its life cycle. (pp. 552–53)
9. Define a fruiting body and name two groups of fungi that typically have fruiting bodies. (pp. 553–54)
10. Name several diseases caused by fungi classified as fungi imperfecti. (p. 554)
11. Describe the makeup of a lichen. (p. 554)

Selected Key Terms

virus (vi′rus)
bacteriophage (bak-te′re-o-fāg′′)
lytic cycle (lit′ik si′kl)
lysogenic cycle (li-so-jen′ik si′kl)
retroviruses (ret′′ro-vi′rus-ez)
monerans (mo-ne′rahnz)
binary fission (bi′na-re fish′un)
endospore (en′do-spor)
facultative anaerobe (fak′ul-ta′′tiv an-a′er-ōb)
cyanobacteria (si′′ah-no-bak-te′re-ah)
protozoans (pro′′to-zo′ahnz)
contractile vacuole (kon-trak′tīl vak′u-ōl)
euglenoids (u-gle′noids)
dinoflagellate (di′′no-flaj′ĕ-lāt)
diatoms (di′ah-tomz)
fungus (fung′gus)
hypha (hi′fah)
mycelium (mi-se′le-um)
sporangium (spo-ran′je-um)
fruiting body (frōōt′ing bod′e)

Further Readings

Ahmadjian, V. 1963. The fungi of lichens. *Scientific American* 208(2):122.

Berg, H. C. 1975. How bacteria swim. *Scientific American* 233(2):36.

Bonner, J. T. 1963. How slime molds communicate. *Scientific American* 209(2):14.

Christensen, C. M. 1965. *The molds and man: An introduction to the fungi*. 3d ed. Minneapolis: University of Minnesota Press.

Echlin, P. 1966. The blue-green algae. *Scientific American* 214(6):75.

Friedman, M. J., and Trager, W. 1981. The biochemistry of resistance to malaria. *Scientific American* 244(3):154.

Litten, W. 1975. The most poisonous mushrooms. *Scientific American* 232(3):90.

Pitelka, D. F. 1963. *Electron-microscopic structure of protozoa*. New York: Pergamon Press.

Simons, K., et al. 1982. How an animal virus gets into and out of its host cell. *Scientific American* 246(3): 58.

Stanier, R. M., et al. 1979. *The microbial world*. 4th ed. Englewood Cliffs, N.J.: Prentice-Hall.

Tiffany, L. H. 1968. *Algae, the grass of many waters*. 2d ed. Springfield, Ill.: Charles C Thomas.

Wollman, E. L., and Jacob, F. 1956. Sexuality in bacteria. *Scientific American* 195(1):22.

26

Plant Kingdom

Chapter concepts

1 Some plants are adapted to an aquatic existence and some are adapted to a terrestrial existence.

2 A study of evolutionary trends among plants shows that full adaptation to life on land developed slowly.

3 Full adaptation is seen in the seed plants that utilize pollen to transport the sperm to the egg. The resulting seed protects the embryo from drying out.

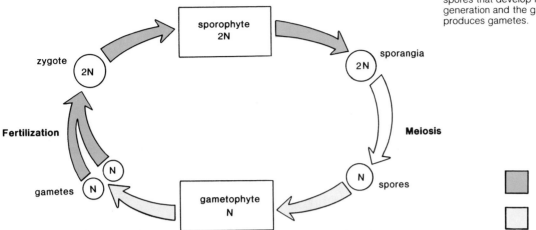

Figure 26.1
Alternation of generations life cycle. In this life cycle, the zygote does not undergo meiosis and instead develops into the sporophyte generation. The sporophyte generation produces haploid spores that develop into the gametophyte generation and the gametophyte generation produces gametes.

P lants can be defined as living organisms that carry on photosynthesis and cannot move about voluntarily. We will include in this kingdom multicellular photosynthetic forms and those unicellular ones that appear to be very closely related to the multicellular forms.

Most plants have a life cycle, termed **alternation of generations,** that includes two phases (fig. 26.1)—the sporophyte generation and the gametophyte generation:

1. The **diploid sporophyte generation** produces spores by meiosis.
2. The **haploid gametophyte generation** produces gametes.

The relative prominence of these generations varies from plant group to plant group, as we will be discussing in this chapter.

Plants first appeared in the seas but eventually became adapted to a land existence. As we survey present-day plants in this chapter, we will attempt to trace the steps by which adaptation to a land existence may have occurred.

Algae

Green algae, brown algae, and red algae are classified as members of the plant kingdom. The term *algae,* as explained previously, refers to aquatic photosynthesizing organisms. The term is old but not very exact. In this text, the blue-green algae, now termed cyanobacteria, are classified in the kingdom Monera; euglenoids, dinoflagellates, and diatoms, being exclusively unicellular, are classified in the kingdom Protista. The algae place in the plant kingdom all have multicellular representatives.

Green Algae

Green algae are believed to be ancestral to the first terrestrial plants because both of these groups possess chlorophylls *a* and *b,* both store reserve food as starch, and both have cell walls that contain cellulose. The green algae are a diverse group that ranges from simple to complex in regard to both structure and sexual reproduction. Some green algae are unicellular, but many types are multicellular.

Figure 26.2
The structure and life cycle of *Chlamydomonas*, a motile green alga. During asexual reproduction, all structures are haploid; during sexual reproduction only the zygote is diploid.

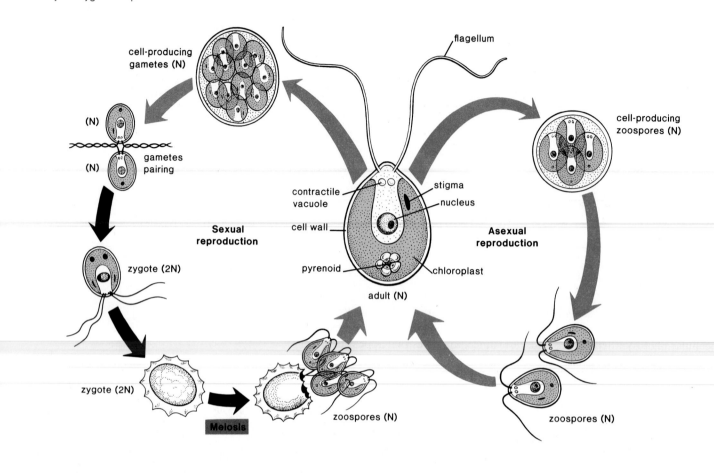

Figure 26.3
A typical green algal colony from the genus *Volvox*. *Left:* the colony produces a definite egg and sperm during sexual reproduction. *Right:* the zygotes develop into daughter colonies that are retained within the mother colony for a while.

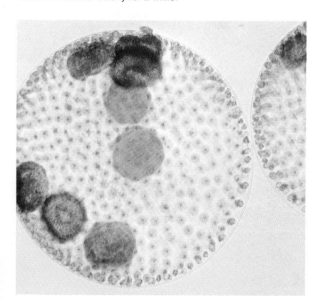

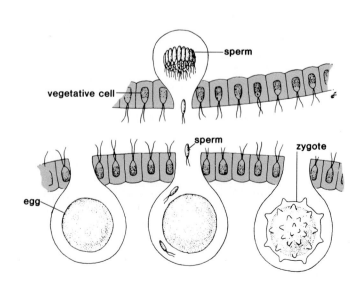

Flagellated Green Algae

The genus *Chlamydomonas* contains unicellular green algae that move by means of a pair of flagella. The single large, cup-shaped chloroplast contains a conspicuous pyrenoid, which is the site of starch production. The stigma, or "eyespot," is a portion of the chloroplast that aids the organism in moving toward the light. There are two small "contractile vacuoles" at the base of the flagella that rhythmically discharge water.

In these organisms, the adult is haploid and the only diploid structure is the zygote. During the asexual portion of the life cycle (fig. 26.2), the interior of the cell divides one or more times to form two to eight zoospores (motile spores) that resemble the parent cell. With the disintegration of the cell wall, the zoospores are released into the water and soon become adult cells like the parent. Sexual reproduction begins when the adult divides repeatedly to produce **isogametes** (gametes that look alike). When gametes from two different strains come into contact, the contents of the two cells join to form a zygote. A heavy wall forms around the zygote, and it becomes a resistant body able to survive until conditions are favorable, at which time the zygote germinates and produces four zoospores by meiosis. Notice that *Chlamydomonas* produces zoospores, or flagellated spores, in both the asexual and sexual life cycles.

The life cycle of *Chlamydomonas* illustrates very nicely the primary differences between asexual and sexual reproduction (table 26.1). *Sexual reproduction* is simply reproduction that involves the use of gametes. Distinct and separate sexes are not required and heterogametes (unlike gametes such as egg and sperm) are not required. Sexual reproduction aids the process of evolution because concurrent recombination of genes may produce a product more suited to the environment than either parent.

A number of **colonial** (loose association of cells) forms occur among the flagellated green algae. *Volvox* is considered the most complex genus among these colonial green algae. A *Volvox* colony is a hollow sphere with thousands of cells arranged in a single layer surrounding a watery interior. The cells of a *Volvox* colony, each one of which resembles a *Chlamydomonas* cell, cooperate in that the flagella beat in a coordinated fashion. Some cells are specialized for reproduction (fig. 26.3). Certain vegetative cells can divide asexually to form a new **daughter colony** that resides within the parental colony. Each female reproductive cell can produce a large, nonmotile egg, and the male reproductive cells produce small flagellated sperm. After fertilization, the zygote divides to form a daughter colony. Daughter colonies stay in the parental colony until it bursts.

Filamentous Green Algae

Filaments are end-to-end chains of cells that form after cell division occurs in only one plane.

Spirogyra (fig. 26.4), which is found in green masses on the surface of ponds and streams, is a genus whose members are filamentous. In *Spirogyra*, the chloroplasts are ribbonlike and spiral within the cell. Asexual reproduction occurs when a filament breaks up and each piece begins to produce new cells. Sexual reproduction occurs when two filaments line up next to one another, and **conjugation tubes** form between their respective cells. The contents of the

Figure 26.4
Members of the genus *Spirogyra* are filaments of cells. *a.* Anatomy of one cell. *b.* Micrograph depicting conjugation between two filaments.

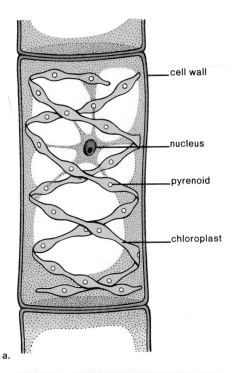

cell wall

nucleus

pyrenoid

chloroplast

a.

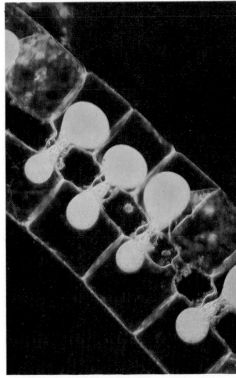

b.

Table 26.1	Life cycles	
Name	**Chromosome number**	**Spores**
Haplontic	Haploid only	Yes
Alternation of generations	Haploid ⇌ Diploid	Yes
Diplontic	Diploid only	No

Figure 26.5

Members of the genus *Oedogonium* are filaments that reproduce both asexually and sexually. During sexual reproduction depicted here, the motile sperm swim to the stationary egg. The zygote is the only diploid portion (*black*) of the cycle, and all the other structures are haploid (*color*).

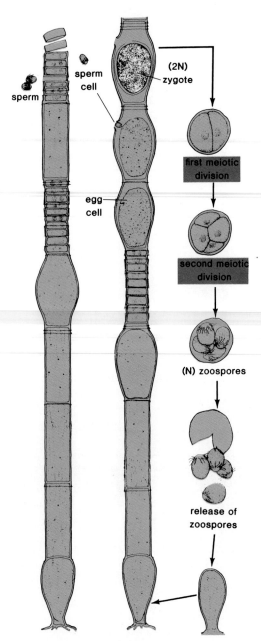

sperm

sperm cell

(2N) zygote

egg cell

first meiotic division

second meiotic division

(N) zoospores

release of zoospores

cells of one filament move into the cells of the other filament, forming 2N zygotes. These zygotes may survive the winter and in the spring undergo meiosis to produce new haploid filaments.

Among the filamentous algae, there are plants that use heterogametes during sexual reproduction. For example, the genus *Oedogonium* contains filamentous algae in which the cells have cylindrical and netlike chloroplasts. Sexual reproduction (fig. 26.5) occurs when an enlarged specialized cell produces an egg, and other short, disklike cells each produce two sperm. The sperm, which look like small zoospores, escape and swim to an egg, after which the zygote is released and enters a period of dormancy. Upon germination, the zygote produces four zoospores, each of which may grow into a filament. Also, any vegetative cell may produce a zoospore asexually and this may develop directly into a filament.

It appears that the development of sexual reproduction was advantageous because it provided these green algae with *(a)* a means to survive unfavorable conditions and *(b)* offspring that had a new combination of genes. The protective coat of the zygote allows it to overwinter. Upon germination, the offspring has a new combination of genes that may be more suited to different environmental conditions. We also note that while some forms have isogametes (*Chlamydomonas*), others have heterogametes (*Volvox* and *Oedogonium*).

Multicellular Sheets

The genus *Ulva* contains green algae that are found in the sea close to shore. *Ulva* is commonly called sea lettuce because of a leafy appearance (fig. 26.6a). Members of this genus may resemble a form ancestral to later plants. *Ulva* is multicellular; and like primitive plants adapted to land, it has a noticeable sporophyte and gametophyte generation. However, both generations look exactly alike. Observation verifies that each haploid zoospore produced by the sporophyte generation develops directly into a gametophyte generation, while gametes produced by the gametophyte generation fuse to give a zygote that develops into the diploid sporophyte generation. While the gametophyte and sporophyte generations are codominant in *Ulva*, in terrestrial plants, one generation is typically dominant over (longer lasting than) the other.

Notice that the life cycles of both *Chlamydomonas* and *Ulva* contain flagellated zoospores that are capable of swimming. This is an adaptation to reproduction in the water.

Seaweeds

Multicellular green algae (such as *Ulva*), red algae, and brown algae are all seaweeds. Their color is dependent on the pigments they contain, which in red and brown algae mask the green color of their chlorophyll. These pigments enable the algae to collect the light they need for photosynthesis. When light strikes seawater, the various wavelengths penetrate to different depths. The accessory pigments in red algae absorb those wavelengths that penetrate deepest; those in brown algae absorb wavelengths that are quickly filtered out of water. Thus it is possible to predict the depth at which these two forms of algae could be found.

Brown Algae

Good examples of brown algae are found in the genus *Fucus* (fig. 26.7), which are often called *rockweeds*. They range from 1 to 3 feet in length and may be attached in great masses to rocks exposed at low tide in the north temperate zone. Air bladders hold the forked branches aloft in the water, and the tips of some of the branches, which are enlarged, contain the sex organs. The life

Evolution and Diversity

a.

Figure 26.6
Members of the genus *Ulva* have a life cycle known
as alternation of generations. In these plants the
sporophyte generation and the gametophyte
generation have the same appearance.

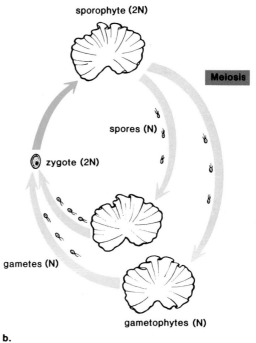

sporophyte (2N)

Meiosis

spores (N)

zygote (2N)

gametes (N)

gametophytes (N)

b.

Figure 26.7
A common representative of the genus *Fucus*.
a. General anatomy. *b.* Enlargement of receptacles
(reproductive structures).

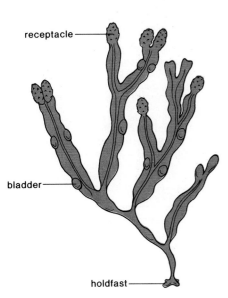

receptacle

bladder

holdfast

a.

b.

cycle of *Fucus* is so advanced that a gametophyte generation is not seen in some species; these have the diplontic cycle in which the adult is always diploid. The largest of the brown algae, the kelps, may be as long as 100 feet.

Red Algae

Like the brown algae, the red algae are multicellular, but they occur chiefly in warmer seawaters, growing both in shallow waters and as deep as light penetrates. Some forms of red algae are filamentous, but more often they are complexly branched; the branches having a feathery, flat, and expanded or ribbonlike appearance. Notably, red algae is used in the production of agar, the gelatinous solid medium on which bacteria and fungi may be grown.

Terrestrial Plants

A land existence offers some advantages to plants. One advantage is the greater availability of light for photosynthesis since water, even if clear, filters out light. Another advantage is that carbon dioxide and oxygen are present in higher concentrations and diffuse more readily in air than in water. Many adaptations, however, are required to live successfully on land (table 26.2). Plants have become adapted in ways that are quite different from those of humans and at the end of this chapter, we will compare the adaptation of humans to that of trees.

There are two main groups of plants adapted to living on land. These two groups—the bryophytes and tracheophytes—may be compared in these ways:

Bryophytes	Tracheophytes
Gametophyte dominant	Sporophyte dominant
No vascular tissue	Vascular tissue

The *dominant generation* is the conspicuous generation in the life cycle of a higher plant (fig. 26.6), the one that lasts longer and is usually considered *the* plant by the layperson.

Table 26.2 Comparison of water environment with land environment

Water	Land
1. The surrounding water prevents the organism from drying out, that is, prevents desiccation.	1. In order to prevent desiccation, the organism must obtain water, provide it to all body parts, and possess a covering that prevents evaporation.
2. The surrounding water buoys up the organism and keeps it afloat.	2. An internal structure is required for a large body to oppose the pull of gravity.
3. The water prevents desiccation and allows easy transport of reproductive units such as zoospores and swimming sperm.	3. The organism may provide a water environment for swimming reproductive units. Alternately, the reproductive units must be adapted to transport by wind currents or by motile animals.
4. The surrounding water prevents the fertilized egg (zygote) from drying out.	4. The developing zygote must be protected from possible desiccation.
5. The water maintains a relatively constant environment in regard to temperature, pressure, and moisture.	5. The organism must be capable of withstanding extreme external fluctuations in temperature, humidity, and wind.

Evolution and Diversity

a.

Figure 26.8
Bryophytes. *a. Marchantia* is a liverwort that can reproduce asexually by means of gemmae—minute bodies that give rise to new plants. As shown here, the gemmae are located in cuplike structures called gemma cups. *b.* Haircup moss with both gametophyte generation (*below*) and sporophyte generation (*above*). Spores are produced in the capsules held aloft by stalks.

b.

Bryophytes

The bryophytes (fig. 26.8) include the liverworts and mosses. Liverworts that have flattened lobed bodies, although widespread and well known, represent but a small fraction of the total number of species. Most species of liverworts are "leafy" and look like mosses, but examination shows that they have a distinct top and bottom surface, with numerous rhizoids (rootlike hairs) projecting into the soil. A moss plant is composed of a stemlike structure with radially arranged leaflike structures. Rhizoids anchor the plant and absorb minerals and water from the soil. Since bryophytes do not have vascular tissue, they *lack true roots, stems, and leaves*. Instead, they are said to have rhizoids, stemlike structures, and leaflike structures.

Figure 26.9
Life cycle in moss. Spores germinate to give
gametophyte generation. The sporophyte
generation occurs after fertilization.

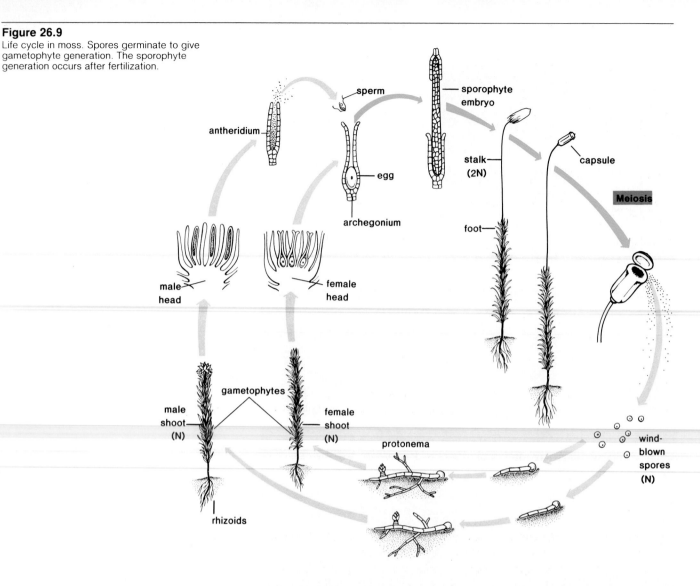

Moss Life Cycle

The moss plant just described is the dominant gametophyte generation that
produces the gametes. This is the more permanent and longer-lasting generation in bryophytes. In some mosses there are separate male and female gametophytes (fig. 26.9). At the tip of a male gametophyte are **antheridia** in
which swimming sperm are produced. After a rain or heavy dew, the sperm
swim to the tip of a female gametophyte where structures called **archegonia**
are found and within which eggs are produced. Antheridia and archegonia are
both multicellular sex organs, and each has an outer layer of jacket cells that
help protect the enclosed gametes from desiccation. After an egg is fertilized,
it is retained within the archegonium and begins development as the sporophyte generation. The *sporophyte generation,* which is parasitic on the gametophyte, consists of a *foot* that grows down into the gametophyte tissue, a
stalk, and upper capsule, or *sporangium,* where meiosis occurs and haploid
spores are produced. The spores are windblown and released in dry weather.

When a spore lands on an appropriate site, it germinates. A single row
of cells grows out and then branches. This algalike structure is called a **protonema.** After about three days of growth under favorable conditions, buds
appear at intervals along the protonema. Each of these sends down rhizoids
and grows upward, giving the generation we call a gametophyte. This com-
pletes the moss life cycle.

Figure 26.10
The growth of moss on rocks contributes to succession, the process by which rocks are eventually converted to fertile soil.

Adaptation

Bryophytes are incompletely adapted to life on land. To prevent water loss, the entire body is covered by a waxy cuticle. The zygote and embryo are also protected from drying out by remaining within the archegonium. The organism is dispersed or distributed to new locations by windblown spores. However, the bryophytes are limited in their adaptations. They lack vascular tissue, and the sperm have to swim in external moisture to reach the egg. It is for these reasons that bryophytes are restricted to moist locations.

Importance

Certain bryophytes and lichens colonize rocks (fig. 26.10) and slowly convert them to soil that can be used for the growth of other organisms.

Sphagnum, bog or peat moss, has commercial importance. This moss has special nonliving cells that can absorb moisture, which is why peat moss is often used in gardening to improve the water-holding capacity of the dirt. In some areas, like bogs, where the ground is wet and acid, dead mosses, especially sphagnum, accumulate and do not decay; this accumulation, called peat, can be used as a fuel.

Tracheophytes

The tracheophytes, with a dominant sporophyte generation, include those plants that are best adapted to a land existence. One advantage of having the sporophyte dominant is that this is the diploid generation; and if a faulty gene is present, it may be masked by a functional gene. In addition, possession of vascular (transport) tissue is a critical adaptation to the land environment.

The tracheophytes have two types of vascular tissue. *Xylem* conducts water and minerals up from the soil, and *phloem* transports organic nutrients from one part of the body to another. Because they have vascular tissue, the specialized body parts of tracheophytes can properly be called roots, stems, and leaves. Further, the strong-walled xylem cells support the body of the plant against the pull of gravity. The tallest organisms in the world are tracheophytes—the redwood trees of California.

There are both nonseed plants and seed plants among the tracheophytes. The first tracheophytes to evolve, including the ferns, are nonseed plants; but later the seed plants, that is, the gymnosperms and angiosperms, evolved.

Figure 26.11

Representative lower tracheophytes. These plants were widespread when the tracheophytes first evolved. *a. Equisetum* (horsetail). *b. Psilotum c. Selaginella* (club moss).

a.

b.

c.

Nonseed Plants

Primitive Tracheophytes

Among the first tracheophytes to evolve are the psilopsids, the club mosses, and the horsetails (fig. 26.11). The psilopsids are of particular interest because they may resemble the common ancestral form of the other tracheophytes. Formerly *Psilotum* was considered to be a psilopsid. Although this assumption

Figure 26.12
Representative fern. *a.* The sporophyte generation
of a bracket fern. *b.* The undersurfaces of the large
fronds are covered with sori, clusters of tiny
sporangia.

a.   b.

is now being questioned and *Psilotum* is considered by some to be a fern, it
is discussed here because it is a primitive genus of living tracheophytes. The
sporophyte body consists of stems with scalelike structures but no leaves. In
these living fossils, there is a horizontal stem (lacking roots) from which rhizoids extend on the underside, while green, photosynthetic upright branches,
with tiny scalelike structures, grow upward. This is the sporophyte generation,
and sporangia are located on the branches. The gametophyte generation is
separate, smaller than the sporophyte, and water dependent. In fact, the life
cycle of *Psilotum* is very close to that of the fern, which is discussed next.

Ferns

Ferns vary in appearance. Most are only a few feet tall, but in tropical rain
forests some are tall, resembling palm trees. The common temperate zone ferns
(e.g., *Pteridium*) often have a horizontal stem (rhizome) from which hairlike
roots project beneath and large leaves, or **fronds** (fig. 26.12), project above.
The fronds are often subdivided into a large number of leaflets. All parts of
the plant contain vascular tissue and therefore a fern has true roots, stems,
and leaves.

Figure 26.13
Life cycle of a fern. Spores germinate to form the
gametophyte generation. The sporophyte
generation occurs after fertilization.

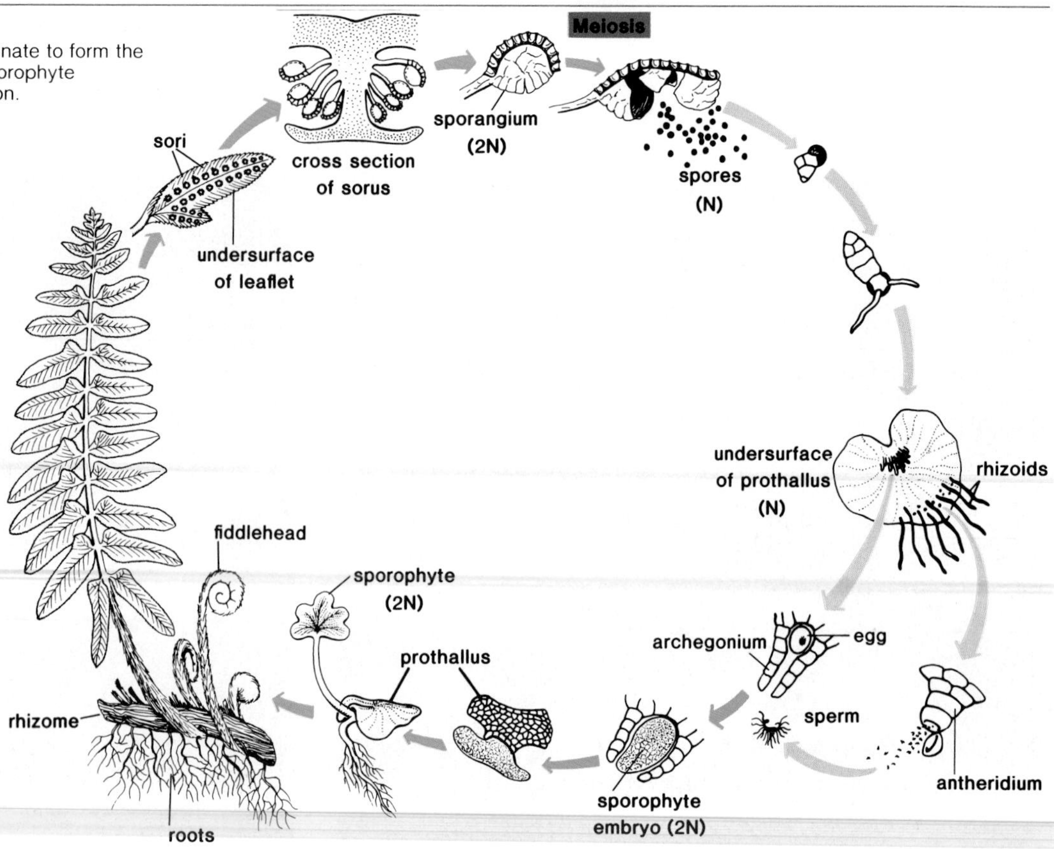

The plant referred to as the fern is the sporophyte generation (fig. 26.13). Sporangia develop in clusters called **sori** (singular **sorus**) (fig. 26.13b) on the underside of the leaflets. Within the sporangia, meiosis occurs and spores are produced. A band of thickened cells breaks open and expels the mature spores that are carried by the wind.

The gametophyte generation is a small, heart-shaped structure called a **prothallus.** Archegonia develop at the notch and at the tip there are antheridia. Spiral-shaped sperm swim from the antheridia to the archegonia where fertilization occurs. A zygote begins its development inside an archegonium, but the embryo soon outgrows the space available there. The young sporophyte becomes visible as a distinctive first leaf appears above the prothallus and the sporophyte's roots develop below it. Often, gametophyte and sporophyte tissues are distinctly different shades of green. The young sporophyte grows and develops into a mature sporophyte, the familiar fern plant.

Adaptation   Ferns are incompletely adapted to life on land due to the water-dependent gametophyte generation. This generation lacks vascular tissue and is separate from the sporophyte generation. Swimming sperm require an outside source of water in which to swim to the eggs in the archegonia. Ferns are most often found in moist environments where water is available for the gametophyte generation.

Importance of Nonseed Tracheophytes
During the Carboniferous period (fig. 26.14), the horsetails, club mosses, and ferns were abundant, very large and treelike. For some unknown reason a large quantity of these plants died and did not decompose completely. Instead they

were compressed and compacted to form the coal that we still mine and burn today. (Oil was formed similarly, but most likely formed in marine sedimentary rocks and included animal remains.)

Seed Plants

Seed plants are fully adapted to a land existence. The primary weakness observed in plant life cycles thus far has been the water-dependent gametophyte generation with its swimming sperm that require outside moisture to swim to the egg. This difficulty has been overcome by the seed plants.

These plants produce **heterospores,** called microspores and megaspores, instead of homospores, or identical spores. Microspores develop into immature male gametophyte generations while still retained within a microsporangium. These are released as **pollen grains** and are either carried by the wind or animals to the vicinity of the female gametophyte. Later, a mature pollen grain contains sperm. Thus, the *pollen grain replaces swimming sperm* in the seed plants.

The megasporangium is found within an ovule. Here a megaspore develops into the female gametophyte generation. While still within the ovule, the female gametophyte generation produces an egg that is fertilized by a sperm. The zygote becomes an embryonic plant enclosed within a seed. Thus, the *seed contains the embryonic plant* of the next sporophyte generation plus stored food. While nonseed plants are dispersed by spores, in seed plants, the seeds serve to distribute the species.

Figure 26.15
Life cycle in seed plants. Notice there is a separate
male and female gametophyte generation.

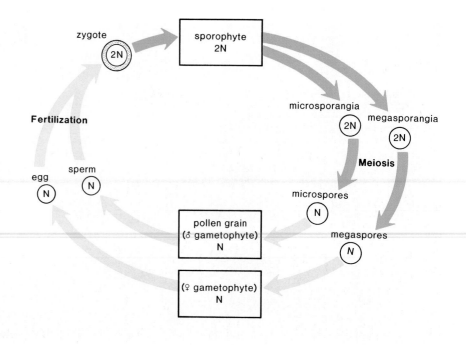

Life Cycle of Seed Plants

Figure 26.15 diagrams the life cycle of seed plants. In this life cycle:

1. The dominant generation is a diploid sporophyte.
2. Meiosis produces microspores within microsporangia and megaspores are produced within megasporangia. Each megasporangium is found within an ovule.
3. The mature male gametophyte generation is the pollen grain and this structure contains a sperm that fertilizes the egg. The female gametophyte generation is retained within the ovule, and within the ovule it produces an egg.
4. Fertilization results in an embryo that lies inside the original ovule. The ovule upon maturation becomes the released seed.
5. The seed contains the new sporophyte generation along with stored food usually within several protective layers.

Gymnosperms (Naked Seed Plants)

The gymnosperms (fig. 26.16) are woody plants that most often bear cones. Of these, pine trees are the most common. They often inhabit colder regions of the globe and therefore their needlelike leaves are modified to withstand low temperatures and harsh winds. The mesophyll lacks air spaces and the stomata are sunken. Thick-walled cells occur beneath the epidermis and the leaf veins are surrounded by an endodermis.

Figure 26.16
Gymnosperm trees. *a.* Hemlock; 1. entire
sporophyte generation, 2. close-up of branches
showing cones. *b.* Pitch pine; 1. entire sporophyte
generation, 2. close-up of branches showing cones.

a.1.

a.2.

b.1.

b.2.

Figure 26.17
Life cycle of pine. The mature sporophyte (pine
tree) has female pine cones that produce
megaspores that develop into female gametophyte
generations and male pine cones that produce
microspores that develop into male gametophyte
generations (mature pollen grains). Following
fertilization, immature sporophyte generations are
present in seeds located on the female cones.

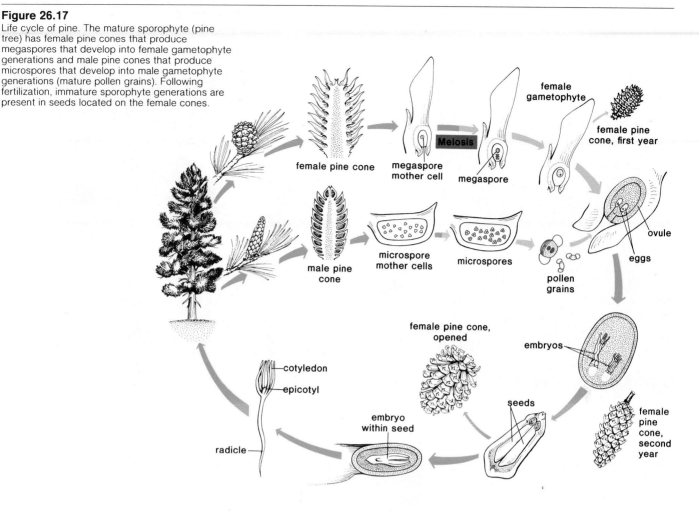

Pine Life Cycle

The sporophyte generation is dominant (fig. 26.17) and its sporangia are located on the scales of the cones. There are two types of cones—male and female.

Typically, each scale of the male (pollen) cone has two or more microsporangia on the underside. Within these sporangia, meiosis produces **microspores,** each of which develops into a pollen grain, the *male gametophyte generation.* A pollen grain has a thickened, protective coat and a pair of balloonlike bladders to the sides. These structures give the pollen grains some buoyancy in the wind. Pine trees release so many pollen grains during the pollen season that everything in the area around them may be covered with a dusting of yellow, powdery pine pollen.

Typically, each scale of the female (seed) cone has two ovules that lie on the upper surface. Within each ovule, there is a megasporangium where meiosis produces four **megaspores.** Only one of these spores develops into a *female gametophyte* that has two to six archegonia, each containing a single large egg lying near the ovule opening.

During pollination, pollen grains are transferred from the male to the female cones by the wind. Once enclosed within the female cone, the pollen grain develops a pollen tube that slowly grows into the ovule. Inside the tube are two sperm, one of which fertilizes an egg inside the ovule. The pollen tube discharges its sperm, and fertilization takes place 15 months after pollination. Notice, then, that fertilization is an entirely separate event from pollination, which is simply the transfer of pollen.

Figure 26.18
Bristlecone pines, the oldest plants in the world.

After fertilization, the ovule matures and becomes the seed that is composed of the embryo, stored food, and a seed coat. The winged seeds are exposed (or naked.)

Adaptation

The reproductive pattern of conifers has several important advantages over reproduction in other plants that have been considered so far. Transfer of pollen grains and growth of the pollen tube eliminate the requirement of surface water for swimming sperm. Enclosure of the dependent female gametophyte inside a cone protects it during its development and shelters the developing zygote as well. Finally, the embryo is protected by the seed and provided with a store of nutrients that support development for the first period of its growth following germination. All of these factors increase chances for reproductive success.

Importance

The conifers supply much of our lumber for building and wood for the production of paper, turpentine, and other products. The wood of a conifer is usually a softwood while that of flowering trees is usually a hardwood. The wood in a pine tree contains only tracheids and has no vessel elements and accompanying support cells.

As mentioned on page 140, it is possible to examine wood to determine the age of a tree. The oldest living thing on earth is believed to be a bristlecone pine (fig. 26.18) that has been dated at 4,600 years old by the Laboratory of Tree Ring Research, University of Arizona.

Figure 26.19
Representative angiosperm trees. *a.* Sweet
chestnut *b.* Wych elm *c.* Red maple *d.* Pear.

a.

b.

c.

d.

Evolution and Diversity

a.

b.

c.

d.

Angiosperms (Flowering Plants)

Angiosperms are the most successful of all the plants. All hardwood trees (fig. 26.19) including all the deciduous trees of the temperate zone and the broad-leaved evergreen trees of the tropical zone, are angiosperms although sometimes the flowers are inconspicuous. All herbaceous (nonwoody) plants common to our everyday experience, such as grasses and most garden plants (fig. 26.20), are flowering plants. Angiosperms are adapted to every type of habitat, including water (e.g., water lilies and duckweed).

The angiosperms are divided into two groups (fig. 7.8); the *monocots* (e.g., lily) and the *dicots* (e.g., buttercup). The monocots are always herbaceous, with flower parts in threes, parallel leaf veins, scattered vascular bundles in the stem, and one cotyledon or seed leaf in the embryo. The dicots may be woody or herbaceous, with flower parts usually in fours and fives, net veins, vascular bundles forming a circle in the stems, and two cotyledons or seed leaves in the embryo.

Figure 26.21
Life cycle of flowering plants. Microspores are produced in the anther and megaspores contained within ovules are produced in the ovary. Each microspore develops into a male gametophyte generation (pollen grain) and each megaspore develops into a female gametophyte generation. After fertilization, the embryo is enclosed within a seed that disperses the species.

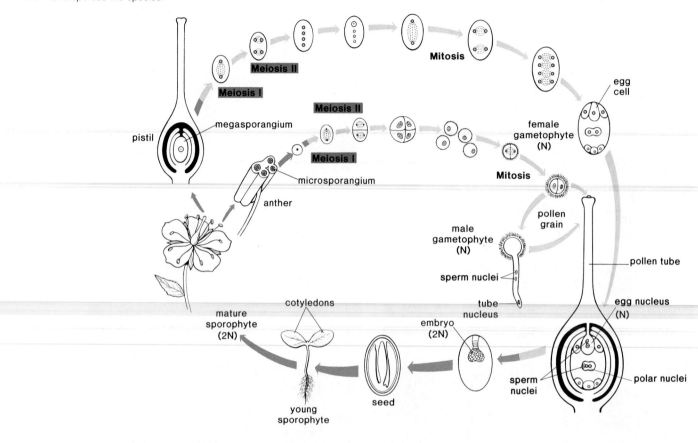

Flower

In angiosperms, the microsporangia and megasporangia are located in the **flower** (fig. 7.24). The flower is advantageous in three ways: it attracts insects and birds that aid in pollination (pollination by wind is also possible); it protects the developing female gametophyte; and it produces seeds enclosed by **fruit.** There are many different types of fruits, some of which are fleshy (e.g., apples) and some of which are dry (e.g., peas enclosed by pods). The fleshy fruits are sometimes eaten by animals, which may transport the seeds to a new location and then deposit them during defecation. Fleshy fruits may also provide additional nourishment for the developing embryo. Both fleshy and dry fruits provide protection for the seeds. Many so-called vegetables are actually fruits; for example, tomatoes, string beans, and squash. Nuts and berries and grains of wheat, rice, and oats are also fruits.

Flowering Plant Life Cycle

Stamens, the pollen-producing portion of the flower, consist of a slender *filament* with an *anther* at the tip. In the anther, meiosis within microsporangia produces microspores (fig. 26.21). Each microspore divides and becomes a binucleated pollen grain, the male gametophyte generation. The pollen grains are either blown by the wind or carried by pollinators (fig. 7.27) to the *pistil*, consisting of the *stigma, style,* and *ovary.* The ovary contains from one to many ovules, depending on the species of plant. Each ovule contains a mega-

Evolution and Diversity

Figure 26.22
Flowering plants are sources of food for the biosphere.

sporangium where meiosis produces one functional megaspore. The latter develops into a multicelled female gametophyte generation. One of these cells is an egg cell.

When pollen grains are transferred to the stigma, each develops a pollen tube that carries two sperm to the ovule. One sperm fertilizes the egg and the other unites with two other nuclei (polar nuclei) of the female gametophyte generation to form endosperm food for the embryo. This so-called double fertilization is unique to angiosperms. The mature ovule, or seed, contains the embryo and food enclosed with a protective seed coat. The wall of the ovary, and sometimes adjacent parts, develops into a fruit that surrounds the seeds. Thus angiosperms are said to have *covered seeds*.

Adaptation

Like the gymnosperms, angiosperms are well adapted to a land environment. Their vascular tissue is more complex and the seeds are enclosed by fruits. Both of these are selective advantages for the angiosperms.

Importance of Angiosperms

The angiosperms are the major producers in most terrestrial ecosystems. They provide the food (fig. 26.22) that sustains most of the animals on land, in-

cluding humans (table 7.B, p. 135). A study of all the plants that have been cultivated has led two authorities, as discussed in the reading below, to suggest that only 12 types of plants stand between humans and starvation. Also, grasses, alfalfa, and clover are used as forage, plant food for livestock.

Humans use angiosperms for many other functions. In the United States much of the wood produced is used for construction and the making of furniture. In developing countries, about half the timber cut is used for fuel. Flax and cotton are a source of natural fibers for making cloth, but cellulose, from any plant, can be treated to yield rayon. Oils are not only used in cooking, they are also, for example, used in perfumes and as medicines. Spices are from various parts of plants: peppercorns are small berrylike structures from a vine; cinnamon comes from the bark of a tree; cloves are dried-up flower buds. Various drugs are taken from angiosperm plants, including morphine and heroin from the juice of the poppy and marijuana from the leaves of the plant *Cannabis*.

Twelve Plants Standing between Man and Starvation

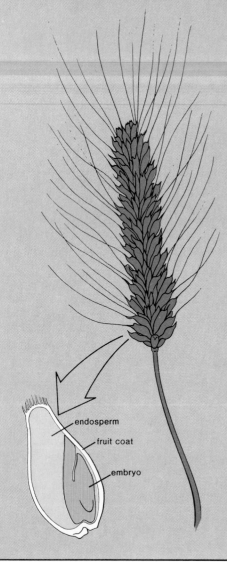

Since we earlier stressed the point that all our food is ultimately derived from plants, it may come as somewhat of a surprise to learn that relatively few species of plants are involved. Of the 800,000 kinds of plants estimated to be in existence only about 3,000 species have provided food, even in the form of nuts, berries, and other fleshy fruits. Virtually all of these food plants are angiosperms, or flowering plants. This is not surprising when you recall that only the angiosperms have seeds enclosed in a carpel [pistil] and hence only they produce true fruits, many of which are used by man for food.

Of the 3,000 plants noted above, only 150 species have been extensively cultivated and have entered the commerce of the world. And of the 150, only 12 species are really important—indeed it can be said that these 12 plants stand between man and starvation. If all 12 or even if a few of these cultivated plants were eliminated from the earth, millions of people would starve.

Three of these all-important species are cereals—**wheat, corn, and rice;** the last alone supplies the energy required by 50 percent of the people of the world. It is a remarkable fact that each of these cereals, or grains, is associated with a different major culture or civilization—wheat with Europe and the Middle East, corn or maize with the Americas, and rice with the Far East. Three of the 12 food plants are so-called root crops—**white,** or **Irish, potato** (not a root but a **tuber,** an enlarged tip of a rhizome, or horizontal underground stem); **sweet potato;** and **cassava,** or **manioc** or **tapioca,** from which millions of people in the tropics of both hemispheres derive their basic food. Two of the 12 are sugar-producing plants—**sugar cane** and **sugar beet.** Another pair of species are legumes—the **common bean** and **soybean,** both important sources of vegetable protein and hence sometimes referred to as the "poor man's meat." The final two plants of this august company are tropical tree crops—**coconut** and **banana.** . . .

endosperm
fruit coat
embryo

Wheat plant with enlarged grain.

Evolution and Diversity

Comparison of Plants

We have seen how plants are adapted to living on land. Not all land plants, however, are *completely* adapted to life on land. Table 26.3 indicates the relative degree of adaptation among bryophyte and tracheophyte plants.

The role of the haploid and diploid stages in the life cycle of various plants may be correlated with their relative adaptation to land (fig. 26.23). Algae, whose life cycle contains only haploid structures except for the diploid zygote, are adapted to a water environment. Mosses with a dominant gametophyte and small, dependent sporophyte have a limited distribution on land. Ferns have a well-developed sporophyte body that has vascular tissue, but they still require very wet conditions for growth of a small, independent nonvascular gametophyte and fertilization by swimming sperm.

Table 26.3 Adaptation summary of terrestrial plants

	Vascular tissue	**Sperm**
Nonseed plants: Spores disperse the species.		
Bryophytes	Both generations lack vascular tissue.	Swimming sperm require a source of outside moisture.
Psilopsida and ferns	The nonvascular gametophyte generation is separate and independent of the vascular sporophyte.	Same
Seed plants: Seeds disperse the species.		
Gymnosperms (naked seeds)	The nonvascular gametophyte generation is retained by and protected from desiccation by the vascular sporophyte.	Pollen grains replace swimming sperm.
Angiosperms (seeds covered)	Same	Same

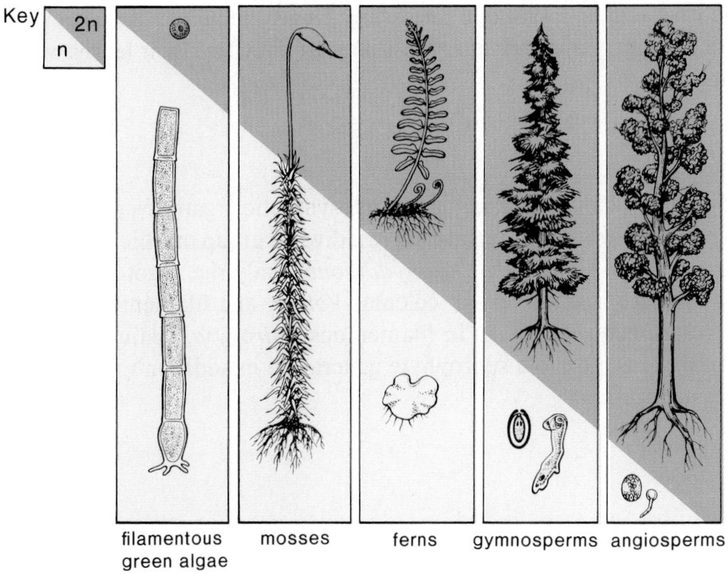

Key

Figure 26.23
The relative importance of the haploid and diploid generation among plants. In most green algae, with the exception of a few—such as those in the genus *Ulva*—only the zygote is diploid. In the rest of the plants depicted, the zygote develops into a diploid sporophyte generation. The haploid generation is then called the gametophyte generation.

filamentous green algae mosses ferns gymnosperms angiosperms

Table 26.4 Comparison of human to tree

	Human	Tree
Protective covering	Skin	Bark or waxy cuticle
Obtain water	Drinking	Absorption by roots
Transport water	Blood vessels	Xylem
Internal support	Skeleton	Woody xylem
Reproduction	Seminal fluid and vaginal secretions provide water for sperm during sexual intercourse.	Pollen grain. Pollen tube allows sperm to reach egg.
Protection of embryo	Internal development in female uterus.	Partial internal development in ovule. Seed coat prevents desiccation.
Constancy of internal environment	Maintains a constant internal environment.	Dormancy during winter, and other unfavorable conditions.

The gymnosperms and angiosperms are widely distributed on land because the large, dominant sporophyte is well adapted to terrestrial life. Furthermore, the delicate spores, gametes, zygotes, and embryos are enclosed within protective coverings produced by the sporophyte plant.

Comparison of Animal and Plant

As indicated in the first paragraph of this chapter, many animals too, are adapted to life on land. Table 26.4 compares the adaptations of *humans* to that of *trees* and shows that they both have similar adaptations. Two adaptations seem to be of particular interest.

Terrestrial organisms need to protect the gametes and zygote from drying out. In humans, swimming sperm are passed directly from male to female during sexual intercourse. In seed plants, however, the pollen grain replaces swimming sperm. Pollination is the process by which pollen grains are brought to the approximate location of the female gamete. This adaptation seems appropriate for plants in which the structures that produce male and female gametes may remain quite distant from one another.

In regard to maintaining a constant internal environment, higher animals have become somewhat independent of the external environment. Terrestrial plants use an entirely different approach. They often become dormant, during which time their metabolic needs are greatly reduced. We are most familiar with dormancy in deciduous plants because they lose their leaves when dormancy sets in.

Summary

The plant kingdom includes multicellular photosynthetic organisms and closely related unicellular ones. The green algae are a diverse group in which different life cycles are seen. In single cell *Chlamydomonas*, only the zygote is diploid and the gametes are isogametes. In colonial *Volvox* and filamentous *Oedogonium*, there are heterogametes. In filamentous *Spirogyra*, conjugation occurs. In *Ulva* there is a diploid sporophyte generation in addition to a haploid gametophyte generation.

Evolution and Diversity

Seaweeds include multicellular green, red, and brown algae. Green algae and brown algae are typically found in cool waters along rocky coasts. Red algae are found farther from the shore in warm tropical waters.

There are two main groups of plants adapted to a land existence: (1) the bryophytes (e.g., mosses) are not well adapted; the nonvascular gametophyte (N) is dominant and requires external water for swimming sperm to reach the egg; (2) the tracheophytes (ferns, pines, and flowering plants) vary in their degree of adaptation. The vascular sporophyte (2N) generation is dominant in the fern, but there is a separate water-dependent gametophyte. Swimming sperm need external water to reach the egg. In both mosses and ferns, the species is dispersed by means of spores.

Gymnosperms and angiosperms have microspores and megaspores. In the pine, these are located in pine cones. In the male cone, the microspores develop into pollen grains, the male gametophyte generation. In the female cone, the megasporangium lying within an ovule, produces a megaspore. Each megaspore develops into a female gametophyte generation. After pollination, the pollen grain develops a pollen tube within which a sperm travels to an egg. Following fertilization, the ovule becomes a windblown seed.

In flowering plants, the anther portions of the stamen produce pollen grains. The ovules are found in the ovary, a portion of the pistil. After pollination and during fertilization, one sperm fertilizes an egg and the other initiates development of endosperm. The ovules become seeds, still enclosed by the ovary, which contributes to the development of a fruit. In gymnosperms and angiosperms, seeds disperse the species.

Gymnosperms and angiosperms are well adapted to life on land. The gametophyte generation is small and protected by the vascular, diploid sporophyte generation. Vascular tissue gives trees an internal skeleton that helps them oppose the force of gravity.

Study Questions

1. Name the two major groups of plants studied in this chapter. (pp. 561, 566)
2. Describe each of the green algae studied, emphasizing their reproductive patterns, both asexual and sexual. (pp. 561–64)
3. Compare the three types of seaweed mentioned in this chapter. (pp. 564–66)
4. Compare and contrast the bryophytes to the tracheophytes in regard to dominant generation and presence of vascular tissue. (p. 566)
5. Draw the diagram of alternation of generations for those terrestrial plants in which the spore disperses the species. (p. 561)
6. Draw the diagram of alternation of generations for those terrestrial plants in which the seed disperses the species. (p. 574)
7. Compare the life cycles of the moss, fern, pine, and flowering plant, emphasizing adaptation to life on land. (pp. 567–81)
8. Name the reproductive parts of the flower and state a function for each part. (pp. 580–81)
9. Give examples to support the statement: The moss and fern are not fully adapted to life on land, but the pine and flowering plant are fully adapted. (p. 583)
10. With reference to table 26.4, contrast the adaptations of a tree and human to the land environment. (p. 584)

Selected Key Terms

alternation of generations (awl''ter-na'shun uv jen''ĕ-ra'shunz)
gametophyte (gam'ĕ-to-fīt)
sporophyte (spo'ro-fīt)
algae (al'je)
isogametes (i''so-gam'ēts)
colonial (ko-lōn'e-al)
filament (fil'ah-ment)
conjugation (kon''ju-ga'shun)
bryophytes (bri'o-fīts)
Tracheophytes (tra'ke-o-fīts)
antheridium (an''thah-rid'y-um)
archegonium (ar''kĕ-go'ne-um)
frond (frond)
sorus (so'rus)
gymnosperm (jim'no-sperm)
angiosperm (an'je-o-sperm'')
heterospores (het'er-o-sporz)
pollen (pol'en)
seed (sēd)
fruit (fro͞ot)

Further Readings

Jensen, W. A., and Salisbury, F. B. 1972. *Botany: An ecological approach*. Belmont, Calif.: Wadsworth.

Muller, W. H. 1979. *Botany: A functional approach*. 4th ed. New York: Macmillan.

Raven, P. H., et al. 1981. *Biology of plants*. 3d ed. New York: Worth.

Rayle, D., and Wedberg, L. 1980. *Botany: A human concern*. 2d ed. Boston: Houghton Mifflin.

Stern, K. R. 1982. *Introductory plant biology*. 2d ed. Dubuque, Iowa: Wm. C. Brown Publishers.

Tippo, O., and Stern, W. L. 1977. *Humanistic botany*. New York: W. W. Norton.

Evolution and Diversity

27

Animal Kingdom

Chapter concepts

1 Animals are classified according to certain criteria such as body plan, symmetry, number of germ layers, and level of organization.

2 There is an increase in complexity of organization when the animal groups are surveyed from those first evolved to those latest evolved.

3 Animals are adapted to their way of life; for example, those adapted to an inactive life may be contrasted to those adapted to an active life, those adapted to an aquatic existence may be contrasted to those adapted to a terrestrial existence.

Figure 27.1
Evolutionary tree of the animal kingdom. On the
basis of embryological and other anatomical
evidence, it is possible to divide the animal
kingdom into the major groups shown. In rare
instances intermediate forms that may be closely
related to certain common ancestors have been
discovered.

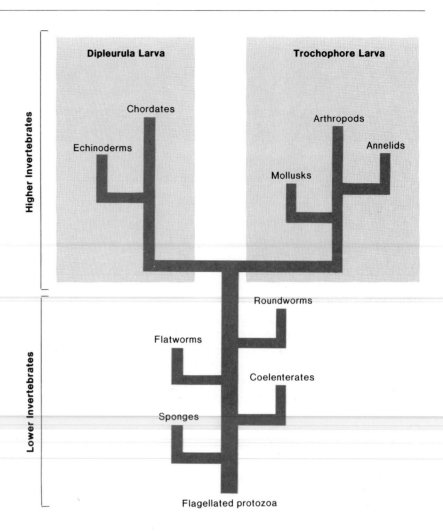

Animals are heterotrophic and must take in food. In contrast to stationary
green plants that absorb energy from the sun and make their own organic
food, animals are nongreen and possess some means of locomotion that enables
them to acquire food.

A predator that actively seeks out and captures food exemplifies best the
animal way of life. Predators have bilateral symmetry, good musculature, and
a well-developed nervous system including sense organs. All of these help the
animal seek prey and escape enemies. Good predators also have a means of
seizing and digesting their food.

All animals must digest their food, carry on gas exchange, excrete waste,
circulate nutrient and waste products to and from cells, coordinate their movements, protect themselves, and reproduce and disperse the species. The more
complex animals have organ systems to carry out these functions; in simple
animals, these functions are sometimes carried out by specialized tissues.

Evolution and Classification

The History of Life table (fig. 23.1) shows that all modern phyla of animals[1]
had evolved by the beginning of the Paleozoic era some 600 million years ago.
The evolutionary tree of animals (fig. 27.1) indicates that animals are believed

[1]The classification system utilized in this text is given on page A-3.

Evolution and Diversity

Table 27.1 Primitive versus advanced

	Most primitive	Primitive	Advanced	Most advanced
Body plan	None	Sac plan	Tube within tube	Tube within tube with specialization of parts
Symmetry	None	Radial	Bilateral	Bilateral with cephalization
Germ layers	None	Two	Three	Three
Level of organization	None	Tissues	Organs	Organ systems
Body cavity	Acoelomate	Acoelomate	Pseudocoelom	True coelom
Segmentation	Nonsegmented	Nonsegmented	Segmented	Segmented with specialization of parts

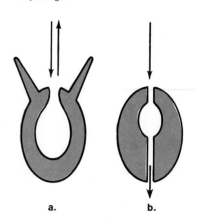

Figure 27.2
Animals have two basic body plans. *a.* Sac plan with only one opening. *b.* Tube-within-a-tube plan with two openings.

a.   b.

to have arisen from flagellated protozoans—perhaps a colonial form whose cells become differented into various types of cells.

After animals became moderately complex, there was a split into two main lines. Indeed, the evolutionary tree of animals resembles a tree with two main branches. The animal phyla located on the main trunk of the tree are referred to as the *lower invertebrates* in this text, and the animals of the upper branches include the *higher invertebrates* and the vertebrates. **Invertebrate** animals lack a dorsal backbone, while **vertebrates** have a backbone made up of vertebrae.

Classification of Animals

A study of the evolution of animals indicates that increased complexity of organization can be related to certain anatomical features of structure (table 27.1). Classification is based on the presence of these features, which are termed advanced because they evolved later in time.

Body Plan

Two body plans (fig. 27.2) are observed in the animal kingdom: the **sac plan** and the **tube-within-a-tube plan.** Animals with the sac plan have only one opening, which is used both as an entrance for food and an exit for waste. Animals with the tube-within-a-tube plan have an entrance for food and an exit for waste. Two openings allow specialization of parts to occur along the length of the tube.

Symmetry

Asymmetry means that the animal has no particular symmetry. **Radial symmetry** means that the animal is circularly organized and, just as with a wheel, it is possible to obtain two identical halves no matter how the animal is sliced longitudinally. **Bilateral symmetry** means that the animal has a definite left and right half so that only one longitudinal cut down the center of the animal will produce two equal halves (fig. 27.3). Radially symmetrical animals tend to be attached to a substrate, or *sessile.* This type of symmetry is useful to these animals since it allows them to reach out in all directions from one center. Bilaterally symmetrical animals tend to be active and to move forward with one anterior end. This end develops a head region (called *cephalization*) that is acutely aware of the environment and aids the animal in its forward progress.

Figure 27.3
Animals have two types of symmetry. *a.* Bilateral symmetry. Notice that only the one longitudinal cut shown will give two identical halves of the animal. *b.* Radial symmetry. Notice that any longitudinal cut, such as those shown, will give two identical halves of the animal.

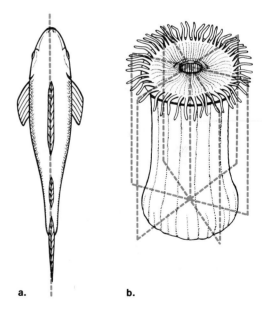

a.   b.

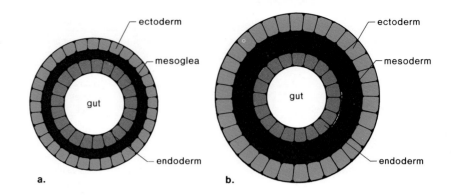

Figure 27.4
Animals have either no germ layers, two germ layers, or three germ layers. *a.* Cross section of animal with two germ layers. In these animals, a packing material called mesoglea is found between the two layers. *b.* Cross section of animal with three germ layers. Endoderm surrounds the gut lumen; mesoderm is the middle layer; and ectoderm is the outer germ layer.

Figure 27.5
Comparison of mesoderm organization. *a.* In acoelomate animals, the mesoderm is packed solidly. *b.* Pseudocoelomate animals have mesodermal tissue inside the ectoderm, but not adjacent to the gut endoderm. *c.* In coelomate animals, there is mesodermal tissue both inside the ectoderm and adjacent to the gut endoderm. True coeloms are body cavities completely lined by mesodermal tissue. Mesenteries hold organs in place within the body cavity.

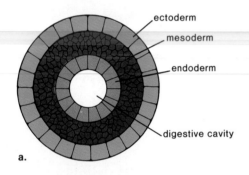

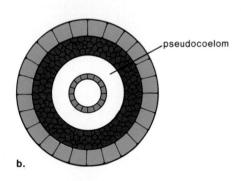

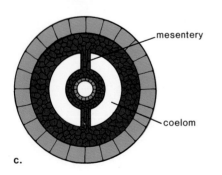

Germ Layers

Although a total of three germ layers is seen in most animals during embryonic development (p. 420), in fact, some animals have two germ layers (fig. 27.4a). Such animals have the tissue level of organization. Animals with three germ layers (fig. 27.4b) have the organ level of organization.

Body Cavity

The body cavity we wish to consider is the space surrounding the digestive system. Some animals do not have such a space (fig. 27.5a). Animals that do have this cavity are either (1) pseudocoelomates, having a **pseudocoelom** (fig. 27.5b), or coelomates, having a true **coelom** (fig. 27.5c). A pseudocoelom is lined with mesoderm only beneath the body wall, but a true coelom is lined with mesoderm both beneath the body wall and around the gut.

Segmentation

Some animals are nonsegmented and some have repeating units called segments. It is easy to tell, for example, that an earthworm (fig. 27.21) is segmented because its body appears to be a series of rings. Segmentation leads to specialization of parts in that the various divisions of the body can become differentiated for specific purposes.

Lower Invertebrates

The lower invertebrates include the sponges, coelenterates, flatworms, and roundworms. By studying the animals in this order, we will observe an increase in complexity that may reflect the order in which these animals evolved. Nevertheless, it is difficult to determine their exact evolutionary relationship.

Sponges

Most sponges are marine and are more abundant in warm ocean water, near the coast. Some sponges grow on rocks and are brightly colored, appearing almost lichen-like when seen at a distance. Sponges are often shaped like vases, with either simple flat walls or convoluted walls containing canals. Regardless, the wall of a sponge is perforated by numerous **pores** surrounded by contractile cells that are capable of regulating their size.

The wall of a sponge is made up of three layers of cells (fig. 27.6). The outer layer is made up of flattened **epidermal cells.** The inner layer contains **collar cells** with flagella whose constant movement produces water currents that flow through the pores into the central cavity and out through the upper opening of the body (called the **osculum**).

Evolution and Diversity

Figure 27.6
Generalized macroscopic and microscopic
anatomy of a sponge. The arrows indicate the flow
of water that is kept moving by the beating of the
flagella of the collar cells, one of which is drawn
enlarged. Notice that because the water enters the
animal by way of pores, the mouth is used as an
exit rather than an entrance. Sponges are classified
according to the chemical makeup of their
spicules, which serve as an internal skeleton.

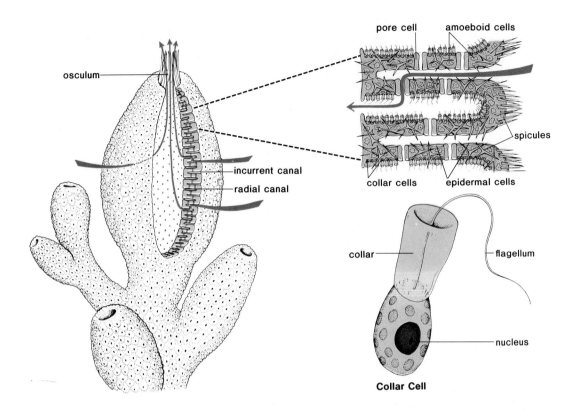

Sponges are **sessile filter feeders.** This means that they remain in one place as an adult and that the food they acquire filters through the pores. Microscopic food particles brought by the water are engulfed by the collar cells and digested by them in food vacuoles or are passed to the amoeboid cells for digestion.

The **amoeboid cells** in the middle layer of the sponge not only act as a circulatory device to transport nutrients from cell to cell, they also produce **spicules,** the needle-shaped structures that serve as the internal skeleton of a sponge, and the sex cells, the egg and sperm. Cross fertilization within the central cavity, where the gametes are released, is usually assured in that a sponge at any one time produces only eggs or sperm. Fertilization results in a zygote that develops into a ciliated larva that may swim away to a new location. Such a larva assures dispersal of the species for the sessile adult sponges. Sponges also reproduce by budding, and this process produces whole colonies of sponges that may become quite large. Like all less-specialized organisms, sponges are capable of **regeneration,** or growth of a whole from a small part. Thus, if a sponge is removed, chopped up, and returned to the water, each piece may grow to a complete sponge.

Figure 27.7
The two body forms of coelenterates. The drawings indicate the manner in which these germ layers surround the central gastrovascular cavity.
a. Polyps, with the oral side uppermost, are usually attached to surfaces. *b.* Medusae, with the aboral side uppermost, are free-swimming.

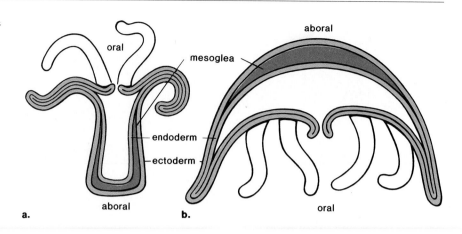

Figure 27.8
Members of the genus *Obelia* have a life cycle that is a type of alternation of generations in that the stationary colonial polyp produces medusae asexually and the medusae produce egg and sperm for sexual reproduction. Following fertilization, the zygote develops into a larva that settles down to become a colonial polyp. Notice that the motile medusae serve to disperse the species in this life cycle.

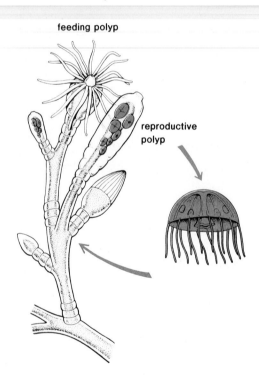

feeding polyp

reproductive polyp

Coelenterates

The body of a coelenterate is often a two-layered sac. (*Coelenterate* means hollow sac). The outer layer, ectoderm, is separated from the inner layer, endoderm, by a jellylike material called **mesoglea** (fig. 27.4). Typically, a ring of tentacles surrounds the mouth region of coelenterates, contributing to their *radial symmetry* (fig. 27.3).

Some coelenterates, referred to as **hydroids** or **polyps** (fig. 27.7), have a tubular shape, with the mouth region directed upward. Others, which have a bell shape with the mouth region directed downward, are called **jellyfishes** or **medusae**. The polyp is adapted to a sessile life, while the medusa is adapted to a floating or free-swimming existence. At one time, both body forms may have been a part of the life cycle of all coelenterates since today we see an alternation of generations[2] of these two forms in certain coelenterates, such as members of the genus *Obelia* (fig. 27.8). When alternation of generations does exist, the polyp stage produces medusae, and the medusae, which produce egg and sperm, disperse the species.

All coelenterates have specialized stinging cells (fig. 27.10) with fluid-filled capsules called **nematocysts.** Each nematocyst contains a long, spirally coiled hollow thread. When the trigger of a stinging cell is touched, the discharged thread, which sometimes contains poison, serves to stun either prey or enemy. Thereafter, the tentacles capture prey and stuff it into the central cavity.

Coelenterates are quite diversified (fig. 27.9). The **Portuguese man-of-war,** whose nematocysts may cause serious or even fatal poisoning in humans, is a colony of polyps suspended from a large medusoid form that serves as a gas-filled float. Many species of jellyfish, such as *Aurelia,* show alternation of generations, but the two generations are not equal—the medusa is the primary stage and the polyp remains quite small and insignificant. **Sea anemones** are solitary polyps with thick walls. They may be brightly colored and look like beautiful flowers. **Corals** are similar to sea anemones, but they have calcium carbonate skeletons. Some corals are solitary, but most are colonial, with either flat, rounded, or upright and branching colonies. The slow accumulation of coral skeletons has formed reefs (p. 762) in the South Pacific, including the Great Barrier Reef along the eastern coast of Australia. An ancient coral reef that now lies beneath Texas is the source of petroleum for that state.

[2]This is not the same as alternation of generations in plants because here both generations are diploid.

a.

Figure 27.9
Coelenterate diversity. *a.* Medusae of *Aurelia*. A thick layer of jelly (mesoglea) gives the medusae buoyancy and accounts for their common name, jellyfish. *b.* Living coral polyps. Each polyp is individual but has a chalky skeleton that is joined to its neighbor's. The polyps feed on zooplankton but may contain symbiotic algae that contribute to their nutrition. *c.* Sea anemone. These animals are called the "flowers of the sea," but despite their appearance they are carnivorous animals.
d. Jellyfish. This large jellyfish has no tentacles around its bell margin. It has eight oral arms whose deep folds take up plankton. *e. Hydra*. The mouth lies between the long tentacles that reach out and grasp food. Asexual reproduction by means of budding is apparent in this micrograph of an animal that is just visible to the naked eye.

b.

c.

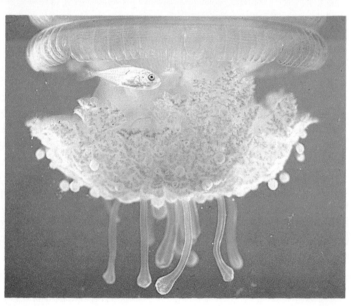

d.

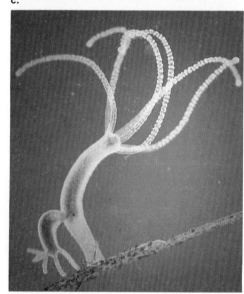

e.

Figure 27.10
Hydra anatomy. Compare the longitudinal section
(*far left*) with the cross section (*lower right*). Before
discharge, the nematocyst is tightly coiled within a
stinging cell (*upper right*) and after discharge, it is
extended.

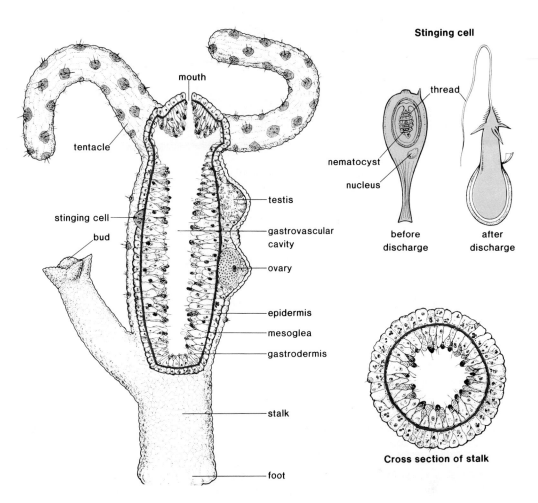

Stinging cell

thread

nematocyst

nucleus

before
discharge

after
discharge

mouth

tentacle

testis

stinging cell

gastrovascular
cavity

bud

ovary

epidermis

mesoglea

gastrodermis

stalk

foot

Cross section of stalk

Hydra

It is customary to study the anatomy of *Hydra* (fig. 27.10), as representative
of coelenterates. Animals that belong to this genus are likely to be found attached to underwater plants or rocks in most lakes and ponds. The body is a
small, tubular polyp about one-quarter inch in length. Although hydras usually remain in one place, they may glide along on their base, or even move
rapidly by means of somersaulting. Hydras, like other animals capable of locomotion, possess both muscular and nerve cells. The nerve cells form a connecting network throughout the mesoglea known as the **nerve net.** The nerve
net makes contact with the outer layer of cells, called the **epidermis,** and the
inner layer, called the **gastrodermis.** These cells contain contractile fibers.

The cells of the gastrodermis secrete digestive juices that pour into the
central cavity. The enzymes begin the digestive process, which is completed
within food vacuoles when small pieces of the prey are engulfed by the cells
of the gastrodermis. Nutrient molecules are passed by diffusion to the rest of
the cells of the body. The presence of the large inner cavity makes it possible

for all cells to exchange gases directly with the surrounding medium. Since this function is carried out by a vascular system in more complex animals, the cavity is known as a **gastrovascular cavity.**

In *Hydra,* the mesoglea not only contains the nerve net. It (the mesoglea) also contains interstitial, or embryonic cells, capable of becoming other types of cells. For example, they can produce the ovary and testes and probably also account for the animal's great regenerative powers. Like the sponges, coelenterates can grow whole from a small piece.

Flatworms

In the evolution of animals, flatworms represent a considerable advance in that they have three germ layers. The presence of mesoderm not only gives bulk to the animal, it also allows for greater complexity of internal structure. Free-living forms have muscles and excretory, reproductive, and digestive organs. The worms lack respiratory and circulatory organs but since the body is flattened, diffusion alone is adequate for the passage of oxygen and other substances from cell to cell. Very important is the presence of bilateral symmetry and good cephalization along with a well-developed nervous system, including sense organs, in free-living forms. This combination makes them efficient predators.

Flatworms are nonsegmented, lack a coelom, and have the sac plan with only one opening. Therefore, if we evaluate them according to table 27.1, we see that they have a combination of primitive to advanced features.

There are three classes of flatworms: one is free living and two are parasitic. Parasites are degenerate forms of the free-living specimen that, of course, exemplifies best the characteristics of the phylum. Thus we will begin with the free-living specimen, the planarian.

Planarians

Freshwater planarians (e.g., *Dugesia;* fig. 27.11) are small (several millimeters to several centimeters), literally flat worms. Some tend to be colorless; others have brown or black pigmentation. Planarians live in lakes, ponds, streams, and springs where they feed on small living or dead organisms, such as worms or crustacea.

Since planarians live in fresh water, water tends to enter the body by osmosis. They have a water-regulating organ that especially rids the body of excess water. The organ consists of a series of interconnecting canals that run the length of the body on each side. The beating of cilia in specialized structures keeps the water moving toward the excretory pores. The beating of the cilia reminded some early investigator of the flickering of a flame and so the excretory organ is called a **flame-cell system** (fig. 27.11b).

Planarians are **hermaphroditic,** which means that they possess both male and female sex organs. The worms practice cross fertilization; the penis of one is inserted into the genital pore of the other. The fertilized eggs hatch in two to three weeks as tiny worms.

Planarians are often used in biology laboratories to illustrate regeneration. If a worm is cut crosswise, it usually grows a new head or tail as is appropriate. Planarians have also been used in so-called memory experiments. In these experiments, planarians were trained to swim mazes and then were cut up and fed to untrained planaria. When the cannibals were subsequently taught the same task, they learned faster than the first set. The exact significance of these experiments is debatable, but they have led to some interesting student speculations as to how best to acquire the knowledge of teachers.

Figure 27.11
Flatworm anatomy as exemplified by a planarian worm. *a.* Photomicrograph of worm. This photo shows the oral side of the animal, the pharynx (pink) leads to the digestive organ, which has been darkly stained to illustrate the manner in which it ramifies throughout the body. *b.* Details of organ anatomy. 1. Excretory organ is composed of canals that have flame cells (*enlarged drawing*) whose beating cilia draw in fluid that is excreted by way of nephridiopores. 2. The nervous system has a ladder appearance because cross fibers stretch between longitudinal fibers that extend the length of the animal. 3. The digestive organ is highly branched, and since the worms are hermaphroditic there are both male and female reproductive organs.

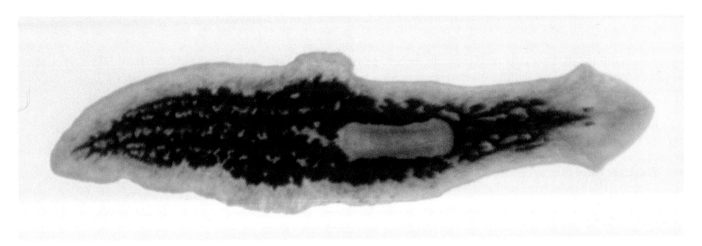

a.

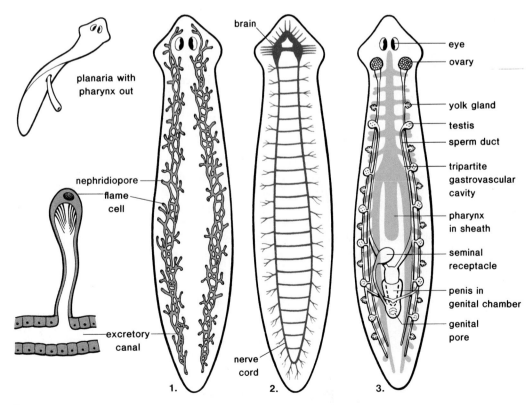

planaria with pharynx out

brain

nephridiopore

flame cell

excretory canal

nerve cord

1.

2.

eye

ovary

yolk gland

testis

sperm duct

tripartite gastrovascular cavity

pharynx in sheath

seminal receptacle

penis in genital chamber

genital pore

3.

b.

Parasitic Flatworms

There are two types of parasitic flatworms: tapeworms (cestodes) and flukes (trematodes). The structure of both these worms illustrates the modifications that occur in parasitic animals (table 27.2). Concomitant with the loss of predation, there is an absence of cephalization; the anterior end notably carries hooks and/or suckers for attachment to the host. The parasite acquires nutrient molecules from the host, and the digestive system is reduced. The presence of mucopolysaccharide coating, called the **glycocalyx,** protects the outer integument against host attack. The extensive development of the reproductive system, with the production of millions of eggs, may be associated with difficulties in dispersing the species. Both parasites utilize a *secondary host,* or intermediate host, to transport the species from main host to main host. The *primary host* contains the sexually mature adult; the secondary host(s) contain(s) the larval stage or stages.

A **tapeworm** has a head region (fig. 27.12), containing hooks and suckers for attachment to the intestinal wall of the host. Behind the head region, called a **scolex,** there is a short neck and then a long series of proglottids. **Proglottids** are segments, each of which contains a full set of both male and female sex organs. Thus, the tapeworm is little more than a reproductive factory. There are excretory canals but no digestive system and only the rudiments of nerves.

After fertilization, the proglottids become nothing but a bag filled with developing embryos (larvae). Mature proglottids such as these break off and, as they pass out with the host's feces, the larvae enclosed by a protective covering are released.

Table 27.2 Free-living versus parasitic worms

	Planarias	Flukes	Tapeworms
Body wall	Ciliated epidermis	Glycocalyx covers integument	Glycocalyx covers integument
Cephalization	Yes. Eyespots and auricles	No. Oral sucker	No. Scolex with hooks and suckers
Nervous connections	Nerves and brain	Reduced	Reduced
Digestive organ	Ramifies	Reduced	Absent
Reproductive organs	Hermaphroditic	Increased in volume	Greatly increased in volume
Larva	Absent	Present	Present

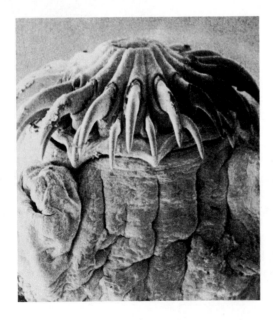

Figure 27.12
Scanning electron micrograph of tapeworm (*Taenia*) scolex. The scolex contains hooks and suckers that permit the animal to cling to the wall of the digestive tract.

Figure 27.13

Life cycle of tapeworm (*Taenia*), showing mature proglottid (*right*) and gravid proglottid (*left*) in detail. The gravid proglottid is little more than a sac of eggs.

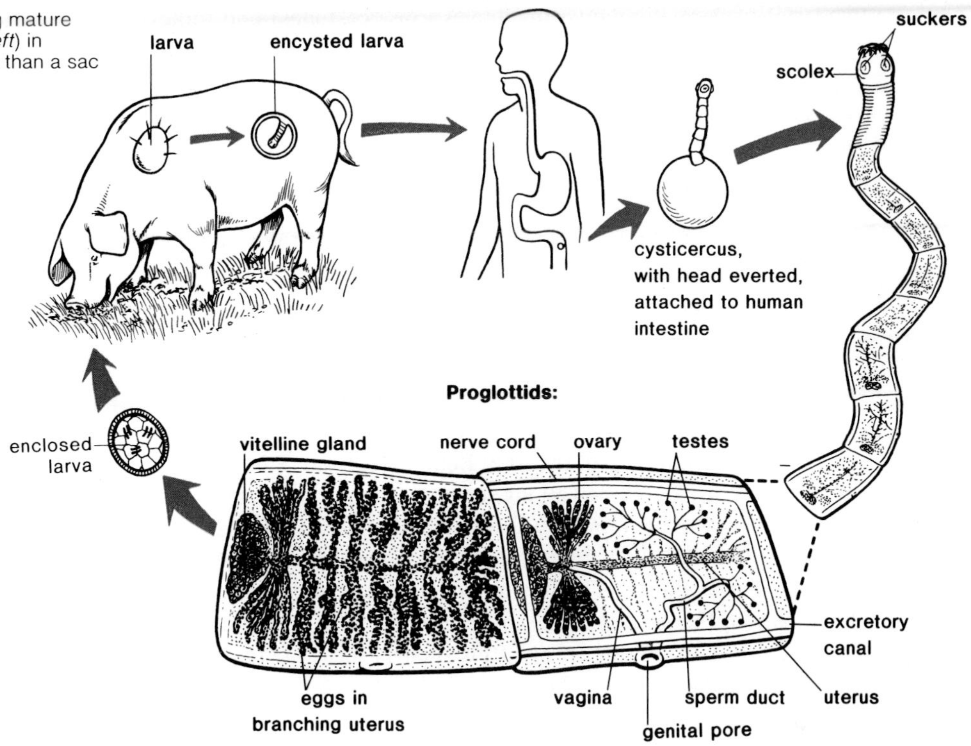

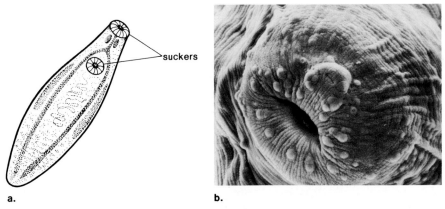

larva

encysted larva

suckers

scolex

cysticercus, with head everted, attached to human intestine

enclosed larva

Proglottids:

vitelline gland

nerve cord

ovary

testes

eggs in branching uterus

vagina

sperm duct

uterus

genital pore

excretory canal

Figure 27.14

Fluke anatomy. *a.* Drawing showing general anatomy. Flukes use suckers for attachment to the host. *b.* Scanning electron micrograph of fluke (*Gorgoderina attenuata*) oral sucker. Note upraised structures, which are believed to be sensory in nature.

suckers

a.

b.

If feces-contaminated food is fed to pigs or cattle, the larvae escape when the covering is digested away. They burrow through the intestinal wall and travel in the bloodstream to finally lodge and encyst in muscle. Here a **cyst** means a small, hard-walled structure that contains a small immature worm. When humans eat raw, or rare, infected meat, the worms break out of the cyst, attach themselves to the intestinal wall, and grow to adulthood. Then the cycle begins again (fig. 27.13).

There are many different types of **flukes** (fig. 27.14a), usually designated by the type of vertebrate organ they inhabit; for example, there are blood, liver, and lung flukes. While the structure may vary slightly, in general the fluke body tends to be oval to elongate with no definite head except that the oral sucker surrounded by sensory papillae is at the anterior end (fig. 27.14b).

Evolution and Diversity

Usually there is at least one other sucker for attachment to the host. Inside, there is a reduced digestive, nervous, and excretory system. There is a well-developed reproductive system, and the adult fluke is usually hermaphroditic although there are exceptions.

A blood fluke causes **schistosomiasis** in Africa and South America. This disease is especially prevalent in areas with irrigation ditches because the secondary host is a freshwater snail. The disease is spread when egg-laden human feces gets into the water and newly hatched larvae enter the snails. Asexual reproduction occurs within the snail and the resulting larvae penetrate human skin to enter blood vessels where they mature.

The Chinese liver fluke requires two hosts: the snail and the fish. Humans contract the disease when they eat uncooked fish. The adults reside in the liver and deposit their eggs in the bile duct, which carries the eggs to the intestine.

Roundworms

Roundworms (nematodes), as their name implies, are rounded rather than flattened worms. They have a smooth outside wall, indicating that they are nonsegmented. These worms, which are generally colorless and less than 5 centimeters long, occur almost anywhere—in the sea, in fresh water, and in the soil—in such numbers that thousands of them can be found in a small area.

Roundworms possess two anatomical features not seen before: a tube-within-a-tube body plan (fig. 27.2) and a body cavity. The body cavity is a *pseudocoelom* (fig. 27.5b), or a cavity incompletely lined with mesoderm. This fluid-filled pseudocoelom provides space for the development of organs, and substitutes for a circulatory system by allowing easy passage of molecules and for a skeleton, by providing turgidity. Worms in general do not have an internal or external skeleton but they do have a so-called *hydrostatic skeleton*, a fluid-filled interior that supports muscle contraction and enhances flexibility.

When roundworms are evaluated according to table 27.1, they are seen to have features associated with advanced animals except that they are non-segmented. Roundworms are thought to be a side branch to the main evolution of animals and may have arisen from a common ancestor that also produced coelomate animals.

Ascaris

Most roundworms are free living, but a few are parasitic. *Ascaris,* a large parasitic roundworm, is often studied as an example of this phylum.

Ascaris (fig. 27.15b) females (20 to 35 centimeters) tend to be larger than males, which have an incurved tail. Both sexes move by means of a characteristic whiplike motion because only longitudinal muscles lie next to the body wall.

The internal organs (fig. 27.15b), including the tubular reproductive organs, lie within the pseudocoelom. Because mating produces embryos that mature in the soil, the parasite is limited to warmer environments. When larvae within their protective covering are swallowed, they escape and burrow through the intestinal wall. Making their way through the organs of the host, they move from the intestine to the liver, heart, and then the lungs. Within the lungs, molting takes place and after about 10 days, the larvae migrate up the windpipe to the throat, where they are swallowed to once again reach the intestine. Then the mature worms mate and the female deposits embryo-containing eggs that pass out with the feces. In this life cycle, as with other roundworms, feces must reach the mouth of the next host; therefore proper sanitation is the best means to prevent infection with *Ascaris* and other parasitic roundworms.

Figure 27.15
Roundworm anatomy. *a.* Photomicrograph of roundworm. *b.* Anatomy of female *Ascaris*. The cross section shows the pseudocoelomate arrangement. There is mesodermally derived muscle tissue inside the epidermis, but no mesodermal tissue adjacent to the intestinal wall.

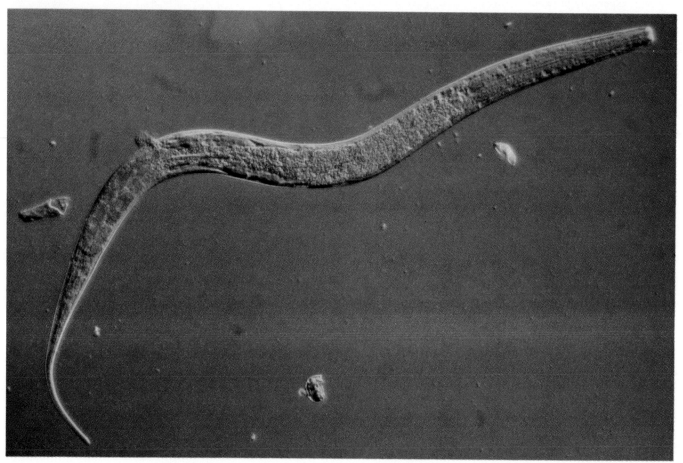

a.

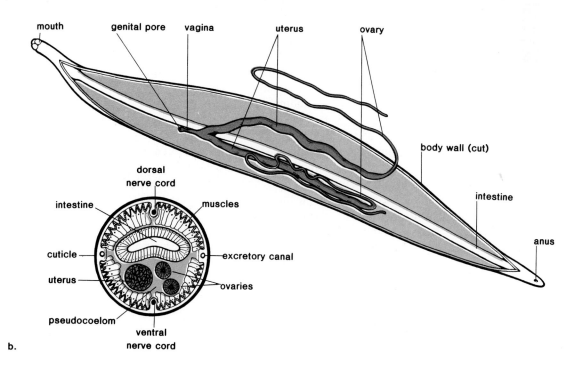

mouth   genital pore   vagina   uterus   ovary

body wall (cut)

intestine

anus

dorsal nerve cord

intestine   muscles

cuticle   excretory canal

uterus   ovaries

pseudocoelom

ventral nerve cord

b.

Other Parasites

Trichinosis is a serious infection of humans caused by *Trichinella*. Humans contract the disease when they eat rare pork containing encysted larvae. After maturation, the female adult burrows into the wall of the small intestine and produces living offspring that are carried by the bloodstream to the skeletal muscles, where they encyst. Since humans are not normally eaten by any other animals, these larvae never reach another host. However, the cycle can be completed if pigs eat infected pig meat or infected rats.

Elephantiasis is caused by a roundworm called the filarial worm, which utilizes the mosquito as a secondary host. Because the adult worms reside in lymph vessels, collection of fluid is impeded and the limbs of an infected human may swell to a monstrous size (fig. 27.16). When a mosquito bites an infected person, it transports larvae to new hosts.

Other roundworm infections are more common in the United States. Children frequently acquire a pinworm infection, and hookworm is seen in the southern states. *Hookworm,* judged by some to be the most important parasitic intestinal worm of humans, is discussed on page 703. An infection by this worm can be very debilitating because the worms feed on blood.

Higher Invertebrates

All of the higher invertebrate phyla have a true coelom (fig. 27.5c). Nevertheless, they can be divided into two groups (fig. 27.1) on the basis of embryological evidence. For example, marine mollusks and annelids share larvae of the **trochophore** type; while echinoderms and certain invertebrate chordates share larvae of the **dipleurula** type (fig. 27.17). A larva is an immature stage that is independent and can feed itself. Also, in the mollusks, annelids, and arthropods, the embryonic blastopore, which is the site of invagination of the endoderm germ layer in the embryo (fig. 27.18), becomes the mouth, while in the echinoderms and chordates, the blastopore becomes the anus and the second opening becomes the mouth. Some authorities refer to the former as the **protostomes** and the latter as the **deuterostomes.**

Figure 27.16
Elephantiasis. An infection with a filarial worm has caused this individual to experience extreme swelling in regions where the worms have blocked the lymph vessels.

Figure 27.17
The (*a.*) trochophore-type larva is characteristic of mollusks, annelids, and arthropods, while the (*b.*) dipleurula-type larva is characteristic of the echinoderms and chordates.

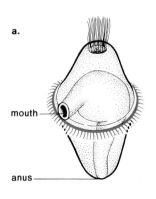

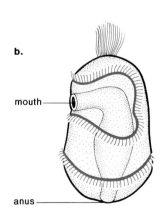

Figure 27.18
Fate of the embryonic blastopore. In the protostomes, the blastopore becomes the mouth. In the deuterostomes, the blastopore becomes the anus.

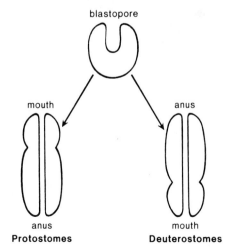

Importance of Higher Invertebrates and Vertebrates

These phyla contain the animals with which we are most familiar chiefly because of their importance to our everyday lives. So-called shellfish, such as clams, oysters (mollusks) and shrimp (an arthropod), and bony fishes, particularly herring, cod, flounder and tuna, are important sources of protein in the diet. In many western countries, however, domesticated animals, particularly cattle and pigs, which are mammals, and chickens and turkeys, which are birds, supply most of the dietary protein.

Mammals, too,—cattle, horses, and water buffalo—have been exploited as draft animals and for transport. The dung is used as fertilizer and in some cultures as fuel and building plaster.

Before the advent of synthetic materials, wool from sheep, a mammal, and silk produced by silkworms were more important than today. Silkworms are insects which have a life cycle that includes complete metamorphosis. The larva spins a cocoon having 1,000 feet of thread. In former days, too, humans made more use of feathers from birds, cow hide for leather goods, and reptilian skin to produce various clothing accessories. Synthetic fur is also in use today but even so mammalian minks and rabbits are raised to supply natural fur.

Some animals transmit diseases to humans and we can only mention a few. For example, among insects mosquitos transmit malaria, elephantiasis and yellow fever; fleas carry plague from rats to humans, and the tsetse fly conveys African sleeping sickness. Snails, which are mollusks, are secondary hosts for flukes, and bats, which are mammals, may carry rabies.

In contrast many higher invertebrates and vertebrates have been useful in biological and medical research. Our knowledge of development has been advanced by the study of echinoderms, particularly sea urchin and amphibian embryos, notably frog embryos. Today mammalian rats are especially bred for many physiological experiments. Rats, molluskan octopi, and rhesus monkeys (primates) have contributed much to behavioral studies. It should never be said, "What use is this animal?" because one never knows how a particular animal might someday be useful to humans.* Adult sea urchin skeletons are now used as molds for the production of small artificial blood vessels and armadillos are used in leprosy research.

*These animals are, of course, important in ecosystems as discussed in chapters 31 and 32. Here we mention only their direct relationship to humans.

Mollusks

Mollusks are a very large and diversified group containing many thousands of living and extinct forms. However, all forms of mollusks have a body composed of at least three distinct parts:

1. **Visceral mass:** the soft-bodied portion that contains internal organs.
2. **Foot:** a strong, muscular portion used for locomotion.
3. **Mantle:** a membranous or sometimes muscular covering that envelops but does not completely enclose the visceral mass. The mantle may secrete a shell.

In addition to these three regions, many mollusks show cephalization and have a head region with eyes and other sense organs.

The division of the body into distinct areas seems to have been a useful evolutionary advance as there are many different types of mollusks, adapted to various ways of life (fig. 27.19). Snails, conches, and nudibranchs are gastropod mollusks that have a ventrally flattened foot. The majority move about by muscular contractions passing along their foot. While nudibranchs, also called sea slugs, lack a shell, most other gastropods have a coiled shell in which

Evolution and Diversity

a.

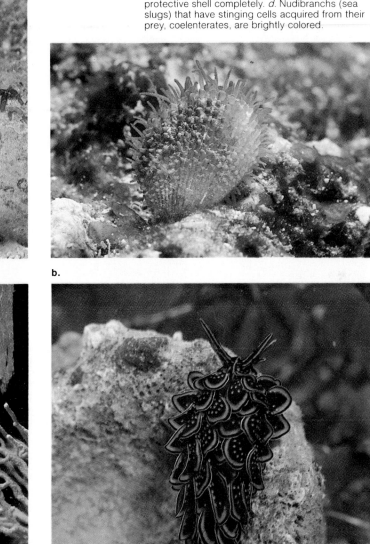

Figure 27.19
Molluskan diversity. *a.* Cowrie, a marine snail with a speckled thick shell. Cowrie shells are so prized that they are worn by royalty in the Pacific islands and used as money in Africa. *b.* A scallop with sensory tentacles extended between the valves. Humans eat only the single large adductor muscle of a scallop. *c.* An octopus moving over the surface of coral in the Pacific Ocean shows that it lacks a protective shell completely. *d.* Nudibranchs (sea slugs) that have stinging cells acquired from their prey, coelenterates, are brightly colored.

b.

c.

d.

the visceral mass spirals. Snails are adapted to life on land. For example, their mantle is richly supplied with blood vessels and functions as a lung when air is moved in and out through respiratory pores.

In cephalopods, including octopuses, squids (fig. 15.5), and nautiluses, the foot has become tentacles about the head. Nautiluses are enclosed in shells, but squids have a reduced and internal shell. Octopuses lack shells entirely. Most cephalopods are fast-moving, predatory animals with well-developed

Figure 27.20

Clam anatomy. Trace the path of food from the incurrent siphon, past the gills, to the mouth, esophagus, stomach, intestine, anus, and excurrent siphon. Locate the three ganglia: pedal, cerebral, and visceral. The heart lies in the reduced coelom. Do clams have an open or closed circulatory system?

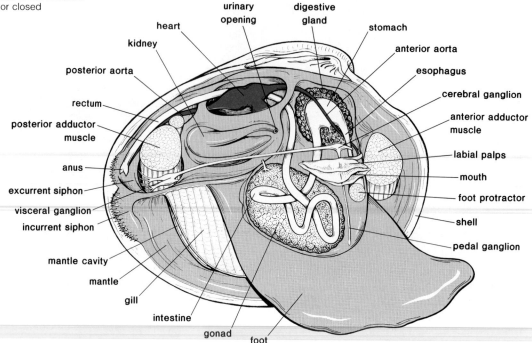

sense organs, including focusing, camera-type eyes that are very similar to those of vertebrates. Aside from the tentacles that seize the prey, cephalopods have powerful jaws and a radula to tear prey apart. Both squids and octopuses can squeeze their mantle cavity so that water is forced out, thus propelling themselves rapidly backward by a sort of jet propulsion. They also possess ink sacs from which they can squirt out a cloud of brown or black ink. This action often leaves a would-be predator completely confused.

Clams, oysters, and scallops are all called bivalves because there are two parts to the shell. Because of the ready accessibility of clams, they are often studied as an example of this phylum. However, it should be noted that a clam is adapted to an inactive life whereas other mollusks such as a squid is adapted to an active life (table 27.3).

Clam

In a clam, such as *Anodonta,* the shell is secreted by the mantle and is composed of calcium carbonate with an inner layer of **mother-of-pearl.** If a foreign body is placed between the mantle and the shell, pearls form when concentric layers of shell are deposited about the particle.

Within the mantle cavity, the gills (fig. 27.20) hang down on either side of the visceral mass, which lies above the foot. **Gills** are vascularized, highly convoluted, thin-walled tissue specialized for gas exchange.

The heart of a clam lies just below the hump of the shell within the pericardial sac, the only remains of the coelom. Therefore, the coelom of the

Table 27.3 Comparison of clam and squid

	Clam	Squid
Food getting	Filter feeder	Active predator
Skeleton	Heavy shell for protection	No external skeleton
Circulation	Open	Closed
Cephalization	None	Marked
Locomotion	Hatchet foot	Jet propulsion
Nervous system	Three separate ganglia	Brain and nerves

clam is said to be *reduced*. The heart pumps blue blood containing blue hemocyanin instead of red hemoglobin into vessels that lead to the various organs of the body. Within the organs, however, the blood flows through spaces, or **sinuses,** rather than vessels. Such a circulatory system is called an **open** circulatory system because the blood is not contained within blood vessels all the time. This type of circulatory system may be associated with an inactive animal because it is an inefficient means of transporting blood throughout the body. An active animal needs to have oxygen and nutrients transported quickly to rapidly working muscles, while an inactive animal is able to survive with a sluggish system for transporting these necessities.

The nervous system (fig. 27.20) is composed of **three pairs of ganglia** (cerebral, pedal, and visceral) which are all connected by nerves. Clams lack cephalization that may be associated with their way of life since they are adapted to slow burrowing in the sand and mud. They have a muscular foot that is compressed and bladelike and is referred to as a **hatchet** foot (class Pelecypoda means "hatchet foot"). The foot projects anteriorly from the shell, and by expanding the tip of the foot and pulling the body after it, the clam moves forward.

The clam is a filter feeder, meaning that it feeds on small particles that have been filtered from the water environment. Particles and water enter the mantle cavity by way of the **incurrent siphon,** a posterior opening between the two valves. Mucus secretions cause smaller particles to adhere to the gills, and cilia action sweeps them toward the mouth. Many inactive animals are filter feeders, since this method of feeding does not require rapid movement.

The digestive system (fig. 27.20) of the clam consists of a mouth, esophagus, stomach, and an intestine, which coils about in the visceral mass and then goes right through the heart before ending in a rectum and anus. The anus empties at an **excurrent siphon,** which lies just above the incurrent siphon. There is also an accessory organ of digestion called a digestive gland. It is readily seen in the clam that the tube-within-a-tube plan does lead to specialization of parts.

There are two excretory kidneys (fig. 27.20), which lie just below the heart and remove waste from the pericardial sac for excretion into the mantle cavity. The clam excretes ammonia (NH_3), a poisonous substance that requires the concomitant excretion of water. Land-dwelling animals tend to excrete a less toxic substance in a more concentrated form.

The male or female gonad (fig. 27.20) of a clam may be found about the coils of the intestine. While all clams have some type of larval stage, only marine clams have a trochophore larva. The presence of the trochophore larva (fig. 27.17a) among some mollusks indicates a relationship to the annelids, some of whose members also have this larval stage.

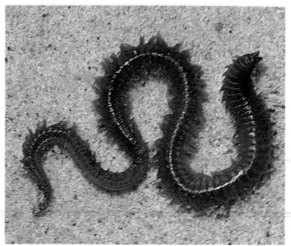

a.

b.

Figure 27.22
Head of *Neanthes*. The worm shows cephalization in that there are antennae and eyes on a definite head region. The presence of well-developed jaws indicates that the worm is a predator.

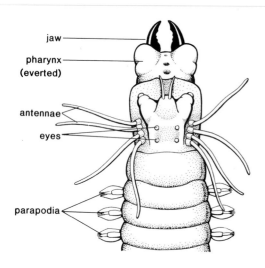

jaw

pharynx (everted)

antennae

eyes

parapodia

Annelids

The primary characteristic of this group of animals compared to the other groups studied is the presence of segmentation (fig. 27.21); obvious rings encircle the body, and the well-developed coelom is even partitioned by membranous septa. Both segmentation and an ample coelom prove to be important advances, facilitating the development of specialization of parts as seen in later phyla.

While we will use the earthworm as an example of this group of animals, marine worms such as sandworms (*Neanthes*) may be more representative. Sandworms are distinguished by the presence of a pair of fleshy lobes, the **parapodia,** on each body segment. These are used not only in swimming but also as respiratory organs where the expanded surface area allows for exchange of gases. Numerous chitinous bristles grow out from the parapodia and hence the name polychetes or "many bristled." These worms are predators. They prey on crustaceans and other small animals, which are captured by a pair of strong chitinous *jaws* that evert with a part of the pharynx when *Neanthes* is feeding. Associated with its way of life, *Neanthes* shows cephalization and has a head region with sense organs including eyes and antennae (fig. 27.22)

Earthworm

The earthworm (*Lumbricus,* fig. 27.23) is terrestrial, but it is not well adapted to life on land because it is always in danger of drying out. Since the body wall and surrounding cuticle must be kept moist for gas exchange, the worm is protected by burrowing in moist soil and certainly never ventures forth on a dry, hot day without dire consequences.

The earthworm lacks obvious cephalization and feeds on leaves or any other organic matter, living or dead, that can conveniently be taken into its mouth along with dirt. Food drawn into the mouth by the action of the muscular pharynx is stored in a crop and ground up in a thick, muscular gizzard. Digestion and absorption occur in a long intestine whose dorsal surface is expanded by a **typhlosole** that allows additional surface for absorption. Notice that the tube-within-a-tube plan has indeed allowed specialization of the digestive system to occur.

Figure 27.23
Earthworm anatomy. Drawing shows the internal anatomy of the anterior part of an earthworm's body. Segmentation is apparent in that there are pairs of setae for each segment; septa divide up the internal coelom; and there are paired nephridia and branch blood vessels in each segment.

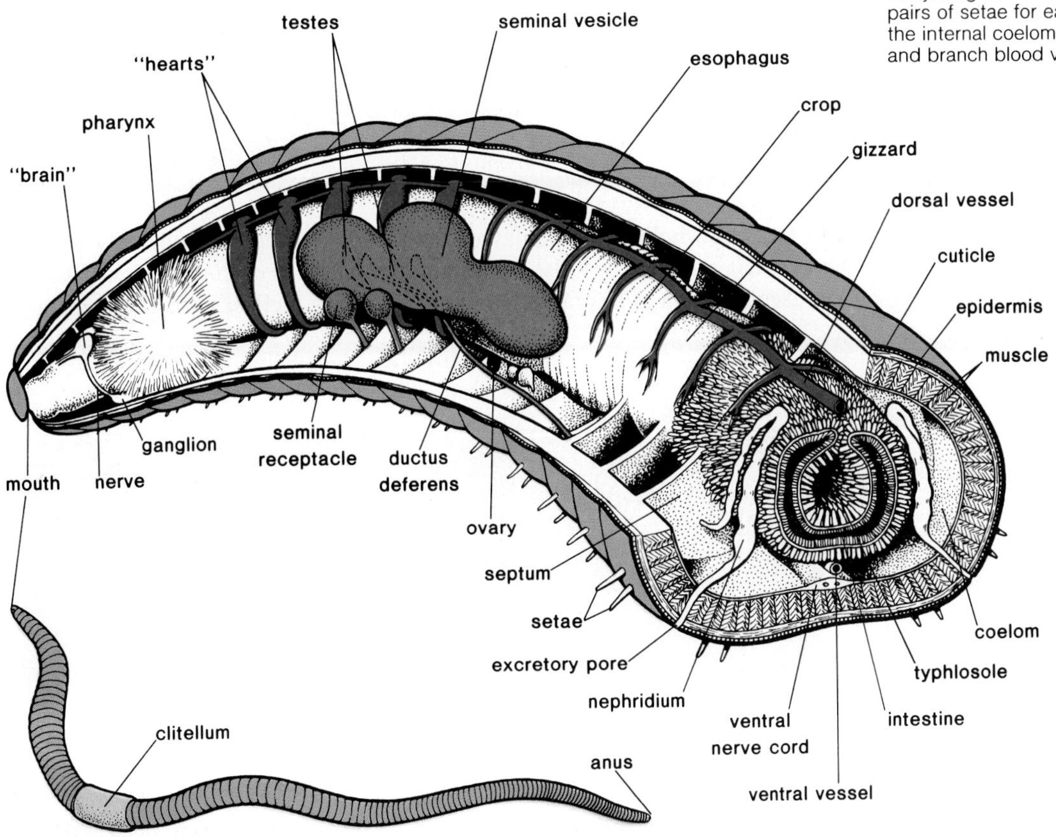

testes seminal vesicle esophagus crop gizzard dorsal vessel cuticle epidermis muscle

"hearts" pharynx "brain" ganglion seminal receptacle ductus deferens ovary septum setae excretory pore nephridium ventral nerve cord ventral vessel intestine typhlosole coelom

mouth nerve clitellum anus

Locomotion in the earthworm is suitable to its way of life, and each segment of the body has four pairs of **setae,** or slender bristles. The setae are inserted into the dirt and then the body is pulled forward. Both a circular and longitudinal layer of muscle in the body wall make it possible for the worm to move and change its shape. Muscular contraction is aided by the fluid-filled coelomic compartments that act as a hydrostatic skeleton.

The nervous system (fig. 27.23) consists of an anterior, dorsal, ganglionic mass, or brain, and a long **ventral solid nerve cord** with ganglionic swellings and lateral nerves in each segment. When invertebrates are compared to vertebrates, it is often said that the former have a ventral solid nerve cord, while the latter have a dorsal hollow nerve cord.

The excretory system consists of paired **nephridia** (fig. 27.23), or coiled tubules, in each segment. Nephridia have two openings: one is a ciliated funnel that collects coelomic fluid and the other is an exit in the body wall. Between the two openings is a convoluted region where waste material is removed from the blood vessels about the tubule.

The earthworm has an extensive *closed circulatory* system. Red blood moves anteriorly in a dorsal blood vessel and then is pumped by five pairs of hearts into a ventral vessel. As the ventral vessel takes the blood toward the posterior regions of the worm's body, it gives off branches in every segment.

The worms are *hermaphroditic,* with a complete set of organs for both sexes. The male organs of an earthworm are the testes, seminal vesicles, and sperm ducts; the female organs are the ovaries, oviducts, and seminal receptacles. Copulation occurs when two worms come to lie ventral surface to ventral surface, with the heads pointing in opposite directions. The **clitellum,** a smooth girdle about the worm's body, secretes mucus, which holds the worms

Table 27.4 Segmentation in the earthworm

1. Body rings
2. Coelom divided by septa
3. Setae on each segment
4. Ganglia and lateral nerves in each segment
5. Nephridia in each segment
6. Branch blood vessels in each segment

together and provides moisture in which the sperm swim from one body to the other. Then the mucus becomes a cocoon from which each worm backs out, releasing both eggs and sperm as they leave. Fertilization results in zygotes (fertilized eggs), which develop directly into miniature earthworms. There is no larval stage.

The annelids show the most obvious segmentation of any phylum of animals. Table 27.4 lists structures that have repeating units illustrating segmentation in earthworms.

Relationships

The fact that mollusks are related to annelids is supported not only by embryological evidence but also by an animal called *Neopilina,* which is somewhat segmented and has molluskan features. Since mollusks are nonsegmented

Figure 27.24

Arthropod diversity. *a.* Cleaner shrimp. As their name implies these shrimp make a living by removing debris and parasites from other sea animals, particularly fish that line up at cleaning stations. *b.* American Lobster. Lobsters typically lie in wait and then spring forward to capture prey by means of well-developed claws. *c.* Giant centipede. Segmentation is obvious in this organism that has a set of appendages on every segment.
d. Dragonflies mating. Dragonflies begin life at the bottom of a pond and after a series of nymph stages they metamorphize into adults that live only a short time. *e.* Walking stick. Looking like sticks helps these animals survive as they walk about on limbs of trees and bushes. *f.* Garden spider. Spiders spin beautiful webs that vibrate when touched by prey, alerting the spider to its prospective meal.

b.

a.

c.

Evolution and Diversity

while annelids are segmented, the existence of this animal seems to suggest that these two groups may be related by way of a common ancestor. The existence of another animal, *Peripatus* (fig. 23.3) indicates that arthropods are related to annelids. Peripatus is segmented like the annelids but also has certain arthropodal characteristics with respect to respiration, circulation, and appendages.

Arthropods

The arthropods have more species (900,000) than any other group of animals and are often said to be the most successful of all the animals. The phylum includes animals adapted to living in water, such as crayfish, lobsters, and shrimp; and animals adapted to living on land, such as spiders, insects, centipedes, and millipedes (fig. 27.24).

d.

e.

f.

Figure 27.25
Scanning electron micrograph showing that
compound eyes are composed of individual units.
Each unit has its own lens and retina.

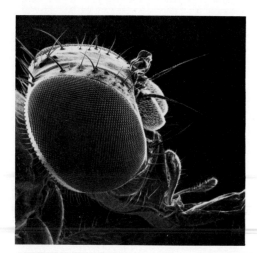

Arthropods have an external skeleton containing **chitin,** a strong, flexible polysaccharide. The skeleton serves many functions such as protection, attachment for muscles, and prevention of desiccation on land. The appendages are also covered by the skeletal material, but they are jointed. The presence of **jointed appendages** is a great advance in the animal kingdom and aids locomotion on land.

An external skeleton is not without difficulties and, since this particular skeleton does not grow larger, arthropods **molt,** or shed the skeleton periodically.

Specialization of parts is readily seen in that the arthropod body is not composed of a series of like segments but rather, due to a fusion of segments, is composed of three parts—head, thorax and abdomen. The head shows good cephalization with sense organs. The sense organs include **antennae** (or feelers) and eyes. The eyes are of two types: **compound** (fig. 27.25) and **simple.** The compound eye is not seen in any other phylum. It is composed of many complete visual units grouped together in a composite structure: each visual unit contains a separate lens and a light-sensitive cell. In the simple eyes, a single lens covers many light-sensitive cells.

The coelom, which is so well developed in the annelids, is **reduced** in the arthropods and composed chiefly of the space about the reproductive system. Instead of a coelomic cavity, there is a **hemocoel,** or blood cavity, consisting of vessels and open spaces where the blood flows about the organs. The dorsal heart keeps the blood moving around in the sinuses. Arthropods, like most mollusks, have an open circulatory system.

Crayfish

Crayfish are in the class Crustacea along with lobsters, shrimps, copepods, and crabs. Figure 27.26a gives a view of the external anatomy of the crayfish, and it can be seen that the head and thorax are fused into a **cephalothorax,** which is covered on the top and sides by a nonsegmented **carapace.** The abdominal segments, however, are marked off clearly.

On the head are a pair of stalked compound eyes and two pairs of antennae. Chitinous jaws and mouthparts are also present. The appendages in the thorax include accessory mouthparts, **pinching claws,** and four pairs of **walking legs;** the abdominal segments are equipped with **swimmerets,** small paddlelike structures. The first pair of swimmerets in the male are quite strong and are used to pass sperm to the female, reminiscent of the squid, which passes sperm by means of specialized tentacles. The last two segments bear the **uropods** and the **telson,** which make up a fan-shaped tail used for swimming backwards.

Ordinarily, the crayfish lies in wait for prey. It faces out from an enclosed spot with the claws extended and the antennae moving about. If a small animal, dead or alive, happens by, it is quickly seized and carried to the mouth. When a crayfish does move about, it generally crawls slowly but may swim rapidly backwards by using heavy abdominal muscles.

Respiration is by means of **gills** (fig. 27.26a), which lie above the walking legs protected by the carapace. Gills, as we have seen, are typical organs of respiration in water-dwelling animals. The crayfish has blue blood containing the pigment hemocyanin, which aids in the transport of oxygen.

Internally, the digestive system (fig. 27.26b) includes a stomach, which is divided into two main regions: an anterior portion called the **gastric mill,** equipped with chitinous teeth to grind coarse food, and a posterior region, which acts as a filter to sort out food according to consistency.

The nervous system (fig. 27.26b) is quite similar to that of the earthworm. There are anterior ganglia from which a solid **ventral nerve cord** passes posteriorly. Along the length of the nerve cord, periodic ganglia give off lateral nerves.

Figure 27.26
Anatomy of the crayfish. *a*. Externally, it is possible to observe the jointed appendages, including the swimmerets, walking legs, and claws. These appendages, plus a portion of the carapace, have been removed from the right side so that the gills are visible. *b*. Internally, the parts of the digestive system are particularly visible. The circulatory system can also be clearly seen. Note also the ventral solid nerve cord.

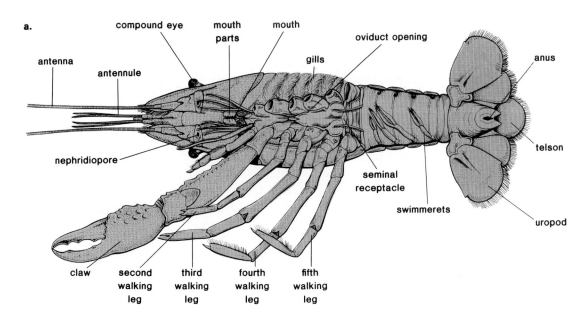

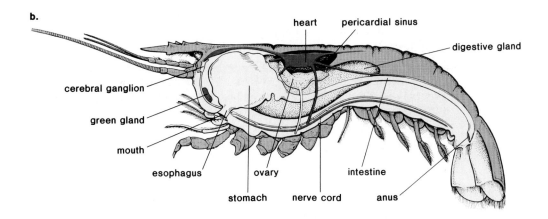

The excretory system (fig. 27.26b) consists of a pair of **green glands** lying in the head region anterior to the esophagus. Each organ possesses a glandular region for waste removal: a bladder and a duct that opens ventrally at the base of the antennae.

The sexes are separate in the crayfish. The white testes of the male are located just ventral to the pericardial sinus. From each side the coiled ductus deferens passes ventrally and opens to the outside at the base of the fifth walking leg. Sperm transfer is accomplished by the modified first two swimmerets of the abdomen. In the female, the ovaries are located in a position similar to that occupied by the testes and the oviducts pass ventrally, opening near the bases of the third pair of walking legs. There is a cuticular fold between the bases of the fourth and fifth pair that serves as a seminal receptacle.

Table 27.5 Comparison of crayfish and grasshopper

	Crayfish	Grasshopper
Locomotion	Legs and swimmerets	Hopping legs and wings
Respiration	Gills	Tracheae
Excretion	Liquid waste by way of green gland	Solid waste by way of Malpighian tubules
Circulation	Blue blood	Colorless blood
Nervous system	Cephalization	Cephalization with tympanum
Reproduction	Modified swimmerets in male	Penis in male, ovipositor in female

Table 27.5 compares the crayfish to the grasshopper to illustrate how one is adapted to the water and the other to the land.

Grasshopper

Insects comprise one of the largest animal groups both in number of species and in number of individuals, perhaps because of the presence of **wings.** Wings enhance the insects' ability to survive by providing a new way of escaping enemies, finding food, facilitating mating, and dispersing the species. Figure 27.24 gives representative examples of insects, of which we will study the grasshopper in detail.

Every system of the grasshopper (*Romalea*) (fig. 27.27a) is adapted to life on land. As a part of the exoskeleton there are **three pairs** of legs, one pair of which is suited to jumping. There are two pairs of wings; the forewings are tough and leathery and when folded back at rest they protect the broad, thin hindwings. The first abdominal segment bears on its lateral surface a large **tympanum** for the reception of sound waves. The posterior region of the exoskeleton in the female has two pairs of projections that form an ovipositor.

The digestive system (fig. 27.27b) is suitable for a grass diet. Digestion begins in the mouth where the mouthparts grind the food and there are salivary secretions. Food is temporarily stored in the crop before passing into the gizzard, which sends finely ground food to the stomach. Here digestion is completed with the aid of enzymes secreted by the gastric caeca.

Excretion is carried out by means of **Malpighian tubules** (fig. 27.27b), which extend out into the hemocoel and empty into the digestive tract. A solid nitrogenous waste is excreted, conserving water.

Respiration occurs when air enters small tubules called **tracheae** (fig. 27.27b) by way of openings in the exoskeleton called **spiracles.** The tracheae branch and rebranch, finally ending in moist areas where the actual exchange of gases takes place. The movement of air through this complex of tubules is not a passive process; air is pumped through by a series of several bladderlike structures (air sacs), which are attached to the tracheae near the spiracles. Air enters the anterior four spiracles and exits by the posterior six spiracles. This mechanism of breathing found in insects and arachnids (e.g., spiders and scorpions) is an adaptation to land that requires a drastic modification of the body. It may even account for the small size of insects since the tracheae are so tiny and fragile that they would be crushed by any amount of weight.

The heart is a slender, tubular organ that lies against the dorsal wall of the abdominal exoskeleton. Blood passes from a dorsal aorta to the hemocoel and circulates through the body spaces, finally returning to the heart again. The blood is colorless and lacks a respiratory pigment since the tracheal system transports gases.

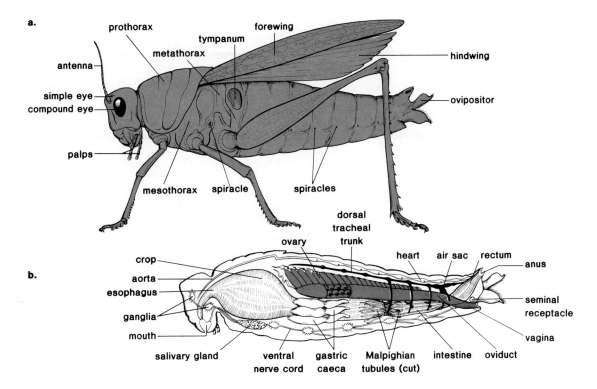

Reproduction is adapted to life on land. The male has two testes and associated ducts that end in the penis. The female has ovaries that occupy the whole dorsal part of the animal, and oviducts that end in the vagina. The sperm received during copulation are stored in the seminal receptacles for future use. Fertilization is internal, usually occurring during late summer or early fall. The female deposits the fertilized eggs in the ground with the aid of her ovipositor.

Insects are land animals that often have larval stages and undergo metamorphosis. **Metamorphosis** means a change, usually a drastic one, in form and shape. Some insects undergo what is called complete metamorphosis, in which case they have three stages of development: **larval stage,** the **pupa stage,** and finally the **adult stage.** Metamorphosis occurs during the pupa stage when the animal is enclosed within a hard covering. The animal that is best known for metamorphosis is the butterfly, whose larval stage is called a caterpillar and whose pupa stage is the cocoon; the adult is the butterfly. Grasshoppers undergo incomplete metamorphosis, which is a gradual change in form rather than a drastic change. The immature stages of the grasshopper are called nymphs rather than larvae and they are recognizable as grasshoppers even though they differ somewhat in shape and form. Metamorphosis is controlled by the same hormones as molting.

Figure 27.28
Representative echinoderms. *a.* Starfish about to feed on a clam, which it can open by utilizing the suction provided by tube feet. *b.* A sea cucumber whose general overall shape resembles the vegetable of this name. *c.* Sea urchins demonstrating their many external spines.

a.

b.

c.

Echinoderms

The echinoderms (fig. 27.28) include only marine animals—starfish, sea urchen, sea cucumber, feather star, sea lily, and sand dollar. The most familiar of these is the starfish, and we will study this representative. Echinoderms are **radially symmetrical** as adults, with a body plan based on **five parts.** Their other unique feature is the **water vascular system,** which is used as a means of locomotion. They also have a calcareous **endoskeleton,** whose projecting spines give the phylum its name, Echinodermata, meaning "spiny skin."

Starfish

The starfish (*Asterias*) sometimes called the sea star, is commonly found along rocky coasts. It has a five-rayed body plan with an **oral** (mouth) and **aboral** (anus) side (fig. 27.29). The oral side is actually the underside and the aboral side is the upper side. On the aboral side there are various structures that project through the epidermis: (1) spinelike projections of the endoskeletal plates; (2) pincerlike structures called **pedicellarie,** which keep the surface free of small particles; and (3) skin gills, which serve for respiratory exchange. The mouth is located on the oral surface, where each of the five arms has a groove lined by little **tube feet.**

Evolution and Diversity

Figure 27.29
Starfish anatomy. Like other echinoderms, starfishes have a water vascular system shown here in color. Water enters the sieve plate and is eventually sent into tube feet by the action of the ampullae. Each arm of a starfish contains digestive glands, gonads, and water vascular system.

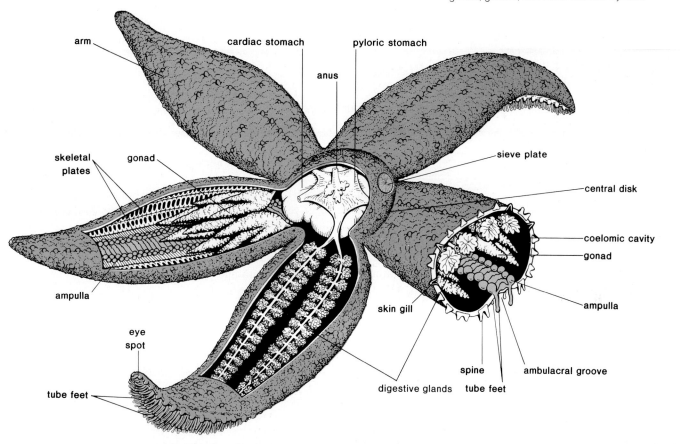

Starfish feed on mollusks. When a starfish attacks a clam, it arches its body over the shell and by the concerted action of the tube feet forces the clam to open. Then it everts a portion of its stomach to digest the contents of the clam.

The mouth of a starfish opens into a narrow esophagus, which in turn leads to an expanded stomach. The stomach has two portions: the saclike cardiac, which can be everted as described, and the narrower pyloric, which is connected to a short intestine. The anus opens on the aboral or upper side of the animal.

Each of the five arms contains a well-developed coelom, a pair of large **hepatic caeca** that secrete powerful enzymes into the pyloric portion of the stomach, and gonads, which open on the aboral surface by very small pores; the nervous system consists of a central nerve ring that supplies radial nerves to each arm. At the tip of each arm is a light-sensitive eyespot.

Coelomic fluid, circulated by ciliary action, performs many of the normal functions of a circulatory system; the water vascular system is purely for locomotion. Water enters this system through a structure on the aboral side called the **sieve plate,** or madreporite. From there it passes through a short canal, called the *stone canal,* to a *ring canal,* which surrounds the mouth. From the ring canal, five *radial canals* extend into the arms along the ambulacral grooves. From the radial canals many lateral canals extend into the tube feet. One canal goes to each tube foot, where it ends in the **ampulla.** When

Figure 27.30

An enlargement of the notochord (*above*) and a diagram of an idealized chordate (*below*). All chordates at some time in their life history have a dorsal hollow nerve cord, pharyngeal gill pouches, and a notochord. The notochord is a supporting rod covered by an inner fibrous sheath and an outer elastic sheath.

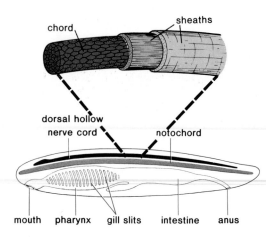

Figure 27.31

Protochordates *a*. Tunicate anatomy. Gill slits are the only chordate feature retained by the adult. *b*. Amphioxus (lancelet) anatomy. This animal retains all three chordate characteristics as an adult

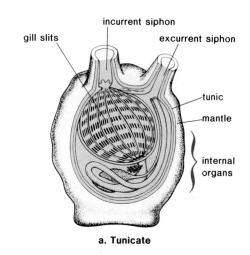

a. Tunicate

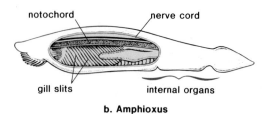

b. Amphioxus

the ampulla contracts, the water is forced into the tube foot, expanding it and giving it suction. By alternating the expansion and contraction of the tube feet, the starfish moves along slowly.

The adult starfish is anatomically unique, but it is believed that echinoderms and chordates share a common ancestor because of similar embryonic development. The dipleurula-type larva is seen in both the echinoderms and the hemichordates (see following). Thus the echinoderms are bilaterally symmetrical as embryos; radial symmetry is found only in the adult.

Chordates

Among the chordates are those animals with which we are most familiar, including human beings themselves. All members of this phylum are observed to have three basic characteristics (fig. 27.30) at some time in their life history:

1. A dorsal supporting rod called a **notochord,** which is replaced by the vertebral column in the adult vertebrates.
2. A **dorsal hollow nerve cord** in contrast to invertebrates, which have a ventral solid nerve cord. By hollow, it is meant that the cord contains a canal that is filled with fluid.
3. **Gill pouches** or **slits,** which may be seen only during embryological development in most vertebrate groups but which persist in adult fish. Water passing into the mouth and pharynx goes through the gill slits, which are supported by gill bars.

Protochordates

Two groups of animals are sometimes called the **protochordates** because they possess all three typical structures in either the larval and/or adult forms as did the first chordates to evolve. They are also called **invertebrate chordates** because they are not vertebrates. These two groups of animals link the vertebrates to the rest of the invertebrates and show how modestly the chordates most likely began.

A **tunicate,** or sea squirt (fig. 27.31a), appears to be a thick-walled, squat sac with two openings, an incurrent siphon and an excurrent siphon. Inside the central cavity of the animal and opening into a chamber that opens to the outside are numerous gill slits, the only chordate feature retained by the adult. The larva of the tunicate, however, has a tadpole shape and possesses the three chordate characteristics. It has been suggested that such a larva may have become sexually mature without developing the other adult tunicate characteristics. If so, it may have evolved into a fishlike vertebrate.

Amphioxus (fig. 27.31b) is a chordate that shows the *three chordate characteristics as an adult*. In addition, segmentation is present as witnessed by the fact that the muscles are segmentally arranged and the nerve cord gives off periodic branches.

Despite its distinctive chordate features, amphioxus is now believed to be a specialized side branch only and is not believed to be in the direct line of vertebrate ancestry. The true ancestor to the vertebrates, which perhaps evolved from a tunicate larva, has never been found as a fossil remain.

Vertebrate Chordates

Vertebrates have all the advanced characteristics listed in table 27.1. They are segmented chordates in which the notochord is replaced in the adult by a *vertebral column* composed of individual vertebrae. The skeleton is internal, and in all the vertebrates there is not only a backbone but also a skull, or cranium, to enclose and protect the brain. In higher vertebrates, other parts of the skeleton serve as attachment for muscles and for protection of internal organs of the chest and abdomen. All vertebrates have a *closed circulatory system* in which red blood is contained entirely within blood vessels. They show good

Evolution and Diversity

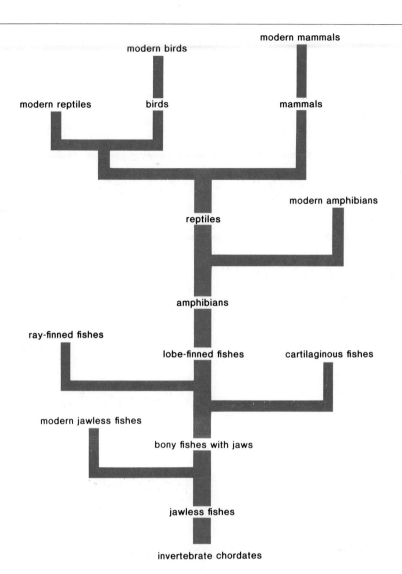

Figure 27.32
Evolutionary tree of vertebrates. Vertebrates evolved in an aquatic environment, but adaptation to a land environment gradually improved as later groups evolved. Mammals and birds possess all the features required for successful adaptation to a terrestrial existence.

cephalization with sense organs; the eyes develop as outgrowths of the brain; and the ears serve as equilibrium devices in aquatic vertebrates plus sound wave receivers in land vertebrates. The kidneys are important excretory and water-regulating organs that conserve or rid the body of water as appropriate.

Two comparisons are often made between invertebrates and vertebrates: (1) invertebrates have a *ventral solid* nerve cord, while vertebrates have a dorsal hollow nerve cord; and (2) invertebrates have an *external* skeleton, while vertebrates have an *internal* skeleton. There are, however, many exceptions among the invertebrates.

Occasionally, the same adaptations are seen in both invertebrates and vertebrates; for example, both insects and birds have wings; both fish and squid have fins. Such structures are called **analogous** rather than **homologous** because, while they serve the same function, the anatomy is completely different, indicating that each evolved independently. Only homologous structures, such as vertebrate limbs, support an evolutionary relationship (p. 507).

As a group, the vertebrates are the dominant animals in the world today. They are found in every habitat from the ocean floor to the mountaintop and in the forest and desert. Included in the group are the three classes of fishes and one class each of amphibian, reptile, bird, and mammal. Figure 27.32 represents an evolutionary tree of the vertebrates and you can see that it is possible to trace the evolution of the vertebrates from fishes to amphibians, to

Figure 27.33
Cartilaginous and bony fishes. *a.* The bull shark is cartilaginous. *b.* Barracuda and (*c.*) freckled grouper are both bony fishes.

a.

b.

c.

Evolution and Diversity

reptiles, to both birds and mammals. All but the fishes are **tetrapods,** meaning that they have four limbs. With one exception, there are no living representatives of the primitive ancestors that connect the classes. These ancestors are known from the fossil record, and when we see the artist's drawing of them they appear quite strange to us.

Fishes (Superclass Pices)

There are three classes of fishes: the jawless fishes, the cartilanginous fishes, and the bony fishes. Living representatives of the **jawless fishes** are cylindrical, up to a meter long, with smooth, scaleless skin and no jaws or paired fins. There are two families of jawless fishes: **hagfishes** and **lampreys.** The hagfishes are scavengers, feeding mainly on dead fish, while some lampreys are parasitic. When parasitic, the round mouth of the lamprey serves as a sucker by which it attaches itself to another fish and taps into its circulatory system.

These first primitive fishes were nonpredaceous. However, with the development of jaws (which are believed to have evolved from the first pair of gill bars) and paired fins among the next two classes of fishes, the possibility for predation exists. Many fishes of the next two groups are predators.

Cartilaginous fishes (Chondrichthyes; fig. 27.33a) are the sharks, rays, and skates, which have skeletons of cartilage instead of bone. The dogfish shark is a small shark often dissected in biology laboratories to show the main features of the vertebrate body. Other sharks are well known to us as vicious predators that attack human swimmers. One of the most dangerous sharks inhabiting both tropical and temperate waters is the hammerhead. The largest of the sharks, the whale sharks, feed on small fishes and marine invertebrates and do not attack humans. Skates and rays are rather flat fishes that live partly buried in the sand and feed on mussels and clams.

Bony fishes (Osteichthyes; fig. 27.33b,c) are by far the most numerous and varied of the fishes. Most of the fishes we eat, such as perch, trout, flounder, and haddock, are a type of bony fish called **ray-finned fishes.** These fishes have a *swim bladder* that aids them in changing their depth in the water. By secreting gases into the bladder or by absorbing gases from it, a fish can change its density and thus go up or down in the water. "Ray-finned" refers to the fact that the fins are thin and supported by bony rays. Another type of bony fish called the **lobe-finned** fish evolved into the amphibians. These fishes not only have fleshy appendages that could be adapted to land locomotion, they also have a lung[1] that is used for respiration. A type of lobe-finned fish called a coelacanth, which exists today, is the only "living fossil" among the fishes.

Fishes are adapted to life in the water. Their streamlined shape, fins, and muscle action are all quite suitable to locomotion in the water. Their bodies are covered by *scales,* which protect the body but do not prevent water loss. Fishes breathe by means of *gills,* respiratory organs that are kept continuously moist by the passage of water through the mouth and out the gill slits. As the water passes over the gills, oxygen is absorbed by the blood and carbon dioxide is given off. The heart of a fish is a simple pump, and the blood flows through the chambers, including a nondivided atrium and ventricle, to the gills only (fig. 27.39). Oxygenated blood leaves the gills and goes to the body proper.

Generally speaking, reproduction in the fishes requires external water; sperm and eggs are usually shed into the water where fertilization occurs, and the zygote develops into a swimming larva that can fend for itself until it develops into the adult form.[2]

---

[1]Actually, the swim bladder of modern-day bony fishes is believed to be derived from an ancient lung.

[2]Some fish, such as sharks, practice internal fertilization and retain their eggs during development. Their young are born alive.

Figure 27.34

Representative amphibian and reptile. *a.* Fire-bellied toad can afford to be so brightly colored because of its poisonous nature. *b.* The Galápagos snake is in the process of eating a lizard.

a.

b.

Evolution and Diversity

Amphibians

The living amphibians include **frogs** and **toads** (Order Anura) and **newts** and **salamanders** (Order Urodela) (fig. 27.34a). These animals have distinct walking legs, each with five (or fewer) toes. This repesents an adaptation to land locomotion. Respiration is accomplished by the use of small, relatively *inefficient lungs* supplemented by gaseous exchange through the skin. Thus, the skin is smooth, moist, and glandular. This is a distinct disadvantage on land because of the danger of drying out; therefore, frogs spend most of their time in or near the water. All amphibians possess two nostrils that, unlike those of most fish, are connected directly with the mouth cavity. Air enters the mouth by way of the nostrils and when the floor of the mouth is raised, air is forced into the lungs. Associated with the development of lungs there is a change in the circulatory system. The amphibian heart has a divided atrium but a single ventricle (fig. 27.39). The right atrium receives impure blood with little oxygen from the body proper and the left atrium receives purified blood from the lungs that has just been oxygenated, but these two types of blood are partially mixed in the single ventricle. Mixed blood is then sent, in part, to the skin where further oxygenation may occur.

Nearly all the members of this class lead an amphibious life—that is, the larval stage lives in the water and the adult lives on the land. The adults must return to the water, however, for the purpose of reproduction. Just as with the fish, the sperm and eggs are discharged into the water and fertilization results in a zygote that develops into the familiar tadpole. The tadpole undergoes metamorphosis into the adult before taking up life on the land.

Amphibians are not fully adapted to life on land. The appendages are not sturdy, the moist skin is a constant threat, and external water is required for reproduction.

Reptiles

The reptiles living today are the **turtles** (Order Chelonia), **alligators** (Order Crocodilia), and **snakes** and **lizards** (Order Squamata) (fig. 27.34b). Reptiles with limbs, such as lizards, are able to lift their bodies off the ground and the body is covered with hard, *horny scales* that protect the animal from desiccation and from predators. Both of these features are adaptations to life on land.

Reptiles have well-developed lungs with a *ribcage* to protect them. When the ribcage expands, the lungs expand and air rushes in. The creation of a partial vacuum establishes a negative pressure that causes air to rush into the lungs. The atrium of the heart is always separated into right and left chambers, but division of the ventricle varies. There is always at least one interventricular septum but it is incomplete in all but the crocodiles, thus permitting exchange of oxygenated and deoxygenated blood between the ventricles in all but the latter.

Perhaps the most outstanding adaptation of the reptiles is the fact that they have a means of reproduction suitable to the land. There is usually no need for external water to accomplish fertilization because the penis of the male passes sperm directly to the female. After *internal fertilization* has occurred, the egg is covered by a protective leathery shell and laid in an appropriate location.

The *shelled egg* made development on land possible and eliminated the need for a swimming larva stage during development. It provides the developing embryo with oxygen, food, and water; removes nitrogen wastes; and protects it from drying out and from mechanical injury. This is accomplished by the presence of *extraembryonic membranes*. These membranes, as we saw in chapter 20, are not a part of the embryo itself and are disposed of after development is complete. There are four membranes: a yolk sac containing nourishing yolk is connected to the digestive tract and provides the embryo with

food; the allantois is attached to the rear of the embryo and is a depository for nitrogenous waste; the chorion lies right next to the shell and carries out gas exchange across the porous shell; and the fluid-filled amnion envelops the embryo, preventing it from drying out and protecting it against mechanical injury. It is an interesting fact that all animals actually develop in water—either external water as in the fish and amphibian or in amniotic fluid as in the reptiles, birds, and mammals.

The reptiles are fully adapted to life on land except for one limitation: they cannot regulate their body temperature. Sometimes animals that cannot maintain a constant temperature, that is, fish, amphibians, and reptiles, are called *cold-blooded*. Actually, however, they take on the temperature of the external environment. If it is cold externally, they are cold internally; and if it is hot externally, they are hot internally. Reptiles try to regulate body temperatures by exposing themselves to the sun if they need warmth or by hiding in the shadows if they need cooling off. This works reasonably well in most areas of the world.

The reptiles at one time were an extremely successful group of animals. Who has not heard of the great dinosaurs that ruled the world long ago? One particular group of reptiles, called the *Ruling Reptiles,* invaded many different habitats, and it is among this group that we find both the dinosaurs and the ancestors to the birds. Some investigators now believe that the dinosaurs were warm-blooded and they even suggest that *Archaeopteryx* (fig. 23.2) was a small dinosaur. According to this controversial theory, modern birds are also dinosaurs. Other investigators continue to maintain that dinosaurs could not regulate their body temperature.

The Ruling Reptiles can be traced back to the *Stem Reptiles,* which also produced a group of animals called therapsids. The *therapsids* are the mammalianlike reptiles whose limbs bend in the familiar mammalianlike manner. The skull also has mammalian features, and the teeth are differentiated like a mammal's. These animals looked something like large dogs and were aggressive predators.

Birds

Birds (fig. 27.35) are characterized by the presence of feathers, which are actually modified reptilian scales. There are many orders of birds including birds that are flightless (ostrich), web-footed (penguin), divers (loons), fish eaters (pelicans), waders (flamingos), broad-billed (ducks), birds of prey (hawks), vegetarians (fowl), shore birds (sandpipers), nocturnal (owl), small (hummingbirds), and song birds, the most familiar of the birds.

Nearly every anatomical feature of a bird can be related to its *ability to fly*. The anterior pair of appendages (wings) has become adapted for flight; the posterior is variously modified, depending on the type of bird. Some are adapted to swimming, some to running, and some to perching on limbs. The breastbone is enormous and has a ridge to which the flight muscles are attached. Respiration is efficient since the lobular lungs form *air sacs* throughout the body, including the bones. The presence of these sacs means that the air circulates one way through the lungs during both inspiration and expiration so that "used" air is not trapped in the lungs. Another benefit of air sacs is that the air-filled, hollow bones lighten the body and aid flying. Birds have a four-chambered heart that completely separates oxygenated from deoxygenated blood.

Birds have well-developed brains, but the enlarged portion seems to be the area responsible for instinctive behavior. Thus, birds follow very definite patterns of migration and nesting.

Birds are fully adapted to life on land and are *warm-blooded* and, like mammals, are able to maintain a constant internal temperature. This may be associated with their efficient nervous, respiratory, and circulatory systems. Also the feathers provide insulation.

Evolution and Diversity

a.

b.

c.

Figure 27.36
Monotreme and marsupial, two rare types of
mammals. *a.* Spiny anteater is a monotreme that
lays a shelled egg. *b.* Koala bear is a marsupial
whose young are born immature and complete
their development within a pouch.

a.   b.

Mammals

The chief characteristics of mammals are the presence of **hair** and **mammary
glands** that produce milk to nourish the young. Human mammary glands are
called breasts.

Mammals are completely adapted to life on land and have limbs that
allow them to move rapidly. In fact, an evaluation of mammalian features
leads us to the obvious conclusion that they lead active lives. The brain is well
developed; the lungs are expanded not only by the action of the ribcage but
also by the contraction of the *diaphragm,* a horizontal muscle that divides the
chest cavity from the abdominal cavity; and the heart is *four chambered.* The
internal temperature is constant and hair, when abundant, helps insulate the
body.

The mammalian brain is enlarged due to the expansion of the foremost
part—the cerebral hemispheres. These have become convoluted and expanded
to such a degree that they hide many other parts of the brain from view.

Mammals are classified according to their means of reproduction: there
are **egg-laying** mammals, mammals with **pouches** for immature embryos, and
placental mammals.

Monotremes   These are the egg-laying mammals represented by the duck-
billed platypus and spiny anteaters (fig. 27.36a). In the same manner as birds,
the female incubates the eggs, but after hatching, the young are dependent
upon the milk that seeps from glands on the abdomen of the female. Thus
monotremes have retained the reptilian mode of reproduction while evolving
hair and mammary glands. The young are blind, helpless, and completely dependent on the parent for some months. The mouth is variously modified among
the monotremes. The **platypus** has a horny, bill-like structure somewhat resembling that of a duck, while the **anteater** has an elongated, cylindrical snout.

a.

Figure 27.37
a. White-tail deer feeding fawn. The white of the tail is exposed when the tail is raised as a sign of imminent danger. *b.* Red fox is a small bristly tailed member of the dog family hunted by humans for sport and fur. *c.* Torpedo-shaped harbor seals are excellent swimmers and divers having short paddlelike flippers and a thick layer of subcutaneous fat (blubber).

b.

c.

Marsupials   Another primitive group of mammals are the **marsupials** (fig. 27.36b) found in large numbers in Australia, such as the **kangaroo** and **koala.** In the Americas, they are represented by the **opossum.** The young in all members of this group are born in a very premature state. Once born, they leave the uterus and crawl to the pouch where each attaches itself to a nipple and continues development for a time.

Placental Mammals   The vast majority of living mammals are the placental type (fig. 27.37). In these mammals, the extraembryonic membranes have been modified for internal development within the uterus of the female. The chorion

Table 27.6 Some major orders of placental mammals

Insectivora (moles, shrews)	Primitive; small, sharp-pointed teeth
Chiroptera (bats)	Digits support membranous wings
Carnivora (dogs, bears, cats, sea lions)	Canine teeth long; teeth pointed
Rodentia (mice, rats, squirrels, beavers, porcupines)	Incisor teeth grow continuously
Perissodactyla (horses, zebras, tapirs, rhinoceroses)	Large, long-legged, one or three toes, each with hoof; grinding teeth
Artiodactyla (pigs, cattle, camels, buffalos, giraffes)	Medium to large; two or four toes, each with hoof; many with antlers or horns
Cetacea (whales, porpoises)	Medium to very large; forelimbs paddle-like; hind limbs absent
Primates (lemurs, monkeys, gibbons, chimpanzees, gorillas, men)	Mostly tree-dwelling; head freely movable on neck; five digits, usually with nails; thumbs and/or large toes usually opposable

From ESSENTIALS OF BIOLOGY, Second Edition by Willis H. Johnson, Louis E. Delanney, Thomas A. Cole, and Austin E. Brooks. Copyright © 1969 and 1974 by Holt, Rinehart and Winston. Reprinted by permission of CBS College Publishing.

Figure 27.38
Breathing mechanisms of vertebrates. Fish breathe by means of gills; amphibians have poorly developed lungs; reptiles have a ribcage and lungs; birds have lungs with air sacs; and humans have well-developed lungs plus ribcage and diaphragm.

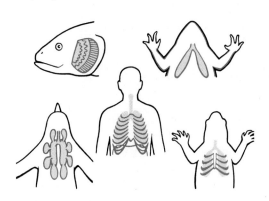

contributes to the fetal portion of the placenta, while a portion of the uterine wall contributes the maternal portion. Here nutrients, oxygen, and waste are exchanged between fetal and maternal blood. These mammals not only have a long embryonic period, they are also dependent on their parents until the nervous system is fully developed and they have learned to take care of themselves.

Placental mammals may be classified into twelve orders, eight of which may be considered major (table 27.6). A study of these reveals that mammals have largely differentiated, or become specialized, according to the mode of locomotion and how they get their food.

The anatomy and physiology of human beings may be used to exemplify vertebrate, and especially mammalian, anatomy and physiology. Chapters 8 through 19 may therefore be used for detailed information regarding vertebrate and mammalian anatomy and physiology.

Comparisons between Vertebrates

Vertebrates, like other animals, are adapted to their way of life. Figure 27.38 shows that fish breathe by means of gills, respiratory organs appropriate to life in the water. Amphibians have small, ineffectual lungs that must be supplemented by the skin as a respiratory organ. Reptiles have more efficient lungs with a ribcage, which not only protects the lungs but helps fill them with air. Birds have lungs expanded by air sacs that allow one-way flow of air, and mammals have highly subdivided lungs surrounded by a ribcage and separated from the abdominal cavity by a diaphragm. These anatomical features make breathing by negative pressure possible in reptiles, birds, and mammals.

Figure 27.39 compares the hearts of the vertebrates. Fishes have a nondivided artrium and ventricle and the heart pumps blood only to the gills. Amphibians have a heart in which there is a right and left atrium but only a single ventricle where oxygenated and deoxygenated blood are partially mixed before being sent, in part, to the skin for further oxygenation. The reptiles have a right and left atrium and a ventricle that has two partial septa. Some mixing of oxygenated and deoxygenated blood is not a serious disadvantage for the cold-blood amphibians and reptiles since their oxygen demands are relatively

Figure 27.39

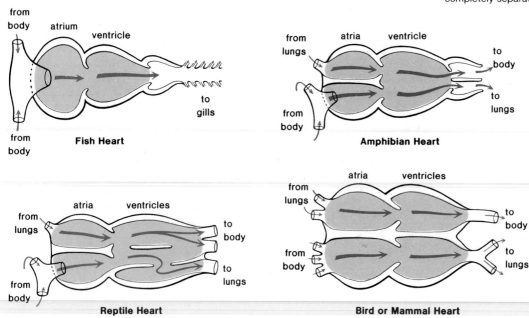

Comparison of hearts of vertebrates. A fish heart has only one atrium and one ventricle; an amphibian heart has two atria but only one ventricle; a reptilian heart has two atria and two ventricles that are not completely separated; but the heart of a bird and mammal has two atria and two ventricles, and all four chambers are completely separated one from the other.

Table 27.7	Comparison of vertebrates				
	Fishes	**Amphibians**	**Reptiles**	**Birds**	**Mammals**
Habitat	Water	Water/land	Land	Land	Land
Heart	Nondivided atrium and ventricle	Two atria and nondivided ventricle	Two atria and partially divided ventricle	Two atria and two ventricles	Two atria and two ventricles
Respiration	Gills	Gills/lungs/skin	Lungs and ribcage	Lungs, ribcage, and air sacs	Lungs, ribcage, and diaphragm
Fertilization	External	External	Internal	Internal	Internal
Egg	Small, no shell, develops externally	Small, no shell, develops externally	Large, with shell, develops externally	Large, with shell, develops externally	Small, no shell, develops internally

low. Birds and mammals have a four-chambered heart in which division of the atria and ventricles is complete and there is no opportunity for mixing to occur. The right side of the heart pumps blood to the lungs and the left side pumps blood to the rest of the body.

A comparison of the eggs of vertebrates shows that fish and amphibian eggs are generally small with little yolk; these eggs are deposited into the water where they may develop into swimming larva. Both reptiles and birds lay a shelled egg with extraembryonic membranes to take over the functions previously performed by external water. The placental mammals have modified membranes that permit internal development during which time the mother provides for the needs of the developing fetus (chap. 20).

Table 27.7 summarizes the vertebrate classes by comparing basic features.

Table 27.8 Classification features

	Sponges	Coelenterates	Flatworms	Roundworms
Body plan	———	Sac	Sac	Tube within a tube
Symmetry	Radial or none	Radial	Bilateral	Bilateral
Germ layers	———	2	3	3
Level of organization	———	Tissues	Organs	Organs
Body cavity	———	———	———	Pseudocoelom
Segmentation	———	———	———	———

Table 27.9 Comparison of lower invertebrates

	Sponge	Hydra	Planaria	Ascaris
Nervous connections	———	Nerve net	Ladder-type	Ring with two nerve cords
Muscles	Fibers	Fibers	Three layers	One layer
Digestion	Intracellular, food vacuoles only	Extracellular and intracellular	Extracellular and intracellular	Extracellular only
Excretion	———	———	Flame cells	———
Reproduction	Asexual by budding	Asexual by budding	Asexual by fission	Sexual by sex organs; separate sexes
	Sexual by egg and sperm—motile larva	Sexual by egg and sperm—motile larva	Sexual by sex organs	

Summary

An examination of the evolutionary tree of animals indicates that they can be divided into three groups. The lower invertebrate phyla form the trunk of the tree. The other phyla are placed on one of two branches according to whether they exhibit the dipleurula or trochophore larva.

Classification of animals considers type of body plan, symmetry, number of germ layers, presence or absence of coelom, and segmentation. Among the lower invertebrates it is possible to observe an increase in complexity regarding these features if the various groups are arranged as in table 27.8. While the flatworms and roundworms have specialized organs, all of the lower invertebrates carry out respiration and circulation by diffusion. The means by which representative animals carry out other animal functions are listed in table 27.9.

All of the higher invertebrates have the tube within a tube body plan, three germ layers, and a true coelom. Embryological evidence indicates that the mollusks, annelids, and arthropods are closely related. The mollusks have a body plan composed of three parts: the foot, mantle, and visceral mass. All mollusks have modified these regions to suit their own way of life and environmental conditions. The mollusks are nonsegmented, however, while the annelids and arthropods are segmented. The annelids show the best segmentation of all the animal phyla while specialization of parts has occurred among the arthropods that also have an exoskeleton with jointed appendages.

Embryological evidence suggests that the echinoderms and chordates are related. The former are unique among the higher invertebrates because of their radial symmetry and locomotion by tube feet. All of the higher invertebrates show complex organ systems that are adapted to their way of life. Table 27.10 summarizes the descriptions that are used to indicate the type of organs found in representative animals.

At some time in their life history all chordates have a dorsal notochord, dorsal hollow nerve cord, and gill pouches. Only amphioxus retains all three features as an adult. The vertebrates are chordates in which the notochord is replaced by a vertebral column in the adult. The evolutionary tree indicates

Table 27.10 Comparison of higher invertebrates

	Clam	Earthworm	Crayfish	Grasshopper	Starfish
Nervous system	Three ganglia joined by nerves	Brain and ventral solid nerve cord	Brain and ventral solid nerve cord	Brain and ventral solid nerve cord	Nerve ring and five cords
Digestion	Filter feeder; intestine goes through heart	Eats dirt; has a crop and gizzard	Predator; has a two-part stomach; anterior portion is a gastric mill	Vegetarian; has a crop and gizzard	Prefers clams; has a two-part stomach, one of which everts to digest clams
Skeleton	Bivalve shell (exoskeleton)	Absent	Chitin exoskeleton	Chitin exoskeleton	Spiny endoskeleton
Excretion	Kidney	Nephridia	Green gland	Malpighian tubules	Utilizes coelom
Circulation	Open, with internal sinuses and blue blood	Closed, with two large vessels and red blood	Open, with hemocoel and blue blood	Open, with colorless blood	Utilizes coelom
Respiration	Gills	Cuticle and body wall	Gills	Tracheae	Skin gills
Reproduction	Sexes separate, marine forms have trochophore larva	Hermaphroditic; clitellum supplies mucus	Sexes separate; male swimmerets (1st pair) are modified	Sexes separate; male has penis; female has ovipositor	Sexes separate; dipleurula larva
Locomotion	Hatchet foot	Setae; hydrostatic "skeleton"	Jointed appendages (walking legs and swimmerets)	Hopping legs and wings	Water vascular system with tube feet

that amphibians evolved from the lobe-finned fishes; reptiles evolved from amphibians; and birds and mammals evolved from reptiles.

Each group of vertebrates shows adaptation to its environment. Fish are very well adapted to life in the water. Adaptation to land life begins poorly in the amphibians, but it is almost complete with the reptiles, which are able to reproduce on land due to a shelled egg with extraembryonic membranes. Only birds and mammals are warm-blooded, however, and able to maintain a constant internal temperature; thus, only they are completely adapted to life on land.

Study Questions

1. Which groups of animals comprise the lower invertebrates? (p. 590) Compare the representatives of these four phyla in regard to body plan, symmetry, germ layers, level of oganization, coelom, and segmentation. (pp. 590–601, 628)
2. Compare the representatives of the four lower invertebrate phyla in regard to nervous conduction, musculature, digestion, excretion, and reproduction. (pp. 590–601, 628)
3. Describe the life cycle and structure of a tapeworm. Compare the anatomy of free-living flatworms with that of the fluke and tapeworm. (p. 597)
4. What biological data are used to divide higher animals into two groups? (p. 601)
5. Compare the clam, earthworm, crayfish, grasshopper, and starfish with respect to nervous, digestive, skeletal, excretory, circulatory, and respiratory systems and means of reproduction and locomotion. (pp. 604–16)
6. Compare the adaptations of the clam to those of the squid to show that the clam is adapted to an inactive life and the squid is adapted to an active life. (p. 605)
7. Compare the adaptations of the crayfish to those of the grasshopper to show that the former is adapted to an aquatic existence while the latter is adapted to a terrestrial existence. (p. 612)
8. Name and describe unique features of echinoderm anatomy and physiology. (p. 614)

9. What three characteristics do all chordates have at some time in their life history? Which invertebrate chordate has these features as an adult? (p. 616)
10. Define a vertebrate and discuss the vertebrate's general anatomy, including type of skeleton, nervous system, circulatory system, and excretory system. (pp. 616–17)
11. Trace the adaptation to a land existence among the vertebrate groups. (pp. 621–24)
12. What are the three types of mammals? What are the different types of placental mammals? (pp. 624–26)

Selected Key Terms

symmetry (sim'ĕ-tre)

cephalization (sef''al-i-za'shun)

germ layers (jerm la'ers)

coelom (se'lom)

segmented (seg-ment'ed)

filter feeder (fil'ter fēd'er)

regeneration (re-jen''er-a'shun)

mesoglea (mes''o-gle'ah)

gastrovascular cavity (gas''tro-vas'ku-lar kav'ĭ-te)

proglottids (pro-glot'idz)

larva (lar'vah)

mantle (man't'l)

nephridia (nĕ-frid'e-ah)

jointed appendage (joint'ed ah-pen'dij)

hemocoel (he'mo-sēl)

Malpighian tubules (mal-pig'ĭ-an tu'būlz)

trachea (tra'ke-ah)

metamorphosis (met''ah-mor'fo-sis)

water vascular system (wah'ter vas'ku-lar sis'tem)

mammary glands (mam'er-e glandz)

Further Readings

Bakker, R. T. 1975. Dinosaur renaissance. *Scientific American* 232(4):58.

Barnes, R. D. 1980. *Invertebrate zoology*. 4th ed. Philadelphia: W. B. Saunders.

Boolootian, R. A., and Stiles, K. A. 1981. *College zoology*. 10th ed. New York: Macmillan.

Buchsbaum, R. M. 1975. *Animals without backbones*. Rev. 2d ed. Chicago: University of Chicago Press.

Hickman, C. P. et al. 1982. *Biology of animals*. 3d ed. St. Louis: C. V. Mosby.

Romer, A. S. 1971. *The vertebrate story*. Rev. ed. Chicago: University of Chicago Press.

Roper, C. F. E. and Boss, K. J. 1982. The giant squid. *Scientific American* 246(4):96.

Russell-Hunter, W. D. 1968. *A biology of lower invertebrates*. New York: Macmillan.

Russell-Hunter, W. D. 1969. *A biology of higher invertebrates*. New York: Macmillan.

Simpson, G. G. 1961. *Principles of animal taxonomy*. New York: Columbia University Press.

Vertebrate structure and function: Readings from *Scientific American,* 1974. San Francisco: W. H. Freeman.

Yange, C. M. 1975. Giant clams. *Scientific American* 232(4):96.

Young, J. Z. 1981. *The life of vertebrates*. 3d ed. New York: Oxford University Press.

28

Human Evolution

Chapter concepts

1 Humans are primates and many of their physical traits are the result of their ancestors' adaptations to living in trees.

2 Humans share a common ancestor with the apes.

3 The first family ancestor of humans was most likely a terrestrial tool-user.

4 Tool use, walking erect, and intelligence all may have evolved together.

5 All human races are classified as *Homo sapiens*.

Figure 28.1

Primate evolutionary tree. Primates are believed to have evolved from an insectivore mammal; monkeys are distantly related to humans, but apes are closely related. Just exactly when the ape-man split occurred is not known. Perhaps humans share a recent common ancestor with only certain of the apes. If this is the case, then the split could have occurred within the past several millions of years. This tree suggests that this is the case.

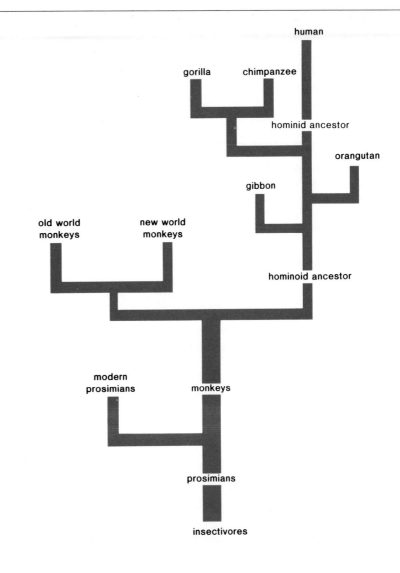

Figure 28.2

The tree shrew is believed to resemble the ancestor of the primates. Primates remained small and insignificant until the great dinosaurs became extinct.

Evolution and Diversity

Primates

Humans are mammals in the order Primates (Table 28.1). **Primates** were originally adapted to an arboreal life in trees. Long and freely movable arms, legs, fingers, and toes allowed them to reach out and grasp an adjoining tree limb. The opposable thumb and toe, meaning that the thumb and toe could touch each of the other digits, were also helpful. Nails replaced claws; this meant that primates could also easily let go of tree limbs.

The brain became well developed, especially the cerebral cortex and frontal lobes, the highest portions of the brain. Also, the centers for vision and muscle coordination were enlarged. The face became flat so that the eyes were directed forward, allowing the two fields of vision to overlap. The resulting stereoscopic (three-dimensional) vision enabled the brain to determine depth. Color vision aided the ability to find fruit or prey.

One birth at a time became the norm; it would have been difficult to care for several offspring, as large as primates, in trees. The period of post-natal maturation was prolonged, giving the immature young an adequate length of time to learn behavior patterns.

Prosimians

As diagrammed in figure 28.1, primates are believed to have evolved from primitive shrewlike **insectivores,** arboreal rat-sized animals with sharp canine teeth. A living tree shrew, which is believed to resemble the primate ancestor is shown in figure 28.2. The first primates were the **prosimians** (fig. 28.3), a term that means premonkeys. The prosimians are represented today by several types of animals, among them the **lemurs,** which have a squirrel-like appearance, and the **tarsiers,** curious monkeylike creatures with enormous eyes.

Table 28.1 Classification of humans and related animals

Phylum chordata
Subphylum vertebrata

Class mammalia*	Monotremes
	Marsupials
	Placental
	Insectivores
	Rodents
	Marine
	Carnivores
	Primates
Order primates	Prosimians
	Lemurs
	Tarsiers
	Anthropoids
	Monkeys
	Apes
	Humans
Superfamily hominoidea	*Dryopithecus*
	Modern Apes
	Humans
Family hominidae	*Ramapithecus?*
	Australopithecus
	Homo habilis
	Homo erectus
	Homo sapiens
Genus homo (Humans)	*Homo habilis?*
	Homo erectus
	Homo sapiens

a.

Figure 28.3
Prosimians. At one time prosimians were widespread, but (a.) lemurs, such as this ring-tail lemur, are found only on the island of Madagascar and (b.) tarsiers adapted to a nocturnal existence are found only in the Philippines and East Indies.

b.

Figure 28.4

Ape diversity. *a.* Of the apes, gibbons are the most distantly related to humans. They dislike coming down from trees, even at watering holes. They will extend a long arm into the water and then drink collected moisture from the back of the hand. *b.* Orangutans are solitary except when they come together to reproduce. Their name means "forest man"; early Malayans believed that they were intelligent and could speak but did not because they were afraid of being put to work. *c.* Gorillas are terrestrial and live in groups in which a silver-backed male such as this one is always dominant. *d.* Of the apes, chimpanzees are most closely related to humans, and the pygmy chimp may be our closest living relative, as discussed in the reading on page 640.

a.

b.

c.

d.

Evolution and Diversity

Table 28.2 Comparison of apes and humans

Feature	Primitive ancestor	Apes	Humans
Brain size	Small brain and skull	Slightly enlarged	Very much enlarged
Face	Sloping brow, heavy eyebrow ridges, and projection of face	Same as primitive	High brow, reduced eyebrow ridges, face flat
	Rectangular-shaped jaw with large molars and long canine teeth	Same as primitive	U-shaped jaw with small molars and shortened canine teeth
Locomotion	Quadrupedal locomotion	Same as primitive	Bipedal locomotion
	Limbs of equal length	Forelimbs elongated	Same as primitive
	Opposable thumb and toe	Same as primitive	Opposable thumb retained

Figure 28.5

Comparison of the skull of *Dryopithecus* with that of modern apes and humans. *Dryopithecus* has primitive features; the common chimpanzee has apelike features, including a low brow, heavy eyebrow ridges, and a face that protrudes; while humans have a high brow, reduced eyebrow ridges, and a flat face.

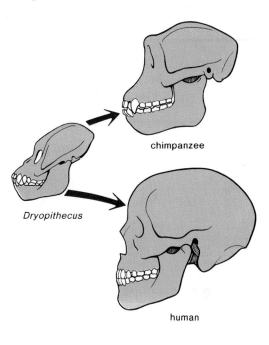

Figure 28.6

Comparison of the jaw of an ape with that of a human. *a.* The ape jaw has a rectangular shape. *b.* The human jaw has a U shape.

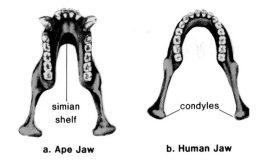

Anthropoids

Monkeys evolved from the prosimians some 60 million years ago. There are two types of monkeys: the **New World monkey,** which has a long, prehensile tail capable of grasping, such as the organ grinder's monkey; and the **Old World monkey,** which does not have a prehensile tail but has a narrow septum between the nostrils, such as the rhesus monkey. The baboon, whose behavior is discussed on page 658, is another Old World monkey.

Although the designation **anthropoid** (Table 28.1) includes the monkeys, humans are more closely related to the apes. There are four types of apes: gibbon, orangutan, gorilla, and chimpanzee (fig. 28.4). The **gibbon** is the smallest of the apes, with a body weight ranging from 12 to 25 pounds. Gibbons have extremely long arms that are quite specialized for swinging between tree limbs. The **orangutan** is large (165 lbs.) but nevertheless spends a great deal of time in trees, while the **gorilla,** the largest of the apes (400 lbs.), spends most of its time on the ground. **Chimpanzees,** which are at home both in trees and on the ground, are the most humanlike of the apes in appearance and are frequently used in psychological experiments.

Hominoids

The fossil record suggests that apes, and perhaps humans too, may have shared an ancestor about 25 million years ago. This ancestor, which may have been similar to *Dryopithecus,* is termed the **hominoid ancestor** because both apes and humans are in the superfamily Hominoidea (table 28.1). Several fossil species from Europe, Africa, and Asia are believed to be *Dryopithecus,* sometimes known as the oak apes because oak leaves have been found near their fossil remains. Table 28.2 lists some of the characteristics of these apes and indicates the manner in which modern day apes and humans have come to differ from the supposed ancestral stock. This ancestor may appear to be apelike, but these features may simply be primitive features. A small brain enclosed by a small skull, with heavy eyebrow ridges and a sloping forehead, may be primitive features rather than apelike ones. The brain size and skull size are increased in apes, but not to the same degree as they are in humans. Humans have developed a high forehead, with only moderate eyebrow ridges (fig. 28.5).

The primitive jaw must have been rectangular shaped with long canine teeth. Although the apes have retained these features, human teeth are smaller, the jaw is U-shaped (fig. 28.6), and the face does not project forward.

Walking on all four limbs, each of which is at an angle to the spinal cord, is a primitive feature. Although their forelimbs are elongated, this placement of the limbs has been retained by apes. In humans, dramatic skeletal changes

Figure 28.7

Comparison of skeleton of "knuckle-walking" ape (a gorilla) with the skeleton of a bipedal human. Note differences in proportions of hindlimbs and forelimbs, in the shape of the rib cage and of the pelvis, and in the curvature of the spine and the angle and position of attachment of the vertebral column to the skull. The head of the ape faces forward when knuckle-walking because the vertebral column attaches to the rear of the skull.

pelvis

pelvis

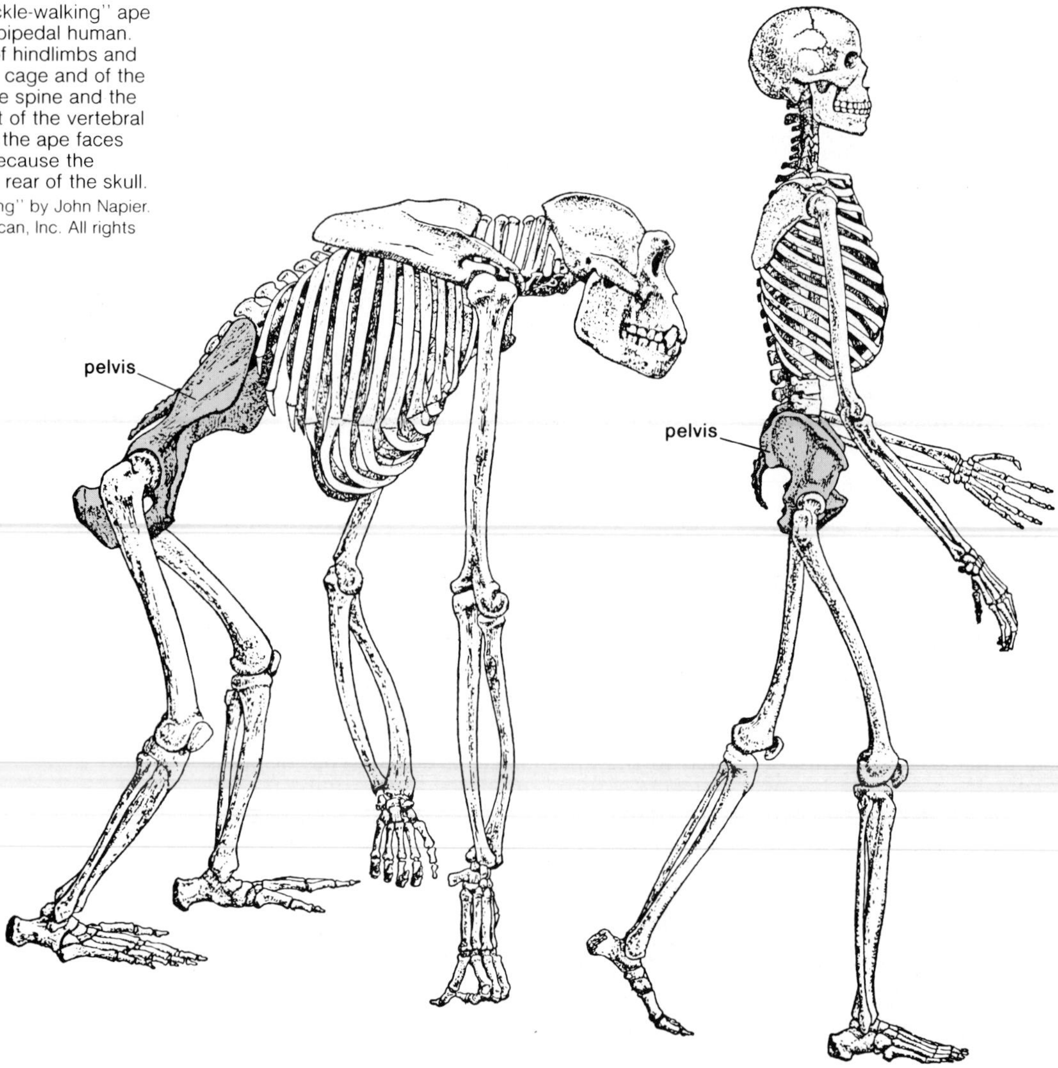

Figure 28.8

The primate hand is capable of grasping objects because of the opposable thumb. In humans, this ability is used to grasp tools.

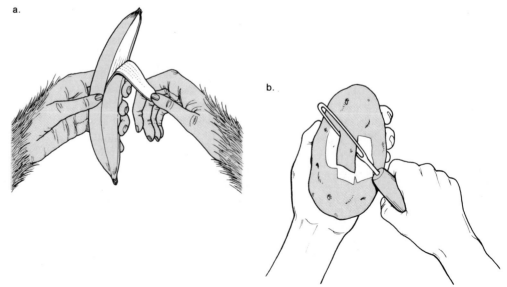

a.

b.

Evolution and Diversity

occurred during the process of developing an erect posture (fig. 28.7). The head, spinal cord, pelvis, and legs are in a nearly straight line. The pelvis is short and upright, rather than long and tilted forward. The spine has an S curve, allowing a better weight distribution and improved balance when the body is upright.

Skeletal changes have also occurred in the fingers and toes. In the primitive condition, the thumb and large toe are both opposable, as seen in apes today. Humans have retained only the opposable thumb (fig. 28.8).

Hominid Ancestor

Hominids (family Hominidae) include living humans and human and humanlike fossils. It is generally agreed that hominids did evolve from an ancestor that has characteristics of both apes and humans. It is also believed that adaptation to a changing environment encouraged the evolution of the first hominid. During the late tertiary period (table 23.1) the climate became progressively cooler and drier. Much of the lush forests were replaced with vast savannas (grasslands with occasional clumps of trees). It is reasoned that during this time, a hominoid evolved into a hominid after coming down out of the trees, standing partially upright, and moving onto the savanna.

For quite some time it has been suggested that the fossil remains of *Ramapithecus* may be those of the first hominids. These remains are found in India and Pakistan, through the Near East and the Balkans to Africa, and are dated from as long ago as 17 to just 8 million years ago. On the basis of only an upper jaw and a handful of molars, it was suggested in the 1960s that the **ramapithecines** were the first hominids. The molars were large and had a thick enamel cover, as do human molars, and since the canine teeth were missing, it was believed that the jaw was U-shaped. However, in 1971 when a lower jaw was properly identified, it became obvious that the ramapithecines had apelike features. In the late 1970s and early 1980s, fossilized skulls were found. The close-set eye sockets, eyebrow ridges, protruding jaw, and flaring cheekbones indicate that the ramapithecines were most likely ancestral orangutans. However, since the ramapithecines lived for as long as 10 million years, it cannot be ruled out that they are also ancestral to the gorilla, chimp, and human. Interesting enough is the observation that the ramapithecines, orangutans, and humans have a thick layer of tooth enamel.

Large molars with a thick layer of enamel are believed to be significant because they suggest that the diet includes not just fruit but also tough morsels of food such as seeds, grass, stems, and roots. These foods are plentiful on the ground but not in trees, and thus the first hominid would most likely have had this dentition. Despite this consideration, the ramapithecines are no longer believed to be the first hominids.

A new theory is that we along with gorillas and modern-day chimpanzees are descended from an ancestor that resembles pygmy chimpanzees. Those who support this contention, which is discussed in the reading on page 640, emphasize that biochemical data show that the DNA of humans and chimpanzees varies less than 2 percent and that this much variation is even sometimes found between two different humans. So it may just be that humans share a recent common ancestor only with certain apes rather than with all the apes. In that case, the split between humans and apes may very well have occurred as late as 4 to 5 million years ago.

The hominid ancestor was most likely a "knuckle-walker" like the great apes (all but the gibbons) are today (fig. 28.9). This type of walking would have allowed retention of an opposable thumb since an opposable thumb does not interfere with knuckle walking. This is important, because with the opposable thumb retained, the evolution of bipedal locomotion (walking on two legs) could have led to manipulation of tools. While tool use has been found in other animals, it is still a distinctly human characteristic.

Figure 28.9

A common chimpanzee knuckle-walking. Most likely, the first hominid walked like this. Knuckle-walking does not interfere with having an opposable thumb and thus the thumb previously used for grasping tree limbs could have eventually been used for grasping tools.

Figure 28.10

Hominid evolutionary tree. Adaptive radiation is evident because this tree indicates that australopithecines and humans coexisted. This is only one possible tree and others have been suggested.

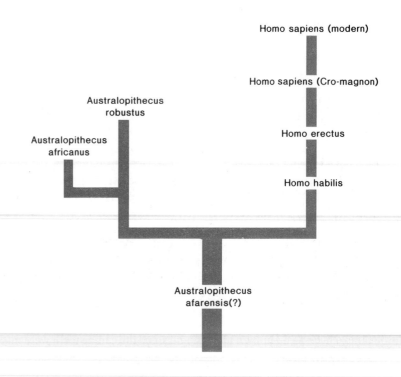

Homo sapiens (modern)

Homo sapiens (Cro-magnon)

Homo erectus

Homo habilis

Australopithecus robustus

Australopithecus africanus

Australopithecus afarensis(?)

Hominids

New ideas regarding speciation, particularly the concept of adaptive radiation (p. 521), have been used to interpret the hominid fossil record. Whereas formerly scientists attempted to place each hominid fossil in a straight line from the most primitive to the most advanced, it is now reasoned that several hominid species could have existed at the same time. Figure 28.10 indicates a possible evolutionary tree for the known hominids.

Australopithecines

The fossil remains of those classified in the genus *Australopithecus* (Southern Apeman) were found in southern Africa and have been dated at about 4 million years ago. The australopithecines (fig. 28.11) were 4 to 5 feet in height, with a brain that ranged in size from 400 to 800 cubic centimeters. The pelvis definitely indicates that they were capable of bipedal locomotion; supporting this contention is the nonopposability of the big toe.

Three species of *Australopithecus* have been identified—*afarensis, africanus, and robustus.* The exact relationship between the three species is in dispute. Because of its small brain size, large canine teeth, and protruding face, it has been suggested that *afarensis* is the more primitive of the three. Moreover, because of certain skeletal limb features indicating that these peoples walked erect, some believe that *afarensis* is also ancestral to humans, as

Figure 28.11
Australopithecus africanus. Some authorities
believe this hominid is directly ancestral to humans,
while others believe that they share a common
ancestor.

indicated in figure 28.10. It may also be noted that the combined characteristics of this fossil indicate that an enlarged brain was *not* needed for bipedal
locomotion to evolve.

Those promoting the suggestion that there is a close relationship between pygmy chimps and humans have found anatomical similarities between
pygmy chimps and *Australopithecus afarensis,* who is affectionately known
as "Lucy." (The fossil is named after the Beatles' song, "Lucy in the Sky with
Diamonds.") The reading on page 640 discusses these anatomical similarities.

Australopithecus robustus, as its name implies, is larger than *africanus.*
Certain anatomical differences may be due to diet. The massive skull and face
and the large cheek teeth of *robustus* indicate a vegetarian diet. The gracile
facial features of *africanus* indicate a more varied diet, possibly including meat.
Facial bones, including eyebrow ridges, facial muscles, and teeth, need not be
as large in meat eaters. If members of this species were meat eaters, it is likely
that they were also hunters.

Lucy's Uncommon Forbearer

Pygmy chimps have anatomical and social characteristics that suggest they are closely related to us.

. . . Along with Sarich [of Berkeley], Zihlman (of University of California, Santa Cruz) and Cramer (of New York University) have become the champions of the bonobo model [which claims that the pygmy chimp, called "bonobo" in the Republic of Zaire where it lives, resembles the common ancestor to chimpanzees, gorillas, and humans] and they have based their claims primarily on studies of the anatomy of living apes and fossilized hominids. Sarich's biochemical data indicate that the two species of chimpanzee split about 2 to 3 million years ago so that the pygmy chimp is fundamentally a chimp; but according to Zihlman, a comparative study of the limb bones reveals that the pygmy chimp is more "primitive" than its common cousin. In other words, it has gone through fewer specialized adaptations and therefore most closely resembles the ancestral condition.

To make her point, Zihlman compares the pygmy chimpanzee to "Lucy," one of the oldest hominid fossils known, and finds the similarities striking. They are almost identical in body size, in stature and in brain size, she notes, and the major differences (the hip and the foot) represent the younger Lucy's adaptation to bipedal walking (Lucy, officially called *Australopithecus afarensis*, has been dated at 3.6 million years, although that date has recently been challenged). These commonalities, Zihlman argues, indicate that pygmy chimps use their limbs in much the same way that Lucy did—and that they inherited those habits from the same ancestor.

In contrast to the shared traits of pygmy chimps and early hominids, Zihlman says, the common chimpanzee has legs that are shorter and arms that are longer, relative to its body size. Zihlman also made estimates of the muscle and tissue mass of an animal like Lucy and compared it to that of modern apes: only the pygmy chimpanzee, with its strong legs, smaller arms and slender chest, closely matches Lucy. What this suggests, according to Zihlman, is that the pygmy chimpanzee is more flexible than other modern apes; a common ancestor would have to be flexible—somewhere between a strict quadruped and a strict biped—in order to evolve simultaneously into man and ape.

The major critics of the pygmy chimpanzee model have been Johanson, discoverer of Lucy, and his co-workers—Tim D. White of Berkeley, C. Owen Lovejoy of Kent State University and Bruce Latimer of the Cleveland Museum of Natural History. They say that Zihlman has gone far beyond the fossil data in her interpretation. Johanson, who is director of the Berkeley-based Institute of Human Origins, argues that the pygmy chimpanzee may actually be a specialized, derived form of ape, whose anatomy is adapted specifically to the isolated jungle habitat that it now inhabits. . . .

Furthermore, Johanson and his colleagues say, the proponents of the pygmy chimpanzee have failed to explain some significant contrasts between bonobos and Lucy. Specifically, they note, the pygmy chimpanzee's teeth are much smaller and less rugged than those of the early hominids; and secondly, the male and female bonobos differ very little in anatomy, whereas the earliest hominids were extremely sexually dimorphic.

Zihlman and Sarich both concede that these differences

exist. But in putting forth the bonobo model, they say, they were not trying to reconstruct the hominid *A. afarensis;* they were trying to reconstruct the common ancestor of chimpanzees, gorillas and *A. afarensis.* "That's one step further back," says Sarich. "We assume that *afarensis* is highly derived in certain features. Specifically it's an open country ape, which makes it much different. The common ancestor was almost certainly not an open country form." Lucy's rugged dentition, they say, could have evolved very rapidly as the species moved from the forests into the savannas and began dealing with heavy, rough food.

Similarly, Zihlman says, the sexual dimorphism of *A. afarensis* could have evolved very rapidly; as the ape-like race emerged from the jungle to take advantage of the resources of the savanna mosaic, she says, the male of the species probably became larger to provide the necessary protection against predation. Sexual dimorphism is one characteristic, Zihlman says, that can evolve rapidly in either direction, so that the extreme male-female differences of *A. afarensis* could easily appear in a brief period of evolutionary time—2 million years—and then disappear (as they have) in modern humans. Another possibility, however, is that *A. afarensis* is not as dimorphic as Johanson and his colleagues maintain; what they interpret as males and females of the same species, Zihlman says, may very well be two different species. . . In any case, he says, in order to make sense of the anatomical differences and similarities between living and long-dead apes, it is necessary to observe living apes in the forest—the ecological setting in which the common ancestor was most likely to have lived. The bonobo is the only true forest dweller among modern apes, and as a result it alone can model what the common forebear was doing 5 million years ago.

Only recently, with the completion of Susman's 18–month study of pygmy chimpanzees in the Lomako Forest of Zaire, has such evidence been available. . . . hours of observation along the Zaire River have confirmed that the bonobo is a unique ape—if not more human, then at least more flexible in its behavior than gorillas, humans or common chimpanzees.

Susman and his colleagues Noel and Alison Badrian have found, for example, that the pygmy chimpanzee is different from the other African apes in terms of sexual behavior; specifically, the females have extraordinarily long periods of sexual receptivity, like humans, and like humans they tend to mate front to front. From an evolutionary point of view, this evidence suggests one of two things: that humans and pygmy chimpanzees, once on their own evolutionary tracks, developed this behavior separately or that such sexual behavior was once an ancestral characteristic.

Susman also discovered that pygmy chimps have a unique style of locomotion. Like modern gorillas they tend to be knuckle-walkers on the ground, yet they seem to be natural bipeds, too, frequently walking upright both on the ground and in the trees. And even more important, Susman says, is that when they are in the trees, the pygmy chimps are by far the most acrobatic of the apes, capable of arm swinging, leaping and diving. This documented behavior is consistent with Susman's own anatomical studies, which have shown that the bonobo's shoulder blade, arm and hand are well adapted for arboreality—the primitive condition, according to Susman . . . .

Susman's group has also gathered extensive data on social and feeding behaviors of living bonobos, and they are beginning to piece together a picture of what the hypothetical 5-million-year-old ancestor might have looked like. "What we have is a unique animal that didn't behave like any animal does today," Susman says. "It climbed trees and probably nested in trees. It weighed 60 pounds. It ate mostly fruit but perhaps some meat. It spent its nights and early mornings in trees, but it probably moved between patches of forest during the day by walking bipedally. It did not have tools or a tool-making hand, and it had a small brain." . .

The rest of the fossils to be mentioned are in the genus *Homo*, which is the genus for all humans, including ourselves.

Homo Habilis

This newly discovered fossil man is dated about 2 million years ago, which may have made him a contemporary of *Australopithecus africanus*. *Homo habilis* is significant for several reasons. (1) Formerly it was believed that humans evolved less than 2 million years ago in the Pleistocene epoch (fig. 23.1). Current evidence seems to suggest that they must have begun evolving in the Pliocene epoch *more* than 2 million years ago. (2) Almost certainly, humans evolved in Africa. (3) Formerly only hominid fossils with a brain capacity of

Figure 28.12
Stone tools used by early hominids (e.g., *Homo habilis*), *Homo erectus,* and *Homo sapiens* (Cro-Magnon). Note the increasing amount of refinement in the tools.

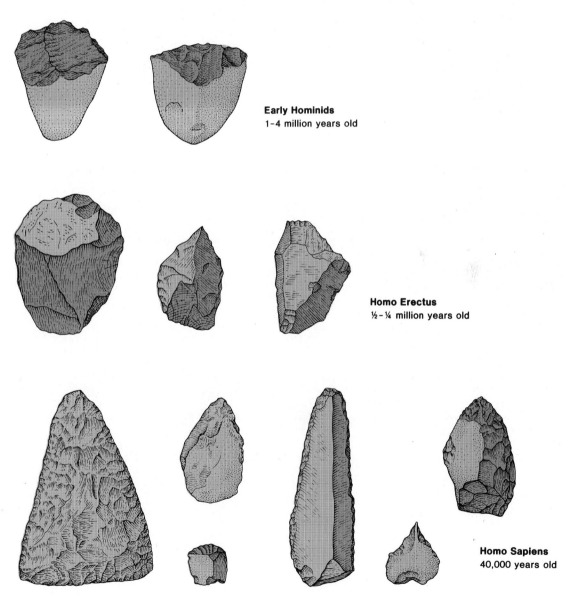

Early Hominids
1–4 million years old

Homo Erectus
½–¼ million years old

Homo Sapiens
40,000 years old

1,000 cubic centimeters were designated humans. *Homo habilis* had a maximum brain capacity of 800 cubic centimeters and yet he has been placed in the genus *Homo* because he not only used tools, he also made them. *Homo habilis* means handyman; he was given this name because of the quality of the tools (fig. 28.12) found with his bones.

At one time it was thought that only primitive people with a brain capacity of more than 1,000 cubic centimeters could have made tools. Since *Homo habilis* has a smaller brain and yet made tools, are we to think that the making of tools preceded the evolution of the enlarged brain? This seems to be an unnecessary question. Increased brain capacity, no matter how slight, would have permitted better tool making; this combination would have been selected because tool making would have fostered survival in a grassland habitat. Thus, as the brain became increasingly larger, tool use would have become more sophisticated.

Homo Erectus

Homo erectus (fig. 28.13) was prevalent throughout Eurasia and Africa during the Pleistocene Age, also called the Ice Age because of the recurrent cold weather that produced the glaciers of this epoch. *Homo erectus* has a brain size of 1,000 cubic centimeters, but the shape of the skull indicates that the areas of the brain necessary for memory, intellect, and language were not well developed.

Figure 28.13
Homo erectus. Evidence indicates that *Homo erectus* was a successful species found throughout Africa and Europe.

Figure 28.14

The striding gait of modern humans. Each limb alternately goes through a stance and swing phase. In the first drawing, the left limb is in the stationary stance phase and the right is beginning the swing phase. First, the knee is bent and extended forward. Then as the knee straightens, the heel and ball of the foot touch the ground. In the last two drawings, the left limb has entered the swing phase and the right limb is entering the stance phase.

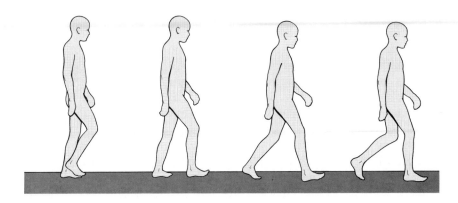

Even so, the grasp, posture, and locomotion of *Homo erectus* were all similar to those seen in modern humans. Humans have a *striding gait* (fig. 28.14), which means that the legs have alternate phases, the stance phase and the swing phase. When one hindlimb is in the stance phase, the other is in the swing phase. During the swing phase, first the knee is bent and then extended forward. The knee straightens, the heel touches the ground, and as the foot follows, that limb enters the stance phase.

Homo erectus made core-type tools (fig. 28.12), meaning that the tool was produced by removal of flakes from all sides of a stone. These people were also hunters but, unlike previous humans, they possessed knowledge of fire and would have been able to cook their meat in order to tenderize it.

Homo Sapiens

Neanderthal

About 100,000 years ago, a distinctive group of humans called **Neanderthal** emerged. The name comes from the Neander Valley in Germany where the first fossil of this type of human was unearthed, but actually they are also found in much of southern Europe, Asia, and Africa. The Neanderthals are considered to be so nearly like modern-day humans, including their brain size, that they are classified *Homo sapiens neanderthal*. Only in western European fossils do the Neanderthals (fig. 28.15) exhibit the massive faces and heavy eyebrow ridges for which they are generally known. In other parts of the world their fossils have a modern appearance.

There is extensive evidence that the Neanderthals carried out ritualistic practices, which is strongly suggestive of abstract thought, and perhaps even had a concept of religion and life after death. Their brain capacity was essentially the same as that of modern humans. Some researchers believe that Neanderthals are simply a variation of modern humans, or that they gave rise to modern humans. Others have suggested, however, that they may have died off, or even that they were killed off by Cro-Magnon.

Cro-Magnon

Cro-Magnon fossils appear in the fossil record some 40,000 years ago. With a cranial capacity of about 1500 cubic centimeters they are generally accepted as modern humans and classified as *Homo sapiens sapiens,* as are present-day humans. Cro-Magnon was such an accomplished hunter that these people have been held responsible by some for the extinction, during the Upper Pleistocene, of many large mammalian animals, such as the giant sloth, mammoth, saber-toothed tiger, and giant ox. A predatory life-style would have encouraged the evolution of intelligence and the ability to speak. The fact that these people preyed on animals larger than themselves (fig. 28.16) would have required cooperation among the hunters, which would have been facilitated by

Evolution and Diversity

Figure 28.15
Homo sapiens neanderthal. It is now believed that Neanderthals generally had a more modern appearance than this drawing depicts them.

Figure 28.16
Socialization may have begun when men organized for the hunt. Since they hunted animals larger than themselves, cooperation was required.

language. Cooperation, in turn, led to socialization and the advancement of culture. Humans are believed to have lived in small groups, the men going out to hunt by day while the women remained at home with the children. The many ways in which hunting may have influenced our behavior is discussed on page 680.

Cro-Magnon is believed to have introduced a new tool industry, evidence of which is generally widespread. The major characteristic of this industry was the production of the blade, a tool with roughly parallel sides (fig. 28.12). Cro-Magnon is known not only for advanced technology but also for artistic ability since these people are believed to have painted the beautiful drawings on cave walls in Spain and France (fig. 28.17) and also, as discussed in the reading on page 649, to have sculpted many small figurines.

If the Cro-Magnon did cause the extinction of many types of animals, this may account for the transition from a hunting economy to an agricultural economy about 12,000 to 15,000 years ago. Technological progress requiring the use of metals and energy sources led, in an amazingly short time (fig. 28.18), to the Industrial Revolution, which began about 200 years ago. After this time, many people typically lived in cities, in large part divorced from nature and endowed with the philosophy of exploitation and control of nature. Only recently have we begun to realize that the human population, like all other organisms with which we share an evolutionary history, should work with rather than against nature. The ecology chapters that follow will stress this theme.

Human Races

All human races (fig. 28.19) are placed in the genus *Homo* and the species *sapiens*. This is consistent with the biological definition of species because it is possible for all types of humans to interbreed and bear fertile offspring. The close relationship between the races is supported by biochemical data showing that differences in amino acid sequence between two individuals of the same race are as great as those between two individuals of different races.

It is generally accepted that racial differences developed as adaptations to climate. Although it might seem as if dark skin is a protection against the

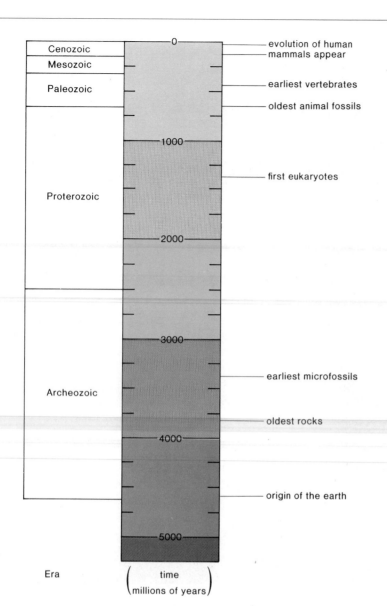

Era	time (millions of years)	
Cenozoic	0	— evolution of human
		— mammals appear
Mesozoic		
Paleozoic		— earliest vertebrates
		— oldest animal fossils
	1000	
Proterozoic		— first eukaryotes
	2000	
	3000	
Archeozoic		— earliest microfossils
		— oldest rocks
	4000	
		— origin of the earth
	5000	

Figure 28.18
Compared to the total history of the earth, humans have been present a very short period of time. When the history of the earth is compared to a 24-hour day, humans evolved a few seconds before midnight.

hot rays of the sun, it has been suggested that it is actually a protection against ultraviolet ray absorption. Dark-skin persons living in southern regions and white-skin persons in northern regions absorb the same amount of radiation. (Some absorption is required for vitamin D production.) Other features that correlate with skin color, such as hair type and eye color, may simply be side effects of pleiotropic genes.

Differences in body shape represent adaptations to temperature. A squat body with shortened limbs and nose retains more heat than an elongated body with longer limbs and nose. Also, the "almond" eyes, flattened nose and forehead, and broad cheeks of the Oriental are believed to be adaptations to the extremely cold weather of the last ice age.

While it has always seemed to some that physical differences might warrant assigning human races to different species, this contention is not borne out by the biochemical data mentioned previously.

Figure 28.19

All human beings belong to one species, but there are several races such as (a.) Negroid, (b.) Mongoloid, (c.) Australoid, (d.) Caucasian, and (e.) American Indian.

b.

a.

c.

d.

e.

Evolution and Diversity

It was a startlingly different world. Vast areas of the Northern Hemisphere were covered with ice. Across the ice-free parts of Europe and Asia, consisting largely of tundra and great treeless steppes, herds of mammoths, bison, reindeer and horses freely roamed. For long periods, winters were cruelly cold, and even in the summertime the average temperature was 12° to 15° C (54° to 59°F). Still, under these difficult conditions, during a period of 25,000 years before the dawn of civilization, the Ice Age Cro-Magnon people not only thrived, but created a surprisingly sophisticated culture that totally belies the popular image of them as savage, club-swinging brutes.

Nowhere is this cultural richness more apparent than in the artworks that these paleolithic hunters left in caves in France and Spain. When the first of these subterranean galleries was discovered in Spain nearly a century ago, Europe's savants, still reeling from the shock of Darwinian evolution, refused to believe that the find was anything more than a hoax. Since then, nearly a hundred richly decorated prehistoric caves have been found in Spain and France, and the existence of paleolithic painting has been established beyond doubt. The ancient artisans also left behind tiny sculptures of exquisite beauty, meticulous carvings on mammoth bone, and other stunning objects. Like the tableaux on the cave walls, some portray paleolithic man's animal neighbors . . . .

The dazzling collection includes, for example, a tiny, 6.4–cm-long (2½ in.) curving sculpture of a horse carved out of a mammoth tusk; it hardly seems possible that this graceful piece, fashioned more than 30,000 years ago, is one of the oldest *objets d'art* ever found. No less remarkable are the voluptuous "Venus" statuettes, some of them coiffed in Stone Age chic, that date back some 27,000 years. Even the wall paintings, some of them on a larger-than-life scale, show a mastery of form and perspective that was not seen again for almost 6,000 years.

These magnificent works reflect far more about Cro-Magnon man than his artistic ability. Indistinguishable from modern man either in brain capacity or physical appearance, he was clearly using his artistic skills to embellish a culture of a richness and complexity that is only beginning to be plumbed by scholars.

What that culture was like remains an enigma. What, for instance, is the significance of the Venus figures, with their exaggerated sexual features? What role did the great cave paintings play in the lives of those ancient people? Whatever the answers, it is clear the art is exceptionally complex, more than simple "hunting magic," as some turn-of-the-century scholars thought. Every indication is that Cro-Magnon man was deeply involved in rituals, ceremonies, myths, pehaps even a kind of religion.

Marshack, a former science writer who has devoted 15 years to paleolithic studies, has suggested even bolder ideas. In his writings, notably *The Roots of Civilization,* he says that what looks like random scribbling on cave walls and even on some artifacts may actually represent many different symbol systems. These could have been used to record the passage of the seasons and astronomical observations and to indicate periods of rituals and ceremonies. If these controversial yet hardly dismissable ideas are correct, Cro-Magnon man may well have been experimenting with the precursors of writing, arithmetic, calendar making and other "civilized" skills.

Marshack asks the key question: "Did these traditions prepare the way for the artistic and symbolic traditions of the civilizations that began to develop not long after the ice melted, about 10,000 B.C.?" No one can say for sure whether paleolithic man did in fact light that intellectual spark. But it is undeniable, as Marshack notes, that the complex art comes from "persons like us, with our brains and our capacity, and that no visitors from space were required to teach them."

A Treasure from the Ice Age

This stone sculpture from the upper Paleolithic culture stage was found in a cave near Menton, France.

Summary

Primates evolved from shrewlike insectivores and became adapted to living in trees, as exemplified by skeletal features, good vision, and even reproduction. The first primates were prosimians, followed by the monkeys, apes, and humans. The latter three are all anthropoids, the latter two are hominoids, and the last are hominids.

Humans and apes share a common ancestor which may possibly have been *Dryopithecus,* a fossil commonly regarded as the first hominoid. The first hominid did, however, live about the same time when the change in weather made it advantageous to dwell on the ground.

Perhaps humans share a recent common ancestor only with pygmy chimpanzees, present day apes that are anatomically similar to *Australopithecus afarensis* (Lucy), another suggested ancestor to humans. Biochemical evidence does indicate that we share an ancestor with chimpanzees as late as 3 million years ago.

One possible hominid evolutionary tree shows *Australopithecus afarensis* as a common ancestor to two other members of this genus and *Homo habilis,* the first fossil to be placed in the genus *Homo. Australopithecus africanus* was a meat eater and a possible hunter; *Australopithecus robustus* was a vegetarian. *Homo habilis,* a fossil dated at least 2 million years ago, may not have been highly intelligent, but he did make tools. Intelligence and making of tools probably evolved together. If *Homo habilis* is indeed considered human, then humans evolved much earlier than previously thought.

Homo erectus had a large brain and walked with a striding gait. He also used fire. *Homo sapiens neanderthal* was not as primitive as formerly thought. The Neanderthals probably evolved directly into Cro-Magnon, the first *Homo sapiens sapiens.* Cro-Magnon was an expert hunter. Hunting promoted language and socialization. In a relatively short time, humans developed an advanced culture that has tended to separate them from other organisms in the biosphere. All human races belong to the same species.

Study Questions

1. Name several primate characteristics still retained by humans. (p. 633)
2. Draw an evolutionary tree that includes all primates. What fossil may have been the hominid ancestor? (pp. 632, 635)
3. What animals mentioned in this chapter, whether living or extinct, are anthropoids? Hominoids? Hominids? Humans? (p. 633)
4. How did adaptations to a grassland habitat influence the evolution of humans? (p. 637)
5. Draw a hominid evolutionary tree. (p. 638)
6. More than one type of hominid existed at the same time. Explain in terms of adaptive radiation. (p. 638)
7. Which came first—tool use, walking erect, or intelligence? (p. 643)
8. Which humans were tool users? Walked erect? Had a striding gait? Used fire? Drew pictures? (pp. 642–46)
9. What evidence do we have that all races of humans belong to the same species? (p. 646) Name several races of humans. (p. 648)

Selected Key Terms

primates (pri′mãts)
arboreal (ar-bo′re-al)
prosimian (pro-sim′e-an)
anthropoids (an′thro-poidz)
hominoids (hom′ĭ-noidz)
Dryopithecus (dri″o-pith′e-cus)
Ramapithecus (ram″ah-pith′e-cus)
hominids (hom′ĭ-nidz)
australopithecines (aw″strah-lo-pith′e-sīnz)
Lucy (lu′se)
Homo habilis (ho′mo hah′bĭ-lis)
Homo erectus (ho′mo ĕ-rek′tus)
Neanderthal (ne-an′der-thawl)
Cro-Magnon (kro-mag′non)
Homo sapiens (ho′mo sa′pe-enz)
races (rãs′ez)

Further Readings

Eckhardt, R. B. 1972. Population genetics and human origins. *Scientific American* 226(1):94.

Hay, R. L., and Leakey, M. D. 1982. The fossil footprints of Laetoli. *Scientific American* 246(2):50.

Holloway, R. I. 1974. The casts of fossil hominid brains. *Scientific American* 231(1):106.

Katz, S., ed. 1975. *Biology anthropology.* Readings from *Scientific American.* San Francisco: W. H. Freeman.

Pilbeam, D. 1984. The descent of hominoids and hominids. *Scientific American* 250(3):84.

Simons, E. L. 1977. *Ramapithecus. Scientific American* 236(5):28.

Strauss, L. G., et al. 1980. Ice-age subsistence in northern Spain. *Scientific American* 242(6):142.

Trinkaus, R., and Howells, W. W. 1979. The Neanderthals. *Scientific American* 241(6):118.

Tullar, R. M. 1977. *The human species.* New York: McGraw-Hill.

Washburn, S. L. 1978. The evolution of man. *Scientific American* 239(3):194.

Weiss, M. L., and Mann, A. E. 1978. *Human biology and behavior: An anthropological perspective.* 2d ed. Boston: Little, Brown.

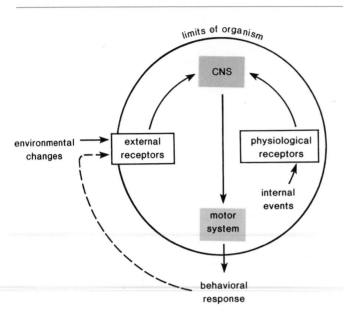

limits of organism

CNS

environmental changes → external receptors

physiological receptors

internal events

motor system

behavioral response

Fig
Diag
stim
caus
the c
direc
beha
be m
even
comr
a cer

Behavior is exemplified by an organism's routine activity. All organisms, even one-celled bacteria and certainly plants, have behavior; but in this chapter we shall concentrate on animal behavior. The particular behavior of an animal is determined by its anatomy and physiology. In other words, for example, frogs can't play the violin and humans can't flick out their tongues to catch flies. Anatomy and physiology are determined by the genetic makeup of an organism; therefore we must recognize the tenet that behavior, or at least the capacity for the behavior, is inherited. Since behavior is inherited, it stands to reason that behavior is adaptive. Behavior that enhances the chances of an organism surviving and reproducing will be that which is most likely passed on to the next generation. Behavior most appropriate to the species' environment will be the behavior to come into existence through the evolutionary process, and populations of the same species living in different habitats will have behavioral differences.

Figure 29.1 tells us that behavior is suitable to the limits of the organism and occurs as a response to a stimulus. Therefore it is dependent on sense receptors, the nervous system, and the musculoskeletal system. But often the organism must be ready or *motivated* to respond, and motivation is most likely dependent on the physiological or internal state of the organism. Thus frogs are more likely to capture passing flies when they are hungry.

The complexity of behavior increases with the complexity of the nervous system. Animals with simple nervous systems tend to respond automatically to stimuli in a programmed way, whereas animals with complex nervous systems are apt to choose behavior that suits the particular circumstance. The first type of behavior, which is inherited, is called **innate** or **instinctive,** while the second type, which requires modification of behavior, is said to involve **learning.** All animals have some instinctive behavior, even humans; but as higher animals with complex nervous systems evolved, the capacity to learn behavioral patterns developed. In order to exemplify the difference in behavior from primarily innate to primarily learned, compare the brief sketch of army ant behavior to that of gull behavior to that of baboon behavior given on pages 656–59.

Army ants are found in the southern hemisphere. They have a life cycle typical of most insects, consisting of egg, larval, pupal, and adult stages. The larval stage is an active, wormlike stage requiring much food, while the pupal stage is a quiet, encapsulated stage that produces the adult by metamorphosis. When the adult army ant first emerges, it is light in color and is called a callow.

A colony of army ants is established after a queen ant has been fertilized on her nuptial flight. Thereafter, the queen produces thousands of eggs every five weeks, and these develop into three different sizes of sterile female workers. Thus, the colony is **polymorphic** because it consists of three different types of individuals. The smallest workers (3 mm), called the nurses, take care of the queen and larvae, feeding them and keeping them clean. The intermediate size workers, constituting most of the population, go out on raids to collect food. The largest workers (14 mm), with huge heads and long, powerful jaws, are called soldiers because they run along the sides and rear of raiding parties where they can best attack any intruders.

Army ants do not have a permanent nest; instead they live in so-called bivouacs, or temporary quarters. In the genus *Eciton,* the most common of the western hemisphere, the ants themselves form the bivouacs when they interlock their legs in such a way as to cover and protect the queen and brood (larvae).

The life of the colony has two phases, a nomadic phase and a resting phase. During the **nomadic phase,** the ants pour out of the bivouac and form several columns, which in turn divide into branches or flanks. This flanking procedure is quite useful, inasmuch as it allows the ants to surround their prey, usually insects, snakes, and lizards. As the ants move along, all wildlife is apt to try to move away as quickly as possible; it has even been reported that army ants have killed human beings. Once they do kill an animal, they tear it apart, and some of the ants carry the parts back to the bivouac so that the nurses, queen, and larvae may be fed. The column may extend some 300 feet from the nest in a two-way stream, with some ants advancing and others returning with the catch.

At night, the entire colony emigrates to a new bivouac, probably along one of the trails laid down that day. In the front and sides of the advancing wave of ants are the soldiers, followed by the medium-sized workers, and finally, in the center, the smallest workers transporting the brood and helping the queen. The next

Sea gulls, or simply gulls, are large (8 to 30 inches), long-winged birds that are white, grey, and black in differing proportions. Generally, the gull's body and tail are white; the back and wings are light or dark grey; and the wing tips are usually black with a contrasting white pattern. The bill is stout and slightly hooked, and the front toes are webbed. Gulls begin life as speckled brown chicks that gradually acquire their characteristic colors in three to four years.

During the winter, gulls form mixed flocks and spend their time eating and resting. They are coastal scavengers that eat anything organic, including garbage and sewage. Because of today's pollution, they are generally increasing in number.

Only when resting does a gull seek out company. Then, as part of a group in which some are always alert, a gull tucks its head under a wing, raises one leg, and dozes.

In the spring, thousands of gulls migrate to a breeding colony on small rocky islands, inaccessible cliffs, or in grassy marshes, according to the species. There the males stake out territories of about fifty square yards and build nests of grass, stems, seaweed, feathers, and discarded food. The males defend their territories and seek mates. Gulls are considered **monogamous** inasmuch as they have a single mate per breeding season; very often mates from the previous year find each other.

Male and female gulls, which look alike and are therefore **monomorphic,** share in incubating the three eggs laid by the female; and when the chicks hatch, both parents help feed and protect the young. Just after hatching, the chick is fed by parent regurgitation when it pecks the parent's bill, but within a month the chick is able to feed itself and has learned what is edible and drinkable.

29

Behavior within Species

Chapter concepts

1 Behavior that occurs as an automatic response to a stimulus is inherited and subject to natural selection in the same manner as anatomy and physiology.

2 Behavioral patterns may be associated with a continuum that ranges from completely innate at one end to completely learned at the other.

3 The internal state of the animal affects the degree to which the animal performs a certain behavior.

4 The behavior of animals that cooperate with one another in a society can also be explained on the basis of evolutionary theory.

Part Six

Behavior and Ecology

The behavior of organisms allows them to interact with their own kind and with other species. These interactions are the framework for ecosystems, units of the biosphere in which energy flows and chemicals cycle. Mature natural ecosystems contain populations that remain constant in size and require the same amount of energy and chemicals each year.

Humans have created their own ecosystem, which differs in that the population constantly increases in size and ever greater amounts of energy and raw materials are needed each year. Since energy is used inefficiently and raw materials are not properly cycled, the human ecosystem is dependent on natural ecosystems to absorb pollutants. Because the natural ecosystems are no longer able to support the human ecosystem in this manner, we must find ways to use energy more efficiently and to recycle materials. Furthermore, the preservation of the natural communities, called biomes, is beneficial to all ecosystems. Preserving the biomes helps to assure the continuance of the biosphere.

Since 1850 the human population has expanded at such a rapid rate that it is doubted by some that there will be sufficient energy and food to permit the same degree of growth in the future. Concomitant with indications that humans desire to preserve the biosphere, the growth rate of the human population has begun to decline.

day the hunting raids begin again. This daily routine of extensive raiding and nightly emigrations continues for about two weeks.

Then rather abruptly the colony begins a **resting phase,** which lasts for about three weeks. During this phase, few workers go on raids, the colony stops emigrating, and marked activity ceases. Then the raids begin again.

Army ants form an insect society in which the members are apparently born to a particular task that they slavishly carry out. When we seek an answer to this regimented type of behavior, we find that it seems to be chemically controlled. The term **pheromone** is used for a chemical released by one organism that controls the behavior of another organism usually of the same species. In the case of army ants, this chemical is most likely released primarily by the queen and larvae. Ants pass a portion of the contents of their crops from one to the other, and in this way the nurses could pass hormones from the queen and larvae to a few workers, who in turn would pass it to others.

Pheromones passed in this manner are believed to trigger the raiding behavior of army ants. The nomadic phase always begins when the eggs have developed into larvae and new adults (callows) have emerged from

pupae. Both of these populations, from two different broods, need food, and they are believed to chemically activate the raiders to begin the process of scouring the countryside. On the other hand, when larvae pupate, the pupae would no longer emit such a hormone, and the signal for food would cease. Pupation, then initiates the resting phase of army ant life.

Baboons

Baboons are a type of Old-world monkey called dog-headed monkeys because they have an elongated muzzle. In particular, we will describe the behavior of the savanna baboons that live in a grassy area of central Africa.

The face of the baboon is characterized by a long muzzle that leads back to small, beady eyes set deep beneath eyebrow ridges. The muzzle is hairless, while the body is covered with long hair except for the buttocks, which range in color from pink to scarlet.

Baboons walk on all four limbs but sit when resting so that the hands remain free for manipulation. They show extreme dimorphism, with the males being much larger and stronger than the females. An average male weighs about 60 pounds and is about 26 inches high when seated, while the female is about 25 pounds and 18 inches high. One interesting explanation for this extreme difference in size is that it is the most economical way to achieve protection while limiting food consumption.

Savanna baboons may be attacked by jackals, hyenas, lions, cheetahs, and leopards, but these animals rarely succeed in killing a baboon because of the ferocity of the baboon male. The males are formidable not only because of their size but also because of their disposition and their long, sharp canine teeth. When necessary they sink their teeth into an attacker and tear out a hunk of flesh. A troop of baboons is very dependent on the males to fight off would-be predators, and observers report that the males rarely fail in this responsibility.

Baboons are vegetarians that live in communal troops that travel three to six miles each day hunting for food and sleeping in trees at night. As the troop moves along, it generally has a definite organization: the dominant males along with new mothers and infants are in the center surrounded by other females. Less dominant males are on the fringes.

The dominant males decide where and when the troop will move. If the troop is threatened, they immediately leave the center and move toward the danger. They cover the troop as it retreats and attack when necessary.

Dominance is very important to the organization of the troop, and it is established along rigid lines. Each individual has a place in the hierarchy; the males know their station in relation to one another, the females know theirs, and there is even a hierarchy among the juveniles. Dominance is established by psychological contests between individuals in which it is determined who will give way to whom. Dominance is obvious in two aspects of troop life: grooming and presenting.

Grooming occurs when one member of the troop picks through the hair of another and removes dirt, parasites, and bits of debris with fingers or teeth. The dominant males are groomed by other members of the troop and, in general, a less dominant animal grooms another.

Presenting occurs when one baboon takes a position with the head down on the ground and the buttocks raised toward another. Presenting demonstrates submission because it puts the presenting animal in a very vulnerable position. Presenting is used to show respect and to prevent aggressive behavior by another.

The establishment of dominance is useful to the troop because it often prevents actual fighting. When a disturbance occurs within the troop, it is usually quieted by a dominant male. The male will try a series of actions including staring, blinking, raising the eyebrows, and yawning to show the canine teeth before coming over to administer a harmless bite on the neck of the offender. Meanwhile, the disturber of the peace may begin to show fear by moving backward, grimacing, screeching, and finally running and presenting just before the bite occurs. In order to prevent all this ruckus, the offender may quickly make a motion that indicates a willingness to present, and the dominant male, being satisfied, turns away.

All of these actions may or may not occur, as the situation warrants, and so we see an absence of stereotyped

behavior. Rather, behavior is modified to suit the occasion.

When females are in estrus and able to conceive, they seek out the company of the dominant males. During this time the female's position in the troop is elevated as the males keep company with her. Following copulation, pregnancy lasts for about six months. The mother takes care of the birth herself and travels with the troop on the day of the birth. The newborn is capable of clinging to the underside of its mother, where it may nurse as convenient. The first day or so, the mother may assist the infant by holding a hand on it. Since this tends to slow her progress, a dominant male may drop back and walk with her, stopping when she stops and generally looking after her.

Males seem to be very interested in the newborn and especially watch over infants and youngsters when the troop is on the move. Should a youngster be stranded when the troop is under attack, a male will quickly snatch it up and return it to its mother.

Many members of the troop want to hold newborn infants, approaching the new mother with lipsmacking and presenting. **Lipsmacking** is used among baboons to indicate friendliness. The mother is very protective of her newborn and generally will not let anyone touch it except a dominant male. She grooms it constantly and holds it to her. The birth elevates the mother's position in the troop, and she is under the protection of the dominant males.

After about three months, the infant begins to play with members of its peer group. Generally, baboons are born in December and thus have a number of playmates of the same age. Peer-group play is probably very important to the proper development of baboons. At five months, the youngster easily rides on the back of its mother, crouching only when afraid. Weaning occurs at one year. During the process of weaning the juvenile shows its distress by screeching and moving from one member of the troop to another. This is the only time the males seem to ignore its distress. After weaning, however, the juveniles spend most of their time with one another and the dominant males see to their safety. Maturity is reached at about four years in females and about seven years in males. The males soon become larger and more adventurous than the females. They may even leave the troop for a time and return when they are bigger and stronger and more able to challenge the position of the dominant males.

Figure 29.2
Pheromone control of insect behavior. Army ants
follow a trail of pheromone even if it causes them
to circle about until they die of exhaustion.

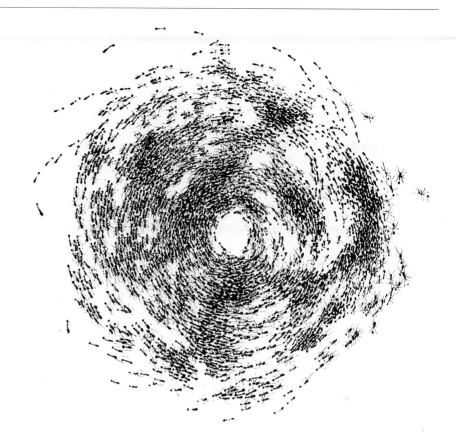

Innate Behavior

In innate behavior, the stimulus appears to trigger a fixed response that does not vary according to the circumstances.

Taxes

Orientation of the body toward or away from a stimulus is termed **taxis** in animals. Animals exhibit a number of different types of taxes; **phototaxis** is movement in relation to light and **chemotaxis** is movement in relation to a chemical. Some insects, for example moths and flies, will fly directly toward a light. Often they orient themselves by shifting the body until the light falls equally on both eyes. If one eye is blind, the animal will move in a spiral, forever trying to find the direction in which the light will be balanced between the two eyes.

Chemotaxes are quite common. Insects are attracted to minute quantities of chemicals, called **pheromones,** given off by members of their species. Army ants are so set on following a pheromone trail that if, by chance, it leads them into a circle, they will continue to circle until they die of exhaustion (fig. 29.2). Vertebrates, too, are sometimes highly responsive to chemicals. A few whiffs of a piece of clothing, and bloodhounds are capable of tracking down a single individual.

Only a few animals are able to orientate themselves by means of *sonar* or *echolocation*. Bats send out series of sound pulses and listen for the echoes

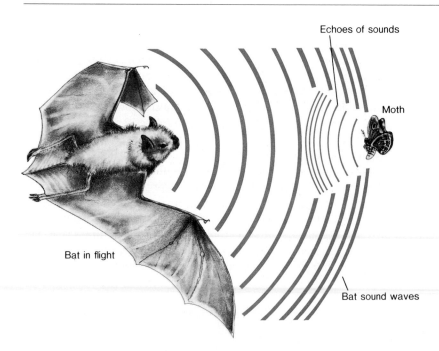

Echoes of sounds

Moth

Bat in flight

Bat sound waves

that come back. The time it takes for an echo to return indicates the location
of both inanimate and animate objects and enables a bat to find its way through
dark caves and to locate food at night (fig. 29.3). Some moths have evolved
the ability to hear the sounds of a bat and they begin evasive tactics when they
sense that a bat is near.

Migration and Homing

Migration, which often occurs seasonally, and **homing,** the ability to return
home after being transported some distance away, have been studied in birds.
Evidence indicates that they use the sun in the day (fig. 29.4) and the stars
at night as compasses to determine direction (north, south, east, or west). Like
bees, they even allow for the east-to-west movement of the sun during the
course of a day. Bees offered a food source can return to their hive no matter
where the sun is located in the sky. Therefore, it is believed that they have an
internal or *biological clock* that tells them the location of the sun according
to the time of day.

Although birds use the sun and stars as compasses, they cannot rely on
these to tell them that home is in a particular direction. In other words, suppose you were blindfolded and then transported away from home. Upon being
set free, you are given a compass to tell direction; how would you know which
direction to select? Some investigators now believe that birds are sensitive to
magnetic lines of force, which are dependent on the earth's magnetic field,
and this allows them to determine the direction of home.

Salmon use chemotaxis to find their way home. Salmon are born in a
tributary of a river but grow to maturity in the open sea. At spawning time,
mature salmon travel back up the river to the same spot at which they were
born. Experiments have shown that the fish appear to return to the spawning
ground by following the chemical scent of their first home.

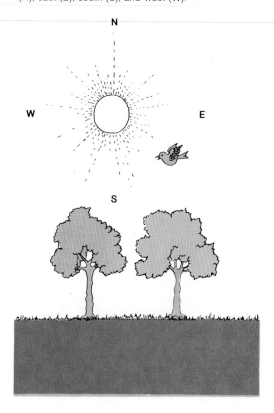

N

W

E

S

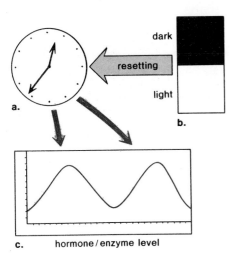

Figure 29.5
Biological clock system. Biological clock systems have three components: (a.) an internal timekeeper, (b.) a means of detecting light/dark periods, (c.) a method of communication that eventually brings about a behavioral response.

dark

resetting

light

a.

b.

c.   hormone / enzyme level

Rhythmic or Cyclic Behavior

Certain adaptive behaviors of animals reoccur at regular intervals. Behavior influenced by a **circadian rhythm** occurs on a daily basis. For example, some animals, like humans, are usually active during the day but sleep at night. Others, such as bats, sleep during the day and hunt at night. Behavior controlled by a **circannual rhythm** occurs on a yearly basis, such as when some birds migrate south every fall. Other rhythms are also known; lunar behavior occurs monthly and tidal behavior occurs every 12.4 or 12.8 hours.

Originally, it was assumed that environmental changes, such as day and night, controlled cyclical behavior in animals. But it is now known that such behavior will occur even when the associated stimulus (daylight or darkness) is lacking. For example, fiddler crabs are dark in color during the day but light in color at night, even when kept in a constant environment. But if the crabs are kept in the constant environment indefinitely, the timing of the daily change tends to drift and become out of synchronization with the natural cycle. For this reason, it has been suggested that rhythmic behavior is under the control of an innate, internal biological clock that runs on its own but is reset by external stimuli. Aside from keeping time, a biological clock must also be able to bring about the change in behavior. In the fiddler crab, for example, it must be able to stimulate the processes that cause the shell to change color. Thus a biological clock system needs (fig. 29.5):

1. a time-keeping mechanism that *keeps time* independently of external stimuli (i.e., a minute is always a minute);
2. a receptor that is sensitive to light/dark periods and *can reset* the clock for circadian rhythms or indicate a change in the length of the day/night for circannual rhythms;
3. a communication mechanism by which the clock *induces* the appropriate behavior.

A review of the discussion concerning flowering (p. 151) shows that the last two components of a biological clock system have been tentatively identified: phytochrome is believed to be the receptor that is sensitive to light and dark periods, and plant hormones are believed to be the means by which flowering is induced. The time-keeping mechanism has not been identified, however.

In animals, experiments with birds suggest that the biological clock resides within the pineal gland. The pineal gland produces **melatonin,** a hormone that lowers body temperature and causes birds to roost, both of which are characteristic of nighttime. Also, there is evidence of neural communication between the eyes and the pineal gland, which is closely associated with the "third eye" present in some lizards, amphibians, and fishes.

Reflexes

Reflexes are simple automatic responses to a stimulus over which the individual appears to have little or no control. When a human knee is hit by a mallet, the lower leg jerks in a characteristic manner. Since reflexes are clearly innate, some investigators believe that it might be possible to explain complex behavior of some lower animals, such as army ants, by a series of reflexes, each one of which acts as a stimulus for the one following.

Behavior and Ecology

Fixed Action Pattern

The term **fixed action pattern** (stereotyped behavior) has been given to complex behavior that occurs automatically as if it were a composite of reflex actions. A stimulus that initiates a fixed action behavior is called a **sign stimulus.** An animal must possess neural mechanisms (called releasing mechanisms) that are sensitive to the sign stimulus in order to respond in the stereotyped manner. As an example, consider the fact that male robins attack a red tuft of feathers in preference to an exact replica of a male robin without the red breast (fig. 29.6). The color red is a sign stimulus that provokes the releasing mechanism, which controls the attacking behavior. This behavior is therefore a fixed action pattern. Animals performing fixed action patterns may seem to be acting in a purposeful manner but, just as with the robin in the previous example, experimentation proves that this is not the case. For example, certain solitary digger wasps dig a hole and then seek out a caterpillar, paralyzing it with a series of stings along the undersurface. The wasp carries the prey to the nest and pulls the prey in. After laying her egg on the side of the caterpillar, she begins to close the burrow. If at this point the experimenter removes the caterpillar and puts it on the ground nearby, the wasp will continue to cover the hole even though the caterpillar is in full view.

Figure 29.6
Behavior as a response to sign stimuli. Male robins will attack a red tuft of feathers rather than an exact replica of a robin without a red breast. This indicates that the color red is a sign stimulus for this behavior.

Figure 29.7
Example of a fixed action pattern among insects. A digger wasp is about to deposit this caterpillar in a previously dug hole. She will lay an egg on the caterpillar before closing the hole.

Figure 29.8
Example of a fixed action pattern among birds. Gulls are more apt to remove broken egg shells with a serrated edge, such as *b*. This shows that an environmental stimulus controls their behavior

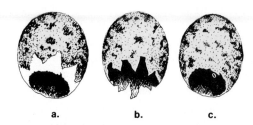

a.   b.   c.

Figure 29.9
Supernormal stimulus. A gull tries to brood an abnormally large egg even though the proper size egg lies just to one side.

Figure 29.10
Feeding behavior among gulls. *a*. Chicks peck at the red spot on a parent's bill when they want to be fed. *b*. In an experiment with laughing gull baby chicks it was shown that with time, chicks peck at a more exact replica of parents' bill (first model). In each example, the top bar represents the pecking frequency of newly hatched chicks; the bottom bar represents the pecking frequency of chicks three-to-five days old.

a.

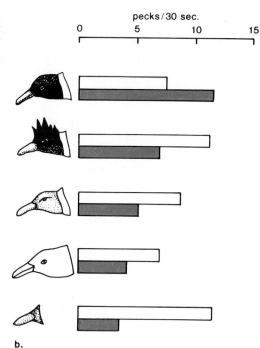

b.

The same sort of approach is typical of army ants. While it was formerly believed to be correct to describe their behavior as we would that of an army of humans, we now know that this is extremely misleading. For example, the army ant soldiers do not purposefully take their places at the forward and side positions of the column; rather, they are pushed and shoved there because they are in the way. Also, the column does not purposefully use a flanking procedure to capture the prey; rather, some of the ants at the front run forward a short distance and then back, and this motion causes the branching of the column. So it is incorrect to think of their behavior as premeditated.

Gulls, too, have fixed action patterns. When a gull is incubating eggs, it will retrieve any egg that rolls out of the nest. The more speckled the egg, the more strongly is the gull stimulated to roll it into the nest and incubate it. But a gull removes egg shells from the nest as soon as the chicks have hatched because empty shells attract predators. Now the stimulus appears to be a white serrated edge; "hollowness" by itself is not effective (fig. 29.8). One curious result of experiments like these has been the discovery of **supernormal stimuli.** For example, parent gulls, if given a choice in size, will prefer to retrieve eggs much larger than normal even if they are then unable to brood them (fig. 29.9).

Fixed action patterns are adaptive and largely inherited. When the cheek of a human baby is touched it will seek the nipple and begin to suck. Gulls automatically retrieve eggs that roll out of the nest and get rid of broken shells. Birds who perform these tasks are more successful parents and therefore have more offspring who also perform these tasks. Thus it is reasoned that behavioral patterns are subject to natural selection as are physical characteristics.

In order to tell if a specific behavior is innate (instinctive), it is customary to determine if it is *(a)* performed by all members of the species in the same manner, and *(b)* performed by animals that have been raised in isolation and/or have been prevented from practicing it.

It is not surprising that some forms of fixed action behavior improve in efficiency after an opportunity to practice has been provided. For example, isolation experiments have been performed to see if songbirds will sing their species' song without having had an opportunity to learn it from another bird. Often they sing a song that is less complicated than the natural song, but they will learn the complete song when they are given the opportunity to hear it, even if many other bird songs are played for them at the same time.

Baby chicks peck at the parent's beak in order to induce the parent to feed them. Experimentation with various models has shown that the chicks are not very discriminatory at first and choose a model that does not resemble the parent to any extent. With time, however, pecking accuracy and efficiency improves and the chicks become progressively more selective and choose a model that more nearly resembles the parent (fig. 29.10).

In the same manner, young chicks at first hide and crouch whenever a shadow passes overhead; but soon they hide and crouch only when a hawk passes over (fig. 29.11). It has been shown by the use of models that chicks do this because they never lose a fear of short-necked birds, while they do learn that long-necked birds pose no danger.

Our discussion of reflex behavior has shown that some behavior may be primarily innate in that the behavior is performed automatically with no forethought. It is fair to assume that animals with simple nervous systems are more apt to rely on this form of behavior, while animals with more complex nervous systems are apt to rely primarily on learned behavior (fig. 29.12). However, a new field of behavior is emerging in which investigators are attempting to determine the degree to which animals, aside from humans, are able to think. There have been some surprising findings as discussed in the reading on page 666.

Figure 29.11
Learning among birds. When this model is pulled to the left, it frightens gulls, presumably because it appears to have a short neck, such as a hawk might have. But when the model is pulled to the right, it does not frighten the birds, presumably because it now appears to have a long neck such as gulls have.

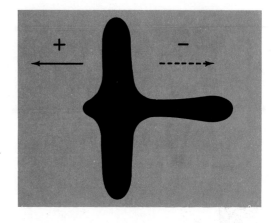

Figure 29.12
Comparative modes of adaptive behavior among invertebrates and vertebrates. While invertebrates rely primarily on types of innate behavioral patterns, vertebrates have increasingly come to rely on learned behavioral patterns.

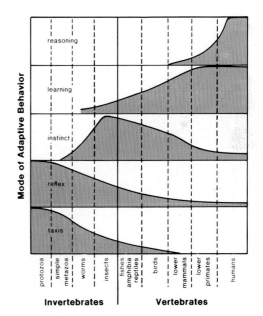

Do Animals Think?

Herbert Terrace, a psychologist, points to a puppet's mouth, and a chimp responds by pointing to its own.

Cognitive psychologists believe their new found acceptance by other scientists is based on their strict adherence to the tenets of experimental observation. And what they have observed has much to say about intelligence in animals. For example:

Working with a pair of young female bottlenose dolphins, psychologist Louis Herman of the University of Hawaii has been able to achieve what many other scientists have not: proof of an animal's understanding not only of vocabulary, but also of communicative grammar and syntax.

Beginning in 1979, the dolphins—which were kept in two 50-foot pools—were taught two artificial languages, each made up of about 35 words. The languages have their own syntactical rules for word order, from which sentences up to five words long can be constructed. Depending on the order of the words, about 1,000 different sentences can be expressed. In one language, the words are made up of electronically generated whistles, created by a computer. In the second language,

the words are formed by hand- and-arm gestures performed by the trainer. ("Ball," for instance, is signaled by a human at poolside, raising his hands quickly above his head and then lowering them.) One dolphin specialized in each language.

The evidence that the dolphins appreciate syntax was found in their ability to differentiate correctly between phrases such as "hoop fetch Frisbee" (in other words, "get the hoop and take it to the Frisbee") and "Frisbee fetch hoop" ("get the Frisbee") and take it to the hoop") Or, "net in basket" ("put the fishing net in the plastic laundry basket") and "basket in net" (the reverse). For research purposes, Herman named the two study dolphins Phoenix (the animal that responds to the trainer's hand gestures) and *Akea Kamai,* which means "Lover of Wisdom" in Hawaiian. "It is a name," remarks Herman, "chosen with hope."

Psychologists R. J. Herrnstein of Harvard, Mark Rilling of Michigan State University and Anthony Wright of the University of Texas have tested pigeons for their ability to make mental discriminations. Some of their most interesting work goes beyond the birds' ability to differentiate between objects; it tests their capacity to form concepts or categories—a higher order of intellectual accomplishment.

To date, the experiments indicate that pigeons can form generalized concepts of such test categories as "human being," "tree," "fish" and "oak leaf." In a typical test, a bird was trained to peck at a small plastic paddle when shown a slide of a human being. It was rewarded with food for a correct answer.

After the bird learned to peck when the picture of a person appeared, and not to peck when photos lacking people were shown, the experiment moved to the test phase. In this stage, the bird was shown a large number of photos— 1,200 in all—which it had never seen. Some of the new test

Behavior and Ecology

transparencies were pictures of human beings. Without food reinforcement, the pigeon correctly pecked when photos of people were shown. It even did so when the new pictures portrayed groups of people. The scientists' conclusion: the pigeons must be able to understand the concept of "people." Similar experiments have also been successfully conducted using trees as a photographic subject. And Wright has achieved the same kind of results using different tests with rhesus monkeys as his subjects instead of pigeons.

Robert Seyfarth and Dorothy Cheney, ethologists from the University of California at Los Angeles, spent more than a year in Kenya studying the vocalizations of vervet monkeys. Their research shows that the animals have at least three distinct alarm cries: a raspy bark for the sighting of a leopard, after which the troop retreats to nearby trees; a short grunt for eagles and hawks, which signals a quick glance skyward and a rush for thick vegetation; and a high chutter for a snake, causing the troop members to rise up on their haunches and survey any tall grass in the vicinity. This sort of *specificity* of communication, "culturally transmitted" from adults to young, is clearly within the province once thought to be reserved only for humans.

The *Schwanzeltanz*, or "waggle dance," with which honey bees tell their sisters back at the hive the precise location of a good nectar source, was first decoded by Karl von Frisch in the 1940s (he was finally awarded the Nobel Prize for the discovery in 1973). More recently, researchers such as James Gould of Princeton University have further investigated the complexity of bee communication. "It is," says Gould, "the second most complex language we know of." (The first is the one you are employing to read this article.) For example, it appears that bee language even has dialects—and that a waggle which in effect means five yards to an

Egyptian bee means 50 yards to an Austrian bee.

During the course of his work, however, Gould has chanced upon other evidence of seemingly intelligent behavior. For example, the time-honored method of training bees to fly from a hive to a desired location is to begin by placing a small container of sugar water very close to the hive and then progressively move it farther away. That allows the bees to collect the sugar and return to the hive before flying out again. The standard practice is to relocate the sugar source 25 percent farther away with each move. If it was 20 yards from the hive, it was moved to 25 yards. If it was 200 yards from the hive, it was moved to 250 yards. Using this method, Gould and his associates have successfully trained bees to fly to sources six miles from the hive.

However, during the course of the training, a curious thing began to happen. "The bees seemed to have figured out that we were moving the source in a regular way," recalls Gould. "When we moved to a new, more distant area, they were already there waiting for us—as if they had been able to figure out the rule we were using to move the sugar water. And it's hard to explain this behavior as innate, because there is nothing I know of in the natural history of flowers that has allowed them to move."

Lastly, consider the prodigious feats of memory accomplished by the Clark's nutcracker, a food-hoarding relative of the crow that lives in the southwestern United States. Biologist Russell Balda of Northern Arizona University found that the small bird spends the late summer collecting seeds from the piñon pine and burying as many as 33,000 of them in caches of four or five seeds each. According to Balda's research, the nutcracker finds its caches months after their creation by an elaborate system in which the creature memorizes nearby landmarks . . . .

. . . . "Now that the legitimacy of animal cognition as a field of inquiry

has been demonstrated," says Columbia's Herbert Terrace, a leading cognitive psychologist, "I see little gained by approaching the subject as one would approach human cognition." Maryland's Hodos is even more emphatic: "Attempts to equate animal and human intelligence may be doomed to failure," he argues. "Intelligence is a human concept. Our ideas about it are still evolving. The same behavior may be viewed as intelligence in one situation and not in another. Therefore, it represents a value judgment as much as a biological property. And what we as humans value may not be what an animal values."

The recent move to study animal intelligence on its own terms, in the words of University of Massachusetts psychologist Alan Kamil, "to refocus attention on animals as animals, with less concern for immediate human relevance," is certainly a healthy sign. "It's high time" observes Kamil, "that we stopped looking at animals as proto-humans."

As the search to understand the nature of animal intelligence continues, the quarry is now being viewed in a new light. "I believe," observes Anthony Wright, "that the role of science should be to discover the underlying processes, the foundations, the basic laws."

Most of the researchers are well aware of the scientific dangers. "If you look through the history of this field," says Mark Rilling, "it's littered with corpses rotting in blind alleys." Rutgers' Beer sounds a note of caution: "Most of the work today is quite sober, but you still have to separate the romance from the hard thinking." To which psychologist Roger Thomas of the University of Georgia adds: "We never really know who is more clever, the animal or the experimenter."

By David Abrahamson (originally published in National Wildlife Magazine). © 1983 David Abrahamson. All Rights Reserved.

Learned Behavior

Learning is a change in behavior as a result of experience. The capacity to learn is inherited and allows an organism to change its behavior to suit the environment. The organism alters its behavior to respond to a specific stimuli in ways that promote its own survival or that of its offspring and/or near relatives.

Kin Recognition

Currently there is a great deal of interest in detecting how organisms learn to recognize close relatives. For instance, tadpoles of American toads can apparently discriminate between siblings and nonsiblings, and Belding's ground squirrels even know the difference between full-siblings and half-siblings.

Kin recognition is beneficial in at least four ways. Parents who can recognize their offspring do not waste time rearing unrelated young; young that recognize their parents avoid the risk of being harmed by a nonparent; giving aid to a sibling rather than a nonsibling increases an animal's fitness (the likelihood of passing on one's genes); and mating with an individual that is to a degree unrelated increases the likelihood of fertile offspring.

Thus far, research has been most extensive in the area of imprinting.

Imprinting

Konrad Lorenz, a famous ethologist, observed that birds become attached to and follow the first moving object they are exposed to. He termed this behavior **imprinting** and suggested that it was a means by which organisms learn to recognize their own species. Ordinarily, the object followed is the mother; however, Lorenz caused chicks to be imprinted to him (fig. 29.13), and then showed that they would choose him over their own mother when given the opportunity.

A variety of animals, in addition to birds, are susceptible to imprinting during a certain critical period of time following birth. Therefore this term may be used in a larger context to refer to any period of time during which a type of learning is apt to be more successful. For example, puppies take to dog training better than older dogs; and adult humans are better at sports they learned as a child.

Habituation

When presented with the same stimuli time and time again, animals will eventually cease to respond. For example, at first baby chicks crouch in fear even when a leaf flutters overhead, but then they learn to disregard these and even long-necked birds. A possible explanation for this is that they have grown accustomed to and habituated to seeing long-necked birds. Habituation is useful because it prevents animals from wasting time on unnecessary responses.

Conditioned Learning

In a type of conditioned learning called **associative learning,** an animal learns to give a response to an irrelevant stimulus. Pavlov's dogs learned to expect food and began to salivate when a bell rang because they had learned to associate the ringing of the bell with food (fig. 29.14). Associative learning can explain behavior that seems out of place. For example, the ringing of a bell in and of itself does not have anything to do with food. Even so, associative

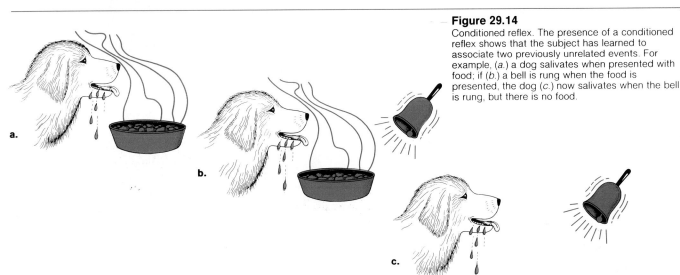

Figure 29.14
Conditioned reflex. The presence of a conditioned reflex shows that the subject has learned to associate two previously unrelated events. For example, (a.) a dog salivates when presented with food; if (b.) a bell is rung when the food is presented, the dog (c.) now salivates when the bell is rung, but there is no food.

learning can be useful; for example, it has been suggested that mothers who fondle their children while reading to them instill in these children a love of learning.

Operant conditioning is **trial-and-error learning.** An animal faced with several alternative choices is *rewarded* for making the proper choice and thereafter learns to make this response repeatedly. While it is possible to also punish an animal for making an improper response, B. F. Skinner, the major proponent of operant conditioning, believes that learning should always be based on positive operant conditioning. Rewarding good behavior can experimentally be shown to be far more lasting than punishing bad behavior. Also, rewarding good behavior has no undesirable side effects, whereas punishment results in anxiety.

Figure 29.15
Latent learning exemplified among insects.
a. Cones were arranged about a nest site of a
wasp that makes an orientation flight. *b.* Wasp
returns to this landmark even if it has been moved
to an inappropriate location. This shows that the
wasp learned to associate the cones with the
location of its nest.

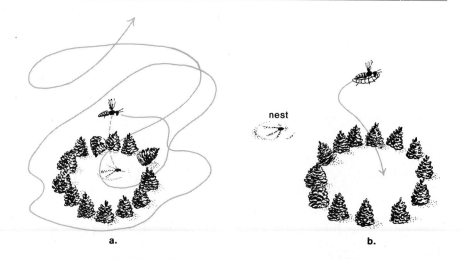

a.

nest

b.

Figure 29.16
Example of insight learning among apes. On the far
left, a chimpanzee is unable to reach a banana. In
the middle sketch, the chimp seems to suddenly
realize a solution to the problem. On the far right, it
stands on boxes to reach the banana.

Figure 29.17
Primates depend on learned behavior. Interaction
between young baboons and elders is essential to
the learning process.

Latent (or **exploratory behavior**) **learning** takes its name from the fact that
this type of learning may not be apparent at the time because it is stored away
for possible future use. Many animals explore new surroundings so that they
are more likely to be able to find their way about should it become necessary.
Insects make "orientation flights" of nest areas so that they can more easily
locate the nest area upon their return (fig. 29.15).

Insight Learning

Insight, or **reasoning,** is the ability to solve a problem by using previous experiences to think through to a solution. To the observer, it seems as if the
animal needed no practice to successfully reach a goal. For example, apes can
devise means to get bananas that are placed out of their arms' reach. They
will pile up boxes or use a pole in order to reach food (fig. 29.16).

Of the three kinds of animals whose behaviors are briefly sketched in
this chapter (army ants, gulls, and baboons), baboons show a marked ability
to modify their behavior. This is consistent with the fact that isolated baboons
are unaware of proper baboon behavior, and we must conclude that learning
(fig. 29.17) plays a large role in the normal behavior of baboons. Therefore,
it takes several months before young baboons are capable of caring for themselves and in the meantime they are dependent on their mothers and other
adults in the troop.

Behavior and Ecology

Figure 29.18
Ring dove mating behavior. Successful
reproductive behavior requires these steps:
(a.) male performs courtship behavior as he bows
and coos; (b.) suitable nesting materials and a site
lead to building of the nest; (c.) incubation of eggs;
(d.) baby chicks are fed "crop milk"; (e.) cycle
begins again.

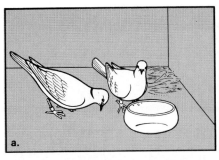

Motivation

As mentioned earlier, certain types of behavior recur periodically in animals.
These types of behavior seem to require an internal readiness before the animal shows the behavior. Thus it is said that the animal is **motivated** to perform
this behavior. Notice that in figure 29.1 motivation would be included in the
diagram as an internal event. Motivated behavior seems to include three stages:

1. An appetitive stage during which the animal searches for the goal
2. A consummatory stage or a series of responses directed at the goal
3. A quiescent stage when the animal no longer seeks the goal

A good example of motivated behavior is the need for food. A hungry animal
goes out to look for food; once food is found, it eats the food; and then, being
satisfied, it no longer seeks food.

Motivated behavior requires the supposition that the animal is made
ready to perform the behavior because of some internal state. A study of reproduction (fig. 29.18) in ringdoves has indeed shown this to be the case. Ringdoves reproduce in the spring but when male and female ringdoves are
separated one from the other, neither shows any tendency toward reproductive
behavior. In contrast, when a pair are put together in a cage, the male begins
courting by repeatedly bowing and cooing. Since castrated males do not do
this, it can be reasoned that the hormone testosterone readies the male for this
behavior. The sight of the male courting causes the pituitary gland in the female to release FSH and LH; these in turn cause her ovaries to produce eggs
and release estrogen into the blood stream. Now both male and female are
ready to construct a nest, during which time copulation takes place. The hormone progesterone is believed to cause the birds to incubate the eggs and while
they are incubating the eggs, the hormone prolactin causes crop growth so
that both parents are capable of feeding their young crop milk.

Figure 29.19
Example of chemical communication. Moth male antennae are capable of detecting female pheromone from miles away.

Reproductive behavior in the ringdove can be explained on the basis of both *external and internal stimuli*. The external stimuli are processed by the central nervous system, which directs the secretion of hormones. Thus the nervous and endocrine systems work together to produce physiological changes that lead to appropriate behavior patterns. Even so, animals with complex nervous systems are more likely to be able to control their behavior due to previous learning regardless of their internal state. For example, human beings can decide whether to engage in sexual behavior even if their testosterone or estrogen blood level is high. This decision is probably based on their previous learning experiences in regard to sexual behavior. Also, human beings do not need a hormonal state to be motivated, as when they voluntarily decide to continue the process of learning.

Societies

A **society** is a group of individuals belonging to the same species that are organized in a cooperative manner. In order to accomplish cooperation, evidence suggests that members of a society have a means of reciprocal communication, a means of overcoming aggression either by territoriality or dominance, and a division of labor.

Communication

Communication by chemical, visual, auditory, or tactile (touch) stimuli often includes a social releaser, a sign stimulus used between members of the same species that causes the receiver to respond in a certain way.

Chemical Communication

The term *pheromone* is used to designate chemical signals that are passed between members of the same species. Pheromones can have either releaser effects or primer effects. A pheromone with a **releaser effect** evokes an immediate behavioral response, while a pheromone with a **primer effect** alters the physiology of the recipient, leading to a change in behavior.

Sex attractants are good examples of pheromones with releaser effects. For example, female moths secrete chemicals from special abdominal glands. These chemicals are detected downwind by receptors on male antennae (fig. 29.19). This signaling method is extremely efficient since it has been estimated that only 40 out of 40,000 receptors on the male antennae need to be activated in order for the male to respond.

It is well known that male dogs and cats are attracted to the opposite sex by means of scent. Experimentation with apes and humans has also suggested that some individuals may be attracted by body odors. Following puberty, human males are reported to secrete about twice as much of a chemical called exaltolide in their urine as compared to females. Females are also reported to produce a vaginal chemical called copulin that can possibly attract certain males.

The members of ant colonies are controlled by numerous pheromones, each one inducing a particular response. Some are used as alarm signals, causing the ants to move about rapidly and to attack all foreign objects. Others are used to mark trails to food or new nesting sites. Still others cause the adults to take care of and nurture the larvae.

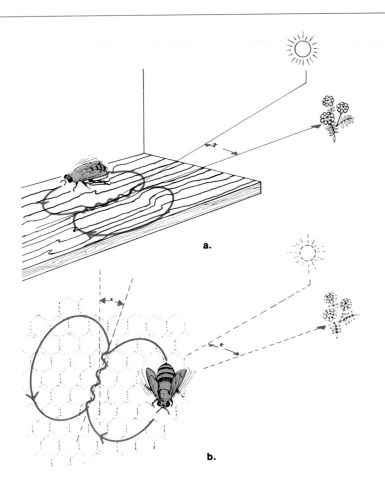

Figure 29.20
Communication among bees. Honeybees do a
waggle dance to indicate the direction of food. *a.* If
the dance is done outside the hive on a horizontal
surface, the straight run of the dance will point to
the food source. *b.* If the dance is done inside the
hive on a vertical surface, the angle of the straight
run to that of the direction of gravity is the same as
the angle of the food source to the sun.

a.

b.

    Ants and honeybees provide an example of a pheromone with a primer
effect. The queen produces a substance that is passed from worker to worker
by regurgitation. This substance prevents the workers from raising other queens
and it also prevents the ovaries of the workers from maturing. Primer effects
have also been seen in mice. Male mice produce a substance that can alter the
reproductive cycle of females. When a new male and female are placed together, this substance can cause the female to abort her present pregnancy so
that she can then be impregnated by her new mate. Similarly, crowding of
female mice causes disturbances or even blockage of their estrous cycles. In
both instances, removal of the olfactory lobes prevents these occurrences.

Visual Communication

The communication of honeybees is believed to be remarkable because the so-
called language of the bees uses not only visual stimuli but other stimuli as
well to impart information about the environment and not about the bee itself.
When a foraging bee returns to the hive, it performs a routine known as the
waggle dance (fig. 29.20). The dance, which indicates the distance and direction of a food source, has a figure-eight pattern. As the bee moves between
the two loops of the figure eight, it buzzes noisily and shakes its entire body
in so-called waggles. *Distance* to the food source is believed to be indicated

Figure 29.21
Ritualized behavior among birds. *a.* When another bird approaches his territory, a male will utter a long call in the oblique posture. The two males may face each other in a threatening manner. Following the upright posture, the invading male will usually retreat. *b.* Similar type of behavior is seen during courtship. The male greets the female with a long call. Male and female may face each other in a threatening manner. However, they then turn away from each other, averting the menacing beak. This is an appeasement act that signifies their willingness to stay together.

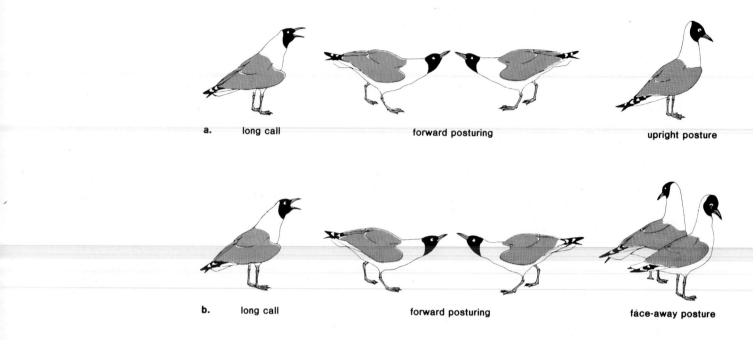

a.   long call   forward posturing   upright posture

b.   long call   forward posturing   face-away posture

by the number of waggles and/or the amount of time taken to complete the straight run. The straight run also indicates the *location* of the food; when the dance is performed outside the hive, the straightaway indicates the exact direction of the food, but when it is done inside the hive, the angle of the straightaway to that of the direction of gravity is the same as the angle of the food to the sun. In other words, a 40° angle to the left of vertical means that food is 40° to the left of the sun. The bees can use the sun as a compass to locate food because their biological clocks (p. 662) allow them to compensate for the movement of the sun in the sky.

Visual communication includes many social releasers. Male birds and fish sometimes undergo a color change that indicates they are ready to mate. Female baboons show that they are in estrus by a reddening of the sex flesh on the buttocks. On the other hand, visual communication also includes defense and courtship patterns (fig. 29.21) that comprise a type of body language. These patterns are **ritualized,** which means that the behavior, which is stereotyped, exaggerated, and rigid, is always performed in the same way so that its social significance is clear. Ritualized behavior is believed to be derived from body movements associated with such activities as locomotion, feeding,

Behavior and Ecology

and caring for the young. Facial expressions of human beings (fig. 29.22) seem to be universal and it has been suggested that they, too, are ritualizations of movements that were first used for biological processes.

Auditory Communication

Because auditory (sound) signals are able to reach a larger audience and can be sent even in the dark, they are sometimes favored even by animals with good vision.

Male crickets have calls and male birds have songs for a number of different occasions. For example, birds may have one song for distress, another for courting, and still another for marking territories. Sound stimuli have been shown to be more important than visual stimuli for birds that live in dense woods where vision is obstructed. In experiments, a male wood thrush attacked models of an unrelated species as long as they were silent. But if the unrelated species' song was played via a loudspeaker, the wood thrush paid the model no heed. Then, too, the wood thrush attacked a model of its own species with vehemency directly proportional to the loudness with which his own species' song was played.

One advantage of auditory communication is that the message can be modified by the sound's intensity, duration, and repetition. In an experiment with rats, an experimenter discovered that an intruder could avoid attack by increasing the frequency with which it made an appeasement sound. Rats isolated from birth could make the sound but had not learned to increase the frequency.

The fact that organisms with good vision often rely on sound communication has been demonstrated in chickens. Hens will react vigorously when they hear a chick peep even if they cannot see the chick. However, they will ignore a chick that is peeping within a soundproof glass container.

Language is the ultimate auditory communication, but only humans have the biological ability to produce a large number of different sounds and to put them together in many different ways. Nonhuman primates have at most only about 40 different vocalizations, each one having a definite meaning, such as the one that means "baby on the ground," which is uttered by a baboon when a baby baboon falls out of a tree. Investigators have overcome the biological inability of chimpanzees to articulate many sounds by having them learn a sign language. Chimpanzees have learned as many as 400 signs of an artificial visual language; however, thus far it has not been possible to demonstrate unequivocally that they are capable of putting the signs together to create new sentences and meanings. It still seems as if humans may possess a communication ability unparalleled by other animals.

Tactile Communication

Ants communicate with one another by means of touch, including jostling, tapping, and licking. For example, a food seeker can receive food by regurgitation when it taps the mouthparts of another ant with its forelegs. Similarly, baby gulls peck at the parent's beak in order to induce the parent to feed them (fig. 29.10).

Grooming, a behavior frequently seen in adult primates, occurs when one animal cleans the coat of another. Grooming helps cement social bonds within a group. A series of experiments with rhesus monkeys has shown that the psychological well-being of primates is enhanced by close bodily contact

Figure 29.22
Ritualized behavior among humans. Eyebrow flashing is a common human trait displayed by (a.) a French woman, (b.) a Balinese man, and (c.) a member of the Woitapmin tribe of New Guinea.

a.

b.

c.

Figure 29.23
Importance of tactile communication. Monkeys prefer a surrogate mother that is covered and warm to a wire mother that nurses when the receptacle contains a bottle.

between mother and offspring and between the offsprings themselves. Monkeys raised in isolation with no mother show marked signs of abnormality. They stare fixedly into space, rock back and forth aimlessly, and even gnaw and bite themselves.

Monkeys raised with a soft terry cloth surrogate mother do better. If given the choice between two surrogate mothers (fig. 29.23), the monkeys cling to the soft terry cloth one in preference to a wire-frame mother that can nurse. But these monkeys are not completely normal, showing signs of fear and aggression when placed with other monkeys, and they usually reject their own young. Monkeys who lack a normal mother can still develop normally, however, if they are allowed to play with peers for as little as 15 minutes per day. The peer-to-peer affectional system is extremely important to normal development in primates.

Competition
Members of the same population compete with one another for resources, including food and mates. **Aggression** is belligerent behavior that helps an animal compete. Thus aggression helps animals establish territories, obtain mates, train and/or defend their young, and maintain status in a group.

Territoriality and Dominance
Territoriality means that a male defends a certain area, preventing other males of the same species from utilizing it. Territoriality spaces animals and thereby reduces aggression. It also avoids overcrowding, ensuring that the young will have enough to eat. Since animals without a territory do not mate, it also has the effect of regulating population density to some extent.

A **dominance hierarchy** exists when animals within a society form a sequence in which a higher ranking animal receives food and a chance to mate before a lower ranking animal. Dominant males lead the group and maintain order; therefore a dominance hierarchy assures that the stronger males are in this position.

When the male defends his territory or engages another in contest to determine dominance, rarely is any blood shed. Rather, the animals have a repertoire of sign signals that comprise a threat *display* or *ritual*. As in figure 29.24, the display often includes postures that make the body appear larger and color changes that make the animal more conspicuous. Many of the sign signals in the display are derived from the normal activity of the animal, but now they are used to convey a social message.

When the animals are facing one another, two opposing responses—*approach response* and *avoidance response*—are simultaneously present in each individual. They are often in *conflict* as to whether to fight or to escape and this conflict causes them to threaten one another rather than to fight outright. Natural selection favors this situation because a contest decided by threat rather than by fighting is more apt to preserve each animal for the purpose of reproducing. The contest result is decided when one animal backs down and flees or submits.

Appeasement, or submission, occurs when an animal actually exposes itself to the attack of another and this gesture prevents further attack. For example, a subordinate baboon of either sex turns away from an aggressor

a.   b.

Figure 29.24
Ritualization of aggression among baboons and humans. *a.* This male baboon is displaying full threat, a mechanism that is used to establish dominance. *b.* The human being is also displaying full threat, a behavioral pattern that is more likely to occur in a dominant rather than a submissive individual.

Figure 29.25
Presenting among baboons. Presenting is an appeasement act because it prevents the more-dominant baboon from attacking a less-dominant one.

and crouches in the sexual presentation posture (fig. 29.25). Investigators studying gull behavior have found that the birds have a whole range of postures from actual fighting to appeasement. In gulls, food-begging behavior in adults is appeasement. Appeasement behavior is believed to cause even greater conflict in the aggressor so that inactivity results.

When an aggressor is in conflict, *redirection* of aggression or *displacement* of aggression may occur. As an example of the first of these, a bird might

Figure 29.26

Aggressive behavior among humans. Human beings often show conflict in regard to aggressive behavior, including (*a.*) vacillation, (*b.*) redirection of aggressiveness, and (*c.*) irrelevant activity.

a.

b.

c.

peck at the ground and a human might bang on a table (fig. 29.26). Displacement of aggression is recognized by the fact that the animal performs an irrelevant activity. A bird might preen its feathers, while a human might pull on his chin.

Courtship is a time during which aggression must be at least temporarily suspended in order for mating to take place. At this time, conflict within the male may cause him to vacillate between aggression and nonaggression. For example, in the spring the male stickleback stakes out a territory and builds a nest; at this time, his body becomes highly colored, including a red belly. Any male attempting to enter the territory is attacked as the owner repeatedly darts toward and nips the intruder. (Experiments have shown that the red belly of the male acts as a sign signal.) On the other hand, the owner entices a female to enter the territory by first darting toward her and then away in a so-called zigzag dance (fig. 29.27). Finally, he leads her to the nest, where she deposits her eggs. Investigators have pointed out that the zigzag dance of the male actually contains the same aggressive movements as when the male darts toward and attacks a trespassing male.

Cooperation

Societies have reached their highest and most complex development among insects and primates. In both of these the individual has discrete tasks that it performs for the benefit of the whole, because to do so increases its own fitness. This statement explains the formation of societies on the basis of evolutionary theory.

There is little or no aggression among the members of an insect society because the society is like a superorganism in which each type of insect has a specific task. For example, in army ants, only the queen lays eggs; the small

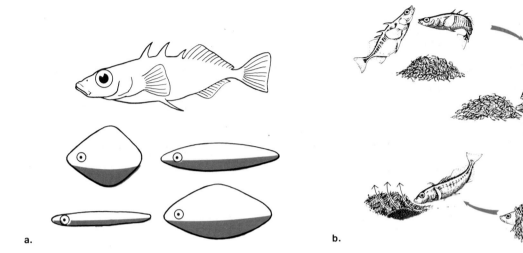

Figure 29.27
Reproductive behavior of stickleback fish. *a.* A red belly on a crude model stimulates females and males more than an exact replica of a fish without a red belly. *b.* The male entices the female to the nest, where he lies flat on his side; the female swims into the nest and lays her eggs when prodded by the male, and the male will later enter the nest to fertilize the eggs. Finally, the male beats his pectoral fins and creates a current of water that flows through the nest. This action supplies oxygen to the developing eggs.

a.

b.

workers, or nurses, care for the queen and larvae; the intermediate-size workers collect food; and the soldiers attack intruders. How does this society increase the fitness of individuals that do not participate in reproduction? The answer lies in the fact that the male parent is haploid and therefore siblings have three-fourths of their genes in common (fig. 29.28). Since offspring have only one-half of their genes in common with their parents, it actually increases the fitness of worker ants to raise siblings rather than offspring.

Among primates, there is a great deal of competition and aggression. At one time, it was believed that humans were more aggressive than other primates but, as discussed in the reading for this chapter on p. 680, this is no longer believed to be the case. In primate societies, aggression is apt to be checked and cooperation is likely fostered whenever (1) the members of the group are relatives who carry at least a portion of the same genes; (2) out-and-out fighting would cause an early death of the participants; and (3) aggression would waste time and energy that might better be spent in reproduction and the care of the young. Therefore, cooperative societies have evolved among the primates.

Altruism

Sometimes it may even seem as if the members of a society perform altruistic acts, acts that increase the fitness of another at the expense of their own fitness. For example, the dominant male baboons sometimes give up their lives in order to defend the troop. However, we must realize that when they are so

Figure 29.28
Sociobiology and the behavior of ants. It is more genetically advantageous for female ants and bees to assist in caring for siblings rather than their own offspring. Offspring possess one-half of the queen's genes, but since the male parent is haploid, siblings share, on the average, three-fourths of their genes.

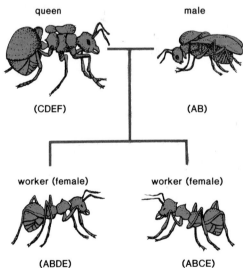

Animals That Kill Their Young

Females charge an adult male in attempt to save infant he has seized and bitten.

In his classic work *On Aggression,* Nobel Laureate Konrad Lorenz argued that man is the only species that regularly kills its own kind. . . . But Lorenz was wrong; since 1963, when his book was published, naturalists have identified dozens of species that kill their own, including lions, hippos, bears, wolves, hyenas, herring gulls and more than 15 types of primates other than man.

In the new perspective, animals are not benign machines that live for the group and kill only to eat. Instead, they are programmed for selfish, even murderous acts when survival and propagation are threatened. This radical shift in thinking is shown most dramatically by studies of India's sacred monkey, the hanuman langur. In 1965, a naturalist wrote that the long-tailed black and gray langurs were "relaxed" and "nonaggressive." Now, a Harvard researcher has shown that the langur society operates more like the House of Borgia, complete with kidnaping, constant sexual

protective of females and offspring, they are actually defending their genes because they are most likely the fathers of these same offspring. Even "mother love" can be explained on the basis that the mother is sure that the offspring carries one-half of her genes, an assurance that a father never has in the same way. Whenever animals act to save a group of related animals, called **kin selection,** behaviorists believe that they are actually protecting their fitness, which may include all of their relatives. The survival of the group is a beneficial side effect to this innate tendency.

Human Society

Humans share many behavioral characteristics with other primate societies. A close mother/offspring relationship, childhood dependency, learning by observing and imitating others, communication by vocal and nonvocal means, occupation of a home range, and dominance hierarchies are all common behaviors for primates.

In-depth study of primate societies has also shown other common traits that were formerly thought to be exclusively human traits. For example, it is now known that chimpanzees occasionally use tools, hunt small prey even though they are primarily vegetarians, and at times kill their own kind. These common behavioral traits support the belief that humans and chimpanzees share a common ancestor.

Humans, however, unlike other primates, began to hunt large game. Adaptations to this way of life can possibly explain why these and other features listed in table 29.1 are typical of humans. It is believed that cooperative hunting and increased intelligence evolved together. As the brain increased in size and complexity, a longer length of time was needed for development after the child was born. Women tended to take care of the children because they were

harrassment, group battles, abandonment of some wounded young by their mothers, and the regular practice of infanticide.

In her new book, *The Langurs of Abu,* Harvard Anthropologist Sarah Blaffer Hrdy, 31, portrays langur life as a "soap opera" that revolves around the struggle between the sexes. As in other species, the strongest males compete for control of each troop. What makes the langurs different is that the winner tries to bite to death the young offspring of his predecessor. The mothers resist the infanticide until the struggle looks hopeless, then pragmatically present themselves to the new ruler for copulation.

Why so brutal a society? Hrdy believes that the answer lies in the theory of sociobiology, which holds that each organism is engaged in a one-against-all struggle to get as many of its genes as possible into the next generation. That explains the sexual aggressiveness of langurs—and males of other species; it usually makes evolutionary sense for males to inseminate the maximum number of females. But why infanticide? Hrdy reasons that the grisly practice evolved among the langurs to solve a problem for the new dominant male. Because of the competition of other males, his reign over a harem or troop is usually short, and his genetic drive dictates that he impregnate the females as quickly as possible. As Hrdy explains: "By eliminating infants in the troop that are unlikely to be his own, a usurping male hastens the mother's return to sexual receptivity and reduces the time that will elapse before she bears his offspring."

Hrdy, who spent 1,500 hrs. observing langur behavior around India's Mount Abu from 1971 to 1975, documented the disappearances of 39 infants around the times of new male takeovers; she estimates that only half of all langurs survive infancy. While males shift constantly among groups, females usually spend a lifetime in one troop and cooperate in warding off danger.

When a new male ascends to power, pregnant females use deceit in an attempt to save their unborn young from his later attack: they demonstrate estrus behavior to the new leader, presumably to trick him into thinking the future offspring are his. But once the new male shows that he is determined to kill the infants, the mothers abandon their young. Though they could gang up on the male or refuse to copulate with him after infanticide, Hrdy notes, it is always in their individual self-interest to break ranks and accept him. Reason: their own male offspring will eventually benefit from the infanticidal trait.

Hrdy's portrait of the langurs is a far cry from the traditional view of animals as social creatures that act to ensure group survival. But as Lorenz's work was, it is in tune with its times. In stressing chaotic individualism at the expense of the group, *The Langurs of Abu* reads like a jungle version of Tom Wolfe's essay on *The Me Decade.*

Table 29.1 Relationship of hunting-related behaviors and early human behavioral traits

Behavioral adaptations	Novel selection pressures
1. Invasion of savannah habitat	
2. Bipedal posture and locomotion	
3. Tool making and use	Selection for locating and capturing large prey:
4. Pursuit of prey animals	1–2–3–4–5–6–7–13–14
5. Sexual division of labor (above and beyond child care)	
6. Cooperation in hunting and food sharing	Selection for skill in coping with carnivorous competitors, including other men:
7. Capture of exceptionally large prey	
8. Highly structured social organization (dominance)	3–8–9–10–11–12–13–14
9. Group territoriality, under some circumstances	Selection for infant care and protection:
10. Possible intense aggression toward strangers	5–6–12–13–14–15–16
11. Killing of competitors, including members of own species	
12. High degree of intelligence	
13. Language	
14. Multiple cultural adaptations	
15. Prolonged maternal care of dependent infant	
16. Prolonged pair bond between mates	

sure the offspring belonged to them. But the extended dependence of the children meant that the women had a greater need for mates. Pair bonding became feasible when continuous receptivity replaced the estrus in humans. The sexual and behavioral differences between the sexes may have led to a division of labor that did not necessarily involve child rearing. Women stayed home while the men went out to hunt.

It is also interesting to note that some investigators believe that the human tendency to kill members of our own kind, including perhaps our tendency to have wars between different groups, also stems from our adaptation to a hunting way of life. On the other hand, it could be that we are only seeking to perpetuate our own genes or the genes of our relatives, as sociobiologists suggest.

The new field of sociobiology supports the tenet that behavior, including social behavior, is subject to natural selection and evolves as an adaptation to the environment. The behavior that increases the likelihood of an individual's genes being passed on to the next generation will eventually be the species' behavior and is thereafter inherited by most individuals. If human behavior is, at least in part, inherited, this might explain any tendency to resist change or alter our behavioral patterns. Nevertheless, we must recognize that a large portion of human behavior is learned behavior. This learned behavior is a part of our culture, and certainly culture, which is not inherited by way of the genes, can evolve much faster than innate tendencies.

Summary

Behavior, or at least the capacity for behavior, is inherited and evolves and is selected for just as are anatomy and physiology. Behavior occurs as a response to a stimulus and therefore must rely on the sensory, nervous, endocrine, and muscular systems of an animal. Behavior is often divided into innate (instinctive) and learned behavior.

Innate behavior includes taxes (orientation toward a stimulus) and reflexes. Rhythmic behavior occurs without the need for an external stimulus and relies on an internal clock. Taxes together with the possession of an internal clock can in some cases offer an explanation for the ability of animals to return to a former location. Fixed action patterns, which may be a series of reflexes, occur as a response to a sign stimulus and are largely inherited, although they often increase in efficiency after practice has been allowed.

Learned behavior includes imprinting, habituation, conditioned learning, latent learning, and insight learning. The motivation of an animal can cause it to respond to a stimulus, unless, as in the case of humans, it can learn not to respond.

Members of a society communicate and compete with one another. However, the members must overcome aggression enough to be cooperative if they are to survive as a group. Territoriality and dominance are two mechanisms by which aggression is controlled. During courtship, aggression is minimized although aggressive actions are still detectable.

Insects and primates, especially, have well-developed societies. Insects have a division of labor and cooperate with one another because to do so increases the fitness of all, even those who do not reproduce. The answer lies in the fact that all siblings have three-fourths of their genes in common. The same explanation holds true for primate societies. Even altruistic acts can be explained by the fact that it increases the inclusive fitness of the individual.

There are many similarities between baboon or chimpanzee behavior and human behavior. This is to be expected since they share a common ancestor. The major ways in which human societal behavior differs from that of other primates can probably be explained by the fact that only humans began to cooperatively hunt large game. Although a portion of our behavior is inherited, the majority of our behavior is learned behavior. Therefore it should be possible for us as a society to change our behavioral habits.

Study Questions

1. Describe in general the behavior of army ants, sea gulls, and baboons. (pp. 656–57)
2. Draw a diagram that indicates that behavior is a response to a stimulus. Include in your diagram the internal readiness of the animal. (p. 655)
3. Give several examples of taxes in animals. Include an explanation for migration and homing. (pp. 660–61)
4. What are the three components of a biological clock system? What might these components be in birds? (p. 662)
5. Define a fixed action pattern. Give several examples, and discuss their improvement with practice. (pp. 663–65)
6. Name four types of learning and give an example of each type. (pp. 668–70)
7. Describe the ringdove reproductive experiment and its significance in regard to motivation. (pp. 671–72)
8. Define a society and name three aspects of behavior that are usually found within a society. (pp. 672–79)
9. Name and give examples of four common means of communication between members of a society. (pp. 672–76)
10. Give both advantages and disadvantages of aggression. How are these two sometimes balanced within a primate society? (pp. 676–78)
11. Support the theory maintained by sociobiologists that much of human reproductive behavior is inherited. (pp. 680–81)

Selected Key Terms

innate (in'nāt)
taxis (tak'sis)
pheromone (fer'o-mōn)
homing (hōm'ing)
circadian rhythm (ser''kah-de'an rith'm)
fixed action pattern (fikst ak'shun pat'ern)
learning (lern'ing)
kin recognition (kin rĕ-kog-ni'shun)
imprinting (im'print-ing)
habituation (hah-bit''u-a'shun)
conditioned reflex (kon-dish'und re'fleks)
insight (in'sīt)
motivated (mo'ti-vāt-ed)
society (so-si'ĕ-te)
waggle dance (wag'l dans)
tactile (tak'til)
territory (ter'i-tor''e)
dominance hierarchy (dom'i-nans hi'er-ar''ke)
courtship (kort'ship)
altruism (al'troo-izm)

Further Readings

Alcock, J. 1979. *Animal behavior: An evolutionary approach.* 2d ed. Sunderland, Mass.: Sinauer.

Bentley, D., and Hoy, R. 1974. The neurobiology of cricket song. *Scientific American* 231(2):34.

Bertram, B. C. R. 1975. The social system of lions. *Scientific American* 232(5):54.

Bonner, J. T. 1983. Chemical signals of social amoebae. *Scientific American* 248(4):114.

Eaton, G. G. 1976. The social order of Japanese Macaques. *Scientific American* 235(4):96.

Eberhard, W. G. 1980. Horned beetles. *Scientific American* 242(5):166.

Eisner, T., and Wilson, E. O., eds. 1975. *Animal behavior.* Readings from *Scientific American.* San Francisco: W. H. Freeman.

Gardner, R. 1972. *The baboon.* New York: Macmillan.

Koenig, W. D., and Stacey, P. B. 1984. Cooperative breeding in the acorn woodpecker. *Scientific American* 251(2):114.

Ligon, J. D., and Ligon, S. H. 1982. The cooperative breeding behavior of the green woodhoopoe. *Scientific American* 247(1):126.

Lore, R., and Flannelly, K. 1977. Rat societies. *Scientific American* 236(5):106.

Manning, A. 1979. *An introduction to animal behavior.* Paper text ed. Reading, Mass.: Addison-Wesley.

Ryker, L. C. 1984. Acoustic and chemical signals in the life cycle of a beetle. *Scientific American* 250(6):112.

Saunders, D. S. 1976. The biological clock of insects. *Scientific American* 234(2):114.

Scheller, R. H., and Axel, R. 1984. How genes control an innate behavior. *Scientific American* 250(3):54.

Topoff, H. R. 1972. The social behavior of army ants. *Scientific American* 227(5):70.

Wiley, R. H. 1978. Lek mating system. *Scientific American* 238(5):114.

Williams, T. C., and Williams, J. M. 1978. Oceanic mass migration of land birds. *Scientific American* 239(4):166.

Wilson, E. O. 1972. Animal communication. *Scientific American* 227(3):52.

Wilson, E. O. 1980. *Sociobiology.* Abridged ed. Cambridge, Mass.: Belknap Press of Harvard University.

30

Chapter concepts

1 Competition, predation, and symbiotic relationships are common modes of behavior between species.

2 Competition leads to diversity of species because no two species occupy the same niche.

3 Predator and prey coevolve and therefore a predator rarely overkills its prey population. Antipredator defenses are extremely varied.

4 Symbiotic relationships usually require a close relationship between two species that also coevolve. Parasitism helps control the size of the host population, while mutualism helps each population increase in size. Commensalism benefits one population without affecting the other.

Figure 30.1
Effects of competition. Competition prevents two species of barnacles from occupying as much of the intertidal zone as possible. The area on the farthest right indicates the area of competition between *Chthamalus* and *Balanus*.

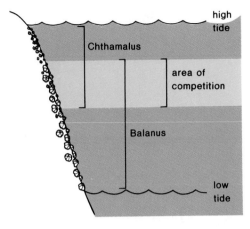

n the preceding chapter we discussed behavior typical of members of the same species; in this chapter we will be discussing interactions between species. These interactions, such as competition and predation, are important in controlling the sizes of populations. Because of this, these interactions are important in achieving a stable community of populations in which all populations are able to maintain a size appropriate to ensuring their continued existence.

Competition

Similar types of species with the same needs are most apt to compete with one another for resources, such as water, food, sunlight, and space. The outcome of competition is often the predominance of one species and the virtual elimination of the other.

A classic example of competition concerns two species of barnacles. Barnacles are attached to rocks in the intertidal zone and an investigator noticed that one species (genus *Chthamalus*) occupied the upper part of the intertidal zone along a Scottish coast, while another species (genus *Balanus*) occupied the lower intertidal zone. It was found that if one species was removed, each species could live in at least a portion of the other's zone (fig. 30.1). Since *Balanus* could occupy almost the entire zone occupied by *Chthamalus,* it must

Figure 30.2

Competition between two squirrels. *a.* Activity of the antelope ground squirrel on a typical summer day. The squirrel emerges from its burrow. Then it suns and grooms itself. If the squirrel's body temperature rises too high, it retreats to a special burrow to cool off. *b.* In the early afternoon, it stays in the shade and feeds on insects, seeds, and dead animals. The squirrel returns to its burrow at night. *c.* Activity of the Mohave ground squirrel during six months of the year. The squirrel emerges from its burrow in March, and the young are born in April. *d.* From May through July it fattens on desert vegetation. In August it again burrows underground where it remains until the following March.

From "Activity of Antelope Ground Squirrel" by George A. Bartholomew, and "Activity of Mohave Ground Squirrel" by Jack W. Hudson. Copyright © 1961 by Scientific American, Inc. All rights reserved.

a.

b.

Behavior and Ecology

be competition that prevents it from doing so. In some manner, *Chthamalus* is better adapted to the upper intertidal zone than is *Balanus*.

A study of desert ground squirrels also illustrates how many organisms deal with competition. Both the antelope ground squirrel and the Mohave ground squirrel (fig. 30.2) possess similar anatomical and physiological adaptations that allow them to cope with high temperatures and lack of water. Even so, their behavior is extremely different. The antelope ground squirrel is active the entire year, but the Mohave ground squirrel emerges from its burrow only during the months of March through August when desert vegetation is at its annual peak. It reproduces and fattens and then returns to its burrow, where it is dormant for the rest of the year. Both squirrels eat seeds as a source of water, but the antelope ground squirrel also eats insects and other animals. This adaptation has not been observed in the Mohave ground squirrel. Perhaps this explains why the Mohave ground squirrel is dormant when the desert is driest. It appears that the antelope ground squirrel is the better competitor because it has a wider range; the Mohave ground squirrel is found only in one corner of the Mohave Desert. It is unclear at this time whether the Mohave ground squirrel will become extinct or not. Perhaps its extended dormancy is an adaptation that permits it to coexist with the antelope ground squirrel.

c.

d.

Figure 30.3

Extinction due to competition with a species better adapted to the environment. *a.* In this experiment, species X survives and Y becomes extinct because the environmental conditions were favorable to X. *b.* In this experiment, the environmental conditions were changed, with the result that Y survives and X becomes extinct.

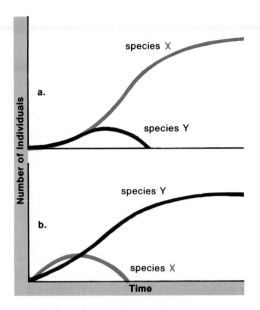

Table 30.1   Aspects of niche

Plants	Animals
Season of year for growth and reproduction	Time of day for feeding and season of year for reproduction
Sunlight, water, soil requirements	Habitat requirements
Contribution to ecosystem	Food requirements
Competition and cooperation with other organisms	Competition and cooperation with other organisms
Effect on abiotic environment	Effect on abiotic environment

Figure 30.4

Representation of niche by use of a hypervolume. There is competition if the hypervolumes of two species overlap. For example, the range of sunlight, temperature, and humidity tolerated by species 1 and species 2 overlap in the shaded area.

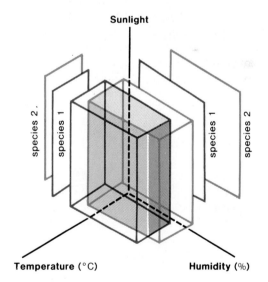

   Behavior and Ecology

Human Intervention

The fact that successful competition can cause one species to increase in size at the expense of another has been inadvertently demonstrated by human intervention. The carp is a fish imported from the Orient that is able to tolerate polluted water. Therefore, this fish is now often more prevalent than our own native fishes. An ornamental tree, the melaleuca, was introduced into Florida and has now invaded the Everglades, where it is drying up the cypress swamps and preventing natural vegetation from surviving. The burro, which originated in Ethiopia and Somalia and is adapted to a dry environment, is now threatening the existence of deer, pronghorn antelope, and desert bighorn sheep in the Grand Canyon.

Exclusion Principle

The fact that one species can cause another to become extinct has been demonstrated in the laboratory. For example, species X and Y both vie for the same resource when grown together in the same container, and eventually one will replace the other entirely. Which population is successful depends on the environmental conditions. In figure 30.3a, the environmental conditions are favorable to X, while in figure 30.3b, the environmental conditions are favorable to Y.

Such experiments have led biologists to formulate and support a **competitive exclusion principle,** which states that no two species can occupy the same niche at the same time. **Niche** is the term that is used to refer to the role a species plays in a community of organisms (this is discussed in chapter 31). To describe a species' niche it is necessary to state all the requirements and activities of the species. Table 30.1 lists factors to be included when describing the niche of a particular plant or animal species. Some investigators have suggested that it is possible to represent a niche as a many-sided figure occupying three-dimensional space (fig. 30.4).

Diversity

Competition leads to diversity of species because similar species will evolve to fill different niches. While it may seem as if several species living in the same area are occupying the same niche, it is usually possible to find slight differences. For example, the six species of monkeys in figure 30.5 have no difficulty living in close proximity because they have different, although sometimes overlapping, habitats and food requirements.

Predation

The ways in which predators are adapted to capturing prey and the ways in which prey are adapted to escaping predators are extremely diverse. Some predators are generalists and attack many different types of prey, while others are specialists, attacking only a few types of prey. Among predatory animals, in particular the specialists, evolution tends to favor those predators most capable of capturing prey. This also holds true for prey that are most successful in escaping predators. In this manner, predators and prey **coevolve,** as observed with both herbivores, animals that feed on plants, and carnivores, animals that feed on other animals.

Figure 30.5
Diversity of monkey species in a tropical rain forest.
All of these monkeys can coexist in a tropical rain
forest because they fill different niches. Each
prefers to live at a different height above ground
and each feeds on slightly different foods.

Red Colobus
Colobus Badius

group size: 50
range size: 90 acres
diet: mostly young leaves, buds
of flowers and leaves,
stalks of mature leaves;
also some insects, fruits,
and flowers.

Blue Monkey
Cercopithecus Mitis

group size: 20
range size: 200 acres
diet: fruits, small
insects,
flowers, and
flower buds;
also a few
young
leaves.

Redtail
Cercopithecus Ascanius

group size: 20
range size: 50 acres
diet: small insects and fruit;
also some flowers and
their buds.

Behavior and Ecology

Gray-cheeked Mangabey
Cercoebus Albigena

group size: 15
range size: 1,000 acres
diet: inner bark of trees, fruits, insects, and other small animals; also a few leaves and flowers.

Black-and-white Colobus
Colobus Guereza

group size: 10
range size: 40 acres
diet: young leaves, mature leaves; also some buds, flowers, and fruits.

L'Hoest's Monkey
Cercopithecus L'Hoesti

group size: 20
range size: unknown
diet: fruits and shoots of herbs, mushrooms, insects.

Figure 30.6
A gerenuk feeding on an acacia. The acacias in
Africa have thorns, but the gerenuk is still able to
feed off them.

Table 30.2   Some poisonous green plants

Plant	Toxin(s)
	Alkaloids
Poison hemlock *(Conium maculatum)*	Atropine*
Jimson weed *(Datura stramonium)*	Scopolamine*
Deadly nightshade *(Atropa belladonna)*	Hyoscyamine*
	Glycosides
Milkweed *(Asclepias Curassavica)*	Calactin
Foxglove *(Digitalis purpurea)*	Digitoxin
Oleander *(Nerium oleander)*	Oleandrin
Wild cherry (leaves and seeds) *(Prunus)*	Glycoside that produces hydrocyanic acid (HCN)
Manioc or Cassava *(Manihot esculenta)*	
Rhubarb (leaves) *(Rheum rhaponticum)*	Oxalic acid

*These toxins, in controlled dosages, have important
medical uses.

Adapted from CONTEMPORARY BIOLOGY, 2nd edition,
by Mary E. Clark. Copyright © 1979 by W. B. Saunders.
Reprinted by permission of CBS College Publishing.

Herbivores

The grassland of all continents support populations of **grazers** that feed on grasses and **browsers** that feed on shrubs and trees. Acacia trees and shrubs occur in tropical and subtropical regions. In Africa and tropical America, where there are browsers, the acacia species are protected by thorns that are often highly developed. Still the browsers (fig. 30.6) are adapted to feed on these plants. In Australia, where there are no natural browsers, most of the species of acacia lack thorns entirely.

The most prevalent terrestrial herbivores are the insects, which have evolved efficient and diverse means of eating plants. In return, plants have evolved mechanisms to discourage predation, such as the sharp spines of the cactus, the pointed leaves of holly, and the tough and leathery leaves of oak trees. Above all, plants produce chemicals called **toxins** that interfere with the normal metabolism of the adult insect (table 30.2) and hormone analogues that interfere with the development of insect larvae. Some insects can still inhabit trees that produce toxins because the insects have evolved detoxifying enzymes or a method of holding the toxin within their bodies so that it is not harmful to them.

The gypsy moth is apparently one insect that is immune to any toxins produced by trees. As the reading on page 698 discusses, the gypsy moth population periodically increases to the point that it dies off only after its food source is depleted. Scientists are now seeking a predator or parasite that could help keep the population in check.

Carnivores

Some carnivores, such as birds of prey, go out alone and seek their prey. While most birds of prey are specialized for hunting, seizing, and killing small terrestrial animals, some, like the osprey and kingfish are specialized for fishing (fig. 30.7). Instead of seeking prey, some solitary predators lie in wait. The

Behavior and Ecology

Figure 30.7
The Eurasian kingfisher is a solitary predator. *a.* The bird drops instantly in a sharp dive towards the water surface. *b.* Through the rising air bubbles, he catches his prey. *c.* With strong strokes, he gains height *d.* to land with his trophy on the nearest branch.

a.

b.

c.

d.

Figure 30.8
Lions are social animals and share their kill with others.

ocotopus hides within a protective shelter until an unsuspecting prey, such as a crab, should happen by. The octopus then quickly catches the prey with its arms and carries it home to be poisoned and eaten. Sperm whales probably lie suspended and hidden within the darkness of the ocean's depth but are ever ready to dart and snap at luminescent shoals of shrimp.

Predators, such as lions and wolves, prey on animals larger than themselves and therefore they often hunt their prey as a group. Lions typically prey on animals that have been separated from their herd. Although only a small number of lions participate in the kill, the food is shared with the rest of the pride (fig. 30.8).

Control of Prey Population Size

A carnivore helps control the size of its prey population. Biologists reasoned that there would be an oscillation in sizes of the predator and prey populations, as illustrated in figure 30.9. The oscillation was expected because it was reasoned that as the number of prey increased, predation would also increase until finally the prey population would suffer a decline. This would be followed by a reduction in the number of predators until the prey population would eventually begin to recover. Then the cycle would begin again. Rarely are such oscillations observed, however. Therefore, it is believed that factors other than predation may have caused the oscillation in the lynx and hare populations observed between 1845 and 1935 (fig. 30.9). For example, the prey population

Behavior and Ecology

Figure 30.9
Example of predator-prey cycling. As the hare increases in number, so does the lynx, but then the hare, followed by the lynx, suffer a dramatic decrease in number. While it may seem as if this cycling is due to "overkill" by the lynx, other factors could be involved. (These data are based on the number of lynx and snowshoe hare pelts received by the Hudson Bay Company in the years indicated.)

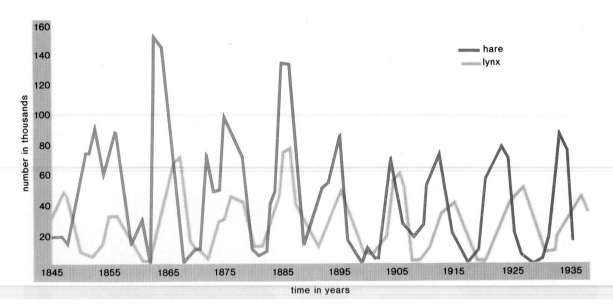

may have been reduced because of some other reason, such as lack of food due to the weather, and this, in turn, may have led to a decline in the predator population.

Usually predators do not overkill the prey population. For example, on Isle Royale, an island in Lake Superior, a population of about 1,000 moose has coexisted with a population of about 24 wolves year after year since 1948. Biologists who have studied this phenomenon tell us that the wolves are able to capture only the old, sick, or very young animals. The inability of the wolves to capture more moose keeps the wolf population in check.

Often predators even have a beneficial effect on the prey population. In 1930, before there were wolves in Isle Royale, the moose overpopulated, over-ate their food sources and subsequently suffered a sharp decline in population. Once the vegetation recovered, the moose population again grew to the point of overexploitation and suffered another decline in the 1940s. The chance introduction of wolves has kept the moose population stable ever since.

Sometimes humans have neglected to take into consideration that predators help keep prey populations in check. For example, in the past coyotes have been indiscriminately killed off in the West without regard to the fact that they help keep the prairie dog population under control. Similarly, when the dingo, a wild dog in Australia, was killed off because it attacked sheep, the rabbit and wallaby populations greatly increased. In contrast, humans formerly kept the burro population in the Grand Canyon within reasonable limits by killing them for meat. However, since a federal law was passed in 1971 forbidding the killing of burros, their numbers have increased until they are now considered destructive pests.

Whenever predator and prey are coadapted, each population is maintained at proper levels, as can be shown by a laboratory experiment involving two protozoans, *Didinium* and *Paramecium*. Didinia prey on paramecia, and

if the paramecia are denied a place to hide, the didinia capture all the paramecia and then they both die out (fig. 30.10). But if debris is provided so that the paramecia can hide, each population remains at a fairly constant level.

Prey Defenses

Prey defenses are varied (table 30.3). Some prey hide from predators by resembling an inedible object, such as a twig (fig. 30.11a), rock, or the background. Decorator crabs camouflage themselves by gathering pieces of algae, hydroids, and sponges, which they attach to hook-shaped setae on their shells (fig. 30.11b).

Bright colors warn a predator to beware that a prey is poisonous (fig. 30.12a). If attacked, a prey may try to disorientate the predator by a sudden movement. Some moths even flash spots that resemble eyes in order to startle the predator. The bombardier beetle sprays its assailant with an irritating chemical (fig. 30.12b). Mice have learned, however, to grab the beetle and stick it in the ground head up so that it cannot spray the irritant. Then the beetle is eaten.

Figure 30.10
Paramecium and *Didinium* interactions. *a.* A didinium (*below*) stalks a paramecium (*above*). *b.* Didinium in the act of engulfing the paramecium. *c.* If, in the experimental situation, the paramecia have no place to hide, the didinia will sometimes kill off all the paramecia and then die off themselves.

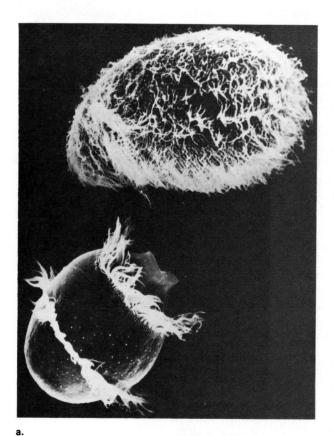

a.

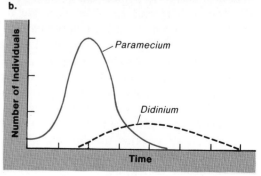

b.

c.

Behavior and Ecology

Table 30.3 Antipredator adaptations

Name	General behavior	Example
Sensory methods	Cryptic coloration and behavior	Blending in with the background
	Startle display	Sudden unexpected noise or visual effect
	Distraction display	Pretending injury
	Death display	Playing dead
Vigilant method	Detection of predator	Alerting others by alarm calling
Escape behavior	Evasion	Running away
Repellent behavior	Having a chemical or mechanical defense	Injuring predator
	Resembling animals that have a chemical or mechanical defense	Mimicry

a.

b.

Figure 30.11
Prey defenses. Some prey use disguises to avoid predators such as (*a.*) the stick caterpillar that resembles a twig and the decorator crab (*b.*), which covers itself with debris.

Figure 30.12

Other prey defenses. *a.* Arrow-poison frogs are so poisonous that natives use them to make their arrows instant lethal weapons. The coloration of these frogs warns others to beware. *b.* A bombardier beetle (held in place by an attached line) responds to an attacker by squirting it with an irritating liquid.

a.

Munch Gypsy, Crunch Gypsy

Due to gypsy moth infestation trees are defoliated in Boxford, Mass.

Untold numbers of trees defoliated, stripped naked of all leaves. Squishy little creatures hanging from branches, crawling over the sides of houses, creating a sickly goo on roads. People rushing to their doctors with rashes on hands, faces, anywhere they may have brushed against the little pests. Neighbors arguing angrily about whether to resort to risky chemicals. Noisy town meetings. Anguished editorials in newspapers.

Episodes from a late-night horror flick? Not at all. More like *cinema verite.* Once again, the Northeast has been infested by gypsy-moth caterpillars in record numbers. . . .

Only 2 in. to 3 in. long when fully grown, the gypsy-moth caterpillar looks harmless enough: a brownish, multilegged strip of fur with telltale pairs of red and blue spots running down its back. But looks are deceptive. Ever since 1869, when it was inadvertently turned loose in Massachusetts by a misguided French naturalist who wanted to cross the European gypsy with the silkworm to produce a disease-resistant hybrid that would eat virtually anything, it has been munching its way across the Northeast. As many as 30,000 caterpillars can infest a single tree,

and each of them can consume five or ten small leaves a day. They seem especially partial to the majestic oak but also eat fruit trees like apple and cherry, the maple and, alas, the already imperiled elm. If nothing else is available, they will nibble away at spruce, hardy pines and hemlocks, even shrubs—more than 500 species in all.

The feast begins in late April or May, when the caterpillars first emerge from their eggs. As they finish off one tree, they swing easily to another on silken threads they secrete. Their vagabond life accounts for the name gypsy. Millions can infest a small wooded patch. As they crunch, dropping excrement and half-eaten leaves, they sound like steady rain. Some homeowners complain that the noise actually keeps them awake. The caterpillars crawl up walls, spread over driveways, drop into plates and glasses at backyard barbecues. Last month Massachusetts officials got a call from a badly flustered woman. So many caterpillars had swarmed across her front door, she said, that she could not get in.

After molting for the last time, in late June and early July, the caterpillars spin the flimsiest of

Behavior and Ecology

b.

cocoons and harden into shell-like pupae, to emerge a week or two later as full-grown moths. Gypsy moths themselves do not eat. But each female lays velvety, tan masses of 100 to 1,000 eggs on tree trunks and buildings, on the undersides of cars, trucks and trailers, in carefully stacked woodpiles. Lighter colored and larger than the male, the female does not fly but attracts the male with a powerful chemical sex lure. By August both parents will have died, but the hardy eggs will survive through the winter, hatching in the spring and starting anew the devastating cycle.

After the first big outbreak, in Medford, Mass., in the late 19th century, New Englanders began battling the gypsy moth by putting out arsenic, soaking egg masses in creosote, burning down whole trees. But the bugs kept spreading. Wafted by winds, hitchhiking on cars and campers, they slowly migrated to at least 21 states, including Florida and California, although so far only pockets of serious infestation have occurred west or south of West Virginia. In the 1950s, scientists thought they finally had the moths under control with DDT. But the pesticide caused so much ecological havoc,

including the death of some of the birds and rodents that are the moth's natural enemies, that DDT has been banned from general use.

Entomologists agree that the moths can never be entirely eliminated. But containment may be possible. Early in the larval stage, when the caterpillars are still small and vulnerable, shorter-lived, milder pesticides like Sevin are useful, though Sevin also kills bees, which are needed for pollinating many fruits and flowers. Warns Cornell University Entomologist Warren Johnson: "By the time you see caterpillars greater than one-half inch in length, the time has passed for the most effective insecticide application." At that stage, however, a simpler tactic may help: encircling tree trunks with sticky bands that trap the caterpillars as they scurry up or down. One difficulty: as the bands become overloaded, the caterpillars evade capture by clambering over the bodies of their trapped brethren.

Lately, subtler forms of warfare have been introduced, including sprays that contain Bt (Bacillus thuringiensis), a bacterium that kills various moth and butterfly larvae. It, too, should be applied early. Another new experimental spray

spreads a virus that afflicts the gypsies with fatal wilt disease, so called because the dying caterpillar shrivels into a kind of inverted-V shape. More diabolical are traps scented with sex lures to attract male moths. Scientists have also been distributing different types of insects—wasps, flies, beetles—that prey on gypsy moths at various stages in their life cycle.

Despite all this lethal ingenuity, the only really good news from the bug battlefront is that most healthy trees can survive two or three onslaughts. Indeed, foresters like to point out that the moths often strengthen the woodland by eliminating sickly specimens. But such Darwinian reassurances are little comfort to suburbanites worried about a favorite elm or oak. By now, about all they can do is keep the tree as healthy as possible—faithful watering and feeding help—gather up and destroy every clutch of moth eggs in sight, and wait until next year.

Figure 30.13
Musk oxen adult males and a juvenile. Musk oxen
adult males are adapted to form a circle and
perform this act to protect females and offspring
from wolves.

Other prey, such as the opossum, play dead. Most likely this behavior removes the social releaser that prompts attack. In contrast, lizards may offer their tail to the predator because its removal does them no harm. They can regrow a new tail.

Flocks of birds, schools of fish, and herds of mammals stick together as protection against predators. Grazing herbivores are constantly on the alert; if one begins to dart away, they all run. Baboons, who detect predators visually, and antelope, who detect predators by smell, sometimes forage together, providing double protection against stealthy predators. Musk oxen males form a circle to protect the females and offspring within (fig. 30.13). This defense is effective against wolves but has also made it easier for humans to shoot and kill the oxen, which are now almost extinct.

One very unusual possible defense is discussed in the reading on cicadas (p. 702). It is believed that the life cycle of these animals protects them from predators. They emerge only every 13 to 17 years and thus far a predator adapted to their life cycle has not evolved.

Mimicry

When one species resembles another species that possesses an overt antipredator defense, it is called **mimicry.** As mentioned previously, an animal that is capable of actively repelling predators is often brightly colored or conspicuous in some other way. This is a form of advertisement that tells would-be predators that they had better keep their distance. Predators tend to avoid these animals; therefore their mimics are protected as if they too possessed the defense.

The monarch butterfly (fig. 30.14) is brightly colored and, if eaten by a bird, sometimes causes the bird to vomit a short time later. This occurs because monarch butteflies feed on milkweeds, which contain poisonous, digitalislike compounds (table 30.2) that make the bird sick. The butterfly larvae are able to collect and store this toxin without enzymatically destroying it. Birds that have had the experience of eating a poisonous monarch butterfly avoid all monarch butterflies in the future.

The viceroy butterfly (fig. 30.14) mimics the monarch butterfly but is not toxic. Birds eagerly eat viceroy butterflies unless they have had previous experience with a poisonous monarch. Then because the butterflies closely resemble each other, birds avoid both types of butterflies. The queen butterfly also mimics the monarch, but it too is poisonous.

A mimic that lacks the defense of the organism it resembles is called a **Batesian mimic;** the viceroy butterfly is a Batesian mimic of the monarch. A mimic that also possesses the same defense is called a **Müllerian mimic;** the queen butterfly is a Müllerian mimic of the monarch butterfly.

Since the monarch, viceroy, and queen butterflies look so much alike, it is expected that each would have a complex courting ritual involving precise pheromones, which would allow the butterflies to recognize members of their own species. The courtship behavior of the queen butterfly has been studied and it includes a series of courtship actions plus several different pheromones.

Wedding Whirs

Cicada sheds its skin for the last time.

Already, from the Carolinas to New York, little holes are appearing in lawns and backyards, hillsides and woodlands. Any evening now, out will pop millions of dark little bugs. They will scamper up almost any upright object—trees, poles, buildings—and soon strike up a joyous racket, marking nuptial rites after being buried alive for 17 years.

They are periodical cicadas (pronounced sih-*kay*-duhs). the world's longest-lived insects. Despite a locust-like appearance, they neither bite nor sting nor devastate vegetation. Entomologists currently count 19 separate ''broods,'' which appear at various times in different parts of the country, some once every 13 years. But all follow roughly the same miraculous life cycle. Growing through five skin-shedding molts and sucking nourishing juices from roots, they emerge with uncanny precision, triggered by some still mysterious internal clock.

In the open, they shed their dry, yellowish skins for the last time. Soon the males strike up their cacophony of ticking, buzzing and shrill whirring sounds. It is all music to the females, who slit open tree bark after they have been impregnated and store their fertilized eggs there. A few weeks later, both parents die. But cicada life goes on as the eggs hatch. The newborn nymphs drop to the ground, burrow, and the age-old cycle starts anew.

Baffled scientists are still unsure why the cicadas behave as they do, but suspect that it may all be a defense against predators like birds. As Entomologist Chris Simon of the State University of New York at Stony Brook writes in *Natural History*, when the cicadas finally emerge, it is in the shadows of dusk. They also gain protection from their monstrous numbers—as many as 1.5 million per acre. Finally, since they appear only once every 13 or 17 years, nature may have endowed them with an unlikely mathematical defense. These are prime numbers, divisible only by themselves, and so parasites would have to live at least as long—a half or a quarter would be improbable—to partake in a 17-year feast.

Symbiosis

Symbiotic relationships (table 30.4) are close relationships between two different species. Here, too, coevolution occurs and the species are closely adapted to one another. The first relationship to be discussed, parasitism, benefits the parasite but harms the host. The next two relationships, commensalism (one species benefits; the other is unaffected) and mutualism (both species benefit), are cooperative relationships.

Parasitism

Parasitism is similar to predation in that the **parasite** derives nourishment from the **host.** Usually, however, the host is larger than the parasite and the parasite does not kill the host. While viruses are the only obligate parasites (p. 537), there are also parasites among bacteria, protista, plants, and animals. Parasites are closely adapted to their host and infect only certain closely related species. Some of the viral infections of humans are listed in table 12.3 and some of the bacterial infections are listed in table 12.4. Malaria is a well-known protozoan infection (fig. 25.15) and athlete's foot is a well-known fungal infection of humans (fig. 25.23). Tapeworms and flukes illustrate typical life cycles of parasitic worms (p. 597).

Just as predators can dramatically reduce the size of a prey population (fig. 30.10) that lacks a suitable defense, so parasites can reduce the size of a host population that lacks a defense. Many thousands of elm trees have died off in this country due to the inadvertent introduction of Dutch elm disease. Figure 30.15 shows a tree-lined street before and after the elms contracted this disease, which is caused by a parasitic fungus. A new method of treating Dutch elm disease consists of inoculating the trees with bacteria that produce a fungus-killing antibiotic. The competition of the bacterium and fungus for existence in the same host tends to keep the population of each in check. Dutch elm disease is carried from tree to tree by a bark beetle. Therefore, attention has also centered in reducing the size of the bark beetle population.

Many animal parasites also use a secondary host not only for dispersal but also to complete their life cycles. For example, tapeworms utilize cattle or pigs and flukes require snails and sometimes fish as a host for their larvae. When one studies the anatomy and life-style of these animals it seems as if they have traded an active predatory life for an inactive secure life. Table 30.5 contrasts certain features of the animal predator with the animal parasite.

The hookworm, which is judged to be the most important parasitic worm, does not require a secondary host. In the New World hookworm, the males are from 5 to 9 millimeters in length and the females are usually about 1 centimeter long. The head is sharply bent in relation to the rest of the body, accounting for its characteristic hooklike appearance. Adult hookworms attach themselves to the intestinal wall (fig. 30.16) of their host and the eggs pass out with the feces. When deposited on moist, sandy soil, the larvae develop and hatch within 24 to 48 hours. After a period of growth and development, the worms then extend their bodies into the air and remain waving

Table 30.4 Symbiosis

	Species 1	Species 2
Parasitism	+	−
Commensalism	+	0
Mutualism	+	+

Key: + = benefits
    − = harmed
    0 = no effect

Figure 30.15
Effects of Dutch elm disease. *a.* Elm-lined Gillet Avenue in Waukegan, Illinois, as it appeared before attack. *b.* Gillet Avenue after the elms were destroyed by the parasitic fungus.

a.

b.

Figure 30.16
Hookworm parasite. Longitudinal section through hookworm attached to intestinal wall. In this position the worms suck blood and tissue fluid from the host.

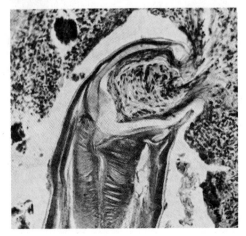

Table 30.5 Predator versus parasite

Predator	Parasite
Well-developed nervous system	Reduced nervous system
Sense organs, such as eyes	Sense organs, such as touch
Fast moving, with protective devices	Locomotion limited
Well-developed muscles	Minimal muscle fibers
Efficient circulatory system	Reduced circulatory system
Protection of offspring	Complicated life cycle

Figure 30.17
Example of social parasitism. Slave-making
Amazon ants invade the colony of a slave species
in order to carry off cocoons. When ants emerge
from the cocoons, they serve as slaves.

Figure 30.18
Example of commensalism. Shark suckers (remora)
do not feed on the shark; rather they feed on
leftovers from the shark's meal.

about in this position until they come into contact with the skin of a suitable
host, such as a human. Penetration usually occurs through the feet. Once in
the blood vessels, the worms are passively carried to the lungs where they
invade the alveoli. From the lungs, the larvae migrate up the trachea to be
swallowed and passed along to the small intestine where they mature. They
attach to the intestinal wall by means of their stout mouth parts and suck
blood and tissue fluid from the host. Symptoms of hookworm infection include
abdominal pains, nausea, diarrhea, and finally, iron deficiency anemia.

Social Parasitism

Social parasitism occurs when one species exploits another species. For example, the cuckoo lays eggs in nests of songbirds and the newly hatched cuckoo ejects its nestmates so that the songbird parents attend only to it. Slave-
making Amazon ants of the species *Polyergus rufescens* raid the ant colonies
of a slave species *Formica fusca* (fig. 30.17). They destroy any resisting
defenders with their mandibles, which are shaped like miniature sabers.
Polyergus ants are so specialized that they can only groom themselves. In
order to eat, they must beg slave workers for food. The slave workers not only
provide food for the slave-making ants but also care for the eggs, larvae, and
pupae of their captors.

Commensalism

In a **commensal** relationship, only one species benefits while the other is neither benefited nor harmed. Often the host species provides a home and/or
transportation for the benefited species. Barnacles, which attach themselves
to the backs of whales and the shells of horseshoe crabs, are provided with
both a home and transportation. Remoras are fish that attach themselves to
the bellies of sharks (fig. 30.18) by means of a modified dorsal fin that acts
as a suction cup. The remoras obtain a free ride and also feed on the remains

Figure 30.19
Example of commensalism. Clownfish live among a
sea anemone's tentacles and yet are not seized
and eaten as prey. The reason why this relationship
is maintained is not known.

of the shark's prey. Epiphytes, such as orchids, grow in the branches of trees
where they can receive light, but they take no nourishment from the trees,
instead, their roots obtain nutrients and water from the air. Clownfish (fig.
30.19) live within the tentacles and gut of a sea anemone and thereby are
protected from predators. Perhaps this relationship is one that borders on mutualism since the clownfish may attract other fish on which the anemone can
live. The sea anemone's tentacles quickly paralyze and seize other fish as prey.

Mutualism

Mutualism is a symbiotic relationship in which both members of the association benefit. Mutualistic relationships often help organisms obtain food or
avoid predation.

Bacteria that reside in the human intestinal tract are provided with food,
but they also provide us with vitamins, molecules we are unable to synthesize
for ourselves. Termites would not even be able to digest wood if it were not
for the protozoans that inhabit their intestinal tract. These organisms digest
cellulose, which termites cannot.

Fungi and usually cyanobacteria live together as lichens (fig. 25.9).
Lichens can grow on rocks most likely because the fungi provide moisture and
dissolve minerals from the rock. The cyanobacteria carry on photosynthesis
and return carbohydrates to the fungi. **Mycorrhizae** (p. 556) are symbiotic
associations between the roots of plants and fungal hyphae. It is now believed

Figure 30.21
A unique mutualistic relationship. *a.* This particular species of orchid resembles a female bee of a particular species. Therefore the male attempts to copulate with it and thereby picks up pollen. *b.* Enlargement of orchid and bee.

a.

b.

that the roots of most plants form these relationships. Mycorrhizal hyphae, which often penetrate the root but may lie just outside the root, increase the solubility of minerals in the soil, improve the uptake of nutrients for the plant, protect the plant's roots against pathogens, and produce plant growth hormones. In return, the fungus obtains carbohydrates from the plant.

Flowers and their pollinators have coevolved since flowering plants first appeared. For this reason, flowers that attract bees or birds are brightly colored and flowers that attract beetles or bats have strong scents. Sometimes the nectaries, which produce a sugar solution, are located at the base of a tubular corolla so that they are accessible only to the moth or bird that has evolved a long sucking tongue, such as the hummingbird (fig. 30.20). Flowers that attract bees have a landing platform and a structure that requires the bee to brush up against the anther and stigma as it moves toward the nectaries. One orchid has evolved a unique adaptation: the flower resembles a female bee and when the male of that species attempts to copulate with it (fig. 30.21), it receives pollen. The coevolution of flowers and their pollinators is of mutual benefit because the pollinator receives food while the flower achieves cross-pollination.

Mutualistic relationships are also common between ants and other organisms. **Leaf cutter ants** keep fungal gardens (fig. 30.22). They are called leaf cutter ants because they gather flowers and leaves, which they cut into pieces and transport to underground nests. After preparing the leaves, they implant them with fungal mycelia, which then grow in profusion. This relationship helps the fungi compete with other fungal species but also provides the ants with a source of food.

Figure 30.23

Mutualistic relationship. Ants sometimes serve to protect aphids. The sap that aphids suck from plants passes through their digestive system so that an ant can feed from the rear of an aphid.

Figure 30.24

Mutualistic relationship. The bull's horn acacia is adapted to provide a home for *Pseudomyrmex ferruginea*, a species of ant that protect the acacia from other insects. *a.* Thorns are hollow and the ants live inside. *b.* Base of leaves has nectaries (openings) where ants can feed. *c.* Leaves have bodies at the tips that ants harvest for larvae food.

a.

b.

c.

Behavior and Ecology

a.

b.

Some ants protect aphids from their predators and in return receive food from the aphids (fig. 30.23). As mentioned previously, aphids remove phloem sap from plants by means of a styletlike mouthpiece. The sap passes through an aphid's body relatively unchanged so that an ant can feed from the rear of an aphid.

In Central America, the bull's-horn acacia is adapted (fig. 30.24) to provide a home for ants of the species *Pseudomyrmex ferruginea.* Unlike other acacias, it has swollen thorns with a soft, pithy interior where ant larvae can grow and develop. In addition to housing the ants, the acacias provide them with food. The ants feed from nectaries at the base of the leaves and eat nodules, called beltian bodies, at the tips of some of the leaves. The ants constantly protect the plant from herbivorous insects because, unlike other ants, they are active 24 hours a day. The plants, on the other hand, have leaves throughout the year, while related acacia species lose their leaves during the dry season.

Cleaning symbiosis (fig. 30.25) is a phenomenon that is believed to be quite common among marine organisms. There are species of small fish and shrimp that specialize in removing parasites from larger fish. The large fish line up at the "cleaning station" and wait their turn, while the small fish feel so secure that they even clean the insides of the mouths of the larger fish. Not everyone plays fair, however, since there are small fish that mimic the cleaners in order to take a bite out of the larger fish, and cleaner fish are sometimes found in the stomachs of the fish they clean.

Summary

Certain interactions are typical between species. When two species compete for the same resources, one usually eliminates or restricts the range of the other. For example, two barnacles, *Chthamalus* and *Balanus,* each prevent

the other from occupying the entire intertidal zone. The antelope ground squirrel possibly restricts the range and activity of the Mohave ground squirrel. In some instances humans have inadvertently imported successful competitors into this country to the extent that the native species have been threatened.

The fact that a species tends to exclude a competing species is in keeping with the competition exclusion principle. This principle states that no two species can occupy the same niche. An organism's niche is defined by the parameters listed in table 30.1.

Predators and prey coevolve. Herbivores prey on plants and carnivores prey on other animals. Plants have physical and chemical attributes that discourage predation. Some carnivores seek out their prey, while others lie in wait. Still others hunt in groups. Although predators control the prey population size, there is usually no oscillation in carnivore and prey population sizes. Instead, both populations are maintained at constant levels. Humans should realize that when they kill off a carnivore predator population, the prey population will tend to increase in size.

Prey defenses are varied, as listed in table 30.3. Sometimes an animal with a successful antipredator defense is mimicked by other animals that have the same defense (Müllerian mimic) or that lack the defense (Batesian mimic). The monarch butterfly has both types of mimics.

Symbiotic relationships are of three types. In parasitism, the host species is harmed. Parasites, which help control the size of their host population, may require a secondary host for dispersal. The anatomy and life-style of an animal parasite may be contrasted to that of an active predator. The hookworm is considered an important animal parasite. In social parasitism, one society exploits another.

In commensalism, one species is benefited but the other is unaffected. Often the host simply provides a home and/or transportation. In mutualism, both species benefit. The two species are often closely adapted to one another, such as flowers and their pollinators. There are also examples of mutualistic species that live together in the same locale, such as ants that keep fungal gardens. But in cleaning symbiosis, the relationship is transitory.

Study Questions

1. Give two examples to show that competition between species results in the elimination or restriction of range of the other. (pp. 686–87)
2. Give examples to show that it is sometimes unwise to bring a competitor into a new area. (p. 689)
3. Define the competition exclusion principle and describe four aspects of an organism's niche. (p. 689)
4. Would it be correct to say that the more species variety there is in an area, the more niches there must be? (p. 689)
5. Give examples to show that plants have defenses against herbivores. (p. 692)
6. Give examples to show that predators have a beneficial effect, helping to stabilize population sizes in an area. (pp. 694–95)
7. Give examples of antipredator defenses that prevent predator populations from overkilling prey populations. (pp. 696–700)
8. Define a mimic and give two examples in relation to the monarch butterfly. (pp. 700–701)
9. What are the three types of symbiotic relationships? Give several examples of each. (pp. 703–4)
10. Describe the life cycle of the hookworm. (pp. 703–9)
11. Give two examples of social parasitism. (p. 704)

Selected Key Terms

competition (kom''pĕ-tish'un)

competitive exclusion principle (kom-pet'ĭ-tiv eks-kloo'zhun prin'sĭ-p'l)

niche (nich)

predation (pre-da'shun)

coevolve (ko-e-volv')

grazers (gra'zerz)

browsers (browz'erz)

prey (pra)

oscillation (os''ĭ-la'shun)

antipredator defense (an''tĭ-pred'ah-tor de-fens')

mimicry (mim'ik-re)

symbiosis (sim''bi-o'sis)

parasite (par'ah-sīt)

social parasitism (so'shal par'ah-si''tizm)

commensalism (kŏ-men'sal-izm)

mutualism (mu'tu-al-izm'')

cleaning symbiosis (klēn'ing sim''bi-ō'sis)

Further Readings

Alcock, J. 1975. *Animal behavior*. Sunderland, Mass.: Sinauer.

Beddington, J. R., and May, R. M. 1982. The harvesting of interacting species in a natural ecosystem. *Scientific American* 247(5):62.

Bekoff, M., and Wells, M. C. 1980. Social ecology of coyotes. *Scientific American* 242(4):130.

Bergerud, A. T. 1983. Prey switching in a simple ecosystem. *Scientific American* 249(6):130.

Emmel, T. C. 1973. *An introduction to ecology and populations*. New York: W. W. Norton.

MacArthur, R. H., and Connell, J. H. 1969. *Biology of populations*. New York: John Wiley.

Prestwich, G. D. 1983. The chemical defenses of termites. *Scientific American* 249(2):78.

Ricklefs, R. E. 1979. *Ecology*. 2d ed. Newton, Mass.: Chiron Press.

Strobel, G. A. 1975. Mechanism of disease resistance in plants. *Scientific American* 232(1):80.

Strobel, G. A., and Lanier, G. N. 1981. Dutch elm disease. *Scientific American* 245(2):56.

Trager, W. 1970. *Symbiosis*. New York: Van Nostrand Reinhold.

Wallace, B., and Srb, A. M. 1974. *Adaptation*. Englewood Cliffs, N.J.: Prentice-Hall.

Wickler, W. 1968. *Mimicry*. New York: McGraw-Hill.

Wicksten, M. 1980. Decorator crabs. *Scientific American* 242(2):146.

Wilson, E. O. 1975. Slavery in ants. *Scientific American* 232(6):32.

Wilson, E. O., ed., 1974. *Ecology, evolution and population biology*. Readings from *Scientific American*. San Francisco: W. H. Freeman.

31

Ecosystems

Chapter concepts

1 Natural ecosystems, which use solar energy efficiently and chemicals that cycle, produce little pollution and waste.

2 The man-made ecosystem, which comprises both country and city, utilizes fossil fuel energy inefficiently and material resources that do not cycle. Therefore there is much pollution and waste.

3 Humans have begun to address the problem of pollution, which affects the quality of air, water, and land.

All living things reside within the **biosphere,** a narrow sphere or shell that encircles the earth. Although most organisms reside near the surface, others are also found a short distance into the air above and into the waters beneath the surface. Each belongs to a **population,** the members of a species that reside in a particular area. Here, various populations interact with each other and the physical environment. Thus an **ecosystem** contains both a **biotic** (living) and **abiotic** (nonliving) environment. The study of **ecology** is defined as the study of the interactions of organisms among themselves and with the physical environment.

Ecosystem Composition

Each organism in an ecosystem has a habitat and a niche. The **habitat** of an organism is its place of residence, that is, the location where it may be found, such as "under a fallen log" or "at the bottom of the pond." The **niche** of an organism is its profession or total role in the community. A description of an organism's niche (table 30.1) includes its interactions with the physical environment and with the other organisms in the community. One important aspect of niche is the manner in which the organism acquires energy and chemicals. In fact, the entire ecosystem has two important aspects: *energy flow and chemical cycling.* These begin when photosynthesizing organisms use the energy of the sun to make their own food. Thereafter, chemicals and energy are passed from one population to another as the populations form food chains.

Food Chains

Essentially, **food chains** are described by telling "who eats whom." Autotrophic green organisms, which are at the start of a food chain, are called the **producers** because they have the ability to change what was formerly inorganic chemicals to organic food. Thus the producers in a food chain produce food. The other two types of populations in the biotic community are the consumers and the decomposers. **Consumers** are heterotrophic organisms that must take in preformed food. **Herbivores** are primary consumers that feed directly on producers; **carnivores** are secondary or tertiary consumers that feed only on consumers; **omnivores** are consumers that feed on both producers and consumers. **Decomposers** are saprophytic organism of decay, such as bacteria and fungi, that break down nonliving organic matter (**detritus**) to inorganic matter, which can be used again by producers. In this way the same chemicals can be used over and over again in an ecosystem. This is not true for energy. As detritus decomposes, all that remains of the solar energy taken up by the producer populations dissipates as heat. Thus energy does not cycle.

Grazing Food Chain

A specific example of an aquatic food chain is described pictorially and diagrammatically in figures 31.1a and 31.1b. Notice that the producer (algae) requires a source of inorganic molecules as nutrient molecules. Utilizing energy from the sun, the algae are able to combine water (H_2O) and carbon dioxide (CO_2) to form the organic molecule glucose. Thereafter, nitrate (NO_3) is used by the producer during the production of amino acids, and phosphate (PO_4) is used during the production of nucleotides.

Figure 31.1

An aquatic food chain. The arrows indicate the steps by which chemicals cycle through a food chain until they are returned once again to the producer. *b.* Pictorial representation of the food chain outlined in *a.*

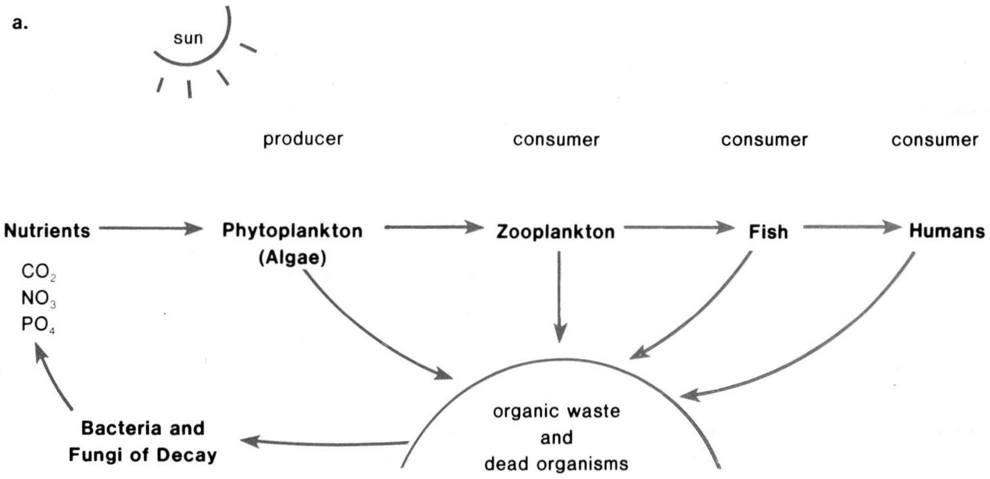

a.

sun

| producer | consumer | consumer | consumer |

Nutrients ⟶ **Phytoplankton (Algae)** ⟶ **Zooplankton** ⟶ **Fish** ⟶ **Humans**

CO_2
NO_3
PO_4

Bacteria and Fungi of Decay

organic waste and dead organisms

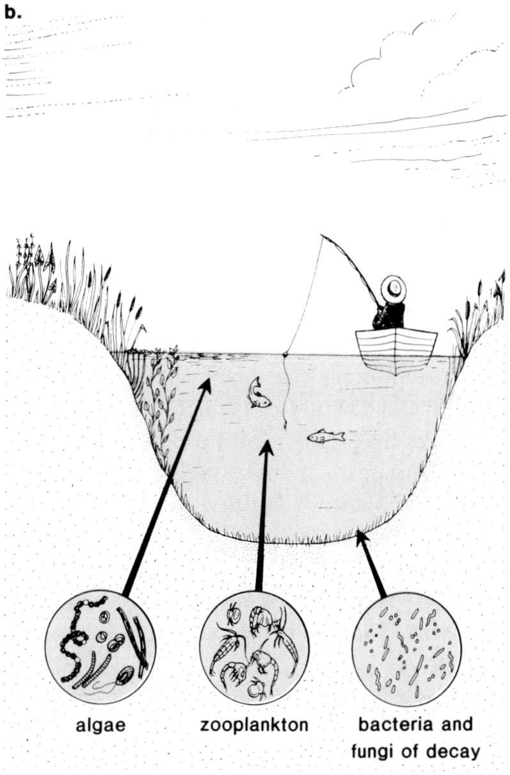

b.

algae   zooplankton   bacteria and fungi of decay

Each of the consumer populations in the food chain (zooplankton, fish, and humans) feed on the preceding population and thus are either directly or indirectly dependent on the producer for their source of organic food. The food chain in figure 31.1 is a grazing food chain because it contains a herbivore population.

Behavior and Ecology

Figure 31.2
A simplified terrestrial food web. Food chains are
interwoven in ecosystems; for example, mice are
food for hawks and snakes.

Detritus Food Chain

Figure 31.1a contains decomposer populations (bacteria and fungi of decay) to illustrate that chemicals cycle, but it does not really show a detritus food chain. A detritus chain occurs when detritus and decomposers are eaten by scavenging animals, such as worms, insect larvae, and snails. Whenever these animals serve as food for fish, the detritus food chain joins the grazing food chain. While the grazing food chain is most important in aquatic food chains, the detritus food chain is most important in terrestrial food chains. For example, it is estimated that only about 10 percent of the annual leaf production is consumed by herbivores; the rest is either degraded by decomposers or eaten by scavengers, such as soil mites, earthworms, and millipedes.

Food Webs

Each food chain represents just one pathway by which chemicals and energy are passed along in an ecosystem. Natural ecosystems have numerous food chains, each linked to form a complex **food web** (fig. 31.2). Since organisms may belong to more than one food chain, energy flow is best described in terms of trophic (feeding) levels. The first level is the producer population, and each successive level is further removed from this population. All animals acting as primary consumers are part of the second level; all animals acting as secondary consumers are part of the third level; and so on. Each succeeding trophic level passes on less energy than was received due to a number of reasons. For example, usable energy is lost at each level because

1. not all members of the previous trophic level become food; of those that do become food, some portions, such as hair and bones, may be uneaten, or if eaten, may be undigestible;
2. some food is used for maintenance and never contributes to growth;
3. energy transformations always result in a loss of usable energy.

To understand that energy is lost due to energy transformations, consider that the conversion of energy in one molecule of glucose to 38 ATP molecules represents only 50 percent of the available energy in a glucose molecule. The rest is lost as heat.

Figure 31.3

Dissipation of energy in an ecosystem. Producers convert inorganic matter to organic food after capturing a small amount of solar energy. When nutrient molecules are used for maintenance, their energy content eventually becomes heat. Energy, therefore, does not cycle, and ecosystems must have a continual supply of solar energy.

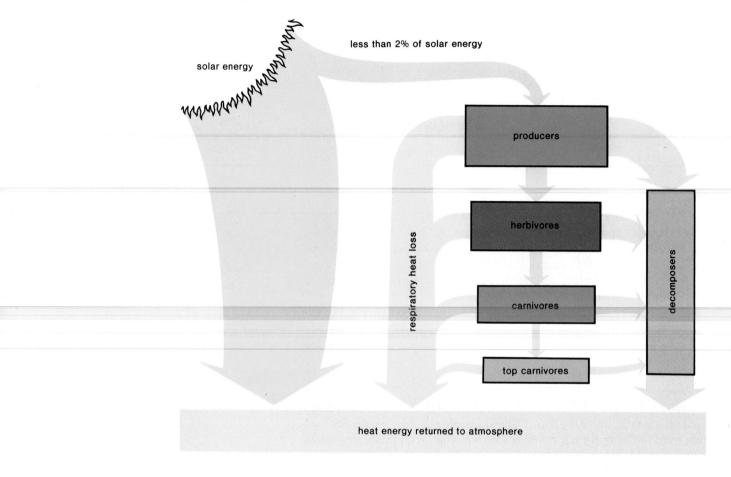

Figure 31.4

Food pyramid. Whether one considers numbers, biomass, or energy content, each trophic level is smaller than the one on which it depends. Therefore the various trophic levels form a pyramid in which the producer population is at the base and the last consumer population is at the peak.

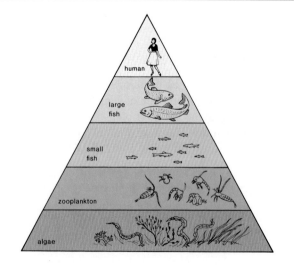

Behavior and Ecology

Eventually, as one population feeds on another and as decomposers work on detritus, all of the captured solar energy that was converted to chemical bond energy by algae and plants is returned to the atmosphere as heat (fig. 31.3). *Thus energy flows through an ecosystem and does not cycle.*

Food Pyramid

Whether one considers the number of individuals, the biomass (weight of living material), or the amount of energy, each succeeding trophic level is generally smaller than the one preceding. To illustrate this, the various trophic levels are often depicted in the form of a pyramid (fig. 31.4). The producer population is at the base of the pyramid. Obviously it must be the largest because it indirectly produces food for all the other populations.

In regard to the pyramid of energy, it is generally stated that only about 10 percent of the energy absorbed by one trophic level can be stored by the next level. About 90 percent is lost for the reasons previously stated. In practical terms, this means that about 10 times the number of people can be sustained on a diet of grain than on a diet of meat.

Stability

Mature natural ecosystems tend to be diverse and stable. *First,* the various populations stay at a constant size for the reasons listed in table 31.1. *Density independent* effects are those forces of nature whose magnitude of influence does not depend on the size of the population affected. For example, the severity of a drought or flood has nothing to do with how many plants or animals there are in a particular area. The severity of *density dependent* effects, several of which were studied in the preceding chapter, does depend on the size of the population affected. For example, two large populations compete to a greater degree for the same resource than do two small populations.

Second, diversity assists maintenance of population size. For example, referring again to figure 31.2, you can imagine that if the rabbit population suffered an epidemic and declined in size, the mice population would increase in size due to decreased competition for food. The increase would mean that the producer population would still be held in check, and the hawks that eat rabbits or mice would still eventually have the same amount of food. It may take a little time for the new balance to come about; a few hawks may have to migrate or starve, but essentially there would be little change in the size of the hawk population. Further, as the rabbits recover, the mice population would decline and eventually there would be exactly the same balance as before the epidemic. In this manner, variation of species at each level of a food pyramid gives stability to an ecosystem. Actually, we see the same principle at work in business; a company diversifies its products so that as demand fluctuates profits will remain the same.

Constancy of population sizes means that most of the solar energy utilized by a biotic community supports stability rather than growth. The same amount of solar energy is required each year in order to maintain a highly diversified community that has little material waste due to the cycling of chemicals.

Chemical Cycles

In contrast to energy, matter does cycle through ecosystems. Because there is normally no input from outside an ecosystem, the same 30 or 40 elements essential to life are used over and over. For each element, the cycling process

Table 31.1 Population control

Density independent effects
Climate and weather
Natural disasters

Density dependent effects
Competition
Predation
Parasitism
Emigration

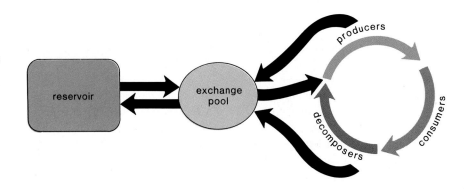

Figure 31.5
Components of a chemical cycle. The reservoir stores the chemical, the exchange pool makes it available to producers, after which it cycles through food chains. Decomposition returns the chemical to the exchange pool once again if it has not already returned by another process.

Figure 31.6
Relationship between photosynthesis and respiration. Animals are dependent on plants for a supply of oxygen and plants are dependent on animals for a supply of carbon dioxide.

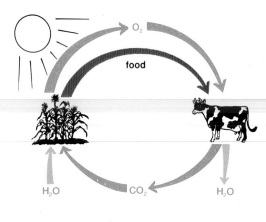

Figure 31.7
Major components of the terrestrial portion of the carbon cycle. In the upper half of the illustration, the solid arrow is photosynthesis; the dotted arrows are respiration; the colored arrow is the burning of fossil fuels, the contribution of humans to the cycle.

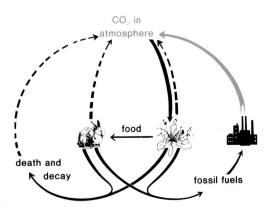

(fig. 31.5) involves (1) a **reservoir**—that portion of the biosphere that acts as a storehouse for the element; (2) an **exchange pool**—that portion of the biosphere from which the producers take and consumers return inorganic nutrients; and (3) the **biotic community**—through which chemicals move along food chains from and to the reservoir.

Carbon Cycle

The relationship between photosynthesis and respiration should be kept in mind in discussing the carbon cycle. Recall that for simplicity's sake this equation in the forward direction represents respiration and in the other direction may be used to represent photosynthesis:

$$C_6H_{12}O_6 + 6\,O_2 \rightleftharpoons 6\,CO_2 + 6\,H_2O$$

The equation tells us that respiration releases carbon dioxide, the molecule needed for photosynthesis. However, photosynthesis releases oxygen, the molecule needed for respiration. From figure 31.6, it is obvious that animals are dependent on green organisms, not only to produce organic food and energy but also to supply the biosphere with oxygen.

In the carbon cycle on land (fig. 31.7), the carbon dioxide released by respiration enters the atmosphere. Animals and animal-like organisms continuously release carbon dioxide into the air, but plants release carbon dioxide only when, as at night, respiration is occurring at a faster rate than is photosynthesis. After organisms die, decomposition also releases carbon dioxide. On the other hand, photosynthesizing organisms take carbon dioxide up from the air.

The carbon cycle also occurs in aquatic communities, but in this case, carbon dioxide is taken up from and returned to water. Carbon dioxide from the air combines with water to give bicarbonate ions (HCO_3^-) that serve as a source of carbon for algae, which produce food for themselves and heterotrophs. Similarly, when aquatic organisms respire, the carbon dioxide they give off becomes bicarbonate ions. The amount of bicarbonate in the water is in equilibrium with the amount in the air.

Reservoirs

Living and dead organisms contain organic carbon and serve as one of the reservoirs for the carbon cycle. When we destroy forests and burn wood, we are reducing this reservoir and increasing the amount of CO_2 in the atmosphere.

Behavior and Ecology

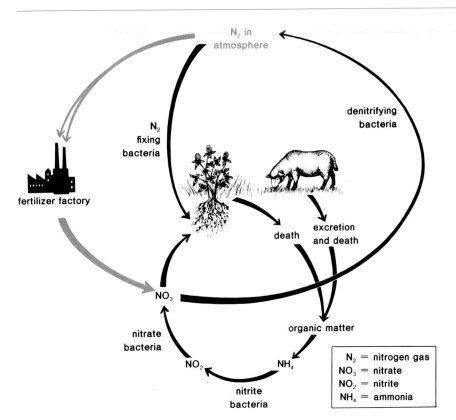

Figure 31.8
Major components of the terrestrial portion of the nitrogen cycle. Three types of bacteria are at work: nitrogen-fixing bacteria convert aerial nitrogen to a form usable by plants; nitrifying bacteria, which include both nitrite and nitrate bacteria, convert ammonia to nitrate; and the denitrifying bacteria convert nitrate back to aerial nitrogen again. Humans contribute to the nitrogen cycle by converting aerial nitrogen to fertilizer.

Diagram labels:

N_2 in atmosphere

denitrifying bacteria

N_2 fixing bacteria

fertilizer factory

death

excretion and death

NO_3

nitrate bacteria

organic matter

NO_2 NH_4

nitrite bacteria

N_2 =	nitrogen gas
NO_3 =	nitrate
NO_2 =	nitrite
NH_4 =	ammonia

In prehistoric times, certain plants and animals did not decompose, by chance, and instead were preserved in **fossil fuels** such as coal, oil, and gas. When humans burn fossil fuels, they not only add CO_2 but also other combustion products to the air. This contributes greatly to air pollution, which is discussed on page 720.

Another reservoir for the carbon cycle is the formation of inorganic carbonate that accumulates in limestone and in carbonaceous shells. Limestone, particularly, tends to form in the oceans. In this way, the oceans act as a sink for excess carbon dioxide. The burning of fossil fuels in the last 22 years has probably released 78 billion tons of carbon, and yet the atmosphere registers only an increase of 42 billion tons.

Nitrogen Cycle

That portion of the nitrogen cycle that involves terrestrial organisms is represented by the diagram in figure 31.8. Aerial nitrogen, the reservoir for the nitrogen cycle, is not usable by most organisms, but there are two types of nitrogen-fixing bacteria that make use of it. One type is free-living in the soil, but the other type infects and lives in nodules on the roots of legumes (fig. 31.9). Here the nitrogen-fixing bacteria convert aerial nitrogen to a form that can be used by plants.

Nitrates, in particular, are taken up by the roots of plants and converted to amino acids. Thereafter, animals acquire this organic nitrogen when they eat plants. Plants and animals die, and decomposition produces ammonia that can be converted to nitrates by **nitrifying** (nitrite and nitrate) **bacteria.** To make the cycle complete, there are some bacteria, the **denitrifying bacteria,** that can convert ammonia and nitrates back to aerial nitrogen again.

Figure 31.9
Nodules on the roots of a legume (in this case, soybean). The bacteria that live in these nodules are capable of converting aerial nitrogen to a source that the plant can use.

Figure 31.10

Balance sheet for the nitrogen cycle. This graph estimates the amount of nitrogen being fixed and denitrified per year. Under nitrogen fixation, notice that that in addition to the ways depicted in figure 31.8, nitrogen is also fixed in the ocean (marine) and in the atmosphere (atmospheric fixation by lightning, for example). Also denitrification not only occurs on land, it also occurs in the sea. The difference between the total gain and total loss represents the rate at which fixed nitrogen is accumulating in the soil and water.

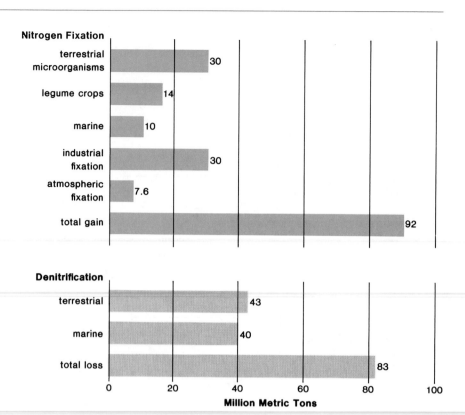

Other Contributions to the Nitrogen Cycle

There are at least two other ways by which nitrogen fixation occurs. Nitrogen is fixed in the atmosphere when cosmic radiation, meteor trails, and lightning provide the high energy needed for nitrogen to react with oxygen. Also, humans make a most significant contribution to the nitrogen cycle when they convert aerial nitrogen to nitrates for use in fertilizers. This industrial process requires an energy input that equals that of the eventual increase in crop yield. The application of fertilizers also contributes to water pollution, as discussed on page 733. Since nitrogen-fixing bacteria do not require fossil fuel energy and do not cause pollution, research is now directed toward finding a way to make all plants capable of forming nodules (fig. 31.9) or, even better, through recombinant DNA research, to possess the biochemical ability to fix nitrogen themselves.

When the amount of nitrogen fixation is compared to the amount of denitrification, it is seen that there is more nitrogen fixation, especially due to fertilizer production (fig. 31.10).

Human Ecosystem

Mature natural ecosystems tend to be stable and to exhibit the characteristics listed in table 31.2. Each population is of a proper size in relation to other populations; the energy that enters and the amount of matter that cycles is appropriate to support these populations. **Pollution,** defined as any undesirable change in the environment that may be harmful to humans and other life, and excessive waste do not normally occur. The man-made ecosystem that replaces natural ecosystems is quite different, however.

Table 31.2 Ecosystems

Natural	Human
Independent	Dependent
Cyclical (except energy)	Noncyclical
Nonpolluting	Polluting
Renewable solar energy	Nonrenewable fossil fuel energy
Conserves resources	Uses up resources

Behavior and Ecology

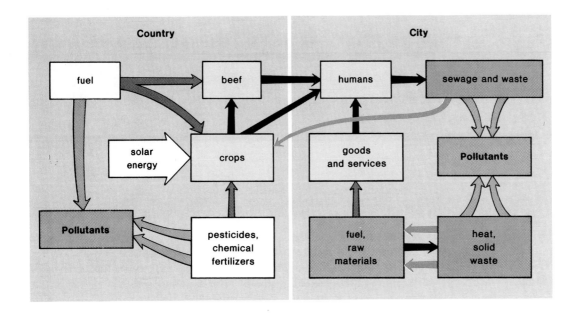

Human beings have replaced natural ecosystems with the man-made ecosystem that is depicted in figure 31.11. This ecosystem essentially has two parts: the *country,* where agriculture and animal husbandry are found, and the *city,* where industry is carried on. Most humans in developed countries live in the city (or suburbs) and thus humans are shown on the right side of the diagram in figure 31.11. This representation of the human ecosystem, although simplified, allows us to see that the system requires two major inputs: *fuel energy and raw materials* (e.g., metals, wood, synthetic materials). The use of these two necessarily results in *pollution and waste* as outputs.

Country

Modern United States agriculture produces exceptionally high yields per acre, but this bounty is dependent on a combination of five variables given here.

1. **Planting of a few genetic varieties.** The majority of farmers specialize in growing one of these. Wheat farmers plant the same type of wheat, and corn farmers plant the same type of corn. This so-called **monoculture agriculture** is subject to attack by a single type of parasite. For example, a single parasitic mold reduced the 1970 corn crop by 15 percent, and the results could have been much worse because 80 percent of the nation's corn acreage was susceptible.

2. **Heavy use of fertilizers, pesticides, and herbicides.** *Fertilizer* production requires a large energy input, and fertilizer runoff contributes to water pollution (p. 733). *Pesticides* reduce soil fertility because they kill off beneficial soil organisms as well as pests, and some pesticides concentrate in food chains (p. 732), eventually producing toxic effects in predators, possibly even humans. *Herbicides,* especially those containing the contaminant dioxin, have been charged with causing reproductive effects and cancer.

Figure 31.12
Planting in straight rows facilitates the use of
modern machines but causes more soil erosion.

3. **Generous irrigation.** River waters are sometimes redirected for the purpose of irrigation, in which case "used water" returns to the river carrying a heavy concentration of salt. The salt content of the Rio Grande River in the Southwest is so high that the government has built a treatment plant to remove the salt. Water is also sometimes taken from aquifers (underground rivers) whose water content can be so reduced that it becomes too expensive to pump out more water. This is the situation that some farmers in Texas are now facing, as discussed in the reading on page 724.

4. **Increase fuel consumption.** Energy is consumed on the farm for many purposes. Irrigation pumps have already been mentioned but large farming machines also are used, to spread fertilizers, pesticides, and herbicides and to sow and harvest the crops (fig. 31.12). It is not incorrect to suggest that modern farming methods transform fossil fuel energy into food energy.

    Supplemental fossil fuel energy also contributes to animal husbandry yields. At least 50 percent of all cattle are kept in **feedlots** where they are fed grain. Chickens are raised in a completely artificial environment where the climate is controlled and each one has its own cage to which food is delivered on a conveyor belt. Animals raised under these conditions often have antibiotics and hormones added to their feed to increase yield.

5. **Loss of land quality.** Evaporation of excess water on irrigated lands can result in a residue of salt. This process, termed **salinization**, makes the land unsuitable for the growth of crops. Between 25 and 35 percent of the irrigated Western croplands are thought to have excessive salinity. Soil erosion is also a serious problem. It is said that we are **mining the soil** because many farmers are not taking measures to prevent the loss of topsoil. The Department of Agriculture estimates that erosion is causing a steady drop in the productivity of land equivalent to the loss of 1.25 million acres per year. Even more fertilizers, pesticides, and energy supplements will be required to maintain yield.

Organic Farming

Some farmers have given up this modern means of farming and instead have begun to adopt organic farming methods. This means that they do not use applications of fertilizer, pesticides, and herbicides. They use cultivation of

Behavior and Ecology

row crops to control weeds; crop rotation to combat major pests; and the growth of legumes to supply nitrogen fertility to the soil. Some farmers use natural predators and parasites instead of pesticides to control insects (fig. 31.13). For the most part, these farmers switched farming methods because they were concerned about the health of their families and livestock and had found that the chemicals were sometimes ineffective.

A study of about 40 farms showed that organic farming for the most part was just as profitable as conventional farming. Crop yields were lower, but so were operating costs. Organic farms required about two-fifths as much fossil energy to produce one dollar's worth of crop. The method of plowing and utilization of crop rotation resulted in one-third less soil erosion. The researchers concluded it would be well to determine how far farmers can move in the direction of reduced agricultural chemical use and still maintain the quality of the product. They noted that a modest application of fertilizer would have improved the protein content of the crop.

City

As figure 31.11 shows, the city is dependent on the country. For example, each person in the city requires several acres of land for food production. Overcrowding in cities does not mean that less land is needed; each person still requires a certain amount of land to ensure survival. Unfortunately, however, as the population increases, the suburbs and cities tend to encroach on agricultural and range land (fig. 31.14). An estimated 875, 000 acres of actual or potential cropland were converted to urban uses annually between 1967 and 1975.

The city houses workers for both commercial businesses and industrial plants. Solar and other renewable types of energy are rarely used; cities currently rely mainly on fossil fuel in the form of oil, gas, electricity, and gasoline. The city does not conserve resources. An office building, with constantly burning lights and windows that cannot be opened, is an example of energy waste. Another example is people who drive cars long distances instead of taking public transportation and who drive short distances instead of walking or bicycling. Materials are not recycled and products are designed for rapid replacement.

The burning of fossil fuels for transportation, commercial needs, and industrial processes causes air and water pollution (p. 727). This pollution is compounded by the chemical and solid waste pollution that results from the

manufacture of many products. Consider that any product used by the average consumer (house, car, washing machine) causes pollution and waste, both during its production and when it is disposed of. Humans themselves produce much sewage that is discharged into bodies of water, often after only minimal treatment.

Table 31.2 lists the characteristics of the human ecosystem as it now exists. Just as the city is not self-sufficient and requires the country to supply it with food, so the whole human ecosystem is dependent on the natural ecosystems to provide resources and absorb waste. Fuel combustion by-products,

Ebbing of the Ogallala

Holding his hat against a swirling dust storm on his farm in Lubbock, Texas, W. E. Medlock leans into a dry future.

There are two documentary images of the Great Plains. The first is a black-and-white photograph of the '30s Dust Bowl, with windblown homesteaders treading the cracked earth. The second: a glossy color shot of the same land 40 years later, showing the lush checkerboard farms of America's breadbasket. Now, as if through a strange reversal in time, the second image threatens to fade into the first. For in another 40 years, the territory could backslide into dust and despair. The Ogallala Aquifer, the vast underground reservoir of water that transformed much of the Great Plains into one of the richest agricultural areas in the world, is being sucked dry.

The aquifer is a California-size deposit of water-laden sand, silt and gravel. It ranges in thickness from 1,000 ft. in Nebraska, where two-thirds of its waters lie, to a few inches in parts of Texas. Although it was first tapped in the 1930s, it has been extensively exploited only since the development of high-capacity pumps after World War II. The Ogallala's estimated quadrillion gallons of water, the equivalent of Lake Huron, have irrigated farms in South Dakota, Nebraska, Wyoming, Kansas, Colorado, Oklahoma, Texas and New Mexico, changing a region of subsistence farming into a $15 billion-a-year agricultural center.

For the past three decades, farmers have pumped water out of the Ogallala as if it were inexhaustible. Nowadays they disperse it prodigally through huge center-pivot irrigation sprinklers, which moisten circular swaths a quarter-mile in diameter. The annual overdraft (the amount of water not replenished) is nearly equal to the yearly flow of the Colorado River. Like all aquifers, the Ogallala depends on rain water for recharging, and only a trickle of the annual local rainfall ever reaches it. Gradually built up over millions of years, the aquifer is being drained in a fraction of that time. The question is no longer if the Ogallala will run dry, but when.

W. B. Criswell has been raising cotton on his 1,700-acre farm in Idalou, Texas, since 1955. Cotton is called the camel of crops because it requires little water, yet Criswell is now in trouble. His water table has dropped 100 ft, since he started farming. Nine years ago, he paid $4 an acre to water his cotton; today

sewage, fertilizers, pesticides, and solid wastes are all added to natural ecosystems in the hope that these systems will cleanse the biosphere of these pollutants. But we have replaced natural ecosystems with our man-made ecosystem and have exploited natural ecosystems for resources, adding ever more pollutants, to the extent that the remaining natural ecosystems have become overloaded.

Natural ecosystems have been destroyed and overtaxed because the human ecosystem is noncyclical and because an ever-increasing number of people wish to maintain a standard of living that requires many goods and

he pays more than $45. "It's like a disease," he says. "You just accept it and go on." Gerald Wiechman farms 6,000 acres and feeds 2,500 head of steer near Scott City in western Kansas. When his farm's first well started pumping, it tapped water at 54 ft. Today he has to go 130 ft. "I've got another 20 years, maybe," he reckons. On the High Plains of eastern Colorado, the water level has dropped as much as 40 ft. since the 1960s. In parts of Oklahoma, it has dipped that much in four years. Texas, the thirstiest of the eight states, has consumed 23% of its Ogallala reserves since World War II.

According to a major study just completed by Camp Dresser & McKee, a Boston engineering firm, 5.1 million acres of irrigated land (an area the size of Massachusetts) in six Great Plains states will dry up by the year 2020. If current trends continue, Kansas will lose 1.6 million irrigated acres. Texas 1.2 million, Colorado 260,000, New Mexico 224,000. Oklahoma 330,000. Yet this drastic estimate, declares Herbert Grubb of the Texas department of water resources, is "20% too optimistic."

"When the water goes," says W. E. Medlock, a stoic, third-generation farmer from Lubbock, Texas, who has lost 47 of his 73 wells in ten years, "we'll just go back to dry-land farming." To the farmers of the Great Plains, those words summon up visions of *The Grapes of Wrath*. Dry-land farming means larger farms with lower yields, fewer workers and probably higher prices in the supermarkets. Cattlemen know that less water means less corn and therefore

smaller herds. Grubb calls such farming the "Russian roulette" of agriculture. Over a ten-year period, he says, dry-land farming will yield two strong harvests, four average ones and four "busts."

Although the cause of the trouble is obvious, the cure is not. Indeed, there may be no fundamental solution to the ebbing of the Ogallala. "We can prolong the supply," concludes John Weeks, a U.S. Geological Survey engineer who heads a five-year U.S.G.S. study of the situation, "but we are mining a limited resource, like gold, and we can't solve the ultimate depletion problem."

Conservation may forestall the end. Farmers can simply use less water. They are already converting from profitable but water-thirsty corn to water-thrifty crops such as wheat, sorghum and cotton. James Mitchell, a cotton farmer from Wolfforth, Texas, has installed an experimental center-pivot sprinkler that, instead of spraying outward, gently drops water directly into the planted furrows, thereby reducing evaporation. Sophisticated laser-guided land graders can now almost perfectly flatten the terrain so that water is not wasted in runoff. Electrodes planted in the fields can measure soil wetness and determine exactly when water is needed. Today, these techniques are rarities, but they may soon be routine. As Kansas Cattle Feeder Harold Burnett puts it: "Water misers" will last longer. But even the stingiest will go under if neighbors are wasteful and the whole aquifer dries up.

Figure 31.15
Modified human ecosystem. In order to cut down
on the amount of lost heat and waste matter, heat
could be used more efficiently and discarded
materials could be recycled.

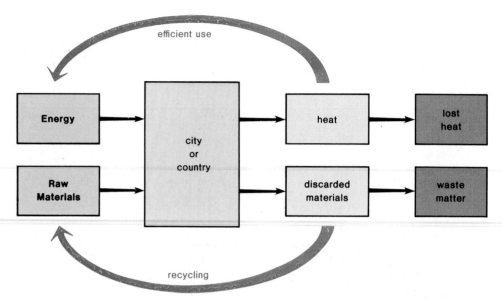

Figure 31.16
Efficient use of resources. Instead of allowing cattle
waste to enter a water supply, it could be sent to a
conversion plant that would produce methane gas.
(The residue remaining could be converted back
into feed for cattle.) Excess heat, which arises from
the burning of the methane gas in order to produce
electricity, can be cycled back to the conversion
plant.

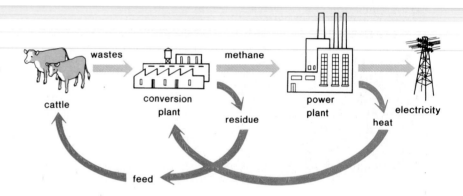

services. But we can call a halt to this spiraling process if we achieve zero
population growth and if we conserve energy and raw materials. Conservation
can be achieved in three ways: (1) use wisely only what is actually needed;
(2) recycle nonfuel minerals such as iron, copper, lead, and aluminum; and
(3) use renewable energy resources (p. 786) and find more efficient ways to
utilize all forms of energy. Figure 31.15 presents a diagrammatic representation of what is needed to maintain the delicate balances of the man-made
and natural ecosystems. As a practical example, consider a plant that was
built in Lamar, Colorado, which produces methane from feedlot animals' wastes
(fig. 31.16). The methane is burned in the city's electrical plant and the heat
given off is used to incubate the anaerobic digestion process that produces the
methane. In addition, a protein feed supplement is produced from the residue
of the digestion process. This system represents a cyclical use of material and

Behavior and Ecology

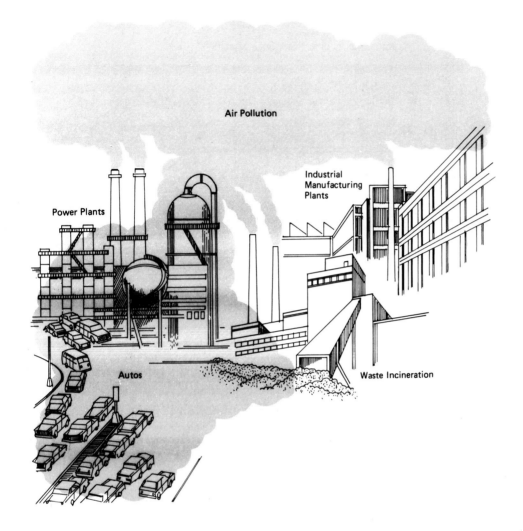

Air Pollution

Power Plants

Industrial
Manufacturing
Plants

Autos

Waste Incineration

efficient use of energy similar to that found in nature. Many other such processes for achieving this end have been and will be devised. However, as long as the human ecosystem on the whole remains inefficient and noncyclical, it will continue to cause pollution.

Pollution

Pollution affects all portions of the biosphere: air, land, and water.

Air Pollution

Pollutants enter the atmosphere from various sources (fig. 31.17), but the burning of fossil fuels contributes greatest to the five categories of primary pollutants: carbon monoxide (CO), hydrocarbons (HC), nitrogen oxides (NO,

Figure 31.18

Components of air pollution. CO (carbon monoxide), HC (hydrocarbons), NO$_x$ (nitrogen oxides), particulates (solid matter), SO$_x$ (sulfur oxides). Transportation contributes most to air pollution because when gasoline burns it gives off CO, a gas that interferes with the capacity of hemoglobin to carry oxygen.

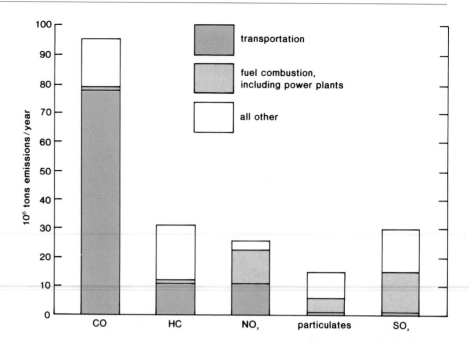

NO$_2$), particulates, and sulfur oxides (So$_2$, SO$_3$). The graph in figure 31.18 compares the sources of these pollutants. It is obvious that modes of transportation, especially the automobile, are the main cause of carbon monoxide air pollution. This chemical combines preferentially with hemoglobin to prevent circulation of oxygen within the body, causing unconsciousness. In New York City traffic, the blood concentration of carbon monoxide has been shown to reach 5.8 percent, a dangerous level when compared to the 1.5 percent that physicians consider safe. The particulates, dust and soot, can collect in the lungs, and nitrogen oxides and sulfur oxides irritate the respiratory tract. The hydrocarbons (various organic compounds) may be carcinogenic.

Unfortunately, these primary pollutants interact with one another, producing pollutants that are even more dangerous.

Photochemical Smog

Photochemical smog results when two pollutants from automobile exhaust—nitrogen oxide and hydrocarbons—react with one another in the presence of sunlight to produce nitrogen dioxide (NO$_2$), ozone (O$_3$), and PAN (peroxylacetyl nitrate). **Ozone** and **PAN** are commonly referred to as oxidants. Breathing ozone affects the respiratory and nervous systems, resulting in respiratory distress, headache, and exhaustion. These symptoms are particularly apt to appear in youngsters; therefore Los Angeles schoolchildren must remain at rest inside the school building whenever the ozone level reaches 0.35 ppm (parts per million by weight). PAN is especially damaging to plants, resulting in leaf mottling and reduced growth.

Normally, warm air near the ground rises, allowing pollutants to be dispersed and carried away by air currents. Sometimes, however, air pollutants, including smog and particulates, are trapped near the earth due to **thermal inversions.** During a temperature inversion, the cold air is at ground level and the warm air is above. (This may occur when a cold front brings in cold air and settles beneath a warm layer). Cities surrounded by hills, such as Los Angeles and Mexico City, are particularly susceptible to the effects of a temperature inversion.

Behavior and Ecology

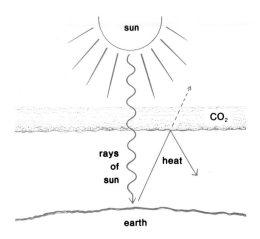

Figure 31.19
Greenhouse effect. The sun's rays can penetrate CO_2 (*wavy line*) and reach the earth's surface. But the resulting heat cannot pass (*dotted arrow*) through a layer of CO_2 and instead is trapped near the earth's surface (*solid arrow*), just as heat is trapped by the glass of a greenhouse. The earth's temperature is forecast to substantially increase due to the accumulation of CO_2 in the atmosphere.

Acid Rain

A side effect of air pollution is now the common occurrence of acid rain. As explained in the reading on page 730, when SO_2 and NO_2 are injected high into the atmosphere by high smokestacks they tend to react with water to produce sulfuric and nitric acid. These acids are eventually deposited on the earth's surface. Here they corrode marble, metal, and stonework and leach metals from the soil. It is the latter action that leads to the destruction of forests and causes lakes to become devoid of most forms of life.

Coal burning supplies most of the sulfur oxides that lead to acid rain. There are several ways to minimize the problem. Coal could be washed prior to burning. Low-sulfur coal could be substituted for high-sulfur coal. Devices called scrubbers could be installed in the tall stacks to prevent sulfur oxide from entering the air. Under development is a new method of burning coal that uses a mixture of coal and limestone. This technique has reduced both sulfur and nitrogen oxide emissions.

The problem of NO_2 emissions is also solvable. Continued use of the catalytic converter in automobiles is important because it reduces the amount of NO_2 given off by about three-fourths. A low NO_2 burner is also being developed for industrial boilers.

The Weather

It is predicted that the earth's average temperature could rise as much as 8°F over the next 100 to 200 years because of CO_2 buildup due to fossil fuel combustion. CO_2 allows the sun's rays to pass through, but absorbs and reradiates heat back toward earth. This may be compared to a greenhouse in that the glass of a greenhouse also allows sunlight to pass through, but traps the heat. This phenomenon also occurs when a car sits in the hot summer sun.

This rise in temperature from the **greenhouse effect** (fig. 31.19) could have a serious impact on agriculture and eventually on sea levels because of melting polar ice. Wind pattern changes would make the midwestern United States, the Soviet Union, and China drier. Favorable climates for the growth of crops would move north, where the soil is not so favorable for agriculture. Even though it is not certain that polar ice would melt, the sea level would rise simply because water expands when it absorbs heat. If the arctic ice should melt substantially, it is predicted that most of the world's cities would be flooded and so would some of our richest farmlands.

The concensus among researchers is that we should begin to prepare for these consequences since it is very unlikely that we will ever curtail fossil fuel combustion enough to prevent their occurrence.

Indoor Air Pollution

While much attention has been given to the quality of outdoor air, it has been pointed out that indoor air is often more polluted. NO_2, traced to gas combustion in stoves, has been found indoors at twice the outdoor level; CO, especially in homes with gas, coal, or wood-heating or cooking stoves, often exceeds the current health standards; hydrocarbons from myriad sources appear in high concentrations; and radioactive radon gas, emitted naturally from a variety of building materials, has been detected indoors at levels that exceed outdoor levels by factors of 2 to 20. A surprising number of household items

Storm over a Deadly Downpour

Monstrous 1,250 ft. smokestack rises above clouds in Sudbury, Ont.

A uniquely modern, postindustrial blight, acid rain is as widespread as the winds that disperse it. In the northeast U.S. and in Canada and northern Europe, it is reducing lakes, rivers and ponds to eerily crystalline, lifeless bodies of water, killing off everything from indigenous fish stocks to microscopic vegetation. It is suspected of spiriting away mineral nutrients from the soil on which forests thrive. Its corrosive assault on buildings and water systems costs millions of dollars annually. It may also pose a substantial threat to human health, principally by contaminating public drinking water. Says Canada's Minister of the Environment John Roberts: "Acid rain is one of the most devastating forms of pollution imaginable, an insidious malaria of the biosphere." . . .

Such natural processes as volcanic eruptions, forest fires and the bacterial decomposition of organic matter produce some of the damaging acidic sulfur and nitrogen compounds. But most experts believe that the current problem is directly traceable to the burning of fossil fuels by power plants, factories and smelting operations and, to a lesser extent, auto emissions. When tall smokestacks vent their fumes, sulfur dioxide, nitrogen oxides and traces of such toxic metals as mercury and cadmium mix with water vapor in the atmosphere. Chemical reactions follow that form dilute solutions of nitric and sulfuric acids—acid rain.

The acidic "depositions," as scientists call them, can come down in almost any form, including hail, snow, fog or even dry particles. On the standard chemical scale for measuring acidity or

Behavior and Ecology

continuously give off formaldehyde, including insulation, particleboard, and plywood; asbestos fibers are often dislodged from building materials, especially due to vibration, abrasion, cutting, grinding, sanding, or aging.

Because of these findings, some have questioned the advisability of tightening up residential buildings to save energy. Reduced ventilation unquestionably increases concentrations of pollutants. Others are calling for research studies so that the proper regulatory legislation may be prepared. For example, it would be possible to modify building codes so that safer building materials are used.

alkalinity, they are defined as having a pH level under 5.6 (a neutral solution has pH 7). Despite the use of tracking balloons and other sophisticated techniques, it is difficult to link acidity with a specific smokestack. But there is little doubt about the damaging effects of acid rain. Absorbed into the soil, it breaks down minerals containing calcium, potassium and aluminum, robbing plants of nutrients. Eventually the acid enters nearby bodies of water, often with a deadly burden of toxic metals that can stunt or kill aquatic life.

As successive rainfalls make the water increasingly acidic, lakes and rivers turn oddly clear and bluish. Their surviving microorganisms are trapped beneath layers of moss on the bottom; the afflicted water cannot support any but the most primitive forms of life. Some areas, rich in alkaline limestone, are able to resist the assault by "buffering" or neutralizing acid precipitation. But much of New York, New England, eastern Canada and Scandinavia is covered with thin, rocky topsoil left by glaciers long ago, and is particularly vulnerable to acid rain.

In New York's Adirondack Mountains, 212 of the 2,200 lakes and ponds are acidic, dead and fishless. Acid rain has killed aquatic life in at least 10% of New England's 226 largest fresh-water lakes. On Cape Cod in Massachusetts, fishery biologists have stopped restocking eight of the area's top ten fishing ponds because the waters are too acidic for young trout to survive in; the onslaught has spurred experimentation with new breeds of acid-resistant fish.

In the mid-'70s, the Quabbin Reservoir in central Massachusetts

became so acidic that it dissolved water conduits and fixtures, producing unhealthy levels of lead in drinking water. Cost since that time for neutralizing chemicals: $1 million annually. In Maine, where the measured acidity of rainfall has increased 40 times in the past 80 years, high levels of toxic mercury, lead and aluminum in acidified streams have killed or deformed salmon embryos. The problem is spreading to other parts of the country. Damage from acid rain has been reported in Minnesota, Wisconsin, Florida and California.

But Canada suffers most severely. Environmental officials project the loss of 48,000 lakes by the end of the century if nothing is done to curb acid rain. Already, 2,000 to 4,000 lakes in Ontario have become so acidified that they can no longer support trout and bass, and some 1,300 more in Quebec are on the brink of destruction. In Nova Scotia, nine rivers used as spawning grounds by Atlantic salmon in the spring no longer teem with fish.

To some extent, environmentalists can blame themselves. In an attempt to reduce industrial smog in 1970, the fledgling Environmental Protection Agency hastily ordered industrial plants to increase the height of their smokestacks. As a result, winds carried pollutants farther afield. The pollutants are now injected so high into the atmosphere that they remain there long enough for the critical chemical reactions to take place. Ironically, the tallest of these smokestacks is in Canada, a 1,250-ft. monster that belches fumes from a nickel smelter in Sudbury, Ont. The area is so bleak and lifeless that U.S. astronauts

practiced moon walking there in the late 1960s. . . .

Most utilities are vociferously opposed to any emission-control program without further research into the causes of acid rain. The industry argues that 1) scientific data on acid rain are still fuzzy, especially in the crucial matter of precisely who is responsible; 2) costs of eliminating sulfur-dioxide emissions by installing expensive "scrubbers" (which collect harmful subtances before they are expelled) are prohibitive; and 3) it is questionable whether the situation is critical enough to justify immediate action. Says Joseph Dowd, general counsel for American Electric Power, which serves 2.5 million customers in the Midwest: "Installing scrubbers could break the economic backbone of the Midwest. And there's no assurance it will improve the acidity of rainfall in the East.". . .

With an increasing number of areas classified by scientists as "sensitive" to acid stress in both North America and Europe, some environmental experts fear that an even more alarming and often irreversible deterioration may take place before corrective measures are taken. The concern of environmentalists is that industrialists will continue to use delaying tactics to put off costly capital improvements necessary to reduce emissions. Says Richard Ayres, chairman of the National Clean Air Coalition, an amalgam of environmentalist groups: "The costs [for cleaning up emissions] aren't trivial. But neither is the damage. A nation that can afford to spend $5 billion a year on video games can afford the same amount to save its lakes and forests."

Air Quality Control

Air quality in the United States is controlled by the *Clean Air Act*. The Environmental Protection Agency (EPA) has established standards for the pollutants listed in figure 32.18. In addition, asbestos, beryllium, mercury, vinyl chloride, and lead are more strictly controlled because they are extremely hazardous to health. Industry is required to use special equipment, such as collectors and scrubbers, to cut down on stack emissions. Automobiles have been equipped with special devices, such as the **catalytic converter** that chemically changes HC and CO to carbon dioxide and water. The catalytic converter, which necessitates the use of nonleaded gasoline, also contributes to fuel economy. The result has been generally improved air quality throughout most of the nation.

Land Pollution

Every year, the United States population discards billions of tons of solid wastes, much of it on land. **Solid wastes** include not only household trash but also sewage sludge, agricultural residues, mining refuse, and industrial wastes. Those solid wastes, containing substances that cause human illness and even sometimes death, are called **hazardous wastes.**

Hazardous Wastes

Hazardous wastes fall into three general categories:

1. **Heavy metals,** such as lead, mercury, cadmium, nickel, and beryllium, can accumulate in various organs, interfering with normal enzymatic actions and causing illness, including cancer.
2. **Chlorinated hydrocarbons,** also called **organochlorides,** include pesticides and numerous organic compounds, such as PCBs (polychlorinated biphenyls), in which chlorine atoms have replaced hydrogen atoms. Research has often found organochlorides to be cancer producing in laboratory animals.
3. **Nuclear wastes** include radioactive elements that will be dangerous for thousands of years. For example, plutonium must be isolated from the biosphere for 200,000 years before it has lost its radioactivity.

Hazardous wastes are often subject to **bioaccumulation** (fig. 31.20). Bacteria in the soil do not break down these wastes, and when other organisms take them up, they remain in the body and are not excreted. Once they enter a food chain, they become more concentrated at each trophic level. Notice in figure 31.20 that the number of dots representing DDT becomes more concentrated as the chemical is passed along from producer to tertiary consumer. Bioaccumulation is most apt to occur in aquatic food chains; there are more trophic levels in aquatic food chains than there are in terrestrial food chains. Humans are the final consumers in both types of food chains, and in some areas, mothers' milk contains a detectable amount of DDT and PCB.

The public has become aware of hazardous dump sites that have polluted nearby water supplies. Chemical wastes buried over a quarter century ago in Love Canal, near Niagara Falls, have seriously damaged the health of some residents there. In other places, manufacturers have left thousands of drums in abandoned or uncontrolled sites where toxic chemicals are oozing out into the ground.

Legislation   In an effort to control the level of toxic wastes, Congress passed the *Resource Conservation and Recovery Act,* which empowers the EPA to track all significant quantities of hazardous waste from wherever it is generated to its final disposal. Government regulations are designed to encourage

Behavior and Ecology

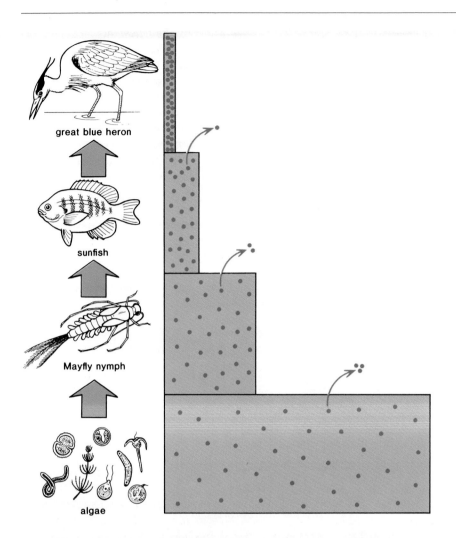

Figure 31.20
Bioaccumulation. A poison (*dots*) that is minimally excreted (*arrows*) becomes maximally concentrated as it passes along a food chain due to the reduced size of the trophic levels. Because of this problem, fishermen are warned against consuming fish from the Great Lakes.

industry to adopt new waste management strategies, including *reduction* (changing a manufacturing process so it does not produce hazardous by-products); *recycling* (reusing waste material); and *resource recovery* (extracting valuable material from waste).

Water Pollution

Surface Waters

Figure 31.21 shows the many ways that humans cause surface water pollution. Some pollutants, such as fertilizers, sewage, and certain detergents, add nitrates and phosphates to freshwater lakes and ponds. This overabundance of nutrients, called *cultural eutrophication,* speeds up the tendency of bodies of water to fill in and disappear. First, the nutrients cause overgrowth of algae. The death of these algae promotes the growth of a very large decomposer population. The decomposers break down the algae but in so doing they use up oxygen. In addition, algae also consume oxygen during the night when photosynthesis is impossible. Both of these cause a decreased amount of oxygen available to fish, ultimately causing the fish to die. The increased amount of life, and dead remains, causes the lake to be more eutrophic leading to a reduction in its size.

Figure 31.21
Water pollution is caused in all the ways shown here. There is much concern of late about Chesapeake Bay because it is dying due to sediments and nutrients that enter the bay, particularly from the Susquehanna River that begins in New York and flows through Pennsylvania.

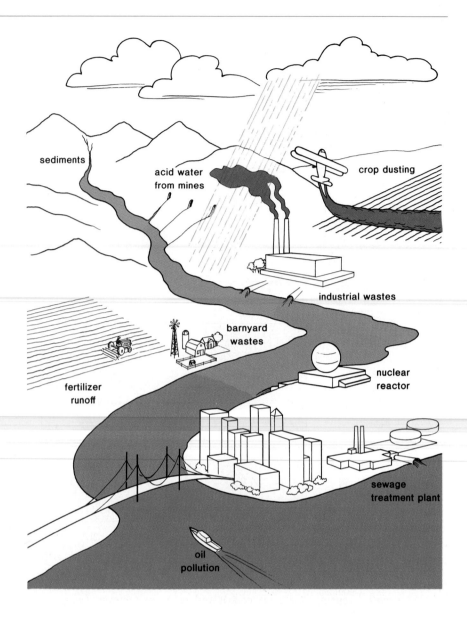

Water Quality Control The *Water Pollution Control Act* empowers the federal government to set minimum water quality standards for rivers and streams. The building of sewage treatment plants has helped clean up U.S. waters. In addition, large industries are no longer permitted to dump chemicals indiscriminately into water or to send their wastes to local sewage treatment plants. This two-pronged attack has helped clean up the Great Lakes, which previously were said to be dying because of pollution. Even so, the presence of toxic waste is still a problem, especially since public sewage treatment plants are not designed to rid water of nonbiodegradable wastes.

Treatment plants capable of not only digesting sewage but also of removing nutrient molecules are very expensive to build. Some communities have devised ingenious ways to accomplish this end, however. Effluent from a sewage treatment plant can be passed through a swampy area or forest before it drains into a nearby waterway. Or the effluent can be used to irrigate crops and/or grow algae and aquatic plants in a man-made shallow pond. Since these products can be used as food for animals, this represents a cyclical use of chemicals.

Figure 31.22
Groundwater pollution is caused in all the ways
shown here. Discontinuance of these means of
disposal for industrial wastes has been difficult to
achieve because citizens do not wish to have
waste disposal plants located near them.

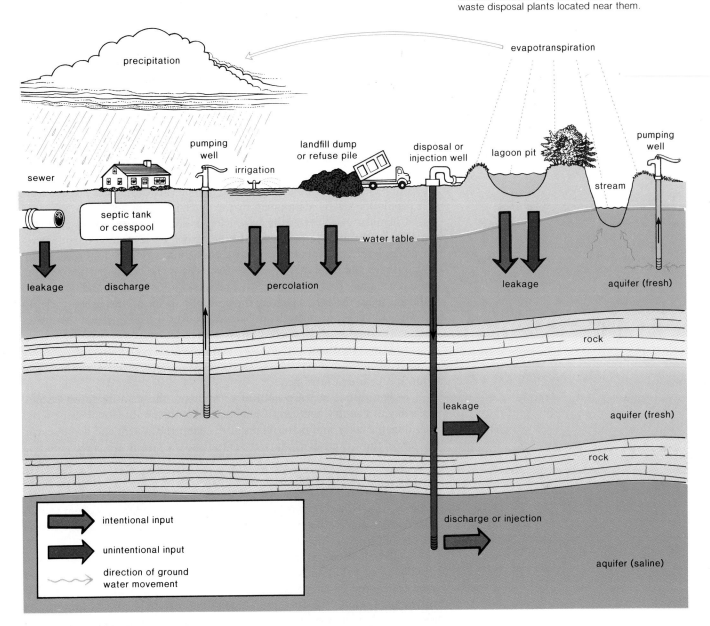

Groundwater

Groundwater is subject to pollution in the ways illustrated in figure 31.22.
Chemicals drain from hazardous waste dumping sites, and bacteria and viruses drain from septic tanks and cesspools into the ground and may eventually reach aquifers (underground rivers). Also, previously industry was
accustomed to running wastewater into a pit. The pollutants could then seep
into the ground. Or pollutants were injected into deep wells from which the
pollutants constantly discharged. Both of these customs have been or are in
the process of being phased out. However, it is very difficult for industry to
find other ways to dispose of wastes. More adequately managed and controlled
waste treatment plants are needed, but because citizens do not wish to live
near waste treatment plants, towns are often successful in preventing their
construction. In the meantime, industries are still employing less approved
methods of dealing with industrial wastes.

Summary

Ecosystems are portions of the biosphere that contain a biotic and an abiotic component. The biotic component consists of populations that interact with each other and with the abiotic component, or physical environment. The populations in an ecosystem form food chains in which the producers produce food for the other populations by being able to capture the energy of the sun, the ultimate source of energy for our universe. Chemicals cycle through the food chain and, in time, the same inorganic molecules are returned to the producer population for conversion to organic food. Energy flows through the ecosystem but does not cycle since there is a loss of useful energy at each trophic level. The food pyramid illustrates that each successive population has a smaller size or biomass because energy does not cycle.

While it is convenient to study food chains, the populations in an ecosystem actually form a food web in which food chains join and overlap with one another. Mature natural ecosystems are stable because population sizes are held in check by density independent and density dependent factors. Further population size is maintained by the diversity of the food web.

Chemical cycles typically have a reservoir and an exchange pool that supplies the chemical to producers. In the carbon cycle, carbon dioxide is removed from the atmosphere by photosynthesis but is returned by respiration. Living things and dead matter are reservoirs as is the ocean, particularly because it accumulates limestone. When humans cut down trees and burn wood and fossil fuels, they are changing the current balance of the carbon cycle. In the nitrogen cycle, nitrogen fixing bacteria convert aerial nitrogen to nitrates and denitrifying bacteria convert nitrates back to aerial nitrogen. Nitrifying bacteria convert ammonia to nitrate, the form of nitrogen most often used by plants. Humans again tap into a reservoir when they convert aerial nitrogen to nitrate for use in fertilizer.

In contrast to mature natural ecosystems, the man-made ecosystem, consisting of country and city, is not stable. Nonrenewable fossil fuel energy is used inefficiently, and material resources enter the system and do not cycle. Because of these excessive inputs, the outputs to the system are much pollution and waste. While in the past we could rely on natural ecosystems to process our wastes, such as sewage, this is no longer feasible because the size of the natural systems has been reduced and the amount of pollution has been steadily increasing. However, it is possible to change the human ecosystem so that it more nearly corresponds to a natural system by using renewable energy sources and by recycling material resources.

Pollution affects all portions of the biosphere. The greatest single contribution to air pollution is caused by the burning of fossil fuels. Acid rain and smog are caused by air pollutants. It is forecast that carbon dioxide buildup will gradually cause the weather to become warmer. While much concern has been focused on outdoor pollution, it appears that indoor pollution may need our attention even more.

Hazardous wastes which are byproducts of various industries have been deposited on the land but have seeped into surface and groundwater. Unfortunately, due to bioaccumulation, these substances return to humans in a concentrated form. While progress has been made in reducing cultural eutrophication, much has to be done in keeping our water supplies free of hazardous wastes.

Study Questions

1. Give an example of an aquatic and a terrestrial food chain. Name the producer, the consumers, and the decomposers in each chain.
   Explain the manner in which chemicals cycle. (pp. 713–15)

2. A pyramid describes the size of the various trophic levels within a food web. Discuss the relationship of size and the fact that energy does not cycle. (p. 717)
3. Describe the carbon cycle. How do humans contribute to this cycle? (p. 718)
4. Describe the nitrogen cycle. How do humans contribute to this cycle? (p. 719)
5. Draw a diagram to represent the man-made ecosystem and discuss its inputs and outputs. (p. 721) Compare the man-made ecosystem to a natural ecosystem. (p. 720)
6. What are the primary components of air pollution? (p. 728) How do they contribute to smog and acid rain? (pp. 728–29)
7. How is air pollution related to the weather? (p. 729)
8. What type chemicals are termed hazardous wastes? Why? (p. 732)
9. What is cultural eutrophication and how might it be prevented? (pp. 733–34)
10. Why do nonbiodegradable poisons concentrate as they go from trophic level to trophic level? (p. 733)

Selected Key Terms

biosphere (bi-o-sfēr)

ecosystem (ek″o-sis′tem)

biotic (bi-ot′ik)

abiotic (ab″e-ot′ik)

habitat (hab′ĭ-tat)

niche (nich)

food chain (fōod chān)

producers (pro-du′serz)

consumers (kon-su′merz)

herbivores (her′bĭ-vorz)

carnivores (kar′nĭ-vorz)

omnivores (om′nĭ-vorz)

decomposers (de-kom-po′zerz)

food web (fōod web)

food pyramid (fōod pir′ah-mid)

nitrifying bacteria (ni′tri-fi″ing bak-te′re-ah)

denitrifying bacteria (de-ni′tri-fi-ing bak-te′re-ah)

cultural eutrophication (kul′tu-ral u″tro-fi-ka′shun)

Further Readings

Batie, S. S., and Healy, R. G. 1983. The future of American agriculture. *Scientific American* 248(2):45.

Bell, R. H. V. 1971. A grazing ecosystem in the Serengeti. *Scientific American* 226(1):86–93.

Brill, W. J. 1977. Biological nitrogen fixation. *Scientific American* 236(3):68–81.

Collier, B. D., et al. 1973. *Dynamic ecology.* Englewood Cliffs, N.J.: Prentice-Hall.

Gosz, J. R., et al. 1978. The flow of energy in a forest ecosystem. *Scientific American* 238(3):93–102.

Moran, J. M., et al. 1980. *Introduction to environmental science.* San Francisco: W. H. Freeman.

Pond, W. G. 1983. Modern pork production. *Scientific American* 248(5):96.

Rasmussen, W. D. 1982. The mechanization of agriculture. *Scientific American* 247(3):76.

Revelle, R. 1982. Carbon dioxide and world climate. *Scientific American* 247(2):35.

Ricklefs, R. E. 1983. *The economy of nature.* 2d ed. Portland, Ore.: Chiron Press.

Scientific American. 1970. 225(3). Entire issue devoted to the biosphere.

Woodwell, G. M. 1978. The carbon dioxide question. *Scientific American* 238(1):34–43.

32

The Biosphere

Chapter concepts

1 All life forms exist in major communities called biomes, which are adapted to climate.

2 Humans have exploited and altered both terrestrial and aquatic biomes to the point that representative examples should be preserved as wilderness areas.

3 Complex biomes, such as forests, come into existence by the process of succession, a series of stages leading to a climax or mature stage.

4 While productivity is highest in the early stages of succession, the climax stage is the most stable. Human need for productivity should not cause the abandonment of stability because only a stable biosphere ensures our continued existence.

5 In order to maintain stability of the biosphere, alteration and pollution of the biomes should be curtailed.

Terrestrial Biomes

Each major type of terrestrial ecosystem contains a characteristic community of populations called a **biome.** The location of the terrestrial biomes to be studied in this text are indicated in figure 32.1. The number of biomes could easily be increased by subdividing these. We should also recognize that the various biomes are somewhat artificially imposed upon the biosphere. Actually, one type of biome gradually becomes another type, and it is not surprising to find within any particular biome a region that does not fit the general description.

Physical conditions, particularly *climate,* determine the biome of an area. Figure 32.2 shows that the various biomes can be related to temperature and rainfall. Deserts are biomes with the least amount of rainfall.

Deserts

Deserts (fig. 32.3) are regions of aridity with less than 20 centimeters of rainfall a year. True deserts, with less than 2 centimeters annually, are infrequent, the Sahara in Africa being the largest. **Semideserts,** however, include about one-third of all land areas. In deserts, the days are hot because lack of cloud cover allows the sun's rays to penetrate easily, but the nights are often cold because heat escapes into the atmosphere.

Perennial flowers, including the succulent cacti, non-succulent shrubs, such as sagebrush, and stunted trees are common North American desert vegetation. Reptiles, exemplified by lizards and turtles; rodents, such as the kangaroo and pack rat; and birds, like woodpeckers, hawks, prairie falcons, and the roadrunner, are common small desert animals. The camel is a large herbivorous animal of African deserts, with a reservoir of fat in its hump that allows it to drink and eat infrequently. Large carnivorous animals of the United States desert are the badger, kit fox, and bobcat.

Plants and animals adapted to the desert have structural and/or behavioral adaptations that allow them to prevent evaporation and withstand heat. Plants, and even some animals, reproduce only when adequate water is available. Like the camel, animals often make use of metabolic water. Both desert plants and animals have protective coverings, but animals also hide beneath rocks or burrow in the earth or venture forth only at night to escape the heat of the sun.

Desertification

Thirty-six percent of the biosphere is expected to be a desert on the basis of annual rainfall. Yet a world survey indicates that 43 percent of the land is actually desertlike. The difference of 7 percent probably represents the extent of desertification caused by human misuse of the land. The true deserts are expanding into the area of semideserts, and the semideserts are expanding into the grasslands. Altogether, a collective area the size of Brazil has undergone reduced productivity. In heavily populated countries, such as Africa and India, true deserts are increasing in size because people living in semiarid areas cut down trees for firewood and allow cattle to graze on the shrubs until the entire area lacks ground cover. In the past 55 years, at least 251,000 square miles of farmland and grazing land have been swallowed up by the Sahara along its southern fringe.

In the midwestern and southwestern regions of this country, overgrazing of grasslands has caused them to become semideserts. Also, water has been diverted from the countryside to the cities so that irrigation of once fertile farmland is no longer possible. This land has become as a desert.

Figure 32.1
Distribution of the major biomes of the world. Each
particular biome has its own biotic makeup, but
humans superimpose on each the same man-made
ecosystem. Only remnants of natural, undisturbed
biomes remain.

- Tropical Forest
- Coniferous Forest
- Deciduous Forest
- Tundra
- Grassland
- Savanna
- Woodland and Chaparral
- Desert

Behavior and Ecology

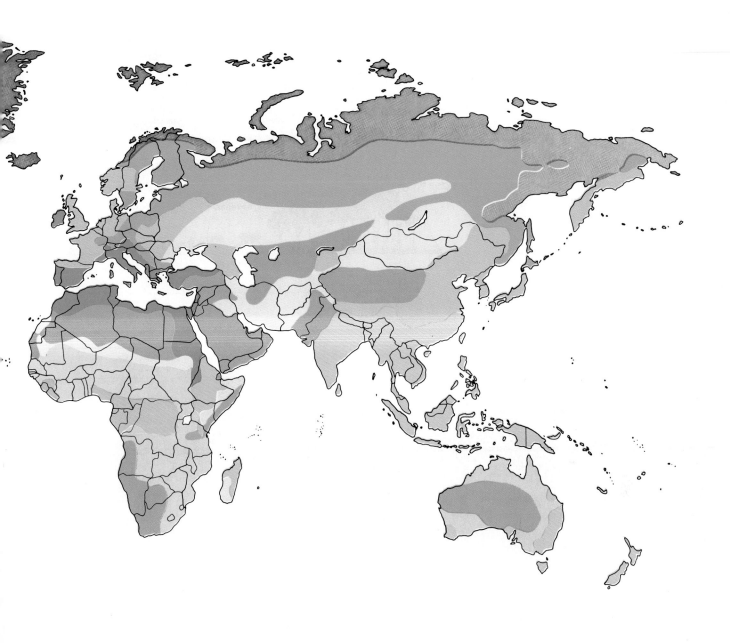

Figure 32.2

Influence of temperature and rainfall on type of biome. Temperature and rainfall determine the biome to a large extent. Deserts occur in regions with minimal rainfall but as the diagram indicates, there are hot deserts and cold deserts. Grasslands occur in regions where there is insufficient water to support trees. The tundra is classified as a grassland.

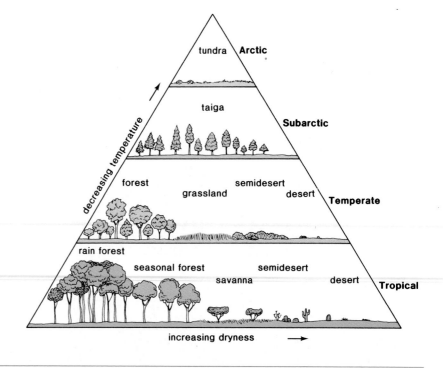

Figure 32.3

Semidesert biome. *a.* Overall view shows a diversity of plants. *b.* Antelope jackrabbit. Note the large ears, a means to rid the body of excess heat. *c.* Beautiful flower of the hedgehog cactus, a plant that stores water in its stem.

a.

b.

c.

Behavior and Ecology

Figure 32.4
Tundra biome. *a*. Overall view illustrates the lack of trees. *b*. Caribou graze on the low-lying vegetation. *c*. Grasses and forbs dominate the biome. This is beargrass.

a.

b.

c.

Grasslands

Grasslands occur where rainfall is greater than 20 centimeters but is generally insufficient to support trees. The extensive root system of grasses allows them to recover quickly from drought, cold, fire, and grazing. Furthermore, these matted roots, which absorb surface water efficiently, prevent invasion by trees.

Grasslands, which occur on all continents (fig. 32.1), are known by various names. We will be discussing the tundra, an arctic grassland; the savanna, a tropical grassland; and the tallgrass and shortgrass prairies, temperate grasslands in this country.

Arctic Tundra

The northernmost biome (fig. 32.4) is dark most of the year but has 24–hour days in the summer. Since precipitation is only about 20 centimeters a year, it could possibly be considered a desert, but water frozen in the winter is plentiful in the summer because so little of it has evaporated. Only the topmost layer of the earth thaws and beneath this the **permafrost** is forever frozen. Trees are not found in the tundra because their roots cannot penetrate the permafrost and cannot become anchored in the constantly shifting soil.

While the ground is covered with shortgrasses and flowering forbs during the summer months, dwarf woody shrubs and frequent patches of lichens and mosses are always present. A few small animals—for example, the arctic

Figure 32.5
Alaskan pipeline. Miles and miles of pipe bring oil
to the contiguous states. There is fear that the
pipeline will contribute to the destruction of the
tundra and taiga biomes.

fox, the snowshoe hare, and the lemming, which resembles a mouse—live in
the tundra the year round. Many migratory birds arrive for the summer, the
arctic tern being the most famous of these. At one time, before its virtual
extermination by human hunters, the musk-ox was a plentiful year-round res-
ident. Now only the large caribou and reindeer migrate to the tundra in the
summer and the wolves follow to prey upon them. Polar bears are common
near the coast.

Plants and animals living in the tundra have adaptations that allow them
to survive the extreme cold. Low-lying plants are typical because it is warmer
near the ground. They have extremely short life cycles, being able to grow and
flower within the short summer. In the winter, smaller animals, like the lem-
mings, burrow, while the large musk-ox has a thick coat and a short, squat
body that conserves heat.

The tundra has been the least altered of all the biomes because of its
unfavorable climate and soil conditions. However, this biome contains reservoirs of minerals and fossil fuels that only now have become profitable and
therefore technologically possible to remove. There is concern that the ecology
of the tundra will be disturbed by such factors as the Alaskan pipeline (fig.
32.5) and that the area may now suffer damaging pollution as a result.

Savanna

The African savanna (fig. 32.6) is a tropical grassland that contains both trees
and grasses and therefore supports populations of browsers and grazers. The
temperature is warm and there is adequate yearly rainfall, but a severe dry
season limits the number of different types of plants. Perhaps the best known

a.

b.

c.

of the trees are the flat-topped acacia trees, which shed their leaves during a drought and remain small due to the limited water supply during the dry season.

The large, always warm African savanna supports the greatest number of different types of large herbivores of all the biomes. Elephants and giraffes are browsers. Antelopes, zebras, wildebeest, waterbuffalo, and rhinoceros are grazers. These are preyed upon by cheetas and lions, whose kill is at times scavenged by hyenas and vultures.

The savanna has been reduced in size due to human encroachment, but those parts that remain, although often misused, have largely retained their usual characteristics. "Conservation by utilization," which means that the native animals are domesticated for dairy and meat products, is favored by farsighted promoters, especially since imported European cattle are susceptible to tsetse fly infection, a constant threat in many parts of Africa.

Figure 32.7
a. Prairie biome. *b.* An abundance of black-eyed
Susans decorate the grassland landscape. *c.* A
sage grouse surveys its prairie habitat.

Prairies

As one travels from east to west across the United States, a tallgrass prairie
(fig. 32.7) gradually gives way to a shortgrass prairie. Although grasses dominate, they are interspersed by flowering forbs.

The lack of trees places a restriction on the variety of animal life. Insects
abound, especially grasshoppers, crickets, leafhoppers, and spiders. Songbirds
and prairie chickens sometimes feed off these, but usually they prefer seeds,
berries, and fruits. Small mammals, such as mice, prairie dogs, and rabbits,
typically burrow in the ground but usually feed aboveground. Hawks, snakes,
badgers, coyotes, and kit foxes capture and feed off these. The largest of the
herbivores, the buffalo and pronghorn antelope, had few enemies until humans
killed them off. Before then large herds of buffalo, in the hundreds of thousands, roamed the prairies and plains, never overgrazing the bountiful vegetation.

The grazers and their predators are specialized for locomotion in all the
grasslands. The same adaptations for leaping, burrowing, and running are seen
in North American animals as well as African animals (fig. 32.8). This is an

Behavior and Ecology

Figure 32.8
Convergent evolution. Similar adaptations in animals of different ancestry living under similar conditions exemplify convergent evolution. This diagram compares North American animals with African animals that live in the grassland biome. *a.* Jackrabbit compared to springhaas. *b.* Prairie dog and ground squirrel compared to African ground squirrel. *c.* Pocket gopher compared to golden mole. *d.* Bison and pronghorn compared to zebra and springbok. *e.* Coyote compared to lion and cheetah.

North America

Africa

a. leaping herbivorous mammals

b. burrowing mammals; feed above ground

c. burrowing mammals; feed underground

d. running herbivorous mammals

e. running carnivorous mammals

Figure 32.9
Taiga biome. This biome stretches round the globe in the northern temperate zone and contains evergreen trees that are adapted to harsh conditions.

Figure 32.10
Coastal redwood trees. The redwoods are the tallest coniferous trees ever to exist and often exceed a height of 300 feet. They are found in the fog belt of the West Coast, from southern Oregon to central California.

Table 32.1 Forest biomes

Biome	Plants
Coniferous forest	Cone-bearing evergreen trees, such as pine and spruce No understory
Temperate deciduous forest	Broad-leaved trees, such as oak and maple Understories
Tropical rainforest	Broad-leaved evergreen trees Multilevel canopy No understory
Tropical seasonal forest	Mixed broad-leaved evergreen and deciduous Rich understories produce jungle

example of convergent evolution—the animals of similar biomes resemble one another because they are suited to the same environment.

Only remnants of the original grasslands remain, and much is now used for farming or for rangeland and pasture. At times, drier areas have undergone desertification, as discussed on page 739.

Forests

The three types of forests discussed here are contrasted in table 32.1. Generally speaking, the evergreen coniferous trees are well adapted to the cold because both the leaves and bark have thick coverings. Also, the needlelike leaves can withstand the weight of heavy snow. The broad leaves of deciduous trees carry on a maximum amount of photosynthesis during the short growing season of the temperate zone. Loss of these leaves and dormancy during the winter protect the trees from the danger of cold weather and heavy snow. Trees living in the moist, warm environment of the tropics are both broad-leaved and evergreen because continuous growth is possible and there is no need for dormancy.

Coniferous Forest

Coniferous forests are found in three locations: in the **taiga,** which extends around the world in the northern part of North America and Eurasia; near mountain tops; and surprisingly enough, along the Pacific coast of North America, as far south as northern California.

The taiga (fig. 32.9) typifies the coniferous forest with its cone-bearing trees, such as pine, fir, and spruce. There is no understory of plants, but the floor is covered by low-lying fungi, mosses, and lichens beneath the layer of needles. Birds harvest the seeds of the conifers, and bears, deer, moose, beaver, and muskrat live around the ponds and along the streams. Wolves prey on these larger mammals. In the mountains, the taigalike forests also harbor the wolverine and mountain lion.

The coniferous forest that runs along the west coast of Canada and the United States contains some of the tallest conifer trees ever in existence, including the coastal redwoods (fig. 32.10). The constant humidity and relatively warm conditions are believed responsible for the unusual growth and development of these trees.

Temperate Deciduous Forests

Temperate forests (fig. 32.11) are found around the world just south of the taiga. They are also found in other areas, such as parts of Japan, Australia,

Behavior and Ecology

Figure 32.11
Deciduous forest. *a.*
in the spring. Notice
through the trees an
forest floor. *b.* Ladys
c. The trees provide
among other herbiv

a.

b.

c.

and South America, where there is a moderate climate and well-defined winter and summer seasons and relatively high precipitation (75 to 150 cm per year).

In North America the trees are bare in winter, but awake from dormancy and start to grow again in the spring. They continue growing in the summer, only to lose their leaves and become dormant again in the fall. In the summer the tops of the trees (oak, birch, beech, and maple) form a canopy open enough to allow sunlight to penetrate to the forest floor, thereby allowing several other layers of growth. Beneath the trees are shrubs, grasses, wild flowers, and finally, mosses and liverworts.

Animal life is abundant. Myriads of insects are food for insectivorous birds, such as the red-eyed vireo and woodpeckers. Mammals, such as squirrels, rabbits, deer mice, and white-tailed deer, make their home in the woods and are preyed upon by foxes and wolves.

Tropical Forests

The largest tropical rain forest (fig. 32.12) is found in the Amazon basin of South America, but such a forest also occurs at the equator in Africa and Australia, where it is always warm and rain is plentiful. The trees are broad-leaved evergreens that form a tall canopy composed of several layers, with characteristic plants and animals in each layer. Woody vines, called **lianas,** reach from the forest floor to the top of the canopy, and **epiphytes,** such as orchids, bromeliads, and ferns, cling to the trees but are not parasitic. Although epiphytes grow on the surface of other plants, they take their nutrients and water from the air. The dense layers of the tropical canopy typically do not allow light to reach the forest floor, and therefore there is no growth here and little litter because decomposition takes place so quickly.

While we usually think of tropical forests as being nonseasonal rain forests, there are tropical forests with wet and dry seasons in India, southeast Asia, West Africa, South and Central America, the West Indies, and northern Australia. Here there are deciduous trees and the canopy does allow light to pass through, helping produce layers of growth beneath the trees. In fact, the tangled mass of growth in tropical seasonal forests is known as jungle.

In both types of tropical forests most animal life is arboreal, living in the trees. There are insects, snakes, lizards, and frogs that spend their whole lives in the treetops and, of interest to us, here we find monkeys in both the Old and New World while apes are found only in the Old World.

Exploitation   The tropical rain forests are being cleared at an ever-increasing rate and no where is this more evident than in Central America and South America. First, logging companies enter the forest to extract valuable hardwoods such as mahogany and tropical cedar. Because these trees are tied to others by vines and lianas, many noncommercial trees are damaged. Then landless peasants arrive to cut and burn the vegetation so that enough inorganic nutrients are provided for a few years of farming. Without further fertilization, the soil soon becomes incapable of sustaining crops because tropical rain forests have nutrient-poor soil—all the nutrients are located in the luxurious living matter. Often the land is now converted to pasture by cattlemen who can acquire the capital to plant pasture grasses since it is known that there will be a market for the beef in the United States and Europe. After about seven to ten years, the effects of overgrazing and torrential rains turn the soil into a wasteland.

Many scientists are extremely concerned about exploitation of the tropical rain forests. They point out that many species will become extinct because there are more kinds of plants and animals here than in any other biome. In regard to trees, for example, researchers have determined that, while there

Behavior and Ecology

Figure 32.12
Tropical rain forest. *a.* This picture shows the forest along a stream. Notice that here the presence of light does allow layers of growth from the ground up. *b.* Bloodflowers provide vibrant color, while (*c.*) the chameleon blends subtly into the tropical background.

a.

b.

c.

are approximately 800 species in all of North America, there are about 2,500 known species of trees in the Malay Penninsula. The same type of data is available for animals:

> For example, a small region at 60° latitude might have as many as ten species of ants, at 40° there may be 50 to 100 species, and within 20° of the equator, 100 to 200 species. Similarly, Greenland has 56 species of breeding birds, New York, 105, Guatemala, 469, and Colombia 1,395.[1]

The reading on page 752 also discusses the great value of the tropical rain forests and tells us of some of the ways that various species are extremely helpful to humans. The great fear is that many species will become extinct before we even know of their existence.

[1]Ricklefs, R. E. 1973. *Ecology*. Portland, Ore.: Chiron Press, p. 701.

The Biosphere

751

The Rush Is on to Study Jungles

Manatees serve humans when they eat unwanted aquatic vegetation. It would behoove us to preserve species that are endangered by destroying the tropical forests.

. . . discoveries are routine in the tropics, where the extraordinary is commonplace and where so much is unknown that it is virtually impossible to do biological fieldwork without finding a new species of animal or plant or uncovering layers of incredibly complex interactions. The grooved hairs on a sloth's back, the pulp of half-eaten figs, even the droppings of fish and mammals contain intricate chapters of natural history. But, for scientists engaged in a variety of tropical research projects, there is not always cause for celebration. The world's rain forests are disappearing so quickly that biologists may run out of time to ferret out their priceless secrets. Now, while there are jungles left, the rush is on to study them, to understand how they work and how they might be saved. The stakes are staggering, the findings fascinating.

Tropical forests, from the monsoon jungles of Borneo to the cloud forests of Costa Rica, make up one of the basic life-support systems of the Earth, affecting not only worldwide rainfall but also the pattern of life's evolution. In a sense all modern living things are creatures of the tropics. Jungles are part of our biological heritage. For some 60 million years tropical forests have been a hothouse for the planet's gene pool, incubators where the great diversity of land plants and animals, including early human ancestors, evolved. Not surprisingly, Charles Darwin and Alfred Russel Wallace independently developed their theories of evolution and natural selection while exploring the tropics. Modern studies of the natural history of tropical forests are helping scientists to learn more about how evolution works.

Unfortunately, many of the discoveries made by researchers in the tropics have been overshadowed by the depressing statistics of deforestation. Fifty million acres of the world's tropical forests—an area about the size of Great Britain—are destroyed each year as forests are cleared for cattle pastures, timber and human settlements. Burgeoning human populations in tropical countries need homes, food and firewood. Tropical governments need the revenue from exports to industrialized nations. And the profits industrialized nations reap from tropical resources are many times greater than their investment in saving them. In the short run, there is more money in turning tropical forests into human settlements and grazing land than in preserving them.

But short-term gains from deforestation are often far outweighed by long-term consequences, and scientists concerned about lasting impacts are proving that the tropical forest is not the menacing, chaotic jungle depicted in movies and folklore. Rather, it is a tightly organized system of dazzling beauty and complexity. Discovery beckons at every turn—and so do potential, uncharted benefits for people . . . .

Tropical plants are getting the attention of biochemists seeking cures for disease and new

Behavior and Ecology

weapons to combat destructive agricultural pests. Already, tropical plants yield L-dopa (a drug that treats Parkinson's disease), curare (a muscle relaxant) and certain steroids used in birth control.

In recent tests, one of the most potent pesticides known has been derived from the East African medicinal plant, *Ajuga remota*. The plant contains hormones that mimic molting hormones within the bodies of insect larvae. . . .

Big trees may seem indestructible, but *Tachigalia* trees kill themselves. For unknown reasons, these huge 150–foot canopy species found in Central America and northern South America have one reproductive event—one flowering in their lives—and then they slowly die. Approximately every four years a group of mature *Tachigalia* flower, fruit and die. "Somehow they count years. They have an internal clock," says plant ecologist Robin Foster of the Smithsonian Tropical Research Institute and Chicago's Field Museum, who discovered this mysterious behavior. Foster also discovered that *Tachigalia versicolor*, a species once listed as rare, is actually one of the more common trees in Panama.

These findings suggest that *Tachigalia* trees—which are fast-growing and which produce good fungus-resistant wood—could be cultivated on plantations and harvested for timber without destroying the natural forest. "Basically it's a tree that naturally times its own harvest," says Foster. "You'd know exactly the year when you should cut them all down."

. . . Various groups are beginning to translate research into short-term technologies for harvesting tropical resources without destroying the remaining forests.

Leading the effort is the Smithsonian Tropical Research Institute in Panama, one of the oldest, most distinguished tropical research centers in the world. It has recently launched a series of projects it calls "Alternatives to Destruction" aimed at sustaining human populations in the tropics without wiping out the forests.

For example, the Institute is trying to cultivate air-breathing fishes. Many natural ponds in the jungle shrink during the tropical dry season, and the crowded conditions create shortages of oxygen in the remaining water. Air-breathing species can cope with this oxygen deficiency; in such fish, a mass of blood vessels on the creature's forehead works much like a human lung, pumping atmospheric oxygen directly into the bloodstream. Should the fish take to artificial ponds, they could provide a much needed native source of protein for people and animals.

Another Institute project is a cooperative effort with the Brazilian National Electric Company to raise manatees—large freshwater mammals whose numbers are dwindling because of overhunting in the upper Amazon. The electric company would like to see more manatees in the reservoirs of its hydro-electric plant. There, aquatic vegetation flourishes and tends to wash down into turbines, causing problems in machinery. Since manatees graze on aquatic vegetation, both the animals and the power company stand to benefit if these creatures can be raised in the reservoirs. Biologists also have an interest in research on raising manatees: they'd like to be able to restock areas where these rare animals have disappeared.

The point of schemes like these is to make it more profitable to save the tropics than to destroy them, thereby creating an economic incentive to prevent their decline. But the success of such efforts depends on knowing more about the complex and often complementary lives of jungle creatures—so that scientists can figure out the best ways to help them flourish.

Figure 32.13
Zones of vegetation change with altitude just as
they do with latitude because vegetation is partially
determined by temperature. Rainfall also
determines vegetation, which is why grasslands
are found at the base of some mountains instead
of a forest as shown here.

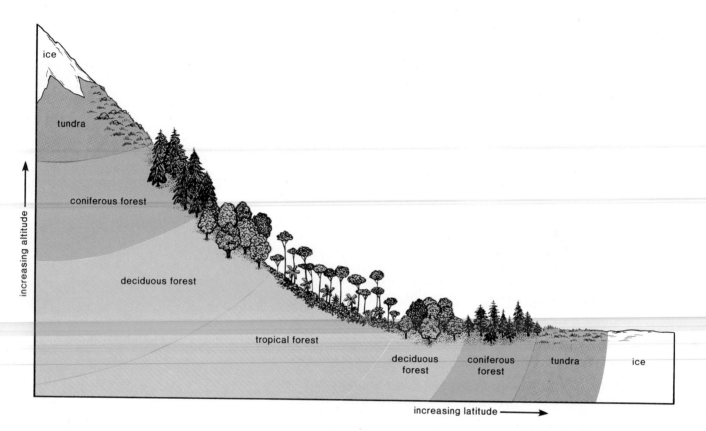

Biomes in the Mountains

If one travels from the Southern to the Northern Hemisphere, it would be
possible to observe first a tropical rain forest, followed by a temperate deciduous forest, and then the taiga and tundra, in that order. One could also observe
a similar series of biomes by traveling from the bottom to the top of mountains
(fig. 32.13). These transitions are largely due to decreasing temperature as
the altitude increases, but they are also influenced by soil conditions and rainfall.

Succession

During the process of succession, a sequence of communities replaces one another in an orderly and predictable way. The complete process is called a **sere,**
and each stage is a **seral stage.**

Behavior and Ecology

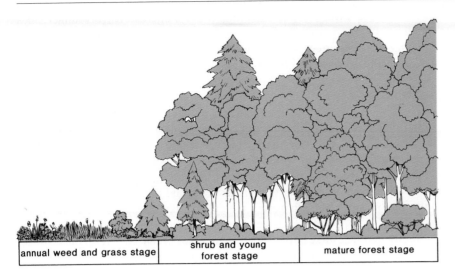

Figure 32.14
Simplified overview of succession as it may occur on abandoned farmland along the East Coast in the United States. For the first few years, annual weeds and grass colonize the area. Then shrubs and trees, such as pines that can tolerate direct sunlight, invade the area. These pave the way for the germination and growth of hardwood saplings that grow to finally dominate the area.

annual weed and grass stage	shrub and young forest stage	mature forest stage

The stages of succession vary from site to site, but figure 32.14 shows a possible succession for abandoned farmland in the eastern United States. The first community, called the **pioneer community,** includes plants such as weeds that are able to colonize disturbed areas because of their ability to survive under harsh conditions, such as limited soil moisture and direct sunlight. This colonization prepares the way for native plants of the area that most likely have a longer growing period. Notice in figure 32.14, how one plant community replaces another until a climax community has been achieved. A different mix of animals would be associated with each of these communities. Whereas previous communities are replaced, the final stage, a **climax community,** is able to sustain itself indefinitely. We have been studying various climax communities in this chapter.

Stability versus Productivity

There is an ecological theory, based on laboratory and field data, that suggests that stability of a biome is related to its complexity. Certainly the climax, or mature, stage of succession is the most stable since it maintains itself with little change. One reason for this might very well be that the last stage is the most complex in terms of species diversity. Thus, if one species or population of a climax community is reduced in size or eliminated, the other populations can compensate for this loss. For example, the deciduous forests in this country did not disappear when both the chestnut and elm trees succumbed to parasitic disease; the other trees simply filled in.

The early stages of succession are obviously unstable, as witnessed by the fact that they are replaced by later stages. However, investigators find that these stages show the most growth and therefore are the most productive.

Knowledge of this relationship between productivity and stability can be utilized by humans when they alter biomes for their own purposes. For example, in forestry the removal of trees places the biome in an earlier successional stage and causes greater productivity (new trees will grow) but also

increases instability (fig. 32.15). Stability can be safeguarded in one of two ways: *(a)* remove only trees of a certain mature age and leave younger trees standing and *(b)* remove trees from certain areas, but leave sections in between standing. Unfortunately, stability is not always safeguarded and many parts of the world that once had forests have them no more.

Aquatic Biomes

Aquatic biomes can be divided into two types: *(a)* inland or fresh water and *(b)* ocean or salt water. An **estuary,** however, where a river flows into the ocean, has mixed fresh and salt water, called brackish water. In these biomes, organisms vary according to whether they are adapted to fresh or salt water, warm or cold water, quiet or turbulent water, and the presence or absence of light. In both salt and fresh water, free-drifting microscopic organisms, called **plankton,** are important components of the biome. **Phytoplankton** are photosynthesizing algae that only become noticeable when they reproduce to the extent that a green scum or red tide appears on the water. **Zooplankton** are animals that feed on the phytoplankton.

Figure 32.16
Stratification of a lake. *a.* In the summer, a large lake has three distinct layers and the nutrients tend to collect in the hypolimnion, which becomes depleted of oxygen. *b.* In the autumn, a turnover occurs that allows mixing. *c.* In the winter, nutrients again collect in the hypolimnion. *d.* In the spring, another turnover occurs.

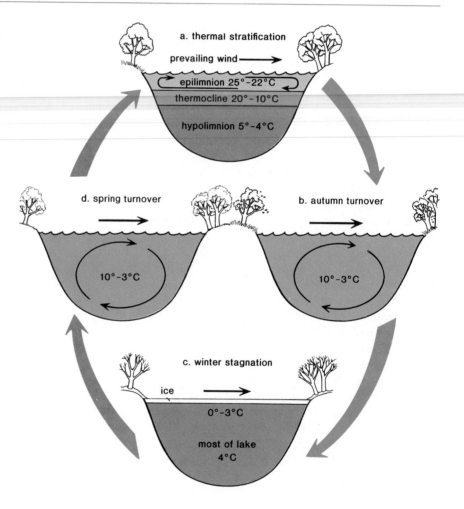

Lakes and Ponds

Lakes, being larger than ponds, have three layers of water that differ as to temperature (fig. 32.16). In summer, the upper surface layer, the *epilimnion,* is warm; the middle *thermocline* experiences an abrupt drop in temperature; and the *hypolimnion* is cold. This difference in temperature prevents mixing; and the epilimnion lacks nutrients found in the hypolimnion, while the hypolimnion lacks oxygen found in the epilimnion. In the fall, as the epilimnion cools and in the spring as it warms, mixing does occur, causing phytoplankton growth to be most abundant at these times.

Lakes and ponds can be divided into three life zones: the **littoral zone** is closest to the shore, the **limnetic zone** forms the sunlit body of the lake, and the **profundal zone** is below the level of light penetration. Aquatic plants are rooted in the shallow littoral zone of a lake. The rest of the organisms are divided into five groups according to their habitats. The *periphyton* (fig. 32.17)

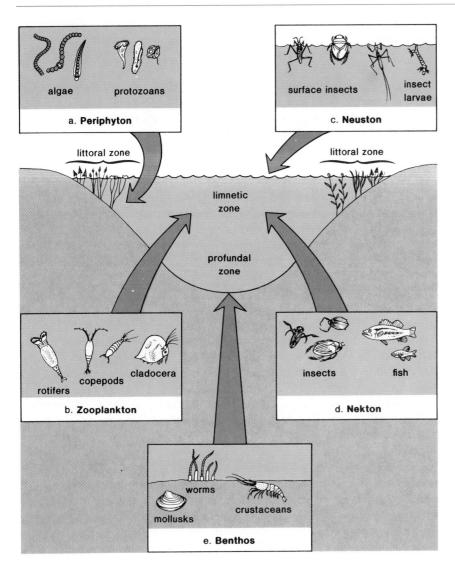

Figure 32.17
Life zones of a lake. *a.* Periphyton occur in the littoral zone. *b.* Zooplankton are in the limnetic zone where they provide food for carnivores. *c.* Neuston are at the water's surface. *d.* Nekton are free swimmers in the limnetic zone where food can be found. *e.* Benthic animals are in the benthos, the bottom of the lake.

are microscopic or near-microscopic organisms, such as algae or protozoans, that cling to plants, wood, and rocks in the littoral zone. The *plankton* of the limnetic zone includes both phytoplankton and zooplankton, such as rotifers, copepods, and water fleas. *Neuston,* insects that live at the water-air interface, include the water strider and water scorpion, animals that can literally walk on water, and the whirligig beetle and mosquito larvae that prefer a location just beneath the surface of the water. Other insects, such as diving beetles, water boatmen, and backswimmers, are a part of the *nekton,* a group of free-swimming organisms. Most nekton are fish. Minnows and killifish are fish that prefer the littoral zone; trout, whitefish, and cisco prefer the profundal zone, particularly the hypolimnion; pike, bass, and gar can tolerate warmer water.

The *benthos* are animals that live on the bottom in the benthic zone. In a lake the benthos include crayfish, snails, clams, various types of worms, and insect larvae. Among the insects that spend a large portion of their life cycle as larvae in the benthic zone are the dragonfly, damselfly, and mayfly. The benthic zone may, however, become so depleted of oxygen at times that only such organisms as sludge worms and midge fly larvae, known as bloodworms, can survive.

Rivers and Streams

At first, rivers and streams have rapidly flowing water as they move down out of the mountains. Here insect larvae and water plants are adapted to clinging to rocks as the water passes by. In intermittent pools, various species of fish, including trout, which prefer cool oxygenated water, may be found. As the river nears the ocean, water flow becomes much slower, plankton can now accumulate, and the community begins to resemble that of a lake or pond.

Water is removed from rivers to grow crops, for use as an industrial coolant, and for various household purposes. At the same time, sewage and pollutants are added to the rivers. With an ever-increasing population, more water is removed and more waste is added to rivers. Since both of these tend to decrease river flow, there is some question if dependable flow can be assured by the year 2000 unless projects are immediately carried out to minimize the use of rivers for these purposes.

The Coast

Rivers flow down to the sea to form estuaries, semienclosed baylike regions. The silt carried by a river forms mudflats, and within the shallow waters, a **salt marsh** (fig. 32.18a) in the temperate zone and a **mangrove swamp** (fig. 32.18b) in the subtropical and tropical zones are likely to develop. Along either side of the estuary community, the seashores reach out along the coast. It is proper to think of sea coasts and estuaries, including mudflats, salt marshes, and mangrove swamps, as belonging to one ecological system.

Estuary

A river brings fresh water into the estuary, and the sea, because of the tides, brings salt water. There is a gradation of salinity, or a gradual increase of salt water, from the river to the sea. Organisms living in an estuary must be able to withstand constant mixing of waters and rapid changes in salinity. Not many organisms are suited to this environment, but for those that are suited there is an abundance of nutrients. An estuary acts as a nutrient trap because the

a.

b.

tides bring nutrients from the sea and at the same time prevent the seaward escape of nutrients brought by the river. Because of this, estuaries produce much organic food.

Although only a few small fish permanently reside in an estuary, many develop there so that there is always an abundance of larval and immature fish. It has been estimated that well over half of all marine fishes develop in the protective environment of an estuary, which explains why estuaries are called the *nurseries of the sea.* Even so, many estuaries are becoming victims of pollution because of construction and development along the seacoast, as discussed in the reading on page 760.

Engulfed in a Rising Sea of Troubles

Farsighted programs are needed to eliminate and prevent further pollution of our coastal areas.

No one can say for sure what the human carrying capacity of America's fragile coastal lands is, but there's no doubt it has already been exceeded many times over. During the past 25 years, development has boomed by 150 percent. A decade from now, more than 75 percent of the entire U.S. population may be living in the coastal belt. On many barrier islands, population density is four times the national average. Right now, counties along the Atlantic and Gulf coasts are growing twice as fast as the rest of the country. More than 40 percent of the wetlands along the Lower 48's seacoast have been destroyed; an additional 300,000 acres are lost every year.

Part of the problem is simply that too many people want to live near the water even though common sense, a steadily rising sea level and meteorological history argue against it. Last fall, Hurricane Frederic [1979] visited $2 billion in damages on the southeastern coast, and weather experts fear the day will come when another, even worse storm kills thousands of people in low-lying areas. With so much at stake, the U.S. Army Corps of Engineers has been trying for decades to "stabilize" eroding beaches with an expensive Maginot Line of jetties, sea walls, groins and dikes that usually just make

matters worse. And still the pressure grows to cover every remaining square foot of coastline with more homes, pavement, hamburger stands, condominiums, motels and trailers.

The trouble does not end there, for the coastal zone is not limited to flood plains, beaches and tidal pools. It extends far out along the ocean floor and it backs well up into the mainland. Across its entire length and breadth, it is an ecosystem in great peril. Channelization projects and denuded streambanks far inland are increasing the load of silt at river mouths faster than the ocean can carry it away. Oyster beds and other marine life are being smothered. Pollution control efforts have not yet stemmed the flow of pesticide residues and toxic chemicals. Consequently, mercury and other dangerous substances are showing up in seal livers and the fat of fish. Poisonous wastes from offshore dumping are being returned by ocean currents as sludge balls that foul the beaches. Oil slicks from spills and wells pose a constant hazard to fish, birds and public beaches. .

These and other disruptions are changing the character of the coast, and in the process they are actually diminishing its value to humans.

Table 32.2 Seashore terminology

Zone	Location	Characteristic
Supralittoral	Above high tide	Rarely wet
Littoral	Intertidal	Half wet, half dry
Sublittoral	Below low tide	Rarely dry

Seashores

Both rocky and sandy shores are constantly bombarded by the sea as the tides roll in and out. Table 32.2 lists the terminology frequently used to divide the seashore into zones according to the length of time they are submerged underwater.

Rocky Shore The rocky shore (fig. 32.19) displays zonation that parallels the tide levels listed in table 32.2. In the *supralittoral zone,* rough periwinkles feed on darkly colored lichens. In the *intertidal zone,* common periwinkles feed on brown algae known as rockweed. Here, too, barnacles and mussels

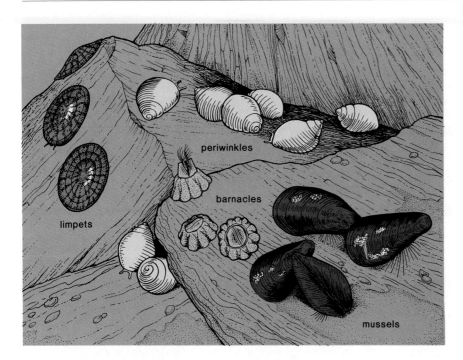

Figure 32.19
Animals typical of a rocky coast in the temperate zone. Limpets, periwinkles, barnacles, and mussels are all adapted to clinging to rocks.

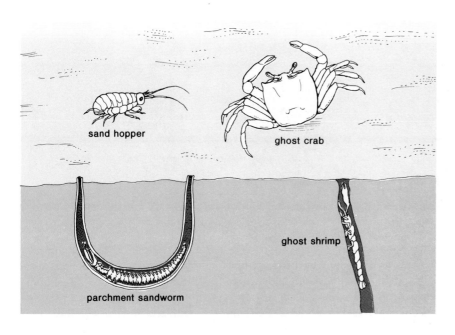

Figure 32.20
Animals typical of a sandy beach in the temperate zone. Sand hoppers, ghost crabs, sandworms, and ghost shrimp are all adapted to burrowing in the sand.

attach themselves to rocks in rows, awaiting the tide before opening their protective coverings to begin filter feeding. At the point where the littoral zone gives way to the *sublittoral zone,* red Irish moss and large brown kelps provide a home for numerous animals. Mussels, sea squirts, sea urchins, and worms all crowd in under the kelp's sturdy holdfasts. Starfish, crabs, and brittle stars prey on these detritus feeders.

Sandy Shore On a sandy beach the dry grains of sand, which are in perpetual motion, do not provide a suitable substratum for the attachment of organisms. Also absent are crevices and seaweeds to protect animals from the burning sun. Therefore, animals that make their home on sandy beaches (fig. 32.20)

either burrow during the day and surface to feed at night or they remain permanently within their burrows or tubes. Ghost crabs and sandhoppers (amphipods) burrow above high tide and feed at night when the tide is out. Sandworms and sand (ghost) shrimp remain within their burrows in the intertidal zone, feeding on detritus whenever possible.

Coral Reefs

Coral reefs (fig. 32.21) are areas of biological abundance found in shallow tropical waters that have a minimum temperature of 70°F. The chief constituents of a coral reef are stony coral animals and calcareous red and green algae. Corals, like sea anemones, have a saclike body with a crown of tentacles about the mouth. The stony corals secrete a calcium carbonate (limestone) exoskeleton; the soft corals secrete only microscopic spicules; and the horny corals secrete only a small amount of limestone but stiffen their bodies with gorgonin, a flexible substance. The horny corals often take the shape of fans

Behavior and Ecology

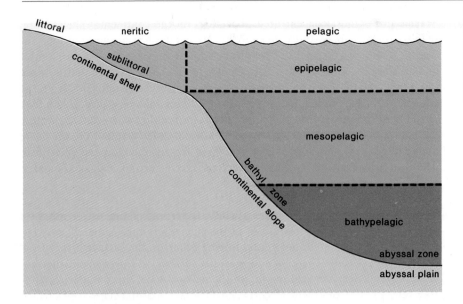

Figure 32.22
Life zones in the ocean. The benthos, comprising the sublittoral, bathyl, and abyssal zones, is found along the ocean floor that is divided into the regions shown. Organisms in the neritic zone live in the shallow waters above the continental shelf; those in the pelagic (open ocean) zone may reside in either the epipelagic, mesopelagic, or bathypelagic zones.

and plumes. Corals do not usually occur individually; rather, they form colonies derived from an individual coral that has reproduced by means of budding.

Microscopic algae live inside the coral. The corals, which feed at night, and the algae, which photosynthesize during the day, share materials and nutrients. Only the top layer of a reef contains living corals. They are like a veneer, growing a few millimeters a year on top of congregate ancestral corals.

A coral reef is densely populated with animal life. There are many types of small fishes (butterfly, damsel, clown, and surgeon), all beautifully colored. In addition, the large numbers of crevices and caves provide shelter for filter feeders (sponges, sea squirts, and fan worms) and for scavengers (crabs and sea urchins). The barracuda and moray eel prey on these animals. Some fish feed on the coral, but the most deadly predator of Pacific coral reefs is the crown-of-thorns starfish, which grows as large as 2 feet across and has from 9 to 21 arms. Along the northeastern coast of Australia, the very existence of the Great Barrier Reef is threatened by a plague of these animals. Some believe that the crown-of-thorns has begun to proliferate because humans have killed off a natural predator of this starfish, the giant triton, for its handsome spriral shell.

Oceans

Approximately three-quarters of our planet is covered by the oceans. Figure 32.22 shows that the depth of the sea is at first shallow and then abruptly deepens as the **continental shelf** gives way to the **continental slope,** which leads to the **abyssal plain..** Water above the continental shelf has an average depth of about 200 meters, while that above the abyssal plain averages about 4,000 to 6,000 meters, or about two to three miles.

Organisms in the ocean occupy three life zones. *Benthos* are animals that reside in the **benthic zone,** which includes the littoral and sublittoral zones of the continental shelf, the bathyl zone of the continental slope, and the abyssal zone of the abyssal plain. Organisms that are found in the sea above the continental shelf are in the **neritic zone** and those in the sea above the abyssal plain are in the **oceanic,** or **pelagic, zone.**

Coastal Zone

Marine life is most concentrated above and on the continental shelf. Here, seaweed, after it is partially decomposed by bacteria, is a source of food for benthic clams, worms, and sea urchins, which are preyed on by starfish, crabs, and brittle stars, all of which are, in turn, eaten by bottom-dwelling fish. More important, the shallow, sunlit neritic waters, which receive nutrients from the sea and estuaries, produce abundant phytoplankton that grow larger than they would in the open sea. Especially in regions of upwelling where surface waters are blown offshore and replaced by cold, nutrient-laden waters from the deep, ample phytoplankton grow and provide food for zooplankton and small fish. These, in turn, are food for the commercial fishes—herring, cod, and flounder.

Open Sea

The *pelagic zone,* which includes all but 10 percent of the sea, is not very productive. In fact, the productivity of the open sea is approximately equal to that of a desert. Whereas horizontal mixing occurs in the surface layer of the ocean due to oceanic currents and seasonal turnovers, a permanent thermocline prevents deep vertical mixing in the pelagic zone. Since nutrient materials tend to fall from the surface to the deep layers of the ocean, the surface becomes nutrient poor. The surface is the only region where photosynthesis takes place because sunlight penetrates the sea only to a level of about 200 to 300 meters.

The pelagic zone of the oceans can be divided into the epipelagic, mesopelagic, and bathypelagic zones (fig. 32.22). Only the epipelagic zone is brightly lit, or euphotic; the mesopelagic zone is in semidarkness; and the bathypelagic zone is in complete darkness.

The greatest amount of life occurs in the **epipelagic zone** (fig. 32.23), but the small size of the phytoplankton and zooplankton accounts for a long food chain, and this contributes to the low fish production in this zone. The phytoplankton includes mostly diatoms, dinoflagellates, and the smaller coccolithophorids. Among the zooplankton, copepods and krill feed on the phytoplankton. There are also carnivorous zooplankton that feed on other zooplankton, such as jellyfishes, comb jellies, wing-footed snails, sea squirts, and worms. The nekton of the epipelagic zone includes herring and bluefish, which are food for the larger mackerel, tuna, and sharks. Flying fishes, which glide above the surface, are preyed upon by dolphin fishes, not to be confused with mammalian porpoises, which are also present. Whales are other mammals found in this zone. Baleen whales strain krill from the water, and the toothed whales feed on the common squid found in the epipelagic zone.

Animals in the **mesopelagic zone,** which are adapted to the absence of light, tend to be translucent, red-colored, or even luminescent. Aside from luminescent jellyfishes, sea squirts, copepods, shrimp, and squid, there are luminescent carnivorous fishes, such as lantern and hatchet fishes. The bathypelagic zone is in complete darkness except for an occasional flash of bioluminescent light. Strange-looking fishes with distensible mouths and abdomens and small, tubular eyes feed on infrequent prey.

The abyssal life zone in the **bathypelagic zone** is inhabited by those animals that live in or just above the cold, dark sea bottom. Because of the cold temperature (averaging 2°C) and the intense pressure (300–500 atmospheres), it was once thought that only a few specialized animals would live in the bathypelagic zone. Yet a diverse assemblage of organisms has been found. Debris from the mesopelagic zone is taken in by filter feeders, such as the sea lilies that rise above the sea floor and the clams and tubeworms that lie burrowed in the mud. Other animals, such as sea cucumbers and sea urchins, crawl around on the abyssal plain, eating detritus and bacteria of decay. They, in turn, are food for predaceous brittle stars and crabs.

Behavior and Ecology

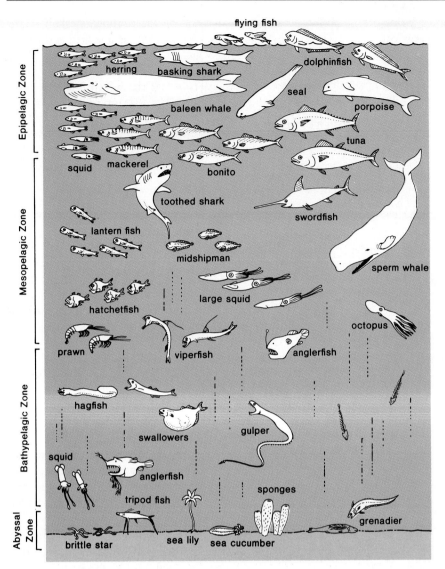

Figure 32.23
Free swimming nekton of the pelagic zone. The epipelagic zone contains the most animals because there are producers here that support various food chains. Only carnivores and scavengers are found in the other zones. The dotted lines represent organic matter that falls from the upper zones into the bottom zone.

Conservation of the Biomes

Very little is left of the original terrestrial biomes. Humans have converted nearly all accessible land to farm or city or even wasteland. Freshwater and marine pollution is also common. It is no wonder, then, that the number of endangered animal species that might become extinct by the year 2000 is several times larger than the sum of all extinct animals since 1900. Biome alteration and hunting account for the extinction of two-thirds of the projected number.

Some of the animals that are now protected by United States law include all types of whales, the American alligator, the peregrine falcon, the California condor, the polar bear, the Florida panther, and the Houston toad. Further, the 1983 Endangered Species Act forbids construction of any federal project that would destroy the habitat of a species listed as imperiled. All types of habitats should be preserved. Even in cities, parks and wildlife areas can be established. Urban renewal can include plans to restore coastlines that were formerly destroyed, and new cities and suburbs can be planned to provide large areas for wildlife. Botanical gardens, zoos, and aquariums can also serve as centers for the preservation of species.

Love Them or Lose Them

Butterflies are fading from many of their former haunts. Several species have been wiped out and several dozen others have been pushed to the brink. Even species still listed in the guidebooks as common are now observed much less frequently than they were only a decade ago. Fortunately for those that remain, butterfly restoration is one task, maybe the only task, in which environmentalists can expect almost instant results.

Here are a few suggestions on how to proceed:

1. Speak out against the application of needless—and often harmful—pesticides in your town and state. There are few good reasons, for example to spray for budworms, gypsy moths and adult mosquitoes. Saving butterflies is one good reason not to.
2. Discourage the spraying and cutting of roadside vegetation. Roadsides provide pockets of prime habitat for butterflies.
3. Let the edges of your lawn grow wild. Perhaps you have a patch—or better yet, a strip—of land that your spouse or your conscience or both have been nagging you to mow. Butterflies are a fine excuse not to.
4. Be tolerant of thistles and nettles, which are among the most valuable food sources for butterflies. You should also think twice before doing away with Joe-Pye weed, ragweed, goldenrod, milkweed, knapweed, dandelions, mallow, marjoram, bugle, wild thyme, clover, meadow sweet, vetch, currant, blueberry and tick trefoil. All are extremely important to butterflies, especially as nectar sources for adults.
5. Provide a variety of habitats. Different butterflies require different conditions. Try to create sunny areas, shaded areas and dappled areas, protection from wind, a shallow pool and an adjacent damp spot for drinking. Naturally, you should avoid using chemical insecticides.
6. Provide the small, bright flowers that butterflies favor. Some of these are: aralia, polyanthus, sweet rocket, honesty, mauve, valerian, pink thrift, catmint, sweet William, the single French marigolds and white alyssum.
7. Cut back shrubs each spring to encourage more blossoms. Also cut back some of the plants used by caterpillars—milkweed, for instance—so that tender new growth will be available for later generations that same season.

It has been suggested that the national park system be expanded and wilderness areas created where plants and animals can live completely undisturbed by humans. A worldwide effort toward the creation of ecological protectorates has been made under the auspices of UNESCO. As of mid-1978, 144 areas in 35 countries had been officially recognized by UNESCO as part of its global network of biosphere reserves. The reserves will help protect biological and genetic diversity while providing scientists with opportunities for basic research. In conjunction with such areas, gene banks can help assure the perpetuation of a wide variety of plants and animals.

Only public education and a concerted effort by everyone can ensure the success of these and other attempts to preserve the diversity of life. For example, the reading on this page lists the things we can all do to preserve butterflies.

Summary

The biosphere can be divided into biomes adapted to climate, especially temperature and rainfall. Deserts are often high-temperature areas with less than 20 centimeters of rainfall a year. Plants and animals living in semideserts need adaptations that protect them from the sun's rays and lack of water. Desertification is a worldwide problem that is reducing the productivity of the biosphere. Grasslands occur where rainfall is greater than 20 centimeters but is insufficient to support trees. The tundra, being the northernmost biome, supports only a limited variety of living things. The savanna, a tropical grassland, in contrast, supports the greatest number of different types of herbivores that all serve as food for carnivores. U.S. grasslands, including the tall grass and shortgrass prairies, became the corn and wheat belt of the country.

Forests require adequate rainfall. The taiga, a coniferous forest, has the least amount of rainfall. The temperate deciduous forest has trees that gain and lose their leaves because of the alternating seasons of summer and winter. The tropical forests, including seasonal ones that experience dry and wet seasons, are the least studied and the most complex of all biomes. Humans are beginning to exploit tropical forests so that many fear that thousands of species could become extinct before we even know of their existence.

There are a series of biomes on mountain slopes that mirror the sequence according to latitude because temperature also decreases with altitude. The term *succession* is applied to the series of stages by which rocks or abandoned land become a climax community. There is an ecological theory that stability is related to complexity, and certainly the climax community is the most complex and stable, as witnessed by the observation that it is not replaced with a later stage. The early stages of succession, however, are the most productive in terms of new growth. When humans alter biomes for their own use, they should balance productivity against stability, never neglecting the latter because only a stable biosphere ensures human existence.

Aquatic biomes are divided into fresh water and salt water. Lakes, especially in the summer, are stratified into layers according to temperature, and only a turnover twice a year restores their productivity. Lakes have various life zones, each with typical organisms. Rivers and streams are characterized by rapidly flowing water that gradually moves more slowly as a river approaches the ocean.

The coastline includes marshes, swamps, estuaries, rocky and sandy beaches. Although the coast is extremely important to productivity of the ocean, it is the region that has suffered the most pollution. Coral reefs, areas of biological abundance, occur in tropical seas. The open seas of the ocean, termed the pelagic zone, can be divided into the epipelagic, mesopelagic, and bathypelagic zones, each having organisms adapted to different environmental conditions. Only the epipelagic zone receives adequate sunlight to support photosynthesis, which limits the oceans to productivity about equal to that of deserts.

Conservation of all the biomes is an ecologically sound endeavor and one that all persons should try to assist in whatever way possible.

Study Questions

1. Arrange the terrestrial biomes discussed in this text in a diagram according to temperature and rainfall. (p. 742)
2. Describe the location, climate, and populations of (1) deserts (p. 739), (2) grasslands (tundra, savanna, prairie) (p. 743), and (3) forests (coniferous, deciduous, tropical) (p. 748).
3. Discuss succession with reference to abandoned farmland. (p. 754-55)
4. Discuss productivity and stability as they relate to succession. (p. 756)
5. Name the terrestrial biomes you would expect to find when going from the base of a mountain to the top. (p. 754)
6. Describe the temperature zones and the life zones of a lake. (pp. 757–58)
7. Describe the coastline biome (including coral reefs) and discuss their importance to the productivity of the ocean. (pp. 758–63)
8. Describe the life zones of the ocean and the organisms you would expect to find in each zone. (pp. 763–64)
9. Discuss the need for conservation of all the biomes and what you personally can do to help this effort. (pp. 765–66)

Behavior and Ecology

Selected Key Terms

biome (bi′ōm)
tundra (tun′drah)
savanna (sah-van′ah)
prairie (prar′e)
taiga (ti′gah)
temperate forest (tem′per-at for′est)
succession (suk-sē′shun)
estuary (es′tu-a-re)
plankton (plank′ton)
littoral zone (lit′or-al zōn)
limnetic zone (lim-net′ik zōn)
profundal zone (pro-fun′dal zōn)
intertidal zone (in″ter-tīd′al zōn)
coral reef (kor′al rēf)
continental shelf (kon″tĭ-nen′tal shelf)
continental slope (kon″tĭ-nen′tal slōp)
abyssal plain (ah-bis′al plān)
benthic zone (ben′thik zōn)
neritic zone (ner-it′ik zōn)
pelagic zone (pe-laj′ik zōn)

Further Readings

Borgese, E. M. 1983. The law of the sea. *Scientific American* 248(3):42.

Brill, W. J. 1981. Agricultural microbiology. *Scientific American* 245(3):198.

Brokaw, H. P., ed. 1978. *Wildlife and America council on environmental quality.* Washington, D.C.: U.S. Government Printing Office.

Cloud, P. 1983. The biosphere. *Scientific American* 249(3):176.

Horn, H. S. 1975. Forest succession. *Scientific American* 232(11):90–98.

Isaacs, J. D., and Schwartzlose, R. A. 1975. Active animals of the deep-sea floor. *Scientific American* 233(4):84.

Pillsbury, A. F. 1981. The salinity of rivers. *Scientific American* 245(1):54.

Richards, P. W. 1973. The tropical rainforest. *Scientific American* 119(6):57–68.

Scientific American. 1969. 221(3). Entire issue devoted to the ocean.

Smith, R. L. 1976. *The ecology of man: An ecosystem approach.* 2d ed. New York: Harper & Row.

———. 1977. *Elements of ecology and field biology.* New York: Harper & Row.

Whittaker, R. H. 1975. *Communities and ecosystems.* 2d ed. New York: Macmillan.

33

Human Population Concerns

Chapter concepts

1 The human population has been undergoing exponential growth since 1850.

2 The very large human population is straining the capacity of the earth to sustain it.

3 There are two points of view regarding the future supplies of nonrenewable resources. There are those who believe that we must conserve what is left and those who believe that technology will ever be able to exploit new sources.

4 Energy is required in order to exploit the environment. Most likely both coal and solar energy will be important sources of energy in the near future.

5 Thus far, the supply of food has kept up with increased population. It may be necessary to institute new agricultural methods if this is to be achieved in the future.

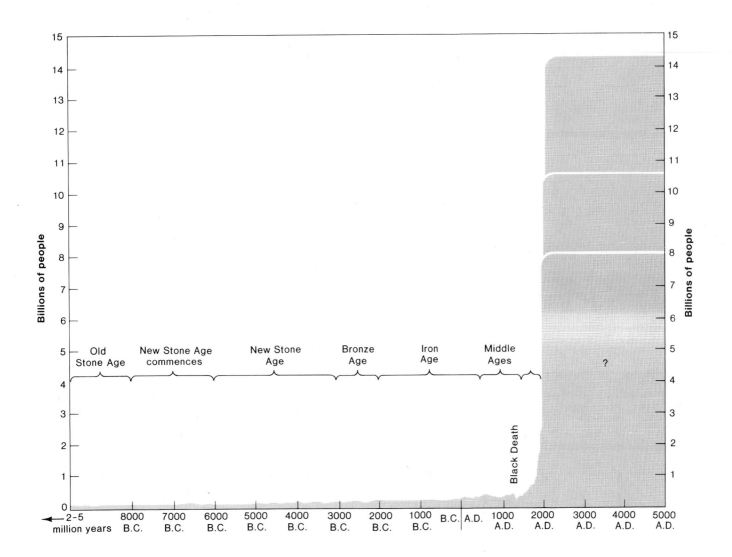

Population Growth

The human growth curve is an **exponential curve** (fig. 33.1). In the beginning, growth of the population was relatively slow, but as a greater number of reproducing individuals were added, growth increased until the curve began to slope steeply upward. It is apparent from the position of 1985 on the growth curve that growth is now quite rapid. The world population increases at least the equivalent of a medium-sized city every day (200,000) and the combined populations of the United Kingdom, Norway, Ireland, Iceland, Finland, and Denmark every year. These startling figures are a reflection of the fact that a very large world population is undergoing exponential growth.

Exponential Growth

Mathematically speaking, **exponential growth,** or geometric increase, occurs in the same manner as compound interest; that is, the percentage increase is added to the principal before the next increase is calculated. Referring specifically to populations, we wish to consider the hypothetical population sizes in table 33.1. This table illustrates the circumstances of world population growth at the moment: the percentage increase has decreased and yet the size of the population grows by a greater amount each year. The increase in size is dramatically large because the world population is very large.

In our hypothetical examples (table 33.1), an initial increase of 2 percent added to the original population size, followed by a 1.99 percent increase, results in the third generation size listed in the last column. Notice that

1. in each instance the second generation has a larger increase than the first generation because the second generation's population was larger than the first;
2. because of exponential growth, the lower percentage increase (i.e., 1.99% compared to 2%) still brings about larger population growth;
3. the larger the population, the larger the increase for each generation.

The percentage increase is termed the **growth rate,** which is calculated per year.

Growth Rate

The growth rate of a population is determined by considering the difference between the number of persons born (birthrate, or natality) and the number of persons who die per year (death rate, or mortality). It is customary to record these rates per 1,000 persons. For example, as shown in table 33.2, Russia

---

Table 33.1 Exponential growth of hypothetical populations

Population size	Percentage increase	Actual increase in numbers	Population size	Percentage increase	Actual increase in numbers	Population size
500,000,000	2.00	10,000,000	510,000,000	1.99	10,149,000	520,149,000
3,000,000,000	2.00	60,000,000	3,060,000,000	1.99	60,894,000	3,120,894,000
5,000,000,000	2.00	100,000,000	5,100,000,000	1.99	101,490,000	5,201,490,000

Table 33.2 1984 World population

Region or country	People (millions)	Birthrate (per 1000)	Death rate (per 1000)	Growth rate (per year)	Population (age 15/64+)(%)	Life expectancy	Urban population	GNP (per capita)
World	4762	28	11	1.7	35/6	61	40	$ 2,800
Developed	1166	16	9	0.6	23/12	73	71	9,190
Developing*	3596	32	11	2.1	38/4	58	32	750
Africa	531	45	16	2.9	45/3	50	29	810
Asia	2782	29	11	1.8	37/4	58	27	970
United States	236	16	9	0.7	22/12	74	74	13,160
Latin America	397	31	8	2.4	39/4	64	65	2,100
Europe	491	14	10	0.3	22/13	73	72	—
USSR	274	20	10	1.0	25/10	69	64	5,940
Oceania	24	21	9	1.3	29/8	70	71	8,700

*These figures include China. Without China, the developing countries birthrate increases to 37 per 1000, the death rate increases to 13, the growth rate increases to 2.4, the life expectancy decreases to 55, and the per capita GNP increases to $940.

Behavior and Ecology

(U.S.S.R.) at the present time has a birthrate of 20 per 1,000 per year, while it has a death rate of 10 per 1,000 per year. This means that Russia's population growth, or simply its growth rate, would be:

$$\frac{20-10}{1,000} = \frac{10}{1,000} = \frac{1.0}{100} = 1.0\%$$

Notice that while birth and death rates are expressed in terms of 1,000 persons, the growth rate is expressed per 100 persons, or as a percentage.

After 1750 the world population growth rate steadily increased until it peaked at 2 percent in 1965, but it has fallen slightly since then to 1.7 percent. Yet there is an ever larger increase in the world population each year because of exponential growth. The explosive potential of the present world population can be appreciated by considering the doubling time.

Doubling Time

Table 33.3 shows that the **doubling time** for a population may be calculated by dividing 70 by the growth rate:

$$d = \frac{70}{gr}$$

d = doubling time
gr = growth rate
70 = demographic constant

If the present world growth rate of 1.7 percent should continue, the world population will double in 40 years.

$$d = \frac{70}{1.7} = 40 \text{ years}$$

This means that in 39 years the world would need double the amount of food, jobs, water, energy, and so on if the standard of living is to remain the same.

It is of grave concern to many individuals that the amount of time needed to add each additional billion persons to the world population has taken less and less time (table 33.4). The world reached its first billion around 1800—some 2 million years after the evolution of humans. Adding the second billion took only about 130 years, the third billion about 30 years, and the fourth took only about 15 years. However, if the growth rate should continue to decline, this trend would reverse itself and eventually there would be zero population growth. Then population size would remain steady. Thus, figure 33.1 shows three possible logistic curves: the population may level off at 8, 11, or 14 billion, depending on the speed with which the growth rate declines.

Table 33.3 Relationship between growth rate and the doubling time of a population

Growth rate%	Doubling time (years)
0.25	280
0.5	140
1.0	70
2.0	35
3.0	23

Table 33.4 Estimated timing of each billion of world population

	Time taken to reach	Year attained
First billion	2–5 million years	About 1800 A.D.
Second billion	Approx. 130 years	1930
Third billion	30 years	1960
Fourth billion	15 years	1975
Projections:		
Fifth billion	12 years	1987
Sixth billion	11 years	1998

Source: Elaine M. Murphy. *World Population: Toward the Next Century* (Washington, DC: Population Reference Bureau, November 1981) page 3.

Figure 33.2

Growth curve for a fruit fly colony. The number of
fruit flies in a laboratory colony were counted every
other day, and when these numbers were plotted,
a sigmoidal growth curve resulted. This type of
curve is expected for a population that is adjusting
to the carrying capacity of the environment.

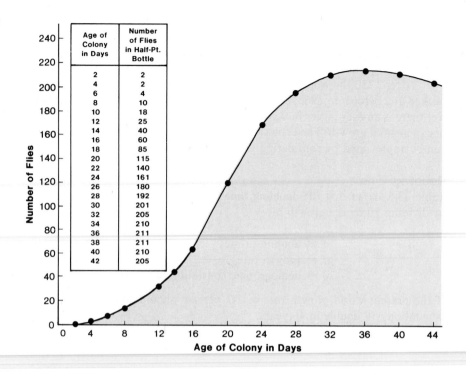

Age of Colony in Days	Number of Flies in Half-Pt. Bottle
2	2
4	2
6	4
8	10
10	18
12	25
14	40
16	60
18	85
20	115
22	140
24	161
26	180
28	192
30	201
32	205
34	210
36	211
38	211
40	210
42	205

Carrying Capacity

Examining the growth curves for nonhuman populations reveals that the populations tend to level off at a certain size. For example, figure 33.2 gives the actual data for the growth of a fruit fly population reared in a culture bottle. At the beginning, the fruit flies were becoming adjusted to their new environment and growth was slow. But then, since food and space were plentiful, they began to multiply rapidly. Notice that the curve begins to rise dramatically just as the human population curve does now. At this time, it may be said that the population is demonstrating its **biotic potential.** Biotic potential is the maximum growth rate under ideal conditions. Biotic potential is not usually demonstrated for long because of an opposing force called **environmental resistance.** Environmental resistance includes all the factors that cause early death of organisms and thus prevents the population from producing as many offspring as it might otherwise have done. As far as the fruit flies are concerned, we can speculate that environmental resistance included the limiting factors of food and space. Also, the waste given off by the fruit flies may have begun to contribute to keeping the population size down.

The eventual size of any population represents a compromise between the biotic potential and the environmental resistance. This compromise occurs at the **carrying capacity** of the environment. The carrying capacity is the maximum population that the environment can support—for an indefinite period.

The carrying capacity of the earth for humans is not certain. Some authorities think the earth is potentially capable of supporting 50 to 100 billion people. Others think we already have more humans than the earth can adequately support.

Behavior and Ecology

Figure 33.3
History of human population. When the size of the
human population is plotted on a log-log scale, it
can be seen that exponential growth occurred on
three ocasions: at the time of the cultural,
ɑgricultural, and industrial revolutions.

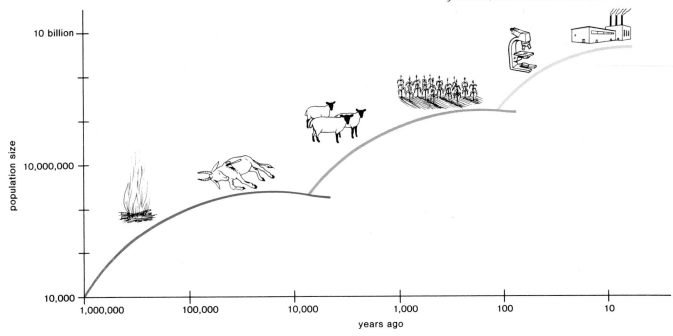

History of the World Population

Figure 33.3 suggests that the human population has undergone three phases
of exponential growth. *Tool-making* may have been the first technlogical advance that allowed the human population to enter a phase of exponential
growth. *Cultivation of plants* and *animal husbandry* may have allowed a second phase of growth; and the *industrial revolution,* which occurred about 1850,
promoted the third phase.

Developed and Developing Countries
When discussing population trends, the countries of the world are sometimes
divided according to whether they are already developed or are presently becoming developed.

Developed Countries
The industrial revolution, which was also accompanied by a medical revolution, took place in the Western world. In addition to European and North
American contries, Russia and Japan also became industrialized. Collectively,
these countries are often referred to as the **developed countries.** The developed
countries doubled their size between 1850 and 1950 (fig. 33.4), largely due to
a decline in the death rate. This decline is attributed to the influence of modern
medicine and improved socioeconomic conditions. Industrialization raised
personal incomes, and better housing permitted improved hygiene and sanitation. Numerous infectious diseases, such as cholera, typhus, and diphtheria,
were brought under control.

Figure 33.4

Size of human population in developed versus developing countries. The population of the developed countries (gray) increased between 1850 and 1950, but the size is expected to increase little between 1975 and 2000. In contrast, the population size of the developing countries (color) increased in the past and is expected to increase dramatically in the future also.

From "The Populations of the Underdeveloped Countries" by Paul Demeny. Copyright © 1974 by Scientific American, Inc. All rights reserved.

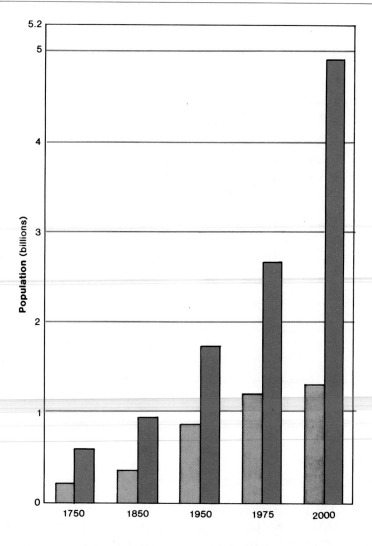

Table 33.5 Analysis of annual growth rates in developed countries

Phase	Birthrate	Death rate	Annual rate
I	High	High	Low
II	High	Low	High
III	Low	Low	Low

The decline in the death rate in the developed countries was followed shortly by a decline in the birthrate. Between 1950 and 1975, populations in the developed countries showed only modest growth (fig. 33.4) because the growth rate fell from an average of 1.1 percent to 0.8 percent.

Demographic Transition   Overall, the growth rate in developed countries has gone through three phases (table 33.5 and fig. 33.5). In Phase I, prior to 1850, the growth rate was low because a high death rate canceled out the effects of a high birthrate; in Phase II, the growth rate was high because of a lowered death rate; and in Phase III, the growth rate was again low because the birthrate had declined. These phases are now known as the **demographic transition.** In seeking a reason for the transition, it has been suggested that as industrialization occurred, the population became concentrated in the cities. Urbanization may have contributed to the decline in the growth rate because, in the city children were no longer the boon they were in the country. Instead of contributing to the yearly income of the family, they represented a severe drain on its resources. It could also be that urban living made people acutely aware of the problems of crowding and for this reason the birthrate declined. Also, some investigators believe that there was a direct relationship between improvement in socioeconomic conditions and the birthrate. They point out that as the developed nations became wealthier (judged by a rise in the Gross National Product [GNP]), as infant mortality was reduced, and as educational levels increased, the birthrate declined.

Behavior and Ecology

Figure 33.5
Time of demographic transition. In the upper graph
it is seen that in the developed countries the
demographic transition occurred in the nineteenth
century. In the lower graph it is seen that the
demographic transition was delayed until the
twentieth century.

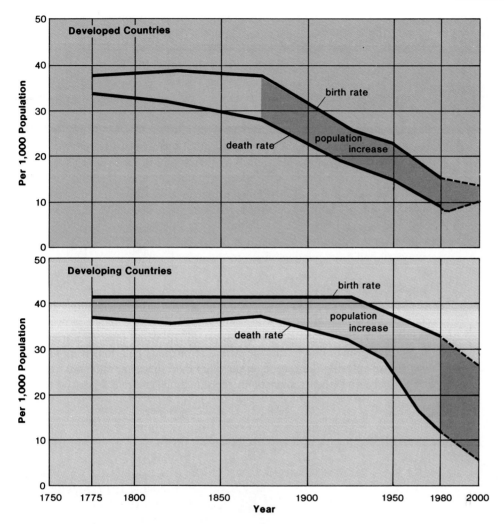

Regardless of the reasons for the demographic transition, it caused the
rate of growth to decline in the developed countries. The growth rate for the
developed countries is now about 0.6 percent and their overall population size
is about one-third that of the developing countries (table 33.2). A few developed countries—Austria, Denmark, East Germany, Hungary, Sweden, West
Germany—are not growing or are actually losing population. As discussed in
the reading on page 781, the United States has a comparatively high growth
rate for a developed country.

Developing Countries
Countries, such as those in Africa, Asia, and Latin America, are collectively
known as **developing countries** because they have not as yet become industrialized. Figure 33.5 indicates that mortality began to decline steeply in these
countries following World War II. This decline was prompted not by socioeconomic development but by the importation of modern medicine from the
developed countries. Various illnesses were brought under control due to the
use of immunization, antibiotics, sanitation, and insecticides. Although the
death rate declined the birthrate did not decline to the same extent (fig. 33.5)

Figure 33.6

Urbanization outside of Mexico City. People from the countryside who lack sufficient funds have built shanties just outside Mexico City.

and therefore the populations of the developing countries began and today are still increasing dramatically (fig. 33.4). The developing countries were unable to cope adequately with such rapid population expansion so that today many people in these countries are underfed, ill housed, unschooled, and living in abject poverty. Many of these poor have fled to the cities where they live in makeshift shanties on the outskirts (fig. 33.6).

The growth rate of the developing countries did finally peak at 2.4 percent during 1960–1965. Since that time, the mortality decline has slowed and the birthrate has fallen. The growth rate is expected to decline to 1.8 percent by the end of the century. At that time, about two-thirds of the world population will be in the developing countries.

Investigators are divided as to the cause of the observed growth rate decreases in the developing countries. Previously, it was argued that this would happen only when these countries enjoyed the benefits of an industrialized society. It has now been shown, however, that countries with the greatest decline were those with the best family-planning programs. From this it may be argued that such programs can indeed help to bring about a stable population size in the developing countries. Nevertheless, it has been found that certain socioeconomic factors have also contributed to a decline in the developing countries' growth rate. Relatively high Gross National Product (GNP), urbanization, low infant mortality, increased life expectancy, literacy, and education all had a dampening effect on the growth rate.

Age Structure Comparison

Lay people are sometimes under the impression that if each couple had two children, zero population growth would immediately take place. However, **replacement reproduction,** as it is called, would still cause most countries today to continue growth due to the age structure of the population. If more young women are entering the reproduction years than there are older women leaving them behind, then replacement reproduction will give a positive growth rate.

Figure 33.7

Age structure diagram for developed countries (*above*) versus developing countries (*below*). Because the population sizes of the developed countries are approaching stabilization, population growth will be modest, but the developing countries will expand rapidly due to their youthful profile.

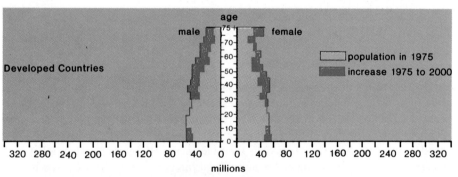

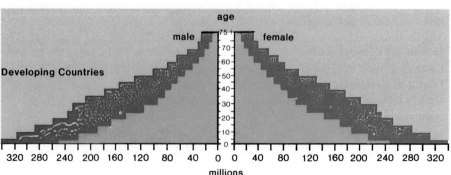

Behavior and Ecology

Reproduction is at or below replacement level in some 20 developed countries, including the United States. Even so, some of these countries will continue to grow modestly, in part because there was a baby boom after World War II. Young women born in these baby boom years are now in their reproductive years and even if each one has less than two children, the population will still grow. It should also be kept in mind that even the smallest of growth rates can add a considerable number of individuals to a large country. For example, a growth rate of 0.7 percent added over 1.6 million people to the United States population in 1984.

Whereas many developed countries have a stabilized age structure (fig. 33.7), most developing countries have a youthful profile—a large proportion of the population is below the age of 15. Since there are so many young women entering the reproductive years, the population will still greatly expand even after replacement reproduction is attained. The more quickly replacement reproduction is achieved, however, the sooner zero population growth will result.

Resource Consumption

Rapid world population growth puts extreme pressure on the earth's resources, physical environment, and social organization. While it might seem as if population increases in the developing countries are of the gravest concern, this is not necessarily the case since each person in a developed country consumes more resources and is therefore responsible for a greater amount of pollution. Environmental impact (EI) is measured not only in terms of the population size but also in terms of the resource used and the pollution caused by each person in the population.

$$EI = \text{population size} \times \frac{\text{resource use}}{\text{per person}} \times \frac{\text{pollution per unit of}}{\text{resource used}}$$

Therefore, there are two types of overpopulation. The first type is due to rapid increased population and occurs mainly in the developing countries. The second type of overpopulation is due to increased resource consumption with its accompanying pollution; this type is most obvious in the developed countries.

Nonrenewable Resources

Nonrenewable resources are those resources whose supply can be used up or exhausted. Figure 33.8a shows a depletion curve for a nonrenewable resource that is consumed at an increasingly rapid rate until a peak of consumption is followed by a decline in consumption as the resource becomes more expensive

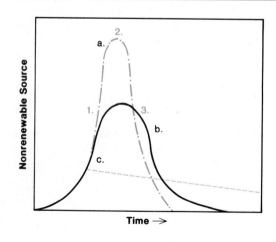

Figure 33.8
Depletion patterns for a nonrenewable resource. *a.* Rapid depletion as the resource is used up quickly: (1) exponential consumption of a resource followed by a peak (2) and decline (3) as the resource becomes difficult to acquire. *b.* Depletion time can be extended with some recycling and less wasteful use. *c.* Efficient recycling extends the depletion curve indefinitely.

Table 33.6 Dates when prices must rise to avoid exhaustion of selected resources

	Based on currently known reserves only (year)	Based on reserves plus prospective reserves (year)
Nonfuel minerals		
Aluminum	2025	2038
Chromium	2048	2095
Cobalt	2004	2016
Copper	2010	2041
Iron	2053	2094
Lead	2000	2016
Manganese	2061	2112
Molybdenum	2014	2480
Nickel	2014	2032
Phosphate rock	2034	2120
Potash	2104	2368
Sulfur	2010	2036
Tin	2003	2030
Titanium	2043	2151
Tungsten	2009	2037
Vanadium	2060	2180
Zinc	1993	2065
Energy		
Coal	2050	3000
Petroleum	1995	2010
Natural gas	2005	2024
Uranium	2005	2030 – 2070

Note: These projections are based on the "standard world case" from the above source and assume that recycling rates remain unchanged and that the rate of growth in demand during the 2015–2025 period remains constant thereafter.

to find and process. The current demand for nonrenewable resources, such as fossil fuels and other minerals, is constantly increasing not only because of population growth but also because of a rise in per capita demand. There are two points of view, called the Malthusian and Cornucopian views,[1] concerning resource availability and consumption. According to the **Malthusian view,** the depletion curve tells us that there are limits to growth and that we are rapidly approaching those limits. Table 33.6 indicates the year in which the price of selected resources will have to rise in order to bring about conservation of presently known and prospective reserves. If conservation of fossil fuels were practiced, the depletion curve could possibly be altered to resemble curve b in figure 33.8. Conservation and recycling of minerals could further change the shape of the depletion curve (fig. 33.8c). Those who hold the Malthusian view believe that, because of exponential consumption, finding new reserves cannot sufficiently extend a depletion curve. Nor do they believe that technology will ever be able to overcome inevitable shortages.

According to the **Cornucopian view,** technology will be able to constantly extend the depletion curve, putting off the day when no further exploitation is possible. Proponents of this view believe that improved technology will enable us to (1) find new reserves, (2) exploit the new reserves, and (3) substitute one mineral or energy resource for another. They point out that while we are sometimes aware of the availability of a resource, we must await the development of a technology in order to exploit it. For example, in the same way that we now utilize offshore drilling to acquire oil, we might one day be able

[1]Malthus was an eighteenth century economist who pointed out that since the size of a population increases geometrically and renewable resources increase only arithmetically, shortages must eventually occur. *Cornucopia* is a Latin word meaning the horn of plenty, a symbol of everlasting abundance.

Behavior and Ecology

The United States now has a population of over 230 million. The geographic center (the point where there are just as many persons in each direction) has moved steadily westward and recently has also moved southward. Another interesting trend is the shifting emphasis from metropolitan areas to nonmetropolitan areas. In the 1970s cities increased by 9.8 percent, but rural small towns increased by 15.8 percent.

The population size of the United States is not expected to level off any time soon for two primary reasons.

1. A baby boom between 1947 and 1964 has resulted in an unusually large number of reproductive women at this time. Thus, although each of these women is having on the average only 1.9 children, the total number of births increased from 3.1 million a year in the mid 1970s to 3.6 million in 1980–81.
2. Many people immigrate to the Unites States each year. In 1981 immigration accounted for 43 percent of the annual population growth. The number of legal immigrants was about 700,000. Even though ordinarily only 20,000 legal immigrants can come from any one country, we give special permission for large numbers of political refugees to enter the United States.

During the 1970s the majority of refugees came from Latin America (e.g., Cuba) and Asia (e.g., Indonesia).

There is also substantial illegal immigration into the United States, although the exact number is not known. Estimates range from 100,000 to 500,000 or more. About 50 to 60 percent of these illegal immigrants come from Mexico, according to most estimates. There has been a continuing effort to stem the tide of illegal immigration to the United States.

Whether the United States can ever achieve a stable population size depends on the fertility rate (the average number of children each woman has) and the net annual immigration. If the fertility rate were kept a 1.8 and immigration limited to 500,000, the population would peak at 274 million in 2050, after which it would decline. Those who favor a curtailment of our population size point out that a fertility rate of 1.8 allows couples a great deal of freedom in deciding the number of children they will have. For example, it means that 50 percent of all couples can have two children; 30 percent can have three children; 10 percent can have one child; 5 percent can have no children; and 5 percent can have four children.

United States Population

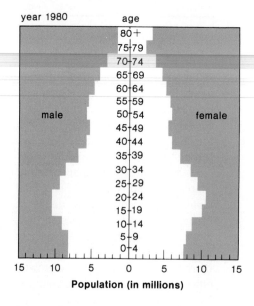

U.S. Age Structure

to develop the means to mine the ocean floor for minerals. Advanced technology might also make it feasible for us to utilize poorer grades of mineral ores. Consider the fact that previously only ores with a minimum of 3 percent copper were mined, whereas now ores with 0.3 percent copper are utilized.

In order to exploit previously inaccessible and/or less concentrated resources, a plentiful supply of energy is required. The Malthusians do not believe that increasing amounts of energy will be available since fossil fuel reserves are being rapidly depleted. The Cornucopians had initially hoped that nuclear power would supply the necessary energy, but because of the many problems associated with nuclear energy, many are now looking forward to the development of an alternate energy source. They feel that, given time, a new and plentiful energy source will be developed.

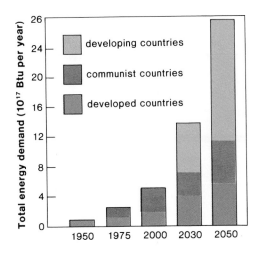

Table 33.7 Energy sources

	Advantages	Disadvantages
Nonrenewable	*Technology well established*	*Finite fuel supply*
Fossil fuels		
Coal, oil shale, and tar sands	Plentiful supply	Surface mining
		Air and water pollution
Petroleum	Cleaner burning	Limited supply
Natural gas	Cleanest burning	Limited supply
Nuclear		
Light water	Fuel availability	Thermal pollution
		Radiation pollution
Renewable	*Infinite Fuel Supply*	*Technology under Development*
Nuclear		
Breeder	Fuel availability	Radiation pollution
		Thermal pollution
		Nuclear weapons proliferation
Fusion	Fuel availability	Radiation pollution
Geothermal	Less pollution	Availability limited
Solar and wind	Nonpolluting	Noncompetitive cost
	Large and small scale possible	
Ocean	Nonpolluting	Applicable only in certain areas
Biomass	Utilizes wastes	Air and water pollution

Energy

A larger world population will require more energy in the years ahead. As figure 33.9 shows, the developing nations are expected to be responsible for most of the increased demand for energy. Between 1956 and 1976, energy consumption in China increased by a factor of 14, in Latin America by a factor of 8, and in Southeast Asia by a factor of 11. This trend will continue through the year 2000 and beyond.

The contrast, the exact opposite trend seems to be occurring in the United States: demand is not increasing and instead could possibly even diminish by almost 25 percent, from 80 quads[2] in 1980 to 62 quads in the year 2000. Surprisingly, it has been suggested that decreased energy consumption need not mean a change in U.S. life-style. Instead it is possible that the GNP could rise even as less energy is used because when less money is spent on energy, more funds can be devoted to growth. Increased energy efficiency in the home, office, and industry is the primary means by which energy has been and will be conserved. For example, as described in the reading on page 784, a new building in Boston is to be kept warm from heat given off by its contents—machinery, lights, people and so forth. It will require 40 percent more electricity to allow storage of this heat in basement water tanks but even so, there will be an overall savings of energy.

Citizens have become increasingly interested in renewable energy sources such as this one, but we are still largely dependent on nonrenewable sources (table 33.7).

Nonrenewable Energy Sources

Petroleum and Natural Gas
Present energy concerns are related to the fact that the depletion curve (fig. 33.8) for petroleum is now in the decline phase, and the depletion curve for natural gas has reached its peak. Supplies of these favored fossil fuels are not

[2]Quadrillion BTUs, the same amount of energy as in 8 billion gallons of gasoline.

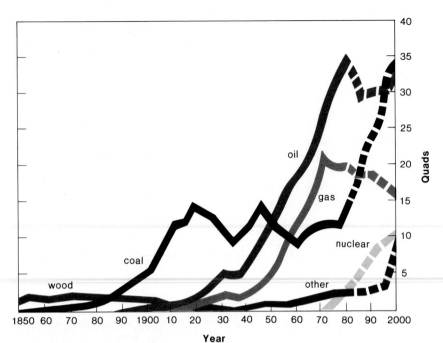

Figure 33.10
United States energy sources past and present.
Around 1850, wood was the favorite energy source
but toward the end of that century, coal became
dominant only to be replaced by oil in the twentieth
century. Now coal is expected to make a
comeback.

Table 33.8 Synthetic fuels

Synfuels	Nonrenewable sources	Renewable sources
Methane (gas)* and methanol (liquid)	Coal and oil shale	Garbage, sewage, plants, woodchips
Oil	Oil shale	Same as above
Ethanol	—	Sugarcane, corn, wheat, etc.

*Called Biogas when produced from renewable resources.

expected to last more than 30 years. Also, the United States has become dependent on foreign supplies of oil, with concomitant financial and political repercussions. Petroleum became the favored fuel during the present generation (fig. 33.10) because it is easily transportable; it is versatile, serving as the raw material for gasoline and many organic compounds; and it is cleaner burning than coal.

Coal and Oil Shale

Coal is in plentiful supply, as is **oil shale,** a misnomer since the "oil" is actually a waxy substance called kerogen that must be treated to produce petroleum. **Tar sands** that contain asphalt, which may be processed into crude petroleum, are also present in the United States and especially in Canada.

Of these three possibilities, it appears that the United States will be depending primarily on the use of coal to make up for decreased use of petroleum; projects utilizing oil shale have actually decreased in number of late. Both coal and oil shale can be used to produce synthetic fuels by means of gasification and liquefaction. Gasification produces **methane,** the main constituent of natural gas, and liquefaction produces **methanol,** a substance that can be added to or converted to gasoline.

Table 33.8 lists the synthetic fuels and the sources from which they can be produced. The term **synthetic fuels** is a catchall phrase for petroleum and petroleum substitutes from unconventional sources. It is sometimes also extended to include hydrogen gas produced by the hydrolysis of water by means of either nuclear or solar energy.

There are environmental drawbacks to using coal and oil shale in place of petroleum. To acquire either one, thousands of acres of land are strip-mined, resulting in huge piles (fig. 33.11) of residue that have the potential of leaching hazardous chemicals into underground aquifers. The burning of coal and/or the production of synfuels add pollutants to the air, with the concomitant effects discussed on page 727.

Nuclear Power

There are three possible types of nuclear energy (table 33.7) that can be used to generate electricity. Thus far, only light-water **fission power plants,** which split uranium-235, have been utilized in this country. There are now approximately 70 operating nuclear plants, and some 65 new ones are being constructed. Many proposed plants have been canceled and there have been no new orders for the past several years. Some predict that the nuclear power industry is dead in this country. The possible reasons for this are numerous; for example, there is little need for new plants because the demand for electricity is not increasing; it can be shown that it is cheaper to build and operate coal-firing plants; and the public is very much concerned about the safety of nuclear power.

Keeping Warm, Boston Style

Transportation building in Boston. This building requires no outside source of heat—it generates heat from people and machines.

Nature, says the second law of thermodynamics, never gives you anything for nothing. Now some engineers in Boston have come close to defying that unbreakable rule. They have produced a building that heats itself without a furnace or conventional fuel and remains warm even during such blustery periods as the recent cold spell, when Boston temperatures plunged to 0°F. The building performs this scientific magic by a cunning engineering stratagem: it recaptures the waste heat of its own machinery, everything from computers to coffeemakers, as well as of the 2,000 people, who will eventually work inside it.

Situated in Boston's theater district, the unusual self-heating structure will open formally at the end of this month, when local temperatures typically hover at 20° or 30° F. Still, the designers of the ingenious heating system, Henry Eggert and Howard McKew of Shooshanian Engineering Associates of Boston, are confident that teeth will not chatter nor pipes freeze. Indeed, they insist, the eight-story, 880,000-sq.-ft., red-brick Transportation Building will stay a comfortable 72°F all year round.

The state-owned structure's basic secret: three swimming-pool-size, 250,000-gal. concrete water

Behavior and Ecology

Figure 33.11
Aerial view of extensive surface mine spoil. Most coal today is surface-mined, a process that reduces the quality of the land unless proper careful measures are taken to restore it as it was before.

tanks resting in the basement. Like giant thermos bottles, these insulated containers can store heat, which can be tapped at will. In daytime, when the building's population is at its peak and office machinery is working full blast, the air in the central core of the building rapidly warms up. (The human body in a 72°F room gives off 250 B.T.U.s per hour, about equal to the heat from a 75–watt light bulb.) This hot air is propelled through a labyrinth of ducts by ventilating fans. Some is mixed with cool air from outside and pumped back into the center of the building to provide fresh air; some is circulated past pipes carrying cooler water from the basement. During this encounter, the water from the basement is initially heated by 10°. Its temperature is raised further by large refrigerator-type compressors called heat pumps, which can add or remove large quantities of heat.

In winter the pumps hike the water's temperature to an optimal 105°. Some of this freshly warmed water is immediately sent off to the building's naturally cooler, outside offices, where it passes through heating coils that warm the air. The heated water can be stored in the tanks for re-circulation through the building at night or on weekends, when the building's population is low and very little machinery is

operating. In summer the system works in reverse, with the heat pumps acting as air conditioners. Rather than being stored in the basement for later use, heat picked up in the building's inner recesses is directed to two cooling towers on the roof, which dissipate the heat into the atmosphere. In all seasons, the building gets a boost from rooftop solar collectors, which are arrays of black-painted pipes that absorb the energy of the sun's rays and provide most of the building's hot water for washrooms.

As added protection against the chill of New England winters, or the heat of its summers, the building has such thermal protection as especially thick walls, recessed windows (for shade) and double-pane glass. It has other features that improve its inhabitability, including a large enclosed atrium, overlooked by interior balconies, and space for restaurants and shops, a rarity in a state building. Says Eggert: "Usually in designing an energy-efficient building, the client is the major stumbling block. But we got the go-ahead to make the building as efficient as possible."

Before the first shovelful of earth was dug, Eggert and McKew subjected their design to various computer tests. For example, they tried to determine whether there would be enough heat in the

building for it to survive a three-day winter weekend. The system breezed through the trials, though innovation does come at a price. The designers acknowledge that the building would require some 45% more electricity to power the fans, sensors, heat pumps and large computer that operate the heating-cooling system. Nonetheless, they figure that when total costs are added up, they will be far ahead of the game. The self-heating scheme should save 740,000 gal. of oil a year (current cost: about $850,000), for a net saving of nearly $400,000 annually. Moreover, at a tab of $9 million, the system's price is about $1 million less than a conventional heating-cooling plant.

Early on, it was decided to forgo an emergency furnace, and some Massachusetts legislators question the wisdom of that decision. (A similar 20–story office in Toronto has taken the timid approach, with a steam-heat backup.) But Site Architect Spiros Pantazi brushes off all fears. Even before the system became fully operational, he set up his own office in the building, oblivious to the chilly temperatures outside. Said he: "It's wonderful. I'm sitting here in my shirtsleeves, and I am comfortable."

In 1979 at Three Mile Island near Harrisburg, Pennsylvania, and in 1982 at the E. Ginna plant, Ontario, New York, there were similar accidents that required the release of radioactive steam into the atmosphere. In both incidents, there was little danger to the public, but the chance still exists for more severe occurrences in the future. Most everyone is also aware of the radioactive waste disposal problem. Spent fuel rods contain low-medium wastes and high-level wastes, depending on the amount and character of radiation they emit. High-level wastes do not lose their radioactivity for hundreds of thousands of years. In 1982, Congress passed the Nuclear Waste Policy Act that establishes a timetable for disposing of nuclear wastes. Most probably, the wastes will be incorporated into glass or ceramic beads, packaged in metal canisters, and buried in stable salt beds, red clay deposits in the center of the ocean, or in stable rock formations.

Nuclear fission breeder reactors do not have the same environmental problems as light-water nuclear power plants. They have little waste because they use plutonium 239, a fuel that is actually generated from what would be reactor waste. Plutonium, however, is a very toxic element that readily causes lung cancer and it is also the very element used to make nuclear weapons. (The chances of a nuclear explosion are much greater with breeder fission reactors than with light-water reactors.) The fear of nuclear weapons proliferation, in particular, has thus far prevented the start-up of a breeder reactor in this country.

For some time it was believed that **fusion power plants** would be free of the environmental problems associated with fission plants, but this no longer seems to be the case. Nuclear fusion requires that two atoms, usually deuterium and tritium, be fused. The heat needed is so great that there is no conventional container for the reaction and scientists are perfecting laser-beam ignition and magnetic containment for the reaction. It is now apparent that since the fusion reaction gives off neutrons that can change uranium to plutonium, the best use of fusion plants would be to provide fuel for breeder reactors. The latest idea is to have hybrid fusion plants that would combine both a fusion and breeder fission plant. But, in any case, the fusion process is still experimental and not expected to be ready for production any time soon.

Renewable Energy Sources

Renewable energy sources are those that are not consumed, regardless of utilization. Because the breeder and fusion reactors produce as well as use fuel, it is possible to include them in this category (table 33.7).

Solar Energy

Solar energy is diffuse energy and before utilization is possible, it must be collected and concentrated. The public rather than the government has provided most of the impetus for solar heating of homes and offices. **Solar collectors,** placed on rooftops, absorb radiant energy but do not release the resulting heat. A fluid within the solar collector heats up and is pumped to other parts of the building for space heating, cooling, or for the generation of electricity. Passive systems are also possible; specially constructed glass can be used for the south wall of a building and building materials can be designed to collect the sun's energy during the day and release it at night.

Solar ponds have been recommended as a means of supplying energy to intermediate-sized communities whose primary need is for low-temperature heat. A solar pond is simply a body of water that is heated by the sun and insulated to prevent heat loss. After the water is heated, it can be pumped into pipes that distribute the heat about the community. Solar ponds are currently in use in Israel but can also be used in temperate zone countries.

The United States government subsidized the building of **Solar One,** the world's largest solar-powered electrical generating facility (fig. 33.12), which

a.

b.

Figure 33.13
Solar energy can be utilized by various community institutions, such as Betatakin National Monument in Arizona.

began operating in 1982 at Daggett, California in the Mohave Desert. A single boiler is located atop a large tower 20 stories high, and a large field of mirrors, called **heliostats,** which are capable of tracking the sun, reflect the sun's rays onto the tower. The water is heated to 500°C and the steam is used to produce electricity in a conventional generator. Systems such as this require much land and cannot be placed just anywhere. For this reason, **photovoltaic (solar) cells** that produce electricity directly, and which have been much improved of late, may be a more promising source of energy. They too can be placed on rooftops (fig. 33.13); or it has even been suggested that cells might

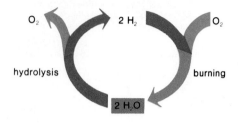

be placed in orbit about the earth where they would collect intense solar energy, generate electricity, and send it back to earth via microwaves.

Solar energy is clean energy. It does not produce air or water pollution. Nor does it add additional heat to the atmosphere since radiant energy is eventually converted to heat anyway. The problem of storage can be overcome in a number of ways, including the use of solar energy to produce hydrogen by means of hydrolysis of water. Hydrogen can be piped in existing pipelines as either a gas or liquid and it can be used for fuel for both automobiles and airplanes. When it is burned, it forms fog, not smog (fig. 33.14).

Other Sources

Two types of renewable energy sources that have been utilized for quite some time are *falling water,* used to produce electricity (hydroelectric plants), and geothermal energy. *Geothermal energy* refers to trapped heat produced by radioactive material deep beneath the surface of the earth. Water, converted to steam by this heat, may be pumped up and used to heat buildings or to generate electricity.

The ocean's tides, waves, currents, and temperature differential can all be used as energy sources. If enclosed coastal basins are dammed, the *tides* can be used to produce electricity as the water flows through specially constructed gates. Various proposals have been made to harness *wave power.* For example, the rolling motion of the waves can be used to rotate vanes that turn a generator. *OTEC* (Ocean Thermal Energy Conversion) projects are based on the temperature difference between warm surface water and cool, deep water, particularly in tropical seas. As liquid propane is piped from the ocean floor to the surface, it changes to a gas and generates electricity before returning to the floor, where it condenses to a liquid again. Utilization of ocean energy is in the pilot stages; it is hoped that full operation can be realized in the next 10 years.

Wind power provides enough force to turn vanes, blades, or propellers attached to a shaft, which, in turn, spins the motor of a generator that produces electricity. The government has allocated a small amount of money for wind research, particularly the promotion of large windmills. Many others believe that it would be best to build numerous small windmills for modest projects. If the latter is done, it is hypothesized by enthusiasts that windmills could supply 75 percent of our energy needs.

Any type of plant material, including garbage, can be fermented to give ethanol that can be added to gasoline, or subjected to bacterial digestion to give methane (table 33.8). The largest gasifier in commercial operation may be a 100–ton-per-day plant at Disney World, which processes solid waste collected from the Magic Kingdom. Other communities have found that one acre of sewage-enriched warm water can produce several tons of water hyacinths a day, enough to yield between 3,500 and 7,000 cubic feet of methane upon bacterial digestion. Some plants even produce an oil that can be used directly in diesel engines. The desert shrub *Euphorbia lathyris* might be able to yield yearly the equivalent of 6.5 barrels of oil per acre, and sunflower seeds are even now being crushed to give fuel for farm tractors.

Food

There is great concern that it will not be possible to provide enough food to feed the population increases that are expected well into the twenty-first century.

Figure 33.15a and 33.15c shows that the *production of food* has increased in the developed countries and the per capita share has also increased. This means that food production has kept pace with population increase in the developing countries where the majority of people can afford the price of

Figure 33.15
Food production in developed versus developing
countries. *a.* and *b.* Food production has increased
at similar rates in the developed and developing
countries over the last three decades, but
population growth has continued at a faster rate in
developing than in developed countries; therefore
(*c.* and *d.*), there has been a per capita gain in the
developed countries but not in the developing
countries.

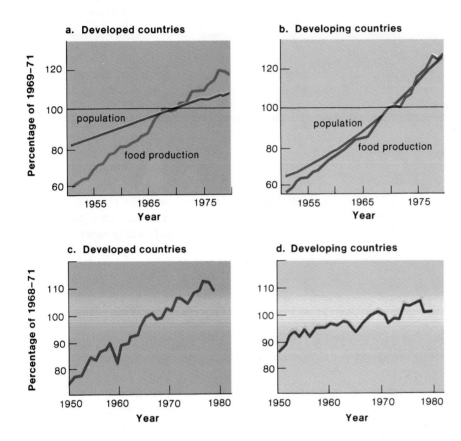

food and are eating better than formerly. Figure 33.15b and figure 33.15d
show that the production of food has increased in the developing countries,
but the per capita share has not substantially increased because increased food
production must be divided by a large population increase. Further, in the
developing countries only a segment of the population can afford the price of
food and are eating well; a larger portion cannot afford the price of food and
are not eating adequately. (Since the per capita share is calculated by dividing
the total production by the total population size, it does not necessarily mean
that all persons are eating an equal amout.) The U.N. Food and Agricultural
Organization estimates that as of 1975 around 450 million people are chron-
ically undernourished. The World Bank estimates that the figure may even be
as high as 1 billion. While some undernourished individuals live in the developed countries, the majority live in the developing countries. In particular,
Africa, with the fastest population growth ever recorded, has had difficulty
providing enough food for its populace. Of the many factors involved in producing enough food, the following are most often discussed.

Land
Up until 1960, growth in food output came largely from expanding the amount
of land under cultivation. But in the 1960s and 1970s the continued increase
in output in both the developed and developing countries came more from
increased yield per acre. Today, most land in both types of countries is already

Figure 33.16
High-yield rice. Hybridization crosses produced this
high-yield rice, one of the types of plants that have
helped developing countries produce more food on
the same amount of land.

being used for one purpose or another and it would be difficult to expand the
amount of farmland. Only in the tropics (sub-Saharan Africa and the Amazon
basin of Brazil) are there still sizable tracts of land not presently utilized that
have enough water to grow crops. However, as discussed previously, many believe that the forests should remain as they are because they lend stability to
the biosphere.

High-yield Plants

The developed countries, especially the United States, have had spectacular
success in increasing yields by utilizing monoculture agriculture and this
method of agriculture, along with its many environmental problems, is now
being exported to the developing countries. The term "Green Revolution" was
coined to describe the introduction and rapid spread of high-yielding wheat
and rice plants (fig. 33.16) that are especially developed to grow in warmer
countries when they are supplied with generous amounts of fertilizer and water.
Research efforts are currently striving toward providing plants with improved
internal efficiency. The new focus is on greater photosynthetic efficiency, more
efficient nutrient and water uptake, improved biological nitrogen fixation, and
genetic resistance to pests and environmental stress.

Fertilizer

The application of fertilizer has contributed greatly to increased yield. When
fertilizer is first applied to soil, there is a dramatic increase in yield, but later,
as more and more fertilizer is used, the increase in yield drops off. This suggests that the developed countries could very well do with less fertilizer, while

Behavior and Ecology

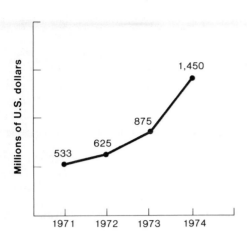

the developing countries need to apply more. Unfortunately, the rising cost of energy has caused fertilizer cost to more than double (fig. 33.17) and developing countries will most likely find it difficult to make or buy adequate amounts of fertilizer.

Legumes, you will recall, increase the nitrogen content of the soil because their roots are infected with nitrogen-fixing bacteria. Research goes forward on the possibility of infecting more plants with these bacteria and even on the possibility of transferring the nitrogen-fixing genes to plant cells themselves by means of recombinant DNA methods.

Energy

High agricultural yields are extremely dependent on both indirect and direct uses of energy. In the United States, fossil fuels have supplied the supplemental energy, as is apparent when one considers that nearly all of the large farm machinery utilizes fossil fuel energy. In Japan, however, much of the supplemental energy required for high yields is provided by human labor. Since we are now entering a time when fossil fuels are in short supply, it would be wise for the developing countries to rely on human energy to increase yield. Keeping people in rural areas close to the source of food would also help solve the problem of transportation and packaging.

Dietary Protein

Providing sufficient calories will not assure an adequate diet; there must also be enough protein in the food eaten. Due to a lack of dietary protein, children in the developing countries often have the symptoms of kwashiorkor in which the entire body is bloated, the skin is discolored, a rash is present, and the hair has an orange-reddish tinge (fig. 9.17).

Plants are low-quality, nutritionally incomplete protein sources because, while they contain some amino acids, they lack others. It is possible, however, to eat a combination of plants so that together they contain an acceptable level of all the essential amino acids. For example, wheat and beans complement each other to give a balance that is comparable to a high-quality protein such as meat and cheese.

In the developed countries most of the grain produced is fed to cattle, pigs, or chickens, which then serve as a source of high-quality protein. In many developing countries most grain must be consumed directly in order to provide sufficient calories. Therefore, beef is not expected to supply the necessary protein. Another possibility is fish consumption. Between 1950 and 1970, fish supplied an increasing portion of the human diet until the catch averaged some

Figure 33.18

World fish catch per capita. The amount of fish caught per capita has remained at the same level for several years due to increased population growth.

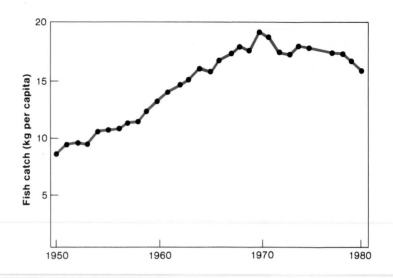

Figure 33.19

Aquaculture methods. *a.* Catfish are raised in ponds and fed commercial fish pellets. *b.* A natural detritus food web supported by animal manure. *c.* Phytoplankton are placed in clear fiberglass cylinders that allow maximum input of solar energy. The phytoplankton that flourish when fertilized with manure serve as food for fish. Methods *b.* and *c.* are more ecologically sound than *a.*

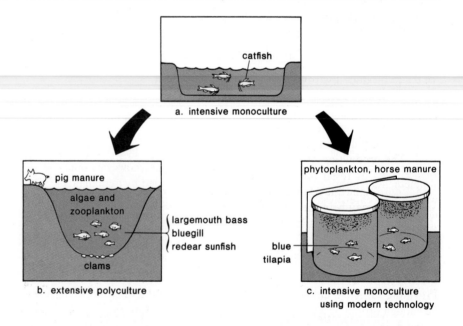

18.5 kilograms per person annually (fig. 33.18). Since that time, although the catch has increased, the per capita share has decreased due to population growth. Also, overexploitation of the ocean's fisheries and pollution of coastal waters may eventually cause a decline in the yearly catch. Therefore, the expansion of aquaculture—the cultivation of fish—is highly desirable. There are some innovative ideas such as the ones described in fig. 33.19 that could make aquaculture an ordinary part of everyone's life, even in the developed countries.

Summary

The human population is expanding exponentially and even though the growth rate has declined, there is a large increase in the population each year. Still, the doubling time has decreased from 35 to 40 years. Since a doubling of the

Behavior and Ecology

population means a doubling of the goods and services needed to sustain the population and since the carrying capacity of the earth is unknown, a decreasing growth rate is welcomed by many.

The history of the world population shows that the developed countries underwent a demographic transition between 1950 and 1975 so that their growth rate is now only 0.6 percent. The developing countries are just now undergoing demographic transition and since it was delayed, their growth rate went as high as 2.4 percent. Now it is declining slowly. Reproduction replacement will not bring about zero population growth even in the developed countries because of a baby boom that occurred after World War II. In the developing countries, where the average age is less than 15, it will be many years before reproduction replacement will mean zero population growth.

Resource consumption is dependent on population size and on consumption per individual. The developing countries are responsible for the first type of environmental impact and the developed countries for the second type. According to the Malthusian view, we are running out of nonrenewable resources and only conservation can help extend their depletion curves. According to the Cornucopian view, we will continue to find new ways to exploit the environment.

Exploitation of the environment is dependent on a plentiful energy source. Among the nonrenewable sources, coal is in ready supply and can be utilized without developing new technologies. Each of the three types of nuclear power has a drawback that has prevented full-scale operation in this country. All of the renewable resources still hold some promise. Solar One is now in operation and solar cells are being perfected. Solar energy can be used to hydrolyze water to produce hydrogen gas or liquid, which can be substituted for fossil fuels. Synthetic fuels is a catchall phrase for petroleum and petroleum substitutes from unconventional sources. Coal, oil shale, and biomass can all be used to produce synthetic fuels. The burning of the synthetic fuels will cause a new set of pollution problems.

The production of food has become increasingly dependent on supplements, such as fossil fuel energy and fertilizer, because the amount of new agricultural land that can be cultivated is limited. Development of new high-yield plants may help the developing countries feed expected increases in population, but there may still be a problem regarding dietary protein. It is doubtful that oceanic fishing can be counted on to supply protein and as yet, aquaculture is not sufficiently developed to do so.

Study Questions

1. Define exponential growth. (p. 771) Draw a growth curve to represent exponential growth and explain why a curve representing population growth usually levels off. (p. 774)
2. Calculate the growth rate and doubling time for a population in which the birthrate is 20 and the death rate is 2 per 1000. (p. 773)
3. Define demographic transition. When did the developed countries undergo demographic transition? When did the developing countries undergo demographic transition? (p. 776)
4. Give at least three differences between the developed countries and the developing countries. (pp. 775–78)
5. Contrast the Malthusian view and the Cornucopian view toward nonrenewable supplies. (pp. 780–81)
6. Draw a typical depletion curve and relate it to the consumption of fossil fuels. (pp. 779, 782)
7. Name the three types of nuclear power and give at least one related drawback to each. (p. 782)

8. Name at least four types of renewable energy resources. (p. 782) What types of fuels might be produced from these sources? (pp. 783–88)
9. Give at least three reasons why intensive monoculture does not seem to be a feasible solution to the food crises of developing countries. (pp. 788–92)

Selected Key Terms

exponential growth (eks″po-nen′shal grōth)
growth rate (grōth rāt)
doubling time (dŭ′b′ling tīm)
carrying capacity (kar′e-ing kah-pas′ĭ-te)
biotic potential (bi-ot′ik po-ten′shal)
environmental resistance (en-vi″ron-men′tal re-zis′tans)
developed countries (de-vel′opt kun′trēz)
demographic transition (dem-o-graf′ik tran-zi′shun)
developing countries (de-vel′op-ing kun′trēz)
age structure diagram (āj struk′tŭr di′ah-gram)
nonrenewable resource (non-re-nu′ah-b′l re′sors)
renewable resource (re-nu′ah-b′l re′sors)
Cornucopian view (kor″nu-ko′pe-an vu)
Malthusian view (mal-thu′se-an vu)
oil shale (oil shāl)
nuclear fission (nu′kle-ar fish′un)
breeder reactor (brēd′er re-ak′tor)
nuclear fusion (nu′kle-ar fu′zhun)
solar energy (so′lar en′er-je)
geothermal energy (je″o-ther′mal en′er-je)

Further Readings

Bebbington, W. P. 1976. The reprocessing of nuclear fuels. *Scientific American* 235(6):30.

Cochran, N. P. 1976. Oil and gas from coal. *Scientific American* 234(5):24.

Cohen, B. L. 1977. Disposal of radioactive wastes from fission reactors. *Scientific American* 236(6):21.

Donaldson, L. R., and Joyner, T. 1983. The salmonid fishes as a natural livestock. *Scientific American* 249(1):50.

Ehrlich, et al. 1977. *Ecoscience: Population, resources, environment.* San Francisco: W. H. Freeman.

Feldman, M., and Sears, E. R. 1981. The wild gene resources of wheat. *Scientific American* 244(1):102.

Flower, A. R. 1978. World oil production. *Scientific American* 238(3):238.

Frejkay, T. 1973. The prospects for a stationary world population. *Scientific American* 228(3):15.

Griffith, E. D., and Clarke, A. W. 1979. World coal production. *Scientific American* 240(1):38.

Gwatkin, D. R. 1982. Life expectancy and population growth in the third world. *Scientific American* 246(5):57.

Hoff, J. E., and Janick, J., eds. 1973. *Food: Readings from* Scientific American. San Francisco: W. H. Freeman.

Keely, C. B. 1982. Illegal migration. *Scientific American* 246(3):41.

Keyfitz, N. 1976. World resources and the world middle class. *Scientific American* 235(1):28.

Long, L., and DeAre, D. 1983. The slowing of urbanization in the U.S. *Scientific American* 249(1):33.

Martin, P. I. 1983. Labor-intensive agriculture. *Scientific American* 249(4):54.

Nef, J. W. 1977. An early energy crisis and its consequences. *Scientific American* 236(5):140.

Scientific American. 1976. 235(3). Entire issue devoted to food and agriculture.

Scientific American. 1978. 239(3). Entire issue devoted to human population.

Scientific American. 1980. 243(3). Entire issue devoted to population and economic concerns.

Swaminathan, M. S. 1984. Rice. *Scientific American* 250(1):80.

Light versus Electron Microscope

In a light microscope, a concentrated beam of light passes through the object on the stage in such a way that the emerging rays of light indicate the areas of light and dark in the object. Very often the prepared specimen has been treated with a stain to enhance these contrasts. A magnified image of the object is achieved when these rays of light are focused by the objective lens; the image is further magnified by means of the ocular lens. Since eyes are sensitive to light rays, the image may be viewed directly by the experimenter.

In an electron microscope, a concentrated beam of electrons passes through a vacuum to bombard the object. The vacuum is created when air is pumped out of the tube transmitting the electrons. The electrons are scattered by the object in such a way as to indicate areas of light and dark. The electrons are focused, and a magnified image is formed as the electrons pass the lenses of the electron microscope. The lenses of an electron microscope are not glass; they are coils of wire about which a magnetic field exists because of an electric current that travels within the wire. This field can focus electrons because they carry a negative charge that makes them sensitive to magnetic fields.

Since eyes are not sensitive to electrons, the image cannot be viewed directly; instead, it is projected onto a screen at the foot of the microscope where the electrons excite the chemical coating of the screen, producing light rays that can be seen by the viewer. A permanent record can also be made on a photographic plate, and this record is an **electron micrograph.**

Appendix

Figure A.1
Light versus Electron Microscope.

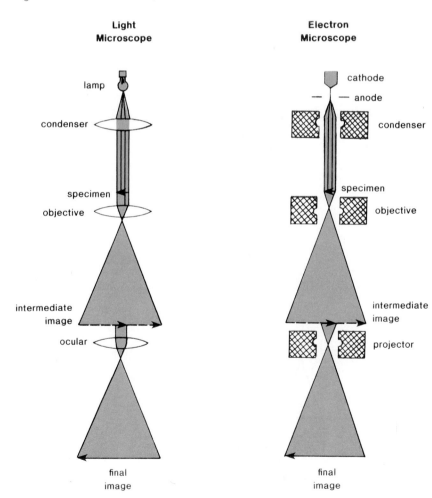

Figure A.2
Periodic Table of the Elements.

Ia																	0
1 H **1.008**	IIa												IIIa	IVa	Va	VIa	VIIa	2 He 4.00
3 Li 6.94	4 Be 9.01												5 B 10.81	6 C 12.01	7 N 14.00	8 O 15.99	9 F 18.99	10 Ne 20.18
11 Na 22.99	12 Mg 24.31	IIIb	IVb	Vb	VIb	VIIb		VIIIb			IB	IIIB	13 Al 26.98	14 Si 28.09	15 P 30.97	16 S 32.06	17 Cl 35.45	18 Ar 39.95
19 K 39.10	20 Ca 40.08	21 Sc 44.96	22 Ti 47.90	23 V 50.94	24 Cr 51.99	25 Mn 54.94	26 Fe 55.85	27 Co 58.93	28 Ni 58.71	29 Cu 63.54	30 Zn 65.37	31 Ga 69.72	32 Ge 72.59	33 As 74.92	34 Se 78.96	35 Br 79.91	36 Kr 83.80
37 Rb 85.47	38 Sr 87.62	39 Y 88.91	40 Zr 91.22	41 Nb 92.91	42 Mo 95.94	43 Tc (99)	44 Ru 101.97	45 Rh 102.91	46 Pd 106.4	47 Ag 107.87	48 Cd 112.40	49 In 114.82	50 Sn 118.69	51 Sb 121.75	52 Te 127.60	53 I 126.90	54 Xe 131.30
55 Cs 132.91	56 Ba 137.34	see below 57-71	72 Hf 178.49	73 Ta 180.95	74 W 183.85	75 Re 186.2	76 Os 190.2	77 Ir 192.2	78 Pt 195.09	79 Au 196.97	80 Hg 200.59	81 Tl 204.37	82 Pb 207.19	83 Bi 208.98	84 Po (210)	85 At (210)	86 Rn (222)
87 Fr (223)	88 Ra (226)	see below 89-103	104 Rf (261)	105 Ha (260)	106 * 263	*newly produced											

57 La 138.91	58 Ce 140.12	59 Pr 140.91	60 Nd 144.24	61 Pm (147)	62 Sm 150.35	63 Eu 151.96	64 Gd 157.25	65 Tb 158.92	66 Dy 162.50	67 Ho 164.93	68 Er 167.26	69 Tm 168.93	70 Yb 173.04	71 Lu 174.97
89 Ac (227)	90 Th 232.04	91 Pa (231)	92 U 238.03	93 Np (237)	94 Pu (242)	95 Am (243)	96 Cm (247)	97 Bk (247)	98 Cf (251)	99 Es (254)	100 Fm (253)	101 Md (256)	102 No (254)	103 Lw (257)

Table A.1 Units of measurement

Unit	Symbol	Seen by
Centimeter	cm = 0.4 inch	Naked eye
Millimeter	mm = 0.1 cm	Naked eye
Micrometer	μm = 0.001 mm	Light microscope
Nanometer	nm = 0.001 μm	Electron microscope

Figure A.3
Six-inch ruler with the equivalent centimeter units.

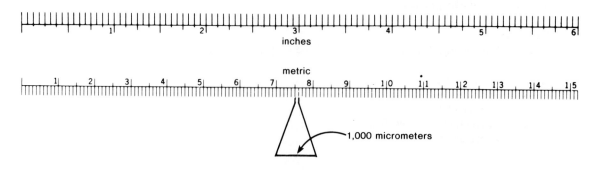

Appendix

A comparison of these two microscopes shows that they both illuminate the object, magnify the object, and produce an image that eventually can be viewed by the observer. The most important difference between these two instruments is not the degree to which they magnify but lies instead in their resolving power. This term is used to indicate the amount of detail that can be distinguished by a microscope. The physical laws of optics tell us that resolving power is dependent on the wavelength of the light or electron beam. The theoretical limit of the resolving power of the light microscope is 200 nm(nanometers), while that of the electron microscope is 0.5 nm. This means that any two structures separated by less than 200 nm in the light microscope and 0.5 nm in the electron microscope will appear as one object. Thus, the electron microscope allows us to see much more.

One drawback to the electron microscope has been that the specimen could be viewed only after drying, because a vacuum is needed to produce the electron beam. We do not know the full extent to which this drying process distorts the true appearance of the specimen. Methods are now being investigated to allow the observation of living materials in their natural state.

Metric System

As mentioned in the text, biologists describe the sizes of objects such as cell organelles in terms of the metric system. The linear units of measurement most often used are given in table A.1. The ruler (fig. A.3) allows you to visualize the relationship of these units.

Classification of Organisms

Kingdom Monera
Unicellular prokaryotic organisms lacking distinct nuclei and membrane-bound organelles. In bacteria, nutrition is principally by absorption, but some are photosynthetic or chemosynthetic. Cyanobacteria are photosynthetic.
Phylum Schizophyta: bacteria
Phylum Cyanobacteria: blue-green bacteria

Kingdom Protista
Typically unicellular, sometimes multicellular, eukaryotic organisms with distinct nuclei and organelles. Nutrition by photosynthesis, absorption, or ingestion.
Phylum Mastigophora: flagellated protozoans
Phylum Sarcodina: amoeboid protozoans
Phylum Ciliophora: ciliated protozoans
Phylum Sporozoa: parasitic protozoans
Phylum Chrysophyta: golden-brown algae
Phylum Euglenophyta: *Euglena* and relatives
Phylum Pyrrophyta: dinoflagellates

Kingdom Fungi
Multinucleate plantlike organisms lacking photosynthetic pigments. Nutrition by absorption.
Division Zygomycota: primitive fungi
Division Ascomycota: sac fungi
Division Basidiomycota: club fungi
Division Deuteromycota: imperfective fungi
Division Myxomycota: slime molds

Kingdom Plantae

Multicellular eukaryotic organisms with rigid cell walls and chlorophyll. Nutrition principally by photosynthesis.

Division Rhodophyta: red algae
Division Phaeophyta: brown algae
Division Chlorophyta: green algae
Division Bryophyta: mosses and liverworts
Division Tracheophyta
    Subdivision Psilopsida: *Psilotum*
    Subdivision Lycopsida: club mosses
    Subdivision Sphenopsida: horsetails
    Subdivision Pteropsida
        Class Filicineae: ferns
        Class Gymnospermae: conifers, cycads, ginkgoes
        Class Angiospermae: flowering plants
            Subclass Monocotyledoneae: grasses, lilies, and orchids
            Subclass Dicotyledoneae: various

Kingdom Animalia

Multicellular organisms without cell walls or chlorophyll. Nutrition principally ingestive with digestion in an internal cavity.

Phylum Porifera: sponges
Phylum Coelenterata: radially symmetrical marine animals
    Class Hydrozoa: *Hydra,* Portuguese man-of-war
    Class Scyphozoa: jellyfish
    Class Anthozoa: sea anemones and corals
Phylum Platyhelminthes: flatworms
    Class Turbellaria: free-living flatworms (planarian worms)
    Class Trematoda: parasitic flukes
    Class Cestoda: parasitic tapeworms
Phylum Nematoda: roundworms (*Ascaris*)
Phylum Rotifera: rotifers
Phylum Mollusca: soft-bodied, unsegmented animals
    Class Amphineura: chitons
    Class Monoplacophora: *Neopilina*
    Class Gastropoda: snails and slugs
    Class Pelecypoda: clams and mussels
    Class Cephalopoda: squids and octopuses
Phylum Annelida: segmented worms
    Class Polychaeta: sandworms
    Class Oligochaeta: earthworms
    Class Hirudinea: leeches
Phylum Arthropoda: joint-legged animals; exoskeleton
    Class Crustacea: lobsters, crabs, barnacles
    Class Arachnida: spiders, scorpions, ticks
    Class Chilopoda: centipedes
    Class Diplopoda: millipedes
    Class Insecta: grasshoppers, termites, beetles
Phylum Echinodermata: marine; spiny, radially symmetrical animals
    Class Crinoidea: sea lilies and feather stars
    Class Asteroidea: starfishes
    Class Ophiuroidea: brittle stars
    Class Echinoidea: sea urchin and sand dollar
    Class Holothuroidea: sea cucumbers

Phylum Hemichordata: acorn worms
Phylum Chordata: dorsal supporting rod (notochord) at some stage; dorsal hollow nerve
  cord; pharyngeal pouches or slits
    Subphylum Urochordata: tunicates
    Subphylum Cephalochordata: lancelets
    Subphylum Vertebrata: vertebrates
        Class Agnatha: jawless fishes (lampreys, hagfishes)
        Class Chondrichthyes: cartilaginous fishes (sharks, rays)
        Class Osteichthyes: bony fishes
            Subclass Crossopterygii: lobe-finned fishes
            Subclass Dipnoi: lungfishes
            Subclass Actinopterygii: ray-finned fishes
        Class Amphibia: frogs, toads, salamanders
        Class Reptilia: snakes, lizards, turtles
        Class Aves: birds
        Class Mammalia: mammals
            Subclass Prototheria: egg-laying mammals
                Order Monotremata: duckbilled platypus, spiny anteater
            Subclass Metatheria: marsupial mammals
                Order Marsupialia: opossums, kangaroos
            Subclass Eutheria: placental mammals
                Order Insectivora: shrews, moles
                Order Chiroptera: bats
                Order Edentata: anteaters, armadillos
                Order Rodentia: rats, mice, squirrels
                Order Lagomorpha: rabbits and hares
                Order Cetacea: whales, dolphins, porpoises
                Order Carnivora: dogs, bears, weasels, cats, skunks
                Order Proboscidea: elephants
                Order Sirenia: manatees
                Order Perissodactyla: horse, hippopotamus, zebra
                Order Artiodactyla: pigs, deer, cattle
                Order Primates: lemur, monkeys, apes, humans
                    Superorder Prosimii: lemurs, tree shews, tarsiers, lorises,
                      pottos
                    Superorder Anthropoidea: monkeys, apes, humans
                      Superfamily Ceboidea: New World monkeys
                      Superfamily Cercopithecoidea: Old World monkeys
                      Superfamily Hominoidea: apes and humans
                        Family Hylobatidae: gibbons
                        Family Pongidae: chimpanzee, gorilla, orangutan
                        Family Hominidae: *Australopithecus,** Homo
                          erectus,** Homo sapiens* neanderthal,* Homo
                          sapiens sapiens*

*extinct

The words in this Glossary are followed by a phonetic guide to pronunciation. The markings in this guide may be confusing at first, because they are unlike the diacritical marks that appear in an ordinary dictionary. This is a simplified system that is standard in medical usage and terminology.

Any unmarked vowel that ends a syllable or stands alone as a syllable is long. In this system, then, the word *play* would be spelled *pla*.

Any unmarked vowel that is followed by a consonant has the short sound. The word *tough*, for instance, would be spelled *tuf*.

If a long vowel does appear in the middle of a syllable (followed by a consonant), then it is marked with the macron (¯), the standard sign for a long vowel. For instance, the word *plate* would be phonetically spelled plāt.

Similarly, if a vowel stands alone or ends a syllable, but should have the short sound, it is marked with a breve (˘).

A

a- (a)   without, lacking.

ab- (ab)   away from, off.

abiotic (ab″e-ot′ik)   not including living organisms, their effects, or products.

abyssal plain (ah-bis′al plān)   the floor of an ocean.

accommodation (ah-kom″o-da′shun)   lens adjustment in order to see close objects.

acetylcholine (as″e-til-ko′lēn)   a neurotransmitter substance secreted at the ends of many neurons; responsible for the transmission of a nerve impulse across a synaptic cleft.

acid (as′id)   a solution in which pH is less than 7; a substance that contributes or liberates hydrogen ions (protons) in a solution.

actin (ak′tin)   one of the two major proteins of muscle; makes up thin filaments in myofibrils of muscle cells. *See* myosin.

action potential (ak′shun po-ten′shal)   the change in potential propagated along the membrane of a neuron; the nerve impulse.

active acetate (ak′tiv as′e-tāt)   an acetyl group attached to coenzyme A; a product of the transition reaction that links glycolysis to the Krebs cycle.

active site (ak′tiv sīt)   the region on the surface of an enzyme where the substrate binds and where the reaction occurs.

active transport (ak′tiv trans′port)   transfer of a substance into or out of a cell against a concentration gradient by a process that requires a carrier and expenditure of energy.

ad- (ad)   next to, at.

adaptation (ad″ap-ta′shun)   the fitness of an organism for its environment, including the process by which it becomes fit, in order that it may survive and reproduce; also the adjustment of sense receptors to a stimulus so that the stimulus no longer excites them.

adaptive radiation (ah-dap′tiv ra″de-a′shun)   the division of a single species into many species, each adapted to a different way of a life.

Addison's disease (ad′i-sonz di-zēz′)   a condition resulting from a deficiency of adrenal cortex hormones.

adenine (ad′e-nin)   a purine (nitrogenous base) that is found in DNA and ATP as well as in other important biological molecules.

adenosine diphosphate (ADP) (ah-den′o-sēn di-fos′fāt)   similar in structure to ATP except that it contains only two phosphate groups.

adenosine triphosphate (ATP) (ah-den′o-sēn tri-fos′fāt)   a compound containing adenine, ribose, and three phosphates, two of which are high-energy phosphates. It is the ''common currency'' of energy for most cellular processes.

adrenal (ah-dre′nal)   endocrine glands of vertebrates located atop the kidneys.

adrenalin (ah-dren′ah-lin)   a hormone produced by the adrenal medulla that stimulates ''fight or flight'' reactions. Also called epinephrine.

aerobic (a″er-ōb′ik)   growing or metabolizing only in the presence of oxygen as in aerobic respiration.

afferent (af′er-ent)   moving toward; for example, neurons that carry nerve impulses to the central nervous system.

afterbirth (af′ter-berth″)   the placenta that is expelled after the birth of a child.

age structure diagram (āj struk′tūr di′ah-gram)   a diagram that depicts the number or percentage of persons at each age level in a population.

agglutination (ah-gloo″ti-na′shun)   clumping of cells, particularly in reference to red cells involved in an antigen-antibody reaction.

agranulocytes (ah-gran′u-lo-sīts″)   white blood cells that do not contain distinctive granules.

albumin (al-bu′min)   a protein found in plasma and egg white.

aldosterone (al″do-ster′ōn)   a hormone, secreted by the adrenal cortex, that functions in regulating sodium and potassium concentrations of the blood.

algae (al′je)   aquatic photosynthesizing organisms that are either unicellular or multicellular, with little structural or functional differentiation; producers in aquatic food chains.

-algia (al′je-ah)   ache, pain.

alkaline (al′kah-lin)   a solution in which pH is more than 7; a substance that contributes or liberates hydroxide ions in a solution; basic; opposite of acidic.

all-, allo- (al, al′o)   other, different.

allantois (ah-lan′to-is)   one of the extraembryonic membranes of reptiles, birds, and mammals; a pouch serving as a repository for nitrogenous waste or as a source of blood vessels to and from the chorion or placenta.

allele (ah-lēl′)   an alternative form of a gene that occurs at a given chromosomal site (locus).

Glossary

allergy (al'er-je) an altered reaction of body tissues to an antigenic substance, which in a nonsensitive person produces no effect.

alternation of generations (awl"ter-na'shun uv jen"ē-ra'shunz) a life cycle that involves two distinct adult forms; particularly in plants where one form is haploid and the other is diploid.

altruism (al'troo-izm) behavior performed for the benefit of others without regard to its possible detrimental effect on the performer.

alveoli (al-ve'o-li) saclike structures that are the air sacs of a lung.

amino acid (ah-me'no as'id) a unit of protein that takes its name from the fact that it contains an amino group (NH_2) and an acid group (COOH).

ammonia (ah-mo'ne-ah) NH_3, a nitrogenous waste product resulting from deamination of amino acids.

amnion (am'ne-on) one of the extraembryonic membranes of reptiles, birds, and mammals; a fluid-filled sac around the embryo.

amoeboid (ah-me'boid) like an amoeba, moving or eating by means of pseudopodia (cytoplasmic outflows).

amphibian (am-fib'e-an) a member of the vertebrate class Amphibia such as frog or salamander.

amylase (am'i-lās) a starch digesting enzyme secreted by salivary glands and the pancreas.

an- (an) without.

anabolism (ah-nab'o-lizm) synthetic chemical reactions in which the product is larger than any reactant.

anaerobic (an-a"er-ōb'ik) growing or metabolizing only in the absence of molecular oxygen as in anaerobic respiration.

analogous (ah-nal'o-gus) similar in function but not in structure; particularly in reference to similar adaptations.

androgen (an'dro-jen) male sex hormone.

anemia (ah-ne'me-ah) a condition characterized by a deficiency of red blood cells or of hemoglobin.

angio-, angium (an'je-o, an'je-um) container, receptacle.

angiosperm (an'je-o-sperm") flowering plant with seeds enclosed in fruit.

annelid (an'ah-lid) segmented worm belonging to the phylum Annelida, such as the earthworm and sandworm.

antagonistic (an-tag"o-nist'ik) opposing another as in counteracting muscles.

ante- (an'te) before, in front of.

anterior (an-te're-or) the head region of an organism.

anther (an'ther) that portion of a stamen in which pollen is formed.

antheridium (an"thah-rid'y-um) male organ in certain nonseed plants where swimming sperm are produced.

anthropoids (an'thro-poidz) higher primates, including only monkeys, apes, and humans.

anti-, ant- (an'ty, ant) opposite, against, counter.

antibody (an'ti-bod"e) a protein produced in response to the presence of some foreign substance in the blood or tissues.

antibody mediated immunity (an'ti-bod"e me'de-āt"ed i-mu'ni-te) immunity dependent upon the activities of B cells.

anticodon (an"ti-ko'don) a "triplet" of three nucleotides in transfer RNA that pairs with a complementary triplet (codon) in messenger RNA.

antidiuretic hormone (ADH) (an"ti-di"u-ret'ik hor'mōn) sometimes called vasopressin, a hormone secreted by the posterior pituitary that controls the rate at which water is reabsorbed by the kidneys.

antigen (an'ti-jen) a foreign substance, usually a protein, that stimulates the immune system to produce antibodies.

antipredator defense (an"ti-pred'ah-tor de-fens') a physiological or behavioral activity or a structural modification that protects an organism from its predators.

aorta (a-or'tah) major systemic artery that receives blood from the left ventricle.

arboreal (ar-bo're-al) organisms that spend their lives in trees.

arch- (arch) primitive, original.

archegonium (ar"kē-go'ne-um) female organ in certain nonseed plants where an egg is produced.

archenteron (ar-ken'ter-on) central cavity or primitive gut in the animal embryo.

arteriole (ar-te're-ōl) a branch from an artery that leads into a capillary.

artery (ar'ter-e) a vessel that takes blood away from the heart; characteristically possessing thick elastic walls.

arthropod (ar'thro-pod) an invertebrate belonging to the phylum Arthropoda that possesses jointed appendages, such as crayfish, lobster, and grasshopper.

asexual reproduction (a-seks'u-al re"pro-duk'shun) a form of reproduction, such as budding or simple fission, that does not involve the use of gametes.

aster (as'ter) short rays of microtubules that appear at the ends of the spindle apparatus in animal cells during cell division.

atom (at'om) smallest unit of matter.

ATP (a te pe) see adenosine triphosphate.

atria (*sing.* atrium) (a'tre-ah) chambers; particularly the upper chambers of the heart that lie above the ventricles.

atrioventricular (a"tre-o-ven-trik'u-lar) a structure in the heart that pertains to both the atria and ventricles; for example, an atrioventricular valve is located between an atrium and a ventricle.

atrioventricular node (a"tre-o-ven-trik'u-lar nōd) a small region of neuromuscular tissue located near the septum of the heart that transmits impulses from the S-A node to the ventricular walls; A-V node.

auditory canal (aw'di-to"re kah-nal') a tube in the external ear that lies between the pinna and the tympanic membrane.

australopithecines (aw"strah-lo-pith'e-sinz) referring to three species of *Australopithecus*, the first recognized hominids.

auto- (aw'to) self.

autonomic nervous system (aw"to-nom'ik ner'vus sis'tem) a part of the peripheral nervous system that controls smooth muscles and glands.

autotroph (aw'to-trof) an organism that is capable of making its food (organic molecules) from inorganic molecules.

A-V node (a-ve nōd) see atrioventricular node.

axon (ak'son) process of a neuron that conducts nerve impulses away from the cell body.

B

bacteriophage (bak-te're-o-fāj") a virus that infects a bacterial cell.

bacterium (bak-te're-um) a microscopic, rod-shaped, round, or spiral unicellular prokaryote other than a cyanobacterium.

base (bās) a solution in which pH is more than 7; a substance that contributes or liberates hydroxide ions in a solution; alkaline; opposite of acidic. Also, in genetics the chemicals adenine, guanine, cytosine, thymine, and uracil that are found in DNA and RNA.

B cell (be sel) a type of lymphocyte involved in antibody mediated immunity. When stimulated, a B cell develops into an antibody-producing plasma cell.

behavior (be-hāv'yor) the way in which an organism acts in response to stimuli.

benthic zone (ben'thik zōn) a region containing those organisms that reside at the bottom of bodies of water.

bi- (bi) two.

bicarbonate ion (bi-kar'bo-nāt i'on) HCO_3^-.

bilateral symmetry (bi-lat'er-al sim'ē-tre) the condition of having a right and left half so that only one vertical cut gives two equal halves.

bile (bīl) a secretion of the liver that is temporarily stored in the gallbladder before being released into the small intestine where it emulsifies fat.

binary fission (bi'na-re fish'un) reproduction by simple cell division that does not involve a mitotic spindle.

binomial system (bi-no'me-al sis'tem) the assignment of two names to each organism, the first of which designates the genus and the second of which designates the species.

bio- (bi'o) life, living.

biogeography (bi"o-je-og'rah-fe) study of the geographic distribution of plants and animals.

biological evolution (bi-o-loj'ē-kal ev"o-lu'shun) see evolution.

biome (bi'ōm) major ecological region that has the characteristics of a climax community.

biosphere (bi'o-sfer) that part of the earth's surface and atmosphere where living organisms exist.

biotic (bi-ot′ik) pertaining to any aspect of life, especially to characteristics of entire populations or ecosystems.

biotic potential (bi-ot′ik po-ten′shal) the maximum population growth rate under ideal conditions.

blast, blasto- (blast, blas′to) embryo.

blastula (blas′tu-lah) an early stage in animal development; usually a hollow sphere of cells about a central cavity called the blastocoele.

bond energy (bond en′er-je) the amount of energy stored in a chemical bond.

Bowman's capsule (bo′manz kap′sūl) a double-walled cup that surrounds the glomerulus at the beginning of the kidney tubule.

breeder reactor (brēd′er re-ak′tor) a nuclear reactor that produces more nuclear fuel than it consumes by converting nuclear wastes into plutonium 239.

bronchi (brong′ki) the two major divisions of the trachea leading to the lungs.

bronchiole (brong′kē-ōl) the smaller air passages in the lungs of mammals.

browsers (browz′erz) animals that feed on higher-growing vegetation such as shrubs and trees.

bryophytes (bri′o-fits) members of the division, Bryophyta, that contains plants lacking vascular tissue and having the gametophyte generation dominant.

budding (bud′ing) a form of asexual reproduction in which the new organism simply grows from the body of the parent and eventually detaches itself.

buffer (buf′er) a substance or compound that prevents large changes in the pH of a solution.

C

calorie (kal′o-re) the amount of heat required to raise one kilogram of water one degree centigrade.

Calvin cycle (kal′vin si′kl) a circular series of reactions by which CO_2 fixation occurs within chloroplasts.

cambium (kam′be-um) meristem tissue found in the stems of plants; vascular cambium and cork cambium.

capillaries (kap′i-ler″es) microscopic vessels located in the tissues connecting arterioles to venules through whose thin walls molecules either exit or enter the blood.

carbaminohemoglobin (kar-bam″i-no-he″mo-glo′bin) hemoglobin that has combined with carbon dioxide.

carbohydrate (kar″bo-hi′drāt) organic compounds with the general formula $(CH_2O)_n$ including sugars, starch, and cellulose.

carbonic anhydrase (kar-bon′ik an-hi′drās) an enzyme found in red cells that speeds up the transformation of carbon dioxide to carbonate.

carcinogen (kar-sin′o-jen) a substance or agent capable of inducing cancerous growth.

cardiac (kar′de-ak) of or pertaining to the heart.

carnivores (kar′ni-vorz) animals that eat flesh.

carrier (kar′e-er) a molecule that combines with a substance and actively transports it through the cell membrane; an individual that transmits an infectious or genetic disease.

carrying capacity (kar′e-ing kah-pas′i-te) the largest number of organisms of a particular species that can be maintained indefinitely in an ecosystem.

cartilage (kar′ti-lij) a flexible connective tissue, usually part of the skeleton, which is composed of cells in a matrix.

catalyst (kat′ah-list) a substance that regulates the speed at which a chemical reaction occurs without affecting the end point of the reaction and without being used up in the reaction. Enzymes are biological catalysts.

cecum (se′kum) a blind pouch below the site where the small intestine is joined to the large intestine to which is attached the appendix.

cell first hypothesis (sel ferst hi-poth′e-sis) a suggestion that only the first cell or cells contained true proteins and nucleic acids.

cell mediated immunity (sel me′de-āt″ed i-mu′ni-te) immunity dependent upon the activities of T cells.

cell membrane (sel mem′brān) a membrane that surrounds the cytoplasm of cells and regulates the passage of molecules into and out of the cell.

cell plate (sel plāt) a structure that forms between two plant cells during telophase and marks the location of the cell membrane and cell wall.

cellular respiration (sel′u-lar res″pi-ra′shun) the reactions of glycolysis, Krebs cycle, and electron transport system that provide energy and the anabolic reactions that use this energy to produce ATP.

cellulose (sel′u-lōs) a polysaccharide composed of glucose molecules; the chief constituent of a plant's cell wall.

cell wall (sel wawl) a relatively rigid structure composed mostly of polysaccharides that surrounds the cell membrane of plants, fungi, and bacteria.

central dogma (sen′tral dog′mah) modern genetic hypothesis in which DNA serves as a template for its own replication and the synthesis of RNA, which in turn directs the order of the amino acids in proteins.

central nervous system (CNS) (sen′tral ner′vus sis′tem) the brain and spinal cord in vertebrate animals.

centriole (sen′tre-ōl) a short, cylindrical organelle in animal cells that contains microtubules in a 9 + 0 pattern and is associated with the formation of the spindle during cell division.

centromere (sen′tro-mēr) a region of attachment for a chromosome to a spindle fiber that is generally seen as a constricted area.

cephalization (sef″al-i-za′shun) the presence of a well-developed head region with sense organs and a definite brain.

cerebellum (ser″ě-bel′um) the part of the vertebrate brain that controls muscular coordination.

cerebral cortex (ser′ě-bral kor′teks) the external layer of the cerebrum that is gray in color and highly convoluted in humans.

cerebrum (ser′ě-brum) the main portion of the vertebrate brain that is responsible for consciousness.

cervix (ser′viks) the lower and narrower section of the uterus that projects into the vagina.

chemical evolution (kem′i-kal ev″o-lu′shun) a gradual increase in the complexity of chemical compounds that is believed to have brought about the origination of the first cell or cells.

chemosynthesis (ke″mo-sin′the-sis) the process of making food by using energy derived from the oxidation of reduced materials in the environment.

chlorophyll (klo′ro-fil) the green pigment found in photosynthesizing organisms that is capable of absorbing energy from the sun's rays.

chloroplast (klo′ro-plast) a membrane-bounded organelle in which membranous grana contain chlorophyll and where photosynthesis takes place.

cholinesterase (ko″lin-es′ter-ās) an enzyme that breaks down acetylcholine, a neurotransmitter substance.

chordate (kor′dāt) a member of the animal phylum Chordata possessing, at some time in the life history, a notochord, dorsal nerve chord, and pharyngeal gill slits.

chorion (ko′re-on) an extraembryonic membrane that forms an outer covering around the embryo in reptiles, birds, and mammals and contributes to the formation of the placenta in mammals.

choroid (ko′roid) the vascular, pigmented middle layer of the wall of the eye.

chrom-, chrome (krōm, krom) colored, pigment.

chromatids (kro′mah-tidz) the two identical parts of a chromosome following replication of DNA.

chromatin (kro′mah-tin) threadlike network in the nucleus that is made up of DNA and proteins.

chromosomes (kro′mo-sōmz) rod-shaped bodies in the nucleus, particularly during cell division, that contain the hereditary units or genes.

cilia (sil′e-ah) hairlike projections that are used for locomotion by many unicellular organisms and have various purposes in higher organisms.

ciliary muscle (sil′e-er″e mus′el) a muscle that controls the curvature of the lens of the eye.

circadian rhythm (ser″kah-de′an rith′m) a regular physiological or behavioral event that occurs on an approximately 24-hour cycle.

citric acid (sit′rik as′id) a metabolite that is designated as the first molecule in the Krebs cycle (citric acid cycle).

class (klas) in taxonomy, the category below phylum and above order.

clavicle (klav'ĭ-k'l) a slender, rodlike bone located at the base of the neck that runs between the sternum and the shoulders.

cleaning symbiosis (klēn'ing sim''bĭ-o'sis) a mutualistic relationship in which one type organism gains benefit by cleaning another that is benefited by having been cleaned of debris and parasites.

cleavage (klēv'ij) cell division of the fertilized egg that is unaccompanied by growth so that numerous small cells result.

clitoris (kli'to-ris) a small, erectile body in females located at the vaginal vestibule where the labia minora meet.

clonal selection (klōn'al sĕ-lek'shun) the ability of an antigen to cause clones of B cells to appear, all of whom produce the same type antibody.

clone (klōn) asexually produced organisms having the same genetic makeup; also DNA fragments from an external source that have been reproduced by *E. coli*.

coacervate (ko-as'er-vāt) a mixture of polymers that may have preceded the origination of the first cell or cells.

cochlea (kok'le-ah) that portion of the inner ear that resembles a snail's shell and contains the organ of Corti, the sense organ for hearing.

codominance (ko-dom'ĭ-nans) the ability of both members of an allelic pair to express themselves so that the phenotype shows both characteristics.

codon (ko'don) a "triplet" of three nucleotides in messenger RNA that directs the placement of a particular amino acid into a polypeptide chain.

coel- (sēl) hollow, cavity.

coelenterates (se-len'ter-āts) the common name for organisms in the phylum Coelenterata, such as *Hydra*, jellyfish, and coral.

coelom (se'lom) a body cavity of higher animals that is lined with mesoderm.

coenzyme (ko-en'zim) a molecule that aids the action of an enzyme, to which it is loosely bound.

coevolve (ko-e-volv') the interaction of two species such that each determines the evolution of the other species.

coitus (ko'ĭ-tus) sexual intercourse during which sperm are passed from one animal to another.

coleoptile (ko''le-op'til) a pointed sheath covering the shoot of grass seedlings.

collecting duct (kŏ-lekt'ing dukt) a tube that receives urine from several distal convoluted tubules.

colon (ko'lon) the large intestine of vertebrates.

colony (kol'o-ne) an organism that is a loose collection of cells that are specialized and cooperate to a degree.

com- (kom) together.

commensalism (kŏ-men'sal-izm) the relationship of two species in which one lives on or with the other without conferring either benefit or harm.

common ancestor (kŏ'mun an'ses-tor) an ancestor to two or more branches of evolution.

community (kŏ-mu'nĭ-te) a group of interacting plant and animal populations inhabiting a particular area.

compact bone (kom'pakt bōn) hard bone consisting of Haversian systems cemented together.

companion cell (kom-pan'yun sel) a specialized cell that lies adjacent to a sieve-tube cell in flowering plants.

competition (kom''pĕ-tish'un) interaction between members of the same or different species for a mutually required resource.

competitive exclusion principle (kom-pet'ĭ-tiv eks-kloo'zhun prin'sĭ-p'l) an observation that no two species can continue to compete for the same exact resources since one species will eventually become extinct.

competitive inhibition (kom-pĕ'tĭ-tiv in''hĭ-bish'un) reduction in rate of a reaction due to the presence of a compound that competes with the enzyme for the reactant(s) so that less of the desired product is produced per unit time.

complement (kom'plĕ-ment) a group of proteins in plasma that produce a variety of effects once an antigen-antibody reaction has occurred.

complementary base pairing (kom''plĕ-men'tă-re bās par'ing) pairing of bases found in DNA and RNA; adenine is always paired with either thymine (DNA) or uracil (RNA) and cytosine is always paired with guanine.

compound (kom'pownd) in inorganic chemistry, a combination of atoms in definite ratios, held together by chemical bonds; in organic chemistry, a substance containing only one kind of molecule, each composed of more than one kind of atom.

conditioned reflex (kon-dish'und re'fleks) a reflex that has been modified by experience.

cones (kōns) bright-light receptors in the retina of the eye that detect color and provide visual acuity; specialized structures composed of scale-shaped sporophylls in conifers.

congenital (kon-jen'ĭ-tal) pertaining to any condition present at birth.

conifer (kon'ĭ-fer) a cone-bearing seed plant, mostly trees, such as pines.

conjugation (kon''ju-ga'shun) a sexual union in which the nuclear material of one cell enters another.

connective tissue (kŏ-nek'tiv tish'u) a type of tissue characterized by cells separated by a matrix that often contains fibers.

consumers (kon-su'merz) organisms of one population that feed on members of other populations in an ecosystem.

continental shelf (kon''tĭ-nen'tal shelf) margin of the continents that forms a broad shallow strip below sea level and declines slowly to a depth of about 100 to 200 meters.

continental slope (kon''tĭ-nen'tal slōp) a boundary of an ocean basin that extends from the edge of the continental shelf to the deep-sea floor at depths of about 4,000 to 5,000 meters.

contractile vacuole (kon-trak'til vak'u-ōl) in certain one-celled organisms, a vacuole that slowly collects water before it is expelled through a pore in the cell membrane.

convergent evolution (kon-ver'jent ev''o-lu'shun) evolution toward the same types of adaptations among different groups of organisms.

copulation (kop''u-la'shun) sexual intercourse in which sperm are passed from one animal to another.

coral reef (kor'al rēf) a structure found in tropical waters formed by the buildup of the skeletons of colonial coelenterates where many and various types of organisms reside.

cork (kork) a tissue made up of dead hollow cells forming a protective covering in woody stems.

Cornucopian view (kor''nu-ko'pe-an vu) the belief that resources are unlimited because technology will be able to overcome any possible shortages.

coronary artery (kor'ŏ-na-re ar'ter-e) an artery that supplies blood to the wall of the heart.

corpus luteum (kor'pus lut'e-um) a body, yellow in color, that forms in the ovary from a follicle that has discharged its egg.

cortex (kor'teks) in animals, the outer layer of an organ; in plants, the tissue beneath the epidermis in certain stems.

cortisol (kor'tĭ-sol) a glucocorticoid secreted by the adrenal cortex.

cotyledon (kot''ĭ-le'don) the seed leaf of the embryo of a plant.

countercurrent exchange (kown''ter-kur'ent eks-chānj) an exchange between two streams of fluid flowing in opposite directions past each other, such as between the ascending and descending limbs of the loop of Henle.

courtship (kort'ship) behavior performed for the purpose of acquiring a mate.

covalent bond (ko'va-lent bond) a chemical bond in which the atoms share electrons.

Cowper's glands (kow'perz glandz) two small structures located below the prostate gland in males.

C₃ photosynthesis (se thre fo''to-sin'thĕ-sis) photosynthesis that exclusively utilizes the Calvin cycle to fix carbon dioxide; so named because the first molecule detected after CO_2 uptake is a C_3 molecule.

C₄ photosynthesis (se for fo''to-sin'thĕ-sis) photosynthesis in which the first detected molecule following CO_2 uptake is a C_4 molecule. Later, this same CO_2 is made available to the Calvin cycle.

cranial nerve (kra'ne-al nerv) nerve that arises from the brain.

creatine phosphate (kre'ah-tin fos'fāt) a compound unique to muscles that contains a high-energy phosphate bond.

creatinine (kre-at'ī-nin) excretion product from creatine phosphate breakdown.

crenated (kre'nāt-ed) an animal cell that has lost water after having been placed in a hypertonic solution.

cretinism (kre'tin-izm) a condition resulting from a lack of thyroid hormone in an infant.

Cro-Magnon (kro-mag'non) the common name for the first fossils to be accepted as representative of modern humans.

crossing over (kros'ing o'ver) the exchange of corresponding segments of genetic material between chromatids of homologous chromosomes during meiosis.

crustacean (krus-ta'shun) a member of the arthropod class Crustacea, such as a crab, shrimp, or copepod.

cultural eutrophication (kul'tu-ral u''tro-fi-ka'shun) enrichment of a body of water causing excessive growth of producers and then death of these and other inhabitants.

Cushing's syndrome (koosh'ingz sin'drŏm) a condition characterized by thin arms and legs and a "moon face" accompanied by high blood glucose and sodium levels due to hypersecretion of cortical hormones.

cyanobacteria (si''ah-no-bak-te're-ah) also called blue-greens or blue-green algae. Photosynthetic prokaryotes that contain chlorophyll and release O_2.

cyclic AMP (sik'lik a em pe) a compound that functions as an intracellular mediator of hormonal action.

cyclic photophosphorylation (sik'lik fo''to-fos''for-i-la'shun) the synthesis of ATP as electrons cycle with Photosystem I beginning and ending with P700 chlorophyll molecules.

cyte-, cyto (sīt, si'to) cell.

cytochrome (si'to-krōm) a heme-containing molecule that carries electrons in electron transport chains in photosynthesis and cellular respiration.

cytokinesis (si''to-ki-ne'sis) division of the cytoplasm of a cell.

cytoplasm (si'to-plazm) the ground substance of cells located between the nucleus and the cell membrane.

D

daughter cells (daw'ter selz) cells formed by division of a parent cell.

deamination (de-am''i-na'shun) removal of an amino group ($-NH_2$) from an amino acid or other organic compound.

deciduous (de-sid'u-us) plants that shed their leaves at certain seasons.

decomposers (de-kom-po'zerz) organisms of decay (fungi and bacteria) in an ecosystem.

degenerate (de-jen'er-āt) an expression used to signify that there can be more than one nucleotide code standing for a particular amino acid.

dehydrogenase (de-hi'dro-jen-as) an enzyme that accepts hydrogen atoms, speeding up the process of dehydrogenation.

dehydrogenation (de-hi''dro-jen-a'shun) a form of oxidation in which hydrogen atoms are removed from a molecule.

deletion (de-le'shun) a chromosome mutation that results in loss of a portion of the chromosome.

demographic transition (dem-o-graf'ik tran-zi'shun) the change from a high birthrate to a low birthrate so that the growth rate is lowered.

denaturation (de-na-tur-a'shun) alternation of the three-dimensional structure of a protein so that it loses its former physical and chemical properties.

dendrite (den'drīt) process of a neuron, typically branched, that conducts nerve impulses toward the cell body.

denitrify (de-ni'-tri-fi) to convert ammonia or nitrate to atmospheric nitrogen as in the denitrifying bacteria.

deoxyribonucleic acid (de-ok''se-ri''bo-nu-kle'ik as'id) *see* DNA.

deoxyribose (de-ok''se-ri'bōs) a five-carbon sugar with one less oxygen than ribose; a constituent of DNA.

depolarization (de-po''lar-i-za'shun) a loss in polarization as when the nerve impulse or action potential occurs.

derm (derm) skin, covering, tissue layer.

dermis (der'mis) the thick skin layer that lies beneath the epidermis.

detoxification (de-tok''si-fi-ka'shun) a chemical process that alters a compound so that it is no longer toxic.

developed countries (de-vel'opt kun'trez) industrialized nations that typically have a strong economy and a low rate of population growth.

developing countries (de-vel'op-ing kun'trez) nations that are not yet industrialized and have a weak economy and a high rate of population growth.

di- (di) two.

diabetes insipidus (di''ah-be'tez in-sip'i-dus) condition characterized by an abnormally large production of urine due to a deficiency of antidiuretic hormone.

diabetes mellitus (di''ah-be'tez me-li'-tus) condition characterized by a high blood glucose level and the appearance of glucose in the urine due to a deficiency of insulin.

diaphragm (di'ah-fram) a sheet of muscle that separates the chest cavity from the abdominal cavity in higher animals. Also, a birth control device inserted in front of the cervix in females.

diastole (di-as'to-le) relaxation of heart muscle.

diatoms (di'ah-tomz) a large group of fresh and marine unicellular organisms having a cell wall consisting of two silica impregnated valves that fit together as in a pill box.

dicot (dicotyledon) (di'kot) a type of angiosperm distinguished particularly by the presence of two cotyledons in the seed.

differentiation (dif''er-en''she-a'shun) the process and developmental stages by which a cell becomes specialized for a particular function.

diffusion (di-fu'zhun) the movement of molecules from an area of greater concentration to an area of lesser concentration.

dihybrid (di-hi'brid) the offspring of parents who differ in two ways; shows the phenotype governed by the dominant alleles but carries the recessive alleles.

dimorphism (di-mor'fizm) having two forms, as when the male and female of a species have a different appearance.

dinoflagellates (di''no-flaj'e lāts) a large group of marine unicellular flagellates that have two flagella; one circles the body while the other projects posteriorly.

dipleurula larva (di-ploor'u-lah lar'vah) a larval form unique to the deuterostomes that indicates that they are related.

diploid (dip'loid) the 2N number of chromosomes; twice the number of chromosomes found in gametes.

disaccharide (di-sak'ah-rid) a sugar such as maltose that contains two units of a monosaccharide.

dissociation (dis-so''she-a'shun) the breaking of a chemical bond when a compound or molecule is put into water, thereby releasing ions.

distal convoluted tubule (dis'tal kon'vo-lūt-ed tu'būl) highly coiled region of a nephron that is distant from Bowman's capsule.

division (di-vizh'un) a taxonomic category applied to plants and fungi that follows kingdom and lies above class.

DNA (de'en-a) (deoxyribonucleic acid) a nucleic acid, found especially in the nucleus where it contains a triplet genetic code.

dominance hierarchy (dom'i-nans hi'er-ar''ke) a system in which animals arrange themselves in a pecking order; the animal above takes precedence over the one below.

dominant allele (dom'i-nant ah-lēl') hereditary factor that expresses itself even when the genotype is heterozygous.

dominant generation (dom'i-nant jen''e-ra'shun) in plants, the most conspicuous and most long-lasting of the two alternating generations.

dormancy (dor'man-se) a period of suspended activity and growth during which life is maintained.

dorsal (dor'sal) toward the back.

double bond (dū'b'l bond) a bond in which two pairs of electrons are shared between two atoms.

double helix (dū'b'l he'liks) a double spiral often used to describe the three-dimensional shape of DNA.

doubling time (dū'b'ling tim) the number of years it takes for a population to double in size.

Down's syndrome (downz sin'drŏm) human congenital disorder associated with an extra 23rd chromosome.

Dryopithecus (dri'o-pith'e-cus) a genus of extinct apes that may have included or resembled a common ancestor to both apes and humans.

ductus deferens (duk'tus def'er-enz) tube connecting epididymis to ejaculatory duct; sperm duct, also called vas deferens.

duodenum (du''o-de'num) the first portion of the small intestine in vertebrates into which ducts from the gallbladder and pancreas enter.

dyad (di'ad) a chromosome having two chromatids held together at a centromere.

E

echinoderm (e-kin'o-derm) a member of the phylum Echinodermata, such as a starfish.

ecology (e-kol'o-je) the study of the relationship of organisms between themselves and the physical environment.

ecosystem (ek''o-sis'tem) a biological community together with the associated abiotic environment.

ecto- (ek'to) outside, external.

ectoderm (ek'to-derm) the outer germ layer of the embryonic gastrula; it gives rise to the skin and nervous system.

effector (e-fek'tor) a structure that allows an organism to respond to environmental stimuli such as the muscles and glands.

efferent (ef'er-ent) moving away; for example, a neuron or nerve that takes impulses away from the central nervous system.

electromagnetic spectrum (e-lek''tro-mag-net'ik spek'trum) radiation of differing wavelengths.

electron (e-lek'tron) a subatomic particle that has almost no weight and carries a negative charge; travels in an orbital, called a shell, about the nucleus.

electron transport system (e-lek'tron trans'port sis'tem) series of metabolic reactions in which electrons are passed along a chain of carrier molecules with the concurrent production of ATP; also called the respiratory chain.

embolus (em'bo-lus) a moving blood clot that is carried through the bloodstream.

embryo (em'bre-o) the developing organism, particularly during the early stages.

emigration (em''i-gra'shun) the deliberate departure of an organism from its home range (area).

emphysema (em''fi-se'mah) a medical condition characterized by an abnormal enlargement of the alveoli within the lungs.

emulsification (e-mul''si-fi'ka'shun) the act of dispersing one liquid in another.

-enchyma (en'ki-mah) tissue.

end-, endo- (end, en'do) within, inside.

endocrine (en'do-krin) secreting internally, particularly hormonal glands whose products are dispersed by the blood.

endocytosis (en''do-si-to'sis) a process in which extracellular material is enclosed within a vesicle and taken into the cell. Phagocytosis and pinocytosis are forms of endocytosis.

endoderm (en'do-derm) an inner layer of cells that line the primitive gut of the gastrula. It becomes the lining of the digestive tract and associated organs.

endodermis (en''do-der'mis) a plant tissue consisting of a single layer of cells that surrounds and regulates the entrance of materials into particularly the vascular cylinder of roots.

endometrium (en''do-me'tre-um) the lining of the uterus that becomes thickened and vascular during the menstrual cycle.

endoplasmic reticulum (en-do-plaz'mic re-tik'u-lum) a complex system of tubules, vesicles, and sacs in cells; sometimes having attached ribosomes.

endospore (en'do-spor) a resistant body formed by bacteria when environmental conditions worsen.

energy of activation (en'er-je uv ak''ti-va'shun) the amount of energy that a molecule must gain to become sufficiently "excited" to enter into a chemical reaction.

energy source (en'er-je sors) a way by which energy can be made available.

environmental resistance (en-vi''ron-men'tal re-zis'tans) sum total of factors in the environment that limit the numerical increase of a population in a particular region.

enzyme (en'zim) a protein catalyst that speeds up a specific reaction or a specific type of reaction.

epi- (ep'i) upon, outer.

epidermis (ep''i-der'mis) the outer layer of cells of an organism.

epididymis (ep''i-did'i-mis) coiled tubules next to the testes where sperm mature and may be stored for a short time.

epiglottis (ep''i-glot'is) a structure that covers the glottis during the process of swallowing.

epinephrine (ep''i-nef'rin) a hormone secreted by the adrenal medulla that acts as a powerful stimulus to the heart; also called adrenalin.

epiphyte (ep'i-fit) nonparasitic plant that grows on the surface of other plants, usually above the ground, such as arboreal orchids and Spanish moss.

epistasis (e-pis'tah-sis) the ability of one gene to mask the genetic expression of another gene located at a different chromosome locus.

epithelial (ep''i-the'le-al) a type of tissue that lines cavities and covers the external surface of the body.

epitope (ep'i-top) that portion of an antigen that stimulates the production of antibodies.

equilibrium (e''kwi-lib're-um) a state of balance; a steady state where forces are equalized.

erythrocyte (e-rith'ro-sit) red blood cell.

esophagus (e-sof'ah-gus) a tube that transports food from the mouth to the stomach.

estrogen (es'tro-jen) a female sex hormone produced by the ovaries, that promotes the secondary sex characteristics.

estuary (es'tu-a-re) an area where fresh water meets the sea; thus, an area with salinity intermediate between fresh water and seawater.

eu- (u) true.

euglenoids (u-gle'noidz) a small group of fresh water unicellular flagellates that are bounded by a flexible pellicle and contain chloroplasts.

eukaryotic (u''kar-e-ot'ik) possessing the membranous organelles characteristic of complex cells.

eustachian tube (u-sta'ke-an tub) an air tube that connects the pharynx to the middle ear.

eutrophication (u''tro-fi-ka'shun) enrichment that causes lakes to fill in. *See also* cultural eutrophication.

evolution (ev''o-lu'shun) genetic changes that occur in populations of organisms with the passage of time, resulting in an adaptation to the environment.

evolutionary tree (ev''o-lu'shun-ar-e tre) a diagram describing the phylogenetic relationship of groups of organisms.

ex-, exo- (eks, ek'so) out of, outside; producing.

excretion (ek-skre'shun) removal of metabolic wastes.

exocrine (ek'so-krin) secreting externally; particular glands with ducts whose secretions are deposited into cavities, such as salivary glands.

exocytosis (eks''o-si-to'sis) a process in which an intracellular vesicle fuses with the cell membrane so that the vesicle's contents are released outside the cell.

exon (eks'on) that portion of a structural gene that has no complementary portion in mRNA.

exophthalmic goiter (ek''sof-thal'mik goi'ter) an enlargement of the thyroid gland accompanied by an abnormal protrusion of the eyes.

expiration (eks''pi-ra'shun) process of expelling air from the lungs; exhalation.

exponential growth (eks''po-nen'shal groth) growth, particularly of a population, in which the total number increases in the same manner as compound interest.

extinction (eks-ting'shun) the demise of all the members of a particular group of organisms.

extraembryonic membranes (eks''trah-em''bre-on'ik mem'branz) in embryology, membranes that are not a part of the embryo but are necessary to the continued existence and health of the embryo.

F

facilitated transport (fah-sil'i-tat-ed trans'port) transfer of a substance into or out of a cell along a concentration gradient by a process that requires a carrier.

facultative anaerobe (fak'ul-ta''tiv an-a'er-ob) a bacterium that can exist whether or not the environment contains oxygen.

family (fam'i-le) a rank in taxonomic classification above genus and below order.

fat (fat) a lipid molecule whose hydrolysis releases three fatty acids and a glycerol molecule.

feces (fe'sēz) indigestible wastes expelled from the digestive tract; excrement.

feedback control (fēd' bak kon-trōl) a system of regulation by which the increase in a product leads to a decrease in its production and vice versa.

femur (fe'mur) the thighbone found in the upper leg.

fermentation (fer''men-ta'shun) anaerobic breakdown of carbohydrates that results in end products such as alcohol and lactic acid.

fertilization (fer''ti-li-za'shun) the union of male and female gametes, often the sperm and egg.

fetus (fe'tus) human development in its later stages following the embryonic stages.

fibrin threads (fi'brin thredz) filaments formed from the protein fibrinogen when the blood clots.

fibula (fib'u-lah) a long slender bone located on the lateral side of the tibia.

filament (fil'ah-ment) a threadlike structure such as the thick (myosin) and thin (actin) filaments found in myofibrils of muscle fibers.

filter feeder (fil'ter fēd'er) an animal that obtains its food, usually in small particles, by filtering it from water.

first filial generation (ferst fil'e-al jen''ē-ra'shun) all of the offspring produced by the sexual reproduction of two individuals (symbol: F_1).

fission (fish'un) *see* binary fission; also nuclear fission.

fixed action pattern (fikst ak'shun pat'ern) a sequence of reflexes that always occurs in the same order and manner.

flagella (flah-jel'ah) slender, long processes used for locomotion by the flagellate protozoans, bacteria, and sperm.

florigen (flor'i-jen) a hypothetical plant hormone responsible for flowering.

follicle (fol'i-kl) a structure in the ovary that produces the egg and particularly the female sex hormone, estrogen.

follicle-stimulating hormone (FSH) (fol'i-kl stim'u-la''ting hor'mon) a gonadotrophic hormone produced by the anterior pituitary that promotes the formation of a follicle and maturation of the egg in the female and seminiferous tubules and maturation of sperm in the male.

follicular phase (fo-lik'u-lar fāz) the period of the ovarian cycle before ovulation when the follicle is maturing.

food chain (food chān) a sequence of organisms, each of which feeds on the previous one to acquire energy and organic building blocks. Includes the producer, various levels of consumers, and the decomposers.

food pyramid (food pir'ah-mid) a diagram that shows the feeding relationship and the energy flow between populations in an ecosystem; photosynthesizers are at the base of the pyramid and the final consumer is at the top.

food web (food web) the complete set of food links between populations in a community.

formed element (form'd el'ē-ment) a cellular constituent of blood.

formula (for'mu-lah) a written designation using atomic symbols to show the fixed proportion of atoms in a compound and/or molecule.

fossil fuel (fos''l fu'el) the remains of once living organisms that are burned to release energy, such as coal, oil, and natural gas.

fossils (fos''lz) any remains of an organism that have been preserved in the earth's crust.

fovea (fo've-ah) a depression, particularly in the retina where the concentration of cones accounts for visual acuity.

frond (frond) the leaf of a fern plant.

fruit (froot) a mature ovary enclosing seed(s).

fruiting body (froot'ing bod'e) a specialized structure found in some fungi in which spores are produced.

fungus (fung'gus) an organism, usually composed of strands called hyphae, that lives chiefly on decaying matter; e.g., mushroom and mold.

furrowing (fur'o-ing) a constriction of the cell membrane that accompanies cytokinesis in animal cells.

G

gamete (gam'ēt) a reproductive cell that joins with another in fertilization to form a zygote; most often an egg or sperm.

gametophyte (gam'ē-to-fīt) the haploid generation that produces gametes in the life cycle of a plant.

ganglion (gang'gle-on) a collection of neuron cell bodies outside the central nervous system.

gastr-, gastro- (gas'tr, gas'tro) stomach, belly.

gastric (gas'trik) of or pertaining to the stomach, such as gastric glands that line the stomach and produce gastric juice that enters the stomach.

gastrovascular cavity (gas''tro-vas'ku-lar kav'i-te) a central cavity, with only one opening, of a lower animal in which digestion takes place and where nutrients are distributed to the cells lining the cavity.

gastrula (gas'troo-lah) a two-layered, later three-layered, animal embryonic stage; each layer is a germ layer.

gastrulation (gas''troo-la'shun) a radical reorganization of the embryo resulting in the formation of primary germ layers and (in chordates) the notochord.

gel (jel) a colloid in which the solid phase is continuous and the liquid phase is dispersed.

gene (jēn) a unit of heredity located at a particular site (locus) on a chromosome. In Mendelian genetics, genes determine phenotypic characteristics. Biochemically, a gene is a segment of DNA that codes for a protein.

gene flow (jēn flo) the movement of genes from one population to another via gametes; increases heterogeneity.

gene pool (jēn pool) the total of all the genes of all the individuals in a population.

genetic drift (je-net'ik drift) evolution by chance processes alone.

genotype (je'no-tīp) the genetic makeup of any individual.

genus (je'nus) a rank in taxonomic classification above species and below family.

geothermal energy (je''o-ther'mal en'er-je) heat energy derived from underground rock formations that have been heated by radioactive elements in the earth.

geotropism (je-ot'ro-pizm) growth in response to gravity; roots show positive and stems negative geotropism.

germ layers (jerm la'ers) primary tissues of an embryo (ectoderm, mesoderm, endoderm) that give rise to the major tissue systems of the adult animal.

gizzard (giz'ard) a very muscular part of a stomach that grinds up food, sometimes with the aid of fragments of stone.

globulin (glob'u-lin) a class of proteins, found particularly in the blood plasma, that contain antibodies.

glomerular filtrate (glo-mer'u-lar fil'trāt) the molecules that pass from the glomerulus to the inside of Bowman's capsule.

glomerulus (glo-mer'u-lus) a cluster; for example, the cluster of capillaries surrounded by Bowman's capsule in a kidney tubule.

glottis (glot'is) slitlike opening between the vocal cords.

glucocorticoids (gloo''ko-kor'ti-koidz) hormones (e.g., cortisol) produced by the adrenal cortex, which regulate gluconeogenesis.

gluconeogenesis (gloo''ko-ne''o-jen'ē-sis) the formation of glucose from amino acids and glycerol.

glucose (gloo'kōs) the most common six-carbon sugar.

glycogen (gli'ko-jen) a polysaccharide that is the principal storage compound for sugar in animals.

glycolysis (gli-kol'i-sis) the metabolic pathway that converts sugars to simpler compounds.

Golgi apparatus (gol'ge ap''ah-ra'tus) an organelle that consists of concentrically folded membranes and functions in the packaging and secretion of cellular products.

gonad (go'nad) an organ that produces sex cells; the ovary, which produces eggs, and the testis, which produces sperm.

gonadotropic (go-nad''o-trōp'ik) a type of hormone that regulates the activity of the ovaries and testes; principally FSH and LH (ICSH).

grana (gra'nah) stacks of flattened membranous vesicles in a chloroplast where chlorophyll is located and photosynthesis begins.

granulocytes (gran'u-lo-sīts) white blood cells that contain distinctive granules.

gray crescent (gra kres'ent)   a portion of cytoplasm in the embryo of a frog that apparently affects differentiation of cells.

grazers (gra'zerz)   animals that feed on low-lying vegetation such as grasses.

growth (grōth)   increase in the number of cells and/or the size of these cells.

growth rate (grōth rāt)   percentage of increase or decrease in the size of a population.

guard cell (gahrd sel)   one of a pair of specialized plant cells forming a stoma.

gymnosperm (jim'no-sperm)   a class of vascular plants whose seeds are not enclosed in an ovary; notable examples are the conifers.

H

habitat (hab'i-tat)   the natural abode of an animal or plant species.

habituation (hah-bit''u-a'shun)   learned behavior that regularly reoccurs without the need for conscious attention.

haploid (hap'loid)   the N number of chromosomes; half the diploid number; the number characteristic of gametes that contain only one set of chromosomes.

Hardy-Weinberg Law (har'de win'berg law)   the observation that in a large, randomly breeding population in the absence of mutation and selection and genetic drift, the frequency of genes does not change from generation to generation.

Haversian canal (ha-ver'shan kah-nal')   a lumen found in compact bone through which blood vessels and nerves pass in order to serve bone cells arranged in concentric rings about the canal.

helix (he'liks)   a spiral shape, such as is found in the double helix of DNA.

hemo- (he'mo)   prefix meaning blood, as in hemocoel, a cavity that contains blood.

hemoglobin (he''mo-glo'bin)   a red iron-containing pigment in blood that combines with and transports oxygen.

hemophilia (he''mo-fil'e-ah)   a genetic disease in which blood clotting is impaired; bleeder's disease.

hepatic (hĕ-pat'ik)   pertaining to the liver.

herbaceous (her-ba'shus)   nonwoody.

herbivores (her'bĭ-vorz)   animals that eat plants.

hermaphroditism (her-maf'ro-di-tizm'')   the state of having both male and female sex organs.

hetero- (het'er-o)   other, different.

heterospores (het'er-o-sporz)   nonidentical spores such as microspores and megaspores produced by the same plant.

heterotroph (het'er-o-trōf)   an organism that cannot synthesize organic compounds from inorganic substances and therefore must acquire food from external sources.

heterotroph hypothesis (het'er-o-trof'' hi-poth'ĕ-sis)   the suggestion that the protocell and first cell(s) were heterotrophs.

heterozygous (het''er-o-zi'gus)   having two different alleles (as *Aa*) for a given trait.

hist- (hist)   tissue.

histology (his-tol'o-je)   study of tissues.

histone (his'tōn)   proteins associated with DNA that apparently have a structural function.

homeo-, homo- (ho'me-o) (ho'mo)   like, similar.

homeostasis (ho''me-o-sta'sis)   the constancy of conditions, particularly the internal environment of birds and mammals: constant termperature, blood pressure, pH, and other body conditions.

homing (hōm'ing)   the ability of an animal to return to a homesite.

hominids (hom'ĭ-nidz)   members of the family of upright, bipedal primates (family Hominidae) that includes modern humans.

hominoids (hom'ĭ-noidz)   members of a superfamily containing humans and the great apes.

Homo erectus (ho'mo ė-rek'tus)   the earliest nondisputed species of humans, named for their erect posture that allowed them to have a striding gait.

Homo habilis (ho'mo hah'bi-lis)   an extinct species that may include the earliest humans, having a small brain but quality tools.

homologous (ho-mol'o-gus)   similarly constructed; homologous chromosomes have the same shape and contain genes for the same traits; homologous structures in animals share a common ancestry.

Homo sapiens (ho'mo sa'pe-enz)   the genus and species designation for present-day humans.

homozygous (ho''mo-zi'gus)   having identical alleles (as *AA* or *aa*) for a given trait; pure breeding.

hormone (hor'mōn)   a chemical secreted in one part of the body that controls the activity of other parts.

host (hōst)   an organism on or in which another organism lives.

humerus (hu'mer-us)   a heavy bone that extends from the scapula to the elbow.

humor (hu'mor)   a fluid found within the chambers of the eye; aqueous humor in the anterior chamber is watery, while the vitreous humor in the posterior chamber is jellylike.

hybrid (hi'brid)   an offspring resulting from the crossing of genetically different strains, populations, or species.

hydro- (hi'dro)   water, fluid; hydrogen.

hydrogen bond (hi'dro-jen bond)   a weak attraction between a hydrogen atom carrying a partial positive charge and an atom of another molecule carrying a partial negative charge.

hydrolysis (hi-drol'i-sis)   the splitting of a bond within a larger molecule by the addition of water.

hydroxide ion (hi-drok'sīd i'on) OH^- ion.

hyper- (hi'per)   much, too much.

hypertonic solution (hi''per-ton'ik so-lu'shun)   one that has a greater concentration of solute, a lesser concentration of water than the cell.

hypha (hi'fah)   one filament of a mycelium that constitutes the body of a fungus.

hypo- (hi'po)   under; lower than normal.

hypothalamus (hi''po-thal'ah-mus)   a region of the brain; the floor of the third ventricle that helps maintain homeostasis.

hypothesis (hi-poth'ĕ-sis)   a scientific theory that is capable of explaining present data and that may be used to predict the outcome of future experimentation.

hypotonic solution (hi''po-ton'ik so-lu'shun)   one that has a greater concentration of water, a lesser concentration of solute than the cell.

I

ICSH (interstitial cell-stimulating hormone) (in''ter-stish'al sel stim'u-lāt-ing hor'mōn)   *see* luteinizing hormone.

immune system (i-mun' sis'tem)   lymphocytes and the organs and tissues that produce lymphocytes or in which lymphocytes mature; the system responsible for immunity.

immunity (ĭ-mu'ni-te)   possessing resistance to certain antigens so that illness does not result; active immunity: ability to produce certain antibodies; passive immunity: antibodies are received by a serum injection.

implantation (im''plan-ta'shun)   the attachment of the embryo to the lining (endometrium) of the uterus.

imprinting (im'print-ing)   the tendency of a newborn animal to follow the first moving object it sees.

incomplete dominance (in-kom-plēt' dom'ĭ-nans)   the inability of either member of an allelic pair to exert dominance so that an intermediate phenotype is produced.

independent assortment (in''de-pen'dent ah-sort'ment)   the random segregation of allelic pairs into gametes during the process of meiosis.

induction (in-duk'shun)   a process by which one tissue controls the development of another, as when the embryonic notochord induces the formation of the neural tube.

inhibitor (in-hib'ĭ-tor)   a substance that combines with an enzyme and prevents it from performing its normal function.

innate (in'nāt)   instinctive, inborn, and not having to be learned.

innominate (ĭ-nom'ĭ-nāt)   one of two hipbones that form the pelvis.

insight (in'sīt)   the use of higher mental abilities and previous experiences to arrive at a new idea or conclusion.

inspiration (in''spī-ra'shun)   the act of breathing in.

instinct (in'stinkt)   genetically innate complex pattern of behavior that requires no previous experience or conditioning.

insulin (in'su-lin)   a hormone produced by the pancreas that regulates carbohydrate storage.

inter- (in'ter)   between.

interferon (in''ter-fer'on) a protein formed by a cell infected with a virus that can increase the resistance of other cells to the virus.

interneuron (in''ter-nu'ron) a neuron that is found within the central nervous system and takes nerve impulses from one portion of the system to another.

interstitial cells (in''ter-stish'al selz) hormone-secreting cells located between the seminiferous tubules of the testes.

intertidal zone (in''ter-tid'al zōn) that portion of a seashore that is alternately wet when the tide is in and dry when the tide is out.

intra- (in'trah) within.

intron (in'tron) that portion of a structural gene that is complementary to mRNA.

inversion (in-ver'zhun) stagnant, nonmoving warm air that covers and traps pollutants beneath it.

invertebrate (in-ver'tē-brāt) an animal that lacks a vertebral column.

ion (i'on) an atom or group of atoms carrying a positive or negative charge.

ionic bond (i-on'ik bond) a chemical attraction between a positive and negative ion.

Islets of Langerhans (i'lets uv lahng'er-hanz) distinctive groups of cells within the pancreas that secrete insulin and glucagon.

iso- (i'so) equal, uniform.

isogametes (i''so-gam'ēts) gametes whose union produces a zygote, but which have a similar appearance.

isolating mechanism (i'so-lāt-ing mek'ah-nizm) a means by which members of two groups of organisms are prevented from reproducing and/or producing fertile offspring.

isomers (i'so-merz) molecules having the same chemical formula but different structural formulas such as glucose and mannose.

isotonic solution (i''so-ton'ik so-lu'shun) one that contains the same concentration of water per volume as does the cell.

isotopes (i'so-tōps) atoms with the same number of protons and electrons but differing in the number of neutrons and therefore in weight.

J

jointed appendages (joint'ed ah-pen'dij-ez) the flexible exoskeleton extensions found in arthropods that are used as sense organs, mouth parts, and locomotion.

K

karyotype (kar'e-o-tīp) the arrangement of all the chromosomes within a nucleus by pairs in a fixed order.

keratinization (ker''ah-tin''i-za'shun) the process by which skin cells acquire keratin, a material that helps make the skin impermeable to water.

kingdom (king'dum) the largest taxonomic category into which organisms are placed: Monera, Protista, Fungi, Plants, and Animals.

kin recognition (kin rē-kog-ni'shun) the ability of an organism to determine those that are related to it.

Krebs cycle (krebz si'kl) a series of reactions found within mitochondria that give off carbon dioxide. Also called the citric acid cycle because the reactions begin and end with citric acid.

L

labium (la'be-um) a fleshy border or liplike fold of skin, as in the labia majora and labia minora of the female genitalia.

lacteal (lak'te-al) a lymph vessel in a villus of the intestinal wall of mammals.

lactic acid (lak'tik as'id) an end product of fermentation (anaerobic respiration) in animals.

lactogenic hormone (lak''to-jen'ik hor'mōn) a hormone secreted by the anterior pituitary that stimulates the production of milk from the mammary glands.

lacuna (lah-ku'nah) a small pit or hollow cavity, as in bone or cartilage where a cell or cells are located.

lamella (lah-mel'ah) a thin leaflike or platelike structure; formed by membrane in cells.

larva (lar'vah) an immature stage in invertebrates, differing significantly from the adult, that is capable of feeding; in aquatic forms, it is capable of swimming.

larynx (lar'ingks) structure that contains the vocal cords; voice box.

lateral (lat'er-al) to the side of.

learning (lern'ing) a change in behavior as a result of experience.

lens (lenz) a clear membranelike structure found in the eye behind the iris. The lens brings objects into focus.

leukocyte (lu'ko-sīt) a white blood cell.

leukoplasts (lu'ko-plasts) colorless plastids in plant cells that function in storage.

lichen (li'ken) fungi and algae coexisting in a mutualistic relationship.

life cycle (līf si'kl) the significant stages in the organism's life from the time of fertilization to the time it reproduces.

ligament (lig'ah-ment) a strong connective tissue that joins bone to bone.

lignin (lig'nin) an organic compound in wood that strengthens cell walls.

limbic system (lim'bik sis'tem) an area of the forebrain implicated in visceral functioning and emotional responses; involves many different centers of the brain.

limnetic zone (lim-net'ik zōn) the sunlit body of a lake.

linkage (lingk'ij) alleles on the same chromosome are linked in the sense that they tend to move together to the same gamete; crossing over interferes with linkage.

lip- (lip) fat or fatlike.

lipase (li'pās) an enzyme that digests or breaks down fats.

lipid (lip'id) a group of organic compounds that are insoluble in water; notably fats, oils, and steroids.

littoral zone (lit'or-al zōn) in a lake the portion closest to the shore; at the seashore, the intertidal zone.

locus (lo'kus) a particular location on a chromosome.

loop of Henle (loop uv hen'le) a U-shaped turn in the kidney tubule of the mammals.

Lucy (lu'se) an affectionate name for the fossil *Australopithecus afarensis*.

lumen (lu'men) the cavity inside any tubular structure, such as the lumen of the gut.

luteal phase (lu'te-al fāz) the period of the ovarian cycle after ovulation when the corpus luteum is active.

luteinizing hormone (LH) (lu'te-in-iz''ing hor'mōn) a gonadotropic hormone of the pituitary that promotes the formation of the corpus luteum; in males (called ICSH) it controls the secretion of the interstitial cells.

lymph (limf) fluid having the same composition as tissue fluid and carried in lymph vessels.

lymphocytes (lim'fo-sīts) white blood cells of two types: T cells are responsible for cell-mediated immunity and B cells are responsible for humoral immunity.

lymphokines (lim'fo-kinz) chemicals secreted by T cells that have the ability to affect the characteristics of monocytes.

lymph vessels (limf ves'elz) vessels that are not a part of the blood circulatory system but nevertheless collect excess tissue fluid (lymph) and return it to systemic veins.

-lysis, lyso- (li'sis, li'so) splitting, breaking open.

lysogenic cycle (li-so-jen'ik si'kl) incorporation of viral DNA into host DNA so that it is passed to any subsequent daughter cells. If and when viral reproduction occurs, the viruses may contain segments of host DNA.

lysosome (li'so-sōm) an organelle in which digestion takes place due to the action of powerful hydrolytic enzymes.

lytic cycle (lit'ik si'kl) reproduction of viruses that leads to a breaking open of the host cell.

M

macro- (mak'ro) large.

macromolecule (mak''ro-mol'ē-kūl) a large molecule composed of many repeating units such as proteins, polysaccharides, and nucleic acids.

macrophage (mak'ro-fāj) an enlarged monocyte that ingests foreign material and cellular debris.

Malpighian tubules (mal-pig'i-an tu'būlz) organs of excretion, notably in insects.

Malthusian view (mal-thu'se-an vu) the belief that resources are limited and that technology will not be able to overcome the advent of eventual shortages.

mammary glands (mam′er-e glandz) milk-producing glands found in mammals.

mantle (man′t'l) fleshy fold that envelops the visceral mass of mollusks.

marsupials (mar-su′pe-alz) mammals in which the immaturely born infant is carried in a pouch.

matrix (ma′triks) the secreted basic material or medium of biological structures, such as the matrix of cartilage or bone.

medulla (mē-dul′ah) the inner portion of an organ; for example, the adrenal medulla.

medulla oblongata (mē-dul′ah ob″long-gah′tah) the lowest portion of the brain that is concerned with the control of internal organs.

medusa (mē-du′sah) a bell-shaped, free-swimming stage capable of sexual reproduction in the life cycle of some sessile coelenterates. Jellyfishes are examples.

mega- (meg′ah) large, female.

meiosis (mi-o′sis) type of cell division that occurs during the production of gametes or spores by means of which the daughter cells receive the haploid number of chromosomes.

melanin (mel′ah-nin) a pigment found in the skin and hair of humans that is responsible for their coloration.

membrane (mem′brān) a thin, pliable layer that is composed of proteins and phospholipids; structural component of many cellular organelles; an outer boundary for the cell and nucleus.

memory cells (mem′o-re selz) cells derived from B cells that are ever present within the body that produce a specific antibody and account for the development of active immunity.

meninges (mē-nin′jēz) protective membranous coverings about the central nervous system.

meniscus (mē-nis′kus) a piece of fibrocartilage that separates the surfaces of bones in the knee.

menopause (men′o-pawz) termination of the menstrual cycle in older women.

menstrual cycle (men′stroo-al si′kl) the female reproductive cycle that is characterized by regularly occurring changes in the uterine lining.

meristem (mer′ĭ-stem) an embryonic plant tissue that always remains undifferentiated and capable of dividing to produce new cells.

meso- (mes′o) middle.

mesoderm (mes′o-derm) the middle germ layer of an animal embryo that gives rise to the muscles, connective tissue, and circulatory system.

mesoglea (mes″o-gle′ah) a jellylike packing material between the ectoderm and endoderm of coelenterates.

mesophyll (mes′o-fil) the middle tissue of a leaf made up of parenchyma cells.

metabolic pathway (met″ah-bol′ik path′wa) a series of enzymes that control the steps by which substrates become an end product.

metabolism (mē-tab′o-lizm) all of the chemical reactions within a cell or organism; sometimes referring to only certain portions as amino acid metabolism.

metamorphosis (met″ah-mor′fo-sis) change in form as when a tadpole becomes an adult frog or as when an insect larva develops into the adult.

micro- (mi′kro) small; male.

microfilament (mi″kro-fil′ah-ment) an extremely thin fiber found within the cytoplasm that is involved in the maintenance of cell shape and movement of cell contents.

micrometer (μm) (mi-krom′ē-ter) one-thousandth part of a millimeter (i.e., 0.001 mm); formerly designated as micron (μ).

microorganisms (mi″kro-or′gan-izmz) organisms so small that it requires a microscope to see them in any detail.

microtubule (mi″kro-tu′būl) an organelle composed of 13 rows of globular proteins; found in multiple units in several other organelles such as the centriole, cilia, and flagella.

microvilli (mi″kro-vil′i) tiny projections from the membrane of a cell; sometimes called a brush border.

mimicry (mim′ik-re) the resemblance of an organism to another that has a defense against a common predator.

mineralocorticoids (min″er-al-o-kor′ti-koids) hormones secreted by the adrenal cortex that influence the concentrations of electrolytes in body fluids.

mitochondrion (mi″to-kon′dre-on) an organelle in which aerobic respiration produces the energy molecule, ATP.

mitosis (mi-to′sis) cell division by means of which two daughter cells receive the exact chromosome and genetic makeup of the mother cell; occurs during growth and repair.

mixed nerves (mikst nervs) nerves that contain both the long dendrites of sensory neurons and the long axons of motor neurons.

mold (mōld) a type of fungus that produces woolly or cottony growth.

molecule (mol′ē-kūl) a chemical unit in which two or more atoms share electrons.

molting (mōlt′ing) shedding all or part of an outer covering; in arthropods, periodic shedding of parts of the exoskeleton to allow increase in size.

monerans (mo-ne′ranz) organisms lacking membranous organelles; for example, bacteria and cyanobacteria.

monoclonal antibodies (mon″-o-klon′al an″ti-bod″ēz) antibodies of one type that are produced by cells that are derived from a lymphocyte that has fused with a cancer cell. These cloned cells produce the same type antibody.

monocot (mon′o-kot) (monocotyledon) a type of angiosperm in which the seed has only one cotyledon, such as corn and lily.

monocyte (mon′o-sit) a large mononuclear leukocyte.

monohybrid (mon″o-hi′brid) the offspring of parents who differ in one way only; shows the phenotype of the dominant allele but carries the recessive allele.

monosaccharide (mon″o-sak′ah-rīd) a simple sugar; a carbohydrate that cannot be decomposed by hydrolysis.

-morph, morpho- (morf, mor′fo) form, shape, structure.

morphogenesis (mor″fo-jen′ĭ-sis) the establishment of shape and structure in an organism.

morula (mor′u-lah) an early stage in development in which the embryo consists of a mass of cells, often spherical.

motivated (mo′tĭ-vāt-ed) physiologically orientated to perform a certain behavior.

motor neuron (mo′tor nu′ron) a neuron that takes nerve impulses from the central nervous system to the effectors.

mucosa (mu-ko′sah) any membrane secreting mucus.

muscle fiber (mus′el fi′ber) muscle cell.

mutagen (mu′tah-jen) an agent, such as a chemical, that increases the rate of mutations.

mutation (mu-ta′shun) a genetic change, most often of the nucleotide sequence in DNA, that is inherited either by daughter cells following mitosis or by an organism following reproduction.

mutualism (mu′tu-al-izm″) a relationship between two organisms of different species that benefits both organisms.

mycelium (mi-se′le-um) a mass of hyphae that make up the body of a fungus.

myelin (mi′ĕ-lin) the fatty cell membranes that cover long neuron fibers and give them a white, glistening appearance.

myo- (mi′o) muscle.

myofibrils (mi″o-fi′brilz) the contractile portions of muscle fibers.

myosin (mi′o-sin) the thick filament in myofibrils made of protein and capable of breaking down ATP.

myxedema (mik″ sĕ-de′mah) a condition resulting from a deficiency of thyroid hormone in an adult.

N

NAD (en a de) a coenzyme of oxidation; a dehydrogenase that frequently accepts hydrogen from metabolites.

NADP (en a de pe) a coenzyme of reduction; a hydrogenase that frequently donates hydrogen atoms to metabolites.

nanometer (nm) (na″no-me′ter) one-thousandth of a micrometer (i.e., 0.001 μm).

natural selection (nat′u-ral sĕ-lek′shun) the process by which better adapted organisms are favored to reproduce to a greater degree and pass on their genes to the next generation.

Neanderthal (ne-an′der-thawl) the common name for an extinct subspecies of humans whose remains are found in Europe, Asia, and Africa.

nematocyst (nem′ah-to-sist) a threadlike structure in stinging cells of coelenterates that can be expelled to numb and capture prey.

nematode (nem′ah-tōd) a member of the phylum Nematoda; a roundworm.

neo- (ne′o) new.

nephr- (nefr) kidney.

nephridia (ně-frid′e-ah) excretory tubules found in invertebrates; notably the segmented worms.

nephron (nef′ron) the anatomical and functional unit of the vertebrate kidney; kidney tubule.

neritic zone (ner-it′ik zōn) the sunlit water containing those organisms that reside above the continental shelves.

nerve (nerv) a bundle of long nerve fibers that run to and/or from the central nervous system.

nerve cord (nerv kord) that portion of the central nervous system that lies posterior to the brain and functions in taking impulses to and from the brain.

nerve impulse (nerv im′puls) an electrochemical change due to increased neurolemma permeability that is propagated along a neuron from the dendrite to the axon following excitation.

neuromuscular junction (nu″ro-mus′ku-lar jungk′shun) the point of contact between a nerve cell and a muscle cell.

neuron (nu′ron) nerve cell that characteristically has three parts: dendrite, cell body, axon.

neurotransmitter substance (nu″ro-trans-mit′er sub′stans) a chemical made at the ends of axons that is responsible for transmission across a synapse.

neurula (nu′roo-lah) the early embryonic stage during which the primitive nervous system forms.

neutron (nu′tron) a subatomic particle that has a weight of one atomic mass unit, carries no charge, and is found in the nucleus.

neutrophil (nu′tro-fil) the most common type of white cell able to phagocytize foreign material.

niche (nich) the functional role and position of an organism in the ecosystem.

nitrifying bacteria (ni′tri-fi″ing bak-te′re-ah) bacteria active in the nitrogen cycle that oxidize ammonia (NH_4^+) to nitrite (NO_2^-) and nitrate (NO_3^-).

noncyclic photophosphorylation (non-sik′lik fo″to-fos″for-i-la′shun) the synthesis of ATP as electrons flow from water to NADP through Photosystems I and II within chloroplasts.

nondisjunction (non″dis-jungk′shun) the failure of homologous chromosomes or chromatids to separate during the formation of gametes.

nonrenewable resource (non-re-nu′ah-b′l re′sors) a resource that can be used up or at least depleted to such an extent that further recovery is too expensive.

notochord (no′to-kord) dorsal supporting rod that exists in all chordates sometime in their life history; replaced by the vertebral column in vertebrates.

nuclear fission (nu′kle-ar fish′un) process in which the nucleus of an atom, typically uranium, is split with the release of neutrons and substantial amounts of energy.

nuclear fusion (nu′kle-ar fu′zhun) a process in which the nuclei of two atoms are forced together to form the nucleus of a heavier atom with the release of substantial amounts of energy.

nucleic acid (nu-kle′ik as′id) a large organic molecule made up of nucleotides joined together; for example, DNA and RNA.

nucleolus (nu-kle′o-lus) an organelle found inside the nucleus; composed largely of RNA for ribosome formation.

nucleotide (nu′kle-o-tīd) a molecule consisting of three subunits: phosphoric acid, a five-carbon sugar, and a nitrogenous base; a building block of a nucleic acid.

nucleus (nu′kle-us) a large organelle containing the chromosomes and acting as a control center for the cell; center of an atom.

O

o-, oo- (o, o′o) egg.

obligate parasite (ob′li-gāt par′ah-sīt) an organism, such as viruses, that always causes disease.

oil shale (oil shāl) rock formations that contain kerogen, a substance that can be refined to give petroleumlike products.

olfactory (ol-fak′to-re) pertaining to the sense of smell.

omnivores (om′ni-vorz) animals that eat both plants and animals.

oncogene (ong′ko-jēn) a gene that contributes to the transformation of a cell into a cancerous cell.

ontogeny (on-toj′e-ne) the developmental history of a single organism.

oogenesis (o″o-jen′e-sis) production of egg in females by the process of meiosis and maturation.

operon (op′er-on) regulatory genes together with the structural genes they regulate.

orbital (or′bi-tal) the path of an electron about the nucleus of an atom; shell.

order (or′der) in taxonomy, the category below class and above family.

organelle (or″gah-nel′) specialized structures within cells such as the nucleus, mitochondria, endoplasmic reticulum.

organic (or-gan′ik) pertaining to any aspect of living matter.

organic soup (or-gan′ik soop) an expression used to refer to the ocean before the origin of life when it contained newly formed organic compounds.

organizer (or′gah-nīz″er) a group of cells of an embryo that influences or directs the differentiation of another group of cells.

organ of Corti (or′gan uv kor′ti) the organ that contains the hearing receptors in the inner ear.

orgasm (or′gazm) physical and emotional climax during sexual intercourse; results in ejaculation in the male.

oscillation (os″i-la′shun) an occurrence that is repeatedly followed by an opposite occurrence.

osmosis (oz-mo′sis) the movement of water from an area of greater concentration of water to an area of lesser concentration of water across a semipermeable membrane.

osmotic pressure (oz-mot′ik presh′ur) pressure generated by the osmotic flow of water.

ossicles (os′ī-k′lz) the tiny bones found in the middle ear: hammer, anvil, and stirrup.

otoliths (o′to-liths) granules that stimulate ciliated cells in the utricle and saccule.

ov-, ovi- (ov, o′vi) egg.

oval opening (foramen ovale) (o′val o′pen-ing) an opening between the two atria in the fetal heart.

ovary (o′var-e) the sex gland in female animals; the base of the pistil in angiosperms.

oviduct (o′vi-dukt) the tube connecting the ovary to the uterus in higher animals.

ovulation (o″vu-la′shun) the discharge of a mature egg from the follicle within the ovary.

ovule (o′vul) structure that contains megasporangium in seed plants where meiosis occurs and the female gametophyte is produced.

oxidation (ok″si-da′shun) the loss of electrons (inorganic) or the removal of hydrogen atoms (organic).

oxidative decarboxylation (ok′si-da′tiv de″kar-bok″si-la′shun) a reaction that involves the release of carbon dioxide as oxidation occurs.

oxidizing atmosphere (ok′si-diz-ing at′mos-fēr) an atmosphere that contains oxidizing molecules such as O_2 rather than reducing molecules such as H_2.

oxygen debt (ok′si-jen det) the amount of oxygen needed to metabolize lactic acid that accumulates during vigorous exercise.

oxytocin (ok′se-to′sin) hormone released by posterior pituitary that causes contraction of uterus and milk letdown.

ozone shield (o′zōn shēld) a layer of O_3 present in the upper atmosphere which protects the earth from damaging U.V. light.

P

pacemaker (S-A node) (pās′māk-er) a small region of neuromuscular tissue that initiates the heartbeat.

palate (pal′at) the roof of the mouth.

palisade cells (pal′i-sād selz) a compact layer of cylindrical cells located in the mesophyll layer near the upper epidermis of a leaf.

pancreas (pan′kre-as) a vertebrate organ located near the stomach that secretes digestive enzymes into the duodenum and produces hormones, notably insulin.

para- (par′ah) alongside of.

parasite (par′ah-sīt) an organism that resides externally on or internally within another organism and does harm to this organism.

parasympathetic nervous system (par''ah-sim''pah-thet'ik ner'vus sis'tem) a portion of the autonomic nervous system that usually promotes those activities associated with a normal state.

parathyroids (par''ah-thi'roidz) hormonal glands located within the thyroid that regulate potassium and calcium metabolism.

parenchyma (pah-reng'ki-mah) relatively unspecialized cells that make up the fundamental tissue of plants.

pelagic zone (pe-laj'ik zōn) the main portion of an ocean containing those organisms that reside in either the upper, middle, or lower levels of the open sea.

pelvis (pel'vis) a bony ring formed by the innominate bones. Also a hollow chamber in the kidney that lies inside the medulla and receives freshly prepared urine from the collecting ducts.

penis (pe'nis) male copulatory organ.

pepsin (pep'sin) a protein-digesting enzyme secreted by gastric glands.

peptide (pep'tid) two to several amino acids joined together by a peptide bond.

peptide bond (pep'tid bond) the bond that joins two amino acids.

peri- (per'e) surrounding.

pericycle (per''i-si'kl) a single layer of tissue next to the endodermis that produces secondary roots.

peripheral nervous system (pě-rif'er-al ner'vus sis'tem) nerves and ganglia that lie outside the central nervous system.

peristalsis (per''i-stal'sis) a rhythmical contraction that serves to move the contents along in tubular organs such as the digestive tract.

peritubular capillary (per''i-tu'bu-lar kap'i-lar''e) capillary that surrounds a nephron and functions in reabsorption during urine formation.

permeable (per'me-ah-b'l) the property of allowing substances to pass through.

PGAL (pe je a el) *see* phosphoglyceraldehyde.

pH (pe āch) a measure of the hydrogen ion concentration; any pH below 7 is acid and any pH above 7 is basic.

phagocytosis (fag''o-si-to'sis) the taking in of bacteria and/or debris by engulfing; cell eating.

pharynx (far'ingks) throat.

phenotype (fe'no-tip) the outward appearance of an organism caused by the genotype and environmental influences.

pheromone (fer'o-mōn) a chemical substance secreted by one organism that influences the behavior of another.

phloem (flo'em) the vascular tissue in plants that transports nutrients; *see* xylem.

phosphoglyceraldehyde (PGAL) (fos''fo-glis''er-al'de-hid) a metabolite in both photosynthesis and glycolysis.

phospholipid (fos''fo-lip'id) lipids containing phosphorus that are particularly important in the formation of cell membranes.

photo- (fo'to) light.

photon (fo'ton) a packet of electromagnetic energy.

photoperiodism (fo''to-pe're-od-izm) a response to light and dark; particularly in reference to flowering in plants.

photosynthesis (fo''to-sin'thě-sis) the process of making carbohydrate from carbon dioxide and water by using the energy of the sun.

photosystem (fo'to-sis''tem) a photosynthetic unit located within the membrane of a thylakoid that contains several hundred molecules of chlorophyll *a* and *b* along with accessory pigments.

phototropism (fo-tot'ro-pizm) a growth response to light in plants.

-phyll (fil) leaf.

phylogeny (fi-loj'ě-ne) the evolutionary history of a particular group of organisms.

phylum (fi'lum) a taxonomic category applied to animals that follows kingdom and lies above class.

-phyte, phyto- (fit, fi'to) plant.

phytochrome (fi'to-krōm) a plant pigment that is involved in photoperiodism in plants.

pinna (pin'nah) outer, funnellike structure of the ear that picks up sound waves.

pinocytosis (pin''o-si-to'sis) the taking in of small, nonpermeable molecules by engulfing them; cell drinking.

pistil (pis't'l) part of the flower that contains a stigma, style, and ovary.

pith (pith) a plant tissue located in the central portion of dicot stems.

pituitary (pi-tu'i-tar''e) a small oval gland that is attached to the hypothalamus of the brain and secretes a number of hormones.

placenta (plah-sen'tah) a region formed from the chorion of the fetus and the uterine lining where nutrients pass from the mother's blood to fetal blood and wastes pass in the opposite direction.

plankton (plank'ton) free-floating microscopic organisms found in most bodies of water.

plasm-, plasmo-, -plasm (plazm, plaz'mo, plazm) viscous material.

plasma (plaz'mah) the liquid portion of blood.

plasma cell (plaz'mah sel) a cell derived from a B cell lymphocyte that is specialized to mass produce antibodies.

plasma membrane (plaz'mah mem'brān) *see* cell membrane.

plasmid (plaz'mid) a circular DNA segment that is present in bacterial cells but is not part of the bacterial chromosome.

plasmolysis (plaz-mol'i-sis) contraction of the cell contents of plant cells due to the loss of water.

plastids (plas'tidz) organelles of plants that are specialized for various functions, including photosynthesis.

platelet (plāt'let) a formed element that is necessary to blood clotting.

pleiotropy (pli-ot'ro-pe) the capacity of a gene to affect a number of different phenotype characteristics.

polar bodies (po'lar bod'es) nonfunctioning daughter cells that have little cytoplasm and are formed during oogenesis.

polar bond (po'lar bond) a covalent bond in which an electron pair is shared unevenly, resulting in a partially positive and partially negative atom.

pollen (pol'en) the male gametophyte generation in seed plants that transports the sperm to the egg.

poly- (pol'e) many.

polymer (pol'i-mer) a large molecule made up of many identical subunits.

polymerization (pol''i-mer''i-za'shun) the formation of a polymer such as a DNA molecule or a protein.

polymorphism (pol''e-mor'fizm) occurrence in a population of two or more distinct forms or genetic types.

polymorphonuclear (pol''e-mor''fo-nu'kle-ar) a white cell with a many-lobed nucleus.

polyp (pol'ip) the sedentary stage in the life cycle of coelenterates; a benign growth.

polypeptide (pol''e-pep'tid) a molecule composed of many amino acids linked together by peptide bonds.

polyploidy (pol'e-ploi'de) several sets of chromosomes; *poly* = many and *ploidy* = sets of chromosomes.

polysaccharide (pol''e-sak'ah-rid) a macromolecule composed of many units of sugar.

polysome (pol'e-sōm) a cluster of ribosomes all attached to the same mRNA molecule and thus all participating in the synthesis of the same polypeptide.

population (pop''u-la'shun) all the organisms of the same species in one place (area/space).

portal system (por'tal sis'tem) a vascular system that begins and ends in capillaries.

prairie (prar'e) a biome characterized by the presence of either short stem or long stem grasses.

predation (pre-da'shun) the killing and eating of one animal by another.

prey (pra) organisms that serve as food for a particular predator.

primates (pri'māts) animals that belong to the order Primates, the order of mammals that includes prosimians, monkeys, apes, and humans.

primitive atmosphere (prim'i-tiv at'mos-fēr) the gases that were found in the atmosphere when the earth first arose.

primitive streak (prim'i-tiv strēk) an elongated mass of cells in bird and mammal embryos that corresponds to the morula stage of other animals.

pro- (pro) before.

producers (pro-du'serz) organisms that produce food and are capable of synthesizing organic compounds from inorganic constituents of the environment; usually the green plants and algae in an ecosystem.

product (prod'ukt) the end result of a chemical reaction.

profundal (pro-fun'dal) that portion of a lake below the level of light penetration.

progesterone (pro-jes'te-ron) a female sex hormone produced by the ovary that helps maintain the secondary sex characteristics and prepares the uterine lining for implantation.

proglottids (pro-glot'idz) the body sections of a tapeworm.

prokaryotic (pro''kar-e-ot'ik) lacking the organelles found in complex cells; such as a bacterium or cyanobacterium.

proprioceptor (pro''pre-o-sep'tor) sensory receptor that assists the brain in knowing the position of the limbs.

prosimians (pro-sim'e-anz) primitive primates such as lemurs, tarsiers, and tree shrews.

prostaglandins (pros''tah-glan'dinz) hormones that have various and powerful effects often within the cells that produce them.

prostate gland (pros'tat gland) a gland in males that is located about the urethra at the base of the bladder; produces most of the seminal fluid.

prot-, proto- (prot, pro'to) first, primary.

protein (pro'te-in) a macromolecule composed of one or several long polypeptides.

proteinoid (pro'te-in-oid) a phase, consisting of polypeptides only, during the chemical evolution of the first cell or cells.

prothallus (pro'thal-us) a small, heart-shaped structure; the gametophyte generation of the fern.

protocell (pro'to-sel) that structure which preceded the true cell in the history of life.

proton (pro'ton) a subatomic particle found in the nucleus that has a weight of one atomic mass unit and carries a positive charge; a hydrogen ion.

protozoans (pro''to-zo'anz) animal-like protists that are classified according to means of locomotion: amoebas, flagellates, ciliates.

proximal convoluted tubule (prok'si-mal kon'vo-lut-ed tu'bul) highly coiled region of a nephron near Bowman's capsule.

pseudo- (su'do) false.

pseudocoelom (su''do-se'lom) a coelom incompletely lined by mesoderm.

pseudopodia (su''do-po'de-ah) projections of cytoplasm characteristic of amoeboid-type cells that function in locomotion and feeding.

pulmonary (pul'mo-ner''e) referring to the lungs.

pulmonary system (pul'mo-ner''e sis'tem) the blood vessels that take deoxygenated blood to and oxygenated blood away from the lungs.

pupa (pu'pah) a dormant stage in the life cycle of insects during which metamorphosis occurs.

pure (pur) see homozygous.

purines (pu'rinz) nitrogenous bases found in DNA and RNA that have two interlocking rings.

pyrimidines (pi-rim'i-dinz) nitrogenous bases found in DNA and RNA that have just one ring.

pyruvate (pi'roo-vat) the end product of glycolysis; pyruvic acid.

R

races (ras'ez) sets of populations occupying particular regions that differ in one or more characteristics from other populations of the same species; subspecies.

radial symmetry (ra'de-al sim'e-tre) regardless of the angle of a cut made at the midline of an organism, two equal halves result.

radioactive atom (ra''de-o-ak'tiv at'om) an atom that spontaneously emits energetic particles by disintegration of the nucleus.

radius (ra'de-us) an elongated bone located on the thumb side of the lower arm.

radula (rad'u-lah) a structure unique to mollusks that aids the process of grinding up food.

Ramapithecus (ram''ah-pith'e-kus) a genus of extinct apes that may be ancestral to humans but are most likely ancestral to orangutans only.

reactant (re-ak'tant) a chemical that undergoes a change during a chemical reaction.

receptor (re-sep'tor) a sense organ specialized to receive information from the environment. Also a structure found in the membrane of cells that combines with a specific chemical in a lock and key manner.

recessive allele (re-ses'iv ah-lel') hereditary factor that only expresses itself when the genotype is homozygous for the factor.

recombinant (re-kom'bi-nant) DNA having genes from two different organisms or gametes carrying recombined chromosomes after crossing over.

recombination (re''kom-bi-na'shun) a recombining of the genes that occurs during sexual reproduction.

rectum (rek'tum) the terminal portion of the intestine.

reducing atmosphere (re-dus'ing at'mos-fer) an atmosphere that contains reducing molecules such as H_2 rather than oxidizing ones such as O_2.

reduction (re-duk'shun) the gain of electrons (inorganic); the addition of hydrogen atoms (organic).

reflex (re'fleks) an inborn autonomic response to a stimulus that is dependent on the existence of fixed neural pathways.

reflex arc (re'fleks ark) the passage of nerve impulses from a receptor to an effector by way of sensory, interneuron, and motor neurons.

regeneration (re-jen''er-a'shun) regrowth of tissue; formation of a complete organism from a small portion.

releaser (re-le'ser) in behavior, a sign stimulus that initiates a behavioral pattern.

renal (re'nal) of or pertaining to the kidney.

renewable resource (re-nu'ah-b'l re'sors) a material needed by organisms that is continually produced in the environment.

replication (re''pli-ka'shun) the duplication of DNA; occurs when the cell is not dividing.

resource (re'sors) anything needed or used by a population or organism.

respiration (res''pi-ra'shun) the inhalation and exhalation of air plus the exchange of oxygen and carbon dioxide across cell membranes. Cellular respiration: the breakdown of glucose with the concomitant accumulation of ATP.

respiratory chain (re-spi'rah-to''re chan) a series of molecules found within mitochondria that pass electrons (sometimes accompanied by hydrogen ions) from one to the other in such a way that the energy of oxidation is captured and ATP is generated.

resting potential (rest'ing po-ten'shal) the voltage recorded from inside a neuron when it is not conducting nerve impulses.

retina (ret'i-nah) the innermost layer of the eyeball that contains the rods and cones.

retroviruses (ret''ro-vi'rus-ez) viruses that contain only RNA and carry out RNA to DNA transcription prior to viral reproduction.

Rh factor (ar'ach fak'tor) a type of antigen on the red cells.

rhizoids (ri'zoidz) rootlike structures that absorb water in certain plants such as the moss and the fern prothallus.

rhodopsin (ro-dop'sin) visual purple, a pigment found in the rods of one type of receptor in the retina of the eye.

ribonucleic acid (RNA) (ri''bo-nu-kle'ik as'id) a nucleic acid important in the synthesis of proteins that contains the sugar ribose; the bases uracil, adenine, guanine, cytosine; and phosphoric acid.

ribose (ri'bos) a five-carbon sugar with the formula $(CH_2O)_5$.

ribosomes (ri'bo-somz) minute particles, found attached to endoplasmic reticulum or loose in the cytoplasm, that are the site of protein synthesis.

ribulose biphosphate (RuBP) (ri'bu-los bi-fos'fat) the molecule that acts as the acceptor for carbon dioxide during the Calvin cycle.

RNA (ar'en-a) see ribonucleic acid.

rods (rodz) dim-light receptors in the retina of the eye that detect motion but no color.

root pressure (root presh'ur) the tendency of water to rise in the xylem of a root.

RuBP (ar u be pe) see ribulose biphosphate.

S

salivary (sal'i-ver-e) pertaining to saliva; a secretion from the glands of the mouth.

salts (sawlts) ionic compounds that are formed by the reaction of an acid with a base (e.g., Na^+Cl^-).

S-A node (es a nod) sinoatrial node; see pacemaker.

saprophyte (sap'ro-fit) a heterotrophic organism such as bacteria and fungi that externally breaks down dead organic matter before absorbing the products.

sarco- (sar'ko) a term meaning flesh; used in reference to skeletal muscle cells (i.e., sarcolemma, sarcoplasma).

sarcolemma (sar''ko-lem'ah) the membrane that surrounds striated muscle cells.

sarcomere (sar'ko-mer) a unit of a myofibril between two Z lines.

savanna (sah-van'ah) a grassland biome that has occasional trees and is particularly associated with Africa.

scapula (skap'u-lah) a broad somewhat triangular bone located on either side of the back.

sclera (skle'rah) white fibrous outer layer of the eyeball.

sclerenchyma (skle-reng'ki-mah) a support tissue in plants made of hollow cells with thickened walls.

scrotum (skro'tum) the sac that contains the testes.

secretin (se-kre'tin) hormone secreted by the small intestine that stimulates the release of pancreatic juice and bile from the gall bladder.

secretion (se-kre'shun) a substance made by a cell or organ that upon release is used by or affects different cells or organs.

seed (sēd) a mature ovule that contains an embryo with food, enclosed in a protective coat.

segmented (seg-ment'ed) the presence of repeating units, such as in segmented animals like the earthworm.

segregation (seg''re-ga'shun) the separation of the members of an allelic pair into different gametes during the process of meiosis.

selection (sĕ-lek'shun) see natural selection.

semen (se'men) the sperm-containing secretion of males; seminal fluid plus sperm.

semicircular canal (sem''e-ser'ku-lar kah-nal') tubular structures within the inner ear that contain the receptors responsible for the sense of dynamic equilibrium.

semilunar (sem''e-lu'nar) a heart valve that consists of three cusps, i.e., the pulmonary semilunar and aortic semilunar valves.

seminal fluid (sem'i-nal floo'id) fluid produced by various glands situated along the male reproductive tract.

seminal vesicle (sem'i-nal ves'i-k'l) a convoluted saclike structure attached to ductus deferens near the base of the bladder in males.

seminiferous tubules (sem''i-nif'er-us tu'būlz) highly coiled ducts within the male testes that produce and transport sperm.

sensory neuron (sen'so-re nu'ron) a neuron that takes the nerve impulse to the central nervous system; afferent neuron.

septum (sep'tum) partition or wall such as the septum in the heart, which divides the right half from the left half.

serum (se'rum) light-yellow liquid left after clotting of the blood.

sessile (ses'il) organisms that lack locomotion and remain stationary in one place, such as plants or sponges.

sex-linked genes (seks-linkt' jēnz) genes found on the sex chromosomes that control somatic traits.

sieve tube cell (siv tūb sel) specialized cells (elements) that form a linear array running vertically through phloem that functions in transport of organic nutrients.

sigmoidal growth curve (sig-moid'al grōth kurv) S-shaped pattern of growth of a population with time.

sinoatrial node (S-A node) (si''no-a'tre-al nōd) see pacemaker.

sinus (si'nus) a cavity, as the sinuses in the human skull and the blood sinuses of some animals with open circulatory systems.

skeletal muscle (skel'ē-tal mus'el) the contractile tissue that comprises the muscles attached to the skeleton; also called striated muscle.

smooth muscle (smooth mus'el) the contractile tissue that comprises the muscles found in the walls of internal organs.

social parasitism (so'shal par'ah-si''tizm) the utilization of one population by another such that the first is harmed and the second is benefited.

society (so-si'ĕ-te) members of a population that are specialized to perform specific duties and cooperate with one another for the good of the group.

solar energy (so'lar en'er-je) energy derived from radiation given off by the sun.

solute (sol'ūt) a substance dissolved in a solvent to form a solution.

solution (so-lu'shun) the mixing of a solute with a solvent to the degree that the solute is not detected by sight.

solvent (sol'vent) a fluid such as water that dissolves solutes.

-soma, somat-, -some (so'mah, so-mat', sōm) body, entity.

somatic (so-mat'ik) pertaining to the body but excluding certain parts; somatic chromosomes exclude the sex chromosomes; somatic nervous system does not include the autonomic nervous system.

somites (so'mits) paired, blocklike masses of mesoderm that appear on either side of the neural tube in higher embryos; become bone and muscle.

sorus (so'rus) a cluster of sporangia found on the underside of fern leaves (plural: sori).

specialization (spesh''al-i-za'shun) the taking on of a particular shape, form, and function.

speciation (spe''se-a'shun) the steps by which one group of organisms gives rise to two different species.

species (spe'shēz) a group of similarly constructed organisms that are capable of interbreeding and producing fertile offspring; organisms that share a common gene pool.

spermatogenesis (sper''mah-to-jen'ĕ-sis) production of sperm in males by the process of meiosis and maturation.

sphincter (sfingk'ter) a muscle that surrounds a tube and closes or opens the tube by contracting and relaxing.

spinal cord (spi'nal kord) neural tube or nerve cord.

spindle (spin'd'l) an apparatus composed of microtubules to which the chromosomes are attached during cell division.

spine (spin) the nerve cord protected by vertebrae.

spongy bone (spun'je bōn) porous bone found at the ends of long bones.

spongy layer (spun'je la'er) the lower layer of the mesophyll of a leaf that carries out gas exchange.

sporangium (spo-ran'je-um) a structure within which spores are produced.

spore (spōr) usually a haploid reproductive structure that develops into a haploid generation; in bacteria, a particularly resistant structure.

sporophyte (spo'ro-fit) the diploid generation of a plant; it produces spores.

stamen (sta'men) part of flower, composed of filament and anther, where pollen grains are produced.

starch (starch) the storage polysaccharide found in plants that is composed of glucose molecules joined in a linear-type fashion.

stereotype (ste're-o-tip) always the same, as in stereotyped behavior.

sterile (ster'il) devoid of living things; particularly following the process of sterilization, which kills even unseen organisms.

steroid (ste'roid) a type of lipid composed of four interlocking rings similar to cholesterol.

stigma (stig'mah) the uppermost part of a pistil.

stimulus (stim'u-lus) any environmental change detected by a receptor.

stomata (sto'mah-tah) openings in the leaves of plants through which gas exchange takes place.

strand (strand) an expression that refers to a polymer of nucleotides found in DNA or RNA.

stratified (strat'i-fid) layered, as in stratified epithelium, which contains several layers of cells.

striated (stri'āt-ed) having bands; cardiac and skeletal muscle are striated with bands of light and dark.

stroma (stro'mah) the interior portion of a chloroplast.

stroma lamellae (stro'mah lah-mel'e) membranous platelike connections between the grana within a chloroplast.

style (stil) the long slender part of the pistil.

subcutaneous (sub''ku-ta'ne-us) a tissue layer found in vertebrate skin that lies just beneath the dermis and tends to contain fat cells.

substrate (sub'strāt) a reactant in a reaction controlled by an enzyme.

succession (suk-sĕ'shun) a series of ecological stages by which the community in a particular area gradually changes until there is a stable community that can maintain itself.

sugar (shoog'ar) a carbohydrate containing a limited number of monosaccharide units; the disaccharide sucrose is table sugar.

sym-, syn- (sim, sin) together.

symbiosis (sim″bi-o′sis) an intimate association of two dissimilar species including commensalism, mutualism, and parasitism.

symmetry (sim′e-tre) the property of having two halves that are mirror images of each other.

sympathetic nervous system (sim″pah-thet′ik ner′vus sis′tem) that part of the autonomic nervous system that generally causes effects associated with emergency situations.

synapse (sin′aps) the region between two nerve cells where the nerve impulse is transmitted from one to the other; usually from axon to dendrite.

synapsis (si-nap′sis) the attracting and pairing of homologous chromosomes during meiosis.

synaptic vesicle (si-nap′tik ves′i-k′l) small vacuoles at the ends of axons that contain a neurohormone.

synovial fluid (si-no′ve-al floo′id) fluid secreted by synovial membrane of a joint.

synthesis (sin′the-sis) to build up, such as the combining together of two small molecules to form a large molecule.

systemic system (sis-tem′ik sis′tem) that part of the circulatory system that serves body parts other than the gas-exchanging surfaces in the lungs.

systole (sis′to-le) contraction of the heart chambers, particularly the left ventricle.

T

tactile (tak′til) communication that requires the touching of others.

taiga (ti′gah) a biome that forms a worldwide northern belt of coniferous trees.

taxis (tak′sis) a movement in relation to a stimulus, such as a phototaxis (a movement oriented to a light source).

taxonomy (tak-son′o-me) the science of naming and classifying organisms.

T cells (te selz) a type of lymphocyte involved in cell mediated immunity.

temperate forest (tem′per-at for′est) a biome characterized by deciduous trees and well-defined understory; found in regions with a moderate climate.

template (tem′plat) a pattern that serves as a mold for the production of an oppositely shaped structure; one strand of DNA is a template for the complementary strand.

tendon (ten′don) a tissue that connects muscle to bone.

territory (ter′i-tor″e) an area or space defended, usually for breeding purposes, by an animal or group of animals against other members of the same species.

testes (tes′tez) the male gonads, the organs that produce sperm and testosterone.

testosterone (tes-tos′te-ron) the most potent androgen.

tetany (tet′ah-ne) severe twitching caused by involuntary contraction of the skeletal muscles due to a lack of calcium.

tetrad (tet′rad) a set of four chromatids resulting from the pairing of homologous chromosomes during meiosis.

thalamus (thal′ah-mus) a mass of gray matter located at the base of the cerebrum in the wall of the third ventricle; the lowest portion of the forebrain.

thorax (tho′raks) chest.

thrombin (throm′bin) the enzyme derived from prothrombin that converts fibrinogen to fibrin threads during blood clotting.

thrombocyte (throm′bo-sit) platelet.

thromboplastin (throm″bo-plas′tin) an enzyme released by platelets that converts prothrombin to thrombin, a necessary step in blood clotting.

thrombus (throm′bus) a blood clot that remains in the blood vessel where it formed.

thylakoid (thi′lah-koid) an individual flattened vesicle found within a granum (pl. grana).

thymus (thi′mus) an organ that lies in the neck and chest area and is absolutely necessary to the development of immunity.

thyroid (thi′roid) a hormonal gland located in the neck region.

thyroxin (thi-rok′sin) the hormone produced by the thyroid that speeds up the metabolic rate.

tibia (tib′e-ah) the shinbone found in the lower leg.

tissue (tish′u) similar-type cells that work together performing a specific function.

tissue fluid (tish′u floo′id) fluid found about tissue cells containing molecules that enter and leave by way of the capillaries.

tone (ton) the continuous partial contraction of muscle; also the quality of a sound.

toxic (tok′sik) poisonous.

toxin (tok′sin) a poison produced by one type organism that harms other type organisms.

trachea (tra′ke-ah) in vertebrates, the windpipe; in insects, the trachea are the air tubes.

tracheids (tra′ke-idz) a component of xylem made of long, tapered nonliving cells.

tracheophytes (tra′ke-o-fits″) members of the division, tracheophyta, that contains plants having vascular tissue and a dominant sporophyte generation.

tracheostomy (tra″ke-os′to-me) a cutting of the trachea in order to allow air to bypass an obstruction within the larynx.

trans- (trans) across, beyond.

transcription (trans-krip′shun) the process that results in the production of a strand of mRNA that is complementary to a segment of DNA.

translation (trans-la′shun) the process involving mRNA, ribosomes, and tRNA that results in a synthesis of a polypeptide having an amino acid sequence dictated by the sequence of codons in mRNA.

transpiration (tran″spi-ra′shun) the evaporation of water from a plant; pulls water from the roots through a stem.

triglyceride (tri-glis′er-id) a neutral fat; so named because it contains one molecule of glycerol and three fatty acid molecules.

triplet code (trip′let kod) sequence of three nucleotides in DNA that serve as a code for some particular amino acid.

trochophore (tro′ko-for) a larval form unique to the deuterostomes that indicates they are related.

trophic level (trof′ik lev′el) the position of species in the food chain; a link in the transfer of energy through an ecosystem.

trophoblast (trof′o-blast) the outer membrane that surrounds the human embryo and, when thickened by a layer of mesoderm, becomes the chorion.

tropic (trop′ik) action brought about by a stimulus, as in phototropic or gonadotropic.

tropism (tro′pizm) a growth response in a nonmotile organism such as the plants.

trypsin (trip′sin) a protein digesting enzyme secreted by the pancreas.

T system (te sis′tem) refers to structural arrangement of the sarcolemma to the sarcoplasmic reticulum in muscles where calcium is stored and released, causing muscle contraction.

tubal ligation (tu′bal li-ga′shun) cutting of the oviducts in females.

tundra (tun′drah) a biome characterized by lack of trees, short growing season, cold temperatures and little rainfall but wet conditions because of little evaporation and the presence of permafrost.

turgor pressure (tur′gor presh′ur) osmotic pressure in plant cells that adds to the strength of the cell.

tympanic membrane (tim-pan′ik mem′bran) eardrum.

U

ulna (ul′nah) an elongated bone found within the lower arm.

umbilical cord (um-bil′i-kal kord) cord connecting the fetus to the placenta through which blood vessels pass.

uracil (u′rah-sil) a nitrogenous base found in RNA.

urea (u-re′ah) primary nitrogenous waste of mammals.

ureter (u-re′ter) tube between kidney and bladder.

urethra (u-re′thrah) tube that takes urine from bladder to outside.

uric acid (u′rik as′id) waste product of nucleotide breakdown.

urinalysis (u″ri-nal′i-sis) a medical procedure in which the composition of a patient's urine is determined.

urinary bladder (u′ri-ner″e blad′der) an organ where urine is stored before being discharged by way of the urethra.

urine (u′rin) waste fluid containing urea, which is made by the kidneys, stored in the bladder, and discharged by the urethra.

uterus (u′ter-us) the organ in females in which the fetus develops.

V

vaccine (vak'sēn) treated antigens that can promote active immunity when administered.

vacuole (vak'u-ōl) a membrane-bounded cavity, usually fluid filled.

vagina (vah-ji'nah) copulatory organ in females.

valves (valvz) an opening that opens and closes, insuring one-way flow only; common to vessels such as the systemic veins and the lymphatic veins and to the heart.

variations (va''re-a'shunz) phenotype differences among members of a population.

vascular (vas'ku-lar) containing or concerning vessels that conduct fluid, for example, vascular bundles in plant stems and vascular cylinders in plant roots contain xylem and phloem.

vascular cylinder (vas'ku-lar sil'in-der) a central region of dicot roots that contains the vascular tissues, xylem and phloem.

vascular tissue (vas'ku-lar tish'u) a transport tissue; circulatory vessels in animals and xylem and phloem in plants.

vas deferens (vas def'er-enz) see ductus deferens.

vasectomy (vah-sek'to-me) cutting of the ductus deferens in males.

vaso- (vas'o) blood vessel.

vasopressin (vas''o-pres'in) secreted by the posterior pituitary; promotes reabsorption of water by the kidneys; also called antidiuretic hormone (ADH).

VD (ve de) see venereal disease.

vein (vān) a blood vessel that takes blood to the heart.

vena cava (ve'nah ka'vah) a primary vein in vertebrates.

venereal disease (VD) (ve-ne're-al di-zēz') an infectious disease that is transmitted through sexual contact; herpes, gonorrhea, and syphilis are common examples.

ventral (ven'tral) the front side of humans; the underside of most animals.

ventricle (ven'tri-k'l) a cavity in an organ such as the ventricles of the heart or the ventricles of the brain.

vertebra (ver'tě-brah) one of the bones found within the backbone of vertebrates.

vessel element (ves'el el'ě-ment) an individual vessel cell in xylem. During development, vessel cells lose their contents and end walls so that they form a continuous pipeline in xylem.

vestigial (ves-tij'e-al) the remains of a structure that was functional in some ancestor but is no longer functional in the organism in question.

villi (vil'i) fingerlike projections that line the small intestine and function in absorption.

virus (vi'rus) a minute infectious particle composed of a coat of protein and a core of nucleic acid.

viscera (vis'er-ah) all the thoracic and abdominal organs of animals.

vitamin (vi'tah-min) usually coenzymes, needed in small amounts, that the body is no longer capable of synthesizing and therefore must be in the diet.

vulva (vul'vah) the external genitalia of the female that lie near the opening of the vagina.

W

waggle dance (wag'l dans) expression used to refer to the movements performed by bees to indicate to other bees the location of food.

water vascular system (wah'ter vas'ku-lar sis'tem) a series of canals that take water to the tube feet of echinoderms allowing them to expand.

X

X chromosome (eks kro'mo-sōm) one of the sex chromosomes, present in duplicate in normal females.

xylem (zi'lem) transport tissue in plants that conducts water and minerals.

Y

Y chromosome (wi kro'mo-sōm) one of the sex chromosomes that must be present in human males.

yolk (yōk) nutrient material in the egg for use by the developing embryo.

yolk sac (yōk sak) one of the extraembryonic membranes within which yolk is found.

Z

zoo- (zo'o) animal; motile.

zoospore (zo'o-spōr) a flagellated or ciliated spore.

zygote (zi'gōt) diploid cell formed by the union of two gametes, the product of fertilization.

Credits

Photographs

Introduction

Opener: Bob Coyle
Fig. I.1a: M. Austerman/ANIMALS ANIMALS
Fig. I.2b: Carolina Biological Supply
Fig. I.2c: Ed Reschke
Fig. I.3: K. G. Preston/PREMAPHOTOS WILDLIFE
Fig. I.4a: Roger Tory Peterson/Photo Researchers, Inc.
Fig. I.4b: Bob Coyle
Page 9 (top): Jack Fields/Photo Researchers, Inc.
Page 9 (bottom): Tom McHugh/Photo Researchers, Inc.
Fig. I.6: Merlin D. Tuttle, Milwaukee Public Museum, and J. Scott Altenbach
Fig. I.7: Clyde H. Smith/Peter Arnold, Inc.
Fig. I.9a: Marilyn Wood/PHOTONATS
Fig. I.9b: Lane Stewart/SPORTS ILLUSTRATED
Page 14: Robert Eckert/EKM-Nepenthe

Part One

Opener: Bruce J. Russell/BioMedia Associates

Chapter 1

Fig. 1.12: Van Bucher/Photo Researchers, Inc.
Fig. 1.18: Gerry Cranhams/Photo Researchers, Inc.
Fig. 1.24: Farrell Grehan/Photo Researchers, Inc.
Page 40: Bob Coyle

Chapter 2

Fig. 2.2a: Bausch and Lomb
Fig. 2.2b: Dr. Ned Feder, National Institute of Arthritis, Diabetes, and Digestive and Kidney Diseases, NIH
Fig. 2.3a: Carl Zeiss, Inc.
Fig. 2.3b: Dr. George E. Palade, Yale University
Fig. 2.4: Karen Anderson/University of Colorado, Boulder
Figs. 2.5, 2.9a, 2.10a, 2.13–2.15a, 2.19c, 2.21 (p. 62 right), 2.21 (p. 63 top, center and bottom): Dr. Keith Porter, University of Colorado, Boulder
Figs. 2.6, 2.21 (p. 62 left): Stephen L. Wolfe
Fig. 2.9c: James A. Lake, Journal of Molecular Biology 105, 131–159 (1976). © Academic Press, Inc., (London) Ltd.
Fig. 2.11: Dr. Andrew Staehelin, University of Colorado, Boulder
Fig. 2.16: Dr. L. K. Shumway, Washington State University
Fig. 2.17: Dr. Jean Paul Revel
Fig. 2.21 (p. 63 top center and bottom center): Biophoto Associates

Chapter 3

Fig. 3.1d: Dr. Daniel Branton
Fig. 3.8: S. J. Singer
Page 75: H. V. Allison Galleries, Inc., New York
Figs. 3.19a & b, 3.20a: L. A. Staehelin and B. E. Hull
Fig. 3.20b: J. H. Troughton and T. B. Sampson

Chapter 4

Figs. 4.7, 4.9–4.11: Eric Grave
Fig. 4.12: G. G. Maul and Academic Press, Inc., from *J. Ultrastr. Res.* 31:375 (1970)
Fig. 4.15: B. A. Palevitz and E. H. Newcomb, University of Wisconsin/Biological Photo Service

Chapter 5

Page 106: Bob Coyle
Fig. 5.15b: H. Fernandes—Moran

Part Two

Opener: Bruce J. Russell/BioMedia Associates

Chapter 6

Fig. 6.2: Gordon Leedale/Biophoto Associates
Fig. 6.6c: Kenneth R. Miller
Page 127: Dr. Raymond Chollet
Fig. 6.13: R. R. Hessler/Scripps Institution of Oceanography

Chapter 7

Figs. 7.3c, 7.6c, 7.7b, 7.9a: Carolina Biological Supply Company
Fig. 7.4: Dr. R. G. Kessel and Dr. C. Y. Shih, From SCANNING ELECTRON MICROSCOPY, Springer-Verlag, Berlin, Heidelberg, New York, 1976
Page 139, Fig. 7.27 (bottom left, top and bottom right): Bob Coyle
Fig. 7.9a: Carolina Biological Supply
Fig. 7.10: Carolina Biological Supply
Fig. 7.11b: J. H. Troughton and F. B. Sampson
Figs. 7.12b, 7.26: J. H. Troughton and L. Donaldson
Fig. 7.15: Dr. B. A. Meylan
Fig. 7.21: Michael Wotton/Weyerhaeuser Company
Fig. 7.29: Michael Godfrey

Part Three

Opener: Al Satterwhite/The Image Bank, Chicago

Chapter 8

Figs. 8.2–8.11: From TISSUES AND ORGANS: A TEXT-ATLAS OF SCANNING ELECTRON MICROSCOPY by R. G. Kessel and R. Kardon. W. H. Freeman and Company, © 1979
Page 173: Robert Eckert/EKM-Nepenthe

Chapter 9

Figs. 9.6, 9.8, 9.12c & d: From TISSUES AND ORGANS: A TEXT-ATLAS OF SCANNING ELECTRON MICROSCOPY by R. G. Kessel and R. Kardon. W. H. Freeman and Company, © 1979
Fig. 9.10: Martin M. Rotker/TAURUS PHOTOS
Fig. 9.13: Nels H. Granholm
Fig. 9.17: J. Somers/TAURUS PHOTOS
Fig. 9.18a & b: WHO
Fig. 9.18c & d: Center for Disease Control, Atlanta, GA
Page 200: Thomas Coard/EKM-Nepenthe

Chapter 10

Fig. 10.1: From: TISSUES AND ORGANS: A TEXT-ATLAS OF SCANNING ELECTRON MICROSCOPY by R. G. Kessel and R. Kardon. W. H. Freeman and Company, © 1979
Fig. 10.13: Tom Ballard/EKM-Nepenthe
Fig. 10.14: Ed Reschke
Fig. 10.17a & b: American Heart Association
Fig. 10.19a & b: Dr. Garrett Lee
Page 223: Bob Coyle

Chapter 11

Fig. 11.2a: Reproduced by permission from "The Morphology of Human Blood Cells" by L. W. Diggs, MD, Dorothy Sturm, and Ann Bell. Copyright 1984 Abbott Laboratories.
Fig. 11.2b: Dr. R. G. Kessel and Dr. C. Y. Shih, From SCANNING ELECTRON MICROSCOPY Springer-Verlag, Berlin, Heidelberg, New York 1976
Fig. 11.2c: Dr. Victor A. Najjar
Fig. 11.2d: Dr. Etienne de Harven and Miss Nina Lampen, Sloan-Kettering Institute, New York
Page 230: Robert Eckert/EKM-Nepenthe
Fig. 11.4a: Ed Reschke
Fig. 11.4b: Dr. Keith Porter, University of Colorado, Boulder
Fig. 11.5: Reprinted from THE STRUCTURE AND ACTION OF PROTEINS by Richard E. Dickerson and Irving Geis. Benjamin/Cummings Publishers, Menlo Park, CA. © 1969 by Dickerson and Geis
Fig. 11.9 (left): Don Fawcett

Fig. 11.10: E. Bernstein and E. Kairinen, *Science,* cover, 27 August 1971, Vol. 173. Copyright 1971 by the American Association for the Advancement of Science
Fig. 11.12a & b: From: TISSUES AND ORGANS: A TEXT-ATLAS OF SCANNING ELECTRON MICROSCOPY by R. G. Kessel and R. Kardon. W. H. Freeman and Company, © 1979
Fig. 11.13a: Pfizer, Inc.
Fig. 11.13b: From: ELECTRON MICROSCOPY OF HUMAN BLOOD CELLS by Tanaka and Goodman, Harper & Row
Fig. 11.14: Almeida, J. D., Cinader, B., and Howatson, A. F. The Structure of Antigen-Antibody Complexes. A Study by Electron Microscopy. *J. Exptl. Med.,* 118: 327–340, 1963 by © of The Rockefeller University Press
Fig. 11.15: Stuart I. Fox

Chapter 12

Fig. 12.6a & b: Dr. Burton D. Goldberg, From *J. Exptl. Med.,* 109(1959): 505 and *Scientific American,* 229(1973):54
Page 253: Bob Coyle
Page 255: Phillip A. Harrington/Schering-Plough Corporation
Fig. 12.9: Dr. Kirk Ziegler
Fig. 12.10a & b: Liepins, A. et al., T-lymphocyte Mediated Lysis of Tumor Cells in the Presence of Alloantiserum. *Cellular Immunology* 36, 331–344, 1978
Fig. 12.11: National Jewish Hospital/National Asthma Center
Fig. 12.12: Sandoz, Ltd., Basel, Switzerland
Fig. 12.13a & b: Professor Nicola Fabris, Ancona, Italy

Chapter 13

Fig. 13.5a–c: Bell Telephone Laboratories
Fig. 13.6: American Lung Association
Fig. 13.7: From: TISSUES AND ORGANS: A TEXT-ATLAS OF SCANNING ELECTRON MICROSCOPY by R. G. Kessel and R. Kardon. W. H. Freeman and Company, © 1979
Page 271: Bruce Russell/BioMedia Associates
Fig. 13.15: Dr. M. W. Jennison, Syracuse University
Figs. 13.16a–c, 13.17a & b: Oscar Auerbach, M. D., Veterans Administration Hospital, East Orange, New Jersey
Fig. 13.18a & b: Martin M. Rotker/TAURUS PHOTOS

Chapter 14

Fig. 14.3a & b: FUNCTIONAL HUMAN ANATOMY, 2nd edition, James E. Crouch. Lea & Febiger, 1972
Fig. 14.9: Dr. R. B. Wilson, Eppiley Institute for Research in Cancer
Fig. 14.12: From: Bloom, W., and Fawcett, D. W.: A TEXTBOOK OF HISTOLOGY, Philadelphia: W. B. Saunders Company, 1966. Photo by Dr. Ruth Bulger

Chapter 15

Fig. 15.4: J. D. Robertson
Fig. 15.6: Linda Bartlett, © 1981
Fig. 15.10: Randy Perkins, from Heuser & Reese, HANDBOOK OF PHYSIOLOGY
Fig. 15.22: Dan McCoy/Rainbow
Page 320: Barbara Burnes/Photo Researchers, Inc.

Chapter 16

Page 335: Mickey Pfleger
Fig. 16.8: NARCO Scientific
Fig. 16.13d: Ed Reschke
Fig. 16.15a: H. E. Huxley
Fig. 16.17c: Bernard Katz

Chapter 17

Fig. 17.5: From: TISSUES AND ORGANS: A TEXT-ATLAS OF SCANNING ELECTRON MICROSCOPY by R. G. Kessel and R. Kardon. W. H. Freeman and Company, © 1979
Page 355: Jacob Steiner/TIME Magazine

Fig. 17.9: Dr. T. Kuwabara
Fig. 17.14: Frank S. Werblen, University of California, Berkeley
Fig. 17.19: Scanning electron micrograph by Robert S. Preston, courtesy of Professor J. E. Hawkins, Kresge Hearing Research Institute, University of Michigan Medical School

Chapter 18

Fig. 18.7: UPI/Bettmann Archive
Figs. 18.8, 18.13, 18.18: From *Clinical Endocrinology and its Physiological Basis* by Arthur Grollman, 1964. Used by permission of J. B. Lippincott Company
Fig. 18.10: From: TISSUES AND ORGANS: A TEXT-ATLAS OF SCANNING ELECTRON MICROSCOPY by R. G. Kessel and R. Kardon. W. H. Freeman and Company, © 1979
Figs. 18.11, 18.14: Lester V. Bergman & Associates, Inc.
Fig. 18.12: F. A. Davis Company, Philadelphia, and Dr. R. H. Kampmeier
Fig. 18.15: Dr. Robert M. Brenner, Oregon Regional Primate Research Center
Page 383: Agence Vandystadt/Photo Researchers, Inc.
Fig. 18.22: Ed Reschke

Part Four

Opener: Claude Edelmann, Petit Format et Guigoz from the book *First Days of Life*/Black Star

Chapter 19

Fig. 19.5a: "Sea Urchin Sperm-Egg Interactions Studied with the Scanning Electron Microscope," Tegner, M. J., and Epel, D., *Science,* Vol. 179, pp. 685–688, Fig. 2, 16 February 1973.
Fig. 19.5b: Dr. Gerald Schatten
Fig. 19.9: Dr. Landrum Shettles
Fig. 19.15a–f: Bob Coyle
Page 408: Charles Lightdale/Photo Researchers, Inc.

Chapter 20

Fig. 20.10a: Lennart Nilsson
Fig. 20.15a–e: Claude Edelmann, Petit Format et Guigoz from the book *First Days of Life*/Black Star
Fig. 20.20: Leonard McCombe/LIFE Magazine © Time, Inc.

Chapter 21

Fig. 21.2a & b: Bob Coyle
Fig. 21.2c, d & f: Tom Ballard/EKM-Nepenthe
Fig. 21.2e: Louis Goldman/Photo Researchers, Inc.
Fig. 21.8: London Express
Fig. 21.9: Wide World Photos, Inc.
Page 452: Library of Congress
Fig. 21.13: Neal T. Nichols/TAURUS PHOTOS
Fig. 21.17: Jill Cannefax/EKM-Nepenthe
Fig. 21.22: Baylor College of Medicine

Chapter 22

Fig. 22.A: Biophysics Department, Kings College London
Fig. 22.B: Original model of the double helix. From J. D. Watson, *The Double Helix,* Atheneum, New York, p. 206. © 1968 by J. D. Watson
Fig. 22.13b: Alexander Rich
Fig. 22.17a: Dr. Pierre Chambon
Fig. 22.C: Dr. G. Steven Martin

Part Five

Opener: Al Szabo/ANIMALS ANIMALS

Chapter 23

Fig. 23.2a: American Museum of Natural History
Fig. 23.2b: Painting by Rudolph Freund, Carnegie Museum of Natural History
Fig. 23.3: Carolina Biological Supply Company
Fig. 23.10: Allan Price/TAURUS PHOTOS
Fig. 23.14: J. A. Bishop and L. M. Cook
Fig. 23.16: USDA
Fig. 23.20a & c: Tui A. DeRoy/Bruce Coleman, Inc.
Fig. 23.20b: F. B. Gill/VIREO
Fig. 23.20d: Miguel Castro/Photo Researchers, Inc.
Fig. 23.20e: F. Erize/Bruce Coleman, Inc.
Fig. 23.20f: Alan Root/Bruce Coleman, Inc.

Chapter 24

Fig. 24.3a: Steven Brooke
Fig. 24.3b: Sidney W. Fox, Institute for Molecular and Cellular Evolution
Fig. 24.4: Dr. Tatiana Evreinova
Fig. 24.5: Professor John M. Hayes

Chapter 25

Fig. 25.1a: Carl Zeiss, Inc., New York
Figs. 25.1b & c, 25.5, 25.24: Biophoto Associates
Fig. 25.2: Parke, Davis & Company
Figs. 25.4a & c, 25.13: Dr. R. G. Kessel and Dr. C. Y. Shih, From SCANNING ELECTRON MICROSCOPY. Springer-Verlag, Berlin, Heidelberg, New York, 1976
Fig. 25.4b: David Scharf/Peter Arnold, Inc.
Fig. 25.7: T. J. Beveridge, University of Guelp/Biological Photo Service
Fig. 25.8: AMSCO/American Sterilizer Company
Page 544: Pfizer, Inc.
Fig. 25.9a: J. Robert Waaland, University of Washington/Biological Photo Service
Fig. 25.9b: Myron Ledbetter/Biophoto Associates
Fig. 25.14: Carolina Biological Supply Company
Fig. 25.17: Gordon Leedale/Biophoto Associates
Fig. 25.18: Eric Grave
Fig. 25.21a & c: Bob Coyle
Fig. 25.21b: Kitty Kohout/Root Resources
Fig. 25.21d: Harold V. Green/Valan Photos
Fig. 25.22a: Philip Zito
Fig. 25.22b: Pat O'Hara
Fig. 25.23a & b: From: Volk and Wheeler: *Basic Microbiology;* 1973. Used with permission of J. B. Lippincott Company
Page 556: Donald H. Marx, Institute for Mycorrhizal Research and Development

Chapter 26

Figs. 26.3, 26.11a & b: Carolina Biological Supply Company
Fig. 26.4: Bruce Russell/BioMedia Associates
Fig. 26.6a: J. Robert Waaland, University of Washington/Biological Photo Service
Fig. 26.7b: John S. Flannery/Bruce Coleman
Fig. 26.8a: Verne Rockcastle
Figs. 26.8b, 26.11c, 26.19b: Biophoto Associates
Figs. 26.10, 26.19c, 26.22 (left): Bob Coyle
Fig. 26.12a: Roger Conant Wilde/Photo Researchers, Inc.
Fig. 26.12b: Foss/Photo Researchers, Inc.
Fig. 26.14: Field Museum of Natural History
Fig. 26.16a (1 & 2): Karlene Schwartz/PHOTO NATS
Fig. 26.16b (1 & 2): Dwight Kuhn
Fig. 26.18: USDA, Forest Service
Fig. 26.19a: Biophoto Associates/NHPA
Fig. 26.19d: Walter Hodge/Peter Arnold, Inc.
Fig. 26.20a–d: Walter Hodge
Fig. 26.22 (top): E. S. Ross
Fig. 26.22 (right): Leonard Lee Rue III

Chapter 27

Figs. 27.9a, c, d, 27.34a: Carolina Biological Supply Company
Fig. 27.9b: Paul Janosi/Valan Photos
Figs. 27.9e, 27.24a: Bruce Russell/BioMedia Associates

Table 9.8: Adapted from TIME Magazine Special Advertising Section on "Eating Well, Looking Fit, Feeling Better." © 1983 TIME Incorporated.

Chapter 10

Figs. 10.5, 10.6: From Van De Graaff, Kent M., HUMAN ANATOMY. © 1984 Wm. C. Brown Publishers, Dubuque, Iowa. All Rights Reserved. Reprinted by permission.
Fig. 10.8: *Functional Human Anatomy* by James E. Crouch, second edition, 1972. Published by Lea & Febiger.
Fig. 10.16: From THE LIVING BODY: A TEXT IN HUMAN PHYSIOLOGY, 4th ed. by Charles Herbert Best and Norman Burke Taylor. Copyright © 1958, 1952, 1944, 1938 by Holt, Rinehart and Winston, Inc. Reprinted by permission of CBS College Publishing.

Chapter 11

Fig. 11.6: From Hole, John W., Jr., HUMAN ANATOMY AND PHYSIOLOGY, 3d ed. © 1978, 1981, 1984 Wm. C. Brown Publishers, Dubuque, Iowa. All Rights Reserved. Reprinted by permission.
Fig. 11.9 (right): From Johnson, Leland G., BIOLOGY. © 1983 Wm. C. Brown Publishers, Dubuque, Iowa. All Rights Reserved. Reprinted by permission.

Chapter 12

Fig. 12.3: From Mader, Sylvia S., BIOLOGY: EVOLUTION, DIVERSITY & THE ENVIRONMENT. © 1985 Wm. C. Brown Publishers, Dubuque, Iowa. All Rights Reserved. Reprinted by permission.
Table 12.2: From LIVING, Health, Behavior and Environment, Fifth Edition by Fred V. Hein, Dana L. Farnsworth and Charles E. Richardson. Copyright © 1970 by Scott, Foresman and Company. Reprinted by permission of the author.

Chapter 13

Fig. 13.13: From Hole, John W., Jr., HUMAN ANATOMY AND PHYSIOLOGY, 3d ed. © 1978, 1981, 1984 Wm. C. Brown Publishers, Dubuque, Iowa. All Rights Reserved. Reprinted by permission.
Fig. 13.14: From A. J. Vander et al.: *HUMAN PHYSIOLOGY*, 1979, McGraw-Hill Book Company.
Table 13.4: Reproduced by permission of the American Cancer Society, Inc.

Chapter 14

Fig. 14.17: Copyright 1978 Time Inc. All rights reserved. Reprinted by permission from TIME.

Chapter 15

Fig. 15.2: Reprinted with permission of Macmillan Publishing Company from ANATOMY AND PHYSIOLOGY, 14th edition by D. C. Kimber, C. Gray, C. Stackpole and L. Leavell. Copyright © 1961 by Macmillan Publishing Company.
Fig. 15.15: From Mader, Sylvia S., BIOLOGY: EVOLUTION, DIVERSITY & THE ENVIRONMENT. © 1985 Wm. C. Brown Publishers, Dubuque, Iowa. All Rights Reserved. Reprinted by permission.
Fig. 15.19: From Hole, John W. Jr., HUMAN ANATOMY AND PHYSIOLOGY, 3d ed. © 1984 Wm. C. Brown Publishers, Dubuque, Iowa. All Rights Reserved. Reprinted by permission.
Fig. 15.25: Reprinted with permission from *Chemical and Engineering News*, Nov. 28, 1977. Copyright 1977 American Chemical Society.
Table 15.6: From Hole, John W., Jr., HUMAN ANATOMY AND PHYSIOLOGY, 3d ed. © 1978, 1981, 1984 Wm. C. Brown Publishers, Dubuque, Iowa. All Rights Reserved. Reprinted by permission.

Chapter 16

Figs. 16.5, 16.6: From Hole, John W., Jr., HUMAN ANATOMY AND PHYSIOLOGY, 3d ed. © 1978, 1981, 1984 Wm. C. Brown Publishers, Dubuque, Iowa. All Rights Reserved. Reprinted by permission.
Fig. 16.17: Adapted from *Nerve, Muscle and Synapse* by Bernard Katz. Copyright 1966 by McGraw-Hill. Used by permission of McGraw-Hill Book Company.

Chapter 17

Figs. 17.4, 17.16: From Hole, John W., Jr., HUMAN ANATOMY AND PHYSIOLOGY, 3d ed. © 1978, 1981, 1984 Wm. C. Brown Publishers, Dubuque, Iowa. All Rights Reserved. Reprinted by permission.
Fig. 17.7: After Carolina Biological Supply Company.
Fig. 17.8: *Functional Anatomy* by James E. Crouch, second edition, 1972. Published by Lea & Febiger.
Fig. 17.12: After Chaffee and Greisheimer, Basic Physiology and Anatomy, 3d ed., 1974.

Chapter 18

Table 18.2: Reprinted from *How To Live With Diabetes* by Henry Dolger, M.D., and Bernard Seeman. By permission of W. W. Norton & Company, Inc. Copyright © 1972, 1965, 1958 by Henry Dolger and Bernard Seeman.

Chapter 19

Fig. 19.14: Data based on Guttmacher, Alan F., Pregnancy, Birth and Family Planning, New York, New American Library, 1973.
Table 19.3: From Mader, Sylvia S., HUMAN REPRODUCTIVE BIOLOGY. © 1980 Wm. C. Brown Publishers, Dubuque, Iowa. All Rights Reserved. Reprinted by permission.

Chapter 20

Fig. 20.16: Courtesy of Carnation Company.

Chapter 21

Fig. 21.14: Courtesy of National March of Dimes.
Table 21.5: From: Antenatal Diagnosis, HEW, 1979, p. I-48.

Chapter 22

Figs. 22.2, 22.3, 22.5, 22.6, 22.11, 22.16, 22.17: From Mader, Sylvia S., BIOLOGY: EVOLUTION, DIVERSITY & THE ENVIRONMENT. © 1985 Wm. C. Brown Publishers, Dubuque, Iowa. All Rights Reserved. Reprinted by permission.
Fig. 22.8: From Johnson, Leland G., BIOLOGY. © 1983 Wm. C. Brown Publishers, Dubuque, Iowa. All Rights Reserved. Reprinted by permission.
Table 22.3: From Volpe, E. Peter, BIOLOGY AND HUMAN CONCERNS, 3d.ed. © 1975, 1979, 1983 Wm. C. Brown Publishers, Dubuque, Iowa. All Rights Reserved. Reprinted by permission.

Chapter 23

Fig. 23.7: Source: General Zoology, 6th ed. Storer et al., McGraw-Hill Book Company, New York.
Fig. 23.8: With permission from JOURNAL OF HUMAN EVOLUTION. Copyright 1972 by Academic Press, Inc. (London) Limited.
Fig. 23.15: Bruce Wallace and Adrian Srb, *Adaption*, 2d. ed., copyright 1964. Reprinted by permission of Prentice-Hall, Englewood Cliffs, N.J.
Fig. 23.19: From Johnson, Leland G., BIOLOGY. © 1983 Wm. C. Brown Publishers, Dubuque, Iowa. All Rights Reserved. Reprinted by permission.
Fig. 23.21: From E. O. Wilson, et al., *Life: Cells, Organisms, Population*, 1977.

Page 515: From BIOLOGICAL SCIENCE: An Inquiry into Life, BSCS Yellow Version, Fourth Edition. Copyright © 1980 Biological Sciences Curriculum Study. Reprinted by permission.

Chapter 24

Table 24.2: From "Chemical Origins of Cells," by Sidney W. Fox; Chemical & Engineering News, vol. 49, Dec. 6, 1971, p. 50.

Chapter 25

Fig. 25.19: From Mader, Sylvia S., BIOLOGY: EVOLUTION, DIVERSITY & THE ENVIRONMENT. © 1985 Wm. C. Brown Publishers, Dubuque, Iowa. All Rights Reserved. Reprinted by permission.

Chapter 26

Figs. 26.2, 26.4: From Mader, Sylvia S., BIOLOGY: EVOLUTION, DIVERSITY & THE ENVIRONMENT. © 1985 Wm. C. Brown Publishers, Dubuque, Iowa. All Rights Reserved. Reprinted by permission.
Figs. 26.3, 26.5: Courtesy of Kendall/Hunt Publishing Company. © 1975 by Kendall/Hunt Publishing Co., Dubuque, IA.

Chapter 27

Figs. 27.4, 27.5: From Mader, Sylvia S., BIOLOGY: EVOLUTION, DIVERSITY & THE ENVIRONMENT. © 1985 Wm. C. Brown Publishers, Dubuque, Iowa. All Rights Reserved. Reprinted by permission.
Figs. 27.6, 27.10, 27.20, 27.26, 27.27, 27.29: After Carolina Biological Supply Company.

Chapter 28

Fig. 28.5: From *Fossil Man* by Michael H. Day. Copyright © 1970 by Grosset & Dunlap, Inc. Copyright © 1969 by The Hamlyn Group. Used by permission of Grosset & Dunlap, Inc. Illustration taken from *Fossil Man* by Michael H. Day. The Hamlyn Group.
Fig. 28.6: From *Fossil Man* by Michael H. Day. Copyright © 1970 by Grosset & Dunlap, Inc. Copyright © 1969 by the Hamlyn Group. Used by permission of Grosset & Dunlap, Inc.
Fig. 28.12: From John Alcock, *Animal Behavior: An Evolutionary Behavior*, Second Edition, 1979. Published by Sinauer Associates, Inc., Sunderland, Massu.

Chapter 29

Fig. 29.1: From *Fundamental Concepts of Biology* by G. E. Nelson, et al. Copyright © 1974 by John Wiley & Sons, Inc.
Fig. 29.3: From Griffin, Donald R., ECHOES OF BATS AND MEN. Copyright © 1959 Doubleday. Reprinted by permission of the author.
Fig. 29.10 (right): From *Behavior Supplement #15* by Dr. J. P. Hailman, published by E. J. Brill, Leiden, Holland.
Fig. 29.12: From V. G. Deither and E. Stellar, *Animal Behavior*, 3rd edition 1970. By permission of Prentice-Hall, Inc., Englewood Cliffs, N.J.
Fig. 29.16: From BIOLOGY TODAY, SECOND EDITION, Copyright © 1972, 1975 by Random House, Inc. Reprinted by permission of CRM Books, a Division of Random House, Inc.
Fig. 29.21: From Volpe, E. Peter, BIOLOGY AND HUMAN CONCERNS, 3d ed. Copyright © 1975, 1979, 1983 Wm. C. Brown Publishers, Dubuque, Iowa. All Rights Reserved. Reprinted by permission.
Fig. 29.27a: From Helena Curtis, Biology 3rd ed. Worth Publishers, New York, 1979, page 904.
Fig. 29.27b: From Curtis & Barnes, INVITATION TO BIOLOGY 3rd ed. Worth Publishers, New York, 1981, page 510.

Table 29.1: From Alcock, J., *Animal Behavior: An Evolutionary Approach* 1975 (First Ed.) Sinauer, page 468.

Chapter 30

Fig. 30.4: Kelly and McGrath: BIOLOGY: EVOLUTION AND ADAPTATION TO THE ENVIRONMENT © 1975 Houghton Mifflin Company. Used with permission.
Fig. 30.5: Adapted from Web of Life. Used with permission of Aldus Books, Inc.
Fig. 30.9: From FUNDAMENTALS OF ECOLOGY, 3rd edition by Eugene P. Odum. Copyright © 1971 by W.B. Saunders Company. Reprinted by permission of CBS College Publishing.
Fig. 30.17: From Mader, Sylvia S., BIOLOGY: EVOLUTION, DIVERSITY & THE ENVIRONMENT. © 1985 Wm. C. Brown Publishers, Dubuque, Iowa. All Rights Reserved. Reprinted by permission.

Chapter 31

Fig. 31.5: From Mader, Sylvia S., BIOLOGY: EVOLUTION, DIVERSITY & THE ENVIRONMENT. © 1985 Wm. C. Brown Publishers, Dubuque, Iowa. All Rights Reserved. Reprinted by permission.
Fig. 31.17: Drawing by Emily Bookhultz.
Fig. 31.22: Adapted from U.S. Environmental Protection Agency, Office of Water Supply and Solid Waste Management Programs, Waste Disposal Practices and Their Effects on Ground Water: Executive Summary (Washington, DC: Government Printing Office 1977.)

Chapter 32

Figs. 32.13, 32.14, 32.19: From Mader, Sylvia S., BIOLOGY: EVOLUTION, DIVERSITY & THE ENVIRONMENT. © 1985 Wm. C. Brown Publishers, Dubuque, Iowa. All Rights Reserved. Reprinted by permission.

Chapter 33

Fig. 33.A: Source: Bureau of the Census.
Figs. 33.1, 33.10, 33.17, 33.18: From Mader, Sylvia S., BIOLOGY: EVOLUTION, DIVERSITY & THE ENVIRONMENT. © 1985 Wm. C. Brown Publishers, Dubuque, Iowa. All Rights Reserved. Reprinted by permission.

Fig. 33.2: Courtesy of Laboratory Studies in Biology: Observations and Their Implications by Chester A. Lawson, Ralph W. Lewis, Mary Alice Burmester and Garret Hardin, W. H. Freeman and Co., San Francisco. Copyright © 1955.
Fig. 33.5: Population Reference Bureau, based on United Nations estimates.
Fig. 33.8: After Cloud, 1969, from Odum: FUNDAMENTALS OF ECOLOGY, 3rd edition, © 1971 by the W. B. Saunders Company, Philadelphia, PA.
Fig. 33.9: From Oak Ridge National Laboratory, operated by Martin Energy Systems, Inc. for the U.S. Department of Energy.
Fig. 33.15: From "The World Food Situation and Global Grain Prospects," by T. N. Barr in SCIENCE, vol. 214, pp. 1087–1095, December 1981. Copyright 1981 by American Association for the Advancement of Science.
Table 33.2: 1983 World Population Data Sheet, Population Reference Bureau, Washington, DC, 1983.
Table 33.6: Ronald G. Fidker and Elizabeth W. Cecelski, "Resources, Environment, and Population: The Nature of Future Limits," Population Bulletin, Vol. 34, No. 3. (Population Reference Bureau, Inc., Washington, DC, 1979.)

Readings

Pages 8–9: Copyright 1982 by the National Wildlife Federation. Reprinted from the January–February issue of International Wildlife Magazine.
Page 40: Reprinted with permission from SCIENCE NEWS, the weekly newsmagazine of science, copyright 1983 by Science Service, Inc.
Page 75: Copyright 1981 Time Inc. All rights reserved. Reprinted by permission from TIME.
Pages 172–73: Copyright 1983 Time Inc. All rights reserved. Reprinted by permission from TIME.
Page 200: Copyright 1982 Time Inc. All rights reserved. Reprinted by permission from TIME.
Pages 222–23: Reprinted with permission from SCIENCE NEWS, the weekly newsmagazine of science, copyright 1983 by Science Service, Inc.
Pages 230–31: Reprinted by permission of SCIENCE 84 Magazine, © the American Association for the Advancement of Science.
Pages 252–53: Reprinted by permission of SCIENCE 84 Magazine, © the American Association for the Advancement of Science.

Pages 254–55: Reprinted with permission from SCIENCE NEWS, the weekly newsmagazine of science, copyright 1983 by Science Service, Inc.
Pages 270–71: Reprinted by permission of SCIENCE 84 Magazine, © the American Association for the Advancement of Science.
Pages 296–97: Reprinted with permission from SCIENCE NEWS, the weekly newsmagazine of science, copyright 1982 by Science Service, Inc.
Pages 320–21: Copyright 1983 Time Inc. All rights reserved. Reprinted by permission from TIME.
Pages 334–35: Reprinted by permission of SCIENCE 84 Magazine, © the American Association for the Advancement of Science.
Pages 354–55: Copyright 1981 Time Inc. All rights reserved. Reprinted by permission from TIME.
Pages 382–83: Copyright 1984 Time Inc. All rights reserved. Reprinted by permission from TIME.
Pages 418–19: Copyright 1981 Time Inc. All Rights Reserved. Reprinted by permission from TIME.
Page 452: Copyright 1978 Time Inc. All rights reserved. Reprinted by permission from TIME.
Pages 496–97: Copyright 1982 Time Inc. All rights reserved. Reprinted by permission from TIME.
Page 582: *Humanistic Botany* by Tippo and Stern. Used by permission of W. W. Norton & Co., Inc.
Pages 640–41: Reprinted with permission from SCIENCE NEWS, the weekly newsmagazine of science, copyright 1983 by Science Service, Inc.
Page 649: Copyright 1978 Time Inc. All rights reserved. Reprinted by permission from TIME.
Pages 680–81: Copyright 1978 Time Inc. All Rights reserved. Reprinted by permission from TIME.
Pages 698–99: Copyright 1981 Time Inc. All rights reserved. Reprinted by permission from TIME.
Page 702: Copyright 1979 Time Inc. All rights reserved. Reprinted by permission from TIME.
Pages 724–25: Copyright 1982 Time Inc. All rights reserved. Reprinted by permission from TIME.
Pages 730–31: Copyright 1982 Time Inc. All rights reserved. Reprinted by permission from TIME.
Pages 752–53: Mary Batten. c/o Purcell Associates. 964 2nd Ave., N.Y., N.Y. 10022.
Page 760: Copyright 1980 by the National Wildlife Federation. Reprinted from April–May 1980 NATIONAL WILDLIFE Magazine.
Page 766: Copyright 1979 by the National Wildlife Federation. Reprinted from the August–September 1979 NATIONAL WILDLIFE Magazine.
Pages 784–85: Copyright 1984 Time Inc. All rights reserved. Reprinted by permission from TIME.

A

Index

Index

Index